BIOLOGY
Evolution, Diversity,
and the Environment

BIOLOGY
Evolution, Diversity,
and the Environment

Second Edition

Sylvia S. Mader

wcb
Wm. C. Brown Publishers
Dubuque, Iowa

Cover Image
The colorful birds pictured on the cover are Scarlet and Red-Blue-and-Green Macaws, the largest of the South American parrots. In addition to an impressive size, these noisy birds have powerful bills and toes that can create a viselike grip. These physical characteristics protect them, and they have few predators. Photo by © Günter Ziesler/Peter Arnold, Inc.

Book Team

John D. Stout *Executive Editor*
Kevin Kane *Editor*
Mary K. Sailer *Designer*
David Lansdon *Assistant Layout Designer*
Renée Pins/Elaine R. Schueller *Art Production Assistants*
Mary Jean Gregory *Production Editor*
Carol M. Schiessl *Photo Research Editor*
Carla D. Arnold *Permissions Editor*
Matt Shaughnessy *Product Manager*

wcb group

Wm. C. Brown *Chairman of the Board*
Mark C. Falb *President and Chief Executive Officer*

wcb
Wm. C. Brown Publishers, College Division

G. Franklin Lewis *Executive Vice-President, General Manager*
E. F. Jogerst *Vice-President, Cost Analyst*
George Wm. Bergquist *Editor in Chief*
Beverly Kolz *Director of Production*
Chris C. Guzzardo *Vice-President, Director of Marketing*
Bob McLaughlin *National Sales Manager*
Craig S. Marty *Manager, Marketing Research*
Marilyn A. Phelps *Manager of Design*
Colleen A. Yonda *Production Editorial Manager*
Faye M. Schilling *Photo Research Manager*

Several chapters in this book contain passages from *Biology* by Leland G. Johnson (Wm. C. Brown Publishers, 1983).

The credits section for this book begins on page 747, and is considered an extension of the copyright page.

Printed in the United States of America
10 9 8 7 6 5 4 3 2

This book is dedicated to my students, who always showed a keen interest in biology and many of whom warmed their instructor's heart by deciding to major in biology.

BRIEF CONTENTS

CONTENTS

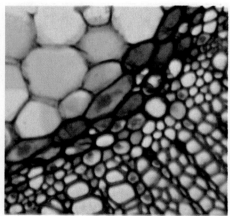

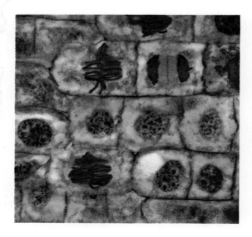

Contents **xi**

READINGS

*B*iology: Evolution, Diversity, and the Environment, a general biology text that covers all areas of modern biology, is designed for freshman biology courses. As indicated by the title, the book has three main themes. The concept of evolution is used to explain the unity and diversity of life. It serves as a background for the study of ecological relationships so that the place of all living things in the biosphere, including humans, can be appreciated. Modern ecological problems are stressed throughout the text since it is extremely important to seek solutions to these problems.

As in the first edition of *Biology*, boxed readings highlight and expand upon topics discussed throughout the text. This edition has many new readings. These were especially chosen or written to better emphasize the text's three themes, to be more in tune with students' interests, and to better stress current concerns. For example, the author's study of the Florida Everglades led to the new reading in chapter 2. Another reading, "New Questions from Fossil Records," considers the recently discovered evidence for comet bombardment every 26 million years.

The text frequently mentions major contributors to the body of biological knowledge. The history of science is even better represented in this edition since an additional number of original contributors are mentioned. As before, certain selected experiments are thoroughly presented.

Organization of the Text

The text is better organized in this edition. There are now four sections: one for overview, and one for each of the text's three themes. All three themes are carried throughout the text, but each has an individual section in which that theme is stressed.

Section 1—Evolution, Diversity, and the Environment: An Overview

Introduction

Part 1 now serves as the introduction to the text. As before, chapter 1 discusses the characteristics of life, introducing important biological concepts. It reviews the scientific method so students can see that the biological view of life is a product of this method. This chapter concludes with an analysis of the place of science in the social sphere and the responsibility of all people to decide how scientific knowledge should be used.

The new second chapter presents the concepts of evolution, diversity, and the environment, and shows how they are related in principle and in the text. Darwin's contribution to the theory of evolution is discussed as a background for showing how taxonomy serves to arrange the diversity of life according to evolutionary relationships. Life is diverse because ecosystems contain numerous niches—various roles that organisms play in the system.

Section 2—Evolution

General Principles

The chapters in parts 2 and 3 show that life is highly organized matter. Life begins when chemical molecules are organized to produce cells, the fundamental units of life. Life's organization comes about and is maintained only by an input of energy. Matter cycles through the biosphere, but energy flows by way of photosynthesis and cellular respiration.

There was much updating of material in these parts. For example, knowledge about cell membrane receptors and the important role that they play in communication between cells has been growing in the past few years. This has led to a revision of our illustration for the cell membrane (fig. 4.4). Instructors will also like the reorganization of the chapter devoted to photosynthesis (chapter 5), and the new figure (5.17) that summarizes photosynthesis. Some may wish to use this figure for a simplified presentation of the subject.

Genetics and Evolution

In parts 4 and 5, coverage of genetics precedes that of evolution. The blueprint of life's organization is passed on through the reproductive process, by which organisms inherit genes from the previous generation. Unequal reproduction by a selected few who are adapted to the environment contributes to evolution, which has resulted in the great variety of organisms on earth.

In part 4 of this edition, new information regarding chromosome structure is included. Research regarding oncogenes and possible gene therapy in humans is proceeding at a rapid pace. Both aspects of molecular genetics are expanded upon in chapter 10.

Part 5 has been reorganized. Coverage of Darwin's contribution to evolutionary history now appears in chapter 2 of the introductory section. The first chapter in this part is entitled "The Evolutionary Process" and replaces "Modern Concepts of Evolution." The principles of evolutionary theory are explained in a more forthright and comprehensible manner. The second chapter, "Evidences for Evolution," now consolidates the data supporting evolution that appeared in the first edition.

Section 3—Diversity

Diversity of Life

Part 6 covers the diversity of life. Each type of organism presently existing has its own evolutionary history and is descended from previously existing organisms. No living group of organisms is ancestral to another living group; nonetheless, presently existing organisms may be representative of ancestors. Acceptance of this principle makes it possible to trace the history of life while reviewing the diversity of life.

In this part, the scientific names of organisms are given in addition to the common names. This aids student appreciation of taxonomic principles. The entire portion of part 6 that is devoted to animals was revised in this edition of *Biology*.

Physiology

A number of body systems that maintain homeostasis have evolved and function in multicellular plants and animals. This concept underlies the discussion of the anatomy and physiology of flowering plants and vertebrates in parts 7 through 10. In each instance, the anatomy and/or physiology of lower animals is compared in an evolutionary context before the anatomy and physiology of mammals is reviewed.

Chapter 19, which discusses plant growth regulators, has been completely updated and rewritten. Also, the comparative zoology portion of each chapter is much improved. Care was taken to provide an interesting and comprehensive introduction to mammalian anatomy and physiology. Chapter 25, "Animal Behavior," is both broader in scope and more pedagogical in this edition, making it more feasible for instructors to include this important and interesting topic in their syllabi.

Section 4—Environment

In parts 11 and 12, both traditional and modern ecology are discussed. The application of ecological principles to current environmental problems is not only an excellent example of the relevance of science, it also places within the grasp of all readers knowledge that is crucial to preserving the biosphere for future generations. Although this important theme is touched upon throughout the entire book, it is emphasized in this section.

This section, particularly praised by reviewers, has been updated. The unique pedagogical approach to modern ecology permits instructors to present this material in an interesting and worthwhile manner.

Comparison of Texts

Biology: Evolution, Diversity, and the Environment and *Inquiry into Life*

Biology: Evolution, Diversity, and the Environment approaches the study of life from the perspective of the three themes in the title—evolution, diversity, and environment. *Biology* explains general principles primarily in reference to forms of life other than the human form and secondarily in reference to the human organism. Some instructors may especially appreciate the frequent application of biological principles to plants and invertebrates. The animal physiology section emphasizes comparative animal physiology in an evolutionary context. Even so, the discussion includes all aspects of human physiology.

As many may know, I am also the author of *Inquiry into Life,* another general biology text. *Inquiry into Life* focuses on human biology—one of the many different ways to approach the study of life. The human biology focus is an emphasis that immediately appeals to many students. General principles are explained primarily in reference to the human organism and secondarily in reference to other forms of life. A significant portion of *Inquiry into Life* is devoted to human physiology. The text also includes plant and animal diversity because human beings cannot understand themselves unless they understand other living things.

Biology: Evolution, Diversity, and the Environment, like *Inquiry into Life,* reviews the entire field of biology. Nonmajors will find either text interesting and stimulating and majors will also acquire the foundation they need in order to proceed to more indepth study. Both *Inquiry into Life* and *Biology: Evolution, Diversity, and the Environment* are highly practical texts. Instructors can choose between the two according to their own preference of emphasis, and knowledge of the students at their particular institution.

Aids to the Reader

Biology was written so that students might enjoy, appreciate, and come to understand the field of biology. The following features were especially designed to assist student learning.

Text Introduction

The introductory section (chapters 1 and 2) sets out the three main themes of *Biology*. The chapters discuss the characteristics of life and, at the same time, lay a foundation upon which the rest of the text depends. This section gives a general overview of the field of biology and provides a specific introduction to the text.

Part Introductions

Each part introduction highlights the topics of the part and relates the topics to those discussed in other parts. The discussion also indicates how the chapters contained in that part contribute to the themes of the text.

Chapter Concepts

Each chapter begins with a list of the biological concepts stressed in that chapter. The way these concepts apply to the specific concerns of the chapter is also shown. The listing constitutes a brief outline that can assist students as they study and review.

Terms

Knowing the definitions of terms in biology is vital to the learning process. Terms appear in bold type the first time they occur in the text; if terms are repeated in later chapters, they may appear in italics. Terms are defined in context, and frequently used terms appear in selected key term lists at the ends of chapters and in the glossary.

Readings

Throughout *Biology,* selected readings reinforce the themes of evolution, diversity, and the environment. Other readings provide insight into the process of science. The purpose of all the selections is to show how information has been discovered and how it is applicable to everyday concerns.

Drawings, Photographs, and Tables

The drawings, photographs, and tables in *Biology* are designed to help students learn basic biological concepts as well as the specific content of the chapters. Often it is easier to understand a given process by studying illustrations, especially when the illustrations are carefully coordinated with the text, as is the case here. The photographs were selected not only to please the eye but also to emphasize specific points in the text. The tables summarize and list important information, making it readily available for efficient study.

Chapter Summaries

At the end of each chapter is a summary. The summary reviews the content of the chapter in a sequential manner and reemphasizes the chapter concepts. The summaries can help students recall the main points and acquire an overall view of the subject matter.

Chapter Questions

Objective questions and study questions are at the close of each chapter. The objective questions allow students to quiz themselves with short fill-in-the-blank objective questions. Answers to these questions appear on the same page. The study questions allow students to test their understanding of the information in the chapter. The thought questions that appeared at the ends of the chapters in the first edition are now located in the Instructor's Manual. These require students to synthesize material in the present chapter and, on some occasions, in previously studied chapters.

Selected Key Terms Lists

A list of key terms now appears at the end of each chapter. Each term is accompanied by its phonetic spelling and a page number indicating where the term is introduced and defined in the chapter. All terms in these lists also appear in the glossary.

Further Readings

The list of readings at the end of each part suggests references that can be used for further study of the topics covered in the chapters of that part. The items listed in this section were carefully chosen for readability and accessibility.

Glossary

The glossary defines many key terms used in the text. In this edition each term is accompanied by its phonetic spelling and is page-referenced to the place in the text it is first introduced and defined. By using this tool, students can review the definitions of the most frequently used terms.

Additional Aids

Student Study Guide

To ensure close coordination with the text, the author wrote the *Student Study Guide* to accompany this text. For each text chapter there is a study guide chapter, which includes a listing of behavioral objectives, a pretest, study exercises, and a posttest. Answers to study guide questions are strategically placed throughout the study guide chapter, giving students immediate feedback.

Instructor's Manual

The *Instructor's Manual* is designed to assist instructors as they plan and prepare for classes using *Biology*. Possible course organizations for semester and quarter systems are suggested, along with alternate suggestions for sequencing of chapters. An outline and a general discussion are provided for each chapter; together these give the overall rationale for the chapter. A large number of objective test questions and several essay questions are also provided for each chapter. Finally, there is a list of suggested films. The appendix contains the addresses of the film suppliers.

Laboratory Manual

The author also wrote the *Laboratory Manual* that accompanies *Biology*. With few exceptions, each chapter in the text has an accompanying laboratory exercise in the manual (some chapters have more than one accompanying exercise in the manual). In this way, instructors will be better able to emphasize a particular portion of the curriculum if they wish. The thirty-two laboratory sessions in the manual were designed to further help students appreciate the scientific method and to learn the fundamental concepts of biology and the specific content of each chapter. All exercises have been tested for student interest, preparation time, safety, and feasibility.

Instructor and Student Aids

To help instructors with laboratory preparation, there is a list of materials at the beginning of each laboratory. To assist the students, a list of learning objectives begins each laboratory. Throughout each exercise, ample space is provided for them to record their observations as the lab proceeds. Students will also benefit from the review questions for each laboratory; answers to the review questions are provided on the same page.

Laboratory Resource Guide

More extensive information regarding preparation can be found in the *Laboratory Resource Guide*. This guide, developed by the author and Dr. Trudy McKee, is a new addition to the package and will assist the instructor in making the laboratory experience a meaningful one for the student. The guide includes suggested sources for materials and supplies, directions for making up solutions and otherwise setting up the laboratory, expected results for the exercises, and suggested answers to all questions in the laboratory manual.

Computerized Ancillaries

wcb StudyPak and wcb QuizPak

Student computer software programs are now available with this edition of *Biology*. wcb StudyPak is an adaptation of the Student Study Guide for use with the Apple® and IBM® PC computers. wcb QuizPak provides students with true-false and multiple choice questions for each chapter in the text. Using these programs in the Learning Resource Center will help students to prepare for examinations.

wcb TestPak and wcb GradePak

wcb TestPak, a computerized testing service, provides instructors with either a mail-in/call-in testing program or the complete test item file on diskette for use with the Apple® and IBM® PC computers. wcb TestPak requires no programming experience. wcb GradePak is a computerized grade management system for instructors. This program tracks student performance on examinations and assignments. It will compute each student's percentage and corresponding letter grade, as well as the class average. wcb GradePak is user friendly: instructors can use it after only twenty minutes of direction.

Acknowledgments

Many people have contributed to making this book possible. In particular, I would like to acknowledge the help of Leland G. Johnson. Some significant ideas came directly from his book *Biology* (Wm. C. Brown Publishers, 1983). Several of the boxed readings are partially or completely derived from *Biology*.

I would also like to thank the following people for their reviews and helpful suggestions during the development of both editions of *Biology:*

First Edition
Karen A. Koos *Rio Hondo College*
A. David Scarfe *Texas A & M University*
David J. Fox *University of Tennessee*
H. W. Elmore *Marshall University*
John D. Cunningham *Keene State College*
William E. Barstow *University of Georgia*
Larry C. Brown *Virginia State University*
A. Lester Allen *Brigham Young University*
Malcolm Jollie *Northern Illinois University*
Carl A. Scheel *Central Michigan University*
Genevieve D. Johnson *University of Iowa*
Eugene C. Bovee *University of Kansas*
Carol B. Crafts *Providence College*
L. Herbert Bruneau *Oklahoma State University*
William H. Leonard *University of Nebraska—Lincoln*
Dean G. Dillery *Albion College*
E. Bruce Holmes *Western Illinois University*
John L. Zimmerman *Kansas State University*
Larry N. Gleason *Western Kentucky University*
Donald R. Scoby *North Dakota State University*
Lester Bazinet *Community College of Philadelphia*

Second Edition
David Ashley *Missouri Western State College*
Jack Bennett *Northern Illinois University*
Oscar Carlson *University of Wisconsin–Stout*
Arthur Cohen *Massachusetts Bay Community College*
Rebecca McBride DiLiddo *Suffolk University*
Gary Donnermeyer *St. John's University (Minnesota)*
D. C. Freeman *Wayne State University*
Sally Frost *University of Kansas*
Maura Gage *Palomar College*
Betsy Gulotta *Nassau Community College*
W. M. Hess *Brigham Young University*
Richard J. Hoffmann *Iowa State University*
Steven J. Loring *New Mexico State University*
Trudy McKee
Brian Myres *Cypress College*
John M. Pleasants *Iowa State University*
Jay Templin *Widener University*

Preparing the manuscript of the text and manuals would have been impossible without the help of the following individuals. My illustrators, Kathleen Hagelston and Anne Green, have produced many fine illustrations which will be enjoyed and appreciated by instructors and students alike. Laura Antler and Mildred Rinehart also helped with final renditions in the previous edition.

The personnel at Wm. C. Brown Publishers contributed greatly to this project. John Stout was always available for counsel and support. Kevin Kane, project editor, provided continual guidance and insight while urging all to achieve their very best. Mary Sailer designed the book with skill, and Carol Schiessl searched diligently for photographs to find just the right ones. Mary Jean Gregory served as production editor and coordinated the efforts of many.

Finally, I wish to express appreciation to my family for their constant support. My sister, Rhetta McMeans, performed clerical duties and my daughter, Karen, acted as a consultant. They, along with my son, Eric, offered daily encouragement.

BIOLOGY

Evolution, Diversity,
and the Environment

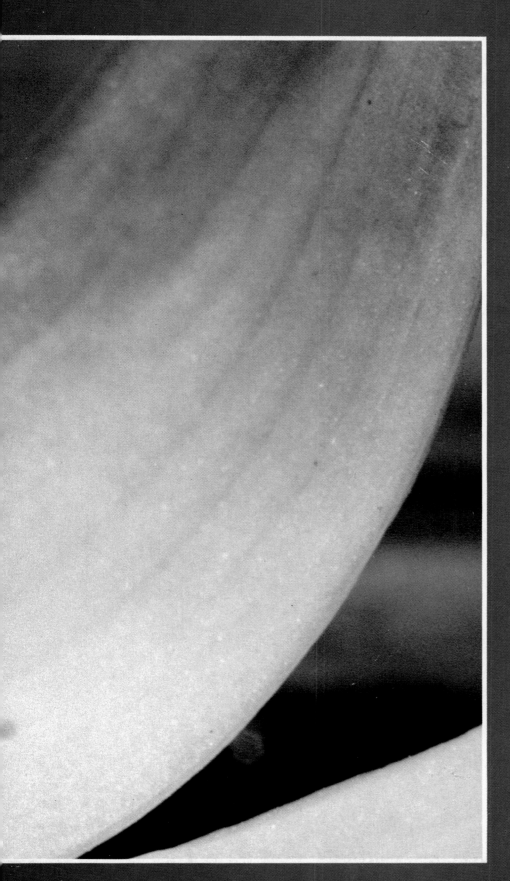

Evolution, Diversity, and the Environment
An Overview

Treat nature in terms of the cylinder, the sphere, the cone; all in perspective.

—PAUL CÉZANNE

Incurvariidae Tegeticula maculata. Yucca moth gathering pollen from the anthers of yucca stamens.

PART 1

Introduction

Despite life's overwhelming diversity, it is possible to identify characteristics that are common to all living things. *Evolution* has produced the great diversity of life we see about us. To understand this *diversity,* we can turn to the science of classification, which groups living things according to their respective organizational patterns. But to truly appreciate life's diversity, we should study ecosystems because each living thing is adapted to a role it plays in the *environment.* Our study of biology emphasizes these three themes, relying on information that has been gathered utilizing the scientific method.

Ridley turtles mating. Although these are marine turtles, the female will come ashore to lay her eggs. When they hatch, the baby turtles will instinctively scurry to the sea.

CHAPTER 1

A View of Life

Concepts

1. Living things share common characteristics: they are organized; they respond, metabolize, reproduce, and are adapted to the environment.

2. The scientific method is the process by which scientists gather information about the material world.

Figure 1.1
There are thousands of species of butterflies and they occur as far north and south as do the plants on which they feed. Coloration is often for the purpose of protection—the bold bands of color we see here break up the body into seemingly unrelated parts.
Photograph by Carolina Biological Supply Company.

L ife. Except for the most desolate and forbidding regions of the polar ice caps, all the earth is teeming with life (fig. 1.1). Without life, our planet would be nothing but a barren rock hurtling through silent space. The variety of life on earth is staggering. Human beings are part of it. So are eagles, whales, houseflies, crabgrass, mushrooms, and giraffes. The variety of living things ranges from unicellular diatoms, much too small to be seen by the naked eye, all the way up to giant redwood trees that can reach heights of three hundred feet or more. This variety seems overwhelming—indeed, this entire book would not be large enough to contain even a list of all the different kinds of living things scientists continue to identify.

Despite the magnitude of this variety, all of these organisms share certain common characteristics. Collectively, these organisms give us insight into the nature of life and help us distinguish living things from nonliving things.

Characteristics of Life

Living Things Are Organized

Both nonliving and living things are composed of chemicals and are entirely controlled by chemical and physical laws, but they are organized differently. Specimens of a nonliving thing generally vary in shape and size, but internally

Figure 1.2

Cells, the smallest of living things, differ in size, shape, and function, as exemplified by this (*a.*) animal nerve cell that conducts nerve impulses and this (*b.*) plant palisade cell that carries on photosynthesis.

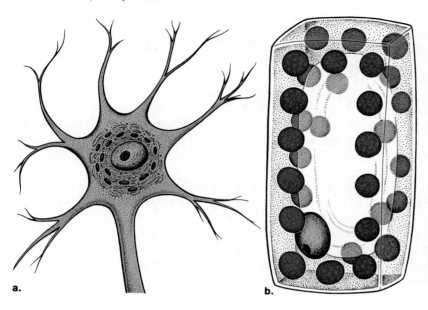

a.   b.

Figure 1.3

The plant *Mimosa* (*a.*) prior to stimulation and (*b.*) after being touched. This sensitivity has been explained on the basis of the collapse of certain cells due to water loss.

a.

b.

each one tends to be rather homogeneous. In contrast, living things are made of different parts, each with a specific structure and function. Thus, the bodies of living things are characterized by a specific complex organization. Even so, there is a great similarity in the basic organization of all living things. At the simplest level of organization, living things are cells (fig. 1.2). The **cell** is the smallest, most basic unit of life. Cells can be broken down into smaller parts, even down to the atoms making up the molecules found in cells, but these smaller parts are not alive. Just as a whole is often greater than its parts, a living cell possesses the various characteristics of life, whereas its parts do not.

Some organisms are single cells. However, most organisms are larger and more complex, having an organization in which a large number of similar cells form tissues. Several types of tissues make up organs that work together within systems.

Living Things Respond

Living things interact with their environment, responding to it in a variety of ways. Even one-celled organisms respond to outside stimuli, such as changes in chemicals, light, or temperature. The more complex the organism, the more complex the ways in which it can respond to its environment. For example, *Mimosa* is a plant that is famous for the ability of its leaves to close quickly when they are touched (fig. 1.3). And a common garden spider is alerted to scurry toward its prey by the vibration the victim causes in the spider's web (fig. 1.4). How does one prove that such behavior is innate and not learned? This question has prompted many scientific investigations.

Figure 1.4

Spiders spin delicate webs in which they capture and store their prey, a source of food to supply them with the energy and molecules they need to continue to exist.

Living Things Metabolize

The organization of the individual must be maintained. This requires a constant supply of nutrient (food) molecules that are broken down to provide a source of building materials and energy not only for repair but also for growth and reproduction. Only plants and plantlife organisms can make their own food by capturing the sun's energy as they photosynthesize. Other organisms must either feed directly on plants or other organisms as do spiders (fig. 1.4).

Homeostasis

Living.things continue to **metabolize** (carry on chemical reactions) as long as their internal environment remains fairly constant. Materials enter and leave the organism in various ways, but its internal chemistry usually remains fairly constant. This is **homeostasis,** or "staying the same." Organisms have various mechanisms and special structures for keeping the internal environment within certain boundaries.

For example, a lizard basks in the sun to raise its internal temperature. If the external temperature rises too high, the lizard will scurry for some shade, and this will keep its internal temperature in the proper range. As another example of homeostasis, consider a biology student who becomes so engrossed in her studies that she forgets to eat lunch. Her liver responds to the emergency by releasing stored sugar into her bloodstream, giving her energy until she eats her next meal.

Living Things Reproduce

Every type of organism has a **life cycle** that involves various stages and takes the organism from creation to death. When a fern develops in a new location, a tiny heart-shaped structure precedes the appearance of the more obvious leafy plant (fig. 1.5). Both structures are the fern. An insect's life involves

Figure 1.5

A fern's life cycle includes (*a.*) a tiny heart-shaped structure and (*b.*) the familiar leafy frond.

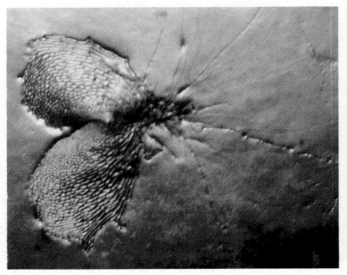

a.

b.

several forms, including egg, larva (e.g., caterpillar), pupa (e.g., cocoon), and finally adult. It is the life cycle, or programmed pattern of development, that is inherited when the hereditary factors, called **genes,** are passed from one generation to another.

The part of the life cycle most essential to the survival of the species is reproduction (fig. 1.6). All living things reproduce, passing on their organization to a new generation, ensuring that, despite the deaths of the individual organisms, the life of the species will continue.

a.

Figure 1.6
Sexual reproduction is common among plants and is the rule among animals. *a.* Each berrylike fruit of a flowering dogwood contains a seed. *b.* Wolf spider carrying its young on its back. *c.* Red fox and offspring called a kit.

b.

c.

Figure 1.7
The same type of environment produces the same type of adaptation in unrelated groups of organisms. Whales, seals, and sea turtles have flippers, while fish and sea skates have flattened bodies.

Figure 1.8
Characteristics of living things.

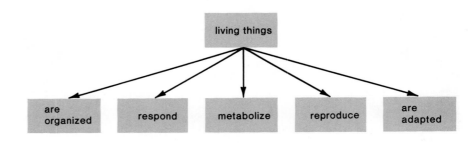

Table 1.1 Living Things	
Characteristic	**Result**
1. Are organized	All composed of cells
2. Respond to stimuli	Interact with the environment
3. Metabolize	Maintain their organization
4. Reproduce	Have offspring
5. Are adapted	Have features suited to the environment

Living Things Are Adapted

The following reading describes how a toucan's bill is adapted for various tasks. In the same way, the features of other organisms are suited to the environment in which they live. The large leafy leaves of the fern (fig. 1.5), for example, are adapted to gathering sunlight on the shady forest floor, and figure 1.7 shows how different types of animals are similarly adapted to locomotion in the sea. In the next chapter we will discuss the evolutionary mechanism by which organisms become adapted to their environment.

Figure 1.8 and table 1.1 summarize the characteristics of living things.

I t's a combination toy and Veg-o-matic: it slices, it dices, it plays catch-the-fruit! The spectacular beak of the toco toucan is a lightweight, nine-inch affair that's as long as the bird's body. Its serrated edges make short work of fruits and insects in South America's woodlands, and its bright colors and sheer size are good for intimidating smaller birds and stealing their eggs. Moreover, the beak seems to provide built-in entertainment: toucans duel with it, knocking beaks like clownish swordsmen, or toss food into the air and catch it again. With toucans, it's the beak that wags the bird.

Big Beak

The colorful toucan, Pteroglossus Bitorguatus.

Scientific Method

Science is one of the most powerful tools human beings have ever had. It stands with religion, law, and art as one of the basic means that the human species has to make sense out of experience. However, unlike religion, law, and art, science considers only the natural world, and it does so in ways that can be tested. Any other way of finding order in our world, no matter how comforting, useful, beautiful, or meaningful to the individual, is not a part of science.

Figure 1.9
Elements of the scientific method.

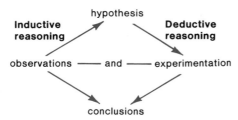

The general approach of scientists, including biologists, is known as the **scientific method** (fig. 1.9). By its use, scientists strive to collect information through systematic observation coupled with equally systematic testing. For all that, however, it is the hallmark of the scientific method that no conclusion is held to be absolutely true. Instead, all conclusions are considered to be the best current explanation of the observations. For instance, a new tool, such as a more powerful microscope, or a new finding, such as a skeleton of a human older than any previously discovered, can change the conclusions reached before these events. And all conclusions are constantly subject to review and replacement by new ones that better explain the observations.

Inductive Reasoning and Hypothesis

The scientific method often begins with **observations** about some particular occurrence. Once observations have been made, it is sometimes possible to formulate a **hypothesis,** which is a tentative explanation that is to be tested. Scientists use a type of reasoning known as **inductive reasoning** to formulate hypotheses. Inductive reasoning proceeds from specific detail to a general statement. For example, biologists observed the feeding behavior of sea gull chicks and noted that a chick only a few days old rotates its head and grasps the parent's bill (fig. 1.10), stroking it downward. Then the parent regurgitates food for the chick. Up to this time, this behavior had been regarded as an instinct, inherited and thus unlearned, but the observing researchers noted that the behavioral pattern was complex enough to suggest that it might require learning. They formed the hypothesis that experience or learning is necessary for chicks to perform the feeding behavior.

Figure 1.10
A young sea gull chick begs for food by grasping the bill of a parent. After the chick pulls the bill downward, the parent regurgitates food for the chick.

Deductive Reasoning

Once the hypothesis is stated, a second type of reasoning, called deductive reasoning, comes into play. **Deductive reasoning** begins with a general statement that infers a specific conclusion. It often takes the form of an "if . . . then" statement: "If the chicks' feeding behavior is learned, then the accuracy with which the behavior is performed will improve over time." Deductive reasoning helps scientists determine the type of experiment and/or observation necessary to support or refute a hypothesis.

Data

To see if the accuracy of the chicks' feeding behavior did improve over time, the scientists collected data in the laboratory. **Data** consists of all the evidence that is collected. This evidence may support or refute a hypothesis. Whenever the nature of the experiment permits, researchers prefer the laboratory setting. Here it is possible to concentrate on just certain phenomena because conditions can be controlled. This is important because all observations and experiments must be repeatable; that is, other scientists must report the same observations under the same circumstances or the data are considered invalid. Laboratory conditions are more easily reproduced than natural conditions. If the chick learning experiment had been conducted in the wild, how could another researcher reproduce the setting, the weather, the availability of food, or other conditions that could affect the behavior of either the chicks or their parents?

Mathematical data are preferred because they are objective and cannot be influenced by the scientist's subjective feelings. In the experiment on gull chick feeding, scientists painted diagrammatic pictures of parent gulls on small cards and recorded the accuracy with which chicks of various ages up to four days were able to strike the "bills" of the drawings. As you can see (fig. 1.11), the researcher chose to display these mathematical data in the form of a graph. Note that all the data are recorded, even if a given observation seems at the time to contradict the hypothesis. For instance, perhaps a chick generally improved over time, even though every so often it didn't perform so well. However, when the whole body of data is analyzed statistically, the overall pattern of behavior becomes clear, and the data show that the accuracy with which the chicks struck the model did improve over time. Therefore, the hypothesis that learning was needed for the chicks to perform the behavior was supported.

However, science is very cautious. Data support a hypothesis but do not actually prove it true. After all, in some instances the original hypothesis may be misleading, and in other instances the observations may be incomplete. And while a hypothesis can never be proven true, it can sometimes be proven false. Suppose a scientist had formulated the hypothesis that only laughing sea gull chicks demonstrate improvement in feeding behavior. When herring gull chicks also show improvement, this particular hypothesis is proven false.

Controlled Experiments

When observations support a hypothesis, researchers often are encouraged to expand their study of the same phenomenon. In our example, scientists wished to determine what caused the observed improvement in feeding behavior. This time they performed an experiment that tested several ideas: did the chicks need to be hungry to show improvement, or was visual experience or the process of hatching necessary to improve accuracy?

Figure 1.11
a. Cards bearing a representation of the head of a laughing sea gull adult were presented to chicks to test chicks' pecking accuracy. (1) Newborn chicks often miss the mark. (2) Two-day-old chicks are more accurate.
b. Data for several chicks recorded in the form of a graph. The graph line shows the median accuracy for the chicks tested.

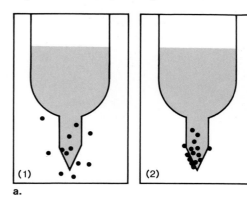

a.

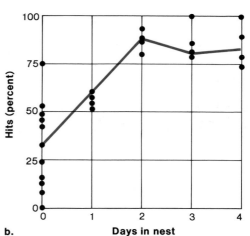

b.

Five groups of chicks were tested, two of which were control groups (a **control** is a sample that goes through all the experiment steps except those being tested):

1. control group reared in the wild;
2. control group reared in the wild;
3. experimental group reared in dark brooder and force-fed;
4. experimental group reared in dark brooder and given no food;
5. experimental group reared in dark brooder but not allowed to hatch normally. (The experimenter broke open the eggs.)

On various days after the chicks hatched, scientists photographed the chicks of all groups pecking at the stuffed head of an adult gull. From the films, the scientists determined the percentage of accurate pecks. The results showed that only the control groups reached a level of over 75 percent accuracy. Since all three experimental groups performed about the same, researchers concluded that visual experience, rather than hunger or hatching, was needed to perform as well as the control groups.

The researchers performed other types of experiments, all of which supported the hypothesis that the feeding behavior of sea gull chicks was in part learned. If related experiments constantly support a hypothesis, the confidence of certainty becomes greater, and scientists then may present the conclusions as if they are factual. Talking and writing about conclusions as if they are factual facilitates communication. However, scientists are aware that the present body of information represents the most accurate available at the moment, but further observations and experiments could lead to changes in the information.

Theories

The ultimate goal of science is to understand the natural world in terms of **concepts,** interpretations that take into account results of many experiments and observations. These concepts are stated as theories. Now in a movie a detective might claim to have a theory about the crime. Or you might say that you have a theory about the won-lost record of your favorite baseball team. But in science the word **"theory"** is reserved for those hypotheses that have been supported by a large number of observations and are considered valid by an overwhelming majority of scientists.

We will see that the theory of evolution is one such conceptual scheme. It enables scientists to understand the history of life, the variety of living things, the anatomy and physiology of organisms, embryological development, and so forth. The theory of evolution is not "just a theory" but has been generally accepted by scientists for almost a hundred years. Some biologists refer to the "principle of evolution," reserving the term "principle" or "law" for those concepts that are now accepted as generally proven. For example, even before we had photographs of the earth taken from space, it was accepted as principle that the earth is (more or less) round. Scientists did not seriously expect to argue with those who believed that the world was flat.

One reason that the scientific method works is that scientific findings are matters of public record. As one scientist examines the findings of another in preparation for research of his or her own, that examination will very likely reveal inaccuracies or inconsistencies in the predecessor's work. The very real possibility of public scrutiny helps ensure the reliability of the information that a scientist reports.

Science and Social Responsibility

There are many ways in which science has improved our lives. The most obvious examples are in the field of medicine. The discovery of antibiotics such as penicillin and the vaccines for polio, measles, and mumps have increased our life spans by decades. Cell-biology research is helping us understand the mechanisms that cause cancer. Genetic research has produced new strains of agricultural plants that have eased the burden of feeding our burgeoning population.

But science has also fostered technologies of grave concern to some of us. Even though we would not like a cancer patient to be denied the benefits of radiation therapy, we are concerned about the nuclear power industry and we do not want nuclear wastes stored near our homes. None of us wish to go back to the days in which large numbers of people died of measles or polio, yet as science has conquered one disease after another, the world's death rate has fallen and the human population has exploded. Few of us are willing to give up technology's gift of the private automobile, but we are concerned about the amount of air pollution that automobiles generate.

All too often we blame science for these undesirable side effects, and we say that our lives are infinitely more dangerous than they were in "the good old days." We think that scientists should label research good or bad and be able to predict whether any resulting technology would primarily benefit or harm humanity. Yet science, by its very nature, is impartial and simply attempts to study natural phenomena. Science does not make ethical or moral decisions. If we wish to make value judgments, we must go to other fields of study to find the means to make those judgments. And these judgments must be made by all people. The responsibility for how we use the fruits of science, including a given technology, must rest with people from all walks of life, not upon scientists alone.

Even though science does not make ethical decisions, scientists can inform the public of the possible consequences of certain actions. For example, what will happen if a community builds a large oil refinery along its coastline? Science can describe the abundant sea life in the area and can suggest the potential effect of the refinery upon that life. Based on previous observations, science may be able to predict the effect of the refinery on the air quality of the town. But science cannot tell the town that the project is a good or a bad thing to pursue. That decision must be reached by all members in the community. And in some cases, such as those dealing with the uses of the world's oceans, the community includes all countries in the world.

Scientists should provide the public with as much information as possible when such issues as nuclear power or genetic engineering experiments are being debated. Then, they, along with other citizens, can help make decisions about the future role of these technologies in our society. All men and women have a responsibility to decide how best to use scientific knowledge so that it benefits the human species and all living things.

Summary

All living things display certain characteristics: they are organized (e.g., they are made up of cells that are, in turn, made up of specialized component parts); they respond to outside stimuli; they maintain their structure by metabolizing nutrient molecules; they reproduce; and they are adapted to the environment.

Scientists gather information about the material world through the process of scientific method. Observations lead to the formulation of hypotheses, or tentative explanations, that are then tested by further observation or experimentation. Experiments are often performed in the laboratory where conditions can be controlled. Objective scientific knowledge is built up gradually, with emphasis on repeatability of experimental results and a constant review and revision of the past conclusions that have been drawn. Inductive reasoning is used to formulate hypotheses. Deductive reasoning helps decide the type of experiments to be performed.

By its very nature, science cannot answer ethical or moral questions. This is the responsibility of all members of society.

Objective Questions

1. In contrast to nonliving things, living things are made of different parts, each with a specific _____ and _____ .
2. The _____ is the smallest, most basic unit of life.
3. Even one-celled organisms respond to outside _____ such as changes in chemicals, light, or temperature.
4. Only plants and plantlike organisms are able to carry on _____ , a process that allows them to make their own food.
5. Homeostasis means "_____ ."
6. As a result of reproduction, the hereditary factors, termed _____ , are passed from one generation to another.
7. Organisms are _____ to their environment—they are suited to living in that particular environment.
8. Inductive reasoning allows scientists to formulate _____ , which are general statements that will be tested by experimentation.
9. Deductive reasoning helps scientists decide what type of _____ to perform.
10. Experimentation helps scientists collect _____ , evidence that will help support or refute a hypothesis.
11. Experiments often have a _____ sample that is subjected to all the steps of the experiment except the ones being tested.
12. In science the word _____ is reserved for those hypotheses that have been supported time and time again by observation and experimentation.

Answers to Objective Questions
1. structure and function 2. cell 3. stimuli 4. photosynthesis 5. "staying the same" 6. genes 7. adapted 8. hypotheses 9. experiments 10. data 11. control 12. theory

Study Questions

1. Name the five characteristics of life and discuss each one.
2. In what manner do scientists support hypotheses? Why can a hypothesis be proven false but not proven true?
3. Contrast and compare inductive and deductive reasoning.

Selected Key Terms

The accent marks used in the pronunciation guides are derived from a simplified system of phonetics. The single accent (') above the line denotes the major stress. Emphasis is placed on the most heavily pronounced syllable in the word. The single accent (,) below the line indicates secondary stress. Each term is page referenced to where it appears in the chapter. Each word is defined in the glossary at the back of the book.

ə abut ᵊ kitten ər further a back ā bake ä cot, cart
au̇ out ch chin e less ē easy g gift i trip ī life
j joke ŋ sing ō flow ȯ flaw ȯi coin th thin t̲h this
ü loot u̇ foot y yet yü few yu̇ furious zh vision

cell ('sel) 7
homeostasis (,hō mē ō 'stā səs) 8
life cycle ('līf 'sī kəl) 8
gene ('jēn) 9
scientific method (,sī ən 'tif ik 'meth əd) 12

observation (,äb sər 'vā shən) 12
hypothesis (hī 'päth ə səs) 12
inductive reasoning (in 'dək tiv 'rēz niŋ iŋ) 12
deductive reasoning (di 'dək tiv 'rēz niŋ iŋ) 13
data ('dāt ə) 13
control (kən 'trōl) 14
concept ('kän ,sept) 14
theory ('thē ə rē) 14

L ife is diverse. Consider, for example, the Florida Everglades (p. 30). Here a saw grass prairie, dotted by islands of hardwood trees, finally gives way to the salt-tolerant mangrove trees at the ocean's edge. Although not immediately apparent, even the seemingly uniform prairie contains numerous types of plants, and the tree islands include various temperate and tropical representatives. At the coast, there is a progression of four types of mangroves, although only the red mangrove extends prop roots into the sea where they protect juvenile shrimps, crabs, and a myriad of fishes.

Inland again, even the casual observer cannot help but be impressed by the abundance of life at a watering hole during the dry season. Here alligators either sun themselves or swim slowly among the many and diverse fishes. These fishes are the prey of such large wading birds as herons and egrets, whose beauty is exhalted by all.

How does diversity such as this come about?

Evolution

The answer to this question about life's diversity was formulated by **Charles Darwin** in the latter half of the nineteenth century. At the age of twenty-two Darwin signed on as naturalist with the H.M.S. *Beagle* (fig. 2.1), a ship that took a five-year trip around the world. The ship sailed in the southern hemisphere, where life is most abundant and varied. Along the way, Darwin encountered forms of life that were very different from those found in his native England.

Figure 2.1
The H.M.S. *Beagle,* the vessel on which Darwin served as naturalist.

CHAPTER 2

Evolution, Diversity, and the Environment

Concepts

The theory of evolution explains the great diversity of life.
a. Classification systems help us order this diversity.
b. The study of ecosystems shows that organisms are adapted to particular roles in the environment.

Figure 2.3
The giant ground sloth and giant armadillo shown here are
now extinct. Therefore, we know them only by their fossil
remains.

Evolutionary Change

Even though it was not his original intent, Darwin (fig. 2.2) began to realize
and gather evidence that life forms change over time and from place to place.

Fossils

Darwin studied **fossils,** which are the remains of ancient animals and plants
that once lived on the earth. For example, on the east coast of South America
he found the fossil remains of a giant ground sloth and armadillo (fig. 2.3).
Were not these extinct (no longer existing) forms related to the living forms
of these animals in some way?

Biogeography

Biogeography is the study of the geographic distribution of life forms on earth.
Darwin could not help but compare the animals of Africa to those of South
America. These tropical continents were occupied by similar animals that were
apparently unrelated. He noted, for example, that the African ostrich and the
South American rhea (fig. 2.4), although similar in appearance, were actually
different animals. Could it be that these two animals had separate lines of
descent because one was native to Africa and the other was native to South
America?

a. **b.**

Figure 2.4
The African ostrich (a.) and the South American rhea (b.) are not related. They look and behave similarly because they have been exposed to the same type of environment.

Figure 2.5
The ancestral form from which the Galápagos finches are descended most likely had a strong beak capable of crushing seeds. Each of the present-day thirteen species of finches has a bill adapted to a particular way of life. For example, (a.) the large tree finch grinds fruit and insects with a parrotlike bill. The small ground finch (b.) has a pointed bill and eats tiny seeds and ticks picked from iguanas. The woodpecker finch (c.) has a stout, straight bill that chisels through tree bark to uncover insects, but because it lacks a woodpecker's long tongue it uses a tool—usually a cactus spine or a small twig—to ferret insects out.

a. **b.** **c.**

Galápagos Islands

The Galápagos Islands are a small group of volcanic islands that lie off the western coast of South America. The few types of plants and animals found here seemed to be slightly different from species Darwin had observed on the mainland. In particular, Darwin became very interested in the birds, now known as Darwin's finches (fig. 2.5).

Darwin's Finches Although the finches on the Galápagos Islands reminded Darwin of a mainland finch, there were many more types. Even today, we can observe ground-dwelling finches that have different-sized beaks, according to the size of the seeds they feed on, and a cactus-eating finch that has a more pointed beak. The beak size of the tree-dwelling finches also varies according to the size of their insect prey. But the most unusual of the finches is a woodpecker-type finch. This bird has a sharp beak to chisel through tree bark, but lacks a woodpecker's long tongue to probe for insects. To make up for this, the bird carries a twig or cactus thorn in its beak with which it pokes into crevices. Once an insect emerges, the finch drops this tool so that it can seize the insect with its beak.

Darwin speculated that all these different species of finches could have descended from one type of mainland finch. In other words, a mainland finch was the **common ancestor** to all the other types. Had speciation occurred because the separate islands allowed populations of birds to evolve independently? Could each type of finch have come into existence on a different island?

Once Darwin had decided that species do change, he turned his attention to the mechanism by which evolution might occur.

Natural Selection

Darwin began to look for an evolutionary mechanism that was nonteleological. A **teleological** mechanism is one in which the end result has been determined from the beginning. A **nonteleological** mechanism is one in which the end result cannot be predetermined. Do certain plants grow in the shade because they strive to or are meant to do so (teleological), or do they grow in the shade because circumstances have brought about this ability over time (nonteleological)?

While Darwin was searching for a response to this question, one of his contemporaries, **Jean Baptiste Lamarck,** proposed a teleological mechanism for evolution. Figure 2.10 contrasts Lamarck's ideas with those of Darwin. Lamarck's mechanism, called the inheritance of **acquired characteristics,** suggests that organisms themselves can bring about physical changes that are then passed on to their offspring. This proposal, however, has not proven viable, whereas Darwin's theory of evolution by means of **natural selection** has been fully substantiated by later investigations. The following are critical to understanding natural selection.

Variations

Individual members of a species vary in their physical characteristics (fig. 2.6). Such variations can be passed on from generation to generation. Darwin was never able to determine the cause of variations, nor how they are passed on. Today, we realize that genes determine the appearance of an organism and that mutations (permanent genetic changes) can cause new variations to arise.

Struggle for Existence

In Darwin's time, a socioeconomist, **Thomas Malthus,** had stressed the reproductive potential of human beings. He proposed that death and famine were inevitable because the human population tended to increase faster than the supply of food. Darwin applied this concept to all organisms and saw that members of any plant or animal population must compete with one another for available resources. Competition must of necessity occur because reproduction produces more members of a population than can be sustained. Darwin

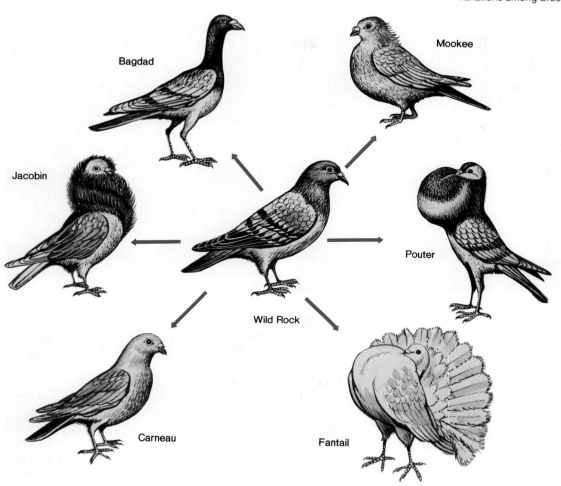

Figure 2.7
Darwin saw artificial selection as a model for evolutionary change in nature. Artificial selection produced these variations among breeds of the domestic pigeon.

Bagdad

Mookee

Jacobin

Wild Rock

Pouter

Carneau

Fantail

calculated the reproductive potential of elephants. Assuming a life span of about one hundred years and a breeding span of from thirty to ninety years, a single female will probably bear no fewer than six young. If all these young survived and continued to reproduce at the same rate, after only 750 years the descendants of a single pair of elephants would number about 19 million! Such reproductive potential necessitates a **struggle for existence** and only certain organisms will survive and reproduce.

Survival of the Fittest

Darwin noted that in **artificial selection,** humans choose which plants or animals will reproduce. This selection process brings out certain traits. For instance, there are many varieties of pigeons, each of which is derived from the wild rock dove (fig. 2.7). In a similar way, several varieties of vegetables can be traced to a single type (fig. 2.8).

Figure 2.8
Humans have been able to develop several types of
vegetables from a single species of *Brassica oleracea*:
(*a*.) Chinese cabbage, (*b*.) brussels sprouts, and
(*c*.) kohlrabi.

a.

b.

c.

In nature, it is the environment that selects those members of the population that will reproduce to a greater degree than other members. In contrast to artificial selection, the natural selection process is not teleological. Selection occurs only because certain members of a population happen to have a variation that makes them more suited to the environment. For example, any variation that increases the speed of a hoofed animal will help it escape predators and live longer; a variation that reduces water loss will help a desert plant survive; and one that increases the sense of smell will help a wild dog find its prey. Therefore, we would expect organisms with these traits to eventually reproduce to a greater extent.

a.

b.

Figure 2.9
Results of selection. *a*. Darwin emphasized that selected individuals survive, whereas those not selected perish. *b*. Modern evolutionists emphasize that selected individuals reproduce to a greater extent than those not selected.

Whereas Darwin emphasized that only certain organisms survived to reproduce, modern evolutionists emphasize that competition results in unequal reproduction (fig. 2.9). If certain organisms can acquire a greater share of available resources and if they have the ability to reproduce, then their chances of reproduction are greater than those of their cohorts.

Evolution, Diversity, and the Environment   23

a. Lamarck's Theory

Early giraffes probably had short necks which they stretched to reach food.

Their offspring had longer necks which they stretched to reach food.

Eventually the continued stretching of the neck resulted in today's giraffe.

Existing data do not support this theory.

b. Darwin's Theory

Early giraffes probably had necks of various lengths.

Competition and natural selection led to survival of the longer-necked giraffes and their offspring.

Eventually only long-necked giraffes survived the competition.

Existing data support this theory.

Adaptation

Natural selection causes a population of organisms and ultimately a species to become adapted to the environment. The process is slow but each subsequent generation will include more individuals that are adapted than the previous generation (fig. 2.10).

If you now reread the first paragraph of this chapter you will immediately realize that the organisms that exist in the Everglades must be adapted or suited to this environment. An ancestor to red mangrove trees did not strive to live in salt water; rather, after generations of natural selection, those features that allow them to live in salt water evolved in the red mangroves.

Origin of Species

Can natural selection account for the origin of new species and for the great diversity of life? Yes, if we are aware that life has been evolving for a very long time. Darwin came to the realization that the earth is very old after studying the work of geologists of his day. One geologist in particular, **Charles Lyell,** had pointed out that our planet's crust had been undergoing gradual change for a very long time, the span of time we now know is approximately 3 billion years.

After the *Beagle* returned home in 1836, Darwin waited over twenty years to publish his ideas. During these intervening years, he collected data, formed hypotheses, and used deductive reasoning to substantiate that evolution does occur. He was prompted to publish his findings after he received a letter from another naturalist, **Alfred Russel Wallace,** who had come to the exact same conclusions about evolution as had Darwin.

Although both scientists subsequently presented their ideas at the same meeting of the famed Royal Society in London in 1858, only Darwin later presented detailed evidence in support of his ideas. He described his experiments and reasonings at great length in his book, *The Origin of Species.* Figure 2.10 and the following listing summarize the theory of evolution as developed by Darwin.

1. **There are inheritable variations among the members of a population.**
2. **Many more individuals are produced each generation than can survive and reproduce.**
3. **Individuals with adaptive characteristics are more likely to be selected to reproduce by the environment.**
4. **Gradually, over long periods of time, a population can become well-adapted to a particular environment.**
5. **The end result of organic evolution is many different species each adapted to specific environments.**

Diversity

The number of different types of organisms in the entire **biosphere,** which is that part of the earth's surface and atmosphere where living organisms exist, is enormous. The number of different trees on the Malay Peninsula alone is 2,500 and 1,395 different species of breeding birds have been observed in Columbia. How can scientists make sense out of all this diversity?

Taxonomy

One way to bring order to life is to classify all the different types of organisms that have been observed. **Taxonomy** is the science of classifying organisms. This science was founded by **Carolus Linnaeus,** who, in the eighteenth century, proposed the classification system we will be describing. In this system, each type organism belongs to a particular **species,** or the lowest level of classification. Several species can be found in one **genus:** several genera in one **family:** several families in one **order** and so forth from order to **class** to **phylum** (plural: phyla) to **kingdom** (table 2.1). (In the plant kingdom, however, the term **divisions** is used rather than phyla.) At any point in the classification system, it is possible to designate supercategories or subcategories.

Table 2.1 Linnean System of Classification

Category	Description		
Species	A type of organism distinguishable from all other types		
Genus	Contains related species	more characteristics in common	characteristics in common are more distinctive
Family	Contains related genera		
Order	Contains related families		
Class	Contains related orders		
Phylum (animals) Division (plants)	Contains related classes		
Kingdom	Contains related phyla		

Table 2.2 Classification of Red Mangrove

Kingdom Plantae
  Division Tracheophyta
    Subdivision Pteropsida
      Class Angiospermae
        Subclass Dicotyledoneae
          Order Myrtiflorae
            Suborder Myrtineae
              Family Rhizophoraceae
                Genus *Rhizophora*
                  Species *mangle*

In order to get a feel for how a particular organism is classified, let us consider the classification of the red mangrove, *Rhizophora mangle*. Notice from table 2.2 that the genus and species levels of classification provide the name for this plant. Likewise, every organism has a two-part, or **binomial, name** based on its genus and species. Only the genus name is capitalized, but both words are italicized. By tradition, Latin is the universal language of taxonomic names.

Since table 2.2 classifies a particular organism, it does not indicate the other divisions in the plant kingdom, the other subdivisions in the division Tracheophyta, nor any grouping within any other level. However, you can verify that the plant kingdom contains other divisions and that each division contains other classes, etc., by consulting the classification system in the appendix.

Evolutionary Tree

Most modern classification systems of organisms are based on evolutionary relationships. The various types of evidence that are used to detect relationships are discussed in chapter 12. Whenever a classification system is based on these evolutionary relationships, it is possible to use the information to construct an evolutionary tree (fig. 2.11). In such a tree, the most closely related groups are twigs on a branch. As greater diversity occurs, we proceed from the branches to the limbs to the trunk. Each juncture of the tree represents a common ancestor or a group of ancestral organisms that gave rise to the lower groups indicated.

Suppose in figure 2.11, for example, that the twigs are genera. The branches, then, are families; the limbs are orders; and the trunk is a class. According to the tree, then, there are two genera in family A; two families in order B; and three orders in class C. This particular tree tells us that genera 3 and 4 shared a recent common ancestor not shared by genera 1 or 2. This makes the organisms in genera 3 and 4 more closely related to each other than they are to 1 and 2. The more closely related two groups are the more characteristics they have in common. This is signified in table 2.1 by the arrow on the side.

Figure 2.11
Evolutionary trees show the relationship between different classification groups. Here a hypothetical class *C* is divided into three orders including order *B*. This order, in turn, contains two families, including family *B*. Since genera *3* and *4* are both in the same family, they are more closely related to each other than they are to genera *1* and *2*.

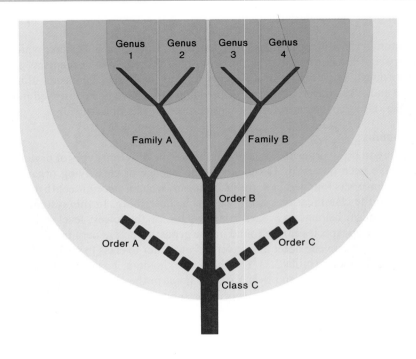

Notice in table 2.1 also that while the implied relationship and the number of common characteristics may decrease as we proceed from lower to higher categories, the characteristics that the organisms do have in common are more distinctive. It is usually easier, for example, to distinguish a plant from an animal than it is to distinguish two different types of animals one from the other.

Classification Systems

While most biologists now recognize five kingdoms of organisms (table 2.3 and fig. 2.12) other classification systems have been proposed and are used. Originally it was customary to utilize only two kingdoms: plants and animals. Plants were stationary photosynthesizers while animals were mobile food gatherers. Difficulties arose because some organisms had characteristics of both the plant and animal kingdoms. It was suggested that there be a separate kingdom, called kingdom Protista, for these organisms.

After this adjustment had been made, however, it was noted that other organisms, like the bacteria, do not have characteristics of either plants or animals. Bacteria are now placed in the kingdom Monera. Yet this four-kingdom setup was still inadequate. Finally, R. H. Whittaker suggested a fifth kingdom, this one comprising the fungi, a group of organisms that have characteristics not shared by other living things. At the present time this five-kingdom system is most often used.

Table 2.3 Classification

Name of Kingdom	Representative Organisms
Monera	Bacteria and cyanobacteria
Protista	Protozoans, unicellular algae of various types
Fungi	Molds and mushrooms
Plantae	Green algae, mosses, ferns, various trees, and flowering plants
Animalia	Sponges, worms, insects, fishes, amphibians, reptiles, birds, and mammals

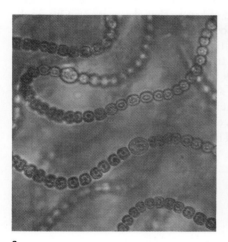
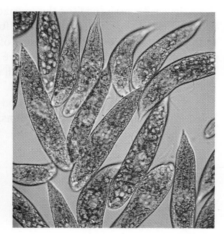

a.

b.

c.

d.

e.

Figure 2.12
Representatives from the five kingdoms recognized in this text: (*a.*) Kingdom Monera: *Nostoc,* a cyanobacterium; (*b.*) Kingdom Protista: *Euglena,* a single-celled organism; (*c.*) Kingdom Fungi: *Mycena,* a mushroom; (*d.*) Kingdom Plantae: rose, a flowering plant; (*e.*) Kingdom Animalia: lynx, a short-tailed cat.

Environment

Classification of organisms is primarily based on anatomical and physiological characteristics. Such characteristics, however, often give us only a general idea of how various organisms function in the wild. Yet it is the very need to specifically function in the biosphere that produces diversity in the first place. Therefore, to make sense out of diversity, we should study the biosphere to determine the many ways organisms satisfy their essential needs of finding food, escaping predators, and reproducing successfully.

Ecosystems

When ecologists go out into the wild, they pick a particular segment of the biosphere to study. For example, they might study the Florida Everglades, which is described in the reading on page 30. Such a segment of the biosphere is termed an **ecosystem.** The Everglades make it possible to observe the various ways in which organisms are adapted to the environment, including both the **abiotic** (physical) environment and the **biotic** (living) environment. Such observations are in the province of **ecology,** which is the study of how organisms relate to the physical environment and to each other.

Food Chains

The organisms in an ecosystem are members of food chains. Essentially, a food chain indicates "who eats whom." For example, in the Florida Everglades, the leaves of a red mangrove tree fall into the sea. When partially decomposed, these leaves become food for the myriad of larvae that are preyed on by small fishes. These fishes, in turn, become food for larger fishes that are eaten by wading birds. Whenever death of any of these organisms occurs, nutrients that can be taken up by the red mangrove trees once more are released. In this way, inorganic nutrients cycle through a food chain and no additional outside source of nutrients is required.

It is possible to analyze this food chain in a more detailed manner (fig. 2.13). The chain consists of **populations,** each of which consists of all the members of a species in a particular region. The chain begins with a photosynthetic

Figure 2.13
Components of a food chain. The producer population provides the food that either directly or indirectly supports all the other populations within an ecosystem.

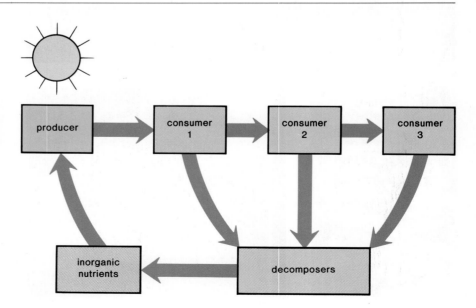

population. When supplied with inorganic nutrients, photosynthesizers are able to harness the energy of the sun to make their own food. Because this population produces food, it is called the **producer.** A producer population is the food source for the next population in the chain. None of the rest of the populations along the chain produce their own food. Therefore, they are termed **consumers.** Each consumer population is the food source for the next population in the chain until the top consumer is reached. Another type of population is also an active participant in the food chain. The bacteria and fungi of decay are the **decomposers,** populations that release inorganic nutrients to the producer.

Niche

The populations that make up the food chains in different ecosystems are adapted to the area. Why is this? First and foremost, we must consider that the abiotic conditions, such as the amount of moisture and the temperature, determine which type of organisms can reside in an area. Also, the biotic environment of the area determines how it is possible for organisms to make their living there. It is said that different ecosystems have different **niches,** which are roles that can be filled or ways in which it is possible for organisms to meet their needs. The more complex an ecosystem, the more niches there are and the greater the diversity.

   Complex ecosystems, especially, are able to maintain themselves. As long as the producer population(s) are provided with a constant supply of solar (sun) energy, they are able to produce food for the consumer populations. The inorganic nutrients they require are continually returned to them by the decomposer populations. When the ecosystem is in equilibrium, the size of each member population remains constant.

Human Ecosystem

These statements are not true of the human ecosystem. Human beings have replaced natural ecosystems with one of their own design. It contains the country, where agriculture is carried on, and the city (fig. 2.14), where businesses and industries are centered. The human ecosystem is not now stable—

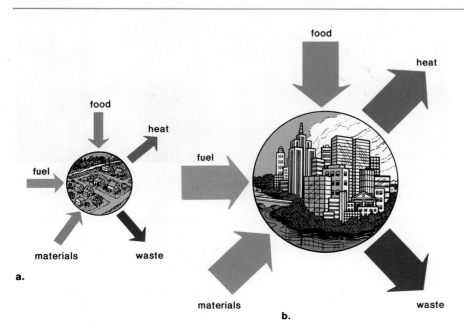

Figure 2.14
As cities grow in size, more materials, fuel, and food are taken from the environment, and more heat and waste are returned to the environment. *a.* Small town. *b.* Large city.

the human population ever increases in size and requires greater amounts of material goods and energy sources each year.

The contrast between instability of the human ecosystem and the stability of the natural ecosystem suggests that it is in the best interests of the human species to preserve what is left of natural ecosystems and to change the human ecosystem to one that is more integrated with natural systems. This is often called "working with nature" rather than against nature. It is certainly possible, and it may prove essential to our survival, to stop working against nature and begin working in harmony with our natural environment. This consideration is explored in the reading that follows.

The Everglades

The Everglades once extended from Lake Okeechobee south to Florida Bay. Now it only encompasses three conservation areas and Everglades National Park.

Originally, the Everglades encompassed the whole of southern Florida from Lake Okeechobee down to Florida Bay (see the accompanying map). Now largely in the Everglades National Park alone do we find the vast saw grass prairie interrupted occasionally by a cypress dome or hardwood tree island. Within these islands, both temperate and tropical evergreen trees grow amongst the dense and tangled vegetation. Mangrove, or salt tolerant trees, are found along sloughs (creeks) and at the shoreline. Only the roots of the red mangrove can tolerate the sea constantly. The prop roots of this tree protect over forty different types of juvenile fishes as they grow to maturity. During the wet season, from May to November, animals are dispersed throughout the region, but in the dry season, from December to April, they congregate wherever pools of water are found. Alligators are famous for making "gator holes" where water collects, and fish, shrimps, crabs, birds, and a host of living things survive until the rains come again. Almost everyone is captivated by the birds that find a ready supply of fish they need for daily existence at these holes. The large and beautiful herons, egrets, roseate spoonbill, or anhinga fill one with awe. These birds once numbered in the millions; now they number only in the thousands. Why is this?

At the turn of the century, settlers began to drain the land just south of Lake

The heart of the Everglades is a vast saw grass prairie interrupted occasionally by hardwood tree islands that contain a variety of tropical and temperate trees.

The Text

The three themes, evolution, diversity, and the environment, are interwoven throughout this text. We begin with the evolutionary theme.

Evolution: Parts 2–5

The evolutionary history of life begins with the first cell(s) and proceeds to multicellular plants and animals that function in organized units termed eco-systems. The text reflects this history in that it first examines the structure, function, and reproduction of the cell. Cellular reproduction is dependent on the passage of genetic material from generation to generation. Evolution is dependent on changes in the genetic material that lead to origination of new life forms. Knowledge of the genetic material has allowed scientists to examine in detail the mechanisms by which evolution occurs.

Okeechobee in order to grow crops on the soil that had been enriched by partially decomposed saw grass. The large dike that now rings the lake prevents the water from taking its usual course: over the banks of Lake Okeechobee and slowly southward. In times of flooding, water can be shunted through the St. Lucie Canal to the Atlantic Ocean or through the canalized Caloosahatchee River to the Gulf of Mexico. In times of drought, water is contained not only in the lake, but also in three so-called conservation areas established to the south of the lake. Water must be conserved for the irrigation of the farmland and to recharge the Biscayne aquifer (underground river) that supplies drinking water for the cities on the east coast of

Florida. Containing and moving the water from place to place has required the construction of over 1,400 miles of canals, 125 water control stations, and 18 large pumping stations. Now the Everglades National Park receives water only when it is discharged artificially from a conservation area. This disruption of the natural flow of water has affected the reproduction pattern of the birds, which is attuned to the natural wet-dry season turnover.

It took considerable human effort and a huge financial investment to control nature and to establish the "Everglades Agricultural Area." Has this attempt to bend nature to human will been worthwhile? The area does, in fact, produce more sugar than Hawaii and a large

proportion of the vegetables consumed in the United States each winter. But this has not been without a price. The rich soil, built up over thousands of years, is disappearing and most likely will be unable to sustain conventional agriculture after the year 2000. It has been suggested that *at that time* we might use the Everglades Agricultural Area for the growth of aquatic plants. Perhaps it should have been decided in the beginning to *work with nature* by growing aquatic plants instead of conventional plants. Then all the canals and pumping stations would have been unnecessary, the water would still flow from Lake Okeechobee to the glades as it had for eons, and the birds yet today would number in the millions.

An alligator beside his "gater hole," which also supplies food for the great egret in the background.

Proproots of the red mangrove provide protective cover for fishes and other sea life during maturation.

Diversity: Parts 6–10

The variety of life seems overwhelming. There are over 800,000 known species of insects alone. However, biologists have been able to categorize all living things into just a few major groups. It is possible to use the information provided by taxonomy and anatomy and physiology to hypothesize the pathway by which higher forms of life may have evolved. An examination of the various systems—circulatory, digestive, excretory, nervous, and reproductive—shows that these systems also have had an evolutionary history. Adaptation to the environment accounts for the modification of these systems in diverse forms of life.

Environment: Parts 11 and 12

Because all living things are members of ecosystems, this part discusses how matter cycles and energy flows through ecosystems. It also examines how various organisms are adapted so as to relate to one another and to the physical environment. Because most human beings live apart from nature, they are apt to forget that they, too, are part of an ecosystem that is dependent on the integrity of the biosphere. They have often disrupted the function of the biosphere, not realizing that it will not always rebound.

Biological Perspective

We live on a planet filled with an amazing and perhaps unique diversity of life. As we come to understand more about the process of evolution that resulted in this diversity, it becomes no less amazing, and we may grow to appreciate it even more. As we see the unity that exists despite the vast diversity, we may come to see ourselves as part of a community of living things. As we learn more about the ecological forces that affect us and the other living things on this planet, perhaps we will decide in favor of preserving this diversity that may be necessary for our survival as a species. It is the purpose of this book, with its themes of evolution, diversity, and ecology, to study present scientific concepts and to explore the problems that face us as members of a complex, evolving ecological web of living things.

Summary

Twenty years after Darwin returned home from his trip around the world on the *Beagle,* he wrote *The Origin of Species.* In this book, he not only presented data in support of organic evolution, he showed that evolution was guided by natural selection. Due to reproductive potential, there is a struggle for existence between members of the same species. Those members that possess variations more suited to the environment will most likely acquire more resources and so have more offspring than other members. Because of this natural selection process there is a gradual change in species composition, which leads to adaptation to the environment.

Classification of the many life forms on earth reflects their evolutionary relationships. It is possible to use classification information to draw evolutionary trees in which it is assumed that the more closely related two groups, the more recently they shared a common ancestor. A species contains organisms of the same type, similar species are in the same genus and thus we proceed from genus to family to order, to class to phylum (or division) to kingdom. Each type organism has a binomial name consisting of its genus and species.

The environment consists of ecosystems, units of the biosphere of convenient size to study. Here we can observe the manner in which organisms relate to the physical (abiotic) environment and themselves (biotic environment). The workings of a food chain show that inorganic nutrients cycle through an ecosystem, but there is also a constant need for solar energy. The diversity of life can be explained in terms of niches, or the roles by which organisms make a living within an ecosystem. Whereas natural ecosystems tend to be stable, the human ecosystem tends to be unstable—a contrast that is well to keep in mind when decisions are made regarding the environment.

These three themes, evolution, diversity, and the environment, are found throughout the text, although each is explored in great detail in a particular section.

Objective Questions

1. Darwin believed that fossil remains showed that extinct animals are _____ to living animals.
2. Darwin observed that African animals and South American animals were similar but unrelated. He reasoned that each must have its own line of _____ .
3. Darwin speculated that all of the Galápagos finches must have descended from a _____ .
4. Darwin's theory of natural selection is _____ , meaning that the organism is unable to predetermine how it shall evolve.
5. Reproductive potential necessitates a _____ and only certain organisms will survive and reproduce.
6. Evolutionary success is judged by _____ success, or the number of an organism's offspring.
7. Animals are first classified into a kingdom and then into a _____ , _____ , _____ , _____ , _____ , and _____ .
8. A food chain always begins with a _____ population.
9. The human ecosystem contains the _____ , where agriculture is carried on, and the _____ , where businesses and industries are centered.
10. This book has three themes: _____ , _____ , and the _____ .

Study Questions

1. What evidence convinced Darwin that evolution occurs?
2. Why are variations and a "struggle for existence" necessary for evolution to occur?
3. What criteria is used to determine the most fit organisms?
4. What is the end result of organic evolution by natural selection?
5. Discuss how organisms are classified.
6. List the types of populations found in an ecosystem and tell how you would expect them to be ordered in a food chain.
7. How is the human ecosystem different from natural ecosystems?
8. In what ways are diversities related to evolution and ecology?

Selected Key Terms

teleological (,tel ē ə 'läj i kəl) 20
natural selection ('nach ə rəl sə 'lek shən) 20
taxonomy (tak 'sän ə mē) 25
species ('spē ,shēz) 25
genus ('jē nəs) 25

family ('fam ə lē) 25
order ('ȯrd ər) 25
class ('klas) 25
phylum ('fī ləm) 25
kingdom ('kiŋ dəm) 25

binomial name (bī 'nō mē əl 'nām) 26
ecosystem ('ē kō ,sis təm) 28
producer (prə 'dü sər) 29
consumer (kən 'sü mər) 29
decomposer (,dē kəm ,pō zer) 29

Suggested Readings for Part 1

Asimov, I. 1960. *Wellsprings of life*. New York: Abelard-Schuman.

Baker, J., and Allen, G. 1971. *Hypothesis, prediction, and implication in biology*. Reading, MA: Addison-Wesley.

Grobstein, C. 1965. *The strategy of life*. San Francisco: W. H. Freeman.

Herbert, S. May 1986. Darwin as a geologist. *Scientific American*.

Himmelfarb, G. 1962. *Darwin and the Darwinian Revolution*. New York: W. W. Norton and Co.

Lack, David. 1978. "Darwin's finches" in *Evolution and the fossil record: Readings from Scientific American*. San Francisco: W. H. Freeman.

Luria, S. E. 1973. *Life: The unfinished experiment*. New York: Charles Scribner's Sons.

Mayr, E. 1977. Darwin and natural selection. *American Scientist* 65:321–27.

Moorehead, A. 1979. *Darwin and the Beagle*. New York: Harper & Row, Publishers, Inc.

Szent-Gyorgyi, A. 1972. *The living state*. New York: Academic Press.

Volpe, E. P. 1981. *Understanding evolution*. 4th ed. Dubuque, IA: Wm. C. Brown Publishers.

Evolution

The general theory of evolution assumes that in nature there is a great, unital, continuous and everlasting process of development. . . .

—ERNST HEINRICH HAECKEL

Bird (penguins), mammal (sea lion), and reptile (Galápagos marine iguana) all adapted for life in the ocean.

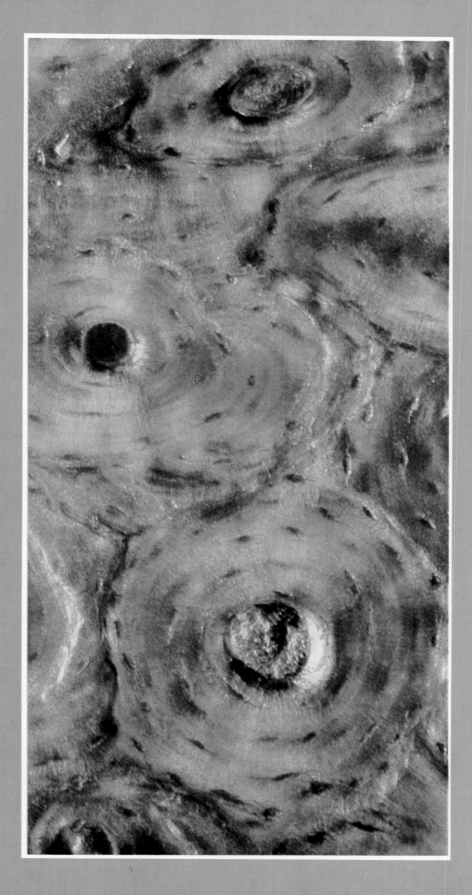

PART 2

Life's Substance

O f the ninety-two naturally occurring chemical elements, six—carbon, hydrogen, nitrogen, oxygen, phosphorus, and sulfur—play a predominant role in living things. Of these, carbon is the basis for the organic molecules that occur in living things. These molecules are organized into cells that display the characteristics of life. Cells (and multicellular organisms) obey chemical and physical laws in the same manner as nonliving matter.

Photomicrograph of compact bone. Bone cells, located in cavities that form rings, are separated from one another by a hard, calcified matrix. The central canal of each system contains blood vessels that bring nourishment to the cells.

CHAPTER 3

Chemistry of Living Things

Concepts

Living things are similarly organized.

    a. The same types of chemicals and similar chemical processes are found in all living things.

    b. Fundamentally, the chemistry of life is the chemistry of water, the most abundant substance, and the chemistry of those carbon molecules that occur in living things.

    c. The life molecules are carbohydrates, proteins, lipids, and nucleic acids.

I n the nineteenth century, many believed that living things were fundamentally different from nonliving things. They felt that life must have a vital force that is neither physical nor chemical in nature. Some of these so-called *vitalists* even thought that living things have their own type of substances and processes not found or duplicated in the nonliving world. Since that time, however, scientific investigation has repeatedly shown that both living and nonliving things obey the same chemical and physical laws (fig. 3.1). Therefore, it is proper for us to first study some basic chemistry as an introduction to our study of life. Later, we will be taking a look at some complex molecules that are associated with living things. We will also discuss how these molecules are arranged into highly organized structures such as membranes and cells. It is at this level that we can begin to see not only the unity of life but also the uniqueness of living things.

Figure 3.1
A bird in flight seems to defy the laws of physics. Actually, it does obey such laws. This Eurasian kingfisher has just risen from the water with a catch in its beak.

Elements and Atoms

All matter, both nonliving and living, is composed of basic substances called **elements.** The oxygen we breathe is an element and so is the gold found in a ring or watch. There are ninety-two naturally occurring elements, but only certain of these are commonly found in living things (table 3.1).

In the early 1800s, scientist John Dalton first proposed that elements actually contain tiny particles called **atoms.** He also pointed out that there is only one type of atom in each type element. Each type atom has been given either an English or a Latin name. One or two letters are used to stand for this name. For example, the letter H stands for the hydrogen atom and the letters Na (for *natrium* in Latin) stand for sodium.

Subatomic Particles

An atom cannot be divided by a chemical means, but even so, each one contains still smaller particles. Physicists have identified a number of **subatomic particles,** but only three will be considered here: protons, neutrons, and electrons. **Protons** and **neutrons** are found in a centrally located **nucleus** (fig. 3.2).

Table 3.1 Common Elements in Living Things*

Element	Symbol	Atomic Number	
Hydrogen	H	1	
Carbon	C	6	
Nitrogen	N	7	These elements make up most biological molecules.
Oxygen	O	8	
Phosphorus	P	15	
Sulfur	S	16	
Sodium	Na	11	
Magnesium	Mg	12	
Chlorine	Cl	17	These elements occur mainly as dissolved salts.
Potassium	K	19	
Calcium	Ca	20	
Iron	Fe	26	
Copper	Cu	29	These elements also play vital roles.
Zinc	Zn	30	

*An atomic table of the elements is given in the appendix.

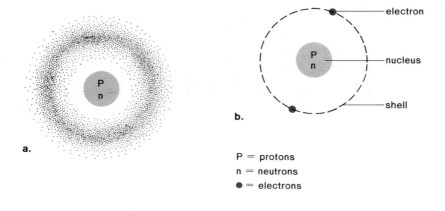

b.

P = protons
n = neutrons
● = electrons

Figure 3.2
Representation of an atom with only one shell. *a.* The nucelus (*center*) contains protons and neutrons; an orbital (represented by the stippled area) around the nucleus may contain electrons. Notice that the first orbital is spherical. *b.* Sometimes only the circumference of the orbital area is represented by a circle, termed *shell,* and each electron is represented as a dot.

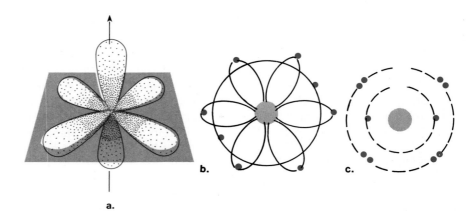

Figure 3.3
Representation of the second shell of an atom. *a.* Three dumbbell-shaped orbitals at right angles to one another, and a spherical orbital are part of the second shell. *b.* All four orbitals of the second shell on a flat plane with the nucleus in the center. *c.* A simple way to show an atom with two shells. The first shell may contain as many as two electrons and the second shell may contain as many as eight.

Figure 3.4
Hydrogen, carbon, and oxygen atoms. Notice that each neutral atom shown has an equal number of protons and electrons.

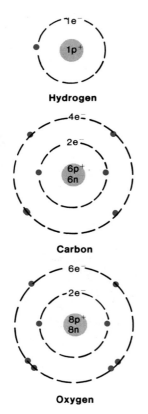

Hydrogen

Carbon

Oxygen

Electrons move about the nucleus in pathways called **orbitals.** These orbitals are grouped into **shells** (fig. 3.3) representing different energy levels. The shell closest to the nucleus has the lowest energy level, and each shell beyond has an ever higher energy level. The first shell of any atom can contain only two electrons, but the second shell can contain up to eight electrons. Although any shells beyond the second can hold more than eight electrons, they are most *stable* when they hold exactly eight electrons.

Protons have a positive ($+$) electrical charge and electrons have a negative ($-$) electrical charge. Each atom has a certain number of protons; this is known as its **atomic number** (table 3.1). An atom is electrically neutral when the number of electrons equals the number of protons.

Figure 3.4 shows one way to depict the structures of hydrogen, carbon, and oxygen. Hydrogen is the smallest atom with only one proton and one electron. Carbon, on the other hand, has an atomic number of six. It has six protons in the nucleus and six electrons in the shells (two electrons in the first shell and four electrons in the second). Each lower-energy shell is filled with electrons before the next higher-energy level shell contains any electrons. Oxygen with an atomic number of eight still has only two shells (two electrons in the first shell and six electrons in the second).

Isotopes

All atoms of an element have the same number of protons (and electrons) but may differ in the number of neutrons. Figure 3.4 indicates only the usual number of neutrons for hydrogen, carbon, and oxygen. But any specific carbon atom, for example, can have more or less than the number indicated. Atoms that have the same atomic number and differ only in the number of neutrons are called **isotopes.** Most isotopes are stable, but a few are unstable and tend to break down to more stable forms. They emit radiation as they break down. These **radioactive isotopes** can be detected by physical or photographic means and are used by biologists as "labels" in biochemical experiments.

^{14}Carbon

Most carbon atoms have six protons and six neutrons (shown as ^{12}C, carbon twelve) and are not radioactive. However, ^{14}C (carbon fourteen) has six protons and eight neutrons and is radioactive. With the release of radioactivity,

^{14}C spontaneously breaks down over a period of time into ^{14}N (nitrogen fourteen). A small amount of ^{14}C is present in all living things. Scientists can determine the age of a fossil because the rate at which ^{14}C breaks down is known. If the fossil still contains organic matter, the relative amounts of ^{14}C and ^{14}N are measured and the age is calculated. However, this method is unreliable for measuring ages in excess of twenty thousand years. The radioactivity becomes so slight that it is difficult to make an accurate determination.

Molecules and Compounds

Living things are not a collection of individual atoms. Instead, the atoms bind together to form units called **molecules**. For example, when two oxygen atoms combine, a molecule of oxygen gas (O_2) is formed. Sometimes a molecule contains different atoms. When a carbon atom combines with two oxygen atoms, a molecule of carbon dioxide gas (CO_2) is formed. A **compound** is a substance made up of molecules that contain different atoms. Therefore, carbon dioxide gas is a compound whereas oxygen gas is not.

Chemical Bonds

Atoms that have eight electrons in their outer shells do not react with other atoms. On the other hand, atoms (with more than one shell) that lack eight electrons in their outer shells do react with other atoms. These observations led to the octet rule: *Atoms (having more than one shell) react with one another to achieve eight electrons in their outer shells.*

The octet rule does not apply to hydrogen, an atom with only one shell. Sometimes hydrogen reacts to have a completed outer shell, and sometimes it reacts in such a way that it is left with only a single proton in the nucleus. Hydrogen's outer shell is complete when it contains two electrons (fig. 3.2).

Ionic Bonding

Ionic bonds form when electrons are transferred from one atom to another. For example (fig. 3.5), sodium (Na), with only one electron in its third shell, tends to be an electron *donor*. Once it gets rid of this electron, the second shell, with eight electrons, becomes its outer shell. Chlorine (Cl), on the other hand, tends to be an electron *acceptor*. Its outer shell has seven electrons, so it needs only one more electron to have a completed outer shell. When a sodium atom and a chlorine atom come together, an electron is transferred from the sodium atom to the chlorine atom. Now both atoms have eight electrons in their outer shells (fig. 3.5).

This electron transfer, however, causes a charge imbalance in each atom. The sodium atom has one more proton than it has electrons, therefore, it has a net charge of $+1$ (symbolized by Na^+). The chlorine atom has one more electron than it has protons, therefore, it has a net charge of -1 (symbolized by Cl^-). Such charged atoms are called **ions.** Ions are attracted to one another by their opposite charges. Thus an ionic compound, such as sodium chloride (Na^+Cl^-), is held together by the attraction between its ions or by **ionic bonds.**

Na^+ and Cl^- are not the only biologically important ions (fig. 3.6). Some, such as K^+, are formed by the transfer of a single electron to another atom; others, such as Ca^{2+} and Mg^{2+}, are formed by the transfer of two electrons.

Ionic bonds are quite strong when a salt like sodium chloride exists as a dry solid. But when such a compound is placed in water, the ions separate as the salt dissolves (fig. 3.10). For example, Na^+Cl^- separates into Na^+ and Cl^-. Ionic compounds are most commonly found in this dissociated (ionized) form in biological systems because these systems are 70 to 90 percent water.

Figure 3.5
In this ionic reaction, sodium donates an electron to chlorine; thereafter, each atom has eight electrons in the outer shell.

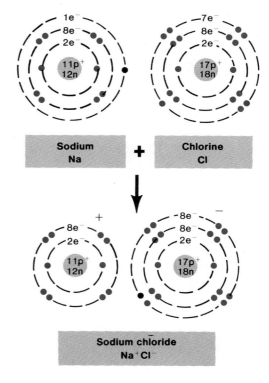

Figure 3.6
Killer whale emerges from the sea. Seawater and blood are strikingly similar in their ion concentrations. We know that life arose in the sea so could it be that blood, the internal environment of whales and of other organisms, is simulated seawater?

Covalent Bonding

Covalent bonding forms when two atoms *share* electrons in such a way that each atom has a completed outer shell. Let us consider the atoms in a water molecule. Oxygen, with six electrons in the second shell, and each hydrogen, with one electron in the first and only shell, must acquire electrons if their outer shells are to be complete. Oxygen requires two more electrons, but each hydrogen requires only one more electron. Under these circumstances, oxygen and each hydrogen share a pair of electrons. In other words, oxygen forms a **covalent bond** with each hydrogen. The pair of electrons being shared circle first the oxygen and then a hydrogen; therefore, they should be counted as belonging to each atom. When this is done for all bonds, oxygen has a completed outer shell of eight electrons, and each hydrogen atom has a completed outer shell of two electrons. Notice in figure 3.7 that a covalent bond may be represented by a short line between atoms (e.g., H—O—H). Sometimes even this notation is absent, and water is simply written as H_2O. The absence of charges (+ or −) indicates that this is a covalent molecule.

Normally, the sharing of electrons between two atoms is fairly equal, and the covalent bond is **nonpolar.** But, in the case of water, the sharing of electrons between oxygen and each hydrogen is not completely equal. The larger oxygen atom dominates the H—O association and attracts the electron pair to a greater extent. This causes the oxygen atom to assume a slight negative charge, showing that it is **electronegative** in relation to the hydrogens. Each hydrogen assumes a slight positive charge, showing that it is **electropositive** in relation to oxygen. The unequal sharing of electrons in a covalent

Figure 3.7
When water forms, oxygen shares a pair of electrons with each of two hydrogen atoms. (Notice below that the covalent bond may be simply represented by short lines between atoms.)

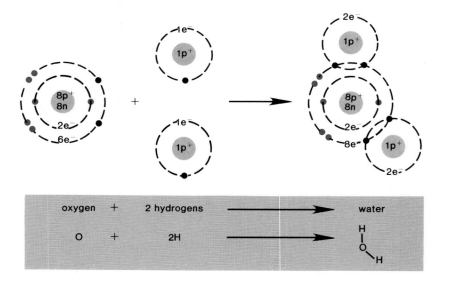

bond is called a **polar covalent bond,** and the molecule itself is a **polar molecule.** The water molecule has an electronegative end and an electropositive end (fig. 3.8*a*), indicating that it is a polar molecule.

Hydrogen Bonding

Polarity within a water molecule causes the hydrogen atoms in one molecule to be attracted to oxygen atoms in other molecules (fig. 3.8*b*). This attractive force creates a weak bond called a **hydrogen bond.** This bond is often represented by a dotted line because a hydrogen bond is easily broken. Hydrogen bonding is not unique to water. It occurs whenever an electropositive hydrogen atom is attracted to an electronegative atom in another molecule. As we shall see, hydrogen bonding between life molecules plays an important role in living systems.

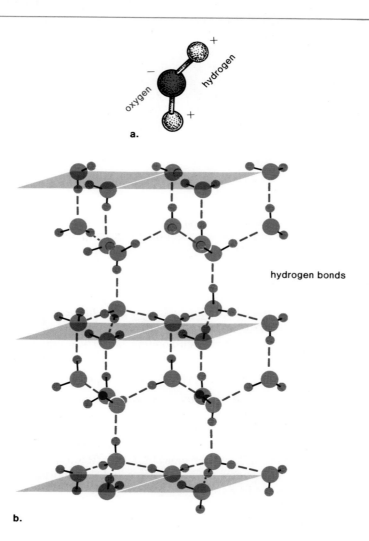

a.

hydrogen bonds

b.

Figure 3.8
a. Representation of a single water molecule. The larger balls stand for oxygen and the smaller balls stand for hydrogen atoms. A water molecule is polar because the electrons are held more tightly by oxygen than by the hydrogens. This makes one end of the molecule electronegative and the other end electropositive.
b. Water molecules are hydrogen-bonded to one another because the electropositive hydrogens are attracted to the electronegative oxygen atoms in other molecules. This is their arrangement in ice; planes have been added to help show the three-dimensional character of the arrangement.

Figure 3.9
a. The temperature of water changes slowly and a great deal of heat is needed for vaporization to occur. *b.* Ice is less dense than liquid water and therefore bodies of water freeze from the top down, making ice skating possible. *c.* The heat needed to vaporize water can help animals in a hot climate to maintain internal temperatures.

a.

Properties of Water

Life evolved in water, and living things are 70 to 90 percent water. The chemical properties of water are absolutely essential to the continuance of life. Water is a polar molecule (fig. 3.8*a*)—the oxygen end of the molecule is electronegative and the hydrogen end is electropositive. Water molecules are hydrogen-bonded one to the other (fig. 3.8*b*). Hydrogen bonds are much weaker than covalent bonds within a single water molecule, but they still cause water molecules to cling together. Without hydrogen bonding between molecules, water would boil at $-80°$ C and freeze at $-100°$ C, making life as we know it impossible. But because of hydrogen bonding, water is a liquid at temperatures suitable for life (figs. 3.11 and 3.12). Water has some other important properties.

1. *Water is the universal solvent and facilitates chemical reactions both without and within living systems.* When a salt, such as sodium chloride (Na^+Cl^-), is put into water, the electronegative ends of the water molecules are attracted to the sodium ions, and the electropositive ends of the water molecules are attracted to the chlorine ions (fig. 3.10). This causes the sodium ions and the chlorine ions to separate and dissolve in water.

   Water is also a solvent for large molecules that contain ionized atoms or are polar molecules. If they are polar, the electronegative ends of the water molecules are attracted to the electropositive ends of these molecules, and the electropositive ends of the water molecules are attracted to the electronegative ends of these molecules. Because of this, these molecules also disperse. When ions and molecules disperse in water, they can move about and collide so that reactions will occur.

b.

c.

Figure 3.10
Sodium chloride dissolves in water because the polarized ends of the water molecules are attracted to the charged Na$^+$ ions and Cl$^-$ ions. *a.* The electronegative end of the water molecule is attracted to the sodium ions. *b.* The electropositive end of the water molecules is attracted to the chlorine ions.

a.   b.

2. *Water molecules are cohesive.* Water flows freely, yet water molecules do not break apart. They cling together because of hydrogen bonding. Water molecules also adhere to (stick to) surfaces, particularly polar surfaces. Therefore, water can fill a tubular vessel and still flow so that dissolved and suspended molecules are evenly distributed throughout a system. For these reasons, water is an excellent transport system both without and within living organisms. One-celled organisms rely on external water to transport nutrient and waste molecules, but multicellular organisms often contain specialized vessels in which water serves to transport nutrients and wastes.

Figure 3.11

A gram of water is solid, liquid, or steam, depending on the number of calories of heat energy added.

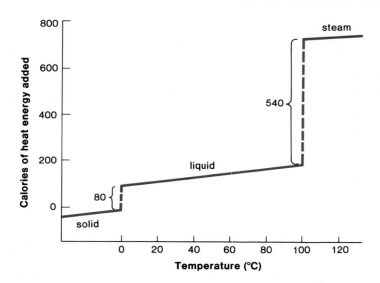

3. *The temperature of liquid water rises and falls more slowly than that of most other liquids.* It requires a calorie of heat energy to raise the temperature of one gram of water one degree (fig. 3.11). This is about twice the amount required for other covalently bonded liquids. The many hydrogen bonds that link water molecules together help water absorb heat without a change in temperature. Because water holds a large amount of heat, its temperature falls more slowly. This property of water is important not only for aquatic organisms but also for all living things. Water protects organisms from rapid temperature changes and helps them maintain their normal internal temperatures.

4. *Water tends to remain a liquid rather than change to ice or steam.* Converting one gram of liquid water to ice requires the loss of 80 calories of heat energy. Converting one gram of water to steam requires an input of 540 calories of heat energy (fig. 3.11). Changing water to steam requires that the hydrogen bonds be broken; this accounts for the very large amount of heat needed for evaporation. This property of water helps moderate the earth's temperature so that it is compatible with the continuance of life. It also gives animals in a hot environment an efficient way to get rid of excess body heat. When an animal sweats, the body heat is used to vaporize the fluid, thus cooling the animal.

5. *Frozen water is less dense than liquid water.* As water cools, the molecules come closer together until they are most dense at 4° C (fig. 3.12), but they are still moving about. Below 4° C, the water molecules cease moving, and hydrogen bonding becomes more rigid but also more open (fig. 3.8). This makes ice less dense than liquid water; this is why ice floats on liquid water. Bodies of water always freeze from the top down. When a body of water freezes on the surface, the ice acts as an insulator to prevent the water below it from freezing. This protects many aquatic organisms so that they can survive the winter.

Table 3.2 summarizes these five properties of water.

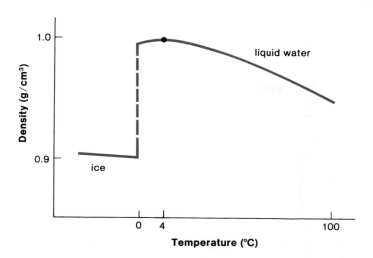

Figure 3.12
Water freezes at 0°C, boils at 100°C, and is a liquid between these temperatures. Water is most dense at 4°C; therefore, ice floats on liquid water.

Table 3.2 Water

Properties	Chemistry	Result
Universal solvent	Polarity	Facilitates chemical reactions
Adheres and is cohesive	Polarity; hydrogen bonding	Serves as transport medium
Resists changes in temperature	Hydrogen bonding	Helps keep body temperatures constant
Resists change of state (from liquid to ice and from liquid to steam)	Hydrogen bonding	Moderates earth's temperature Evaporation helps bodies remain cool
Less dense as ice than as liquid water	Hydrogen bonding changes	Ice floats on water

Acids and Bases

Living things are sensitive to the balance of acids and bases in the environment and within their own systems. If the proper balance is not maintained, it can lead to the death of organisms. To understand why acids and bases have such an effect on biological systems, it is necessary to discuss another property of water. Just as Na^+Cl^- can dissociate, so can water molecules. When water molecules dissociate, they release an equal number of hydrogen ions (positively charged hydrogen) and hydroxide ions (one oxygen and one hydrogen, with a negative charge).

$$H-O-H \rightleftharpoons H^+ + OH^-$$

This equation is written with a short forward arrow and a long backward arrow because few water molecules actually remain dissociated.

Acids are molecules that release hydrogen ions when they dissociate. A strong acid, such as hydrochloric acid (HCl), almost completely dissociates when put into water.

$$HCl \rightarrow H^+ + Cl^-$$

Bases are molecules that release hydroxide ions when they dissociate. A strong base, such as sodium hydroxide (NaOH), almost completely dissociates when put into water.

$$NaOH \rightarrow Na^+ + OH^-$$

Figure 3.13

The pH scale showing the relative concentrations of H^+ and OH^-. Notice that this particular representation should be read in this manner: At pH 7, for every 10,000,000 H^+ ions, there are 10,000,000 OH^- ions.

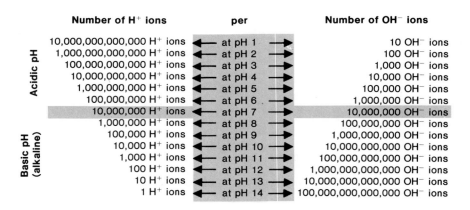

Number of H⁺ ions	per	Number of OH⁻ ions
10,000,000,000,000 H⁺ ions	at pH 1	10 OH⁻ ions
1,000,000,000,000 H⁺ ions	at pH 2	100 OH⁻ ions
100,000,000,000 H⁺ ions	at pH 3	1,000 OH⁻ ions
10,000,000,000 H⁺ ions	at pH 4	10,000 OH⁻ ions
1,000,000,000 H⁺ ions	at pH 5	100,000 OH⁻ ions
100,000,000 H⁺ ions	at pH 6	1,000,000 OH⁻ ions
10,000,000 H⁺ ions	at pH 7	10,000,000 OH⁻ ions
1,000,000 H⁺ ions	at pH 8	100,000,000 OH⁻ ions
100,000 H⁺ ions	at pH 9	1,000,000,000 OH⁻ ions
10,000 H⁺ ions	at pH 10	10,000,000,000 OH⁻ ions
1,000 H⁺ ions	at pH 11	100,000,000,000 OH⁻ ions
100 H⁺ ions	at pH 12	1,000,000,000,000 OH⁻ ions
10 H⁺ ions	at pH 13	10,000,000,000,000 OH⁻ ions
1 H⁺ ions	at pH 14	100,000,000,000,000 OH⁻ ions

(Acidic pH / Basic pH (alkaline))

pH Scale

Acids and bases affect the relative concentration of hydrogen and hydroxide ions. Biologists use the pH scale because it indicates the relative concentrations of H^+ and OH^- (fig. 3.13). The pH scale range is from 0 to 14. The pH of pure water is 7; this is neutral pH.

> pH 7
> $[H^+] = [OH^-]$[1]

Acids lower the pH; any pH below 7 indicates that there are more hydrogen ions than hydroxide ions.

> pH 7 → pH 0
> $[OH^-]$ is less than $[H^+]$
> $[H^+]$ is greater than $[OH^-]$

Bases increase the pH; any pH above 7 indicates that there are fewer hydrogen ions than hydroxide ions.

> pH 7 → pH 14
> $[OH^-]$ is greater than $[H^+]$
> $[H^+]$ is less than $[OH^-]$

Notice in figure 3.13 that a pH change of one unit involves a tenfold change in hydrogen ion concentration. Thus pH 6 has ten times the $[H^+]$ of pH 7, and pH 5 has one hundred times the $[H^+]$ of pH 7.

Buffers

Most organisms maintain a pH of about 7; for example, human blood has a pH of 7.4. A much higher or lower pH causes illness. Normally, pH stability is possible because organisms have built-in mechanisms to prevent pH change. Buffers are the most important of these mechanisms. A buffer is a chemical, or a combination of chemicals, that takes up either excess hydrogen ions or excess hydroxide ions. A buffer also releases hydrogen or hydroxide ions should they be in short supply. It is possible to overcome an organism's buffering ability, but usually buffers keep the pH within normal limits despite many biochemical reactions that either release or take up hydrogen or hydroxide ions.

[1]Square brackets are conventionally used to denote concentrations.

Figure 3.14
The pH of various solutions depicted on the glass
electrode of a pH meter, an instrument that automatically
measures pH.

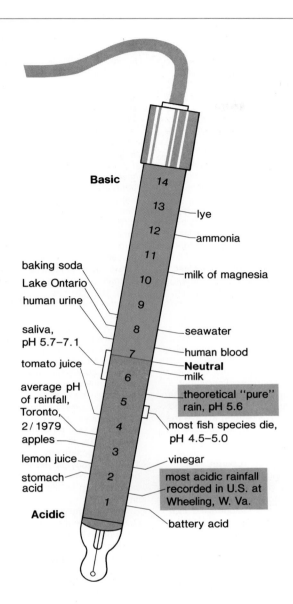

Basic

14
13 — lye
12 — ammonia
11
10 — milk of magnesia

baking soda
Lake Ontario
human urine

9

saliva,
pH 5.7–7.1

8 — seawater

7 — human blood
Neutral
— milk

tomato juice

6

average pH
of rainfall,
Toronto,
2/1979

5 — theoretical "pure" rain, pH 5.6

4 — most fish species die, pH 4.5–5.0

apples

3

lemon juice — vinegar

stomach
acid

2 — most acidic rainfall recorded in U.S. at Wheeling, W. Va.

1

Acidic

— battery acid

Acid Rain

Figure 3.14 indicates the pH of some common solutions, including rain. Acid
rain, also discussed on page 716, is caused by the air pollutants sulfuric acid
and nitric acid. Coal-fired power plants and gasoline-fueled cars add most of
these pollutants to the atmosphere. Our industrialized society depends primarily on consumption of such fossil fuels as coal and petroleum (gasoline is
derived from petroleum). This leads to detrimental environmental effects: lakes
that no longer support life and the possible death of the famed Black Forest
in Germany as described in the reading on the following page. In order to
minimize these effects, can we reduce our consumption of fossil fuels or learn
to depend on alternate and cleaner sources of energy? Concerned citizens and
scientists are working on these questions.

Fast Cars and Sick Trees

The magnificent spruce rise like giant pillars, their clustered tops dimming even a cloudless sky. Underfoot, a carpet of needles cushions footpaths that vanish mysteriously from sight in the forest darkness. The air is cool and moist, carrying a scent of enchantment, as if a fairy tale might suddenly come to life. This is the legendary Black Forest, West Germany's most beloved woodland, and a source of inspiration for such great cultural figures as Goethe, Beethoven and Hesse. For more than three centuries, it has been admired, protected and lovingly tended.

But now, the 2,300-square-mile forest in the south of Germany is sick, perhaps even dying, afflicted by a disease that is still not completely understood. One major culprit is believed to be acidic fallout from the sulfur dioxide gas emitted by Europe's coal-fired industries. Recently, however, scientists have begun to suspect still another villain, one that many Germans ironically treasure as much as the Black Forest itself—the automobile.

In a country without emission controls on its cars or speed limits on its *autobahns,* motor vehicles spew out more than a million tons of nitrogen oxides (NOX) a year, creating a deluge of nitric acid and other pollutants that may also be killing trees. As a result, a bitter battle is now being waged between those who are fighting to save the forests and those who want to preserve the freedom of the roads.

. . . In the late 1960s and early 1970s, Scandinavian researchers reported that their lakes were dying because of acid falling from the sky. Soon after, biologists discovered lifeless acidified lakes in the United States' Adirondack Mountains, and reports of forest damage began to trickle in from Poland, East Germany and Czechoslovakia, where coalburning industry is especially dirty. Scientists examined the evidence and hypothesized that air pollution, in particular acid rain, was responsible for the plight of the lakes and forests.

The notion helped explain the observation that trees at high altitudes suffer the most damage. Fog often shrouds the upper slopes of mountains in the Black Forest and other woodlands, says Schröter. "The fog carries acid. The damage is very heavy in the fog zone."

Many German scientists now believe that there is enough evidence linking the decline of the Black Forest and acid rain to warrant immediate action. "The only way to solve the problem of the forests is to reduce industrial and car emissions," says senior forester Wolfgang Tzschupke of the Baden-Württemberg forest administration. "It must be done in a short time. We don't know how much longer the forests can last."

Tzschupke and like-minded colleagues now have a powerful ally in West Germany's Greens Party. Campaigning on a save-the-forests platform in the March 1984 national elections, the five-year-old party won 27 seats in the 498-seat Parliament, establishing it as the nation's third-largest political party. But despite its success in turning the plight of the Black Forest into a national *cause célébre,* Germany's environmental movement has made slow progress in pressuring the government to enact strict emission controls.

Economic hardships and cultural traditions pose . . . barriers. The Bonn Parliament originally planned to ban leaded gasoline and impose American-style emission controls in 1986. But the nation's auto manufacturers protested, claiming that they would be at a serious disadvantage in European markets compared to competitors in France and Italy, where a gradual six-year reduction in emissions, agreed to by all the countries of the European community, will not begin until 1988. Germany did, however, institute tax breaks for "environment friendly" cars—tax savings of up to $680 that will go into effect on July 1.

With unemployment still high, the Germans have also been less willing to crack down on sulfurous power-plant emissions. Last summer, in a rare emergency session of Parliament, the government allowed the start-up of a new coal-burning plant that lacked pollution control equipment. Claiming that 3,500 jobs depended on the facility, the government did grant environmentalists one concession: it limited the plant to two-percent sulfur coal, instead of the originally planned three-percent.

The government is also resisting strong pressure to impose speed limits, a simple measure that might significantly help the

View of distressed tree in Black Forest with auto in the foreground.

Black Forest. According to scientists at the Heidelberg Institute for Energy and Environment Research, a maximum speed of 62 miles per hour would lower auto emissions all the way down to the levels expected when emission controls finally go into effect. In the land of the *autobahn,* however, voting for speed limits is tantamount to committing political suicide for many members of Parliament. "If there is one freedom that people here are prepared to defend tooth and claw, it is the freedom of being able to hit the gas pedal," explained a recent editorial in the *Rundschau,* a Frankfurt newspaper.

While the politicians balk at even a limited war on airborne pollutants, the grim warnings of German foresters have managed to loosen a generous supply of money for research. Several million dollars from federal and state governments will finance the first comprehensive scientific investigation of the Black Forest and the far more damaged woodlands in other sections of West Germany. In one experiment, researchers will spread lime and potassium over 10,000 acres in the Black Forest to see if trees and soil can be protected from acid rain. If the experiment works, it will be the first good news Germans have heard about their beloved forest in years. Meantime, the situation continues to look grim. Warns Hansjochen Schröter: "Some areas are so damaged they may be gone in five years."

Figure 3.15

This figure illustrates that many different molecules can be formed with carbon and hydrogen atoms. Carbon-to-carbon bonds bring about (a.) long chains, (b.) branched chains, and (c.) ring compounds. Carbon-to-carbon bonds (d.) can be single covalent bonds or double covalent bonds.

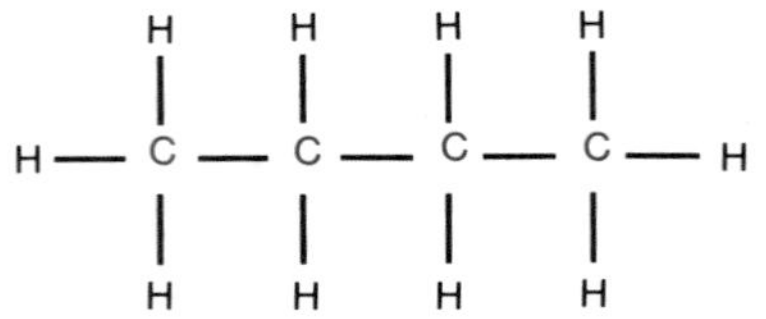

a. Straight chain

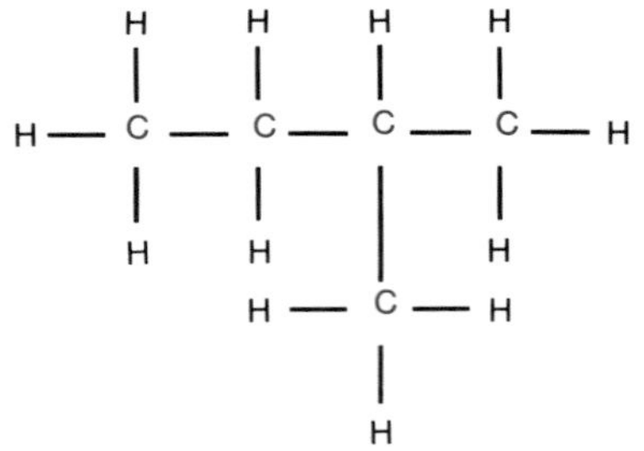

b. Branched chain

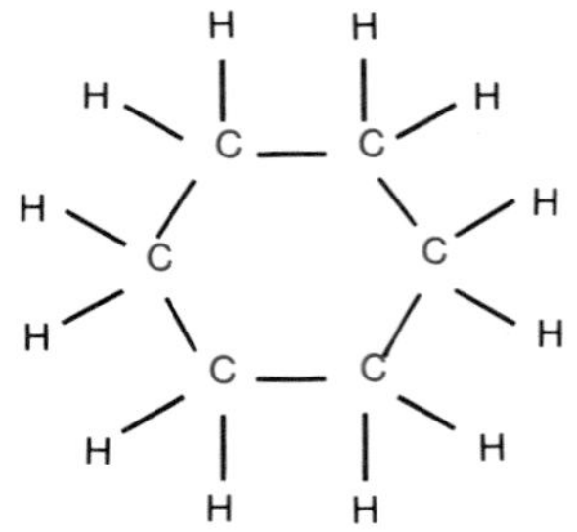

c. Carbon ring

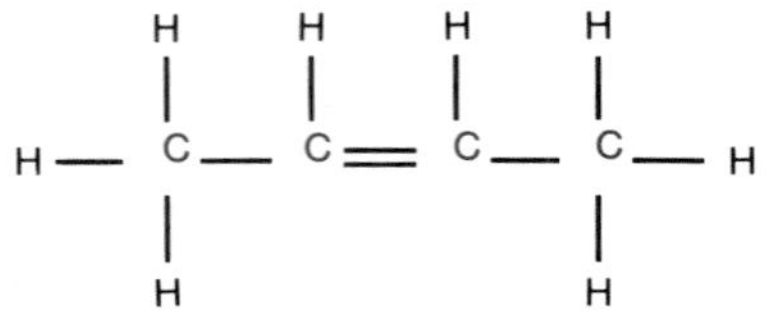

d. Chain with double bond

Figure 3.16

Isomers have the same formula but different configurations. Both of these compounds have the formula $C_3H_6O_3$. a. In glyceraldehyde, oxygen is double-bonded to an end carbon. b. In dihydroxyacetone, oxygen is double-bonded to the middle carbon.

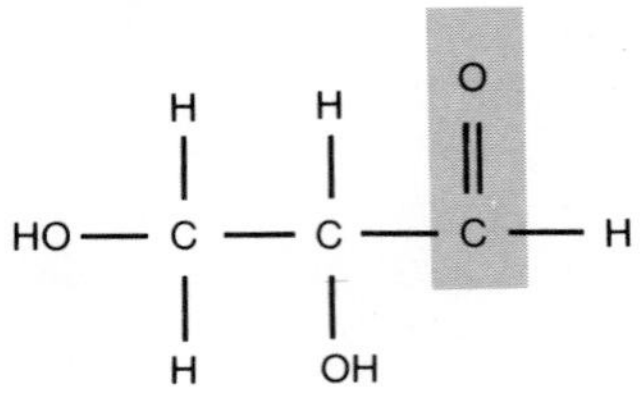

a.

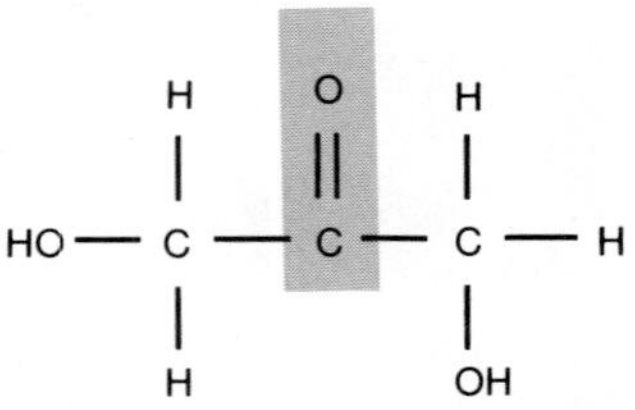

b.

Life Molecules

The most common elements in molecules unique to living organisms are frequently referred to as CHNOPS (table 3.1). Of these, carbon (C) is most fundamental to the chemistry of the life molecules, and its properties account for their great variety (fig. 3.15). The carbon atom has four electrons in its outer shell, so it can form covalent bonds with as many as four other atoms. Usually, carbon bonds to hydrogen, oxygen, nitrogen, or another carbon atom. Carbon-to-carbon bonding makes possible carbon chains of various lengths and shapes. Long chains containing fifty carbon atoms are not unusual in living systems. Even much longer chains (containing other atoms, such as nitrogen and oxygen) are found. Carbon-to-carbon bonding sometimes results in numerous ring compounds of biological significance. Carbon can share more than one pair of electrons with another atom, giving what is called a **double covalent bond.**

Another factor contributing to the diversity of the life molecules is the presence of isomers. **Isomers** are molecules that have identical chemical formulae because they contain the same numbers and kinds of atoms, yet they are different molecules because the atoms in each isomer are arranged differently. For example, figure 3.16 shows two compounds with the formula $C_3H_6O_3$. Each of these molecules is assigned its own chemical name because the molecules differ structurally.

Life molecules are **macromolecules** (large molecules). Macromolecules are often **polymers**; that is, they are chains of unit molecules linked together. **Monomers** are the smaller molecules that form individual units of polymer chains.

Monomers are joined together into polymers by *condensation reactions* that remove the components of water molecules (fig. 3.17). The reverse of a condensation reaction is **hydrolysis** (the addition of water molecules), which separates the monomers from one another. This pattern of condensation and hydrolysis is essential to many syntheses (building up) of macromolecules as well as to the breakdown of macromolecules.

The principal macromolecules in cells are polysaccharides (large carbohydrates), lipids, proteins, and nucleic acids. The first three are discussed here in some detail. Nucleic acids are discussed more fully in a later chapter. Table 3.3 is a simplified listing that briefly describes these macromolecules.

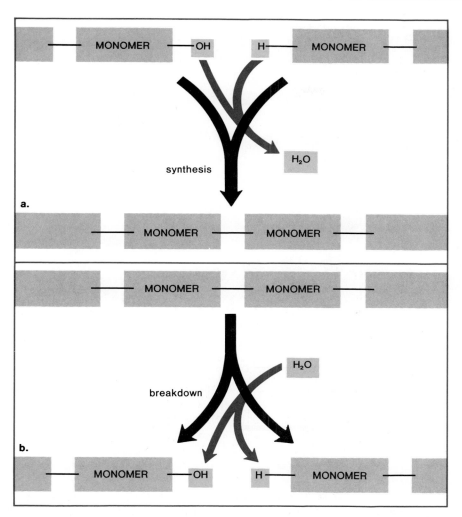

Figure 3.17
a. In cells, synthesis often occurs when monomers are joined together by condensation (removal of H_2O).
b. Breakdown occurs when the monomers in a polymer are separated by hydrolysis (the addition of H_2O).

Table 3.3 Simplified Table of Macromolecules

Selected Macromolecules	Unit Molecules	Common Atoms	An Important Function in Cells
Polysaccharides			
Starch and glycogen	Glucose	C,H,O	Short-term storage of energy
Cellulose	Glucose	C,H,O	Plant cell walls
Lipids			
Neutral fats	Glycerol + 3 fatty acids	C,H,O	Long-term storage of energy
Phospholipids	Glycerol + fatty acids + phosphate	C,H,O,P	Cell membrane structure
Steroids	Ring compounds + hydrocarbon chain		Various
Proteins			
	Amino acids	C,H,O,N	Enzymes are globular proteins.
Nucleic Acids			
	Nucleotides	C,H,O,N,P	Genes are a nucleic acid.

Figure 3.18
a. Starch and glycogen are polymers of glucose.
b. Glucose is a six-carbon sugar that can be expressed in the ways shown. *c.* Starch granules in a plant cell.
d. Glycogen granules in a liver cell.

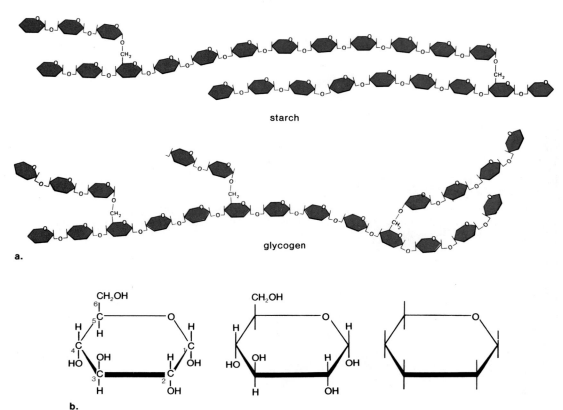

starch

glycogen

a.

b.

Polysaccharides

Polysaccharides (many sugars) function as short-term energy-storage compounds in both plants and animals and as structural compounds in plants.[2] Polysaccharides and the sugar units from which they are constructed belong to the class of compounds called carbohydrates. **Carbohydrates** usually contain carbon and the components of water—hydrogen and oxygen—in a 1:2:1 ratio. The general formula for any carbohydrate is $(CH_2O)_n$; the n subscript stands for whatever number of these groupings there are. For example, the carbohydrate formula $C_6H_{12}O_6 = (CH_2O)_6$.

Starch and Glycogen

Both starch and glycogen (fig. 3.18) are long chains of the six-carbon sugar **glucose.** There are many isomers of glucose; for example, mannose, fructose, and galactose are also six-carbon sugars. But in this text, $C_6H_{12}O_6$ will refer to glucose only, the most biologically important six-carbon sugar.

[2]Cellulose forms plant cell walls, and chitin is a major component of the arthropod exoskeleton. Only cellulose is discussed in this chapter; chitin, which has a similar but more complicated structure, is discussed on page 542.

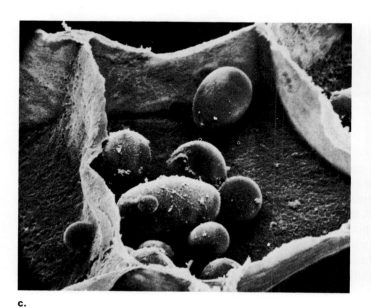

c.

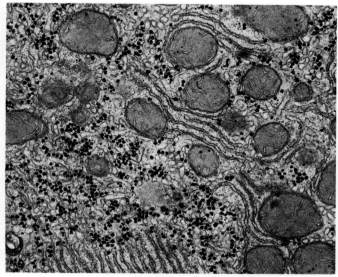

d.

The structures of **starch** and **glycogen** differ only slightly. Glycogen is characterized by many side branches; that is, chains of glucose that go off from the main chain. Starch has few of these chains.

Plant cells store extra carbohydrates as complex sugars or starches. For instance, when leaf cells are actively producing sugar by photosynthesis, they store some of this sugar in the form of starch in their own cells. Sugars are transported to storage organs, especially in roots and modified stems, where they are synthesized to starch in specialized storage cells.

Animal cells store extra carbohydrates as glycogen, sometimes called "animal starch." After a human eats, the liver stores glucose as glycogen. Then, between eatings, the liver releases glucose to keep the normal blood concentration of glucose near 0.1 percent.

Cellulose

Cellulose contains glucose molecules joined together by a slightly different type of linkage than that found in starch and glycogen. The long chains that result are not branched; instead, they are joined together to form units that arrange themselves in fibrils large enough to be seen in a photomicrograph. Layers of

Figure 3.19

a. Cellulose is also a polymer of glucose molecules. Compare the linkage in cellulose to the linkage in starch and glycogen, which is shown in figure 3.18a. b. A portion of a plant cell wall showing the arrangement of cellulose fibrils, each of which contains many cellulose polymers.

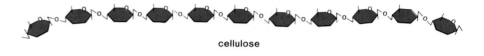

cellulose

a.

b.

Figure 3.20

Cows can eat grass because they have a rumen, a part of the stomach where microorganisms digest cellulose.

Figure 3.21

Structure of glycerol. When glycerol reacts with three fatty acids, the hydrogen atoms indicated are removed.

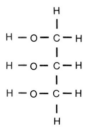

cellulose fibrils make up plant cell walls (fig. 3.19). The cellulose fibrils are parallel within each layer, but the layers themselves lie at angles to one another for added strength.

Most animals cannot digest cellulose because they lack the enzymes needed to break the linkage between the glucose molecules. Cattle and sheep, exceptions to this generalization, can only digest cellulose because they have a special stomach chamber, the rumen, where bacteria break this linkage. It makes sense, then, that cattle should be range fed (fig. 3.20) rather than grain fed. Unfortunately, cattle are often kept in feedlots (fig. 29.10) where they are fed grain. This is ecologically unsound because it requires fossil fuel energy to grow the grain, process it, transport it, and finally feed it to the cattle. This supply of fossil fuel could be saved for other purposes. Also, because range-fed cattle move about, they produce leaner meat than do feedlot-fed cattle. This is another advantage for humans as some nutritionists believe this less fatty meat is more healthful.

Other scientists suggest that cellulose might be used as an energy supply. Special microorganisms would break down indigestible plant material into alcohol; this alcohol would be added to gasoline to produce gasohol. This way, the production of gasohol would not require the use of grain. Presently, yeast fermentation of grain produces much of the alcohol used for gasohol.

Lipids

A variety of organic compounds are classified as **lipids.** Many of these are insoluble in water because they lack ionized atoms and are nonpolar. The most familiar lipids are those found in the neutral fats, such as lard, butter, and oil. Within cells, neutral fats are long-term energy-storage compounds. Phospholipids and steroids are important lipids found in living things. For example, they are components of cell membranes.

Neutral Fats

Neutral fats arise when **glycerol** (fig. 3.21) joins with three **fatty acids** to form a **triglyceride.** Fatty acids consist of a long hydrocarbon chain (one that contains carbon atoms with hydrogens attached) and an acidic **carboxyl group** ($C\!\!\stackrel{\displaystyle O}{\diagdown_{OH}}$) at one end (fig. 3.22).

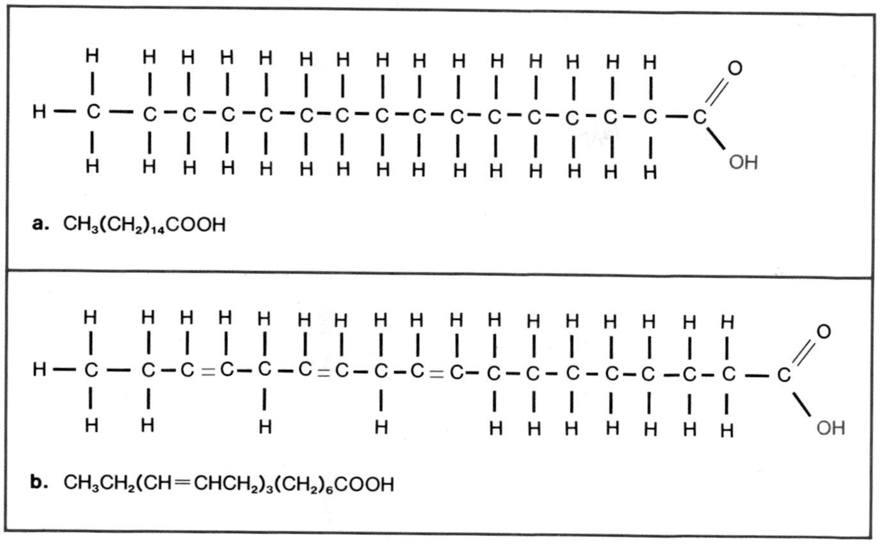

Figure 3.22
a. A saturated fatty acid contains as many hydrogens as possible. It is unnecessary to write out the entire chain; it may be represented by the shortened version given beneath the longer one. *b.* An unsaturated fatty acid does not contain as many hydrogen atoms as possible because some carbon atoms are double-bonded. The OH groups are removed when fatty acids react with glycerol.

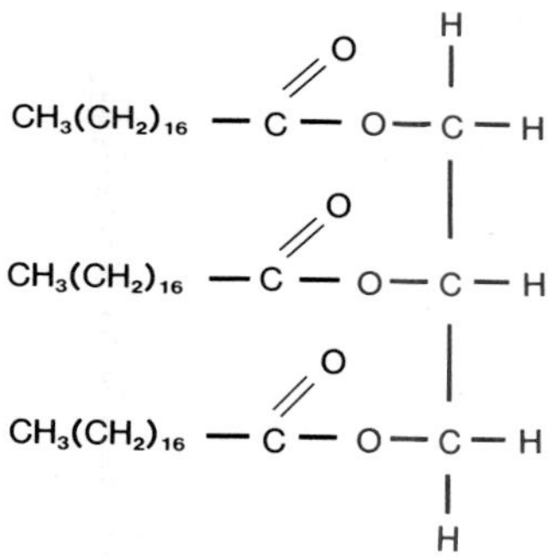

a. $CH_3(CH_2)_{14}COOH$

b. $CH_3CH_2(CH=CHCH_2)_3(CH_2)_6COOH$

Figure 3.23
Structure of a neutral fat. Notice that a neutral fat has no polarity; there are no negatively or positively charged atoms.

Most fatty acids in cells contain 16 to 18 carbon atoms per molecule. Fatty acids are either saturated or unsaturated. The carbon atoms in **saturated fatty acids** have as many hydrogen atoms as possible, but unsaturated fatty acids do not. Instead, **unsaturated fatty acids** have double bonds, thus some of the carbon atoms carry fewer hydrogen atoms than do saturated fatty acids.

Saturated fatty acids are found in lard and butter; unsaturated fatty acids are found most often in vegetable oils, accounting for the liquid nature of these oils. Vegetable oils are hydrogenated (hydrogens are added) to make margarine, but polyunsaturated margarine still contains a large number of unsaturated, or double, bonds. There is some evidence that ingestion of saturated fats contributes to cardiovascular disease, the most common illness in the United States. Therefore, we are often advised, particularly by the advertising media, to consume polyunsaturated fats.

Structure and Function A neutral fat arises when one glycerol molecule condenses with three fatty acids. Figure 3.23 gives the structure of a neutral fat; the portion derived from glycerol is shown in color, and the portion derived from fatty acids is shown in black. Organisms use neutral fats as long-term energy-storage compounds because a lot of energy is made available when fatty acids break down. For example, there are more calories in a pat of butter placed on a potato than there are in the potato itself.

Phospholipids

Phospholipids contain a phosphate group.

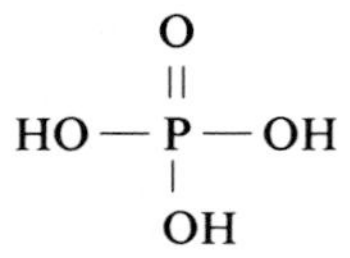

$$\text{HO} - \overset{\displaystyle \overset{\text{O}}{\|}}{\underset{\displaystyle \underset{\text{OH}}{|}}{\text{P}}} - \text{OH}$$

Essentially phospholipids are constructed as are neutral fats, but the third hydrocarbon chain ends with a phosphate group or a group that contains both phosphate and nitrogen. This group can ionize; therefore, this chain forms what is called the *head* of the molecule, while the other two chains are the *tails*. When phospholipid molecules are placed in water, they form a sheet in which the polar heads face outward and the nonpolar tails face each other (fig. 3.24). This property of phospholipids contributes to the structure of membranes.

Steroids

Steroids are lipids that have entirely different structures than neutral fats. Each one contains four fused carbon rings that vary primarily according to the type of carbon chain attached (fig. 3.25). Cholesterol is an important steroid, as are several of the vertebrate hormones, including sex hormones.

Proteins

It would be difficult to overemphasize the importance of proteins in living systems. They have many functions, but we will discuss only a few in this chapter. Later chapters will acquaint you with other protein functions.

Figure 3.24
Structurally, phospholipids have a polar head and nonpolar tails (*left and center*), and in water (*right*) the molecules arrange themselves as shown. The polar heads are attracted to water; the nonpolar tails are not.

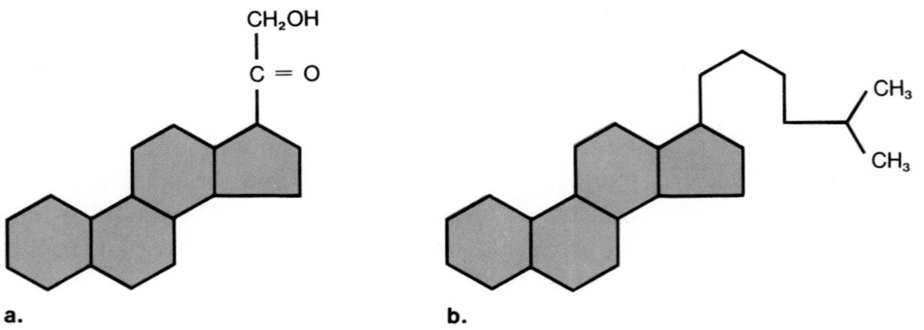

Figure 3.25
All steroids have four fused carbon rings, but the attached carbon chain is different in each one. *a*. This attached chain is found in cortisone. *b*. This attached chain is found in cholesterol.

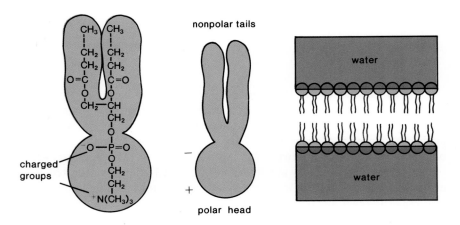

a.

b.

Proteins are polymers made up of **amino acids** linked together. An amino acid joined to an amino acid joined to an amino acid, and so forth, results in a protein. Proteins range from 50 to 500 amino acids long, thus proteins are indeed macromolecular. About twenty different common amino acids are found in cellular proteins. Proteins differ one from the other because the order of these amino acids differs. A certain protein always has exactly the same sequence of amino acids. This remarkable constancy is absolutely necessary to the particular structure and function of a protein.

Amino acids have an **amino group** (NH_2) and an **acidic carboxyl group** (COOH) bonded to a carbon atom (fig. 3.26), which also bonds to an "R" group, the Remainder of the molecule. "R" groups range in complexity from a single hydrogen to complicated ring compounds. The twenty different amino acids in cells differ from one another by their "R" groups.

A **peptide** is two or more amino acids joined together, and a **polypeptide** is many amino acids joined to form a long chain of amino acids. Figure 3.27a shows that two amino acids are joined by a condensation reaction between the acid group of one and the amino group of another. The resulting covalent bond between two amino acids is called a **peptide bond.** There are many peptide bonds in a polypeptide (fig. 3.27b). The atoms associated with the peptide bond share the electrons unevenly because the oxygen is electronegative, making the hydrogen electropositive. Hydrogen bonding, represented by the dotted lines in figure 3.27b, is therefore a frequent occurrence in polypeptides.

Figure 3.26
All amino acids have an amino and an acidic carboxyl group, but the "R" group can vary. There are approximately twenty different amino acids; therefore, there are approximately twenty different "R" groups. Two different ways to indicate amino acid structure are shown in (a.) and (b.).

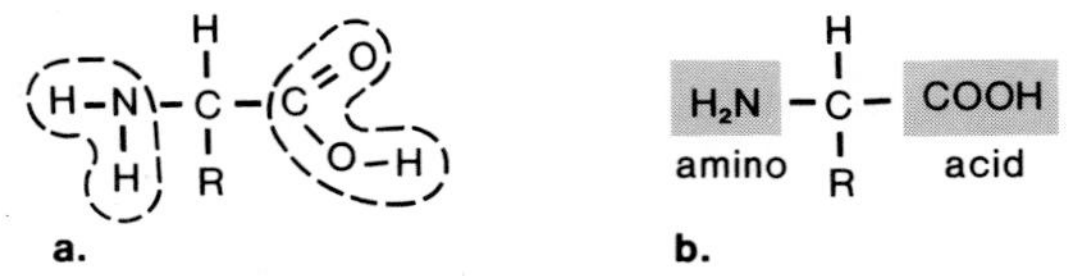

a.

b.

R = remainder of molecule

Figure 3.27
a. When one amino acid joins another, a condensation reaction occurs between the carboxyl group of the first and the amino group of the second. The resulting bond is called a peptide bond, and the resulting molecule is a peptide. b. A polypeptide chain has a backbone of repeating N—C—C—N atoms and side chains of "R" groups. Hydrogen bonding (dotted lines) between peptide linkages is common.

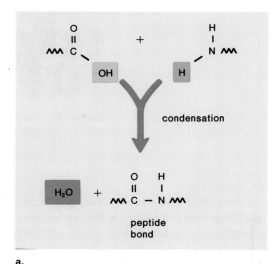

a.

b.

Figure 3.28
Proteins have four possible levels of structure. The primary structure is the particular sequence of the amino acids. The secondary structure is often an alpha helix. A bending and twisting of the coiled polypeptide chain gives the tertiary structure. A quaternary structure is possible for proteins that have more than one type polypeptide chain.

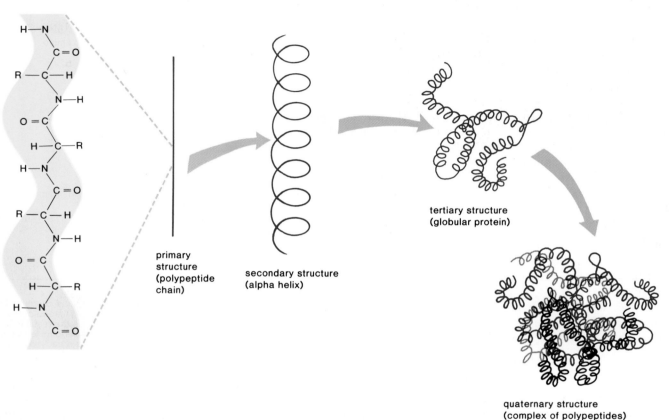

primary
structure
(polypeptide
chain)

secondary structure
(alpha helix)

tertiary structure
(globular protein)

quaternary structure
(complex of polypeptides)

Levels of Protein Structure

Proteins commonly have three levels of structure (fig. 3.28), although some have a fourth level. The **primary structure** is the sequence of the amino acids joined together by peptide bonds. Since these amino acids differ only by their "R" groups, it is correct to say that proteins differ from one another by the particular sequence of the "R" groups.

The **secondary structure** of a protein comes about when the polypeptide chain takes a particular orientation in space. One common arrangement of the chain is the alpha helix, or right-handed coil, with 3.6 amino acids per turn. Hydrogen bonding between amino acids, in particular, stabilizes the alpha helix.

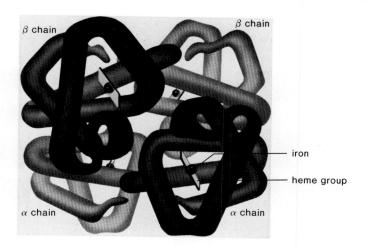

β chain   β chain

iron

heme group

α chain   α chain

Figure 3.29
The hemoglobin molecule is a globular protein that contains four polypeptide chains (2α and 2β chains arranged as shown). Each chain is attached to a nonprotein heme group represented as a plane. The small ball in each plane represents an atom of iron.

The **tertiary structure** of a protein is dependent on the final arrangement of the polypeptide chain. In **fibrous proteins** such as keratin the chains are long and stiff. Keratin makes up the nails, claws, and hair of animals. In **globular proteins** the helix bends and twists into its final shape. This shape is dependent on various interactions between the "R" groups.

Some proteins have more than one type of polypeptide chain, each with its own primary, secondary, and tertiary structures. These separate chains are arranged and bonded to each other in specific ways, yielding a more complex level of protein structure termed the **quaternary structure.** Hemoglobin is a good example of protein that has both a tertiary and quaternary structure (fig. 3.29).

Function of Proteins

Proteins can have a structural or metabolic function or a combination of both. Myosin, one of the main proteins in muscle (fig. 3.30), has such a dual function as will be discussed in chapter 24. Many globular proteins are **enzymes,** which are organic catalysts that speed up chemical reactions in organisms. Enzymes make "warm chemistry" possible. In a chemical laboratory it is often necessary to heat a reaction flask to bring about a reaction; in only slightly warm cells, chemical reactions take place very quickly because a specific enzyme is present for each and every reaction. Enzymes are discussed more fully in chapter 5.

Nucleic Acids

Nucleic acids are macromolecules that control the structure and functions of a cell. **DNA (deoxyribonucleic acid)** makes up the hereditary factors called genes, and **RNA (ribonucleic acid)** works with DNA to control protein synthesis. The structure and functions of these molecules will be discussed in detail in chapter 10; here we discuss them only briefly.

Figure 3.30
Most athletes have well-developed muscle cells containing many protein molecules. A major controversy today is the use of steroids to promote muscle buildup.

Figure 3.31

Generalized nucleotides. Each nucleotide contains phosphate, a pentose sugar, and a nitrogenous base. The base may have a double ring, as shown in (a.) or a single ring, as shown in (b.).

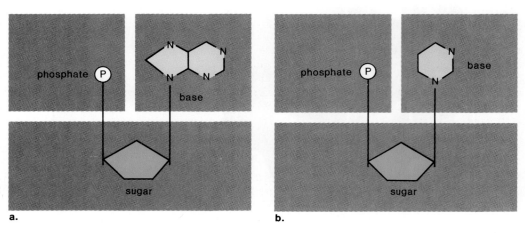

a.

b.

Figure 3.32

Polymer of nucleotides. When nucleotides join, the sugar and phosphate molecules form a backbone and the bases project to the side.

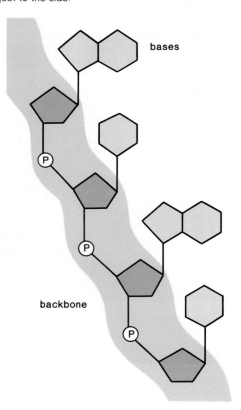

Figure 3.33

ATP is a nucleotide with three phosphate units; two of the phosphate bonds are high-energy bonds, indicated by wavy lines.

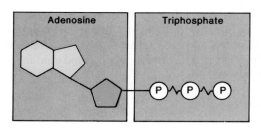

Nucleic acids are polymers of nucleotides. Every **nucleotide** is a molecular complex having three parts: phosphate (or phosphoric acid), a five-carbon (pentose) sugar, and a nitrogenous base. The base has either a single ring or a double ring. Figure 3.31 shows generalized nucleotides because no specific type of single- or double-ringed base is shown. The phosphate is represented simply as Ⓟ .

When nucleotides join together, they form a chain in which the backbone is made up of phosphate-sugar-phosphate-sugar, etc., with the bases projecting to one side of the backbone (fig. 3.32). Such a chain is called a *strand*. The nucleic acid DNA is double-stranded; the two strands are held together by hydrogen bonding between the bases (fig. 10.6). RNA is single-stranded.

ATP

ATP (adenosine triphosphate) is a nucleotide that is a carrier of energy in cells (p. 93). ATP contains the five-carbon *sugar* ribose, three phosphate groups (triphosphate) instead of one, and a base called adenine. It is customary to draw the molecule with the three phosphate groups on the right (fig. 3.33). ATP is the energy molecule because the triphosphate unit contains two high-energy bonds, represented in figure 3.33 by wavy lines. When these bonds are broken, energy is made available to cells.

Summary

All matter is composed of atoms, which contain the subunits protons, neutrons, and electrons. The atomic number signifies the number of protons (having a plus charge) located in the nucleus and the number of electrons (having a negative charge) located in orbitals that form the shells of the atom. The number of neutrons (bearing no charge) located in the nucleus can vary, resulting in isotopes, some of which are radioactive.

Atoms react or form bonds with one another to achieve a full complement of electrons (usually eight) in the outer shell. In ionic reactions there is a transfer of electrons from one atom to another. The atom that gives away electrons becomes positively charged, while the atom that receives electrons becomes negatively charged. The ionic bond is the attraction that ions have for one another. In covalent reactions the atoms share electrons because they require additional electrons to fill their outer shells. The covalent bond is represented by a short line drawn between atoms.

Water contains atoms that are covalently bonded, but since the oxygen atom attracts electrons more strongly than the two hydrogen atoms do, the molecule is polar. Further, the electropositive hydrogen atoms are attracted to electronegative oxygen atoms of other water molecules. This causes water molecules to be hydrogen-bonded to one another. Polarity and hydrogen bonding account for the properties of water. Water facilitates chemical reactions and is an excellent transport medium. The temperature of water tends to remain stable, and a large amount of heat is required to make water vaporize. Ice is less dense than liquid water, thus a pond freezes from the top down.

When water dissociates, an equal number of H^+ and OH^- ions result. On the pH scale, the pH of water is 7; this is neutral pH. Acids increase the $[H^+]$ and produce a lower pH. Bases increase the $[OH^-]$ and produce a higher pH. Living things are sensitive to pH changes and contain buffers that usually keep the pH constant.

The chemistry of carbon accounts for the molecules unique to living things. Carbon compounds have variety because carbon can bond to as many as four other atoms, forming long chains or ring compounds. The macromolecular life molecules (polysaccharides, lipids, proteins, and nucleic acids) are made up of smaller molecules joined together following condensation reactions. Macromolecular compounds can be broken down by hydrolysis reactions.

Polysaccharides are formed when many molecules of glucose or other sugars join together. Starch in plants and glycogen in animals are used for short-term energy storage. Cellulose makes up plant cell walls.

Lipids are usually insoluble in water. Neutral fats hydrolyze to give a glycerol and three fatty acid molecules. Fatty acids are saturated, as in butter, or unsaturated as in oils. Phospholipids have a polar head that is soluble in water and two nonpolar tails (hydrocarbon chains) that are not.

Protein molecules have at least three levels of structure. The primary level is the sequence of amino acids joined by peptide bonds. The secondary level is often an alpha helix, a right-handed coil held in place by hydrogen bonding. In globular proteins (as are enzymes), the tertiary structure occurs when the helix twists and turns on itself. Hemoglobin illustrates that a protein sometimes contains more than one type of polypeptide. Such a protein has a quaternary structure.

Nucleic acids are made up of nucleotides bonded so that phosphate and sugar molecules make a backbone and the bases project to the side. DNA and RNA are both nucleic acids. ATP is an important nucleotide that has three phosphate groups and two high-energy phosphate bonds.

Objective Questions

1. All matter is composed of basic substances called _____ .
2. The subatomic particles, called _____ , move about the nucleus in pathways called orbitals.
3. The two primary types of bonding are _____ and _____ .
4. Water molecules are held together by a special type of bonding, termed _____ bonding.
5. Acids contain more _____ ions than bases, but they have a _____ pH.
6. Starch, glycogen, and cellulose are all _____ , or polymers of glucose molecules.
7. A neutral fat hydrolyzes to give one _____ molecule and three _____ molecules.
8. The primary structure of a protein is the sequence of _____ ; the secondary structure is very often an alpha _____ ; the tertiary structure is the final _____ of the protein.
9. All _____ are proteins and function to speed up chemical reactions.
10. A nucleic acid is made up of _____ joined together.

Study Questions

1. Name all subatomic particles studied; tell their charges and locations in an atom.
2. Draw an atomic structure for carbon having six protons and six neutrons.
3. Draw an atomic representation for the molecule $Mg^{+2}Cl_2^-$. Using the octet rule, explain the structure of this compound.
4. Tell whether CO_2 (O—C—O) is an ionic or covalent compound. Why does this arrangement satisfy all atoms involved?
5. Name five properties of water and relate them to the structure of water, including its polarity and hydrogen bonding between molecules.
6. Name the types of macromolecules important in living cells and state the unit molecule(s) for each.
7. Name three different polysaccharides and state a function for each.
8. Name three types of lipids and state a function for each.
9. Explain how a phospholipid molecule differs from a neutral fat.
10. Describe the four levels of protein structure. Give an example of a fibrous protein and a globular protein.
11. Name two nucleic acids important in cell structure and function. Name the important nucleotide that is the "energy molecule" in cells.

Selected Key Terms

element ('el ə mənt) 39
atom ('at əm) 39
isotope ('ī sə ‚tōp) 40
molecule ('mäl i kyül) 41
compound (käm 'paủnd) 41
hydrolysis (hī 'dräl ə səs) 52
carbohydrate (‚kär bō 'hī ‚drāt) 54
lipid ('lip əd) 56

fatty acid ('fat ē 'as əd) 56
steroid ('stiər ‚òid) 58
protein ('prō ‚tēn) 59
amino acid (ə 'mē ‚nō 'as əd) 59
deoxyribonucleic acid ('dē 'äk si‚rī bō nủ klē ik 'as əd) 61
ribonucleic acid (‚rī bō nủ klē ik 'as əd) 61
nucleotide ('nü klē ə ‚tīd) 62

Figure 4.1
Using this compound microscope that he built himself (*a.*),
Robert Hooke was able to view dead cork cells (*b.*).

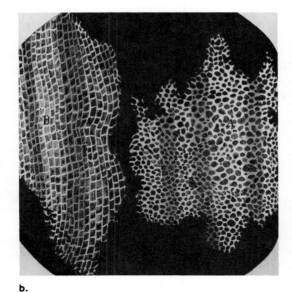

a.   **b.**

CHAPTER 4

Cells
Their Structure and Function

Concepts

1. All organisms are composed of cells, units of structure and function.
   a. Every cell is surrounded by a cell membrane. Membranes also make up various organelles.
   b. The cell membrane regulates the entrance of molecules into and the exit of molecules out of the cell.

2. Increase in complexity often accompanies the evolutionary process.
   a. Eukaryotic cells contain various organelles that are not found in prokaryotic cells.
   b. In eukaryotic cells, endoplasmic reticulum, the Golgi body, and vesicles, including lysosomes, are membranous organelles that function in the production, transportation, and secretion of cell products.
   c. Microfilaments and microtubules, cilia and flagella, and centrioles are concerned with movement within the cell and/or movement of the cell itself.
   d. Eukaryotic cells perhaps evolved when small prokaryotes became residents in large prokaryotes. The former have become organelles.

T o understand how organisms grow, reproduce, and interact with their environments, it is necessary to learn about cell structure and function. Just as matter is composed of atoms or groups of atoms, living things consist of cells, or aggregates of cells, and cell products.

The study of cells did not begin until the development of **light microscopes** in the seventeenth century. A light microscope magnifies the image of an object when rays of light pass through the object and are focused by one or more of the microscope's lenses. Anton van Leeuwenhoek is famous for having observed tiny one-celled living things that no one had seen before, but it was Robert Hooke that first identified cells within a tissue (fig. 4.1). The tiny chambers he observed in the honeycomb structure of cork reminded him of the rooms or *cells* in a monestary. Naturally, then, he referred to the boundaries of these chambers as *walls.*

Although these early microscopists had seen cells, it was more than a hundred years later that Matthias Schleiden published his theory that all plants are composed of cells, and Theodore Schwann published a similar proposal concerning animals. Soon thereafter, the German Rudolph Virchow stated that "every cell comes from a preexisting cell." By the middle of the nineteenth century, biologists clearly recognized the fundamental tenet of the **cell theory:** all organisms are composed of self-reproducing elements called cells.

Modern Microscopes

Even after the formulation of the cell theory, little was known about the fine details of cell structure until the advent of the **transmission electron microscope** in this century. Although modern light microscopes are useful for certain studies, their use in cell research is limited. The problem is not magnification; it is **resolution,** or the ability to discern close objects. A light

Table 4.1 Measurements Used in Cell Biology*

Unit	Symbol	Seen by
Centimeter	cm = 0.4 inch	Naked eye
Millimeter	mm = 0.1 cm	Naked eye
Micrometer	μm = 0.001 mm	Light microscope
Nanometer	nm = 0.001 μm	Electron microscope

*Cell structures are so small they are measured by divisions of the metric system not used in everyday life.

Figure 4.2
This 3-D image of a kidney glomerulus was produced by a scanning electron microscope and a computer. The microscope scans the object from various angles and these views are stored in the computer. Then they are reassembled by the computer to give a lifelike picture on a TV screen.

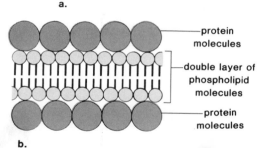

Figure 4.3
When adjoining cell membranes were first viewed with the electron microscope (a.) investigators formulated the unit membrane model of membrane structure, which is diagrammatically illustrated in (b.). This hypothesis proposed that membrane protein molecules formed a layer above and below the lipid bilayer. Later this hypothesis was replaced by the fluid mosaic model illustrated in figure 4.4.

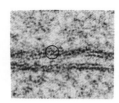

a.

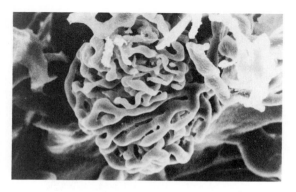

protein molecules

double layer of phospholipid molecules

protein molecules

b.

microscope achieves a resolving power and produces a magnification that allows the human eye to see objects that are 0.2 μm apart (table 4.1). This is an improvement of about 500 times over the capability of the human eye, but even greater resolution is required to make out the fine details of cells.

The electron microscope has a greater resolving power than the light microscope because it uses electrons, not light, to create an image of the object. Electrons moving through space have a shorter wavelength than light waves. As a result, electrons achieve a greater resolving power. Using an electron microscope, it is possible to project on a screen, and so view, objects that are as close as 0.5nm apart. Thus, the electron microscope has a resolving power about 200,000 times greater than that of the human eye.

When biological material is prepared for viewing with the light or electron microscope, it is often fixed (preserved), sliced thinly, and treated to enhance the contrast between light and dark areas. The exact procedures differ, but the result is that lighter areas tend to permit light or electron waves to pass through, while the darker areas tend to deflect or absorb them. Unfortunately, the possibility always exists that preparation procedures may distort the resulting image.

The **scanning electron microscope** often views only the surface of an object. A beam of electrons scans the object, detecting surface variations that cause the beam electrons to scatter and, in turn, emit secondary electrons. Scattered and secondary electrons are then used to produce an image on a television screen. Here, too, the specimen is treated: the surface is coated with a substance like gold, which is a good emitter of secondary electrons.

Micrographs are photographs of materials as viewed with microscopes (fig. 4.2). The appearance of micrographs varies depending on the microscope in use.

Membrane

The transmission electron microscope has added greatly to our knowledge of membrane, which cannot be viewed in detail with the light microscope. **Membrane** encloses a cell and at times forms various structures within cells. Here we will concentrate on cell (plasma) membrane structure and function, although the discussion applies in general to all membrane.

Cell Membrane Structure

When the entire membrane is viewed with the transmission electron microscope, it looks somewhat like a sandwich about 7.5 to 8.0 nm thick: two dark layers are separated by a lighter central area (fig. 4.3). Based on biochemical and biophysical research, investigators of the 1930s believed that the two outer layers consisted of sheets of protein extending over an inner bilayer of phospholipid molecules (p. 58), orientated with their polar heads pointed outward and their nonpolar tails pointed inward. Since it was believed that all membranes had this basic structure, the model was termed the **unit membrane model.**

Later, **freeze-fracture technique** showed that membrane has a more complex and changeable structure. When cells are frozen and then fractured, the fracture splits the membrane in two, exposing its interior. Electron microscopy reveals the presence of globular particles within the membrane. In contrast to the unit membrane model, it now seems that most proteins actually lie within the lipid bilayer (fig. 4.4). There are some proteins that lie just inside the cell and these are apparently attached to those that are embedded in the membrane. This model is called the **fluid mosaic model.** The proteins within

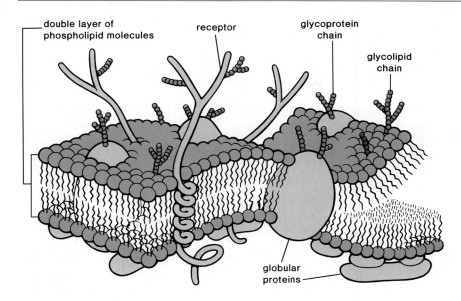

double layer of phospholipid molecules

receptor

glycoprotein chain

glycolipid chain

globular proteins

Figure 4.4
Fluid mosaic model of a cell membrane. Protein molecules are embedded in and project to either side of a double layer of phospholipid molecules. Note that some of these are receptors that specifically bind with certain molecules acting as chemical messengers, thus influencing the activity of the cell. The right side of the illustration shows the effect of fracturing a frozen membrane.

the membrane create a constantly changing pattern as they move laterally within the lipid bilayer, which has the consistency of light oil. Different types of proteins are found in the membranes of different cells. Some of these are carriers that promote the movement of molecules into and out of the cell, and some are receptors for molecules that influence the activity of the cell.

Short chains of sugars are attached to the outer surface of some protein and lipid molecules. There is evidence that these **glycoproteins** and **glycolipids** allow cells to recognize one another and/or a cell as belonging to a particular individual. If this is the case, their presence may explain in part why a patient's system sometimes rejects an organ transplant.

Cell Membrane Function

The cell membrane forms a boundary between the outside of the cell and the inside of the cell, often called the **cytoplasm.** The cell membrane allows only certain molecules to enter and exit the cytoplasm freely; therefore, the cell membrane is said to be **selectively permeable.** In general, small molecules have less difficulty crossing a membrane than do large molecules. Noncharged molecules, particularly some lipids, easily move through the membrane, but charged (polar) molecules often are repelled by like charges in the membrane.

Some small molecules pass through the membrane by diffusion. **Diffusion** is the movement of molecules from the area of greater concentration to the area of lesser concentration. This process can be demonstrated by placing a tablet of dye in a water-filled container (fig. 4.5). As the tablet dissolves, the dye will move slowly away from the tablet into the surrounding water until the water is colored evenly. Diffusion is a physical process that occurs whenever there is a concentration gradient, that is, a decreasing amount of a substance from one area to another. Molecules, such as gases (CO_2 and O_2) and water, that freely pass through a membrane do so by simple diffusion.

Osmosis

Osmosis is the diffusion of water across a cell membrane. It occurs whenever there is an unequal concentration of water on either side of a selectively permeable membrane. When discussing osmosis, a solution is considered in

Figure 4.5
Diffusion demonstration. After a tablet of dye is placed in a beaker of water, the water is colored as the dye moves away from the original area of the tablet.

Figure 4.6
Osmosis demonstration. *a.* A thistle tube covered at the broad end by a selectively permeable membrane contains a solute and a solvent. The beaker contains only solvent. *b.* The solute is unable to pass through the membrane, but the solvent passes through in both directions. *c.* There is a net movement of solvent toward the inside of the thistle tube. This causes the solution to rise in the thistle tube until a back pressure develops, preventing any further net gain of solvent.

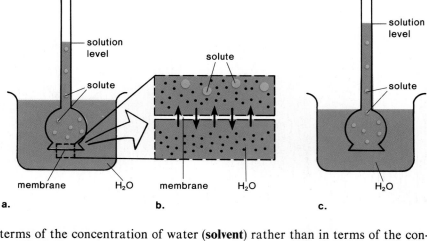

terms of the concentration of water (**solvent**) rather than in terms of the concentration of substance (**solute**). For example, figure 4.6*a* represents a thistle tube covered with a selectively permeable membrane. The tube contains a protein suspension, and the beaker contains distilled water. Because of the presence of protein (solute), there is a lesser concentration of water (solvent) inside the tube than there is outside the tube. Because the membrane is selectively permeable, water can cross the membrane but protein cannot. Under these conditions there will be a net movement of water to the inside of the tube: some water molecules will exit from the tube but more will enter the tube; that is, water will move from the area of greater concentration to the area of lesser concentration (fig. 4.6*b*). Theoretically, water will continue to enter until there is an equal concentration of water on both sides of the membrane. In effect, however, once water enters, a "back pressure" builds up that prevents any further net gain of water. Water molecules now enter and exit the thistle tube in about equal numbers. At this time, it is possible to determine the amount of **osmotic pressure,** the amount of force that must be exerted to prevent a net flow of water across a selectively permeable membrane.

Since cytoplasm contains proteins and salts in addition to water, cells exhibit osmotic properties. Osmosis is of great importance to cells because it can cause them to either gain or lose water to the extent that their existence is threatened. We will examine osmosis in both animal and plant cells.

Osmosis in Animal Cells When animal cells, such as red blood cells (table 4.2; fig. 4.7), are placed in an **isotonic** solution (equal concentration of water on both sides of the membrane), there is no net movement of water and the cells retain their normal appearance. When red cells are placed in a **hypertonic** solution (lesser concentration of water outside the cell), the cells shrink because water molecules have left the cell. When red cells are placed in a **hypotonic** solution (greater concentration of water outside the cell), the cells expand to bursting because water has entered the cell.

Osmosis in Plant Cells A plant cell exhibits osmotic properties that are more difficult to detect because in addition to a cell membrane a cell wall surrounds the cell. The **cell wall,** composed largely of cellulose (p. 55), lies external to the cell membrane and maintains the plant cell's shape and rigidity (fig. 4.8). This cell wall is porous, allowing all types of molecules to pass through freely; the rest of the plant cell, however, responds to hypertonic and hypotonic solutions.

Table 4.2 Effect of Osmosis on Cells

Tonicity of Solution	Description	Cells
Isotonic*	Equal concentration of water on both sides of cell membrane	No change in appearance
Hypertonic*	Lesser water and greater solute concentration outside the cell than inside	Shrink; develop plasmolysis
Hypotonic*	Greater water and lesser solute concentration outside the cell than inside	Swell; develop turgor pressure

*Iso means "same as," *hyper* means "more than," and *hypo* means "less than."

Figure 4.7
Effect of tonicity on red blood cells. *a.* Cells appear normal in an isotonic solution. *b.* Cells shrink in a hypertonic solution. *c.* Cells swell in a hypotonic solution.

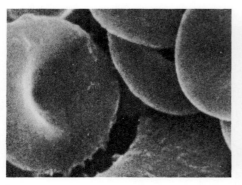

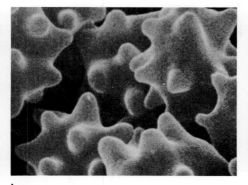

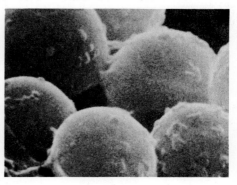

a.

b.

c.

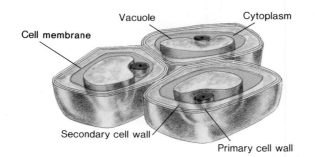

Cell membrane

Vacuole

Cytoplasm

Secondary cell wall

Primary cell wall

Figure 4.8
All plant cells are surrounded by a primary cell wall in addition to the cell membrane. Sometimes, as shown here, the cells produce a secondary wall that lies inside the primary wall. The secondary wall is impregnated with lignin, a hardening material.

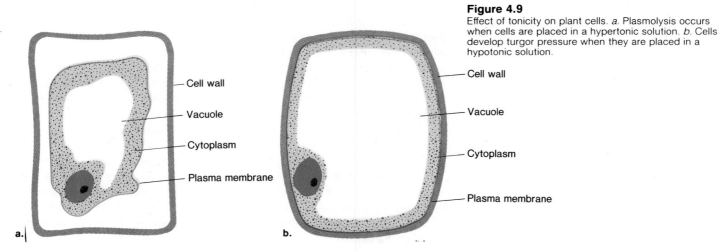

Cell wall

Vacuole

Cytoplasm

Plasma membrane

Cell wall

Vacuole

Cytoplasm

Plasma membrane

a.

b.

Figure 4.9
Effect of tonicity on plant cells. *a.* Plasmolysis occurs when cells are placed in a hypertonic solution. *b.* Cells develop turgor pressure when they are placed in a hypotonic solution.

When plant cells are placed in a hypertonic solution, a large central vacuole loses water and becomes smaller (fig. 4.9). The cell membrane pulls away from the cell wall; the plant cell is now said to have undergone **plasmolysis.** When plant cells are placed in a hypotonic solution, the large central vacuole gains water; the cell membrane pushes against the cell wall, which prevents the cell from bursting. The plant cell is now said to exhibit turgor pressure. **Turgor pressure** keeps a nonwoody plant in its upright position. In the absence of turgor pressure a plant will wilt.

Figure 4.10
Facilitated transport. A carrier protein speeds the rate at which the solute crosses a membrane in the direction of decreasing concentration. Facilitated diffusion occurs only when there is a concentration gradient across the membrane.

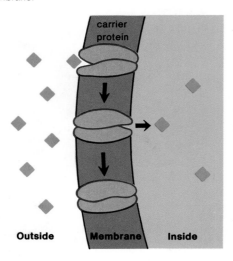

Figure 4.11
Active transport. Active transport occurs when cells collect and concentrate ions or molecules. The carrier protein transports the solute against its concentration gradient. Thus, active transport requires an expenditure of energy.

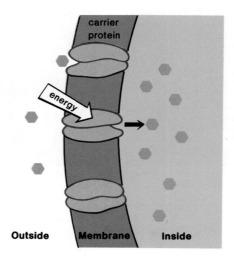

Transport

Most solutes (ions and molecules other than water) do not simply diffuse across a cell membrane; rather, they are transported by carriers. **Carriers** are membrane proteins (or a group of proteins) that are *specific* because they bind only to the particular molecule they transport. Two types of transport are believed to occur in cells: facilitated transport and active transport. In **facilitated transport** the carrier functions to increase the rate with which a specific solute molecule can enter or exit the cell by diffusion. Facilitated transport (fig. 4.10) is involved in carrying various sugars, amino acids, and nucleotides into or out of a cell according to concentration gradients. Because the solute is still traveling down its concentration gradient (from a greater to a lesser concentration), no significant energy input is required.

In **active transport** (fig. 4.11) a solute is moved across a membrane *against* its concentration gradient (from a lesser to a greater concentration). This is the mechanism by which a solute is concentrated either inside or outside a cell. For example, in humans iodine collects in the cells of the thyroid gland; sugar is completely absorbed from the gut by the cells lining the digestive tract; and sometimes sodium (Na^+) is almost completely withdrawn from urine by cells lining the kidney tubules. As described in the next reading, when the chambered nautilus achieves buoyancy by emptying its shell chambers, active transport precedes osmosis.

Active transport requires considerable energy because, in contrast to diffusion, a solute is moving from the area of lesser concentration to the area of greater concentration. Energy may be used to induce a change in the shape of the carrier protein(s) so the molecule can be transported to the other side of the membrane, where it is released. The cycle would end when the carrier protein(s) return(s) to its/their original shape.

Chambered Nautilus

The chambered nautilus, a mollusk that takes its name from its multichambered shell, still occurs today in deep tropical oceans, but was much more prevalent some 65 million years ago. Along with closely related forms, it dominated the oceans. The advantage these animals must have had was the ability to achieve buoyancy. Thus they were no longer confined to the ocean floor. Recently, scientists discovered that as the nautilus grows, it empties liquid from each newly formed chamber of its spiral shell. This makes it light enough to float in water.

Research has been conducted into the puzzle of how the nautilus empties its chambers. A complex tubular organ spirals in the same manner as the shell. This organ contains blood vessels. Water from each newly formed chamber makes its way to these blood vessels, but by what means? The blood pressure within these vessels should cause water to *leave* the blood vessels, not enter them!

Electron micrographs have provided a possible explanation. They show that the wall of the tubular organ has a folded tissue lining containing many small grooves (*a*). If the cells making up this tissue use active transport to carry solutes from the chamber liquid to the grooves, this would establish a concentration gradient so that water would follow passively by osmosis. This process is called *local osmosis* because it depends on the buildup of solutes by active transport in an isolated area, such as the grooves in question. The cells about these grooves do indeed contain many mitochondria (p. 78) that could supply the energy needed for active transport of the solute into the grooves (*b*.) so that water could follow passively by means of osmosis. After being withdrawn, the water moves from the grooves of the tubular organ to collecting channels that funnel it to the blood vessels.

Although the nautilus can empty its chambers, it never attempts to fill them. This observation is consistent with the hypothesis that local osmosis accounts for how the chambered nautilus empties the chambers of its shell.

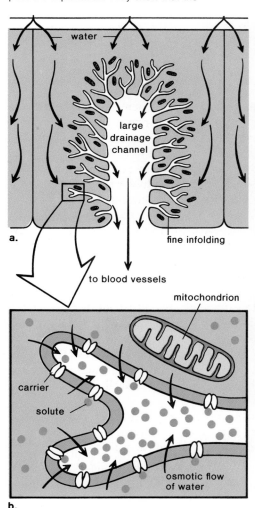

a.

b.

c.

a. *Folded tissue of the tubular organ that spirals within shell of the chambered nautilus.* b. *Enlargement of portion of (a.).* c. *Chambered nautilus.*

Figure 4.12
a. Animal cell structure. *b.* Plant cell structure.

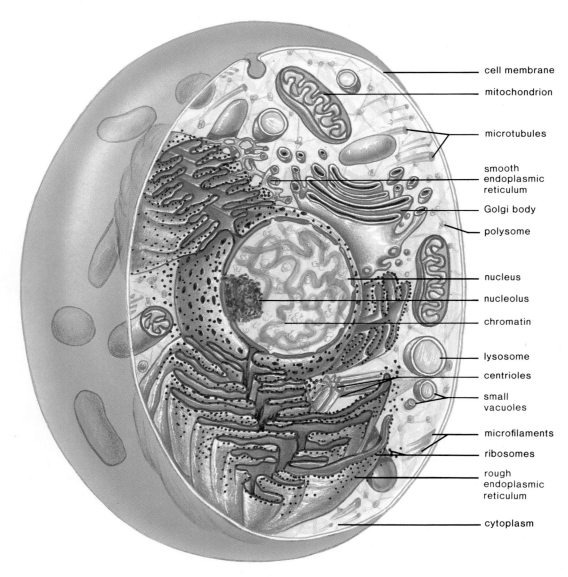

cell membrane

mitochondrion

microtubules

smooth endoplasmic reticulum

Golgi body

polysome

nucleus

nucleolus

chromatin

lysosome

centrioles

small vacuoles

microfilaments

ribosomes

rough endoplasmic reticulum

cytoplasm

a. Animal cell

Eukaryotic Cells

All cells are placed in one of two classes. Bacteria and cyanobacteria are **prokaryotic,** meaning before the nucleus. All other cells—the cells of plants, animals, protists, and fungi—are **eukaryotic,** meaning true nucleus (fig. 4.12). The two types of cells can be distinguished partly by the amount of membrane

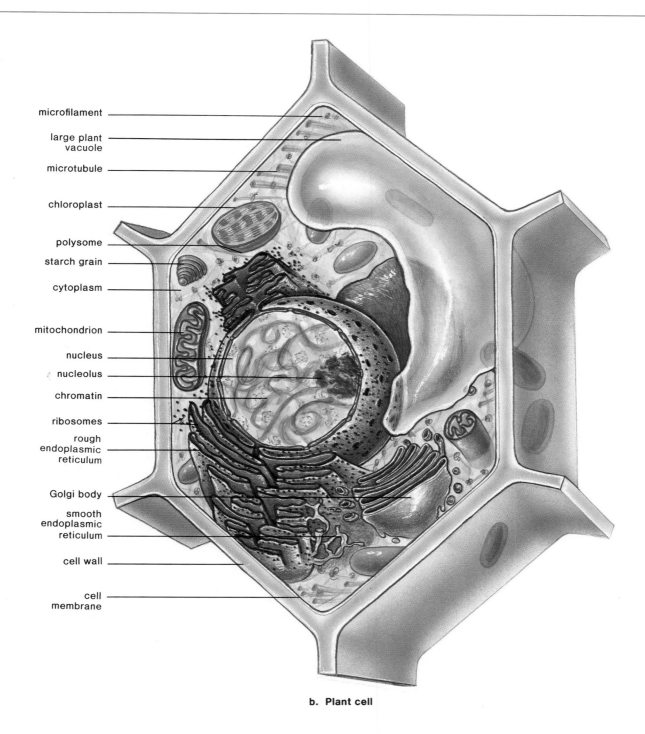

microfilament

large plant
vacuole

microtubule

chloroplast

polysome

starch grain

cytoplasm

mitochondrion

nucleus

nucleolus

chromatin

ribosomes

rough
endoplasmic
reticulum

Golgi body

smooth
endoplasmic
reticulum

cell wall

cell
membrane

b. Plant cell

they contain. For example, prokaryotic cells have nuclear regions that are not surrounded by membrane, whereas eukaryotic cells have a **nucleus,** a nuclear region enclosed by a double layer of membrane called the **nuclear envelope.** Similarly, prokaryotic cells lack the other membranous organelles that will be discussed in this chapter.

Organelles (table 4.3) are small structures, often membranous, that perform specific functions so cells can grow and reproduce. The nucleus is concerned with both these processes. In order to grow, each cell must be able to synthesize and even break down, when necessary, the macromolecules that make up the cell. The first set of organelles discussed in the following are involved with these functions. Cells require energy to synthesize macromolecules and to carry on other activities, such as active transport. The second set of organelles is concerned with energy supply to cells. The third set of organelles specializes in maintaining cell shape and in assisting cellular movement.

Nucleus

The nucleus is the largest organelle in the cell (fig. 4.12). The outer layer of the nuclear envelope is continuous with the membrane making up the endoplasmic reticulum. As discussed in chapter 7, the DNA-containing chromosomes are found within the nucleus as are RNA-containing nucleoli (singular; nucleolus). Because DNA and RNA are involved in controlling protein synthesis, including enzyme synthesis, the nucleus is the **control center** of the cell. DNA, the genetic material, can also reproduce itself and mutate. During cellular reproduction, a copy of the genetic material is passed on to each new cell. During organismal reproduction, the genetic material is passed on to the offspring, some of whom may have inherited a mutation. Such mutations are the raw material for the process of evolution.

Table 4.3 Eukaryotic Organelles (Simplified)

Name	Structure	Function
Cell membrane	Bilayer of phospholipid and globular proteins	Passage of molecules into and out of cell
Nucleus	Nuclear envelope surrounding Chromosomes (DNA) and nucleoli	Cellular reproduction and control of protein synthesis
Nucleolus	Concentrated area of RNA in the nucleus	Ribosome formation
Endoplasmic reticulum	Folds of membrane forming flattened channels and tubular canals	Transport by means of vesicles
Rough	Studded with ribosomes	Protein synthesis
Smooth	Having no ribosomes	Lipid synthesis
Ribosome	Protein and RNA in two subunits	Protein synthesis
Golgi body	Stack of membranous saccules	Packaging and secretion
Vacuole and vesicle	Membranous sacs	Containers of material
Lysosome	Membranous container of hydrolytic enzymes	Intracellular digestion
Mitochondrion	Inner membrane (cristae) within outer membrane	Cellular respiration
Chloroplast	Inner membrane (grana) within outer membrane	Photosynthesis
Microfilament	Actin or myosin proteins	Movement and shape of cell
Microtubule	Tubulin protein	
Centriole	9 + 0 pattern of microtubules	Organization of microtubules
Cilium and flagellum	9 + 2 pattern of microtubules	Movement of cell

Membranous Canals and Vacuoles

Endoplasmic Reticulum

Within the cytoplasm and extending to the nuclear envelope is a ramifying system of flattened channels and tubular canals known as the **endoplasmic reticulum (ER).** Some portions of ER, termed **rough ER** (fig. 4.13), are studded with ribosomes, and some portions, termed **smooth ER,** lack ribosomes. **Ribosomes** are the sites for protein synthesis. Although only 20 nm in diameter, each has two subunits containing proteins in addition to ribosomal RNA. Ribosomes are not only found on rough ER, they may also be found in the cytoplasm where they gather in groups known as **polysomes.**

While rough ER (RER) is a site for protein synthesis, smooth ER (SER) is a site for lipid synthesis. In addition to serving these functions, the ER is a major means by which materials get from one part of the cell to another. After synthesis, substances enter the channels of the ER, which change shape to form vesicles (small membranous sacs). In some cases these vesicles move through the cytoplasm toward the cell membrane, but often they move toward a Golgi body.

Golgi Bodies

Cell biologist Camillo Golgi observed the presence of an organelle at the turn of the century, but was unable to decipher its detailed structure and function. We now know that this organelle, a **Golgi body** (fig. 4.14), is a stack of a half dozen or more flattened membranous sacs called **saccules.** It has also been determined that Golgi bodies are involved in the packaging and secretion of macromolecules. The saccules contain enzymes that can add sugar molecules, phosphate groups, and fatty acids to protein molecules.

Figure 4.13
Rough endoplasmic reticulum. *a.* Electron micrograph. *b.* Diagrammatic drawing. Note the presence of ribosomes (small dots).

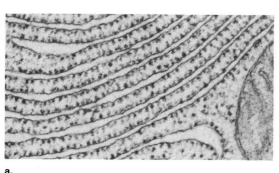

a.

ribosomes

b.

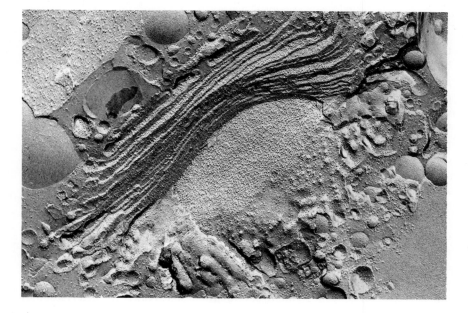

Figure 4.14
Golgi body structure. A frozen cell was fractured (split) and then subjected to scanning electron microscopy. In this case, the treatment reveals the interior of the saccules and shows that vesicles appear along both sides of the Golgi body and at the rims of the saccules.

Figure 4.15

Golgi body function. The Golgi body receives vesicles from the endoplasmic reticulum and thereafter forms at least two types of vesicles: lysosomes and secretory vesicles. Lysosomes contain hydrolytic enzymes that can break down large molecules. Sometimes lysosomes join with vesicles, bringing large molecules into the cell. Thereafter, any nondigested residue is voided at the cell membrane. The secretory vesicles formed at the Golgi body also discharge their contents at the cell membrane.

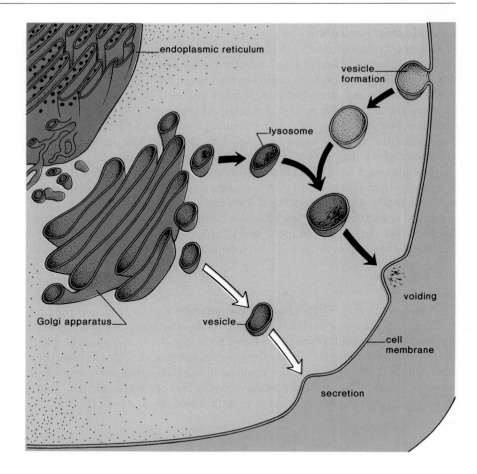

Note in figure 4.15 that one side of a Golgi body faces the nucleus and the other side faces the cell membrane. Microscopic observations along with biochemical evidence suggest that the Golgi body receives protein-filled vesicles from the endoplasmic reticulum at its inner face. After the proteins are processed, they are packaged into vesicles at the outer face. Thereafter, the vesicles move to different locations in the cell. Some proceed to the cell membrane where their contents are released. This process, called **exocytosis,** is the manner in which secretion occurs. During secretion a cell product leaves a cell.

Lysosomes

Some vesicles formed at a Golgi body function within the cell. Chief among these are **lysosomes,** which contain hydrolytic, or digestive, enzymes that break down macromolecules and therefore need to be segregated from the rest of the cytoplasm (so they do not break down the cell itself). Almost all animal cells contain lysosomes, as do some plant cells, some fungi, and algae. Lysosomes are roughly spherical in shape and average about 500 nm in diameter, but they may range from 50 nm to several μm in size. Newly formed lysosomes sometimes fuse with **phagocytic vesicles.** Phagocytic vesicles form when the cell membrane folds around a small quantity of material and then pinches off

Table 4.4 Passage of Molecules Into and Out of Cells

Name	Direction	Requirements	Examples
Diffusion	Toward lesser concentration	———	Lipid-soluble molecules Water Gases
Transport Facilitated	Toward lesser concentration	Carrier	Sugars and amino acids
Active	Toward greater concentration	Carrier plus energy	Sugars, amino acids, and ions
Exocytosis	Toward greater concentration	Vacuole release	Secretion of substances
Endocytosis	Toward greater concentration	Vacuole formation	Phagocytosis of substances

Figure 4.16
An electron micrograph showing two types of lysosomes. The darker ones are residual bodies containing nondigested material; the lighter one is an autodigestive vacuole containing a mitochondrion.

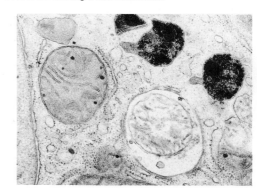

(fig. 4.15). This type of vesicle formation is sometimes called **endocytosis** (table 4.4) because it brings materials into the cell. When a phagocytic vesicle fuses with a lysosome, the vesicle's contents are exposed to the digestive enzymes contained in the lysosome. If there is any nondigestible material, the lysosome becomes a residual body. Residual bodies carry out excretion when they fuse with the cell membrane and release their contents.

Lysosomes sometimes digest cells and parts of cells. For example, fig. 4.16 shows mitochondria that are enclosed within lysosomes. Because lysosomes can digest cell parts, they are sometimes referred to as "suicide bags," meaning that their contents, if released, could completely digest the cell in which they occur. This capability may be important in certain types of transformations, such as a tadpole losing its tail to become a frog.

Cell Vacuoles

Most mature plant cells have **vacuoles** (large membranous sacs), and some animal cells and various unicellular organisms contain certain types of vacuoles. However, vacuoles are much more prominent in plant cells; they occupy a greater proportion of the space in mature plant cells than they do in animal cells. The vacuolar fluid, or cell sap, is a storage site containing relatively high concentrations of stored materials, such as sugars, acids, salts, and pigments. The vacuoles of many plant cells contain water-soluble pigments that contribute to the color of plant parts.

Energy-Related Organelles

Chloroplasts and mitochondria are two important energy-related organelles in cells. **Chloroplasts** contain pigments, notably chlorophyll, that capture the sun's energy so carbohydrates (e.g., glucose) can be synthesized. During photosynthesis chloroplasts transform solar energy into the chemical bond energy of organic molecules. An overall equation for photosynthesis is:

$$\text{energy} + \text{carbon dioxide} + \text{water} \longrightarrow \text{carbohydrate} + \text{oxygen}$$

Here the word "energy" stands for solar energy, since sunlight provides the initial source of energy that powers photosynthesis.

Photosynthesis is the process by which algae and plants make their own food. Eukaryotic algae and plant cells make their own food within chloroplasts. (Cyanobacteria are prokaryotes and do not have chloroplasts. They do have chlorophyll enclosed within saccules, but these saccules are not enclosed within a chloroplast.)

All eukaryotic cells, whether photosynthetic or not, have **mitochondria,** where cellular respiration occurs. Mitochondria contain enzymes that break down energy-rich nutrient organic molecules to build up ATP, the energy molecule of cells (p. 115). An overall equation for cellular respiration is the opposite of that for photosynthesis.

$$\text{carbohydrate} + \text{oxygen} \longrightarrow \text{carbon dioxide} + \text{water} + \text{energy}$$

Here "energy" stands for ATP, the molecule used as an energy source within cells. All plant and animal eukaryotic cells have mitochondria since they all need a supply of ATP. (Prokaryotic cells also need energy and do have respiratory enzymes, but these enzymes function within the cytoplasm and are not enclosed by membrane.)

As we shall discuss further in the next part, chloroplasts and mitochondria are the organelles that make possible the flow of energy from the sun to **autotrophs,** organisms that make their own food, and to **heterotrophs,** organisms that must take in preformed food.

Chloroplasts

Chloroplasts are a type of plastid. Plastids, organelles that function in the synthesis and storage of food reserves, are of two types: leucoplasts (colorless plastids) and chromoplasts (colored plastids). **Leucoplasts** manufacture and store starch and eventually develop into specialized starch-storing bodies. **Chromoplasts,** similar to the plant vacuoles discussed before, often account for the color of plant parts. **Chloroplasts** are chromoplasts containing, in addition to carotenoids and other pigments, a large quantity of the green pigment chlorophyll.

In higher plants chloroplasts are often oval in shape and measure about 2–4 μm by 1–5 μm. They are bounded by two membranes separated from one another by a small space. These boundary membranes enclose an interior compartment called the **stroma.** Here the inner membrane forms **grana.** In an electron micrograph a granum looks like a stack of coins (fig. 4.17), but each "coin" is actually a saccule. As we shall see in chapter 5, the organization of pigments and enzymes within the membrane of each saccule facilitates the process of photosynthesis.

Mitochondria

Mitochondria are usually elongated, measuring about 0.3–1.0 μm by 5–10 μm. The electron microscope reveals that each mitochondrion is bounded by a double membrane, and the outer and inner membranes are separated by a small space. The innermost compartment of a mitochondrion is called the **mitochondrial matrix.** Within the matrix, the inner membrane's surface area is greatly increased by infoldings called **cristae** (fig. 4.18). These cristae vary in shape from organism to organism and even from cell to cell within the same organism. The respiratory enzymes are arranged in an assembly-line fashion along the membrane, and this arrangement facilitates the process of cellular respiration.

Mitochondria frequently are called the *powerhouses of the cell* because they produce ATP, the type of molecular energy that all cells require for cellular activities. They are often scattered throughout the cell, but in some cases large numbers are concentrated in regions where the demand for energy is great; for example, near the cell membrane in cells that engage in active transport, as in the tubular organ in the chambered nautilus (p. 71).

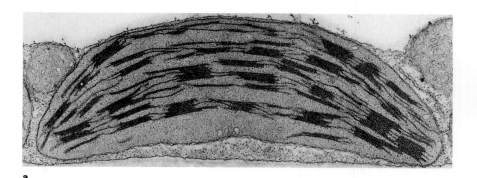

a.

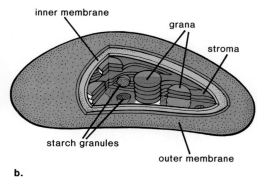

Figure 4.17
Chloroplast structure. *a.* Electron micrograph. *b.* Drawing in which the outer and inner membranes have been cut away to reveal the grana.

inner membrane

grana

stroma

starch granules

outer membrane

b.

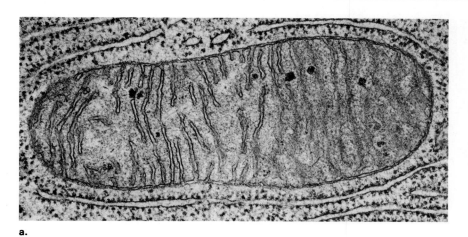

a.

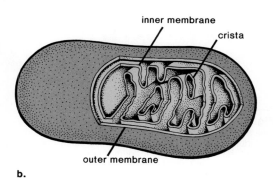

Figure 4.18
Mitochondrial structure. *a.* Electron micrograph.
b. Drawing in which the outer membrane and portions of the inner membrane have been cut away to reveal cristae.

inner membrane

crista

outer membrane

b.

Other Organelles

Cytoskeleton

Several types of filamentous protein structures form a cytoskeleton (fig. 4.19), which supports and suspends the organelles already discussed. The cytoskeleton includes microfilaments and microtubules as well as a newly discovered, irregular, three-dimensional network, or lattice (termed the **microtrabecular lattice**), in the cytoplasm. The lattice is dynamic and its exact shape is constantly changing. While the lattice itself is protein rich, the spaces between the strands are water rich.

    Microfilaments are long, extremely thin fibers (approximately 4–7 nm in diameter) that usually occur in bundles or other groupings. Microfilaments have been isolated from a number of cells. When analyzed chemically, their composition is similar to that of *actin* or *myosin,* the two proteins responsible for muscle contraction.

Figure 4.19
Cytoskeleton of cell. *a.* Electron micrograph of
microtubules and microfilaments within the cytoskeleton,
which also contains tiny, slender strands of the
microtrabecular lattice. *b.* Drawing of cytoskeleton.
c. Detailed structure of a microtubule. *d.* Detailed
structure of a microfilament.

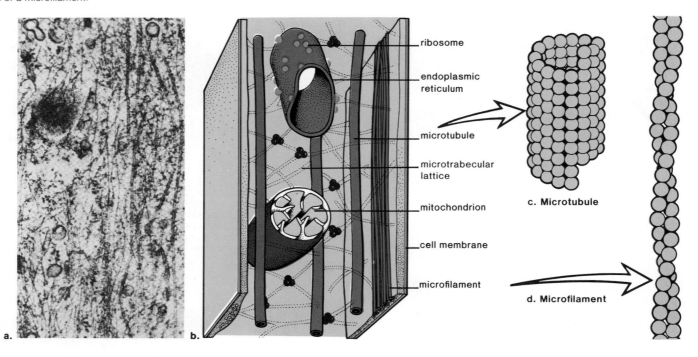

ribosome

endoplasmic
reticulum

microtubule

microtrabecular
lattice

mitochondrion

cell membrane

microfilament

c. Microtubule

d. Microfilament

Figure 4.20
Cilia and flagella move either by undulation or stroking.
a. Undulation, during which waves move either from the
base of the structure to its tip or in the opposite direction.
b. Stroking, during which the beat has two phases. In the
effective stroke, the structure remains fairly stiff as it
pushes back through the water. This is followed by a
recovery stroke, in which the structure bends and returns
to its initial position. The large arrows indicate the
direction of water movement in these examples.

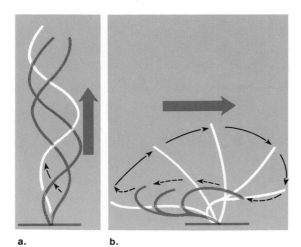

a.   b.

Microtubules are shaped like thin cylinders and are several times larger
than microfilaments (about 25 nm in diameter). Each cylinder contains thirteen rows of *tubulin,* a globular protein, arranged in a helical fashion. Aside
from existing independently in the cytoplasm, microtubules are also found in
certain organelles, such as cilia, flagella, and centrioles.

Cilia and Flagella

Cilia and **flagella** are cell structures associated with cell movement. Our remarks here pertain only to eukaryote flagella; some bacteria (prokaryotes)
have flagella, but as their structure and movement are entirely different from
the flagella of eukaryotes, they will be discussed on page 267.

Some single-celled eukaryotic organisms use cilia or flagella to move independently. For example, paramecia (p. 272) move by means of cilia; sperm
are propelled by flagella. Cells in multicellular organisms also move materials
along their surfaces by means of cilia.

Cilia and flagella differ in length. Cilia are about 10–20 μm in length,
while flagella are much longer (100–200 μm). Movement of each, however,
is similar. Cilia and flagella can move like an oar; in the effective stroke the
cilia and flagella pull backwards; during recovery, they bend forward (fig. 4.20).
Or cilia and flagella can undulate, moving like a whip. Cells are actually able
to coordinate the beats so that some of these structures are moving forward
while others are pushing backward.

Cilia and flagella in eukaryotes are membrane-bound cylinders (about
0.2 μm in diameter) enclosing a matrix area. In the matrix are nine microtubule doublets arranged in a circle around two central microtubules (fig. 4.21).

Figure 4.21
This drawing contrasts the 9 + 2 pattern of microtubules in a cilia or flagella and the 9 + 0 pattern of microtubules in the basal body. *a.* Portion of entire structure. *b.* Cross section of 9 + 2 pattern. *c.* Electron micrograph of basal body, showing the 9 + 0 pattern.

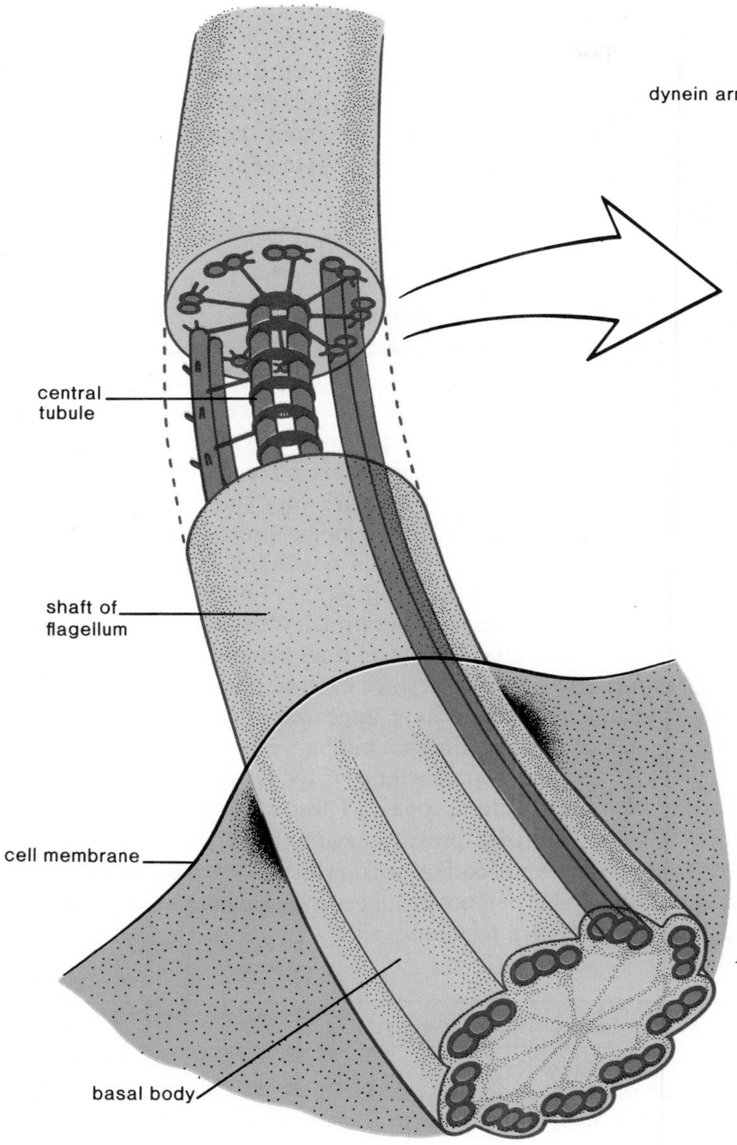

This is called the 9 + 2 pattern of microtubules. Each doublet has pairs of arms projecting toward a neighboring doublet and spokes extending toward the central pair of microtubules. Recent evidence indicates that cilia and flagella move when the microtubule doublets slide along one another. The claw-like arms and spokes seem to be involved in causing this sliding action, which requires ATP energy.

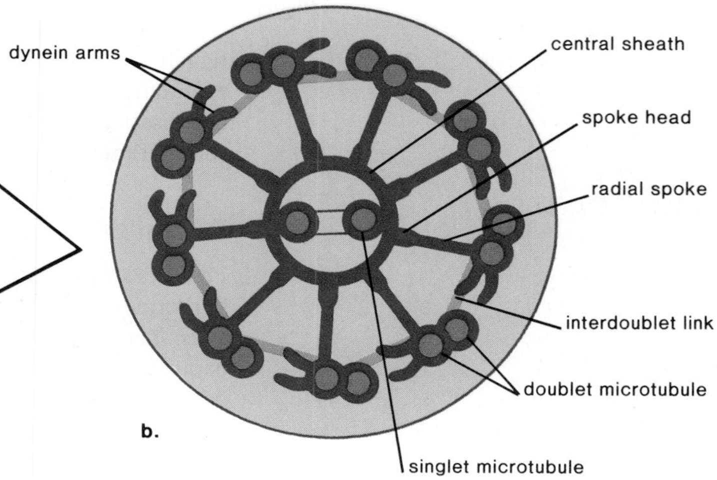

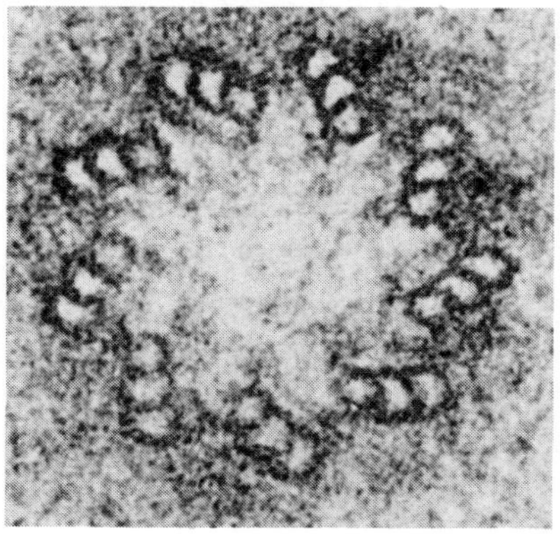

Each cilium and flagellum has a basal body lying in the cytoplasm at its base. **Basal bodies,** short cylinders with a circular arrangement of nine microtubule triplets called the 9 + 0 pattern, are believed to organize the structure of cilia and flagella.

Centrioles

Centrioles are short cylinders (0.15–0.20 μm by 0.3–0.5 μm) with a 9 + 0 pattern of microtubule triplets. There are usually two pairs (fig. 4.12a) lying to one side of the nucleus in certain eukaryotic cells, such as lower fungi, lower plants, and animal cells. The members of each pair are at right angles to one another.

Centrioles give rise to basal bodies that direct the formation of cilia and flagella. Centrioles may be involved in other cellular processes that use microtubules, such as movement of material throughout the cell or appearance and disappearance of the spindle apparatus (p. 134) in animal cells. Their exact role in these processes is uncertain, however.

Evolution of Eukaryotic Cells

Fossil evidence indicates that the prokaryotic cell evolved first, before the eukaryotic cell. Table 4.5 compares prokaryotic cells to typical plant and animal cells. You'll notice that prokaryotes, represented only by bacteria and cyanobacteria (fig. 4.22), lack the various types of organelles we have been discussing. They do have genetic material, respiratory enzymes, and chlorophyll (if the prokaryote is photosynthetic), but these are not enclosed in organelles. In fact, in some respects prokaryotes, which are very small, seem like individual organelles. Is it possible that, in days gone by, prokaryotes became the organelles of today's eukaryotic cells? Such a possibility has been proposed and is termed the **endosymbiotic theory** (fig. 4.23). This theory states that some prokaryotes came to reside *inside* other prokaryotes, establishing a symbiotic relationship known as mutualism (of mutual benefit).

Fossil records indicate that prokaryotes evolved about 3.5 billion years ago, while eukaryotes are thought to have appeared about 1.5 billion years ago. Surely during this interval of time (2 billion years), some prokaryotes became photosynthetic like the cyanobacteria of today, and some carried on cellular respiration like most bacteria today. (These are labeled aerobic prokaryotes in fig. 4.23). Undoubtedly, prokaryotes interacted with one another, and by chance a very large amoebalike prokaryote may have phagocytized other types. If so, a eukaryotic cell possessing such organelles as chloroplasts and/or mitochondria could have resulted.

Table 4.5 Comparison of Prokaryotic and Eukaryotic Cells

	Prokaryotic	Eukaryotic	
		Animal Cell	*Higher Plant Cell*
Cell membrane	Yes	Yes	Yes
Cell wall	Yes*	No	Yes
Nuclear envelope	No	Yes	Yes
Mitochondria	No	Yes	Yes
Chloroplasts	No	No	Yes
Endoplasmic reticulum	No	Yes	Yes
Ribosomes	Yes, small	Yes, large	Yes, large
Vacuoles	No	Yes, small	Yes, large central
Lysosomes	No	Yes, usually	No, usually
Centrioles	No	Yes	No
Flagella	Yes*	Yes	No
Cilia	No	Yes	No

*Composition is different from eukaryotic structure.

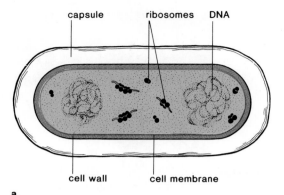

capsule   ribosomes   DNA

cell wall   cell membrane

a.

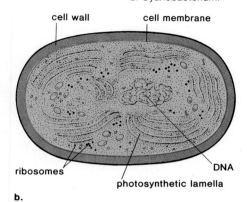

cell wall   cell membrane

ribosomes   DNA

photosynthetic lamella

b.

Figure 4.22
Generalized drawing of prokaryotic cells. Notice the lack of discrete membrane-bound organelles. *a.* Bacterium. *b.* Cyanobacterium.

Figure 4.23
Simplified endosymbiotic theory. *a.* Mitochondria may be derived from aerobic prokaryotes that were phagocytized by a larger amoebalike prokaryote. *b.* Similarly, chloroplasts may be derived from photosynthesizing prokaryotes that were phagocytized by a larger amoebalike prokaryote.

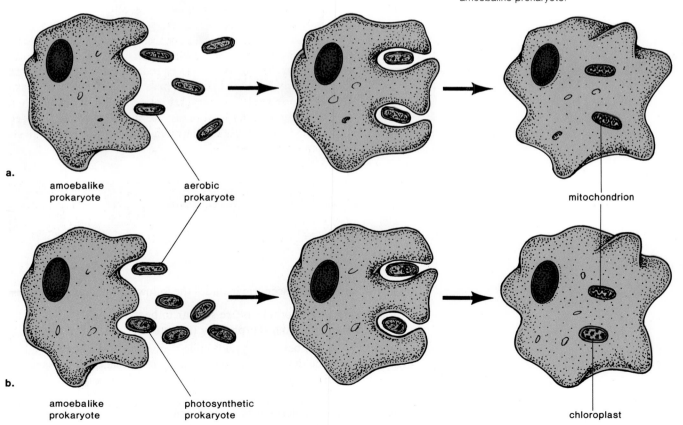

a.

amoebalike prokaryote   aerobic prokaryote   mitochondrion

b.

amoebalike prokaryote   photosynthetic prokaryote   chloroplast

Figure 4.24
Summary of eukaryotic organelles in animal and plant cells.

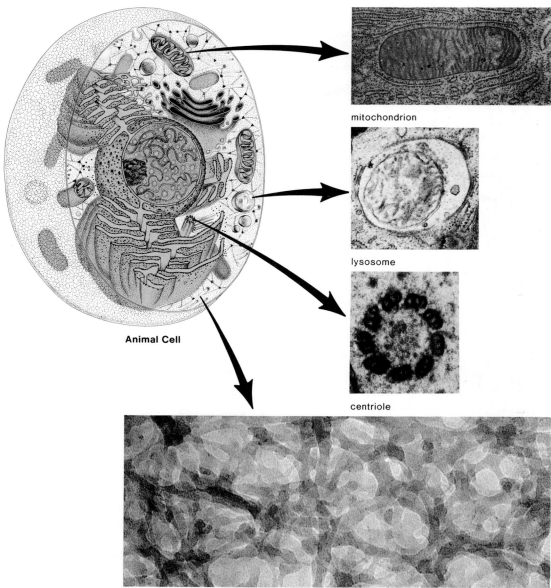

mitochondrion

lysosome

centriole

Animal Cell

microtrabecular lattice

The evidence for this scenario includes the following four considerations.

1. Mitochondria and chloroplasts are similar to bacteria and cyanobacteria in size and morphology (structure).

2. Both organelles are bounded by a double membrane—the outer membrane may be derived from the phagocytic vesicle, and the inner one may be derived from the plasma membrane of the original prokaryote.

3. Mitochondria and chloroplasts contain genetic material and therefore are capable of self-reproduction. Their DNA resembles that of prokaryotes.

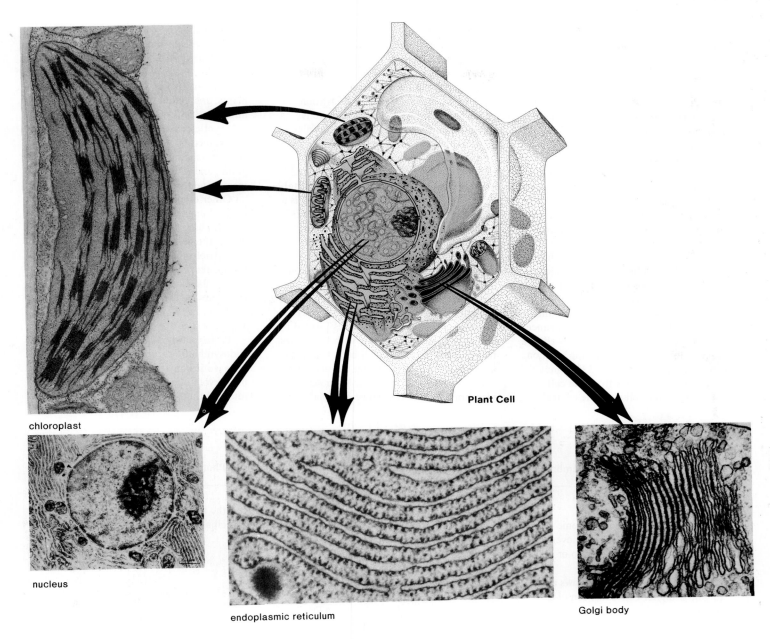

chloroplast

nucleus

endoplasmic reticulum

Plant Cell

Golgi body

4. Mitochondria and chloroplasts are capable of carrying on protein synthesis. Their ribosomes resemble those of prokaryotes.

Although many biologists currently support the endosymbiotic theory, alternative theories have been proposed. For example, perhaps the prokaryotic cell membrane became infolded to enclose copies of DNA, thereby forming several double-walled entities within a single cell. These entities could have evolved into the eukaryotic nucleus, mitochondrion, and chloroplast. This sort of scheme also fits the available data, but it is not possible at this time to determine which (if either) hypothesis is correct.

Summary

The cell theory states that all organisms are composed of self-reproducing units called cells. Electron microscopes make it possible to study cell structure in detail. Micrographs support the fluid mosaic model of membrane structure: globular proteins are embedded in a phospholipid bilayer and also are found next to the inner side of this layer. Short chains of sugars, termed glycoproteins and glycolipids, are attached to the outer surface of protein or lipid molecules, respectively. Most likely, these give cells an identity and the ability to interact with other cells.

The cell membrane is selectively permeable so it controls the entrance and exit of molecules into and out of the cytoplasm. A few types of molecules simply diffuse across a membrane. The diffusion of water is termed osmosis. Animal and plant cells retain their normal appearance in an isotonic solution. Due to water loss, animal cells shrink in hypertonic solution; entrance of water causes the animal cells to swell in a hypotonic solution. In a hypertonic solution, the central vacuole shrinks and the plant cell membrane withdraws from the cell wall; in a hypotonic solution, the central vacuole swells and the cell membrane presses against the cell wall.

Most solutes must be transported across the cell membrane by means of carrier proteins, which are specific. In facilitated transport a solute is still undergoing diffusion according to its concentration gradient; in active transport a solute is moving against its concentration gradient. Active transport requires ATP energy.

Eukaryotic cells have a number of membranous organelles (fig. 4.24), while prokaryotic cells lack these organelles (table 4.5). The endoplasmic reticulum (ER) is a system of flattened channels and tubular canals. Rough ER, which has ribosomes, is a site for protein synthesis; smooth ER is a site for lipid synthesis. Vesicles from the ER apparently transport substances to a Golgi body where they are processed. A Golgi body, a stack of saccules, is involved in packaging (vesicle formation) and secretion (exocytosis). Phagocytic vesicles formed by endocytosis sometimes fuse with lysosomes, bodies formed at the Golgi body that also are capable of autodigestion because they contain hydrolytic enzymes. Vacuoles, membranous sacs that store materials, are more prominent in plant cells than in animal cells.

Chloroplasts and mitochondria permit a flow of energy from the sun to all organisms. Photosynthesis occurs in chloroplasts, which are unique to algae and plants, and cellular respiration occurs in mitochondria, which are found in all eukaryotic cells. These structures are surrounded by a double membrane; in chloroplasts the inner membrane forms grana, and in mitochondria the inner membrane forms cristae.

Other organelles, including microfilaments, microtubules, cilia, and flagella, have to do with movement either within the cell or of the cell itself. Two pairs of centrioles lie to one side of the nucleus in those eukaryotic cells that typically have cilia or flagella. Cilia and flagella contain microtubules arranged in a 9 + 2 pattern, while the centrioles have microtubules arranged in a 9 + 0 pattern.

Prokaryotic cells are quite small, and the endosymbiotic theory proposes that the eukaryotic cell arose when a large prokaryote ingested other prokaryotes and retained them as "partners."

Objective Questions

1. The fluid-mosaic model of membrane structure suggests that _____ molecules drift about within a _____ bilayer.
2. If a cell is placed in a hypertonic solution, there will be a net movement of water _____ (into or out of) the cell.
3. _____ transport allows a cell to gain a higher concentration of a molecule than is outside the cell.
4. Eukaryotic cells have a true _____ .
5. Protein synthesis occurs at the _____ .
6. Lysosomes contain _____ enzymes.
7. Vesicles derived from endoplasmic reticulum make their way to the _____ , an organelle that functions in packaging and secretion.
8. Both plant and animal cells have _____ , where glucose products provide energy for ATP formation.
9. Photosynthesis takes place within _____ .
10. _____ contain thirteen rows of the protein tubulin and are found in cilia and flagella.

Answers to Objective Questions
1. protein, lipid 2. out of 3. Active 4. nucleus 5. ribosomes 6. digestive 7. Golgi body 8. mitochondria 9. chloroplasts 10. microtubules

Study Questions

1. Describe the current model for the structure of membrane.
2. Name the various means by which substances enter and exit cells. Which of these requires ATP energy?
3. Define the terms isotonic, hypotonic, and hypertonic and tell how they affect plant and animal cells.
4. Define the words prokaryotic and eukaryotic and relate these definitions to the two main categories of cells.
5. Describe the structure and function of endoplasmic reticulum; the Golgi body; lysosomes; vacuoles; plastids (in particular, the chloroplasts), mitochondria, microfilaments and microtubules, cilia and flagella, and centrioles.
6. List and explain three differences between the organelles found in animal cells and those found in higher plant cells.
7. Describe the endosymbiotic theory.

Selected Key Terms

resolution (‚rez ə 'lü shən) 65
cytoplasm ('sīt ə ‚plaz əm) 67
diffusion (dif 'yü zhən) 67
osmosis (äz 'mō səs) 67
isotonic (ī sə 'tän ik) 69
hypertonic (‚hi pər 'tän ik) 69
hypotonic (‚hī pə 'tän ik) 69
active transport ('ak tiv trans 'pōərt) 70

prokaryotic (‚prō ‚kar ē 'ät ik) 72
eukaryotic (‚yü ‚kar ē 'ät ik) 72
organelle (‚òr gə 'nel) 74
exocytosis (‚ek sō sī'tō səs) 76
endocytosis (‚en də sī 'tō səs) 77
cilia ('sil ē ə) 80
flagella (flə 'jel ə) 80

Suggested Readings for Part 2

Avers, C. J. 1981. *Cell biology.* 2d ed. New York: D. Van Nostrand.

Baker, J. J. W., and Allen, G. E. 1981. *Matter, energy, and life.* 4th ed. Reading, MA: Addison-Wesley.

Berns, M. W. 1977. *Cells.* New York: Holt, Rinehart & Winston.

Bretscher, M. S. October 1985. The molecules of the cell membrane. *Scientific American.*

Capaldi, R. A. March 1974. A dynamic model of cell membranes. *Scientific American* (offprint 1292).

Dustin, P. August 1980. Microtubules. *Scientific American* (offprint 1477).

Frieden, E. July 1972. The chemical elements of life. *Scientific American.*

Grivell, L. A. March 1983. Mitochondrial DNA. *Scientific American.*

Jensen, W. A. 1970. *The plant cell.* 2d ed. Belmont, CA: Wadsworth.

Lazarides, E., and Revel, J. P. May 1979. The molecular basis of cell movement. *Scientific American* (offprint 1427).

Loewy, A. G., and Siekevitz, P. 1970. *Cell structure and function.* 2d ed. New York: Holt, Rinehart & Winston.

Mahler, H. R., and Cordes, E. H. 1971. *Biological chemistry.* 2d ed. New York: Harper & Row, Publishers, Inc.

Osborn, M., and Weber, K. October 1985. The molecules of the cell matrix. *Scientific American.*

Porter, K. R., and Bonneville, M. A. 1973. *Fine structure of cells and tissues.* 4th ed. Philadelphia: Lea and Febiger.

Porter, K. R., and Tucker, J. B. March 1981. The ground substance of the living cell. *Scientific American* (offprint 1494).

Satir, B. October 1975. The final steps in secretion. *Scientific American* (offprint 1328).

Unwin, N. February 1984. The structure of proteins in biological membranes. *Scientific American.*

Weinberg, R. A. October 1985. The molecules of life. *Scientific American.*

PART 3

Energy and Life

L iving things require not only matter but energy as well. Solar energy captured by photosynthesizing organisms is passed on to other organisms in the form of nutrient organic compounds. Energy from these compounds is converted to ATP (adenosine triphosphate) energy during cellular respiration. Both photosynthesis and cellular respiration are necessary processes for the continuance of life. Solar energy is used in the synthesis of nutrient molecules, and ATP is the most common energy molecule used by cells.

Great blue heron feeding at dawn. With steadfast eyes, the heron marks the position of a fish and then seizes its prey with a direct and quick movement of its sharp bill.

CHAPTER 5

Photosynthesis

Concepts

1. The process of photosynthesis provides food for most living things.
   a. Photosynthesis uses the portion of sunlight that is absorbed by chlorophyll. Chlorophyll appears green because it does not absorb green light well.
   b. Photosynthesis consists of two sets of reactions. The first produces the NADPH and the ATP needed by the second reaction, when carbon dioxide is reduced and converted to a high energy molecule.

2. The energy of the sun is continually needed by photosynthesizers to produce more food.
   Food energy is first converted to ATP energy; ATP is used whenever cells require an energy source.

3. Living things are dependent on enzymatic reactions.
   Cellular metabolism, including photosynthesis and cellular respiration, depends on enzymes that speed up reactions and on coenzymes that participate in the reactions.

All living things require food as a source of unit molecules for the synthesis of cellular macromolecules. Food is a source of the energy that drives the synthesis and organization of macromolecules and makes possible special functions, such as nerve conduction and muscle contraction.

In general, there are two mechanisms by which organisms acquire food. **Autotrophic** organisms make their own food, while **heterotrophic** organisms must take in food that is already made. Autotrophic organisms, such as terrestrial plants and aquatic algae, need only carbon dioxide, water, and the energy of the sun to synthesize carbohydrates and other nutrient compounds. These compounds serve as food not only for autotrophs but also either directly or indirectly for most heterotrophs (fig. 5.1). **Herbivores** eat plants directly. **Carnivores** eat herbivores and may be eaten by other carnivores. Eventually, all organisms die and decay. When they do, inorganic chemicals such as CO_2 and H_2O are made available to autotrophs again. In this manner inorganic chemicals have cycled time and time again through living organisms since life began on earth.

Energy

Unlike chemicals, energy does not cycle. Each organism uses some portion of its food as an energy source. Eventually most of the energy content of any particular organic food molecule is dissipated as heat, and only inorganic chemicals remain. The continual energy of sunlight is essential for plants and algae to make more organic food. Without energy from the sun and the capability of autotrophs to capture this energy within organic food molecules, life would cease to exist.

The Laws of Thermodynamics

Thermodynamics is the study of energy relationships and exchanges. **Energy** is defined as the capacity to bring about change or to do work. For example, work has been done when a macromolecule is synthesized or when a bear captures a fish.

The two laws of thermodynamics explain why energy does not cycle. The First Law of Thermodynamics says that *energy can neither be created nor destroyed.* In other words, the same amount of energy has been present and will continue to be present in our universe from its beginning to its end. Most often, however, we do not wish to consider the amount of energy in the entire universe. We usually consider the amount of energy in a piece of the universe, termed a system. For a biologist, a system might be an organ, an organism, or a community of organisms. Whatever system is examined, all other matter outside that system is considered to be the *surroundings.*

The First Law of Thermodynamics tells us that energy cannot be created within a system; rather, energy must come from the surroundings. For example, the electric company doesn't create energy; it simply converts the energy of falling water, nuclear energy, or fossil fuel energy into electrical energy. Similarly, plants do not create chemical energy; they convert solar energy into chemical energy. In the same way, a system cannot destroy energy,

although it can lose energy to its surroundings. *Usable energy* is concentrated and easily employed to do work. Heat is often *nonusable energy* because it is too diffuse to be easily employed to do work.

The Second Law of Thermodynamics says that *one usable form of energy cannot be completely changed into another usable form;* the system always loses some energy as heat. Muscles change the chemical-bond energy within ATP to the mechanical energy of contraction, *but not completely;* some chemical-bond energy becomes heat, which is not usable for contraction. Eventually, all usable forms of energy are converted to heat, which is lost to

Figure 5.2

During photosynthesis, autotrophs use the energy of the sun to transform inorganic chemicals to organic nutrient molecules, which become food for most heterotrophs. Eventually, the energy content of food is lost as heat, but with death and decay, the chemical content becomes available once more to photosynthesizers.

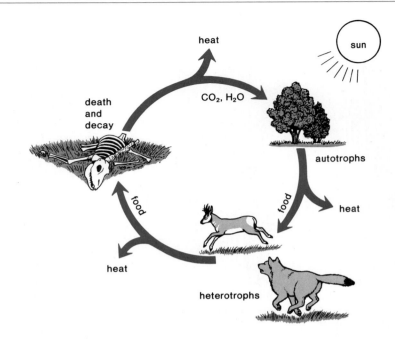

Figure 5.3

Two organelles are necessary for energy transformations in cells; photosynthesis occurs within chloroplasts, and cellular respiration occurs within mitochondria. Photosynthesis captures the energy of the sun and transforms it to chemical energy in carbohydrates, and cellular respiration changes this nutrient molecular energy to ATP, the energy molecule of cells. In the meantime, the by-products of cellular respiration, CO_2 and H_2O, are incorporated once again into carbohydrates within chloroplasts.

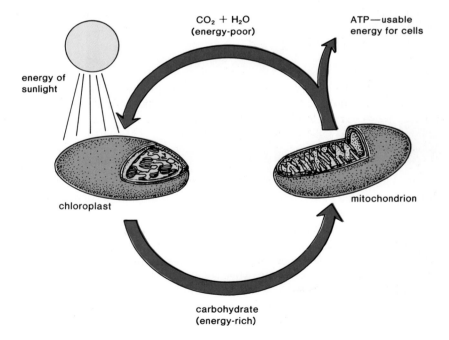

Figure 5.4
ATP, the energy molecule in cells, has two high-energy phosphate bonds (indicated by wavy lines). When cells require energy, the last phosphate bond is broken.

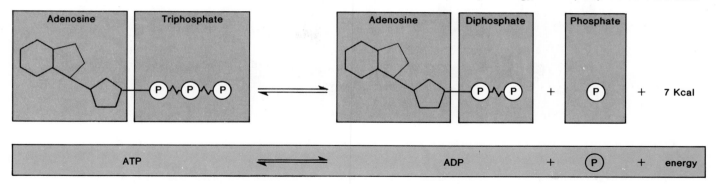

surroundings. Thus, in figure 5.2 food energy is transformed as it is passed from one organism to another. Eventually, it is all converted to heat and dissipated. A continual input of usable energy is required to keep organisms alive. This input of usable energy comes from the sun when photosynthesizing organisms manufacture food.

Energy Transformation

In this chapter and the next we will be concerned with two primary energy transformations. During **photosynthesis** the energy of the sun is converted to the chemical-bond energy within carbohydrates (for example, glucose). In **cellular respiration** this energy is converted to the chemical-bond energy within ATP molecules (fig. 5.3).

ATP

When cells require energy, a certain kind of energy must be available. Cells depend on the molecule called **ATP (adenosine triphosphate)** as an energy source, just as our society has come to depend on electricity to light our homes and to run our electric appliances.

ATP (fig. 5.4) is a nucleotide composed of the base adenine, the sugar ribose (together called adenosine), and three phosphate groups. The wavy lines in the formula for ATP indicate *high-energy phosphate bonds;* when these bonds are broken, an unusually large amount of energy is released. ATP is the energy currency of cells; when cells "need" something, they "spend" ATP. ATP is the common energy currency for use in chemical work, mechanical work (movement), and transport work in living cells. When energy is required for these processes, the end phosphate group is removed from ATP, breaking the molecule down to **ADP** (adenosine diphosphate) and ℗ (phosphate).

Notice in figure 5.4 that ATP can be re-formed from ADP and ℗ whenever enough energy is available. The first step in photosynthesis uses light energy to form ATP.

Energy to Grow Food

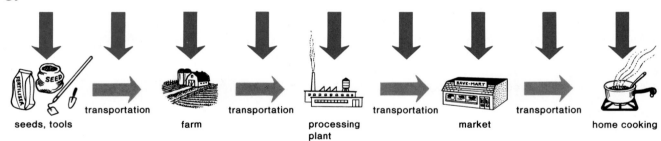

seeds, tools — transportation — farm — transportation — processing plant — transportation — market — transportation — home cooking

a.

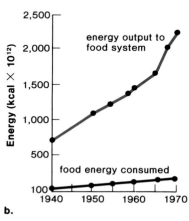

b.

a. *Energy input needed to grow food is represented by colored arrows.* b. *Energy use in food system, 1940 through 1970, compared to the caloric content of food consumed.*

In the United States sunlight provides part of the energy for food production, but it is greatly supplemented by fossil fuel energy. Even before planting time, there is an input of fossil fuel energy for the production of seeds, tools, fertilizers, and pesticides, and for their transportation to the farm (a.). At the farm, fuel is needed to plant the seeds, to apply fertilizers and pesticides, and to irrigate, harvest and dry the crops. After harvesting, still more fuel is used to process the crops to make the neatly packaged products we buy in the supermarket. Most of the food we eat today has been processed in some way. Even farm families now buy at least some of their food from supermarkets in nearby towns.

Since 1940 the amount of supplemental fuel used in the American food system has increased far more sharply than the caloric content of the food we consume, as shown in (b.). This is partially due to the trend toward producing more food on less land. High-yielding hybrid wheat and corn plants require more care and thus about twice as much supplemental energy as the traditional varieties of wheat and corn. Cattle kept in feedlots and fed grain that has gone through the whole production process need about twenty times the amount of supplemental energy as do range-fed cattle. Our food system has been labeled energy intensive because it requires such a large input of supplemental energy.

The intensive use of fossil fuel energy to grow and provide food in the United States is a matter for concern because the supply of fossil fuel is limited and its cost has risen tremendously. This, in turn, affects the cost of farming and of the food produced. What can be done? First of all, devote as much land as possible to farming and animal husbandry. Plant breeders could sacrifice some yield to develop plants that would need less supplemental energy. And we could depend more on range-fed cattle. If cattle are kept close to farmland, manure can substitute, in part, for chemical fertilizer. Biological control, the use of natural enemies to control pests, would cut down on pesticide use. Solar and wind energy could be used instead of fossil fuel energy, particularly on the farm. For example, wind-driven irrigation pumps are feasible.

Finally, of course, consumers could help matters. We could overcome our prejudice against fruits and vegetables that have slight blemishes. We could consume less processed food. We could eat less meat and buy cheaper cuts of beef, which are more likely to have come from range-fed cattle. And we could avoid using electrically powered gadgets when preparing food at home.

Since the United States food system is so energy-intensive, it is doubtful that needy countries abroad could ever duplicate this system. Indeed, if we are concerned about feeding the hungry of the world we should cut back on our own use of supplemental energy to make more available for use by underdeveloped countries.

Enzymes and Coenzymes

Cellular metabolism, including photosynthesis and cellular respiration, requires the participation of enzymes. **Enzymes** are globular protein molecules that increase the rates at which cellular reactions occur. Each enzyme is specific; that is, it increases the rate of only one particular kind of reaction or, in some cases, just one specific reaction. When an enzyme catalyzes (speeds up) a reaction, the reactant(s) that participate in the reaction are called the enzyme's **substrate(s).** Enzymes are often named for their substrate(s) (table 5.1).

Energy of Activation

Molecules frequently will not react with one another unless they are activated in some way. For example, wood does not burn unless its temperature is raised high enough to start the reaction. In the laboratory, the reaction between two substances can be accelerated if heat is applied, if the solution in which they are dissolved is stirred, or if the concentration of the substances is increased. This raises the frequency of collisions between the molecules of the reactants and, therefore, the overall reaction rate.

Enzymes decrease the **energy of activation** needed to produce a reaction (fig. 5.5) and thus speed up that reaction. They have specific regions called **active sites** where the substrates are brought together so that they can react.

Table 5.1 Enzymes Named for Their Substrate

Substrate	Enzyme
Lipid	Lipase
Urea	Urease
Maltose	Maltase
Ribonucleic acid	Ribonuclease
Lactose	Lactase

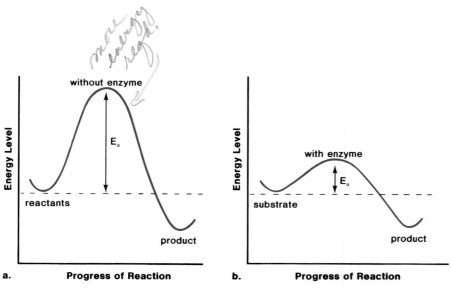

a.   **Progress of Reaction**   b.   **Progress of Reaction**

Figure 5.5

a. Energy of activation (E_a) that is required for a reaction to occur when an enzyme is not available. b. Required energy of activation (E_a) is much lower when an enzyme is available. Notice that the energy level of the entire system is always lower after a reaction.

Figure 5.6
An enzyme (a.) has an active site where (b.) the
substrates and enzyme fit together in such a way that the
substrates are oriented to react. Following the reaction,
the enzyme assumes its prior configuration (c).

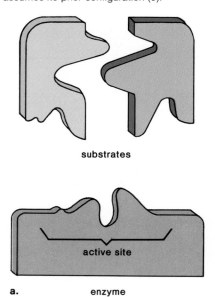
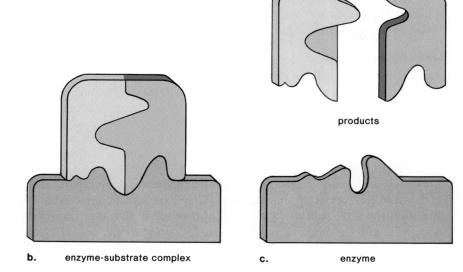

substrates

products

active site

a.   enzyme   **b.**   enzyme-substrate complex   **c.**   enzyme

An enzyme's specificity is caused by the *shape* of the active site, where the
enzyme and its substrate(s) fit together in a specific way, much as the pieces
of a jigsaw puzzle fit together (fig. 5.6). The substrates fit onto the active site
in such a way that they become oriented in the proper manner for the reaction
to take place. Sometimes the active site assumes a different configuration as
the reaction occurs. After the reaction has been completed, the product or
products are released and the active site returns to its original state. What we
have said can be summarized in the following manner.

$$E + S \rightarrow ES \rightarrow E + P$$

where E = enzyme, S = substrate, ES = enzyme substrate complex, and
P = product.

Rate of Enzyme Activity

The rate of an enzyme's activity is judged by the amount of product produced
per unit time. An enzyme's activity increases with an increase in *substrate
concentration*. Increasing the substrate concentration increases the frequency
of collision between substrate molecules and enzymes. As a result, substrate
molecules fill empty active sites for a larger proportion of the time. Hence,
each enzyme molecule participates in more reactions per unit time. When the
concentration of substrate is so great that the enzyme's active sites are almost
continuously filled with substrate, the enzyme's rate of activity will increase
no further. Maximum velocity has been reached.

    An increase in *temperature* generally results in an increase in enzyme
activity. As the temperature rises, the movement of enzyme and substrate molecules increases. An increase in movement increases the frequency of effective
contacts between these molecules. If the temperature increases beyond a certain point, however, enzyme activity eventually levels out and then declines

rapidly because the enzyme is denatured at high temperatures. During **denaturation** an enzyme's shape changes so that its active site can no longer bind substrate molecules efficiently.

A change in pH can also affect enzyme activity. Each enzyme has an optimal *pH* at which its reaction is best activated. You recall that the tertiary structure of a protein is dependent on interactions, such as hydrogen bonding, between "R" groups (fig. 3.28). A change in pH can alter the ionization of these side chains and disrupt the normal interactions, so that denaturation eventually occurs. Again, the enzyme is unable to combine efficiently with its substrate.

Coenzymes

Many enzymes contain a nonprotein portion in addition to a protein portion. These nonprotein components, which must be present for enzymes to be active, are called cofactors. A cofactor may be a metallic ion, such as calcium, potassium, magnesium or zinc ions, or a specific small molecule, in which case they are called **coenzymes.** Coenzymes often participate directly in the enzyme's reaction. Sometimes coenzymes can even transfer electrons, atoms, or molecules from one substrate to another.

Coenzyme research is a good illustration of how basic research can sometimes help resolve questions in seemingly unrelated areas. The field of nutrition is an example. **Vitamins** are relatively small organic molecules that are required in trace amounts in the diets of animals, including humans, for health and physical fitness. By the eighteenth century, specific dietary remedies for particular health problems were recognized. Lime juice, for example, was known to prevent or cure scurvy. Even though more and more dietary requirements were discovered, it was not until the 1930s that biochemists discovered that the role of many vitamins is to serve as components in the synthesis of coenzymes. For example, the B-vitamin group includes niacin, which is involved in the synthesis of a coenzyme important in cellular metabolism. Several other vitamins likewise are precursors from which other coenzymes are synthesized.

NAD and NADP

NAD (nicotinamide adenine dinucleotide) and **NADP** (nicotinamide adenine dinucleotide phosphate) are two coenzymes of particular interest. NAD and NADP are known as coenzymes of **oxidation-reduction** because they transfer electrons from one molecule to another in many different kinds of reactions. When electrons are removed from a substrate, we say the substrate has been **oxidized.** When electrons are added to a substrate, we say the substrate has been **reduced.**

The concept of oxidation-reduction developed in relation to inorganic reactions. For example, when magnesium reacts with oxygen, it is oxidized and loses electrons. Conversely, oxygen is reduced because it gains electrons.

$$Mg + \tfrac{1}{2}O_2 \rightarrow Mg^{++}O^{--}$$

Later, the terms oxidation-reduction were applied to ionic reactions, whether or not oxygen was involved. For example, when sodium reacts with chlorine, the sodium is *oxidized* (loses an electron) and the chlorine is *reduced* (gains an electron).

$$Na + Cl \rightarrow Na^+Cl^-$$

Figure 5.7

NAD is serving as a coenzyme of oxidation-reduction in this reaction. The curved arrow indicates that as NAD becomes reduced, the substrate becomes oxidized, or, conversely, as NADH becomes oxidized, the substrate becomes reduced. (*left to right*) Two electrons and one hydrogen ion are transferred to NAD. The other hydrogen ion appears in the medium. The substrate, therefore, has been oxidized in this reaction. (*right to left*) The substrate accepts hydrogen atoms and becomes reduced.

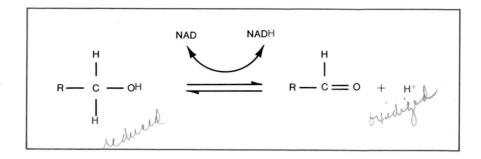

Figure 5.8

This squirrel is a herbivore. It feeds directly on plant material produced by a photosynthesizer. Carnivores, such as a hawk that may feed on this squirrel, are still dependent, although indirectly, on food produced by photosynthesizers.

Oxidation-reduction also occurs in covalent reactions. In covalent reactions the electron is often accompanied by a hydrogen ion. Therefore, the organic molecule is oxidized when it loses hydrogen atoms ($H^+ + e^-$) and reduced when it gains hydrogen atoms. Thus when NAD or NADP accepts electrons it becomes NADH or NADPH, as illustrated in figure 5.7.

Photosynthesis

Photosynthesis is an energy transformation in which energy from the sun in the form of light is converted to the energy within carbohydrate molecules; for example, glucose. Photosynthesis takes place in cells that "capture" the energy of the sun. The end products of photosynthesis provide food for photosynthesizing organisms and, indirectly, for the organisms that feed on them (fig. 5.8). We will begin our discussion of photosynthesis with the energy source—light.

Light

The sun's energy can be analyzed in terms of the **electromagnetic spectrum,** which includes radiation ranging from gamma rays and X rays, with very short wavelengths, to radio waves, with very long wavelengths (fig. 5.9). Photosynthesis utilizes that portion of the electromagnetic spectrum known as **visible light;** the other portions are not utilized.

Electromagnetic radiation comes in discrete packets called **quanta.** The energy of one quantum is inversely proportional to the wavelength of the radiation; that is, radiation of short wavelengths has a higher energy level quantum than does long wavelength radiation. High-energy quanta, such as those of short-wavelength ultraviolet radiation, are dangerous to cells because they can break down organic molecules. Low-energy quanta, such as those of infrared radiation, are ineffective in cells because they only increase the vibrational or rotational energy of molecules. Quanta of visible light, however, have an amount of energy sufficient to promote electrons to higher orbits around atomic nuclei in pigments such as chlorophyll. These considerations explain why *visible* light is utilized by autotrophs. It is energetic enough for photosynthesis but does not harm cells.

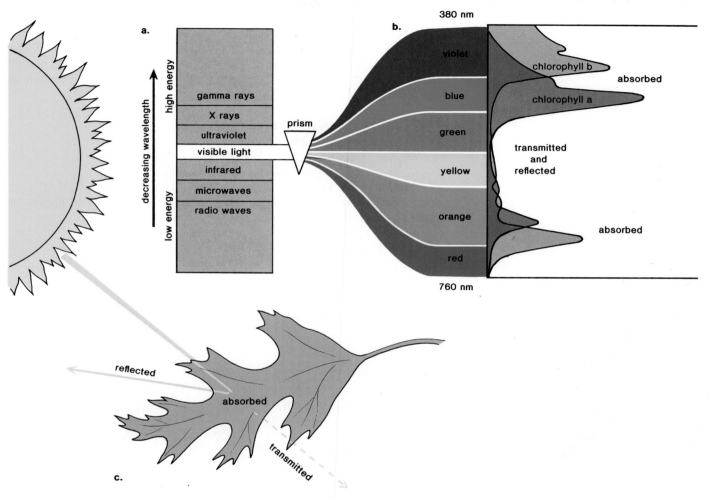

Figure 5.9
a. Radiant energy from the sun is categorized according to the wavelengths of the electromagnetic spectrum. *b.* Visible light contains various colors of light, some of which are absorbed by chlorophyll. *c.* Leaves appear green to the human eye because the color green is largely reflected or transmitted by chlorophyll.

Visible light is made up of a number of different wavelengths of radiation. When white light is passed through a prism, it is divided into these wavelengths, which we often describe in terms of colors. These colors are actually our own interpretation of what we see when specific wavelengths within white light strike our eyes. Figure 5.9 shows that chlorophyll absorbs certain wavelengths of visible light. This is called its **absorption pattern.** Chlorophyll does not absorb green-light wavelengths to any extent. Therefore, they are reflected to our eyes and we see chlorophyll as a green pigment.

Chloroplast Structure

The site of photosynthesis is the **chloroplast,** an organelle found in the cells of green plants (fig. 5.10). The chloroplast is membranous. A double membrane or envelope surrounds a large central space called the **stroma.** The stroma contains many different soluble enzymes that function to incorporate CO_2 into organic compounds. In the stroma, membrane also forms the **grana**. Each granum is a stack of flattened sacs or disks called **thylakoids.** Grana are connected to each other by **stroma lamellae.** Chlorophyll is found within the membranes of the grana, making them the energy-generating system of the chloroplast. Each thylakoid, being saclike, contains a space. Movement of ions from this space across the thylakoid membrane is believed to be important in the photosynthetic process.

Overall Equation for Photosynthesis

An overall equation for photosynthesis

$$H_2O + CO_2 \xrightarrow{\text{light}} (CH_2O) + O_2$$

was known by the beginning of this century. In this equation, (CH_2O) represents carbohydrate. Multiplying this equation by six shows that glucose is often an end product of photosynthesis:

$$6\ H_2O + 6\ CO_2 \xrightarrow{\text{light}} C_6H_{12}O_6 + 6\ O_2$$

Figure 5.10
Leaves are the primary photosynthetic organs of a plant. (*top*) The cells in a leaf contain many chloroplasts, the organelles that carry on photosynthesis. (*bottom*) The stroma within a chloroplast contains grana, each a stack of flattened membranous sacs called thylakoids. The electron micrograph shows stacking of thylakoids to form a granum.

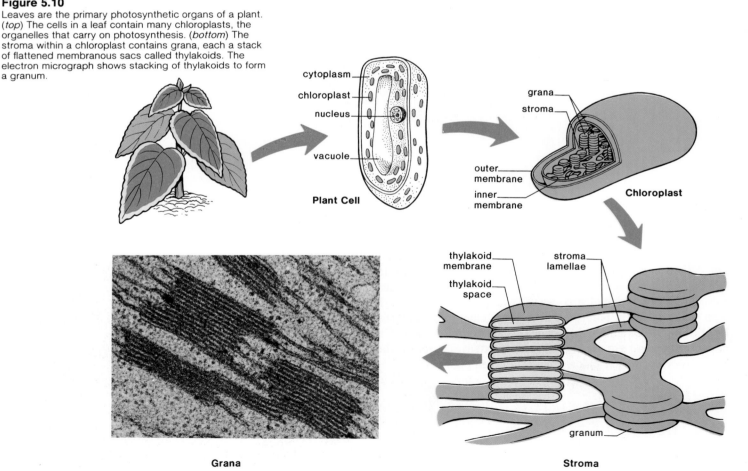

Grana

Stroma

In 1905 a researcher, F. F. Blackman, proposed that the process of photosynthesis involves two different types of reactions. Blackman said that there is a **light reaction,** which requires light, and a **dark reaction,** which does not require light. He suggested this because a rise in temperature could still increase the rate of photosynthesis even when maximum absorption of light had been achieved (fig. 5.11). Blackman's work was an important contribution because it directed the course of future research toward two different aspects of the photosynthetic process.

Today, however, the terms light and dark reaction are being phased out because they do not adequately point out the most important difference between the two parts of the photosynthetic process. The first part captures the energy of sunlight and uses it to produce NADPH and ATP, molecules that are used to reduce CO_2 to a carbohydrate during the second part of the photosynthetic process.

Capturing the Energy of Sunlight

Capturing sunlight energy is the portion of photosynthesis that requires the participation of two photosystems called **Photosystem I** and **Photosystem II.** The freeze-fracture method of preparing thylakoid membrane shows two types of particles (fig. 5.12). Smaller particles that appear in the outer half of the membrane are believed to be locations for Photosystem I, while the larger

Figure 5.11
Blackman's experimental results. Photosynthetic rate is at first dependent only on the amount of light available. After maximum light (dotted line) is available, however, the rate depends on the temperature. Blackman therefore concluded that photosynthesis has two sets of reactions: the rate of the "light" reactions is dependent on light intensity and the rate of the "dark" reactions is dependent on temperature.

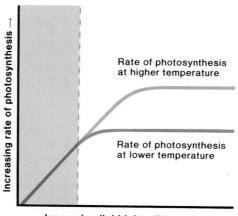

Figure 5.12
Thylakoid membrane. *a.* Thylakoids within grana. *b.* Enlarged thylakoid membrane shows particles in the inner half and outer half of membrane. *c.* Electron micrograph following freeze-fracture of thylakoid membrane.

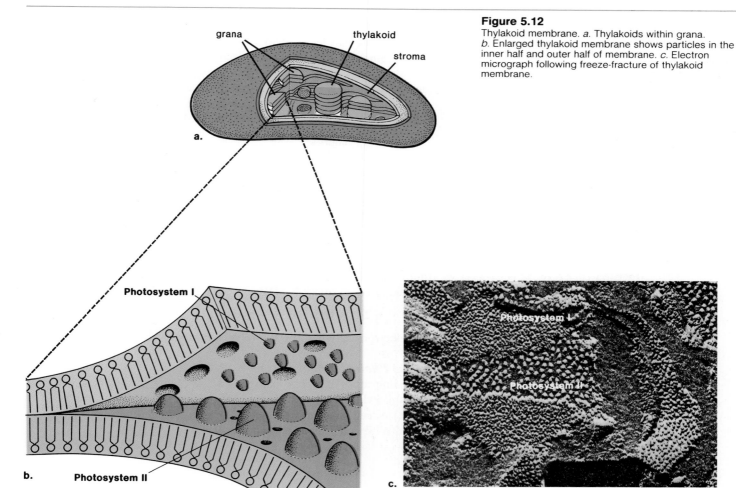

Figure 5.13.
The antenna complex in each photosystem is a collection of pigment molecules. Light energy is funneled from molecule to molecule through the antenna to the reaction center molecule.

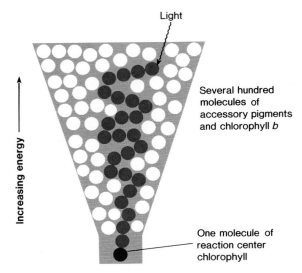

Light

Several hundred molecules of accessory pigments and chlorophyll *b*

One molecule of reaction center chlorophyll

Increasing energy

particles that appear in the inner half of the membrane are believed to be locations for Photosystem II. Each photosystem consists of a pigment complex plus an electron acceptor and donor. The pigment complex contains several hundred molecules of chlorophyll and also carotenoid pigments. The pigment complexes have been termed **antennae** because just as television antennae are aimed to pick up signals, so the leaves of a plant turn to allow the pigment complexes to pick up solar energy.

Photosynthesis begins when the pigments absorb light energy and funnel it to **reaction centers** within the antennae of both Photosystem I and Photosystem II (fig. 5.13). The antenna within Photosystem I has a reaction-center chlorophyll *a* that absorbs maximum light at a wavelength around 700 nm and is therefore called P700 (P stands for pigment). The antenna within Photosystem II has a reaction-center chlorophyll *a* that absorbs slightly shorter wavelengths maximally and is called P680. In figure 5.14, you can see that each photosystem probably absorbs sunlight at the same time. However, it is easiest to describe events as if they occur in a sequential manner and as if they begin with Photosystem II.

1. As energy is received by P680, its electrons become so highly charged that a few actually leave the chlorophyll molecule. P680 would soon disintegrate if it did not receive replacement electrons. It receives these electrons from water, which splits in the following manner.

$$H_2O \rightarrow 2H^+ + 2e^- + \tfrac{1}{2}O_2$$

This freed oxygen is the oxygen gas given off during photosynthesis.

2. The electrons that leave P680 are received by an acceptor molecule before being passed down an **electron transport system.** This system consists of a series of carriers, some of which are cytochrome molecules. (For this reason, the system is sometimes referred to as the **cytochrome system.**) As the electrons pass "downhill" from one cytochrome to another, energy is made available for ATP formation.

$$ADP + P + energy \longrightarrow ATP$$

Just how this occurs is still under investigation. In 1978 British biochemist Peter Mitchell received a Nobel Prize in Chemistry for his theory, which is described briefly in the reading on page 104.

3. After the electrons leave the cytochrome system, they are in a lower energy state and may be received by P700, the reaction-center chlorophyll *a,* in an antenna within Photosystem I. Here, prior to their arrival, light energy has been absorbed by the pigment complex and funneled to P700, energizing its electrons so that some have left to be eventually taken up by NADP, as shown in the following reaction.

$$NADP + 2e^- + H^+ \longrightarrow NADPH$$

The electrons arriving at P700 replace those that have been lost.

Noncyclic Photophosphorylation

The preceding series of events (paragraphs 1–3 above) is known as noncyclic photophosphorylation because (1) it is possible to trace electrons in a one-way (noncyclic) direction from water to NADP and (2) sunlight energy (photo) has caused ATP production (phosphorylation).

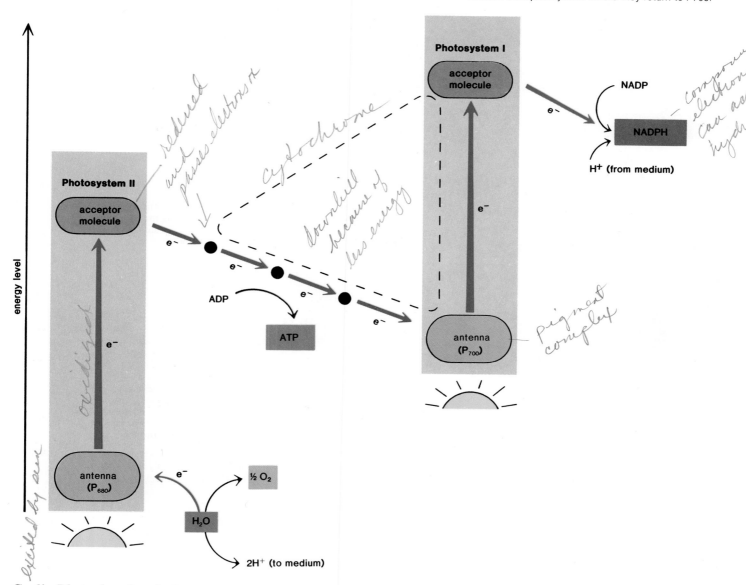

Cyclic Photophosphorylation

At times electrons leave P700 and then return to it instead of being taken up by NADP (fig. 5.14). Before arriving, they pass down the electron transport system, and ATP is produced.

This is called cyclic photophosphorylation because (1) it is possible to trace electrons in a cycle from P700 to P700 and (2) sunlight energy (photo) has caused ATP production (phosphorylation). Cyclic photophosphorylation

ATP Formation in Chloroplasts

In 1978 the British biochemist Peter Mitchell received a Nobel Prize for his hypothesis explaining how ATP production is linked to the electron transport system. According to Mitchell's chemiosmotic theory, hydrogen ions (H^+) collect inside each thylakoid space because they are placed there by certain carriers of the electron transport system. This establishes a hydrogen ion concentration gradient (a.). When hydrogen ions flow down this gradient across the membrane where there is an ATP-synthesizing system, ATP is formed. Andre Jagendorf of Cornell University showed that this would indeed cause ATP production in a famous experiment (b.) in which he soaked chloroplasts in a solution at pH 4. Once the inside of the chloroplast was judged to be at this pH, the external pH was increased. Thereafter, ATP formation occurred as the hydrogen ions flowed down the concentration gradient.

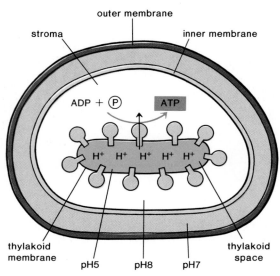

a.

a. Diagram representing a chloroplast. H^+ has been transported by membrane carriers into the thylakoid space. When H^+ flows down this established concentration gradient, ATP synthesis occurs.

b. Jagendorf's "acid-bath" experiment, showing that an artificially generated pH gradient can cause ATP formation. The entire experiment was conducted in the dark.

a. From Alberts, et. al., Molecular Biology of the Cell. Copyright © Garland Publishing Company. Reprinted by permission. b. From Albert L. Lehninger, Biochemistry, Second Edition, Worth Publishers, New York, 1975, pages 612–613. Reprinted by permission.

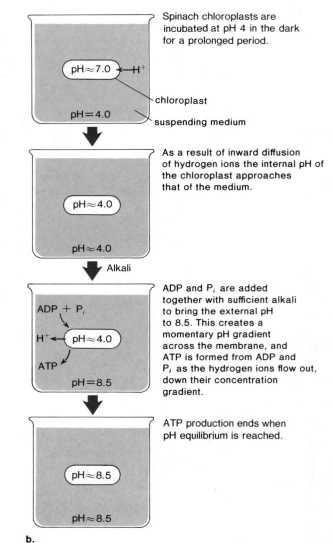

Spinach chloroplasts are incubated at pH 4 in the dark for a prolonged period.

As a result of inward diffusion of hydrogen ions the internal pH of the chloroplast approaches that of the medium.

ADP and P_i are added together with sufficient alkali to bring the external pH to 8.5. This creates a momentary pH gradient across the membrane, and ATP is formed from ADP and P_i as the hydrogen ions flow out, down their concentration gradient.

ATP production ends when pH equilibrium is reached.

b.

is believed to occur whenever CO_2 is in such limited supply that carbohydrate is not being produced. At these times, NADPH buildup prevents noncyclic photophosphorylation from occurring.

Cyclic photophosphorylation provides an independent means by which ATP can be generated. But it does not result in the release of oxygen or in NADPH production. Perhaps this form of photophosphorylation evolved before the noncyclic form, simply as a means to make ATP.

Reducing Carbon Dioxide

The NADPH and ATP produced by noncyclic photophosphorylation are now used to reduce carbon dioxide. Look again at the overall equation for photosynthesis on page 100. Notice that carbohydrate (CH_2O) contains hydrogen atoms, whereas carbon dioxide (CO_2) does not. When carbon binds to hydrogen atoms, it has accepted electrons plus hydrogen ions. Therefore, carbon has been reduced. The reduction process is a building-up, or a synthetic, process because it requires the formation of new bonds. Hydrogen atoms and energy are needed for reduction synthesis, and these are supplied by NADPH and ATP.

Reduction of CO_2 occurs in the *stroma* of the chloroplast. It is a several-step process that is catalyzed by a series of soluble enzymes, with a specific enzyme for each reaction.

C_3 Pathway

When the radioactive isotope carbon fourteen (^{14}C) became available for use in biological research during the 1940s, it became possible to analyze the reactions involved in CO_2 incorporation. Melvin Calvin and his colleagues in California carried out a series of very elegant experiments on CO_2 incorporation, often called **CO_2 fixation.** In Calvin's experiments (fig. 5.15) an illuminated suspension of cells of the unicellular green alga *Chlorella* was exposed to CO_2 that had a radioactive carbon atom ($^{14}CO_2$). The algae were killed by

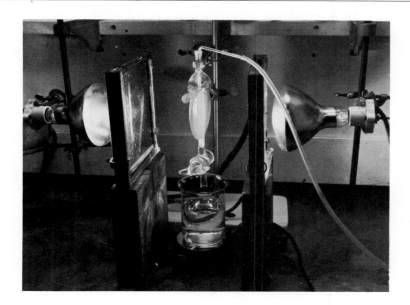

Figure 5.15
When Melvin Calvin and colleagues performed their classic experiments on carbon dioxide fixation, they used the apparatus shown here. Algae were contained in the flat flask, which was illuminated by two lamps. $^{14}CO_2$ was added with a syringe. The algae were killed when the stopcock was opened and the algae fell into the beaker containing alcohol.

dropping the cells into boiling alcohol. This treatment instantly stopped all chemical reactions in the cells. Carbon-containing compounds were then extracted from the cells and analyzed. Calvin and his colleagues were able to trace the reaction series, or pathway, through which CO_2 is incorporated by measuring the amount of radioactive carbon in compounds after ever-increasing periods of exposure to $^{14}CO_2$. The first radioactive compounds detected after a very short exposure to $^{14}CO_2$ were three-carbon compounds; therefore, the whole pathway is often called the **C₃ pathway,** although it is also known as the **Calvin-Benson cycle.**

A later discovery showed that carbon dioxide is taken up by a five-carbon sugar, ribulose bisphosphate (RuBP), and that the resulting six-carbon molecule immediately breaks down to two C_3 molecules of phosphoglycerate (PGA) (fig. 5.16). The intermediate six-carbon molecule is present for such a short time before it splits into two PGA molecules that it has not been possible to isolate or identify it.

Each molecule of PGA then undergoes reduction in two steps to PGAL (phosphoglyceraldehyde).

energy hydrogen

ATP   ADP + Ⓟ

NADPH   NADP

acid (OX) PGA ⟶ ⟶ PGAL *energy rich sugar / reduced form / (reduced) aldehyde*

The reactions that reduce PGA to PGAL use up the NADPH and some of the ATP formed in the thylakoid membrane during noncyclic photophosphorylation (fig. 5.17). These reactions also represent the reduction of carbon dioxide and its conversion to a high energy molecule. In other words, PGAL contains more hydrogen atoms than does PGA. PGAL is the immediate photosynthetic product of the Calvin-Benson cycle.

Some PGAL is used within the chloroplast to re-form ribulose bisphosphate.

5 PGAL ⟶ 3 RuBP

3 ATP   3 ADP

Figure 5.16
A simplified version of the Calvin-Benson cycle. RuBP accepts carbon dioxide (CO_2) forming a six-carbon molecule (C_6) that immediately breaks down to two PGA molecules. These two molecules are then reduced to two PGAL, one of which represents the net gain of the cycle. PGAL can be combined with another PGAL to give glucose-6-phosphate that can be metabolized to other organic molecules.

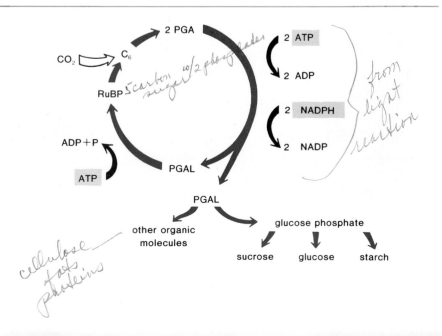

This reaction also utilizes some of the ATP produced by photophosphorylation. Altogether, the ATP and NADPH consumed in converting CO_2 to the level of a hexose (a six-carbon sugar) results in about a 30 percent energy efficiency rate for photosynthesis; that is, about 30 percent of the light energy absorbed is stored.

For each turn of the Calvin-Benson cycle, there is a net gain of one PGAL. Therefore, PGAL is sometimes said to be the end product of photosynthesis. However, since two PGAL combine to give glucose-phosphate, this molecule, or simply glucose, is often called the end product of photosynthesis. Within the leaves of higher plants glucose is converted to the disaccharide sucrose for transport to other parts of the plant and to starch for storage, particularly in the roots. Table 5.2 summarizes the function of each participant in photosynthesis, and figure 5.17 represents an overview.

C_4 Pathway — *gives greater yield (in tropical regions)*

The mechanism of CO_2 incorporation found in Calvin's classic experiments on the green alga *Chlorella,* where the radioactive carbon from CO_2 is first found in C_3 molecules, was subsequently found in many other plants. Recent discoveries have shown that in some plants, particularly those growing in hot climates, the first detectable products of CO_2 incorporation are C_4 molecules. The C_4 pathway requires an extra set of reactions to capture CO_2. In the long run, however, this allows plants to be more efficient, as discussed in the reading on the next page.

Table 5.2 Thylakoids versus Stroma

	Participant	Role
Thylakoids	Sunlight	Provides energy
	Chlorophyll	Absorbs energy
	Water	Donates electrons and releases oxygen
	ADP + $\textcircled{P}$	Forms ATP
	NADP	Becomes NADPH
Stroma	RuBP	Takes up CO_2
	CO_2	Reduced to PGAL
	ATP	Provides energy for reduction
	NADPH	Provides electrons for reduction
	2 PGAL	Becomes glucose phosphate

Figure 5.17
Photosynthesis contains two sets of reactions. Noncyclic photophosphorylation, which takes place in the thylakoid membrane, produces the NADH and ATP required for carbon dioxide reduction that takes place in the stroma.

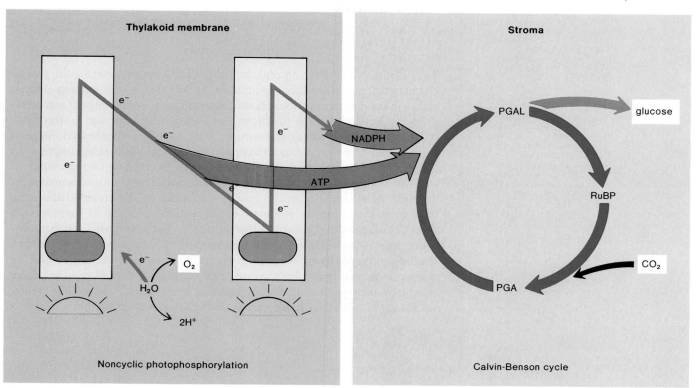

Increasing Crop Yields: C_3 versus C_4 Photosynthesis

Some newly proposed methods of increasing crop yields would not require the expenditure of more supplemental energy. For example, temperate-zone plants, such as wheat, alfalfa, and potatoes, do not take up CO_2 as efficiently as do certain tropical-zone plants, such as corn and sugarcane. The difference lies with the first enzyme of the Calvin-Benson cycle. This enzyme can catalyze one of two reactions.

1. $CO_2 + RuBP \rightarrow C_3$
2. $O_2 + RuBP \rightarrow CO_2 + H_2O$

The first reaction is a normal part of photosynthesis, but it occurs in temperate-zone plants only when CO_2 concentration is high. The second reaction occurs in temperate-zone plants when CO_2 concentration is low. The second reaction is called photorespiration because oxygen is taken up and CO_2 is given off to the environment. Photorespiration accounts for the lower yield in temperate-zone plants since it does not lead to carbohydrate production. How is it avoided by certain tropical-zone plants?

Because of the anatomical structure of certain tropical-zone plants, they avoid photorespiration when CO_2 concentration is low. Photosynthesis occurs only in certain cells, called the bundle-sheath cells (p. 382). CO_2 concentration is always high in the bundle-sheath cells because other cells of the leaf specialize in capturing CO_2 and passing it to the bundle-sheath cells. CO_2 is captured in these cells by PEP, a common organic molecule found in most cells. This reaction occurs whether CO_2 concentration is high or low.

3. $CO_2 + PEP \rightarrow C_4$

Plant Products

Plants have the enzymatic capability to convert glucose phosphate by means of metabolic pathways to all the various types of organic molecules they require.

Thus plants, which have no organic dietary requirements, have a far greater biochemical capability than do animals. In addition to the macromolecules shown in the diagram, various plants also synthesize a wide range of secondary products that have various functions. For example, lignin is used to strengthen plant cell walls, and flower pigments help attract pollinating insects. Some of these products protect plants from predation, either by poisoning insects or by interfering with their life cycle, as discussed in the reading on page 606.

Secondary plant products are often of great importance to humans; for example, quinine is used in the treatment of malaria, and natural rubber can be used to make rubber products. When distilled, the sap of Euphorbia, a desert shrub product, resembles petroleum in that it contains the same hydrocarbons. Melvin Calvin and others propose that _Euphorbia_ could be cultivated commercially in "petroleum plantations" on land that is too arid for other commercial crops. Perhaps we could reduce our dependence on foreign oil not by turning to coal, but by learning more about, and taking greater advantage of, the remarkable energy conversions of photosynthesis.

These C_4 molecules are then transported into the photosynthesizing bundle-sheath cells where CO_2 is released.

It is possible to make a distinction between the two groups of plants according to the first molecule detected following CO_2 uptake. In temperate-zone plants, C_3 is always the first molecule detected following CO_2 uptake (see reaction (1)). In the tropical-zone plants under discussion, C_4 is the first molecule detected (see reaction (3)). Therefore, it is now customary to speak of C_3 *versus* C_4 photosynthesis. Scientists are now exploring the possibility of transforming temperate-zone plants into C_4 photosynthesizers since C_4 photosynthesis is much more efficient. If this can be accomplished, agricultural yields will be greatly increased without supplemental energy.

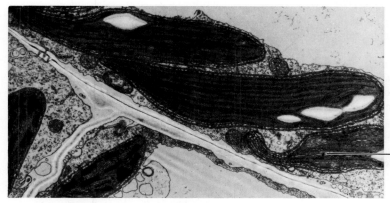

bundle-sheath cells

Corn plant cells capture CO_2 and pass it to the bundle-sheath cells. Notice that the bundle-sheath cells have stroma lamellae but lack grana. Could this be because they specialize in carrying out the Calvin-Benson cycle?

Summary

Autotrophic plants and algae can make their own food through the process of photosynthesis, but heterotrophic animals must take in preformed food. As food is used by organisms, it is gradually converted to inorganic chemicals and heat in accordance with the Second Law of Thermodynamics. This says that one form of usable energy cannot be changed completely into another usable form without some energy being lost as heat. This explains why energy does not cycle back to photosynthesizers and why life cannot continue to exist without sunlight for photosynthesis.

Photosynthesis and cellular respiration require the use of enzymes and coenzymes. Enzymes have active sites where their substrates fit, and this lowers the energy of activation. NAD and NADP are important coenzymes of oxidation and reduction.

Photosynthesis uses visible light, that portion of the electromagnetic spectrum containing radiant energy available in quanta that best activate chlorophyll while not damaging cells. Chlorophyll is found in antennae of Photosystem I and Photosystem II, present in the thylakoid membranes of grana within chloroplasts. When chlorophyll *a* in Photosystem II, known as P680, absorbs light, it loses energized electrons, which are passed down an electron transport system, also called the cytochrome system. As the electrons are passed from one cytochrome to another, energy is made available for ATP formation. Chlorophyll *a* in Photosystem I, known as P700, also absorbs light and loses energized electrons. These are passed to NADP, which also picks up hydrogens from the medium to become NADPH.

Replacement electrons pass to P680 as water splits to release oxygen. (This is the oxygen given off by photosynthesis.)

Noncyclic photophosphorylation is the movement of these electrons from P680 down the cytochrome system with concomitant buildup of ATP, to P700, and on to NADP. In cyclic photophosphorylation, of less importance, the electrons leave P700 and return to it by way of the electron transport system.

Photosynthesis continues within the stroma of chloroplasts where CO_2 is taken up and reduced—made to combine with hydrogen atoms. CO_2 first combines with RuBP to produce a six-carbon molecule that immediately breaks down to two C_3 PGA molecules. Reduction of PGA to PGAL requires the NADPH and some of the ATP produced by photophosphorylation. PGAL is used to regenerate RuBP. Thus a cycle known as the Calvin-Benson cycle is present. PGAL is also used by the plant to make glucose phosphate, a molecule that can be converted to all the organic compounds the plant needs.

Objective Questions

1. Life requires a continual supply of energy. Ultimately, this energy comes from the _____ .
2. To synthesize molecules, cells make use of the chemical energy found in the high energy bonds of _____ .
3. Every reaction that occurs in a cell requires a(n) _____ .
4. NAD and NADP are coenzymes of _____ .
5. The photosystems are located within the _____ membranes of the grana.
6. Chlorophyll molecules are located within the _____ of the photosystems.
7. Lying between Photosystem I and Photosystem II, the _____ system converts ADP to ATP.
8. In addition to ATP, noncyclic photophosphorylation produces the _____ needed by the Calvin-Benson cycle.
9. The Calvin-Benson cycle takes up _____ and reduces it to a carbohydrate.
10. The Calvin-Benson cycle is found in the _____ of the chloroplast.

Study Questions

1. State the two laws of thermodynamics and relate them to the continual need for solar energy to produce food during photosynthesis.
2. Explain the structure of ATP and why it is the energy currency for cells.
3. In what manner do enzymes decrease the energy of activation so that a reaction may more easily occur?
4. Name and explain the manner in which at least three factors can influence the speed of an enzymatic reaction.
5. Define oxidation and reduction. What is the coenzyme of oxidation-reduction in cells?
6. What portion of the electromagnetic spectrum is involved in photosynthesis?
7. Explain the structure of chloroplasts and identify the sites of photophosphorylation and the Calvin-Benson cycle.
8. State the events of noncyclic and cyclic photophosphorylation as if they were sequential. What happens to the NADPH and ATP produced?
9. What reaction in the Calvin-Benson cycle represents the reduction of carbon dioxide?
10. Name the participant molecules in photosynthesis and state their functions.

Selected Key Terms

autotrophic (,ȯt ə 'trō fik) 90
heterotrophic (,het ə rə 'trō fik) 90
herbivore ('ər bə ,vōr) 90
carnivore ('kär nə ,vōr) 90
adenosine triphosphate
  (ə 'den ə ,sēn trī 'fäs ,fāt) 93
photosynthesis (,fōt ō 'sin thə səs) 93
cellular respiration
  ('sel yə lər ,res pə 'rā shən) 93
enzyme ('en ,zīm) 95
substrate ('səb ,strāt) 95
active site ('ak tiv 'sīt) 95
coenzyme (,kō 'en ,zīm) 97
vitamin (,vīt ə mən) 97
oxidation (,äk sə 'dā shən) 97
reduction (ri 'dək shən) 97
chloroplast ('klōr ə ,plast) 100

P hotosynthesis is the means by which food is formed for most living things. Cellular respiration is the means by which energy is extracted from food (fig 6.1). Thus photosynthesis and respiration are woven together as two necessary processes for the continuance of life. In photosynthesis light energy is used to reduce carbon dioxide so that nutrient compounds can be synthesized. In cellular respiration these compounds undergo oxidation and release energy for ATP formation. In photosynthesis reduction occurs when carbon accepts electrons and hydrogen ions. In cellular respiration oxidation occurs when electrons and hydrogen ions are removed from carbon compounds (fig. 6.2). In this chapter we primarily consider the oxidation of **glucose,** but later we briefly discuss the oxidation of other types of organic compounds.

Figure 6.1
Which of these organisms is carrying on cellular respiration—the gerenuk or the acacia tree? Both are because both plants and animals require a source of ATP energy that can be used by cells.

CHAPTER 6

Cellular Respiration

Concepts

1. All living things require a source of energy.
   a. Cellular respiration is the means by which chemical energy within nutrient compounds is converted to ATP (adenosine triphosphate) energy.
   b. Aerobic cellular respiration requires oxygen and results in the maximum number of ATP molecules.
   c. Anaerobic cellular respiration does not require oxygen and results in a minimum number of ATP molecules.

2. Living things are dependent on enzymatic reactions.
   a. Both aerobic and anaerobic cellular respiration require metabolic pathways, each consisting of a series of enzymatically catalyzed reactions.
   b. Cellular respiration can be compared to photosynthesis; both differences and similarities exist.

3. Living things exchange molecules with the physical environment.
   a. Photosynthesis and cellular respiration require that organisms exchange molecules with the environment. There are no specialized regions for exchange in unicellular organisms as there are in complex organisms.
   b. Complex organisms are able to maintain an internal environment that stays relatively constant.

Figure 6.2
The relationships between photosynthesis and cellular respiration. Notice that the energy of the sun (light energy) is eventually transformed to ATP energy.

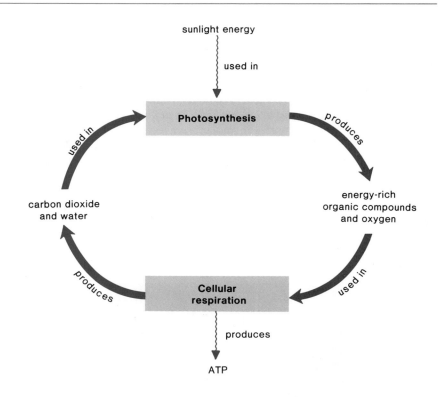

Glucose Oxidation

Burning, a combustion process discussed in the reading on page 114, occurs in one step and immediately gives off a great deal of heat. In contrast, the oxidation of glucose in cells does not occur in one step. A means has evolved by which oxidation occurs gradually, and a portion of the energy released is stored in ATP molecules. Several pathways, each a series of enzyme-catalyzed reactions, are involved in the oxidation of glucose. The first pathway, termed **glycolysis,** begins with glucose and ends with two molecules of pyruvic acid. **Pyruvic acid** is important in cellular respiration because it can be metabolized in different ways (fig. 6.3).

One of the ways in which pyruvic acid can be metabolized following glycolysis is called **anaerobic respiration,** or **fermentation,** which occurs when oxygen is not available. In alcoholic fermentation, carried out by many microorganisms (yeast, for example), pyruvic acid is converted to ethanol (ethyl alcohol). In lactic fermentation, carried out by some microorganisms and some animal cells, pyruvic acid is converted to lactic acid. During heavy exercise, when oxygen supply is insufficient, muscle cells acquire energy by means of lactic fermentation. Anaerobic respiration is believed to be older, in an evolutionary sense, than oxygen-requiring respiration because the primitive earth is believed to have had no available oxygen. We discuss anaerobic respiration in detail later in this chapter.

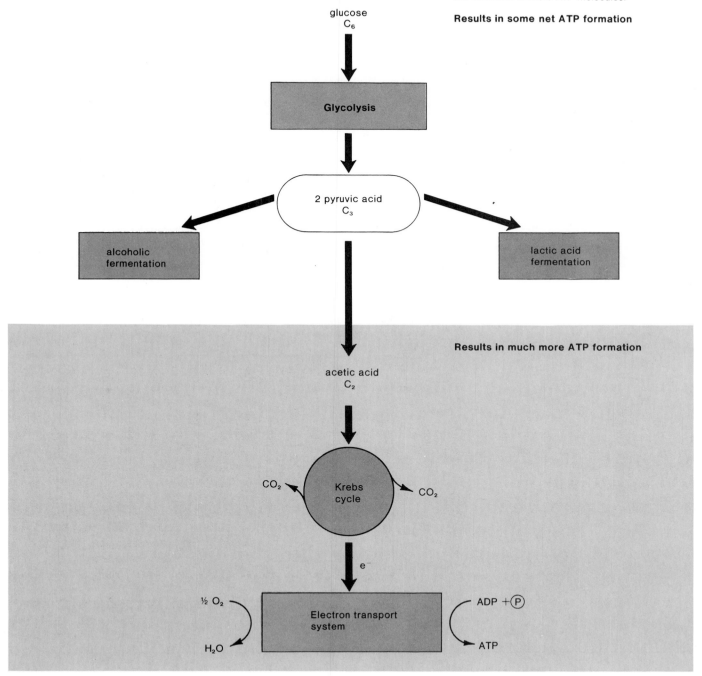

Figure 6.3
Overview of glucose metabolism. Glycolysis results in two
molecules of pyruvic acid. When oxygen is not available,
alcoholic or lactic acid fermentation occurs. When oxygen
is available pyruvic acid is eventually completely broken
down to carbon dioxide and water. Most of the ATP
available from cellular respiration is formed at the electron
transport system; therefore, aerobic respiration results in
the formation of more ATP molecules.

Results in some net ATP formation

glucose
C_6

Glycolysis

2 pyruvic acid
C_3

alcoholic
fermentation

lactic acid
fermentation

Results in much more ATP formation

acetic acid
C_2

CO_2 Krebs
cycle CO_2

e^-

½ O_2 ADP $+\textcircled{P}$

Electron transport
system

H_2O ATP

CO$_2$ in the Atmosphere

Just as in cellular respiration, combustion of fossil fuels, wood, or any other organic material results in the end products carbon dioxide, water, and energy. Human devices often capture the heat energy given off by combustion and use it to do work. Heat energy powers our automobiles, produces electricity, and runs our industrial plants.

This is not the case with carbon dioxide, another important by-product of fossil fuel combustion. Since 1958 the atmospheric concentration of CO$_2$ has risen almost 6 percent (see figure), and it is expected to continue to increase as more fossil fuels are burned by an ever-growing human population. Originally some believed that the concentration of CO$_2$ in the atmosphere would remain about constant despite the amount that combustion of fossil fuels added. It was reasoned that autotrophs would simply take it up as they photosynthesized to a greater degree. This has not proved to be the case, perhaps, in part, because vegetation has been removed to make way for more buildings, roads, and parking lots. If the current increase rate of CO$_2$ in the atmosphere continues, a doubling could occur toward the middle of the next century. The National Academy of Sciences has predicted that this could affect the weather.

Despite its low atmospheric concentration (0.03 percent), CO$_2$ plays a critical role in regulating the earth's surface temperature. CO$_2$ allows the sun's rays to pass through the atmosphere but absorbs and reradiates heat back toward the earth. This may be compared to a greenhouse since the glass of a greenhouse also allows sunlight to pass through, then traps the heat inside the structure. This *greenhouse effect* of CO$_2$ in the atmosphere could cause the average annual temperature to rise as much as 3° to 8° (37° to 46° C). At the least, this would affect agriculture since the favorable climate for growing temperate zone crops would move northward where the soil is not as good. It would also mean that the Soviet Union would have a better climate for agriculture than would the United States. At worst, it might cause polar ice to melt so that the sea level, which has remained the same for thousands of years, would rise to the extent that most of the world's cities would be flooded. In fact, Columbia University researchers recently reported that they have observed a 35 percent decrease in the average summer area of Antarctic pack ice between 1973 and 1980.

Some authorities predict that the greenhouse effect will never materialize because combustion also releases fine carbon particles that prevent the sun's rays from reaching the earth in the first place. Since an increase in the amount of particles could possibly cause the weather to grow colder, the particles are said to have a *refrigerator effect*. Most researchers believe, however, that eventually the greenhouse effect will overcome the refrigerator effect and the average annual temperature will begin to rise.

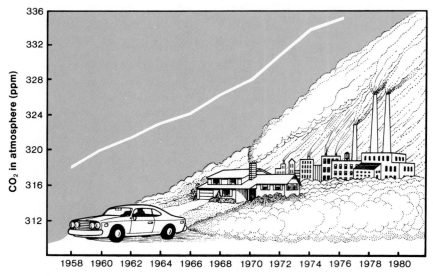

Average concentration of atmospheric carbon dioxide.

When oxygen is available, **aerobic respiration** occurs. Pyruvic acid molecules enter the mitochondria and undergo **decarboxylation,** or removal of carbon in the form of carbon dioxide. A two-carbon compound, acetic acid, remains. This acetic acid enters the **Krebs cycle,** a pathway named for Sir Hans Krebs, who discovered the cyclical nature of the reactions involved. The organic molecules in the Krebs cycle also undergo decarboxylation. Decarboxylation accounts for the carbon dioxide given off during cellular respiration.

As oxidation occurs during aerobic respiration, the electrons are taken, especially by the coenzyme NAD, to an electron transport system located in the mitochondrial membranes. As enzymes in this system pass the electrons from one carrier to another, *energy becomes available for ATP formation.* The electron transport system finally donates electrons to oxygen, which then combines with hydrogen ions to form water.

$$\tfrac{1}{2}O_2 + 2e^- + 2H^+ \longrightarrow H_2O$$

This accounts for the formation of water, called metabolic water, during cellular respiration.

With the synthesis of ATP, the cell now has the form of energy needed to do work such as synthesis, transport, and movement. The structure of ATP and the manner in which it is formed and broken down are discussed on page 93.

Aerobic Respiration

Aerobic cellular respiration takes place in the presence of oxygen and proceeds until glucose has been broken down to carbon dioxide and water.

$$C_6H_{12}O_6 + 6O_2 \rightarrow 6CO_2 + 6H_2O + ATP$$

In the discussion that follows, aerobic cellular respiration is divided into two stages: the first stage takes place outside the mitochondria; the second takes place inside the mitochondria.

Stage One

The glycolytic pathway, or simply glycolysis, is that portion of cellular respiration that occurs outside the mitochondria. **Glycolysis** is the breakdown of glucose to the end product pyruvic acid. Each molecule of glucose results in two molecules of pyruvic acid.

First, two ATP molecules are used to phosphorylate glucose so that it is activated and made ready for subsequent reactions. As hydrogen-carbon bonds are broken, electrons are removed from the molecules of the pathway and are picked up by NAD, which becomes NADH (fig. 5.7). The energy of oxidation results in high energy phosphate bonds attached to certain substrates. Now these can be used to produce ATP.

$$R \sim \text{\textcircled{P}} + ADP \longrightarrow ATP + R$$
R = rest of molecule

This is called **substrate phosphorylation** because the substrate has provided the high energy bond for ATP production.

The steps in glycolysis can be better understood by examining the series of reactions depicted in figure 6.4. In this figure each of the molecules involved is represented by the number of carbon atoms it contains.

When the ATP molecules that were used to start the reaction are subtracted, glycolysis results in a net gain of two NADH and two ATP molecules.

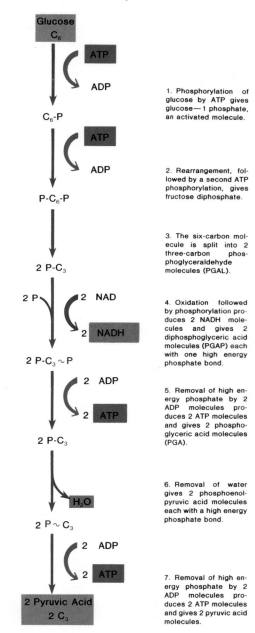

Figure 6.4
Glycolysis is a metabolic pathway that begins with glucose and ends with pyruvic acid. Net gain of ATP molecules can be calculated by subtracting expended ATP molecules from those produced.

1. Phosphorylation of glucose by ATP gives glucose—1 phosphate, an activated molecule.

2. Rearrangement, followed by a second ATP phosphorylation, gives fructose diphosphate.

3. The six-carbon molecule is split into 2 three-carbon phosphoglyceraldehyde molecules (PGAL).

4. Oxidation followed by phosphorylation produces 2 NADH molecules and gives 2 diphosphoglyceric acid molecules (PGAP) each with one high energy phosphate bond.

5. Removal of high energy phosphate by 2 ADP molecules produces 2 ATP molecules and gives 2 phosphoglyceric acid molecules (PGA).

6. Removal of water gives 2 phosphoenolpyruvic acid molecules each with a high energy phosphate bond.

7. Removal of high energy phosphate by 2 ADP molecules produces 2 ATP molecules and gives 2 pyruvic acid molecules.

Figure 6.5
During the transition reaction, pyruvic acid undergoes oxidative decarboxylation to acetic acid, which then combines with Co A to give active acetate. Carbon dioxide is released.

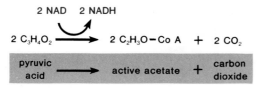

Stage Two

Stage two of cellular respiration, which takes place within the mitochondria, consists of a transition reaction plus the Krebs cycle and an electron transport system, sometimes called the respiratory chain. The **transition reaction** (fig. 6.5), which connects glycolysis to the Krebs cycle, is an oxidation reaction. In this reaction oxidation also involves decarboxylation, so that as pyruvic acid becomes acetic acid, carbon dioxide is given off. The two-carbon acetic acid compound is transferred to coenzyme A, and this combination may be called **active acetate.** Since the transition reaction occurs for each pyruvic acid molecule, it occurs twice for each glucose molecule and produces two molecules of carbon dioxide, two molecules of active acetate, and two NADH molecules.

Krebs Cycle

The next step in cellular respiration is the **Krebs cycle** (fig. 6.6), a metabolic pathway that begins and ends with a six-carbon organic compound called citric acid. For this reason the cycle is also known as the **citric acid cycle.** The Krebs cycle is simply a series of reactions within the mitochondria that always involve the same organic compounds. In figure 6.6, the compounds are only designated by the number of carbon atoms they contain.

The Krebs cycle begins when active acetate reacts with a C_4 compound to give citric acid. Next, *decarboxylation* releases the two carbon atoms derived from active acetate. Consequently, two molecules of carbon dioxide are given off. More important, *oxidation* has also occurred, and the electrons are picked up primarily by NAD. A few are taken by **FAD,** which is another coenzyme of oxidation-reduction but one that is used infrequently compared to

Figure 6.6
The Krebs cycle is a metabolic pathway that begins and ends with citric acid. The cycle occurs twice for each glucose molecule.

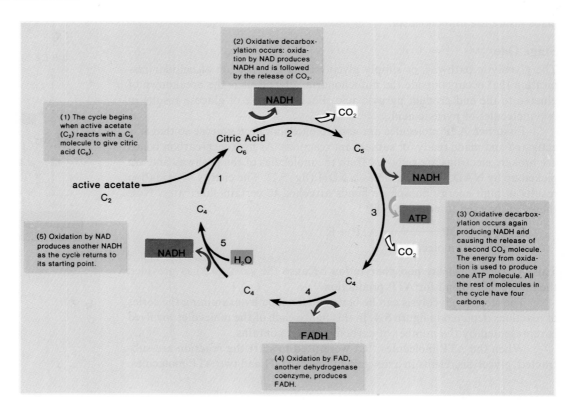

NAD. The energy of oxidation is sufficient to allow the formation of one ATP molecule during the cycle. The cycle occurs twice per glucose molecule and produces four CO_2, six NADH, two FADH, and two ATP.

Respiratory Chain

The **respiratory chain** is an electron transport system found within the mitochondria. It includes cytochrome molecules that pass electrons from one to another (fig. 6.7). Notice that when each cytochrome accepts an electron, it carries one less positive charge than before.

$$\text{cytochrome}^{+++} + e- \rightleftharpoons \text{cytochrome}^{++}$$

Because the respiratory chain contains cytochrome molecules, it is often also called the **cytochrome system.**

Oxygen is the final acceptor for electrons and hydrogen ions at the end of the chain; therefore, *water* is an end product of cellular respiration. As electrons pass down the chain, ATP is formed. ATP formation by way of the respiratory chain is called **oxidative phosphorylation** because it is the passage of electrons down the chain to oxygen that provides the energy for ATP synthesis.

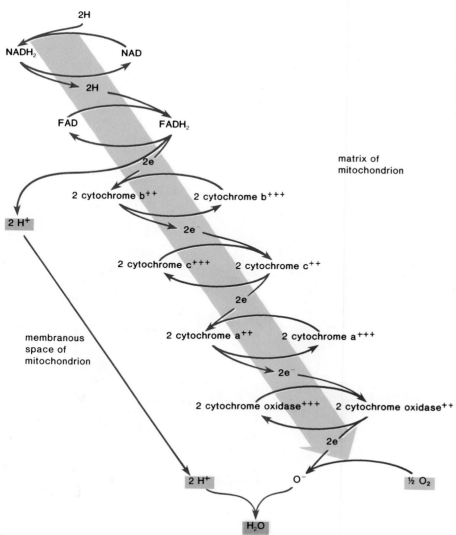

Figure 6.7
Respiratory chain is responsible for ATP production in mitochondria. The carrier molecules that make up this chain are located in the inner membrane of the mitchondrion (fig. 6.9). As electrons (first accompanied by hydrogen ions) pass down the respiratory chain from one carrier to another, energy is released. This energy is used to pump H^+ ions into the membranous space between the inner and outer membranes of the mitochondrion. Later the H^+ gradient is used to bring about the synthesis of ATP as described in the reading on page 121.

Table 6.1 Summary of ATP Produced by Cellular Respiration

	Direct	By Way of Respiratory Chain
Glycolysis	2 ATP	2 NADH = 6 ATP*
Transition reaction		2 NADH = 6 ATP
Krebs cycle	2 ATP	6 NADH = 18 ATP
		2 FADH = 4 ATP
Subtotal	4 ATP	34 ATP
Grand total		38 ATP*

*The numbers in this column and the total number of ATP are usually less because the chain does not always produce the maximum possible number of ATP per NADH.

If electrons enter the chain attached to NAD, the chain may produce as many as three ATP molecules for every two electrons. FAD transport produces at most only two ATP molecules because the electrons enter further down along the chain of carriers. As table 6.1 indicates, the respiratory chain accounts for most of the ATP produced during aerobic cellular respiration. Since the chain is present in mitochondria, this makes them the powerhouses of the cell. For the sake of discussion, it is usually calculated that the chain produces thirty-four of a possible total thirty-eight ATP molecules per glucose molecule. These numbers are for discussion purposes because the chain produces a varying amount of ATP, usually less than thirty-four ATP per glucose molecule.

If thirty-eight ATP should be produced during cellular respiration, this represents an efficiency of about 40 percent. In other words, about 40 percent of the energy available from glucose breakdown has been converted to ATP energy. Figure 6.8 summarizes cellular respiration and indicates the relationship among the three pathways involved.

Figure 6.8
Cellular respiration contains three subpathways: glycolysis, Krebs cycle, and respiratory chain. This overview shows a theoretical maximum of thirty-eight ATP for the breakdown of one glucose molecule.

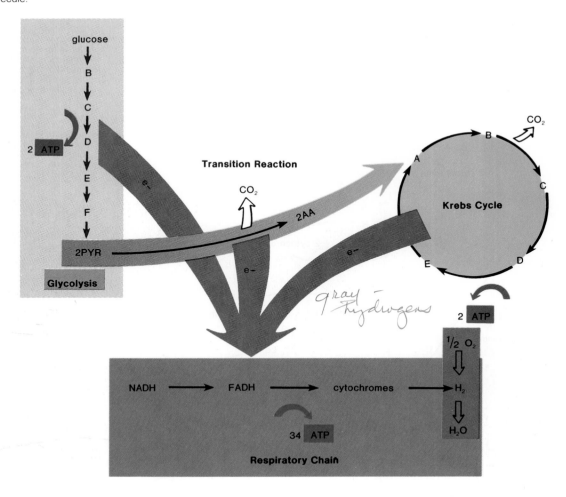

Cellular Organization

We have delayed consideration of cellular organization to avoid interruption of the sequence of reactions in glucose oxidation. We have stated that glycolysis occurs in the cytoplasm outside the mitochondria and that the Krebs cycle and electron transport system are located within the mitochondria. Now we will explore how these pathways are structurally arranged.

Cytoplasmic Organization

Consider that glycolysis (fig. 6.4) occurs in a stepwise manner with each step being controlled by an enzyme. Wouldn't it be appropriate for these enzymes to be arranged in a sequential manner? It now appears that the enzymes are held in proper order by the microtrabecular lattice (fig. 4.19). This structural arrangement organizes the glycolytic pathway and increases the efficiency of the cell.

Mitochondrial Organization

The outer mitochondrial membrane is smooth, while the inner membrane is folded (fig. 6.9). These folds, called **cristae**, increase the inner surface area greatly and it is here that we find the electron carriers of the respiratory chain. The space within the inner membrane, called the **mitochondrial matrix**, contains a background substance that is slightly granular and considerably more dense than the cytoplasm. Most of the Krebs cycle enzymes are located in this gelatinlike substance.

Figure 6.9
Mitochondrial structure. The Krebs cycle occurs in the matrix background substance. The respiratory chain is located within the inner membrane, and ATP formation occurs at the F_1 factor particles. *a.* Drawing of longitudinal section of mitochondrion. *b.* Electron micrograph of inner membrane fragment (EP = Electron transport Particle). *c.* Drawing of particles.

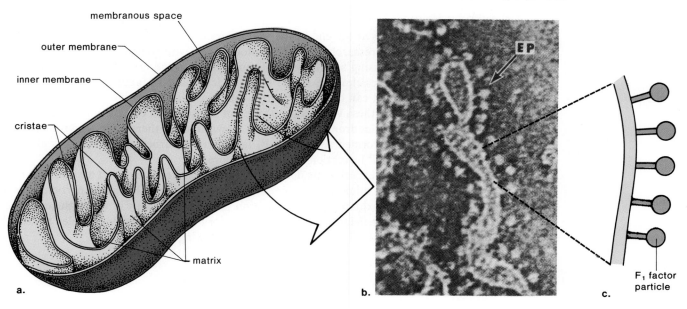

membranous space

outer membrane

inner membrane

cristae

matrix

a.

EP

b.

F_1 factor particle

c.

Figure 6.10
Experiment to determine function of F_1 factor particles. Sonication (high-frequency sound treatment) produces cristae vesicles. Vesicles with F_1 factor particles perform both electron transport and ATP formation. Vesicles lacking F_1 factor particles perform electron transport but not ATP formation. Reconstituted vesicles can perform both functions once more.

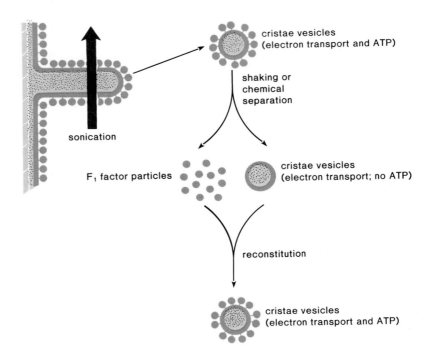

ATP Formation

Detailed examination of the inner mitochondrial membrane and its cristae reveals a number of knoblike particles attached to the matrix side of the inner membrane (figs. 6.9 and 6.10). Are these particles, called F_1 factors, involved in phosphorylation of ADP to produce ATP?

Ephraim Racker and his colleagues have shown that F_1 factor particles are involved in oxidative phosphorylation. ATP formation in the mitochondria is called *oxidative phosphorylation* because oxygen is the final receptor for the electrons at the end of the respiratory chain.

These researchers treated mitochondria with chemicals or with sonic vibrations causing small pieces of material from the cristae to break off and form little vesicles (fig. 6.10). These vesicles have F_1 factor particles on their surfaces and can carry on electron transport and oxidative phosphorylation. However, if these vesicles are shaken or treated with certain enzymes, the F_1 factor particles drop off the vesicles. Now, although there is electron transport, ATP is not formed. But if the F_1 factors are returned to the surfaces of the vesicles, ATP formation occurs once more.

These experiments demonstrated that electron transport assemblies are in the inner mitochondria membrane itself and that F_1 factor particles are involved in oxidative phosphorylation. The proposed way in which the energy derived from electron transport (respiratory chain) is used or "coupled to" the production of ATP is discussed in the following reading.

One of the unsolved problems of cell biology is how the energy derived from the respiratory chain is used in the production of ATP.

$$ADP + \textcircled{P} \rightarrow ATP$$

Just as with chloroplasts, Peter Mitchell (1978 Nobel Chemistry Prize winner) has proposed that his chemiosmotic hypothesis could account for ATP production in mitochondria. According to this proposal, hydrogen ions (H^+) collect [in the space] between the outer and inner membranes of mitochondria because they are placed there by certain carriers of the electron transport system. This establishes a hydrogen ion concentration gradient. When the hydrogen ions flow down their gradient across the membrane at the location of F_1 factor particles, ATP is synthesized.

The evidence for a chemiosmotic scheme in mitochondria is not as strong as it is for chloroplasts. Experiments have shown, though, that hydrogen ions are indeed transported across mitochondrial membrane during electron transport. Thus there is some experimental support for the chemiosmotic thesis. While we understand many aspects of cellular oxidations and energy conversions, there are important gaps in our knowledge when we come to the actual energy "payoff" at ATP synthesis. This provides an incentive for future work in cellular energy conversions.

Adapted from *Biology* by Leland G. Johnson (Wm. C. Brown Publishers, 1983).

ATP Formation in Mitochondria

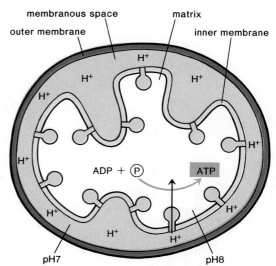

Diagram representing a mitochondrion. H^+ has been transported by membrane carriers into the space between the outer and inner membrane. It is hypothesized that when H^+ flows down this established concentration gradient, ATP synthesis occurs.
From Alberts, et. al., Molecular Biology of the Cell.
Copyright © Garland Publishing Company.
Reprinted by permission.

Anaerobic Respiration

Anaerobic cellular respiration, also called *fermentation,* takes place in the cytoplasm and involves the conversion of glucose to alcohol and carbon dioxide (in yeast and plants) or the conversion of glucose to lactic acid (in most animals) (fig. 6.3). Anaerobic cellular respiration does not require oxygen and allows the glycolytic pathway (see fig. 6.4) to keep operating when oxygen is not available.

The glycolytic pathway will run as long as it has a supply of "free" NAD; that is, NAD that can pick up electrons. We have seen that NAD normally passes electrons to the respiratory chain and thereby becomes "free" of them. However, if the chain is not working because of lack of oxygen, NADH passes electrons to pyruvic acid in either of the following reactions.

For animal cells:

NADH   NAD

pyruvic acid ⟶ lactic acid

For yeast and plant cells:

NADH   NAD

pyruvic acid ⟶ alcohol + CO_2

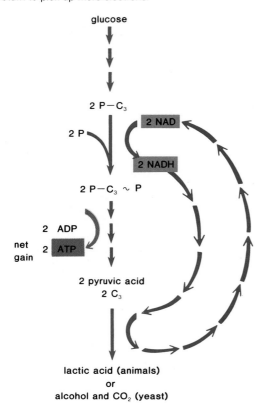

Figure 6.11
Fermentation consists of the glycolytic pathway and one additional reaction in which the end product of the glycolytic pathway, pyruvic acid, accepts hydrogen atoms and becomes reduced. This "frees" NAD so that it can return to pick up more electrons.

NAD is now capable of oxidizing another molecule of PGAL (fig. 6.11), and the glycolytic pathway can keep operating.

Notice that fermentation consists of the glycolytic pathway plus one additional reaction in which the end product of the glycolytic pathway, pyruvic acid, accepts electrons and becomes reduced. The Krebs cycle and respiratory chain have no part in fermentation. When oxygen is once again available in animals, lactic acid can be reconverted to pyruvic acid and metabolism can proceed as usual.

Fermentation is an impractical process for two reasons. First, it produces only two ATP per glucose molecule compared to a much larger number for aerobic respiration. Second, it results in a buildup of molecules, which are toxic to cells; for example, in humans, lactic acid causes muscles to cramp and become fatigued.

The Metabolic Pool and Biosynthesis

Cells can oxidize other molecules besides glucose to release energy. A fat molecule breaks down to glycerol and three fatty acids when it is used as an energy source. Glycerol is easily converted to PGAL (fig. 6.12) and thereby can enter the glycolytic pathway. The fatty acids are converted to active acetate and thereafter enter the Krebs cycle. If a fatty acid contains eighteen carbons, it breaks down to give nine sets of two-carbon active acetates. Calculation shows that these nine active acetates would produce 216 ATP. For this reason, fats are indeed an efficient form of stored energy because there are three long-chain fatty acids per fat molecule.

Amino acids can enter the Krebs cycle at various places, as figure 6.12 also indicates, depending on the length of the "R" group. This group determines the number of carbons left following **deamination,** or removal of the amino group. When the amino group is removed, ammonia (NH_3) is formed.

It should be pointed out that the substrates making up these pathways can also be used as starting molecules for synthetic reactions. In other words, compounds that enter the pathways are oxidized to substrates that can be withdrawn for synthesis. In this way, cells are provided with a **metabolic pool** that allows one type molecule to be converted to another. For example, carbohydrate intake can lead to fat accumulation when PGAL molecules are converted to glycerol molecules and active acetate units are joined together to form fatty acids. Fat synthesis will follow. This explains why it is that one gets fat from eating candy, ice cream, and cake.

Some participants in the Krebs cycle can be converted to amino acids. Plants are able to synthesize all of the amino acids. Animals, however, lack some of the enzymes necessary for synthesis of all amino acids. Humans (and white rats), for example, can synthesize ten of the common amino acids but cannot synthesize the other ten. Amino acids that cannot be synthesized by animals must be supplied by their diets; they are called the *essential amino acids.* The nonessential amino acids are the ones that can be synthesized. It is quite possible for animals to eat large quantities of protein and still suffer from deficiency if their diets do not contain adequate quantities of the amino acids that cannot be synthesized.

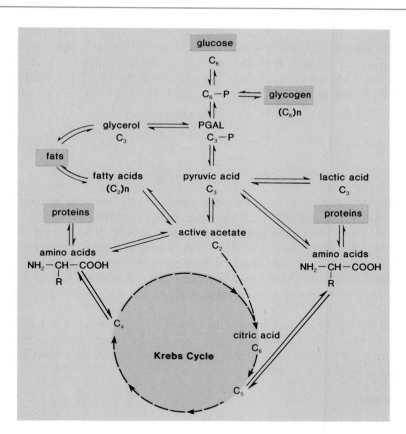

Figure 6.12
The metabolic pool. Organic compounds are broken down to substrates that can be used for synthetic reactions.

Comparison of Cellular Respiration and Photosynthesis

Differences

Whereas only certain cells carry on photosynthesis, all cells, including photosynthesizing cells, carry on cellular respiration, either aerobic or anaerobic. The cellular organelle for cellular respiration is the mitochondrion, while the cellular organelle for photosynthesis is the chloroplast.

The overall equation for aerobic cellular respiration is the opposite of that for photosynthesis:

$$C_6H_{12}O_6 + 6O_2 \underset{\text{photosynthesis}}{\overset{\text{cellular respiration}}{\rightleftharpoons}} 6CO_2 + 6H_2O + \text{energy}$$

The reaction in the forward direction represents cellular respiration and the energy stands for ATP. The reaction in the reverse direction represents photosynthesis, and the energy is the energy of the sun.

Obviously, cellular respiration is the breaking down of glucose, while photosynthesis is the building up of glucose. Table 6.2 summarizes these differences.

Table 6.2 Cellular Respiration and Photosynthesis

Cellular Respiration	Photosynthesis
Oxidation	Reduction
Releases energy	Requires energy
Requires oxygen	Releases oxygen
Releases carbon dioxide	Requires carbon dioxide

Figure 6.13
Simplified overview of the linked processes of photosynthesis and cellular respiration.

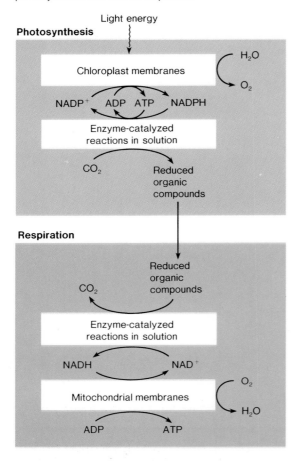

Similarities

Both photosynthesis and cellular respiration are metabolic pathways within cells and therefore consist of a series of reactions (steps) that the overall reaction does not indicate. For example, both pathways utilize this overall reaction, but in the opposite direction. For photosynthesis (fig. 5.16), read from left to right, and for cellular respiration (fig. 6.4), read from right to left.

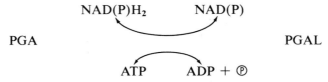

Figure 6.13 gives an overview of the metabolic steps that take place within chloroplasts and mitochondria. Both make use of cytochrome systems located within the inner membrane to generate a supply of ATP. Both depend on enzyme-catalyzed reactions in solution within these organelles. Certain of these reactions make use of oxidation-reduction coenzymes, such as NAD in cellular respiration and NADP in photosynthesis. In photosynthesis, the PGA → PGAL reaction shown previously takes place within the stroma of the chloroplast.

In summary, we want to note that figure 6.13, like figure 6.2, indicates steps that take place in order for light energy to be eventually converted to ATP energy for cellular use. All living things are ultimately dependent on the energy of the sun.

Organisms

If you look again at the overall equation for photosynthesis and cellular respiration (p. 123), you will realize that in order for organisms to carry out these reactions, they must interact with the environment. Figure 6.14 illustrates that unicellular organisms make these exchanges directly with the external environment while complex organisms have an internal environment with which their cells make exchanges.

Eukaryotic Unicellular Organisms

Chlamydomonas (fig. 6.15*a*) is an aquatic autotrophic unicellular organism. The carbon dioxide and water required for photosynthesis diffuse directly from the watery environment into the organism's cup-shaped chloroplast. *Chlamydomonas* stays relatively near the surface in order to absorb the light energy needed to make glucose from these inorganic substances. Just as carbon dioxide diffuses directly from the environment into the chloroplast, so oxygen, the by-product of photosynthesis, diffuses directly out of the chloroplast into the environment.

An amoeba (fig. 6.15*b*) is a heterotrophic unicellular organism that also lives in the water. Amoebas take in preformed organic food by means of phagocytosis. Following vesicle formation, as illustrated in figure 4.15, digestive enzymes break down the food into molecules that enter the cytoplasm. Among these molecules is the glucose needed for cellular respiration. After glycolysis, pyruvic acid molecules enter the mitochondria. Gas exchange occurs directly with the water; oxygen diffuses into the cell and mitochondria, and carbon dioxide diffuses from the mitochondria out of the cell. (The same is true for *Chlamydomonas* whenever these organisms carry on cellular respiration.)

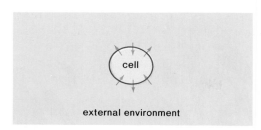

a.

external environment

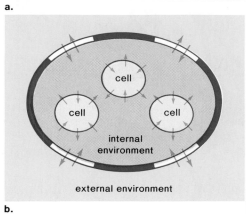

b.

internal environment

external environment

Figure 6.14
Comparison of exchanges made by (*a.*) unicellular organism and (*b.*) multicellular organism. In the former, cell exchanges are made directly with the external environment. In the latter, cell exchanges are made with the internal environment of the organism. The organism makes exchanges with the external environment at specialized regions.

Figure 6.15
Unicellular organisms that exchange materials directly with the external environment. *a. Chlamydomonas* is an autotrophic photosynthesizer. Note the large chloroplast to one side of the single cell. *b.* An amoeba is a heterotroph that engulfs debris and other cells by means of pseudopodia, the long extensions of its single-celled body.

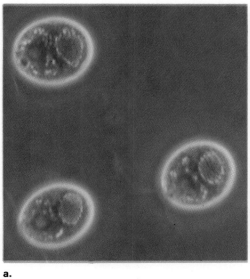

a.

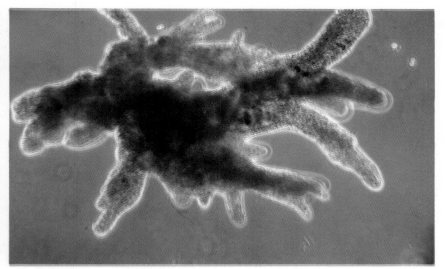

b.

Figure 6.16
Complex organisms have specialized regions where exchanges take place. The leaves of a terrestrial plant, such as a thistle plant, serve as regions of gas exchange and the roots function to take up water and mineral ions. In complex terrestrial animals, such as goldfinches, food is absorbed from the gut, gases are exchanged in the lungs, and water and ions are excreted by the kidneys. The eyes act as receptors.

Figure 6.17
Cheetahs communicate with one another, as do zebras. Whenever the cheetahs kill zebras, there is a flow of energy from one population to the other.

Complex Organisms

In contrast to unicellular organisms, each individual cell of a complex organism does not make exchanges directly with the environment. Instead, the organ systems have specialized regions where a particular type of exchange takes place. For example, in figure 6.16, the root of the plant is specialized to take up water from the soil and the plant's leaves are specialized to take up carbon dioxide from the air. The water must be transported from the roots up the stem to the cells of the leaf in order for photosynthesis to take place. In an animal, the glucose needed for cellular respiration is absorbed from the digestive tract and must be transported in the blood to the cells. Similarly, oxygen enters the body at the lungs and is transported to the cells. Each cell takes the nutrients it needs from, and excretes its wastes into, an internal environment that is nourished and cleansed by blood. The wastes are eventually excreted by the kidneys.

The functions carried out by the organ systems of an animal are coordinated by the nervous and hormonal systems. The nervous system has specialized parts, termed sense organs, that make animals aware of their environment. This awareness is very important because it enables animals to seek and acquire food, and, when necessary, avoid becoming food for other animals. At these times, animals may communicate with others of their own kind (fig. 16.17).

The organ systems of a complex animal all contribute to the maintenance of a relatively constant internal environment, a state termed **homeostasis.** Each cell must make exchanges with this environment; therefore, plants and animals are indeed multicellular and must be studied and understood at the cellular level.

Summary

Three pathways are involved in aerobic cellular respiration. The first of these, glycolysis, occurs in the cytoplasm and is the breakdown of glucose to two molecules of pyruvic acid. During glycolysis, oxidation occurs by the removal of electrons, which are picked up by NAD. The energy made available is enough to result in the net gain of two ATP. The next two pathways occur in the mitochondria. First, there is a transition reaction that converts each pyruvic acid molecule to active acetate as carbon dioxide is released and electrons are received by another NAD. Each active acetate enters the Krebs cycle, a series of reactions that begins and ends with citric acid. During each turn of the Krebs cycle, two more carbon dioxide molecules are released, and three NADH, one FADH, and one ATP are formed.

NADH and FADH carry electrons to the respiratory chain, an electron transport system located in the inner membrane of the mitochondria. For each two electrons that enter by NADH, a maximum of three ATP are formed. For each two electrons that enter by FADH, a maximum of two ATP are formed. Formation of ATP by means of the respiratory chain is called oxidative phosphorylation because the electrons that move down the chain eventually are accepted by oxygen, which then combines with hydrogen ions to become water. Oxidative phosphorylation is associated with F_1 factor particles located on the inner mitochondrial membrane. If the theoretical maximum of thirty-eight ATP are formed per glucose molecule (table 6.1), about 40 percent of the total energy in a glucose molecule is made available. This represents about 40 percent of the energy available from the glucose molecule.

Anaerobic cellular respiration, also called fermentation, consists of glycolysis plus one more reaction, during which pyruvic acid accepts electrons and hydrogen ions and becomes the end products lactic acid (in animals) or carbon dioxide and alcohol (in yeasts and plants). By this means, NAD, which received electrons from a reaction during glycolysis, becomes free of them and can return to accept more electrons. Fermentation results in a net gain of only two ATP and causes a buildup of the end products mentioned, which are toxic.

The substrates of glycolysis and the Krebs cycle form a metabolic pool. By means of this pool, carbohydrates, fats, and amino acids can either be broken down or synthesized.

Cellular respiration is the opposite of photosynthesis, as diagrammed in figure 6.13. However, both photosynthesis and cellular respiration involve cytochrome systems, enzyme-requiring metabolic pathways, and coenzymes of oxidation-reduction.

Photosynthesis and cellular respiration require that organisms exchange gases with the environment. Organisms also get nutrients from the environment and deposit wastes into the environment. The entire cell is the exchange boundary in unicellular organisms such as *Chlamydomonas* (an autotroph) and an amoeba (a heterotroph). All exchanges occur at specialized boundaries in complex multicellular plants and animals. This helps facilitate the maintenance of a constant internal environment.

Objective Questions

1. The two types of cellular respiration are _____, which does not require oxygen, and _____, which does require oxygen.
2. Glycolysis is the break down of glucose to two molecules of _____.
3. The Krebs cycle begins and ends with _____.
4. The respiratory chain is located on the _____ of the mitochondria.
5. The respiratory chain is an electron transport system that produces _____, energy molecules needed by the cell.
6. The final acceptor for hydrogen molecules at the end of the respiratory chain is _____.
7. Carbon dioxide formation should be associated with the _____, one of the three major pathways involved in aerobic cellular respiration.
8. The immediate acceptor of hydrogen molecules during anaerobic cellular respiration is _____.
9. Anaerobic cellular respiration only results in a net gain of _____ ATP.
10. Fatty acids are broken down to _____ molecules which enter the Krebs cycle.

Answers to Objective Questions

1. anaerobic, aerobic 2. pyruvic acid 3. citric acid 4. cristae 5. ATP 6. oxygen 7. Krebs cycle 8. pyruvic acid 9. two 10. active acetate

Study Questions

1. Give several examples to show that photosynthesis and aerobic cellular respiration are reverse processes.
2. Define glycolysis, the Krebs cycle, and the respiratory chain.
3. What are the net results for each of the metabolic pathways listed in question 2?
4. Show that a theoretical maximum of 38 ATP may result from the breakdown of one glucose molecule.
5. State where the Krebs cycle, the respiratory chain, and oxidative phosphorylation are believed to occur within the mitochondria.
6. Define anaerobic cellular respiration. Why is this process biologically wasteful and potentially harmful?
7. Discuss the metabolic pool concept. Explain the manner in which carbohydrates can be converted to fat.
8. Contrast the manner in which unicellular and complex multicellular organisms exchange molecules with the environment.

Selected Key Terms

glucose ('glü ,kōs) 111
glycolysis (glī 'käl ə səs) 112
pyruvic acid (pī ,rü vik 'as əd) 112
anaerobic respiration (,an ə 'rō bik ,res pə 'rā shən) 112
fermentation (,fər mən 'tā shən) 112

aerobic respiration (,ā 'rō bik res pə rā shən) 115
Krebs cycle ('krebz 'sī kəl) 115
citric acid cycle (,si trik ,as əd ,sī kəl) 116
respiratory chain ('res pər ə ,tör ē 'chān) 117

cytochrome system ('sīt ə ,krōm 'sis təm) 117
oxidative phosphorylation ('äk sə ,dāt iv ,fäs ,fór ə 'lā shən) 117
mitochondrion (,mīt ə 'kän drē ən) 119
cristae ('kris ,tē) 119
matrix ('mā triks) 119
metabolic pool (,met ə 'bäl ik 'pül) 122

Suggested Readings for Part 3

Baker, J. J. W., and Allen, G. E. 1981. *Matter, energy, and life*. 4th ed. Reading, MA: Addison-Wesley.

Becker, W. M. 1977. *Energy and the living cell: An introduction to bioenergetics*. New York: Harper & Row, Publishers, Inc.

Cronquist, A. 1982. *Basic botany*. 2d ed. New York: Harper & Row, Publishers, Inc.

Dickerson, E. March 1980. Cytochrome and the evolution of energy metabolism. *Scientific American*.

Hinkle, P. C., and McCarty, R. E. March 1978. How cells make ATP. *Scientific American*.

Miller, K. R. October 1979. The photosynthetic membrane. *Scientific American*.

Rayle, D., and Wedberg, H. L. 1980. *Botany: A human concern*. Boston: Houghton Mifflin.

Sheeler, P., and Bianchi, D. E. 1980. *Cell biology: Structure, biochemistry, and function*. New York: John Wiley.

Stryer, L. 1981. *Biochemistry*. 2d ed. San Francisco: W. H. Freeman.

Wolfe, S. L. 1980. *Biology of the cell*. Belmont, CA: Wadsworth.

Continuance of Life

All species of organisms, whether unicellular or multicellular, are capable of reproducing themselves. When they reproduce, they pass gene-bearing chromosomes on to their offspring. When unicellular organisms reproduce by dividing, each daughter cell has the same genetic makeup as the parent cell. In sexually reproducing organisms, each parent contributes one half of the total number of chromosomes to the offspring. Therefore, each offspring receives a different combination of genes. It is sometimes possible to determine the chances of an offspring receiving particular combinations of chromosomes and/or genes. If the genetic makeup of human parents is known, for example, it is sometimes possible to determine the chances of a child inheriting a genetic disease.

Genes, now known to be constructed of DNA, control not only the metabolism of the cell but also, ultimately, the characteristics of the individual. DNA contains a code for the sequence of amino acids in proteins that are synthesized at the ribosomes. The step-by-step procedure by which the DNA code is transcribed and translated to assure the formation of a particular protein has been discovered.

Pupa of monarch butterfly. The life cycle of a butterfly includes metamorphosis, a transformation from the wormlike larva to the flying insect. Metamorphosis occurs within the enclosed pupa, the structure shown here. When ready the pupa will open and the butterfly will emerge to fly away.

CHAPTER 7

Cell Reproduction

Concepts

Reproduction is a characteristic of life. Both organisms and cells reproduce.

a. Eukaryotic cell reproduction involves nuclear division and cytoplasmic division.

b. Mitotic nuclear division results in two daughter nuclei, each of which has a full complement of chromosomes and genes.

c. Mitosis is necessary for the growth and repair of multicellular eukaryotes and for asexual reproduction of various eukaryotes.

d. Meiotic nuclear division results in four daughter nuclei, each of which has half the number of chromosomes and genes of the mother cell nucleus.

e. Meiosis is a necessary component of sexual reproduction in both plants and animals, and a source of genetic variation.

f. Cell division in prokaryotes does not involve mitosis.

Cell division is necessary to the life of any organism. When the organism consists of a single cell, cell division permits reproduction. Even in a multicellular organism, cell division is involved in the production of cells that allow reproduction. At all times, cell division is necessary for growth and repair. The human embryo (fig. 7.1), or any embryo, begins as a single cell that repeatedly divides to produce the millions of cells needed for the development of organs. Thereafter, human cells differ in their capacity to divide. Some cells, such as skeletal muscle cells and nerve cells, do not usually divide at all. Others, like liver cells, will divide if tissue has been lost due to injury, surgery, or disease. Others, for example those that produce skin cells or blood cells, divide constantly. In the adult human, in fact, millions of cells must divide every second simply to maintain the life of the body.

During cell division, the genetic material (DNA) is distributed to the newly formed cells. Because of this, each and every cell contains genes that control the metabolism and the characteristics of the cell. In eukaryotic cells, these genes are enclosed in a nucleus.

Figure 7.1
Already this human embryo is made up of millions of cells. Due to the process of nuclear division, each cell contains the same complement of genes, copies of the very ones that were inherited from its parents.

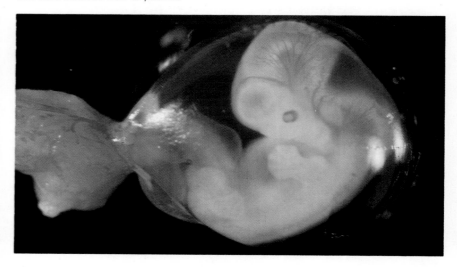

Cell Division

Nucleus

The eukaryotic cell contains two distinct regions: the cytoplasm and the nucleus, which is often centrally located (fig. 7.2). The nucleus is enclosed by two membranes collectively called the nuclear envelope. The outer membrane of the nuclear envelope is continuous with the endoplasmic reticulum. There are numerous **pores,** or openings, in the nuclear envelope that probably allow substances to pass between the nucleus and the surrounding cytoplasm. Some electron micrographs show granular material in the pores, which may be material caught moving through these openings.

The nucleus contains a background substance known as **nucleoplasm.** Within the nucleoplasm, one or more **nucleoli** are usually found. Ribosomal RNA (rRNA) is synthesized and stored in the nucleoli. This RNA is necessary for the formation of ribosomes (p. 75). Most important, the nucleoplasm also contains the threadlike **chromatin** where the DNA of the cell is located. In nondividing cells the chromatin is dispersed, but during cell division it condenses, forming the highly coiled **chromosomes.**

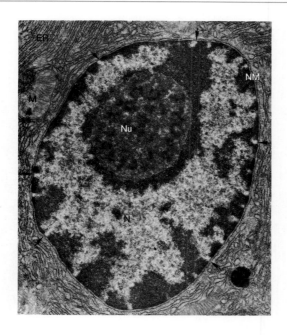

Figure 7.2
An electron micrograph of a nucleus that has a clearly defined nucleolus (Nu) and irregular patches of chromatin scattered throughout the nucleoplasm. The chromatin contains DNA and condenses to form chromosomes prior to cell division. (N = nucleus; Nu = nucleolus; M = mitochondria; ER = endoplasmic reticulum. Nuclear pores are indicated by the arrows.)

Figure 7.3

Steps in karyotype preparation. A karyotype shows the pairs of chromosomes that occur in the nucleus just prior to nuclear division. Only at this time are the chromosomes distinctly visible.

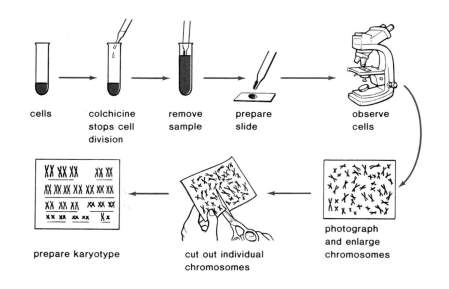

cells   colchicine stops cell division   remove sample   prepare slide   observe cells

photograph and enlarge chromosomes

cut out individual chromosomes

prepare karyotype

Figure 7.4

a. Karyotype of a male. Note the pairs of autosomes, numbered from 1 to 22, and one pair of sex chromosomes, X and Y. b. Drawing of an enlarged chromosome.

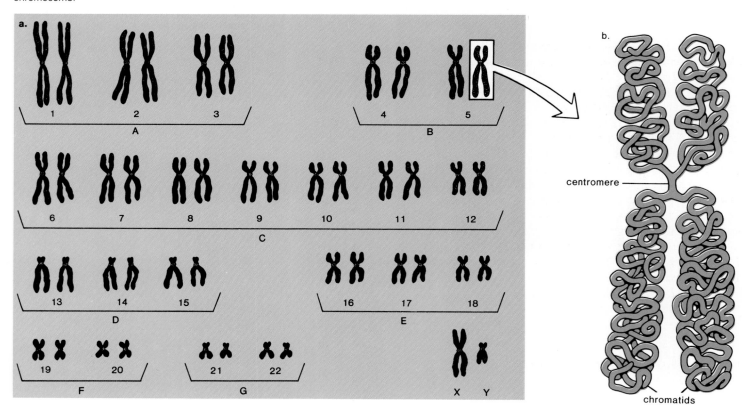

Chromosomes

Each species has a characteristic number of chromosomes. For example, corn cells have twenty chromosomes, mice cells have forty chromosomes, and human cells have forty-six chromosomes. To clearly view the chromosomes so that they can be counted, a cell can be treated and photographed just prior to dividing, as described in figure 7.3. The chromosomes can then be cut out of the photograph and arranged in pairs. The members of a pair have the same size and general appearance and, therefore, are called a **homologous pair.** The resulting display of paired chromosomes is called a **karyotype.** A human karyotype is shown in figure 7.4.

Although both male and female humans have twenty-three pairs of chromosomes, the chromosomes of one pair are of unequal length in the male. The larger chromosome in this pair is called the X chromosome and the smaller one is called the Y chromosome. Females have two X chromosomes in their karyotype. These X and Y chromosomes are called the **sex chromosomes** because they contain genes that determine sex. All of the other chromosomes are known as **autosomes.** Humans have twenty-two pairs of autosomes and one pair of sex chromosomes.[1]

Just prior to division every chromosome is composed of two identical parts called **chromatids,** as illustrated in figure 7.4b. The two chromatids are genetically identical; that is, they contain the genes that control the same traits. The chromatids are constricted and seem to be attached to each other at a region called the **centromere.**

Mitosis

Mitosis is the division of the eukaryotic nucleus in which the nuclei of the daughter cells acquire the same number and kinds of chromosomes as in the mother cell nucleus. The mother cell is the dividing cell, and the daughter cells are the resulting cells. Figure 7.5 is an overview of cell division involving mitosis. Each nucleus in the diagram has four chromosomes. (In determining the number of chromosomes, only the number of independent centromeres is counted.) Figure 7.5 illustrates that a cell prepares for mitosis by replication of the genetic material contained within each chromosome. *Replication,* the process by which DNA makes a copy of itself, is described in detail in chapter 10. Because of replication, every chromosome in the mother cell nucleus contains duplicate chromatids. During mitosis, these *sister chromatids* separate so that both daughter nuclei receive one of each. This ensures that both daughter nuclei receive a copy of every chromosome rather than two copies of one chromosome and none of another. Different series of genes are on different chromosomes. Therefore, for each daughter cell nucleus to receive a full complement of genes, it must receive a copy of every chromosome.

Mitosis (fig. 7.6) is a continuous process that is arbitrarily divided into four stages for convenience of description. These are prophase, metaphase, anaphase, and telophase. In between cell divisions, the cell is said to be in interphase.

[1]Sex chromosomes in other animals are discussed on page 175.

Figure 7.5
A simplified overview of mitosis. The chromosomes are represented diagrammatically, making it possible for us to see that the daughter cell nuclei have the same number of chromosomes as the mother cell nucleus.

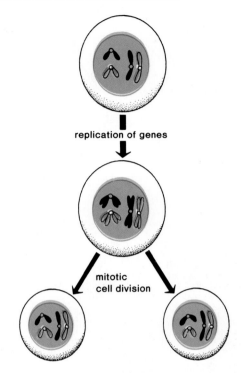

replication of genes

mitotic cell division

Figure 7.6

Mitosis has four stages, excluding interphase and daughter cells. In these drawings, the daughter cells are shown in late interphase (following duplication of chromosomes) since it is assumed that each will be dividing again.

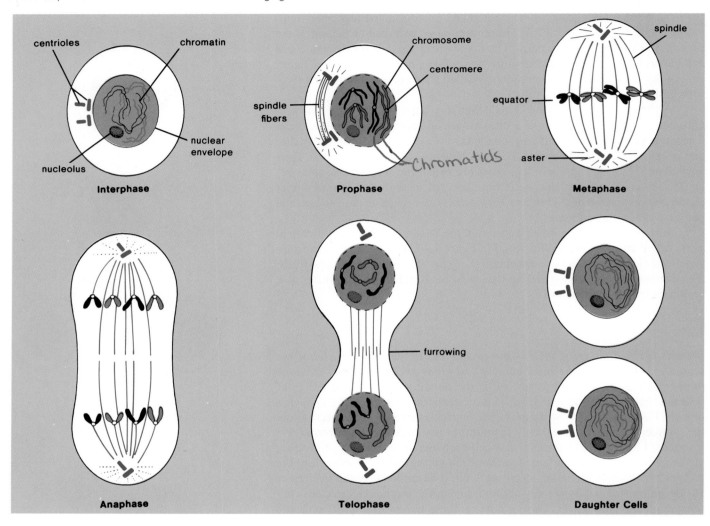

Interphase — centrioles, chromatin, nuclear envelope, nucleolus

Prophase — chromosome, centromere, spindle fibers, Chromatids

Metaphase — spindle, equator, aster

Anaphase

Telophase — furrowing

Daughter Cells

Figure 7.7

Micrograph of prophase. The chromosomes are now distinct and the nuclear envelope has disappeared.

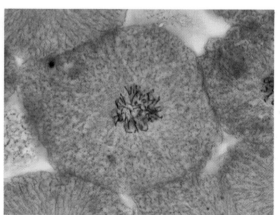

Stages of Mitosis

Prophase

It is apparent during prophase that nuclear division is about to occur because the diffuse chromatin shortens and thickens and forms distinct chromosomes. As the chromatin is condensing, the chromosomes are becoming highly coiled structures (fig. 7.4b). The pairs of centrioles, formerly located to one side of the nucleus, begin moving away from one another. **Spindle fibers** appear between the separating pairs of centrioles. These centrioles also give off short radiating fibers called **asters.** As the spindle appears, the nuclear envelope breaks down and the nucleoli disappear. Figure 7.7 is a micrograph of an animal cell at the time of late prophase. The chromosomes are randomly placed even though the spindle now appears to be fully formed.

Function of Centrioles The entire spindle apparatus is shown in figure 7.8. It consists of asters, spindle fibers, and centrioles. Both the asters and the spindle fibers are composed of microtubules. It is known that microtubules are capable of assembling and disassembling (fig. 7.9), which would account for the

Figure 7.8
A drawing of the spindle apparatus from an animal cell at the time of prometaphase. Some of the microtubules are attached to the chromosome centromeres. Other microtubules seem to extend from pole to pole. The centrioles are surrounded by short microtubules, forming the aster.

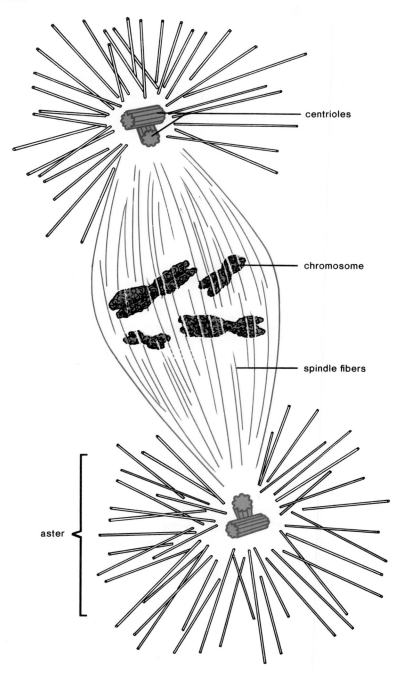

centrioles

chromosome

spindle fibers

aster

Figure 7.9
Microtubules can assemble at one end or disassemble at the other. During assembly, protein dimers join together, and during disassembly the protein dimers separate from one another.

assembly end

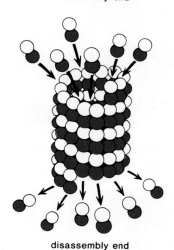

disassembly end

appearance and disappearance of asters and spindle fibers. Does the location of centrioles at the **poles** (ends) of the apparatus signify that they are organizing centers for spindle assembly? While some investigators continue to believe that they are, there is strong evidence against this assumption. Even though plant cells lack centrioles, they still produce spindles during mitosis. (Such spindles have no asters and are less focused at the poles, however.) Also,

Figure 7.10
Metaphase in a whitefish (animal) embryonic cell.

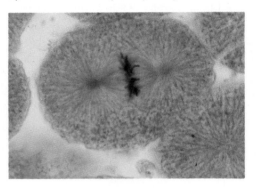

Figure 7.11
Anaphase in a whitefish (animal) embryonic cell.

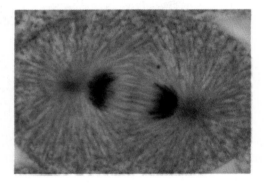

Figure 7.12
Telophase in a whitefish (animal) embryonic cell.

when the centrioles of dividing animal cells are destroyed by a laser beam, the spindle continues to function normally. This growing body of evidence seems to suggest, then, that centrioles are not organizing centers for formation of the spindle. It may be that their location at the poles simply ensures that each daughter cell will have a pair of centrioles.

Metaphase

During metaphase, the chromosomes become aligned so that they are at right angles to the spindle apparatus. They are attached at the region of their centromeres to the equator (center) of the spindle (fig. 7.10). Notice that this arrangement orients the sister chromatids so that one chromatid is roughly parallel to each of the poles (fig. 7.6). The chromosomes are now ready to be divided.

Anaphase

During anaphase, the centromeres divide. This allows the sister chromatids to separate. They then move toward opposite poles of the spindle (fig. 7.11). *Once separated, the chromatids are called chromosomes.* Separation of the sister chromatids ensures that each daughter cell will receive a copy of each type of chromosome and thus have a full complement of genes.

Movement of the Chromosomes The mechanism of chromosomal movement during anaphase is the subject of much investigation and speculation. Two types of spindle fibers have been identified. The so-called *polar fibers* extend from the poles to the equator of the spindle where they overlap one another. These fibers elongate during anaphase, causing the spindle apparatus to increase in length. During elongation, the fibers are pulling away from each other and creating a *push* that moves the chromosomes toward the poles. Since ATP is required, it suggests that the microtubules making up the polar fibers are sliding past one another.

The chromosomes themselves are not attached to the polar fibers. Each is attached to the second type of spindle fibers, the *centromeric fibers,* that extend from the region of the centromere to each of the poles. These fibers get shorter and shorter as the chromosomes move toward the poles, and eventually they disappear. Undoubtedly disassembly (fig. 7.9) is the mechanism by which these fibers shorten, providing a *pull* that moves the chromosomes toward the poles.

Telophase

During telophase (fig. 7.12) a nuclear envelope reforms around the chromosomes in each daughter cell and the chromosomes uncoil to become indistinct chromatin once more. Most likely uncoiling is necessary in order for DNA to carry on its functions during interphase. Also, nucleoli reappear as RNA synthesis resumes.

Cytokinesis

Cytokinesis, or cytoplasmic cleavage, often accompanies mitosis. In animal cells, a gradual indentation of the membrane between the two daughter nuclei, termed **furrowing,** divides the cytoplasm. The equator of the spindle apparatus marks where furrowing will occur. Here a bundle of microfilaments slowly constricts the cell, forming first a narrow bridge (fig. 7.13), and then finally completely separating the cell into the two daughter cells.

Cytokinesis occurs by a different process in plant cells. The rigid cell wall that surrounds plant cells does not permit cytokinesis by means of furrowing. Instead, vesicles largely derived from the Golgi body travel down the polar spindle fibers to the region of the equator. These vesicles fuse to form a membranous structure, which is the early **cell plate** (fig. 7.14). The vesicles also release substances that signal the formation of plant cell walls. These walls are later strengthened by the formation of cellulose fibrils.

Importance of Mitosis

Mitosis ensures that each daughter cell receives the same number and kinds of chromosomes as the mother cell; thus, mitosis ensures that each daughter cell is *genetically identical* to the mother cell.

As mentioned earlier, mitosis is involved in the growth and repair of multicellular organisms. Growth by an increase in the number of cells, rather than by the enlargement of one cell, seems to be dictated by a physical principle. As a sphere increases in size, the volume increases to a greater degree than does the surface area. This means that large cells have more volume per surface area than do small cells. This consideration seems to put a limit on cell size because molecules enter and exit a cell by way of the cell's outer surface, the cell membrane. Cell division provides adequate surface area for the cell to acquire nutrients and excrete wastes at a rate sufficient to maintain its existence.

Figure 7.13
Electron micrograph of a human cell in advanced stage of cytokinesis. Note the remnants of spindle fibers in the furrow region.

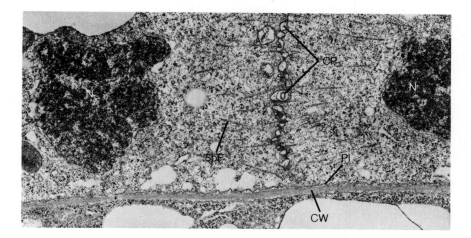

Figure 7.14
Cytokinesis in a plant cell involves the formation of a cell plate, as illustrated in this electron micrograph. The nuclear envelopes of the daughter cells have already reformed. CP = cell plate; CW = cell wall; Pl = plasma membrane; N = nucleus; SpF = spindle fiber.

Figure 7.15
Longitudinal section of an onion root tip. Many of the cells are undergoing cell division because this portion of the onion contributes to root growth. Compare these plant cells to the animal cells in figures 7.10 through 7.12. Note that these plant cells have no asters.

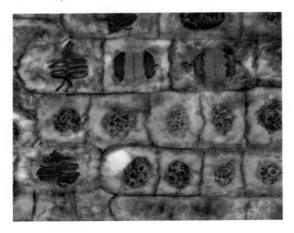

Figure 7.16
The cell cycle contains the four phases shown. DNA replication takes place during the *S* phase and growth occurs during the *G* phases. The diagram indicates the proportional length of time spent in each phase in mammalian cells that complete the cycle in about twenty hours. Interphase consists of all phases except mitosis—a process that usually takes one to two hours. The cell cycle stops when a cell becomes specialized (differentiated).

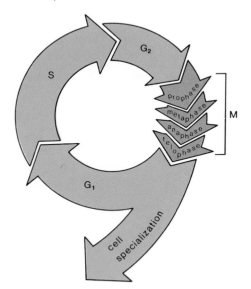

Mitotic cell division is also a common means by which some single-celled eukaryotes reproduce. Many protozoa, for example, an amoeba, reproduce by simply dividing. Multicellular organisms also occasionally reproduce asexually by means of mitotic cell division. As an example, consider that some plants, such as violets and strawberries, send out runners from which new plants develop. There are even some animals, such as sponges and hydras, that reproduce by budding. A bud forms on the parent organism and develops into a completely new organism.

The Cell Cycle

By the 1870s microscopy could give detailed and accurate descriptions of the movement of chromosomes during mitosis. But there were no adequate microscopic techniques for studying the detailed activities of the cell during the period between divisions. Because there was little visible activity between divisions, this period of time was dismissed as a resting stage and termed **interphase.** When it was discovered in the 1950s that DNA synthesis occurs during interphase, the **cell cycle** concept was formulated.

Cells grow and divide during a cycle that has four phases (fig. 7.16). The entire cell division phase, including both mitosis and cytokinesis, is termed the *M phase* (M = mitotic). The period of DNA synthesis is termed the *S phase* (S = synthesis) of the cycle. There are two other phases that fall in between these two. The period of time following the M phase before the S phase begins is termed the G_1 *phase,* and the period of time following the S phase before the M phase begins is termed the G_2 *phase.* At first not much was known about these phases and they were thought of as G = gap phases. Now we know that during the G_1 phase, the cell grows in size, and the cellular organelles double in number. During the G_2 phase, synthesis of various enzymes and structural proteins necessary for mitosis takes place. Therefore, some biologists now prefer the term G = growth phases. In any case, it is now obvious that *interphase* consists of the G_1, S, and G_2 phases (fig. 7.16).

Normally, cells mature and differentiate as cell specialization takes place (fig. 7.16). In these cells, the cell cycle stops in the G_1 phase and DNA synthesis never takes place. If DNA synthesis during the S phase does take place, the cell is committed to undergoing mitosis and completing the cycle. Cancer cells never specialize and keep on undergoing the cell cycle repeatedly. As discussed on page 213, it has been discovered that cancer cells mistakenly produce growth factors, chemicals that presumably stimulate their continual entrance into the S phase.

Meiosis

In sexual reproduction the **zygote,** the first cell of the new organism, is produced by the fusion of two **gametes** (sex cells). Therefore, the gametes, often specialized as a sperm and an egg, must carry traits from one generation to the next. As described in the reading on page 143, in 1868 Charles Darwin attempted unsuccessfully to explain this phenomenon. Since that time we have discovered that the gametes contribute chromosomes to the new individual. Each gamete contains half of the normal complement of chromosomes, and **meiosis** is the special type of nuclear division that halves the chromosome number. If meiosis did not occur, the total number of chromosomes in a new organism would double every time two gametes fused to form a zygote.

The number of chromosomes in the body cells is designated as the **diploid** or **2N number** and the number of chromosomes in the gametes is designated as the **haploid** or **N number** of chromosomes. Thus, meiosis involves the

gametes (N)

Meiosis

zygote (2N)

Mitosis

adult mammal (2N)

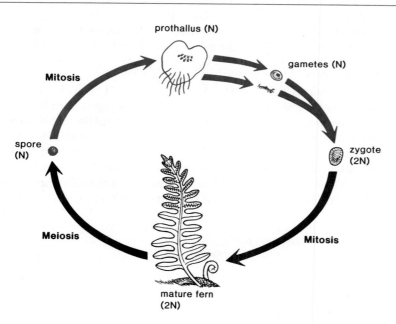

prothallus (N)

Mitosis

gametes (N)

spore (N)

zygote (2N)

Meiosis

Mitosis

mature fern (2N)

reduction in the number of chromosomes from the diploid to the haploid number. For example, a human body cell contains forty-six chromosomes (fig. 7.4). This is the diploid number for humans. These forty-six chromosomes contain twenty-three pairs of homologous chromosomes. Human gametes contain twenty-three chromosomes, one from each of these pairs. This is the haploid number for humans.

Meiosis is part of the life cycle in all sexually reproducing organisms. *In most animals,* meiosis immediately precedes formation of the gametes (fig. 7.17). Plants have life cycles quite different from animals; they alternate between two different generations. The *sporophyte* generation, often a 2N generation, produces haploid spores. In plants, meiosis occurs during the production of spores (fig. 7.18). Each spore develops into a haploid generation called the *gametophyte,* which produces gametes. Meiosis, then, is not directly involved in gamete production in plants.

Figure 7.19
A simplified overview of cell divisions involving meiosis I and meiosis II. The chromosomes are represented diagrammatically, making it possible for us to see that the daughter cell nuclei have half the number of chromosomes as the mother cell nucleus.

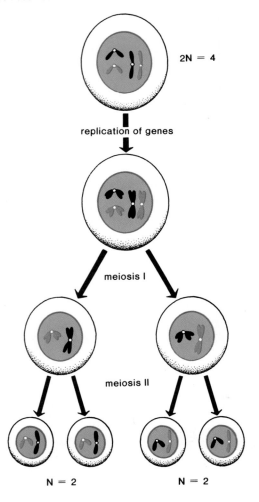

Overview of Meiosis

Meiosis, which requires two nuclear divisions, produces four daughter nuclei, each having one of each kind of chromosome and thus half the total number of chromosomes present in the mother cell nucleus. Figure 7.19 presents an overview of meiosis indicating its two nuclear divisions, **meiosis I** and **meiosis II.** Prior to meiosis I, replication has occurred. Therefore, each chromosome consists of duplicate chromatids.

During meiosis I, the homologous chromosomes of each pair come together and line up side by side. (The mechanism that attracts corresponding chromosomes to each other is still unknown.) This so-called **synapsis** results in **tetrads,** associations of four chromatids that stay in close proximity during the first two phases of meiosis I. Now an important phenomenon called **crossing over** may occur (fig. 7.20). During crossing over, chromatids of the homologous chromosome pairs exchange segments. This produces new combinations of genes on the chromatids. The homologous chromosomes of each pair then separate. This separation assures that one chromosome of every homologous pair reaches each gamete.[2] The separation process has no restrictions; either chromosome of a homologous pair may occur in a gamete with either chromosome of any other pair.[3]

At the completion of the first cell division, as illustrated in figure 7.19, the chromosomes in the nucleus of each daughter cell still consist of duplicate chromatids. During meiosis II the chromatids separate, so that the nucleus in each of four daughter cells has the haploid number of chromosomes.

Stages of Meiosis

The stages of meiosis I are diagrammed in figure 7.21. During *prophase I,* the homologous chromosomes of each pair undergo synapsis, forming tetrads. The nuclear envelope and nucleolus do not disappear until the end of prophase I. At *metaphase I,* tetrads line up at the equator of the spindle. During *anaphase I,* the homologous chromosomes of each pair separate, and the chromosomes (each still composed of two chromatids) move to the poles of the spindle. Each pole receives one-half the total number of chromosomes. In some species there is a *telophase I* stage at the end of the first division. If there is

[2]See Mendel's Law of Segregation, p. 152.
[3]See Mendel's Law of Independent Assortment, p. 157.

Figure 7.20
During crossing over, pieces of chromosomes are exchanged between chromatid pairs. *a.* Chromatid pairs before crossing over has occurred. *b.* Chromatid pairs after crossing over. Notice the change in chromosome structure.

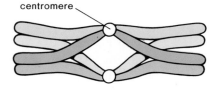

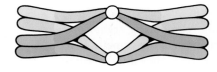

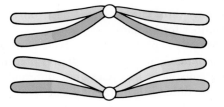

a.

b.

c.

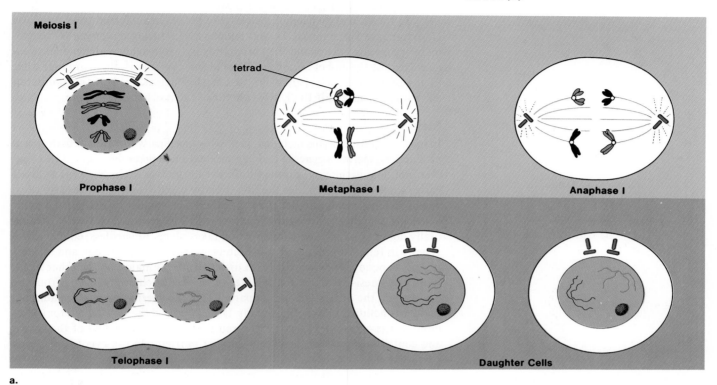

Meiosis I

Prophase I

tetrad

Metaphase I

Anaphase I

Telophase I

Daughter Cells

a.

Meiosis II

Prophase II

Metaphase II

Anaphase II

Telophase II

Daughter Cells

b.

a telophase, the nuclear envelope reforms and the nucleoli reappear. The chromosomes may also uncoil. Also, this stage may be accompanied by cytokinesis, so that daughter cells are formed. In other species the telophase I stage is omitted at the end of the first division. In any case, there is no DNA replication between meiosis I and meiosis II.

In the second division of meiosis, the phases are referred to as prophase II, metaphase II, anaphase II, and telophase II. During *anaphase II,* centromere separation occurs and the chromatids part to become independent chromosomes. At the end of *telophase II,* following cytokinesis, there are four cells. Each of these four cells is haploid; that is, the nucleus has half the number of chromosomes and half the DNA content of the mother cell nucleus.

Meiosis in Humans

In most animals, humans included, meiosis is specifically involved in both sperm production (**spermatogenesis**) and egg production (**oogenesis**) (fig. 7.22). However, not only is the final product different in the two sexes, the process itself varies somewhat. In females the first meiotic cell division produces two cells, one of which is much smaller than the other. The smaller cell is called a **polar body,** and remains attached to the larger cell. The second meiotic division actually does not occur unless fertilization takes place. Regardless, complete oogenesis in females produces from each original cell only one egg and at least two nonfunctional polar bodies that disintegrate. In males, on the other hand, spermatogenesis results in four viable sperm from each original cell. From the drawing on the next page, it is apparent that the sperm and egg are adapted to their functions. The sperm is a tiny, flagellated cell that is adapted for swimming to the egg. The egg is a large cell that awaits the arrival of the sperm and contributes most of the cytoplasm and nutrients to the zygote.

Figure 7.22
Spermatogenesis produces four viable sperm, whereas complete oogenesis produces one egg and two polar bodies. Polar bodies are nonfunctional and eventually disintegrate. In humans, sperm and egg have twenty-three chromosomes each; therefore, the zygote has forty-six chromosomes.

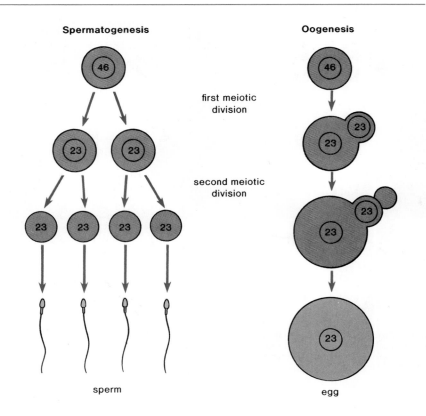

Pangenesis

The egg and sperm cells are often referred to as *germ cells* because they are the beginnings, or germs, of new individuals. The germ cells represent the only connecting thread between successive generations. Accordingly, the mechanism of hereditary transmission must operate across this slender connecting bridge. In 1868, Charles Darwin proposed an ingenious theory to account for the transmission of traits by the germ cells. Darwin thought that all organs of the body sent contributions to the germ cells in the form of particles, or gemmules. These gemmules, or minute delegates of bodily parts, were supposedly discharged into the bloodstream and became ultimately concentrated in the gametes. In the next generation, each gemmule reproduced its particular bodily component. The theory was called *pangenesis,* or "origin from all," since all parental bodily cells were supposed to take part in the formation of the new individual.

Darwin's fanciful explanation of inheritance, however, has no foundation of factual evidence. Indeed, many observations argue against it. In 1909 W. E. Castle of Harvard University provided one of the earliest convincing observations refuting the theory of pangenesis. Castle successfully removed the ovaries of an albino guinea pig and grafted in their place the ovaries of a black guinea pig. The albino foster mother was then mated to an albino male. This pair produced several litters of young, and all the offspring were black. Ordinarily when albinos are crossed only albino offspring are produced. It is obvious, then, that the eggs produced by the transplanted ovaries (from the black guinea pig) were not changed, or instructed otherwise, by their new residence in the albino foster mother guinea pig.

Notwithstanding Darwin's simplistic view, the mechanism of inheritance does reside in the gametes. Although the relatively large egg cell and the comparatively small sperm cell are as strikingly unlike in appearance as any two cells can be, they are equivalent in certain important components—their nuclei and, more particularly, their *chromosomes.* The chromosomes are, in fact, the vehicles for transmitting the blueprints of traits from parents to offspring. Deoxyribonucleic acid (DNA) [is] . . . the *chemical carrier* of the hereditary information.

From Volpe, E. Peter, *Biology and Human Concerns,* 3d ed. © 1975, 1979, 1983 Wm. C. Brown Publishers, Dubuque, Iowa. All Rights Reserved. Reprinted by permission.

Darwin recognized that the egg and sperm must have a means of carrying information to the next generation, but did not realize such information resides in the genes located on the chromosomes.

Figure 7.23

Cell division in bacteria. The single chromosome is attached to the membrane and is in the process of replicating. As the cell begins to elongate, the replicated chromosomes are pulled apart by the elongation. The elongated cell divides in the middle, resulting in two new cells, each with a copy of the chromosome.

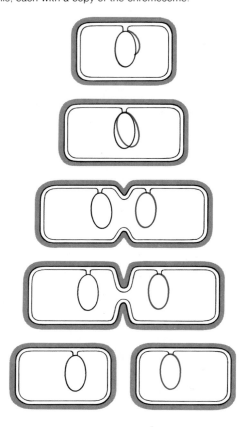

Importance of Meiosis

Meiosis maintains a constant *chromosome number* generation after generation. Because the gametes have half the number of chromosomes, the zygote, following fertilization, has the full number of chromosomes for the species.

Meiosis also helps assure that the zygote will have a different combination of genes than has either parent. As a result of independent assortment, the chromosomes are distributed to the gametes in various combinations. Only two possible combinations can be illustrated in figure 7.21, but the total number of possible combinations is actually 2^N where N is the haploid number of chromosomes. In humans, where N = 23, the number of possible chromosome combinations produced by meiosis is a staggering 2^{23} or 8,388,608! And this does not even consider the variations that are introduced due to crossing over.

Sexual reproduction involving meiosis is therefore one means by which genetic variation is produced. As we will see in a later chapter, genetic variation is the raw material for the evolutionary process by which organisms change and become adapted to their environment.

Evolution of Mitosis

Cell Division in Prokaryotes

Cell division in prokaryotes is termed **binary fission** because the two daughter cells are identical in content and size. Prokaryotes have a single circular chromosome that is not enclosed within a nucleus. This single chromosome replicates during interphase and both copies become anchored to a fold in the cell membrane, as shown in figure 7.23 for a typical bacterial cell. As the cell membrane elongates, the chromosomes are separated and pushed apart. When the cell is approximately twice its original length, the cell membrane indents until the cytoplasm is divided into two approximately equal portions. Cell wall growth between the two cells then takes place.

Evolution of the Spindle

Observation of nuclear division in primitive eukaryotes, such as dinoflagellates and fungi, makes it possible to suggest steps by which mitosis may have evolved. At first, division may have taken place in a manner similar to that described for prokaryote division, except that the separating chromosomes were attached to the nuclear envelope instead of the cell membrane. The purpose of the first involvement of microtubules may have been to support the nuclear envelope. Later, the microtubules became attached to the chromosomes themselves. As microtubules began to take a more active role in separating the chromosomes, the nuclear envelope became less important. Indeed, we have seen that the nuclear envelope disappears completely in the most advanced eukaryotes.

Figure 7.24
Mitotic cell division contrasted to meiotic cell division.

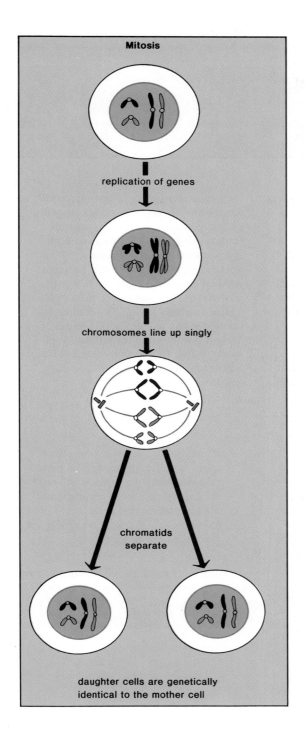

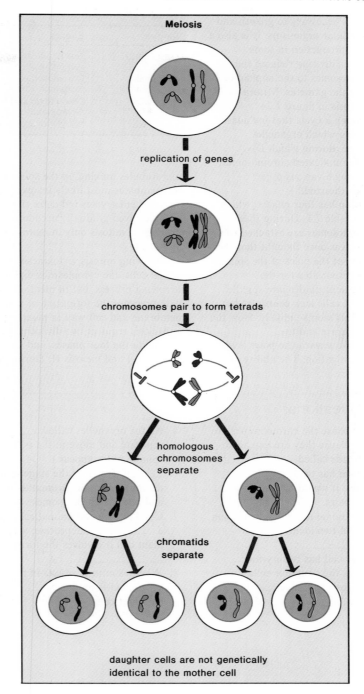

Summary

Cell reproduction involves mitosis and meiosis. Mitosis is necessary to growth and repair of multicellular organisms. It is also a part of asexual reproduction in some organisms. Meiotic division reduces the number of chromosomes to the haploid or N number found in the gametes. Mitosis is compared to meiosis in figure 7.24.

Cells go through a cycle that includes mitosis; G_1, during which organelles duplicate; S phase, during which DNA replicates to give duplicated chromosomes; and G_2, during which various other molecules are synthesized.

Mitotic division has four phases, which are described in table 7.1. During these phases, the chromosomes are attached to spindle fibers. It was once thought that the centrioles located at the poles of the spindle apparatus were responsible for the organization of microtubules into spindle fibers. Since plant cells lack centrioles but form a spindle, this seems unlikely. How the chromosomes separate and the chromosomes move toward the poles is also a matter for investigation. The ability of

microtubules making up the spindle fibers to disassemble is most likely involved. A survey of eukaryotes indicates that this ability evolved gradually accounting for the presence of mitosis only in more complex organisms.

Following mitosis, cytokinesis occurs. In animal cells, the cytoplasm is divided by a furrowing process, but in plant cells a cell plate precedes the formation of a new membrane and cell wall in plant cells.

Meiosis requires two divisions, each containing the four phases, and designated as meiosis I and meiosis II. During meiosis

I, crossing over occurs before anaphase I when homologous chromosomes separate; during meiosis II, chromatids separate. Gamete production in animals requires spermatogenesis and oogenesis. Spermatogenesis in males produces four viable sperm, while oogenesis in females produces one egg and two or three polar bodies. In plants, meiosis is not directly involved in the production of gametes. Meiosis is involved in the production of spores, which divide and mature to become a gametophyte generation that produces gametes.

Table 7.1 Stages of Mitosis

Stages	Events
Prophase	Nuclear envelope and nucleoli disappear; spindle appears. Each chromosome is composed of a pair of chromatids.
Metaphase	Chromosomes are attached to spindle fibers and are lined up at the equator.
Anaphase	Chromatids separate and move to the poles of the spindle. Now each chromatid is called a chromosome.
Telophase	Spindle disappears, nuclear envelope reforms, and nucleoli reappear.

Objective Questions

1. During interphase the chromosomes are not visible because they are extended into fine threads called _____ .
2. If an organism has twelve chromosomes, it would have _____ homologous pairs.
3. Every chromosome just prior to division is composed of two identical parts called _____ .
4. If the mother cell has twenty-four chromosomes, the daughter cells following mitosis will have _____ chromosomes.
5. As the organelles called _____ separate and move to the poles, the spindle fibers appear.
6. _____ is the stage of mitosis during which the chromatids separate and become chromosomes.
7. Cytokinesis in an animal cell occurs by a _____ process, while in a plant cell it involves the formation of a _____ .
8. Whereas mitosis results in two daughter cells, meiosis produces _____ daughter cells.
9. During anaphase I of meiosis, the _____ separate. This means that eventually the gametes will have the haploid number of chromosomes.
10. Meiosis ensures that the zygote will have a _____ combination of genes than has either parent.

Study Questions

1. Describe the makeup of the eukaryote nucleus.
2. Define chromatin, chromosomes, chromatids, and centromere.
3. Draw a diagram to illustrate that mitosis results in daughter cells with the same chromosome number as the mother cell.
4. State the stages of mitosis and describe the events of each stage.
5. Compare and contrast cytokinesis in animal cells and plant cells.
6. Discuss the importance of mitosis and its various functions in eukaryotes.
7. Describe what happens in each portion of the cell cycle, including the various phases of interphase.
8. Draw a diagram to illustrate that meiosis results in four daughter cells, each with the haploid number of chromosomes.
9. Compare and contrast metaphase of mitosis with metaphase I and metaphase II of meiosis.
10. Compare and contrast spermatogenesis with oogenesis.
11. Discuss the importance of meiosis in the life cycle of animals and plants.
12. Describe cell division in prokaryotes and discuss the steps by which mitosis may have evolved.

Selected Key Terms

chromatin ('krō mət ən) 131
chromosome ('krō mə ‚sōm) 131
homologous pair (hō 'mäl ə gəs 'paər) 133
karyotype ('kar ē ə ‚tīp) 133
autosome ('ȯt ə ‚sōm) 133
chromatid ('krō mə təd) 133
centromere ('sen trə ‚mir) 133
mitosis (mī 'tō səs) 133

cytokinesis (‚sīt ō kə 'nē səs) 137
zygote ('zī ‚gōt) 138
meiosis (mī' ō səs) 138
diploid ('dip ‚lȯid) 139
haploid ('hap ‚lȯid) 139
spermatogenesis (‚spər ‚mat ə 'jen ə səs) 142
oogenesis (‚ō ə 'jen ə səs) 142

CHAPTER 8

Patterns of Inheritance

Concepts

1. Inheritance is controlled by genes.
   a. Mendel's Law of Segregation states that organisms contain two factors for each trait and that these segregate so that the gametes have only one factor for each trait.
   b. Mendel's second law, the Law of Independent Assortment, states that each pair of factors segregates independently of all the other pairs.
   c. Mendel's laws can be used to predict the proportions of offspring types resulting from specific crosses.
   d. Interactions between different genes may alter the expected inheritance patterns.

2. An organism's characteristics are the result of both genes and environment. It is sometimes difficult to determine the relative importance of these influences.

Genetics is the branch of biology that deals with the inheritance of biologically expressed traits. In sexually reproducing organisms, an offspring is likely to have characteristics that cause it to resemble both parents. In the midnineteenth century, researchers favored a blending theory to explain this observation. They thought the hereditary material was a fluid, perhaps contained in the blood of animals and in the sap of plants. The combination of the parents' characteristics in the offspring was thought to be caused by the blending of these fluids. This hypothesis did not permit them to design fruitful experiments. They were attempting to study the transmission of a large number of characteristics simultaneously, and this deterred them from determining the mechanics of inheritance.

As we shall see, an entirely different approach was taken by an Austrian monk named Gregor Mendel (fig. 8.1). Mendel experimented with many plants and some animals, but he achieved his best results with garden peas in the early 1860s. In the garden pea, each flower (fig. 8.2) produces both male and female gametes. The pollen grains, containing the male gametes, are produced

Figure 8.1
Mendel working in his garden. Mendel selected twenty-two varieties of the edible pea *Pisum* for his genetic experiments. He grew and tended the plants himself. For each experiment, he observed as many offspring as possible. For example, for a cross that required him to count the number of round seeds to wrinkled seeds, he observed and counted a total of 7,324 peas.

within the anther of the stamens. The female gametes, or eggs, are produced within the ovary of the pistil. Pollination occurs when pollen is deposited on the stigma, the top portion of the pistil. Following pollination, the male gametes usually fertilize the eggs within the ovary. Mendel sometimes allowed the plants to self-pollinate so that the male and female gametes of the same flower produced offspring. At other times he performed cross-pollination so that male and female gametes from different flowers produced offspring.

Mendel's use of pea plants as his experimental material was a good choice—not only can the plants be self-pollinated or cross-pollinated at will, but they are easy to cultivate and have a short generation time.

Mendel performed careful experiments, collected a large amount of data, and then analyzed the data in a quantitative manner. In this way he formulated hypotheses that are still valid today. The significance of Mendel's conclusions is indicated by the fact that even today biologists refer to the study of certain patterns of inheritance as **Mendelian Genetics.** This chapter reviews Mendel's work and shows how it is applicable to the study of inheritance in all types of organisms. Mendel's tenets have been modified over time to explain many other types of inheritance, some of which are also discussed in this chapter.

Figure 8.2
Pea flower anatomy. Self-pollination normally occurs within the flowers of a pea plant. Whenever Mendel wanted to cross two different plants, he removed the pollen-producing anthers of a plant and brushed the stigmas with the anther from another plant. Male gametes within the pollen grains then fertilized the eggs within the ovules. Once the ovules had developed into seeds (peas), they could be planted in order to observe the results of the cross. The enlarged pod shows all possible shapes and colors of peas in a cross involving these characteristics.

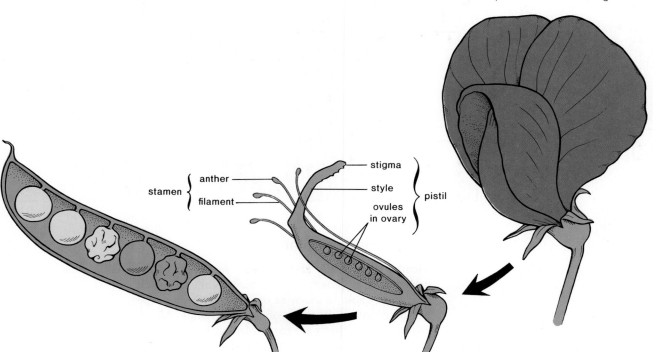

stamen { anther
filament

stigma
style
ovules
in ovary } pistil

Table 8.1 Traits and Characteristics Studied by Mendel

Trait	Characteristics	
stem length	tall	short
pod shape	inflated	constricted
seed shape	round	wrinkled
seed color	yellow	green
flower position	axial	terminal
flower color	purple	white
pod color	green	yellow

Single-Trait Inheritance

Mendel was not the first to study the results of cross-pollination in garden peas, but his approach was different. While others chose to study many traits, he at first chose to examine one. A **trait** is any aspect of the individual that can be described in terms of characteristics. Also, Mendel concentrated on characteristics that were easily distinguishable. The traits and characteristics that Mendel studied are listed in table 8.1.

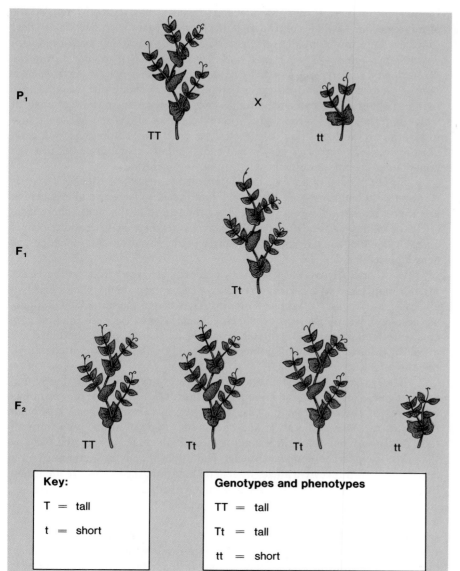

P₁

X

TT

tt

F₁

Tt

F₂

TT

Tt

Tt

tt

Key:

T = tall

t = short

Genotypes and phenotypes

TT = tall

Tt = tall

tt = short

Figure 8.3
Mendel formulated the Law of Segregation after doing an experiment such as this. When tall pea plants were crossed with short plants, the first generation of offspring were all tall. When these plants were allowed to self-pollinate, however, one of every four plants was short in the next generation. The actual ratio Mendel obtained was 787 tall to 277 short, or 2.84:1.

Before Mendel began his experiments, he first made sure his parental plants (called the P₁ generation) bred true. He observed that when these plants self-pollinated, the offspring were like one another and like the parent plant. In the case of stem length, for example, when the plants self-pollinated, a tall parent always had tall offspring; a short plant always had short offspring. Then Mendel cross-pollinated the P₁ plants; he dusted the pollen of tall plants on the stigma of short plants, and vice versa. In both cases, the offspring (called the F₁, or first filial, generation) resembled the tall parent only (fig. 8.3). Mendel then allowed the F₁ plants to self-pollinate.

Regardless of which particular trait Mendel studied, he always observed a 3:1 ratio in the F₂ (second filial) generation. For example, there were approximately three tall plants for every short plant.

Figure 8.4
Diagrammatic representation of a homologous pair of
chromosomes showing that each allelic pair is located on
homologous chromosomes at a particular gene locus.

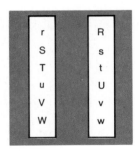

Segregation

In interpreting these results, Mendel stated what is now called **Mendel's Law of Segregation.** Mendel concluded that characteristics such as tall or short stems are determined by discrete factors. Mendel stated that *each organism contains two factors for each trait and that the factors segregate during the formation of gametes so that each gamete contains only one factor from each pair of factors.* Mendel drew these conclusions even though he was unaware that his factors were located on chromosomes. He was also unaware of meiosis, which is responsible for the segregation of chromosomes (and Mendel's factors) prior to the formation of gametes. As we discussed on page 140, each gamete eventually receives only one of each kind of chromosome because the homologous chromosome pairs separate during meiosis I.

Figure 8.3 shows how we now interpret the results of Mendel's experiment on stem length in peas. Each pea plant is said to have two **alleles,** which are alternate types of a gene, that control the length of the stem. There is an allele for tallness *(T)* and an allele for shortness *(t).* One of these alleles occurs on each homologous chromosome at a particular location that is called the **gene locus** (fig. 8.4).

In Mendel's cross, the original P_1 parents were true breeding; therefore, the tall plants had two alleles for tallness *(TT)* and the short plants had two alleles for shortness *(tt).* When an organism has two of the same kind of allele, as these had, we say that the organism is **homozygous.** Because the first parents were homozygous, all gametes produced by the tall plant contained the allele for tallness *(T),* and all gametes produced by the short plant contained the allele for shortness *(t).*

After cross-pollination all the individuals of the resulting F_1 generation had one allele for tallness and one for shortness *(Tt).* When an organism has two different alleles at a gene locus, we say that it is **heterozygous.** Although the plants of the F_1 generation had one of each type of allele, they were all tall. When only one allele in a heterozygous individual is expressed, we say that this allele is the **dominant allele.** The allele that is not expressed in a heterozygote is a **recessive allele.** In genetic notation, the alleles are identified by letters; the dominant allele with an uppercase (capital) letter and the recessive allele with the same letter, but lowercase (small).

Genotype and Phenotype

It is obvious from these results that two organisms with different allele combinations for a trait may have the same outward appearance (*TT* and *Tt* peas are both tall). For this reason, it is necessary to distinguish between the alleles present in an organism and the appearance of that organism.

The word **genotype** refers to the alleles an individual receives at fertilization. Genotype may be indicated by means of letters or by means of short descriptive phrases, as in table 8.2. Genotype *TT* is called **homozygous dominant** and *tt* is called **homozygous recessive.** Genotype *Tt* is called heterozygous, which is sometimes also referred to as a hybrid.

The word **phenotype** refers to the physical appearance of the individual. Homozygous dominant and heterozygous individuals both show the dominant phenotype and are tall, while the homozygous recessive individual shows the recessive phenotype and is short.

We will now digress from Mendel's work with peas to show that his results are applicable to other organisms.

Table 8.2 Genotype versus Phenotype

Genotype	Genotype	Phenotype
TT	Homozygous (pure) dominant	Tall plant
Tt	Heterozygous (hybrid)	Tall plant
tt	Homozygous (pure) recessive	Short plant

Single-Trait Cross

When solving a genetics problem, it is first necessary to know which characteristic is dominant. For example, the following key indicates that unattached earlobes are dominant over attached earlobes.

Key: E = unattached earlobes
e = attached earlobes

Suppose a homozygous man with unattached earlobes marries a woman who has attached earlobes. What type of earlobes will the children have? In row 1 following, P_1 stands for the parental generation, and the letters on this row are the genotypes of the parents. Row 2 shows that the gamete of each parent has only one type of allele for earlobes; therefore, all the children (F_1 offspring) will have a heterozygous genotype that will result in unattached earlobes.

P_1	EE	$\times$	ee
Gametes	E		e
F_1		Ee	

If these children marry someone with the same genotype as themselves, will their children have unattached earlobes?

P_1	Ee	$\times$	Ee
Gametes	E,e		E,e

Whereas previously there was only one possible type of gamete for each parent, now there are two possible types. When designating the gametes, it is necessary to keep in mind that whereas the individual has two alleles for each trait, *each gamete carries only one allele for each trait.* Thus, in this example, the gametes for an individual with the genotype Ee could contain either an E or an e, designated simply as E,e—the comma is used to separate the two possible types of gametes.

When solving genetic problems, it should be kept in mind that no two letters in a gamete may be the same. This is true of single-trait crosses as well as for multitrait crosses. Practice Problems 1 will help you learn to designate gametes.

Practice Problems 1

1. For each of the following genotypes, give all possible gametes.
   a. *WW*
   b. *Ww*
   c. *Tt*
   d. *TT*
2. For each of the following, state whether it represents a genotype or a gamete.
   a. *D*
   b. *GG*
   c. *P*

*Answers to problems appear on page 169.

Figure 8.5

Representation of a cross between two human beings heterozygous for type of earlobe. Because unattached earlobes are dominant to attached earlobes, an offspring of this cross has one chance in four of having attached earlobes.

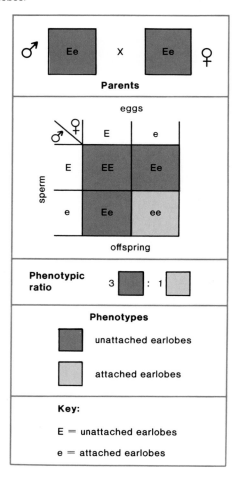

Punnett Square

Once the gametes have been designated, there are two possible ways to calculate the results. The first is to use a **Punnett square.** This simple method was introduced by a prominent poultry geneticist, R. C. Punnett, in the early 1900s. Figure 8.5 shows how the results of the cross under consideration ($Ee \times Ee$) may be determined by using a Punnett square. All possible kinds of male gametes are placed along the left side of the square and all possible kinds of female gametes are placed across the top of the square (or vice versa). It is assumed that all possible types of gametes are present in equal proportions. The ways in which these gametes may combine are determined by filling in the cells of the square. The results show that the expected proportions of offspring are 1 *EE* to 2 *Ee* to 1 *ee* for the genotypes and, phenotypically, three individuals with unattached earlobes to one individual with attached earlobes.

These are only expected proportions, not absolute numbers. For example, if a large number of heterozygous individuals produced a large number of offspring, say 200, then approximately 150 would have unattached earlobes and approximately 50 would have attached earlobes. In terms of genotypes, approximately 50 would be *EE*, about 100 would be *Ee*, and the remaining 50 would be *ee*.

Probability

Another method of calculating the expected ratios is to realize that *the chance or probability of receiving a particular combination of genes is simply the product of the individual probabilities.* In the cross just considered,

$$Ee \times Ee$$

the offspring have an equal chance of receiving *E* or *e* from each parent. Therefore,

> Probability of $E = 1/2$
> Probability of $e = 1/2$

and

> Probability of $EE = 1/2 \times 1/2 = 1/4$
> Probability of $Ee = 1/2 \times 1/2 = 1/4$
> Probability of $eE = 1/2 \times 1/2 = 1/4$
> $3/4 =$ have unattached earlobes
> Probability of $ee = 1/2 \times 1/2 = 1/4$
> $1/4 =$ have attached earlobes

Even when calculating results by means of the Punnett square, the most useful interpretation in humans is to consider the chances that a child will have a particular trait. The Punnett square in figure 8.5 also tells us that in the case under consideration, each child has three chances out of four, or a 75 percent chance, to have unattached earlobes, and one chance out of four, or a 25 percent chance, to have attached earlobes. One should also realize that *"chance has no memory"*; for example, if two heterozygous parents already have a child with attached earlobes, the next child still has one out of four chances of having attached earlobes. While this may not seem important as far as earlobes is concerned, it is important when one is calculating the chances of a child inheriting a genetic disease. (Human genetic diseases are discussed in the reading on page 156.)

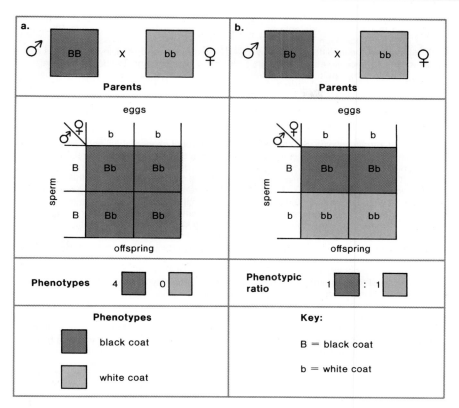

Figure 8.6
Representation of a test cross to determine if the individual showing the dominant trait is homozygous or heterozygous. *a.* Since all offspring show the dominant characteristic, the individual is most likely homozygous, as shown. *b.* Since the offspring show a 1:1 ratio, the individual is heterozygous, as shown.

The cross we have just considered (*Ee* × *Ee*) can be called a **monohybrid cross** because only one trait is being considered and each of the individuals is heterozygous. Whenever a heterozygote reproduces with a heterozygote and one allele is dominant to the other, a 3:1 phenotypic ratio is expected among the offspring.

Test Cross

If a plant, animal, or person has the dominant phenotype, it is not possible to determine by inspection whether the organism is homozygous dominant or heterozygous. However, if the plant or animal is crossed with a homozygous recessive, the results may indicate the correct genotype. In a certain strain of mice, for example, black coat *(B)* is dominant over white coat. Therefore, a black-coated mouse could be *BB* or *Bb*. Figure 8.6 shows the expected results when a homozygous dominant and a heterozygote each reproduce with a white mouse. In the first case, there are only black offspring, but in the second, there is a 50-percent chance that each offspring will have a white coat. A white-coated offspring indicates that the genotype of the black-coated mouse is *Bb*.

The odds of producing a recessive phenotype are greatest when a heterozygote is crossed with the homozygous recessive. Therefore, in the **test cross,** an individual with a dominant phenotype is crossed with an individual having the recessive phenotype. The results of the test cross allow one to determine whether the individual is heterozygous or homozygous dominant.

Human Genetic Diseases

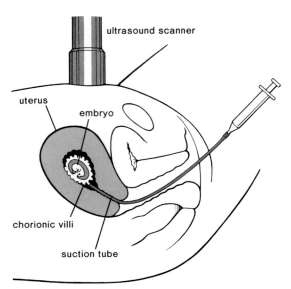

Chorionic villi sampling procedure. In this procedure the physician removes a sampling of embryonic cells from the chorionic villi by means of a suction tube inserted into the uterus by way of the vagina.

Over the years, considerable progress has been made in the study of the inheritance of many human characteristics. A list of Mendelian traits in humans compiled in 1978 includes 736 known dominants, 753 suspected dominants, 521 recessives, 596 suspected recessives, and 107 on the X chromosome (these so-called X-linked genes are discussed in the next chapter). Most of these identified traits involve various genetic diseases because there is much more medical incentive to study inherited diseases than to study the inheritance of normal human characteristics.

If a genetic disorder is inherited as a dominant allele of a single allelic pair and one of the parents is heterozygous for the characteristic while the other is homozygous recessive, each child has a 50:50 chance of either developing the characteristic or escaping it. Huntington's disease is one of the best known of the dominant conditions. It causes the victim literally to "go to pieces" in mind and body, until insanity and finally death occur. Because the condition does not develop until a person has reached the age of thirty or forty, the individual may have already had children, thus passing on genes that cause the disease in offspring before diagnosis is made.

There are many well-known recessively inherited disorders among humans: cystic fibrosis is a disorder affecting the function of the mucous and sweat glands; PKU (phenylketonuria) is an essential liver enzyme deficiency; Cooley's anemia is a blood disorder; and Tay-Sachs disease results in nervous system degeneration, primarily in infants. In these cases, parents who appear to be normal may be carriers of the defective allele. If both parents carry the defective allele, each child will run a 25 percent risk of manifesting the disease, a 25 percent chance of full normality, but will have a 50 percent chance of receiving a single defective gene and being a carrier. Should these carriers marry other carriers, they run the same risk as the parents of transmitting the disease to the next generation.

Sickle-cell anemia is inherited as a pair of incomplete dominant alleles. This condition derives its name from the fact that the blood cells in the affected individual are often sickle-shaped and have a limited ability to transport oxygen. Sickle-cell anemia is caused by an abnormal hemoglobin, Hb^S, that is less soluble than the normal hemoglobin, Hb^A. The sickle-cell hemoglobin molecules tend to crystallize to the point that they pile up and form hairlike rods that cause the cells to become sickle-shaped. Persons with the genotype $Hb^S Hb^S$ have sickle-cell anemia and suffer from major ills such as bone and kidney damage and ulceration of the lower legs. Persons with the genotype $Hb^A Hb^S$ have the sickle-cell trait, but have few outward

Practice Problems 2

1. In rabbits, if B = dominant black allele and b = recessive white allele, which of these genotypes could a white rabbit have? *Bb, BB, bb*
2. In horses, trotter *(T)* is dominant over pacer *(t)*. A trotter is mated to a pacer and the offspring is a pacer. Give the genotype of all horses.
3. In humans, freckles is dominant over no freckles. A man with freckles is married to a woman with freckles, but the children have no freckles. What chance did each child have for freckles?
4. In pea plants, yellow peas is dominant over green peas. Give the genotype of all plants that could possibly produce green peas (Y = yellow, y = green).

*Answers to problems appear on page 169.

symptoms unless the oxygen content of the air decreases or the carbon dioxide content rises.

Color blindness is an X-linked[1] genetic disorder that causes little medical difficulty, but hemophilia, the bleeder's disease, and muscular dystrophy cause untold suffering. While hemophiliacs bleed externally after an injury, they also suffer from internal bleeding, particularly around joints that are in frequent motion. Costly administration of the missing clotting factor can control hemophilia but there is no specific therapy for muscular dystrophy, a wasting away of the muscles. The most common form of muscular dystrophy—Duchenne type—is also an X-linked recessive disorder.

Carrier tests are now available for a large number of potential genetic diseases. Blood tests can identify carriers of Cooley's anemia and sickle-cell anemia. By measuring enzyme levels in blood or skin cells, doctors can detect carriers of some inborn metabolic errors, such as Tay-Sachs disease.

If a woman is already pregnant, it is possible to collect and test the cells of the developing baby. A new method called chorionic villi sampling (see the illustration) allows physicians to collect these cells as early as the fifth week. The doctor inserts a long, thin tube through the vagina and into the uterus. With the help of ultrasound, which gives a picture of the uterine contents, the tube is placed between the lining of the uterus and the chorion, a membrane that surrounds the developing baby. Then suction is used to remove a sampling of the chorionic villi, treelike extensions of the chorion. These cells have the same chromosomal and genetic makeup as the baby's and therefore can be tested for genetic defects. For example, a karyotype will show if the baby will be born with Down's syndrome (formerly termed mongolism), a form of mental retardation that occurs more frequently in children born to mothers over forty years of age. Persons with Down's syndrome have three #21 chromosomes instead of two.

Two other procedures can be done to collect cells for testing. Amniocentesis can be done at about the sixteenth week of pregnancy. In amniocentesis, a long needle is used to withdraw a small amount of amniotic fluid and fetal cells. After the cells are cultured, they can be subjected to chromosomal analysis and tests that detect biochemical deficiencies. In fetoscopy, a tiny periscope with a cold light source is inserted through a small incision in the uterus. This allows the physician to see small portions of the fetus and take skin and blood samples for testing purposes. Currently, this is the only method of acquiring blood samples to test for sickle-cell anemia and Cooley's anemia, for example.

[1]X-linked inheritance is explained on pages 176–79.

Multitrait Inheritance

Mendel performed a second series of experiments in which he crossed true-breeding plants that differed in two traits. For example, he crossed tall plants having inflated pods with short plants having constricted pods. The F_1 generation plants always had both dominant characteristics; therefore, he allowed the F_1 plants to self-pollinate. Among the F_2 generation he always achieved an approximate 9:3:3:1 phenotypic ratio. For example, for every plant that was short with constricted pods, he had about nine plants that were tall with inflated pods, three that were tall with constricted pods, three that were short with inflated pods, and one that was short with constricted pods (fig. 8.7). Mendel saw that these results were explainable if all possible combinations of factors were passed on to the offspring. Therefore the pairs of factors must segregate independently of one another. Mendel's second genetic principle, called the **Mendel's Law of Independent Assortment,** states that *the members*

Figure 8.7

Mendel formulated the Law of Independent Assortment
after doing an experiment such as this. When tall pea
plants with inflated pods were crossed with short plants
with constricted pods, the first generation of offspring
were all tall with inflated pods. When these plants were
allowed to self-pollinate, however, nine of every sixteen
plants were tall with inflated pods, three were tall with
constricted pods, three were short with inflated pods, and
one was short with constricted pods.

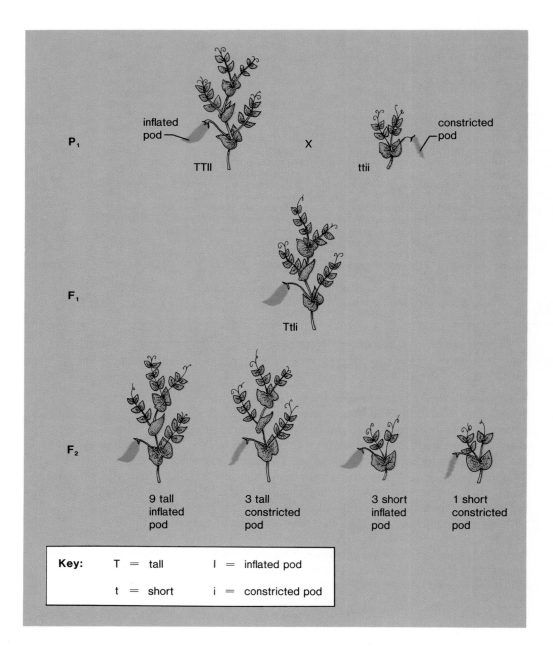

P_1

inflated pod — TTII   X   ttii — constricted pod

F_1

Ttli

F_2

9 tall
inflated
pod

3 tall
constricted
pod

3 short
inflated
pod

1 short
constricted
pod

Key: T = tall   I = inflated pod

t = short   i = constricted pod

of each pair of factors segregate independently of all other pairs. We now recognize that all possible combinations of alleles located *on different chromosomes* occur in the gametes because homologous pairs of chromosomes separate independently during meiosis (p. 171).

Let us use Mendel's Law of Independent Assortment to designate the gametes for one of the genotypes given in figure 8.7. Let us consider *TtIi*. Only one letter of each kind may appear in a gamete; for example, there may be either a *T* or a *t*. The letter *T* or *t* may be present with either one of the other letters. Therefore the possible types of gametes for this individual are *TI, Ti, tI, ti*. Depending on the genotype, there may be fewer possible types of gametes. For example, the genotype *TtII* has only two possible types of gametes: *TI* or *tI*.

Practice Problems 3

1. For each of the following genotypes, give all possible gametes.
   a. *ttGG*
   b. *TtGG*
   c. *TtGg*
   d. *TTGg*
2. For each of the following, state whether a genotype or a gamete is represented.
   a. *TT*
   b. *Tg*
   c. *IiCC*
   d. *TW*

*Answers to problems appear on page 169.

Probability

Mendel realized that if the inheritance of one trait was independent of the other, then the proportion of combined characteristics could be calculated because the *chance that two or more independent events will occur together is the product of their chances of occurring separately*. Using the results from each of two single-trait crosses,

> The chance of tallness = 3/4
> The chance of inflated pods = 3/4
> The chance of shortness = 1/4
> The chance of constricted pods = 1/4

The probabilities for the two-trait cross are, therefore,

> The chance of tallness and inflated pods = 3/4 × 3/4 = 9/16
> The chance of tallness and constricted pods = 3/4 × 1/4 = 3/16
> The chance of shortness and inflated pods = 3/4 × 1/4 = 3/16
> The chance of shortness and constricted pods = 1/4 × 1/4 = 1/16

Again, since all possible male gametes must have an equal opportunity to fertilize all possible female gametes to even approximately achieve these results, a large number of offspring must be counted.

Figure 8.8

Two-Trait Cross

The fruit fly, *Drosophila melanogaster* (fig. 8.8), less than one-half the size of a housefly, is a favorite subject for genetic research because it has easily recognizable mutant characteristics. "Wild-type" flies have long wings and gray bodies. Mutant flies exist that have vestigial (short) wings and ebony (black) bodies. The key for a cross between the wild type and the mutant fly is

L = long wing	G = gray body
l = short wing	g = ebony body

The cross is

P_1	*LLGG*	×	*llgg*
Gametes	*LG*		*lg*
F_1		*LlGg*	

In the P_1 cross, only one type of gamete is possible for each fly; therefore, all the F_1 will have the same genotype *(LlGg)* and the same phenotype (long wings and gray bodies). This genotype is called a **dihybrid** because the individual is heterozygous in two respects: wing length and body color.

If a dihybrid *(LlGg)* reproduces with a dihybrid, each fly has four possible types of gametes: *LG, Lg, lG,* and *lg.* The Punnett square (fig. 8.9) shows the expected results among sixteen offspring if all possible sperm have fertilized all possible eggs. Notice that nine offspring have a gray body and long wings, three have a gray body and short wings, three have a black body and long wings, and one has a black body and short wings. The 9:3:3:1 phenotypic ratio is always expected when a dihybrid is crossed with a dihybrid and simple dominance is present.

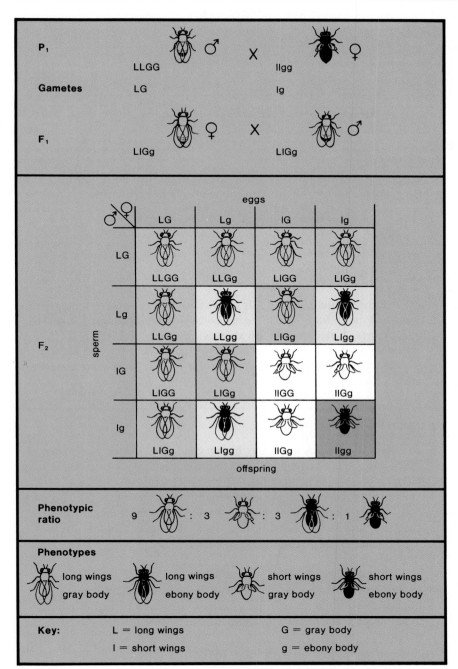

Figure 8.9
Mendel's Law of Independent Assortment, as revealed in the inheritance of two pairs of alleles in the fruit fly. A pure long-winged fly with a gray body mated with a short-winged, ebony-bodied fly produces all long-winged gray-bodied flies in the first generation (F₁). When these F₁ flies are inbred, a second generation (F₂) is produced that displays a phenotypic ratio of 9:3:3:1. What are the four phenotypes?

Figure 8.10

Representation of a test cross to determine if an individual showing two dominant traits is heterozygous or homozygous for both traits. In this instance the individual must be a dihybrid because some of the offspring show the recessive characteristics.

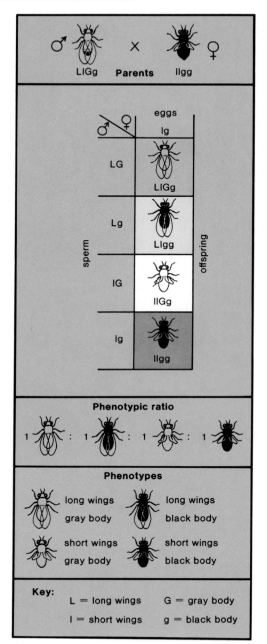

Table 8.3 Common Crosses Involving Simple Dominance

Cross	Phenotypic Ratio
monohybrid × monohybrid	3:1 (dominant to recessive)
*monohybrid × recessive	1:1 (dominant to recessive)
dihybrid × dihybrid	9:3:3:1 (9 dominant; 3 mixed; 3 mixed; 1 recessive)
*dihybrid × recessive	1:1:1:1 (all possible combinations in equal number)

*Called a test cross because it can be used to test if the individual showing the dominant gene is homozygous or heterozygous. For a definition of all terms, see the glossary.

Test Cross

A plant or animal that shows two dominant traits can be tested for the dihybrid genotype by crossing it with the recessive in both traits.

P_1	$LlGg$	×	$llgg$
Gametes	LG		lg
	Lg		
	lG		
	lg		

The Punnett square (fig. 8.10) shows that the expected phenotypic ratio is 1 gray body with long wings; 1 gray body with short wings; 1 black body with long wings; 1 black body with short wings; or 1:1:1:1.

Table 8.3 lists all the crosses we have studied thus far that show a frequently observed phenotypic ratio. When these types of crosses are done, these ratios are observed.

Practice Problems 4

1. In horses, B = black coat, b = brown coat; T = trotter, t = pacer. A black pacer mated to a brown trotter produces a black trotter offspring. Give all possible genotypes for this offspring.
2. In fruit flies, long wings (L) is dominant over short wings (l), and gray body (G) is dominant over black body (g). In each instance, what are the genotypes of the parental flies, if a student gets these results:
   a. 1:1 long to short wing (all gray body)
   b. 1:1:1:1
   c. 3:1 gray to black body (all long wings)
3. In humans, short fingers and widow's peak are dominant over long fingers and continuous hairline. A dihybrid is married to a dihybrid. What is the chance that any child will have the same phenotype as the parents?

*Answers to problems appear on page 169.

Beyond Mendel's Laws

While the study of Mendel's laws is helpful, we know today that they are an oversimplification. There are many exceptions, some of which are discussed in the following material and in the next chapter.

Multiple Alleles

All of the examples discussed thus far have concerned genes that have two alleles. It is possible, however, for a gene to have three or more alleles, although a diploid individual can have a maximum of two alleles per trait. One of the best known examples of a three-allele gene is the one that controls A-B-O blood types in humans. The phenotypes listed in table 8.4 are produced by combinations of three different alleles: I^A, I^B, and i^O. Both I^A and I^B are fully expressed in the presence of the other. On the other hand, both I^A and I^B are dominant to i^O and only the genotype $i^O i^O$ produces a person with type O blood.

An examination of possible matings between different blood types sometimes produces surprising results. For example,

P_1	$I^A i^O$	$\times$	$I^B i^O$	
F_1	$I^A I^B$,	$i^O i^O$,	$I^A i^O$,	$I^B i^O$

Thus from this particular mating every possible phenotype (AB, O, A, B blood type) is possible.

Blood typing can sometimes aid in paternity suits. A blood test of the supposed father can be used to determine whether he *might* be the father but not that he definitely *is* the father. For example, a man with blood type A could be the father of a child with blood type O, in which case the father's genotype must be $I^A i^O$. On the other hand, a man who has blood type AB could not possibly be the father.

It might be noted here that the blood factor called Rh is inherited separately from A, B, AB, or O type blood. It is possible for an individual to be Rh^+ (Rh positive) or Rh^- (Rh negative), meaning in the first case that an Rh factor is present on the red cells, and in the second that an Rh factor is not present. Rh factor inheritance is another example of a trait controlled by multiple alleles. For convenience, however, it is often treated as if it were controlled by a single allelic pair in which simple dominance prevails, Rh^+ being dominant over Rh^-.

Incomplete Dominance

Although Mendel never became aware of the fact, actually in many instances one allele is not dominant over the other allele. Instances of this occurrence are known as incomplete dominance. One type of incomplete dominance might better be termed **codominance.** For example, we have just seen that both I^A and I^B are both expressed equally in regard to blood type. Another example of codominance occurs in shorthorn cattle. When a red bull is crossed with a white cow (or vice versa), the offspring are neither red nor white but roan, with an intermingling of both red and white hairs. To symbolize that the two alleles have equal effect on the offspring, the following key is suggested:

$H^R H^R$ = Red
$H^W H^W$ = White
$H^R H^W$ = Roan

This key clearly indicates that these two genes are alleles and that both contribute equally to the phenotype of the roan offspring. A Punnett square (fig. 8.11) shows that if a roan bull is crossed with a roan cow, the calf has a 25 percent chance of being red or white and a 50 percent chance of being roan.

Table 8.4 Blood Groups	
Phenotypes Blood Types	**Possible Genotypes**
A	$I^A I^A$, $I^A i^O$
B	$I^B I^B$, $I^B i^O$
AB	$I^A I^B$
O	$i^O i^O$

I, i = immunogen gene

Figure 8.11
Inheritance of the roan phenotype in cows. Since neither H^R nor H^W are dominant over the other, three phenotypes are possible among the offspring when a roan bull reproduces with a roan cow.

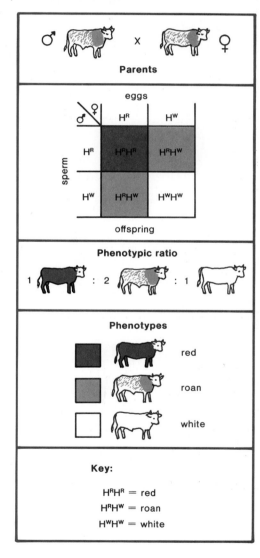

Key:

$H^R H^R$ = red
$H^R H^W$ = roan
$H^W H^W$ = white

Figure 8.12
Snapdragon flower. Pink-flowered snapdragons, which
result from crossing red-flowered and white-flowered
snapdragons, illustrate intermediate inheritance.

Another type of incomplete dominance might better be termed **intermediate inheritance.** For example, if red-flowered snapdragons are crossed with white-flowered snapdragons (fig. 8.12), all the offspring have pink flowers. Following self-fertilization, we again see a 1:2:1 phenotype ratio; only this time it is 1 red:2 pink:1 white-flowered plant.

Modifier Genes

Modifier genes affect the phenotypic expression of alleles that are at a different gene locus from the modifier gene. One example of this occurs in the genetic determination of coat pattern in cattle. The basic spotted pattern in Holstein cattle (fig. 8.13) is determined by a single pair of alleles, but the relative amounts of black and white are determined by modifying genes. Also,

Figure 8.13
Holstein cattle. The amount of white and black present in the coats of Holstein cattle is determined by modifier genes.

in humans there may be only two alleles for eye color: *B* for brown and *b* for "blue" eyes. The presence of modifying genes, however, could explain why there are also shades of gray and green in addition to brown and blue eyes. The modifying genes would affect the amount of pigment actually deposited in the iris.

Epistatic Genes

Epistatic genes interfere with the expression of genes that are at different gene loci from the epistatic gene. However, epistasis ("covering up") works much the same as ordinary dominance. For example, in certain dogs:

	Epistatic
B = black	*W* = prevents color
b = brown	*w* = does not prevent color

If only the alleles *B* and *b* were involved in coat color, one would expect that all of the dogs under consideration would have colored coats. A dog with the genotypes *BB* or *Bb* would have a black coat and a dog with the genotype *bb* would have a brown coat. But this turns out to be the case only if the recessive genotype *ww* is also inherited. If a dog should happen to inherit a dominant *W*, being either *WW* or *Ww*, the coat will be white regardless of the genotype for coat color.

A similar situation occurs in humans and other animals. If an individual inherits the allelic pair *(aa)* that causes albinism, the inability to produce the pigment melanin, it does not matter what genotype for eye color and hair color is inherited. This genotype cannot be expressed and the individual will be an albino, lacking pigment in all parts of the body (fig. 8.14).

Figure 8.14
Albinism is due to the inheritance of two recessive genes. Regardless of any other inherited genes affecting coloration, the individual lacks the ability to produce pigment. Albino Negro and father.

Pleiotropy

A **pleiotropic gene** affects several different phenotypic characteristics. This can occur because the same biochemical pathway is present in all developing organs and because various pathways interact. A gene that affects a single pathway can then affect many characteristics as they develop.

Medical syndromes are often due to pleiotropic genes. For example, persons with Marfan's syndrome (fig. 8.15) have characteristic skeletal, eye, and cardiovascular defects. All of these are due to the inability to produce normal connective tissues. In *Drosophila* there is a single gene locus that affects eye color, body proportions, wing size, wing vein arrangement, body hairs, size and arrangement of bristles, shape of testes and ovaries, viability, rate of growth, and fertility.

Polygenic Inheritance

Sometimes a particular trait is apparently controlled in a quantitative manner by several gene loci. Often in these cases it is possible to find a range of phenotypes from one extreme to another, with most individuals falling somewhere in between, so that the observed phenotypes are distributed according to a bell-shaped or normal curve. Human height, for example, is determined by many allelic pairs, and the distribution of human heights follows a bell-shaped curve (fig. 8.16).

Ear length in corn also seems to be controlled by several gene loci. When plants that produce short ears are crossed with plants that produce long ears, the offspring are intermediate in length. But if the latter self-fertilize, the offspring show lengths that vary following a bell-shaped curve. Investigators have suggested that these results could be explained by assuming that ear length is controlled by three gene loci. Plants that produce short ears have the alleles *aabbcc*. Plants that produce long ears have the alleles *AABBCC*. When these two types of plants are crossed, the F_1 generation *(AaBbCc)* will have ears of an intermediate length, but then self-fertilization will produce offspring that show continuous variation in ear length.

Figure 8.16
In polygenic inheritance, the distribution of the possible phenotyes follows a bell-shaped curve. The curve shows that few individuals have the extreme phenotypes and most have the average phenotype.

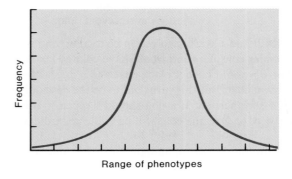

Range of phenotypes

This type of inheritance is considered quantitative because each dominant allele seems to add a definite and precise increment; for example, *AABbcc* will have the same phenotype as *AaBbCc*. Many of the economically important characteristics of agricultural plants and animals are inherited in a quantitative fashion. Examples of this are milk production in dairy cattle and corn yield. For this reason, those involved in breeding agricultural plants and animals must have a thorough understanding of quantitative genetics.

Practice Problems 5

1. A woman with blood type B gives birth to a child with blood type B. Could a man with blood type A be the father of this child?
2. In humans, $H^A H^A$ are persons with normal hemoglobin; $H^A H^S$ are persons with sickle-cell trait; and $H^S H^S$ are persons with sickle-cell anemia. If two carriers ($H^A H^S$) reproduce, what are the chances that a child will have sickle-cell anemia? *25%*
3. Breeders of dogs note a range of coat colors among the offspring. Is this an example of epistatic inheritance or polygenic inheritance? If coat color were controlled by three pairs of genes, how many different shades would be possible?

*Answers to problems appear on page 169.

Heredity and Environment in Genetic Expression

One of the most basic, and also most often forgotten, principles of genetics is that the genotype and the environment interact to produce the phenotype. It is sometimes difficult to determine how much of a phenotype is determined by heredity (nature) and how much by environment (nurture).

Geneticists attempt to assess the relative contribution of genetic factors by using a concept called heritability. Heritability is the proportion of the phenotype variation that is due to genetic factors. If all the variation is due to genetic causes, the heritability is 1 (100 percent). If all of the observed phenotypic variation is due to the environment, the heritability of that variation is zero. All traits demonstrate heritability factors that fall between 0 and 1.

In economically important plants and animals, the heritabilities of different characteristics are estimated from careful breeding studies. In humans, of course, this cannot be done, but the heritabilities of some abnormalities have been estimated from the rate at which they occur among relatives of a person with the particular disease. Some examples of heritabilities that have been estimated in this manner are certain kinds of epilepsy (.4), clubfoot (.8), and harelip (.7). This suggests that clubfoot and harelip are more likely to be due to inheritance than is epilepsy. Other abnormalities are most likely due entirely to environmental effects (fig. 8.17).

Figure 8.17
Children with shortened, deformed limbs are characteristic of those born to women who took the sedative thalidomide in the early months of pregnancy. These children have a congenital defect (a defect not due to heredity), which cannot be passed on to the next generation.

Summary

Mendel established two laws that help explain certain patterns of inheritance. The Law of Segregation says that each organism possesses an allelic pair for each trait, but the gametes have only one allele for each trait. The Law of Independent Assortment says that pairs of alleles (as long as they are associated with different chromosomes) segregate independently of one another so that every possible combination of alleles may occur in the gametes.

Certain terminology and conventions are used to indicate the genotype of individuals and their gametes. The same alphabetic letter is used for both alleles at a given locus. A capital letter indicates a dominant allele and a lowercase letter indicates a recessive allele. A homozygous dominant individual is indicated by two capital letters and a homozygous recessive individual is indicated by two lowercase letters. The genotype of a heterozygous individual is indicated by a capital and a lowercase letter. The gametes of all individuals have only one letter of each type, either capital or lowercase as appropriate. All possible combinations of letters indicate all possible gametes.

In doing an actual cross, it is assumed that all possible types of sperm fertilize all possible types of eggs. The results may be expressed as a probable phenotypic ratio or, in some cases, the chances of an offspring showing a particular phenotype. The results of some crosses may be determined by simple inspection, but certain others that commonly reoccur are given in table 8.3.

There are many exceptions to Mendel's laws, some of which are discussed in this chapter. Sometimes a gene has more than two alleles, although each individual has only any two of these. The inheritance of blood type in humans is a common example of a trait controlled by multiple alleles. There are three alleles, I^A, I^B, and i^O, which determine human blood types. These genes also demonstrate incomplete dominance since I^A and I^B are both expressed when present together. Intermediate inheritance is seen in snapdragons, where the cross of a red-flowered plant and a white-flowered plant results in pink-flowered plants.

Modifier genes can affect the degree to which other genes are expressed, and epistatic genes mask the presence of other genes. For example, an albino has no color regardless of the alleles inherited for color of hair, eyes, etc. Pleiotropic genes have multiple effects, such as those seen in human medical syndromes. Sometimes a number of gene loci affect the same trait and the various phenotypes fit a bell-shaped curve. Often each dominant allele adds a small increment to the phenotypic characteristic; therefore, this is also called quantitative genetics. Height in humans is an example of polygenic inheritance.

Both the genotype and the environment contribute to an organism's phenotype. The relative contributions of each, however, are difficult to resolve.

Objective Questions

1. Whereas an individual has two genes for every trait, the gametes have _____ gene for every trait.
2. The recessive allele for the dominant gene W is __ *W* __ .
3. Mary has a widow's peak and John has a continuous hairline. This would be a description of their _phenotype_
4. W = widow's peak and w = continuous hairline; therefore, only the phenotype _widows peak_ could be heterozygous.
5. Two heterozygotes, each having a widow's peak, already have a child with a continuous hairline. The next child has what chance of having a continuous hairline? _25%_

6. In a test cross, an individual having the dominant phenotype is crossed with an individual having the _recessive_ phenotype.
7. How many letters are required to designate the genotype of a dihybrid individual? _4_
8. If a dihybrid is crossed with a dihybrid, how many offspring out of sixteen are expected to have the dominant phenotype for both traits? _9_
9. How many different phenotypes among the offspring are possible when a dihybrid is crossed with a dihybrid? _4_
10. According to Mendel's Law of Independent Assortment, a dihybrid can produce how many types of gametes having different combinations of genes? _4_

Match the following numbered examples to the type of inheritance (lettered list).
11. albinism _____
12. roan cows _____
13. ABO blood type _____
14. human height _____
15. pink snapdragons _____
   a. polygenic inheritance
   b. multiple alleles
   c. epistatic gene
   d. codominance
   e. intermediate inheritance

Answers to Objective Questions
1. one 2. w 3. phenotype 4. widow's peak
5. 25 percent 6. recessive 7. four 8. nine
9. four 10. four 11. c 12. d 13. b 14. a
15. e

Study Questions

1. Define these terms: allele, dominant allele, recessive allele, genotype, homozygous, heterozygous, phenotype.

2. Mendel first performed a single-trait cross. Describe his experiment and tell which of Mendel's laws is associated with this experiment.

3. What results do you expect when a monohybrid is crossed with a monohybrid? When a monohybrid is crossed with a recessive?

4. If a Punnett square gives these results: *AA, Aa, Aa,* and *aa,* then what proportion of the offspring are *AA, Aa,* and *aa?*
5. Mendel also did two-trait crosses. Describe his experiment and tell which of his laws is associated with such an experiment.
6. What results do you expect when a dihybrid is crossed with a dihybrid?

When a dihybrid is crossed with a recessive in both traits?
7. In peas, yellow seed is dominant over green seed and long stem is dominant over short stem. When a dihybrid for these characteristics is crossed with a dihybrid, what is the expected phenotype for nine plants out of every sixteen plants? for three plants? for another three plants? for one plant?

8. Define multiple alleles, incomplete dominance, modifier gene, epistatic gene, pleiotropic gene, and polygenic inheritance. Give an example of each of these.
9. State the various blood types in humans based on A-B-O and Rh and give all possible genotypes for each.

Additional Genetics Problems

1. If a man homozygous for widow's peak (dominant) marries a woman homozygous for continuous hairline (recessive), what are the chances the children will have widow's peak? Will have a continuous hairline?
2. John has unattached earlobes like his father, but his mother has attached earlobes. Give the genotype of each of the three.
3. In humans, the allele for short fingers is dominant over that for long fingers. If a person with short fingers who had one parent with long fingers marries a person with long fingers, what are the chances for each child to have short fingers?
4. In rabbits, black color is due to a dominant allele *B* and brown color to its recessive allele *b*. Short hair is due to the dominant allele *S* and long hair is due to its recessive allele *s*. In a cross between a homozygous black, long-haired rabbit and a brown, homozygous short-haired one, what would the F₁ generation be like? The F₂ generation? If one of the F₁ rabbits reproduced with a brown, long-haired rabbit, what kinds of offspring, and in what ratio, would you expect?
5. In radish plants, the shape of the radish produced may be long (S^LS^L), round (S^RS^R), or oval (S^LS^R). If oval is crossed with oval, what proportion of offspring would also have oval radishes?

6. A cross is made between parents with genotypes *AABB* and *aabb.* If there are thirty-two offspring, how many of them would be expected to exhibit both dominant characteristics?
7. A man of blood type A and a woman of blood type B produce a child of type O. What are the genotypes of the man, woman, and child? If the couple were to have more children, what possible blood types could be produced?
8. In horses, black coat *(B)* is dominant to brown coat *(b),* and being a trotter *(T)* is dominant to being a pacer *(t).* A black horse who is a pacer is crossed with a brown horse who is a trotter. The offspring is a brown pacer. Give the genotype of all these horses.

Selected Key Terms

trait ('trāt) 150
allele (ə 'lēl) 152
gene locus ('jēn 'lō kəs) 152
homozygous (,hō mə 'zī gəs) 152
heterozygous (,het ə rō 'zī gəs) 152
dominant allele ('däm ə nənt ə 'lēl) 152

recessive allele (ri 'ses iv ə 'lēl) 152
genotype ('jē nə ,tīp) 152
phenotype ('fē nə ,tīp) 152
Punnett square ('pən ət 'skwar) 154
monohybrid cross (,män ō 'hī brəd 'krós) 155

test cross ('test 'krós) 155
codominance (,kō 'däm ə nəns) 163
epistatic gene (,ep ə 'stat ik 'jēn) 165
pleiotropic gene (,plī ə 'trōp ik 'jēn) 166

CHAPTER 9

Chromosomes and Genes

Concepts

1. Inheritance is controlled by genes that are located on chromosomes.
   a. Genes are located in linear arrays along chromosomes.
   b. Alleles linked on the same chromosome tend not to segregate independently during meiosis.
   c. Both sexual and nonsexual characteristics are controlled by the sex chromosome.

2. Chromosomal mutations affect inheritance patterns.
   Chromosomal mutations involve changes in chromosome structure and in chromosome number.

Figure 9.1
Each type chromosome has a characteristic banding pattern as illustrated in this giant chromosome from the salivary gland of a fruit fly. Giant chromosomes are large because they contain many chromatids that have not separated. It is speculated that each light-dark double band may be a single gene locus.

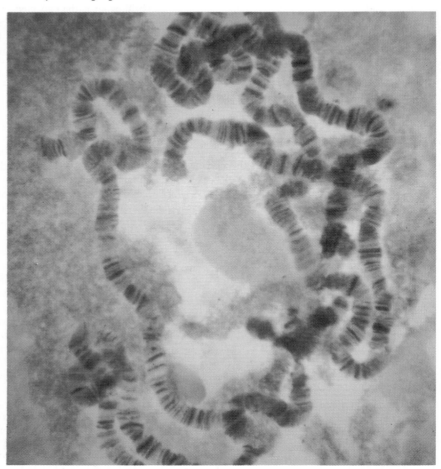

It was not always known that the genes are located on the chromosomes (fig. 9.1). Mendel, himself, never associated his factors with any particular part of a cell. While it was known during Mendel's time that organisms are made up of cells, the behavior of chromosomes during gametogenesis had been little studied (fig. 9.2). But in 1902, Walter Sutton was studying gametogenesis in the grasshopper at Columbia University when it occurred to him that Mendel's factors must be located on the chromosomes in the nucleus. He pointed out that (a) homologous chromosomes of each pair segregate during meiosis; (b) all possible combinations of chromosomes occur in the gametes; and (c) each parent gives one of each type of chromosome to the offspring. Because of this, the offspring has *homologous chromosomes;* that is, two chromosomes of the same kind. Altogether, then, chromosome behavior can account for Mendel's Law of Segregation and Law of Independent Assortment if we assume that the Mendelian factors are located on the chromosomes, as illustrated in figure 9.2.

Figure 9.2
Segregation and independent assortment during meiosis.
The homologous chromosomes of each pair separate
(segregate) during meiosis I. Separation by independent
assortment results in all possible combinations of alleles.
This shows that Mendel's Law of Segregation and Law of
Independent Assortment hold because of the manner in
which meiosis occurs.

Key:

W = allele W on chromosome

w = allele w on homologous chromosome

S = allele S on another chromosome

s = allele s on homologous chromosome

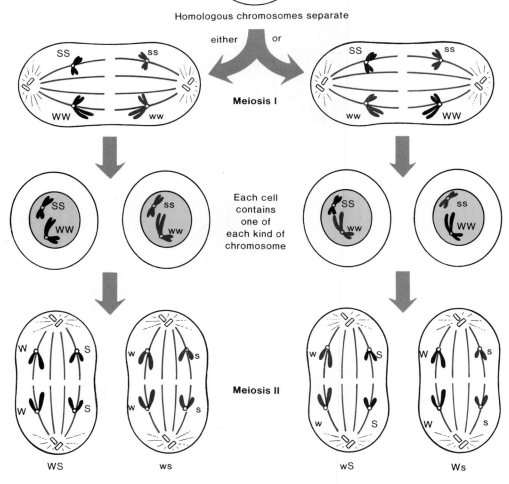

Replication

Tetrads are present in mother cell.

Homologous chromosomes separate

either   or

Meiosis I

Each cell contains one of each kind of chromosome

Meiosis II

WS   ws   wS   Ws

Four different gametes are possible.

What's in a Chromosome?

When early investigators decided that the genes were on the chromosomes, they had no idea of chromosome composition. By the mid-1900s, it was known that chromosomes are made up of both DNA and protein. Only in recent years, however, have investigators been able to produce models suggesting how chromosomes are organized.

A eukaryotic chromosome is more than 50 percent protein. Many of these proteins are concerned with DNA and RNA synthesis, but a large proportion, termed **histones,** seem to play primarily a structural role. A human cell contains forty-six chromosomes, and the length of the DNA in each is about five cm. Therefore, a human cell contains at least two meters of DNA. Yet all of this DNA is packed into a nucleus that is about five μm in diameter. The histones seem to be responsible for

packaging the DNA so that it can fit into such a small space. The packing unit, termed a *nucleosome*, gives chromatin a beaded appearance in certain electron micrographs.

The accompanying drawing shows that DNA is wound around a core of histone molecules in a nucleosome. To fully appreciate this packing unit, you must realize that the DNA strand is twisted on itself even as it winds about the histone core. Also, notice how DNA stretches between the nucleosomes at the location of H_1 histone molecules. Whenever the H_1 molecules make contact, chromatin would shorten. Indeed, if the entire structure should then twist as shown, even more compactness could be achieved. No doubt still more folding processes occur as chromatin condenses to form the chromosomes.

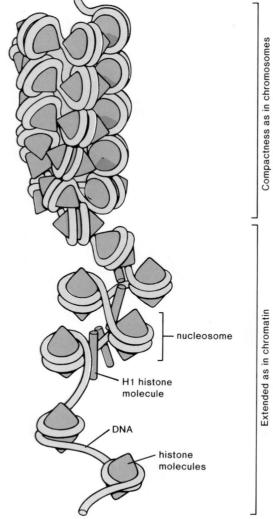

nucleosome

H1 histone molecule

DNA

histone molecules

Compactness as in chromosomes

Extended as in chromatin

Chromosome versus chromatin structure.

Having concluded that the genes are on the chromosomes, Sutton realized that there must be many alleles on each chromosome. For example, humans have only twenty-three pairs of chromosomes, but certainly have more than twenty-three traits. He suggested that the alleles on each chromosome are linked together (fig. 9.3), and do not assort independently but instead are inherited together.

Gene Linkage

The presence of a **linkage group** in which alleles are on the same chromosome is suspected when a cross yields an unexpected phenotypic ratio. For example, it has been observed that when homozygous tall, nonhairy-stem (*DDHH*) tomatoes are crossed with dwarf, hairy-stem (ddhh) tomatoes, the F_1 plants have tall and nonhairy stems (*DdHh*). A 9:3:3:1 ratio is expected when these dihybrid plants are crossed, but a 3:1 ratio was actually observed (fig. 9.4*a*). Approximately three plants had tall, nonhairy stems for every one that had a dwarf, hairy stem. This reduction in the number of phenotypes occurs because the alleles for stem height and hairiness are linked together in the same chromosome. Further, the results indicate that *D* is on the same chromosome as *H*, and *d* is on the same chromosome as *h*.

Crossing Over

Although the cross in figure 9.4 gave an approximate 3:1 ratio, a few plants among the offspring had recombinant characteristics: some had tall, hairy stems, and some had dwarf, nonhairy stems. How can these plants be accounted for? The answer is that linkage does not hold 100 percent of the time

Figure 9.3
Diagrammatic representation of homologous pairs of chromosomes within the nucleus. The letters *Aa, Bb*, etc., stand for allelic pairs. Genes *ABCDEF* are one linkage group, and genes *abcdef* are another. What other linkage groups are present?

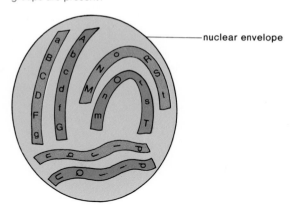

nuclear envelope

Figure 9.4
Two-trait cross in tomatoes. *a*. The expected 9:3:3:1 ratio among the offspring does not occur because of gene linkage. Notice that only two phenotypes are found among the offspring. *b*. Due to crossing over, all four possible phenotypes are present among the offspring, although usually the recombinant ones appear much less frequently.

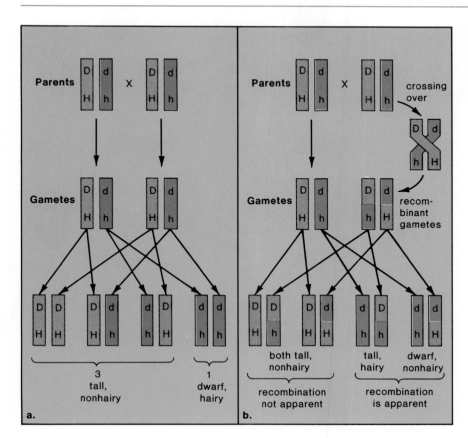

Figure 9.5

A simplified gene map showing the relative positions of some of the genes on *Drosophila* chromosome 2, as calculated by the frequency of crossovers. The chance of crossing over increases as the distance between genes increases. For example, there is a greater chance of crossing over between the genes for feelers and body color than between the genes for feelers and legs.

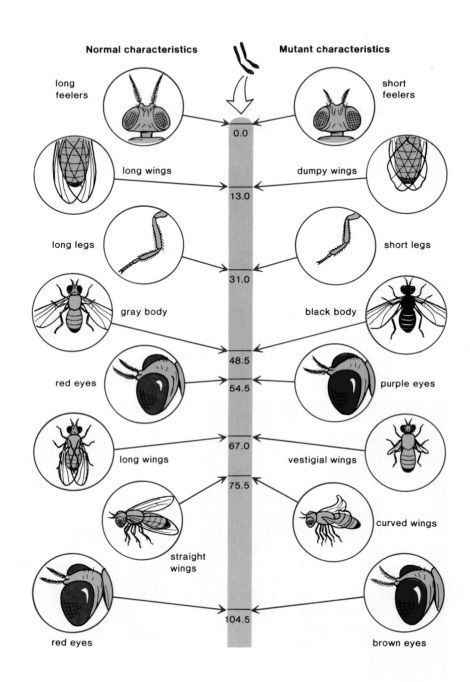

Normal characteristics

Mutant characteristics

long feelers — 0.0 — short feelers

long wings — 13.0 — dumpy wings

long legs — 31.0 — short legs

gray body — 48.5 — black body

red eyes — 54.5 — purple eyes

long wings — 67.0 — vestigial wings

straight wings — 75.5 — curved wings

red eyes — 104.5 — brown eyes

because of **crossing over.** When homologous pairs of chromosomes come together during meiosis, the chromatids may exchange portions (fig. 7.20) by a process of breaking and then reassociating. The gametes that receive these recombined chromosomes are termed recombinant gametes. Figure 9.4*b* shows how recombinant gametes produced recombinant phenotypes in our example.

Chromosome Mapping

It has been possible to map chromosomes in lower organisms by studying the crossover frequency of linked alleles. Alleles distant from one another are more likely to be separated by crossing over than alleles that are close together. Thus the crossover frequency indicates the distance between two gene loci on a chromosome. Each percentage of crossing over is taken to mean a distance of one chromosome unit. Using these frequencies, then, it is possible to indicate the order of the gene loci on the chromosome (fig. 9.5).

The study of the inheritance of human disorders has helped map human chromosomes, but laboratory investigations involving the chromosomes themselves have been even more helpful. For example, human and mouse cells can be fused together in tissue cultures, producing cells with an inordinately large number of chromosomes. Because of this, mitosis is not orderly and the daughter cells do not acquire the full complement of chromosomes present in the mother cell. Eventually some of the daughter cells contain only a few human chromosomes. These chromosomes can then be analyzed for the presence of certain genes. For example, **DNA probes** (DNA from another source that contains a known gene) can be used to determine whether the chromosome under study carries a particular gene. This works because the probe will match up with any portion of the DNA in the cell that is similar to itself. The manner in which this can happen will be discussed in the following chapter.

Genes on Sex Chromosomes

Sex Chromosomes

Many animals, including *Drosophila,* have the same sex determination system as humans: an organism with two X chromosomes is a female, and an organism with one X chromosome and one Y chromosome is a male. But different sex determination patterns are found in other animals. In grasshoppers there is no Y chromosome; instead, the male simply has one X chromosome and the female has two X chromosomes. In birds, moths, and butterflies, the situation is just the reverse: the male has two X chromosomes, while the female has one X chromosome. Thus the Y chromosome may or may not be present, depending on the species.

The parent that forms two chromosomally different sex cells determines the sex of the offspring. Thus in humans the father determines the sex because he produces two types of sperm—those that contain an X and those that contain a Y. Although the mathematical chances of any couple having a boy or

Figure 9.6

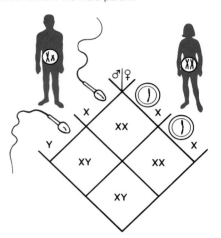

Figure 9.6
In this Punnett square, the sperm and eggs are shown as carrying only a sex chromosome. (Actually, they also carry twenty-two autosomes.) The offspring are either male or female, depending on whether they receive an X or Y chromosome from the male parent.

girl are 50:50 (fig. 9.6). Researchers have been surprised to find that on occasion fathers produce XX sons. About one of every twenty thousand men is infertile because he has two X chromosomes instead of an X and a Y. Apparently, crossing over during spermatogenesis placed male-directing genes on the father's X chromosome. In these cases, the father's X chromosome can produce a son.

Sex Linkage

The sex chromosomes are believed to carry genes that determine the sex of the individual. One such gene may have been identified in humans, as discussed in the reading "Active Y Chromosomes and Inactive X Chromosomes."

In addition to sex-determination genes, sex chromosomes carry genes having effects on traits other than the sex of the individual. These are called sex-linked genes. A gene locus that occurs only on the Y chromosome is easy to detect because if the father has a particular allele, all of the sons but none of the daughters will have the allele and the characteristic it controls. In this way, it was possible to determine that an allele causing hairy ear rims in humans is on the Y chromosome. The Y chromosome has been found to carry very few sex-linked genes; the vast majority of sex-linked genes have alleles only on the X chromosome.

X-linked Characteristics

Thomas H. Morgan and his colleagues, who did their work at Columbia University in the early 1900s, were the first to associate a specific gene with a certain chromosome. In one of their many genetic experiments on the fruit fly, *Drosophila melanogaster,* they crossed red-eyed females and white-eyed males.

		♀		♂
1. $P_1 \times P_1$		red-eyed	×	white-eyed
2. F_1		red-eyed		red-eyed

From these results they knew that red eyes are dominant over white eyes. They then proceeded to cross the F_1 flies.

		♀		♂
3. $F_1 \times F_1$		red-eyed	×	red-eyed
4. F_2		red-eyed		1:1 red to white

They noticed that while there was an approximate 3:1 ratio among all F_2 flies, only the males had white eyes. Further experimentation led the investigators to conclude that in *Drosophila* a gene locus for eye color is located only on the X chromosome, and the Y chromosome bears no gene locus for eye color.

Figure 9.7 illustrates this explanation of Morgan's F_2 results. Notice that X-linked alleles are symbolized by the superscripts attached to the X chromosomes.

$$X^R = \text{red eye}$$
$$X^r = \text{white eye}$$

It is evident that females have three possible genotypes, but males have only two possible genotypes because no superscript is added to the Y chromosome.

$X^R X^R$ = red-eyed female	$X^R Y$ = red-eyed male
$X^R X^r$ = red-eyed female	$X^r Y$ = white-eyed male
$X^r X^r$ = white-eyed female	

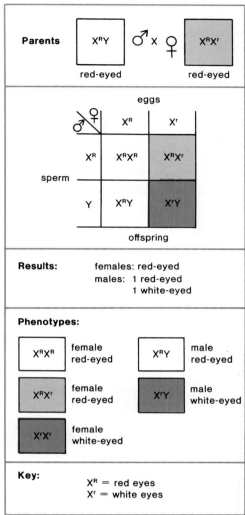

A recessive allele carried on the X chromosome will express itself in males because there is no dominant allele to suppress it. This can explain why no females had white eyes among the F_2 flies in Morgan's experimental cross (fig. 9.7). Females can be heterozygous; thus when an X^r was received from the female parent, it was offset by an X^R from the male parent. However, male offspring received a Y from the male parent and thus had to express an X^r received from the female parent. Since the F_1 female parent was $X^R X^r$, the F_2 males were $1:1$, red eye to white eye.

It is important when solving sex-linked problems to keep track of the X and Y in the manner shown here and to record the results for males and females separately.

X-linked Characteristics in Humans

As discussed in the reading on page 156, several genetic diseases in humans are inherited as sex-linked characteristics. One interesting example of a sex-linked trait in humans is a recessive allele that results in hemophilia, "bleeder's disease." This disease has been found in European royal families related

Figure 9.8

A simplified pedigree showing the X-linked inheritance of hemophilia in European royal families. Because Queen Victoria was a carrier, each of her sons had a 50 percent chance of having the disease and each of her daughters had a 50 percent chance of being a carrier. This pedigree shows only the affected individuals. Many others are unaffected, such as the members of the present British royal family.

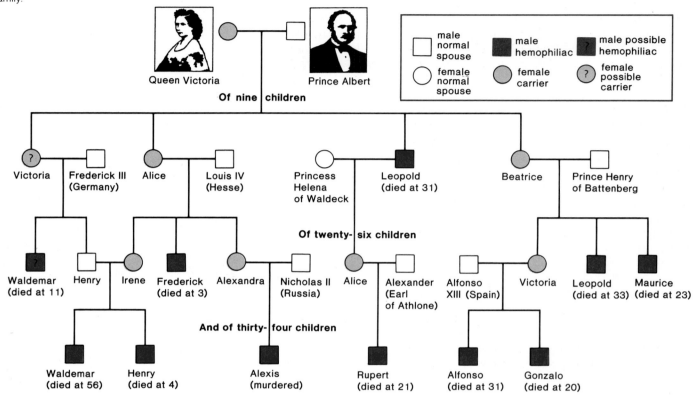

to Queen Victoria of Great Britain, and it has resulted in or contributed to the early death of male members of several royal families (fig. 9.8). Alexis, the last heir to the Russian throne before the Russian Revolution, was a hemophiliac.

A more common and less troublesome affliction is red-green color blindness—a recessive X-linked characteristic. In reference to the statements following, consider these possible crosses involving color blindness, in which X^B = normal color vision and X^b = color blindness.

1. $X^B X^B \times X^b Y$
2. $X^B X^b \times X^B Y$
3. $X^B X^b \times X^b Y$
4. $X^b X^b \times X^b Y$

Notice that these crosses would substantiate the following ways to recognize a recessive X-linked trait.

1. *More males than females are afflicted.* Crosses 2, 3, and 4 would produce color-blind sons, whereas only crosses 3 and 4 would produce color-blind daughters. Does a color-blind son receive the recessive allele from his mother or from his father? Since a son must receive a Y from his father, a color-blind allele must come from his mother.

2. *For a female to express the characteristic, her father must also have it.* Her mother must show it or be a carrier. Only crosses 3 and 4 can produce color-blind daughters; in cross 3 the mother is a carrier and is not color blind herself. In cross 4 the mother is color blind.

3. *The characteristic often skips a generation from grandfather to grandson* as shown in figure 9.9c. This occurrence can be related to cross 1 above. This cross can produce heterozygous daughters who appear to be normal, but who can pass on the color blindness allele to their sons. Individuals who do not show a genetic defect, but who are capable of passing on an allele for a genetic defect, are called **carriers.** Therefore, the color-blind grandfather in figure 9.9c has passed the characteristic to his grandson by way of his carrier daughter.

Pedigree Charts Pedigree charts (fig. 9.9) are constructed to show the inheritance of a genetic condition within a human family. Such charts are a great help in deciding whether a phenotype is controlled by a dominant, recessive, or sex-linked allele. For example, if only one parent shows the characteristic, yet all or several of the children show it, then it must be dominant. Or if two individuals do not show the characteristic, but their children do, then it must be determined by a recessive allele. On the other hand, if the characteristic appears primarily in males and passes from grandfather to grandson, then it must be sex-linked recessive. Notice that if a characteristic is controlled by a dominant allele, no individual can be a carrier.

Practice Problems

1. When the dihybrid *AaBb* was crossed with the homozygous recessive *aabb,* the phenotypic results were approximately 1:1. What ratio was expected? What may have caused the observed ratio?
2. Which *Drosophila* cross below will produce white-eyed males? In what ratio?
   a. $X^R X^R \times X^r Y$
   b. $X^R X^r \times X^R Y$
3. A woman is color blind. What are the chances that her sons will be color blind? If she is married to a man with normal vision, what are the chances that her daughters will be color blind? Will be carriers?
4. Both parents are right-handed (R = right-handed, r = left-handed) and have normal vision. Their son is left-handed and color blind. Give the genotype of each of the three persons.
5. Determine if the characteristic possessed by the darkened squares (males) and circles (females) is dominant, recessive, or sex-linked recessive. Assume the trait is controlled by a single allelic pair.

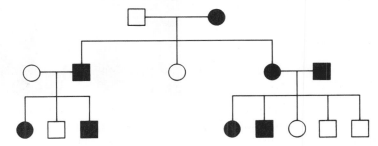

*Answers to problems appear on page 187.

Figure 9.9
Pedigree charts indicate the phenotypes of family members of several generations. Females are indicated by circles and males by squares; afflicted individuals are in black and carriers are in color. The pattern of inheritance often reveals whether a characteristic is (a.) autosomal dominant; (b.) autosomal recessive, or (c.) sex-linked recessive. Notice that only (b.) and (c.) have carriers. If a characteristic is dominant, an individual with the dominant allele must show the characteristic.

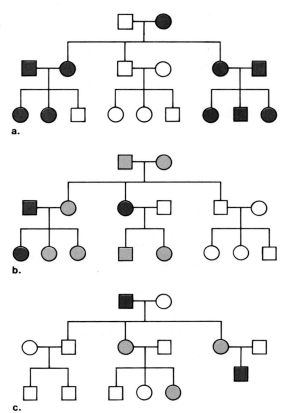

a.

b.

c.

Key:
○ normal female
□ normal male
◉ carrier female
▨ carrier male
● afflicted female
■ afflicted male

Active Y Chromosomes and Inactive X Chromosomes: The Functions of Sex Chromosomes in Body Cells

Although the Y chromosome in mammals has few gene loci, it has a powerful effect on sex determination. One allele that has been found on the Y chromosome is called H-Y$^+$. This allele directs the synthesis of a specific kind of protein molecule called the H-Y antigen. It appears that in the absence of the H-Y$^+$ allele and H-Y antigen, the fetus becomes a female. In other words, the embryo is basically female and will automatically develop into a female unless the H-Y antigen promotes the development of testes instead of ovaries. Following this, the sex hormones promote the maturation of male or female organs plus the other characteristics that distinguish males from females.

The H-Y antigen is present in the membranes of virtually all cells of a male, but not in those of a female. It is called an antigen because females produce antibodies against it. To test for maleness, it is possible to suspend a sample of white blood cells in a solution that contains some of these antibodies. If the cells carry H-Y antigen, indicating that the person is a male, the antibodies bind with them.

Aside from identifying gene loci on the sex chromosomes that cause a person to be male or female, biologists were interested in how males manage with only one X chromosome while females have two X chromosomes. That question turned out to have a rather unexpected answer.

Years ago, M. L. Barr observed a consistent difference between nondividing cells taken from female and male mammals, including humans. Females have a small, darkly staining mass of condensed chromatin present in their nuclei (a.). Males have no comparable spot of chromatin in their nuclei. This darkly staining spot in female nuclei is now called a Barr body.

In 1961, Mary Lyon, a British geneticist, proposed that the Barr body is a condensed, inactive X chromosome.

Her hypothesis has been tested and found to be valid; X chromosomes are inactivated in the cells of female embryos, but which one of the two is determined by chance. About 50 percent of the cells have the one X chromosome and 50 percent have the other X chromosome. The female body, therefore, is a mosaic with "patches" of genetically different cells. Heterozygous females usually escape the effect of harmful genes carried on X chromosomes because although these alleles are probably active in half the cells, the other body cells have active normal alleles.

The female calico cat (b.) provides phenotypic proof of the Lyon hypothesis. In these cats, an allele for black coat color is on one X chromosome and a corresponding allele for orange coat color is carried on the other X chromosome. The patches of black and orange in the coat can be related to which of these X chromosomes is in the Barr bodies of the cells found in the patches.

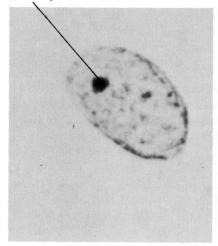

Barr body

a. *Barr body in cell.* b. *The coat of a calico cat is orange-yellow and black. The alleles for coat color occur at a gene locus on the X chromosome. Presumably, the black patches occur where cells have Barr bodies carrying the allele for orange-yellow, and the orange-yellow patches occur where cells have Barr bodies carrying the alleles for black coat.*

a.

b.

Figure 9.10

Baldness is a sex-influenced characteristic. The individual's sex hormones influence whether the characteristic will be present or not.

Figure 9.11

Another sex-influenced trait is the relative lengths of the index finger and fourth finger of the hand. The hand on the right is that of a female and the hand on the left is that of a male.

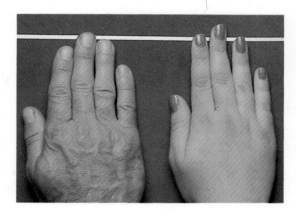

Sex Influence

Some genes not located on the X or Y chromosomes are expressed differently in the two sexes, probably due to the hormonal differences between the sexes. This is called **sex influence.** An example of a sex-influenced characteristic in humans is pattern baldness (fig. 9.10), which is far more common among men than women. The allele for pattern baldness is dominant, but the male hormone testosterone is necessary for its expression. Another example of sex-influenced inheritance in humans involves the length of the index finger compared to that of the fourth finger. An index finger equal to or longer than the fourth finger is dominant in females but recessive in males (fig. 9.11).

Chromosomal Mutations

Gene mutations consist of alterations in a single gene (p. 210), and chromosomal mutations are alterations in the structure and number of chromosomes.

Changes in Chromosome Structure

Figure 9.12 gives examples of structural chromosomal mutations. **A deletion** occurs when a segment of a chromosome is lost. Deletions usually are lethal when homozygous (both members of a pair of chromosomes affected), although exceptions have been detected in corn and other organisms. Even when heterozygous (only one member of a pair of chromosomes affected), a deletion often causes abnormalities. An example of a heterozygous deletion is the cri du chat (cat's cry) syndrome in humans. Afflicted individuals meow like a kitten when they cry and, more importantly, tend to have a small head with malformations of the face and body, and mental defectiveness, which usually causes retarded development. Chromosomal analysis shows that a portion of chromosome number 5 is missing, while the other number 5 chromosome is normal.

Figure 9.12

Chromosomal mutations. The arrows show points of chromosome breakage. In deletions, the broken chromosome fragment does not reattach and is lost. In duplication, a broken segment from one chromosome attaches to its homologous chromosome. An inversion involves breakage and reattachment to the same chromosome in a reversed position. A translocation is a transfer of a chromosome fragment to a nonhomologous chromosome. Radiation damage to cells increases the frequency of these mutations because radiation causes greatly increased chromosome breakage.

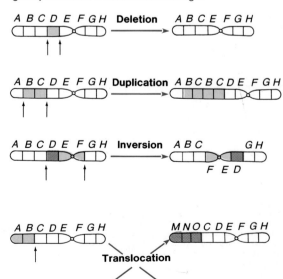

Figure 9.13
Bar eyes (narrow eyes) in *Drosophila* is due to a duplication of a gene locus. *a*. The X chromosome in flies with normal eyes has only one bar segment. *b*. The chromosome in flies with bar eyes has two bar segments.

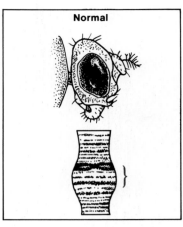

Normal

a.

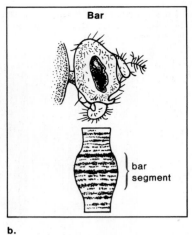

Bar

bar segment

b.

A **duplication** is the presence of a chromosome segment more than once in the same chromosome. An example of the phenotypic expression of a duplication is the bar-eyed phenotype in *Drosophila* (fig. 9.13*b*). This duplication reduces the number of facets in the insect's eye and results in development of small, bar-shaped eyes instead of normal round eyes.

Both deletions and duplications may arise following chromosome breakage. Chromosomes may be broken by radiation, various chemicals, or even viruses. The way in which chromosome fragments rejoin or fail to rejoin following breakage results in deletions or duplications.

An **inversion** occurs when a segment of a chromosome is turned around 180°. Inversions may cause reproductive problems when the organism is heterozygous for the inversion because crossing over during meiosis can lead to lethal chromosome disorganizations.

Translocations result from the interchange of blocks of genes between two nonhomologous chromosomes. Translocation heterozygotes usually have reduced fertility, again due to problems with pairing during meiosis.

Changes in Chromosome Number

Polyploidy
When an organism has more than two sets of chromosomes, the condition is called **polyploidy.** Polyploid organisms are named according to the number of sets of chromosomes that they have: triploids (3N) have three of each kind of chromosome; tetraploids (4N) have four sets; pentaploids (5N) have five sets; etc. In animals, where sex determination is based on a single pair of chromosomes, polyploidy can result in the formation of abnormal gametes. But in plants, polyploidy does not interfere with the reproductive process as long as the number of chromosome sets is even, for example 4N, 6N, 8N, etc.

Many ornamental plants are bred as polyploids because they tend to have larger leaves and flowers (fig. 9.14). In some cases, triploids are purposely developed because the plants are sterile and so will not put any energy into seed production. In these cases, the plants can be propagated asexually from cuttings.

Aneuploidy
While polyploidy involves the number of sets of chromosomes, **aneuploidy** is an excess or deficiency of an individual chromosome in an otherwise normally diploid cell. If a diploid individual has one extra copy of a chromosome, the

Figure 9.14
Polyploid plants are larger than plants with the usual number of chromosomes. The flower at left is an ordinary (diploid) Easter lily. The Easter lily at right is a tetraploid, and is, therefore, a polyploid.

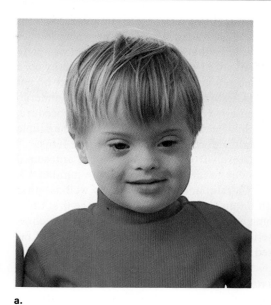

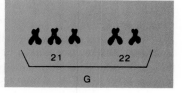

a.

b.

Figure 9.15
The classical eye and facial features of a child with Down's syndrome (a.) are due to (b.) the presence of three number 21 chromosomes among the chromosomes designated by the letter G. Compare to the karyotype in figure 7.4a.

individual is called a trisomic for that chromosome. The condition itself is a *trisomy* and is designated as 2N + 1. If a diploid individual is missing one chromosome, the individual is called a monosomic. The condition is called *monosomy* and is indicated by 2N − 1.

A well-known example of aneuploidy in humans is **Down's syndrome** (fig. 9.15a). This syndrome is sometimes called "mongolism" because the eyes of the person seem to have an oriental-like fold, but the term is no longer preferred. Other characteristics of this syndrome are short stature; stubby fingers; a wide gap between the first and second toes; a large, fissured tongue; a round head; a palm crease (the so-called simian line); and mental retardation that is sometimes severe.

Figure 9.16

Nondisjunction can occur during meiosis I if the members of a chromosome pair fail to separate or during meiosis II if the chromatids fail to separate completely. In either case, the abnormal eggs carry an extra chromosome. Nondisjunction can also occur during spermatogenesis.

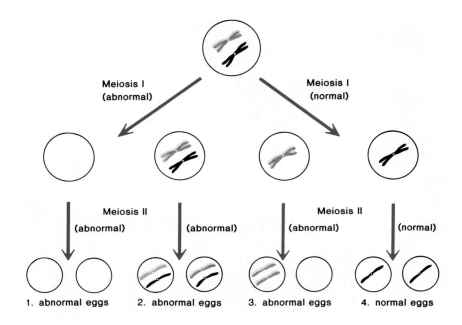

Meiosis I (abnormal)   Meiosis I (normal)

Meiosis II (abnormal)   (abnormal)   Meiosis II (abnormal)   (normal)

1. abnormal eggs   2. abnormal eggs   3. abnormal eggs   4. normal eggs

Persons with Down's syndrome usually have three number 21 chromosomes because the egg had two number 21 chromosomes instead of one. Either the chromosome pair or the chromatids failed to separate completely and instead went into the same daughter cell (fig. 9.16). Either of these occurrences is called **nondisjunction.** It would appear that nondisjunction is most apt to occur in an older female since children with Down's syndrome are usually born to women over age forty. If an older woman wishes to know whether or not her unborn child is affected by Down's syndrome, she may elect to undergo chorionic villi sampling or amniocentesis (p. 157). Following this procedure, a karyotype can reveal whether the child has Down's syndrome. If so, the couple may elect either to allow the pregnancy to continue or to abort the fetus.

There are other diseases that result from aneuploidy in the autosomal chromosomes. Patau's syndrome is due to an extra chromosome number 13. This syndrome is characterized by harelip and cleft palate, as well as other serious defects. Most infants with Patau's syndrome die during the first three months of life. Edward's syndrome occurs when an infant has an extra chromosome number 18. This disease is associated with many malfunctions of internal organ systems, and affected infants live an average of only six months.

Abnormal Sex Chromosome Numbers There are a number of abnormalities resulting from aneuploidy in the sex chromosomes. Nondisjunction of the sex chromosomes during oogenesis may produce an egg that has either no X chromosome or two X chromosomes. When the first of these is fertilized by an X-bearing sperm, a female with **Turner's syndrome** may be born. These XO individuals have only one sex chromosome—an X; the O signifies the absence of the second sex chromosome. The ovaries never become functional, regressing to ridges of white streaks. Because of this, these females do not undergo puberty, do not menstruate, and lack breast development (fig. 9.17*b*). Generally, these individuals have a stocky build, a webbed neck, and a learning disability that interferes with spatial pattern recognition.

When an egg having two X chromosomes is fertilized by an X-bearing sperm, a **superfemale** having three X chromosomes results. While it might be supposed that the XXX female would be especially feminine, this is not the

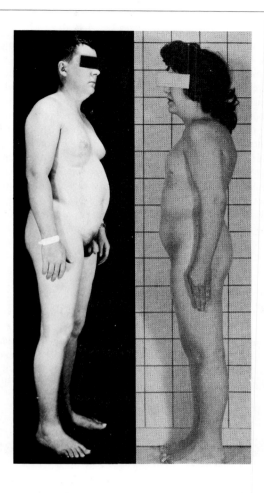

Figure 9.17
a. A male with Klinefelter's syndrome (XXY), which is marked by immature sex organs and development of the breasts. *b.* A female with Turner's syndrome (XO), which is marked by a bull neck, short stature, and immature sexual features.

case. Although there is in some cases a tendency toward learning disabilities, most superfemales have no apparent physical abnormalities, and many are fertile and have children with a normal chromosome count.

When an egg having two X chromosomes is fertilized by a Y-bearing sperm, a male with **Klinefelter's syndrome** results. This individual is male in general appearance, but the testes are underdeveloped, and the breasts may be enlarged (fig. 9.17*a*). The limbs of these XXY males tend to be longer than average, body hair is sparse, and many show learning disabilities.

XYY males may also result from nondisjunction during spermatogenesis. Afflicted males are usually taller than average, suffer from persistent acne, and tend to have barely normal intelligence. At one time, it was suggested that these men were likely to be criminally aggressive, but it has been shown that the incidence of such behavior among them is no greater than that among normal XY males.

Examination of these abnormal sex chromosome patterns indicated how sex is determined in humans. In *Drosophila* and other insects, sex is dependent on the number of X chromosomes present. For example, XY is a male, but so is XO, and XXY is a normal female. At first investigators thought that sex in humans might follow this same pattern, but this is not the case. Turner's syndrome (XO) produces a person who is more female than male. Klinefelter's syndrome (XXY) produces a person who is more male than female. Clearly, the Y chromosome is not neutral in humans. Rather, it seems to bear genes that cause maleness, as discussed in the reading for this chapter on page 180.

Summary

Genes are located in linear arrays on chromosomes. Alleles that are on the same chromosome are linked together and tend to be inherited together. A genetic cross involving linked alleles always has a number of phenotypes lower than expected among the offspring. This shows that Mendel's Law of Independent Assortment does not hold for linked alleles.

Linked alleles are sometimes separated due to crossing over during meiosis. This accounts for a small percentage of offspring with recombinant characteristics in crosses involving linked genes. These percentages can be used to map chromosomes, although other methods have been utilized to map human chromosomes.

Genes that are on the sex chromosomes are called sex-linked genes. Most sex-linked alleles are on the X chromosome. Males are more apt to show a recessive X-linked characteristic than are females. Other differences are also noted between male and female inheritance of these traits.

Sex-influenced inheritance is the differential expression of autosomal genes due to hormonal differences between the sexes. These differences cause an allele to be expressed in one sex and not expressed in the other.

Chromosomal mutations involve a change in the structure of chromosomes and a change in the number of chromosomes inherited. Changes in chromosomal structure include deletions, which occur when a segment of a chromosome is lost; duplications, which occur when a part of a chromosome is repeated; and translocations, which involve an exchange of material between nonhomologous chromosomes.

Changes in chromosome number include polyploidy, in which the individual has multiple sets of chromosomes; and aneuploidy, in which the individual has $2N - 1$ or $2N + 1$ number of chromosomes. A number of examples of aneuploidy, some involving the sex chromosomes, have been identified in humans.

Objective Questions

1. The allele A is on the same chromosome with the allele b. Therefore, these two alleles are _____ .

2. What ratio do you expect among the offspring from a dihybrid cross in which the two gene loci involved are linked (the dominant alleles are both on the same chromosome, and the recessive alleles are on its homologue)? _____

3. During crossing over, two _____ exchange portions.

4. On the *Drosophila* chromosome, the gene for length of feeler is thirty-one units from the gene for length of leg, and 104 units from a gene for color of eyes. The gene for length of feeler is more likely to undergo crossing over with the gene for _____ .

5. Sex-linked genes are found on the _____ chromosomes.

6. Do sex-linked genes determine the sex of the individual? _____

7. If a male is color blind, he inherited the allele for color blindness from his _____ .

8. What is the genotype of a female who has a color-blind father but a homozygous normal mother? _____

9. In a pedigree chart, it is observed that although the children have a characteristic, neither parent has it. The characteristic must be inherited as a _____ gene.

10. Down's syndrome is an example of _____ ploidy.

11. Nondisjunction can occur during meiosis I if the homologous chromosomes fail to _____ .

12. An XXY male will have those characteristics that accompany _____ syndrome.

Study Questions

1. Draw a hypothetical pair of chromosomes and indicate the genes on this pair by using appropriate letters. Define these terms in reference to your drawing: gene loci, linkage group, chromosome map.
2. Explain why linked alleles do not obey Mendel's Law of Independent Assortment.
3. Explain the use of crossover frequencies to map chromosomes.
4. What are sex-linked genes? X-linked genes?
5. Support this statement: "Males always inherit an X-linked characteristic from the female parent."
6. List three possible ways to recognize an X-linked recessive trait.
7. Name two sex-influenced traits in humans and explain this phenomenon.
8. Name four types of chromosomal mutations involving changes in chromosome structure.
9. Name two types of chromosomal mutations involving changes in chromosome number.
10. Discuss these chromosome mutations in humans: Down's syndrome, Turner's syndrome, superfemale, Klinefelter's syndrome, XYY males.

Additional Genetics Problems

1. A hemophiliac is married to a homozygous normal woman. What are the chances that the sons will be hemophiliacs? that the daughters will be hemophiliacs? that the daughters will be carriers?
2. In *Drosophila,* long wings are dominant over vestigial (short) wings. Give the genotypes for these flies.

   ♂
   a. Homozygous long wing, white eye
   b. Heterozygous long wing, red eye

   ♀
   c. Heterozygous long wing, heterozygous red eye
   d. Homozygous long wing, white eye

3. Cross *2b* by *2c* and give the phenotypic results.
4. Imagine that in humans, dimples is on the same chromosome as blunt fingers (both dominant) and further that the two dominants are linked together. Cross a dihybrid by a dihybrid using the method illustrated in this text. What are the chances that a child will have no dimples and long fingers?
5. Crossing over would produce what recombinant gametes in #4?

Answers to Additional Genetics Problems

1. 0 percent, 0 percent, 100 percent
2. a. *LLX^RY*   c. *LLX^RY*
   b. *LlX^rY*   d. *LlX^rX^r*
3. Females: 6 long wings, red eyes
   Males: 3 long wings, red eyes
   2 short wings, red eyes
   1 short wings, red eyes
   3 long wings, white eyes
   1 short wings, white eyes
4. 25 percent
5. *Db, dB*

Answers to Practice Problems

1. 1:1:1:1; gene linkage 2. b.; 1:1 3. 100 percent, 0 percent, 100 percent 4. *RrX^BX^b, RrX^BY, rrX^bY* 5. Dominant autosomal

Selected Key Terms

linkage group ('lin kij grüp) 173
crossing over ('krò siŋ 'ō vər) 175
carrier ('kar ē ər) 179
deletion (di 'lē shən) 181
duplication (,dü pli 'kā shən) 182

inversion (in 'vər zhən) 182
translocation (,trans lō 'kā shən) 182
polyploidy ('päl i ,plòid ē) 182
aneuploidy ('an yü ,plòid ē) 182
Down's syndrome ('dau̇nz 'sin ,drōm) 183

nondisjunction (,nän dis 'jənk shən) 184
Turner's syndrome ('tər nərz 'sin ,drōm) 184
Klinefelter's syndrome ('klin ,fəl tərz 'sin ,drōm) 185

CHAPTER 10

Molecular Genetics

Concepts

1. The hereditary material must be able to perform three functions: replication, control of the cell, and mutation.
   a. The double-stranded nature of DNA means that each strand can serve as a template during replication.
   b. DNA controls the cell by controlling protein synthesis. Four nucleic acid bases in varied order provide a triplet code for twenty amino acids.
   c. Mutations are changes in the DNA code.
2. Cellular functions are regulated in various ways.
   Some genes are structural genes coding for functional polypeptides, and some are regulatory genes that control the activity of the structural genes.
3. Science has developed techniques to control life functions.
   It is now possible to manipulate the genes of both prokaryotes and eukaryotes.
4. Cancer is characterized by uncontrollable growth.
   a. Cancer develops when normal genes mutate to oncogenes.
   b. Oncogenes code for growth factors that stimulate the cell to grow and divide.

We have called the genes units of heredity. But what exactly is the hereditary material and how does it work? How does it carry on these following three necessary functions? The hereditary material must be able to

1. replicate; make copies of itself that may be passed on from cell to cell and from generation to generation.
2. control the activities of the cell; direct the development of phenotypic differences among individuals and between species.
3. mutate or change; with evolution and the origination of new species as subsequent events.

Hereditary Material

In the mid-1900s it was known that chromosomes contained both DNA and protein (p. 172), but it was uncertain which of these was the hereditary material.

DNA Is the Genetic Material

An experiment utilizing a virus known as the T2 bacteriophage (or phage, for short) helped determine that DNA is the genetic material.

Viruses have specific shapes and specific host cells. The T viruses are categorized as complex viruses because they have both a head and a tail (fig. 10.2). They are designated T1–T7 (the T simply means "type") and are termed **bacteriophages** because they attack bacteria.

Figure 10.1
Computer-generated representation of DNA, supercoiled as it is in a nucleosome. Blue spheres are carbon atoms, cyan are nitrogen, red are oxygen, and yellow are phosphorus. (Hydrogen atoms are not shown.)

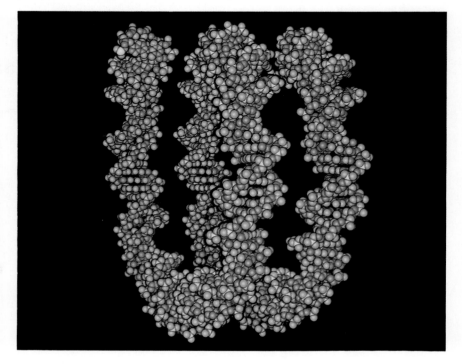

Viruses are **obligate parasites** that reproduce only within the cell they invade. They are tiny particles (5–200nm) composed of just two parts: an outer coat of protein and an inner core of nucleic acid; often, as in bacteriophages, the nucleic acid is DNA. Experimenters wished to determine which part of a virus, the protein or the DNA, enters the bacteria and takes over the machinery of the bacterial cell, causing it to produce more viruses.

Hershey and Chase Experiment

A. D. Hershey and Martha Chase first cultured bacteria and viruses in radioactive sulfur (^{35}S) until they obtained a batch of viruses with labeled protein coats. They also cultured bacteria and viruses in a radioactive phosphorus (^{32}P) until they had another batch of viruses with labeled DNA. Then they allowed each batch to attack new bacteria (fig. 10.2) to determine whether the labeled protein or the labeled DNA entered the cell. They found that only viral DNA entered the cell, while the original protein coats remained on the outside of the cell. Once the infection had started, the empty protein coats could be removed without affecting the viral reproduction. It was clear that DNA is the substance inserted by the T2 bacteriophage to take over the metabolism of the cell. In other words, DNA is the genetic material.

Figure 10.2
Life cycle of a T virus. A T virus is a complex virus with a head and a tail. Even so, it is composed of just a protein coat and inner core of DNA. Experimenters labeled the coat with ^{35}S and the DNA with ^{32}P and allowed the viruses to attack bacteria. Later, they found only ^{32}P in the cell, yet the cell produced many new viruses. From this they knew that DNA is the genetic material.

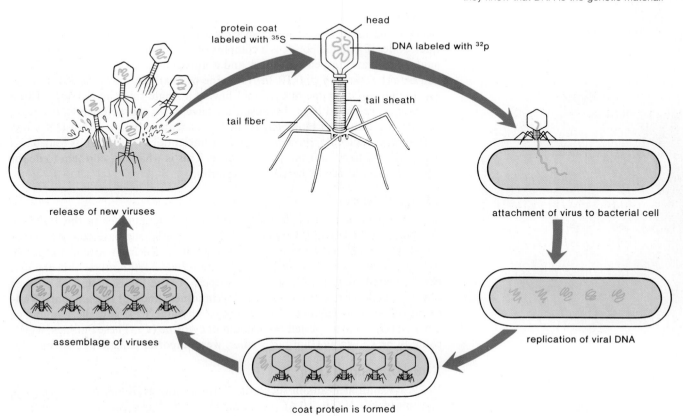

protein coat labeled with ^{35}S

head

DNA labeled with ^{32}p

tail sheath

tail fiber

release of new viruses

attachment of virus to bacterial cell

assemblage of viruses

replication of viral DNA

coat protein is formed

Figure 10.3

All nucleotides found in DNA contain phosphate, the pentose sugar, deoxyribose, and a base. *a*. The bases adenine and guanine are double-ringed purine bases. *b*. The bases cytosine and thymine are single-ringed pyrimidine bases.

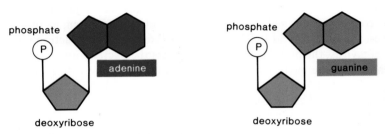

a. DNA nucleotides with purine bases

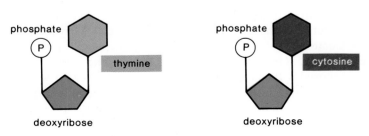

b. DNA nucleotides with pyrimidine bases

Figure 10.4

An X-ray diffraction photograph of DNA taken by Rosalind Franklin. The crossing pattern of dark spots in the center of the picture indicates that DNA is helical. The dark regions at the top and bottom of the photograph show that the purine and pyrimidine bases are stacked on top of one another and are 0.34 nm apart.

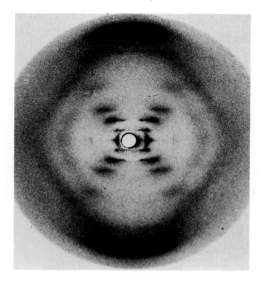

DNA Structure

Nucleotides

Prior to the mid-1900s it was also known that DNA contains four different nucleotides. Each nucleotide is a complex of three united subunits: phosphoric acid (phosphate), a pentose sugar, and a nitrogen base (fig. 3.31). The bases can be either **purines** (fig. 10.3*a*) (adenine or guanine) having a double ring, or **pyrimidines** (thymine or cytosine) having a single ring (fig. 10.3*b*). (These structures are called bases because they have basic characteristics that raise the pH of a solution.) However, at the time scientists believed that DNA was a very small molecule, probably only four nucleotides long, with its four bases arranged in a fixed, unchanging sequence. This is what caused them to doubt that DNA could be the hereditary material.

Chargaff's Rules

With the development of new chemical analysis techniques in the 1940s, it was possible for Erwin Chargaff to analyze in detail the base composition of DNA. He found that instead of being fixed, it varied from species to species just as one would expect for hereditary material. Furthermore, this chemist demonstrated that regardless of the species (1) the total amount of purines always equaled the total amount of pyrimidines, and (2) the amount of adenine (A) always equaled the amount of thymine (T), and the amount of guanine (G) always equaled the amount of cytosine (C). These findings, which came to be known as Chargaff's rules, were important to later research.

X-ray Diffraction Data

In 1951 Rosalind Franklin succeeded in preparing an X-ray diffraction photograph of DNA (fig. 10.4). In that same year, James Watson and Francis Crick began attempting to build a model of DNA structure. After several

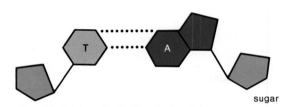

a. thymine (T) is paired with adenine (A)

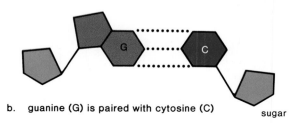

b. guanine (G) is paired with cytosine (C)

Figure 10.5
Complementary base pairing. Watson and Crick determined that a purine is always paired or hydrogen bonded with a pyrimidine. (Hydrogen bonding is represented by dotted lines.) Thymine (*T*) (*a.*) is always bonded with adenine (*A*), and (*b.*) guanine (*G*) is always bonded with cytosine (*C*).

unsuccessful attempts to unravel the structure of DNA, Franklin's photograph finally provided Watson and Crick with the necessary clues. The cross-pattern of X-ray reflections in Franklin's photograph told them that DNA is helical. The black areas at the top and bottom of the photograph meant that the purines and pyrimidines are regularly stacked next to each other at a distance of 0.34 nm. It became clear from other data as well that the DNA helix contains two strands, not three or more, as some chemists had proposed.

Watson and Crick Model

Using Chargaff's rules, Watson and Crick reasoned that within DNA a purine is always bonded to a pyrimidine; *A* is hydrogen bonded to *T;* and *G* is hydrogen bonded to *C* (fig. 10.5). This is now called **complementary base pairing.**

The model that Watson and Crick proposed looks somewhat like a twisted ladder (fig. 10.6). The sugar-phosphate backbone of each polynucleotide strand makes up the *sides* of the ladder; the hydrogen-bonded bases make up the rungs or *steps* of the ladder. The ladder is twisted because DNA is a helix. The Watson and Crick model is known as the **double-helix model.**

DNA Replication

The Watson and Crick model suggested that DNA could replicate by means of complementary base pairing. Further experimentation proved this was indeed the case. The replication process is termed semiconservative because each old DNA strand serves as a template for a new DNA strand (fig. 10.7). A **template** is most often a mold used to produce a shape opposite to itself. Replication requires the following steps:

1. The two strands that make up DNA become enzymatically separated or "unzipped" (i.e., the weak hydrogen bonds between the paired bases break).
2. New complementary nucleotides, always present in the nucleus, move into place by the process of complementary base pairing.
3. The complementary nucleotides become enzymatically joined together. Now a new backbone has formed.
4. When the process is finished, two complete DNA molecules are present, identical to each other and to the original molecule.

Figure 10.6
The DNA double helix resembles a twisted ladder. Sugar-phosphate backbone, symbolized in this drawing by S-P, et cetera, make up the sides of the ladder, and the hydrogen-bonded bases make up the rungs or steps of the ladder.

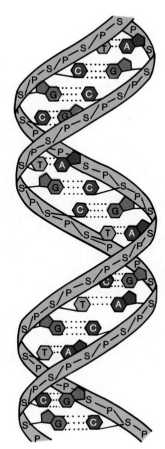

Figure 10.7
Replication of DNA. The two strands of a DNA molecule
separate and each serves as a template for the formation
of a new complementary strand.

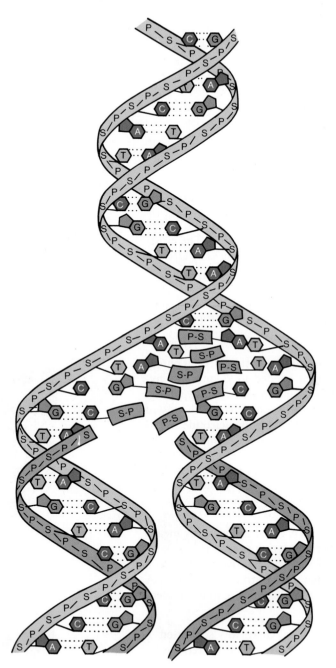

Region of parental DNA helix. (Both backbones
are light).

Region of replication. Parental DNA is unzipped
and new nucleotides are pairing with those in
parental strands.

Region of completed replication. Each double
helix is composed of an old parental strand (light)
and a new daughter strand (dark). Notice that
each double helix is exactly like the other one.

This replication process is described as **semiconservative** because each double strand of DNA contains one old strand and one new strand. Although DNA replication can be easily explained, it is in actuality an extremely complicated process involving many steps and enzymes. There are enzymes that assist the unwinding process, some that join together the nucleotides, and others that assist the rewinding process, just to mention a few. On occasion, errors are made that cause a change in the DNA and, in this way, a mutation arises.

Scientists were surprised to discover that the time required for replication of DNA in human cells is only about twenty times longer than in bacterial cells, even though a human chromosome is much longer than a bacterial one. The explanation for this is that DNA replication in eukaryotic cells is initiated at hundreds of sites, almost simultaneously, along each chromosome, whereas bacterial chromosome replication proceeds from a single site.

Genetic Expression

Inborn Errors

An English physician, Archibald Garrod, suggested in the early 1900s that there is a relationship between inheritance and metabolic diseases. Garrod observed patients who suffered from genetic diseases related to the metabolic pathway shown in figure 10.8. In this pathway, three genetic diseases are now known. In **phenylketonuria** (PKU), phenylpyruvic acid accumulates in the body and spills into the urine because the enzyme needed to convert phenylalanine to tyrosine is missing. If the condition is not treated, a continued accumulation of phenylpyruvic acid in the brain can cause mental retardation. *Albinism*

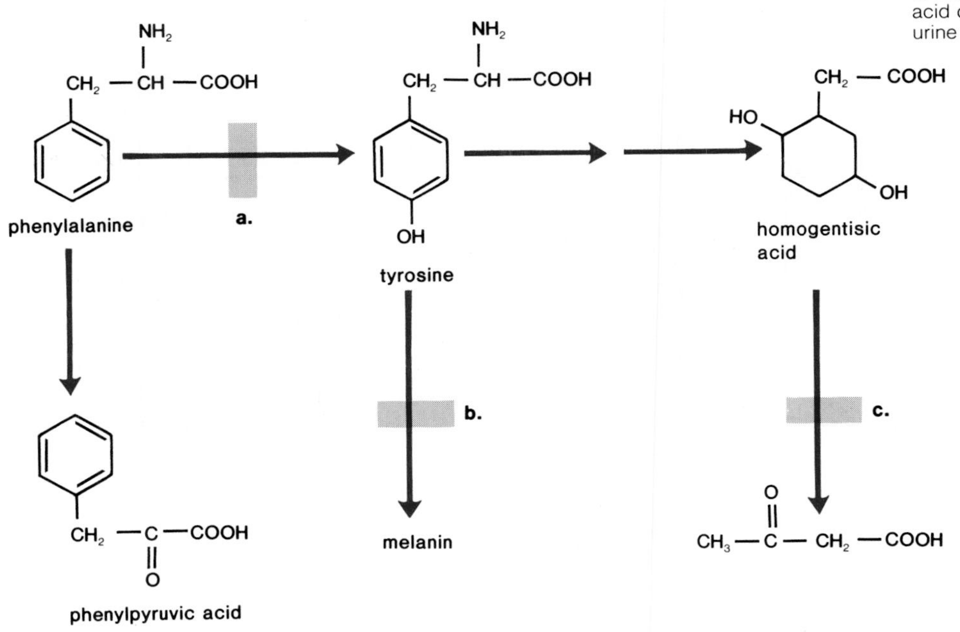

Figure 10.8
Metabolic pathway by which phenylalanine is converted to other metabolites. *a.* If the enzyme that converts phenylalanine to tyrosine is missing, phenylalanine is converted to phenylpyruvic acid instead, and the accumulation of this substance leads to PKU (phenylketonuria). *b.* If the enzyme that converts tyrosine to melanin is missing, albinism results. *c.* If homogentisic acid cannot be metabolized, alkaptonuria results (the urine turns a dark color).

results because tyrosine cannot be converted to melanin, the natural pigment in human skin. *Alkaptonuria* results if the enzyme needed to metabolize homogentisic acid is missing. Because these diseases are inherited, Garrod proposed that each was due to an inherited factor that controlled a metabolic regulator. He referred to these biochemical hereditary diseases as **"inborn errors of metabolism."**

One Gene–One Enzyme Hypothesis

It was many years before Garrod's remarkably insightful hypothesis was verified experimentally. In 1951 G. W. Beadle and E. L. Tatum performed a series of experiments on *Neurospora crassa,* the red bread mold, which reproduces by means of spores. Normally, the spores produce mold that grows on minimal medium because the mold is capable of producing all the enzymes it needs. In their experiments, Beadle and Tatum used X rays to induce mutations in the mold. As described in figure 10.9, the X-rayed spores could no longer produce mold capable of growing on minimal medium. However, growth was possible on medium enriched by certain metabolites. In the hypothetical example given, the mold can grow only when supplied with enriched medium that includes all metabolites or C and D alone. Since C and D are a part of this hypothetical pathway

$$A \xrightarrow{\ 1\ } B \xrightarrow{\ 2\ } C \xrightarrow{\ 3\ } D$$

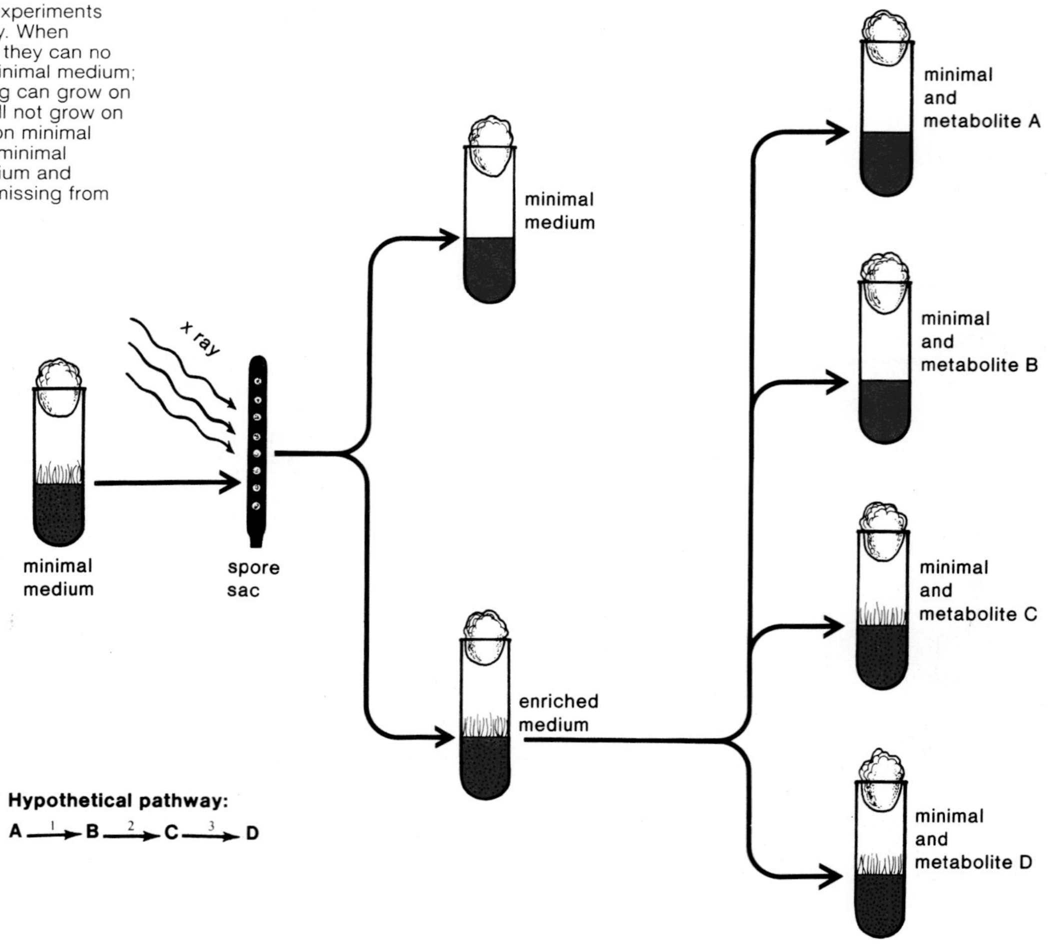

Figure 10.9

A flow diagram illustrating the *Neurospora* experiments that led to the one gene–one enzyme theory. When spores containing haploid DNA are x-rayed, they can no longer produce mycelia that can grow on minimal medium; however, the mycelia produced after x-raying can grow on enriched medium. The mycelia produced will not grow on either minimal medium or metabolite A nor on minimal medium and metabolite B, but will grow on minimal medium and metabolite C and minimal medium and metabolite D. This shows that enzyme 2 is missing from the hypothetical pathway.

Hypothetical pathway:

$$A \xrightarrow{\ 1\ } B \xrightarrow{\ 2\ } C \xrightarrow{\ 3\ } D$$

in which the numbers are enzymes and the letters are metabolites, it is obvious that the mold lacks enzyme 2. Since this is the only enzyme lacking, it may be reasoned that the X-ray treatment of the spore affected only one gene. Beadle and Tatum, on the basis of data such as these, proposed that each gene specifies the synthesis of one enzyme. This is called the **one gene–one enzyme hypothesis.** It is now known that the relationship is more complex than this.

One Gene–One Polypeptide Hypothesis

In 1957, V. M. Ingram reported that the hemoglobin in people with sickle-cell anemia differs from normal hemoglobin at a single amino acid. Normal human hemoglobin, designated as Hb^A, is made up of two alpha (α) polypeptide chains and two beta (β) polypeptide chains. Sickle-cell hemoglobin (Hb^S) differs from Hb^A only in one amino acid in the B chain: valine is present in one position where glutamic acid normally occurs. Unfortunately, this one substitution causes the hemoglobin molecule to be less soluble and to precipitate out of solution, especially when environmental oxygen is low. At those times, the Hb^S molecules stack up into long, semirigid rods that push against the cell membrane and distort the red cell into the sickle shape (fig. 10.10).

The discovery that sickle-cell hemoglobin differs from normal hemoglobin by only one amino acid indicated that *genes control the amino acid sequences of polypeptide chains* whether they are part of enzymes or of other types of proteins. Further, since the α chain is normal in persons with sickle-cell anemia, there must be a separate gene for both α and β chains. This caused Ingram to suggest that the one gene–one enzyme hypothesis be renamed the **one gene–one polypeptide hypothesis.**

Figure 10.10
Sickle-cell anemia in humans. *a.* The first seven amino acids found in the normal and in the abnormal β chain. The substitution of a single amino acid (valine substituted for glutamic acid) at the sixth position results in sickle-cell anemia. *b.* Photomicrographs of normal (*left*) and sickled (*right*) red blood cells.

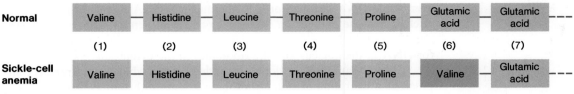

Normal	Valine	Histidine	Leucine	Threonine	Proline	Glutamic acid	Glutamic acid	---
	(1)	(2)	(3)	(4)	(5)	(6)	(7)	
Sickle-cell anemia	Valine	Histidine	Leucine	Threonine	Proline	Valine	Glutamic acid	---

a.

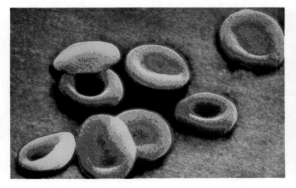 

b.

Figure 10.11
Like DNA, RNA is a polymer of nucleotides. RNA, however, is single-stranded; the pentose sugar is ribose, and uracil replaces thymine as one of the pyrimidine bases. The drawing at left shows a more detailed RNA structure than the drawing at right.

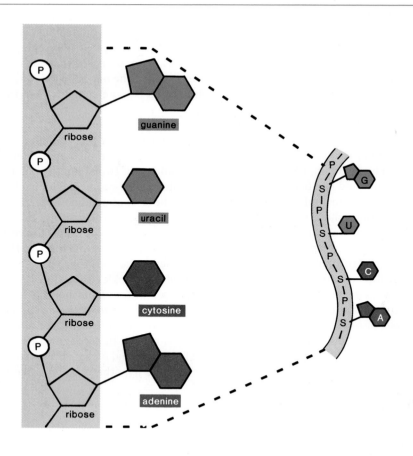

Table 10.1 DNA Compared to RNA

	DNA	RNA
Function	Genes; controls protein synthesis	Helper to DNA; involved in protein synthesis
Sugar	Deoxyribose	Ribose
Bases	Adenine, guanine, thymine, cytosine	Adenine, guanine, uracil, cytosine
Strands	Double-stranded with base pairing	Single-stranded
Helix	Yes	No

Protein Synthesis

The fact that DNA controls the production of proteins may at first seem surprising when we consider that genes are located in the nucleus of eukaryotic cells, while enzymes are synthesized at the ribosomes in the cytoplasm. However, while DNA is found only in the nucleus, RNA exists within both the nucleus and the cytoplasm. Modern biochemical genetic research indicates that a type of RNA called **messenger RNA** (mRNA) serves as a go-between for DNA in the nucleus and the ribosomes in the cytoplasm.

Recall that DNA is a double-stranded helix in which each strand is composed of a backbone of sugar-phosphate groups with four different bases—adenine (A), cytosine (C), guanine (G), and thymine (T)—attached to the sugars. RNA is similar to DNA (fig. 10.11) in structure, except RNA is single-stranded; the sugar is ribose; and the base uracil (U) replaces the base thymine (table 10.1). Because complementary base pairing is possible between the two nucleic acids, DNA can serve as a template for the production of mRNA. The same rules of base pairing hold as in replicating a copy of DNA, except adenine attracts uracil instead of thymine.

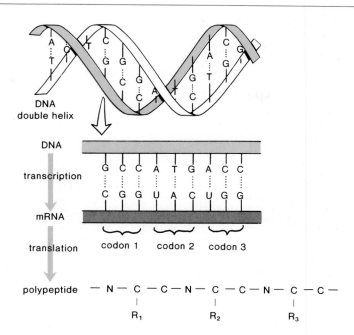

DNA
double helix

DNA

transcription

G C C A T G A C C

C G G U A C U G G

mRNA

translation

codon 1   codon 2   codon 3

polypeptide — N — C — C — N — C — C — N — C — C —
                    | | |
                    R₁   R₂   R₃

Figure 10.12
Transcription and translation in protein synthesis. Transcription occurs when DNA acts as a template for mRNA synthesis. Translation occurs when the sequence of the codons found in mRNA determines the sequence of the amino acids in a polypeptide.

After its formation, mRNA moves from the nucleus to the ribosomes in the cytoplasm, where it dictates the sequence of amino acids in a polypeptide. This concept, often called the "central dogma" of modern genetics, can be diagrammed as follows.

DNA ⟶ mRNA ⟶ polypeptide
        transcription   translation

The diagram indicates that the control of polypeptide synthesis has two parts: transcription and translation. During the process of **transcription,** complementary mRNA is made in the nucleus, and during **translation** its message is used to produce the correct order of amino acids in a polypeptide (fig. 10.12).

Code of Heredity

DNA provides mRNA with a message that directs the order of amino acids during protein synthesis. What is the nature of the message? The message cannot be contained in the sugar-phosphate backbone because it never changes. The order of the bases in DNA and mRNA can and does change however. Therefore, it must be the bases that contain the message. The order of the bases in DNA must code for the order of the amino acids in a polypeptide. Can four bases provide enough combinations to code for twenty amino acids? If the code were a duplicate one (any two bases stand for one amino acid), it would not be possible to code for twenty amino acids (table 10.2). But if the code were a triplet code, the four bases would be able to supply sixty-four different triplets, far more than is needed to code for twenty different amino acids. It should come as no surprise, then, to learn that the code is a **triplet code.**

Table 10.2 Number of Bases in Code

Number of Bases in Code	Number of Amino Acids Coded for
1	4
2	16
3	64

Table 10.3 Three-Letter Codons of Messenger RNA, and the Amino Acids Specified by the Codons

AAU AAC	] Asparagine	CAU CAC	] Histidine	GAU GAC	] Aspartic acid	UAU UAC	] Tyrosine
AAA AAG	] Lysine	CAA CAG	] Glutamine	GAA GAG	] Glutamic acid	UAA UAG	] (Stop)*
ACU ACC ACA ACG	] Threonine	CCU CCC CCA CCG	] Proline	GCU GCC GCA GCG	] Alanine	UCU UCC UCA UCG	] Serine
AGU AGC	] Serine	CGU CGC CGA CGG	] Arginine	GGU GGC GGA GGG	] Glycine	UGU UGC	] Cysteine
AGA AGG	] Arginine					UGA UGG	(Stop)* Tryptophan
AUU AUC AUA	] Isoleucine	CUU CUC CUA CUG	] Leucine	GUU GUC GUA GUG	] Valine	UUU UUC	] Phenylalanine
AUG	Methionine					UUA UUG	] Leucine

*Stop codons signal the end of the formation of a polypeptide chain.
From Volpe, E. Peter, *Biology and Human Concerns*, 3d ed. © 1975, 1979, 1983 Wm. C. Brown Publishers, Dubuque, Iowa. All Rights Reserved. Reprinted by permission.

To crack the code, artificial RNA was added to a medium containing bacterial ribosomes and a mixture of amino acids. Comparison of the bases in the RNA with the resulting polypeptide allowed investigators to decipher the code. Each sequential three-base unit of a messenger RNA is called a **codon.** All sixty-four codons have been determined (table 10.3). Sixty-one triplets correspond to particular amino acids, and three others stand for chain termination.

The DNA code is degenerate—many amino acids are designated by more than one codon. Codons specifying the same amino acid often differ only in the third nucleotide; this sometimes permits DNA to mutate without altering the amino acid sequence of the proteins encoded by the DNA. Therefore, it is possible that degeneracy is a protective device that maintains the constancy of the code.

An analysis of mutations in viruses, bacteria, and higher organisms has thus far shown that the code is essentially *universal.* There are exceptions, but for the most part, the same codons stand for the same amino acids in all living things. Although a few differences have been found, it still appears that all living things have a common evolutionary background. Certainly variances in the code and in the sequence of amino acids in proteins can be used to indicate how distantly related are two species.

Transcription

Messenger RNA

Transcription is the transfer of information from DNA to messenger RNA. The process of transcription allows the formation of a mRNA that contains a sequence of bases complementary to DNA. A segment of the DNA helix unravels; complementary RNA nucleotides pair with DNA nucleotides of one of the strands. When these RNA nucleotides are joined together, an mRNA molecule results. This molecule carries a sequence of bases that are triplet codons complementary to the DNA triplet code (fig. 10.13).

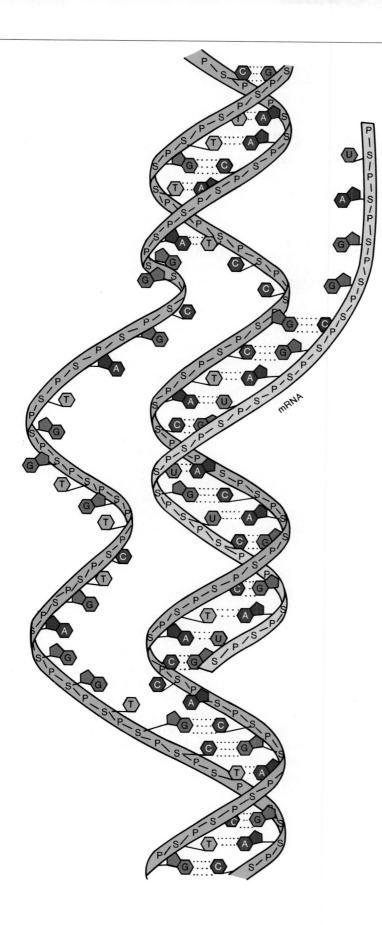

mRNA

Figure 10.13
Transcription takes place in the nucleus. (*below*) DNA (color) unzips and complementary RNA nucleotides lie adjacent to the DNA nucleotides. When they join as shown, mRNA (gray) is formed. Every three bases in mRNA is a codon that corresponds to the DNA code. (*above*) mRNA has disengaged itself from DNA and is ready to begin its process of leaving the nucleus. The sequence of codons is the message that mRNA takes to the cytoplasm.

Translation

Translation is the synthesis of proteins based on messenger RNA codons. It requires the involvement of two other types of RNA: transfer RNA (tRNA) and ribosomal RNA (rRNA).

Ribosomal RNA

Ribosomal RNA is sometimes called structural RNA because it was once believed that ribosomes were like an inert workbench on which amino acids were assembled. High-energy utilization by ribosomes, however, suggests that **ribosomes** probably play an important role in coordinating protein synthesis. Ribosomes are composed of two subunits, each with its characteristic rRNA and proteins. The larger of the two subunits contains at least thirty different proteins, and the smaller unit contains at least twenty different proteins. Ribosomal RNA is transcribed from DNA at the nucleolus; ribosomal proteins are manufactured in the cytoplasm but emigrate to the nucleolus where ribosomes are assembled.

Transfer RNA

Small molecules of tRNA are located in the cytoplasm; each molecule transfers a particular amino acid from the cytoplasm to a ribosome. Molecules of tRNA are shaped in such a way that one end bears a specific sequence of three bases. These bases are referred to as **anticodons** because each one is complementary to an mRNA codon. The other end of a tRNA molecule bears one of the twenty amino acids. (Each tRNA molecule is transcribed from DNA and then, due to intramolecular binding of complementary bases, the anticodon is exposed.)

Since the DNA code is degenerate, there may be more than one type of tRNA molecule for each amino acid. Only one type of amino acid is attached to each type of tRNA. The order in which the tRNA-amino acid complexes come to their complementary codons at the ribosomes determines the sequence of amino acids in a polypeptide. The making of a protein is accomplished codon by codon, as enzymes construct peptide bonds between amino acids.

Polypeptide Synthesis

Polypeptide synthesis involves three steps—initiation, elongation, and termination. During the *initiation* process, a ribosome becomes attached to a messenger RNA. First, the smaller subunit binds to mRNA; then, the larger subunit joins to the smaller subunit, giving a complete ribosomal structure. *Elongation* occurs as the polypeptide chain grows in length. Figure 10.14 shows the process of elongation sometime after initiation. A ribosome is large enough to accommodate two codons; at one codon a tRNA is just about to leave, and at the other codon a tRNA–amino acid complex has just arrived. The bond holding the chain to the first tRNA molecule is enzymatically broken, and the chain immediately becomes attached by a peptide bond to the newly arrived amino acid. The ribosome then moves laterally so the next mRNA codon becomes available to receive the next tRNA-amino acid complex. In this manner, the peptide chain grows and the primary structure of a polypeptide comes

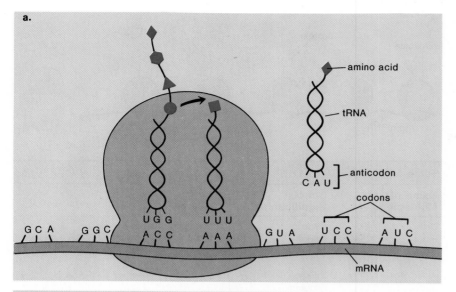

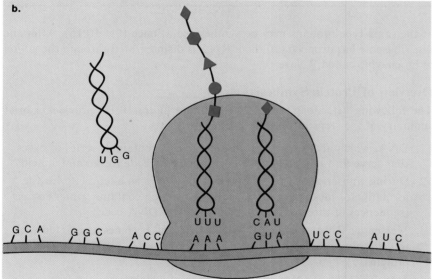

Figure 10.14
Translation. During translation a ribosome moves along an mRNA. A codon on the mRNA attracts an anticodon on the tRNA. When a codon pairs with an anticodon, the tRNA brings an amino acid to the ribosome. *a.* In this diagram, translation is already in progress and two tRNA molecules are in place on a ribosome. Notice that the first of these with the anticodon UGG bears a polypeptide chain, while the second having the anticodon UUU bears an amino acid. The arrow indicates that the chain will be passed from the first to the second tRNA. *b.* In this diagram, the ribosome has moved to the right and the tRNA with the anticodon UGG has departed. The tRNA with the anticodon UUU is now in the first position on the ribosome. When the ribosome moves again, the polypeptide chain will be passed to the newly arrived tRNA–amino acid complex with the anticodon CAU. In this way, during translation, a polypeptide chain is synthesized.

about; the secondary and tertiary structures (p. 60) occur after termination, as the predetermined sequence of amino acids within the polypeptide chain interact with one another.

Termination of polypeptide synthesis occurs at a specific nucleotide sequence on the messenger RNA where the last tRNA and completed polypeptide are liberated from the ribosomal complex. The ribosome dissociates into its two subunits and falls off the messenger. Several ribosomes (collectively called a **polysome**) may move along one mRNA at a time; therefore, several

Figure 10.15
Diagram of a polyribosome (polysome). Several ribosomes move along an mRNA at a time. Ribosomes function independently, enabling several polypeptides to be made at the same time.

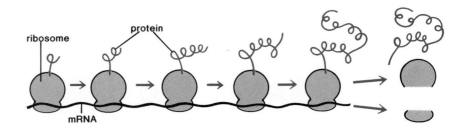

Table 10.4 Steps in Protein Synthesis

Name of Molecule	Special Significance	Definition
DNA	Code	Sequence of bases in threes
mRNA	Codon	Complementary sequence of bases in threes
tRNA	Anticodon	Sequence of three bases complementary to codon
Amino acids	Building blocks	Transported to ribosomes by tRNAs
Protein	Enzyme	Amino acids joined in a predetermined order

of the same type proteins may be synthesized at once (fig. 10.15). After the last ribosome has dropped off, the mRNA is disintegrated through the action of an enzyme called RNase.

Overview of Protein Synthesis

The following list, along with table 10.4 and figure 10.16, provides a brief summary of the steps involved in protein synthesis.

1. DNA, which always remains in the nucleus, contains a series of bases that serve as a *triplet code* (every three bases code for an amino acid).
2. During transcription one strand of DNA serves as a template for the formation of messenger RNA (mRNA), which contains *triplet codons* (sequences of three bases complementary to DNA code).
3. Messenger RNA goes into the cytoplasm and becomes associated with the *ribosomes* that are composed of ribosomal RNA (rRNA) and proteins.
4. Transfer RNA (tRNA) molecules, each of which is bonded to a particular amino acid, have *anticodons* that pair complementarily to the codons in mRNA.
5. As the ribosome moves along mRNA, a newly arrived tRNA-amino acid complex receives the growing polypeptide chain from a tRNA molecule that is about to leave the ribosome. During translation, therefore, the linear sequence of codons of the mRNA determines the order in which the tRNA molecules and their attached amino acids arrive at the ribosomes and thus determines the *primary structure* (linear sequence of amino acids) of a protein.

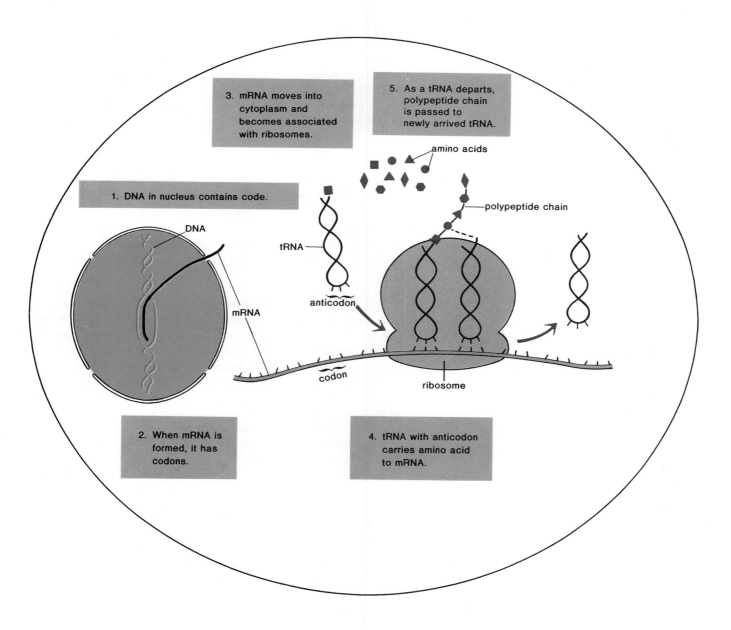

Figure 10.16
Summary of protein synthesis. Transcription occurs in the nucleus (color) and translation occurs in the cytoplasm (white). During translation, the ribosome moves along an mRNA. In the diagram, as the ribosome moves to the left, a tRNA bearing an amino acid will come to the ribosome. Thereafter, the polypeptide chain will be passed to this tRNA–amino acid complex. Each time the ribosome moves, a tRNA departs.

3. mRNA moves into cytoplasm and becomes associated with ribosomes.

5. As a tRNA departs, polypeptide chain is passed to newly arrived tRNA.

amino acids

1. DNA in nucleus contains code.

polypeptide chain

DNA

tRNA

mRNA

anticodon

codon

ribosome

2. When mRNA is formed, it has codons.

4. tRNA with anticodon carries amino acid to mRNA.

Table 10.5 Participants in Regulatory Models

Participants	Action
Operon	Genes that code for enzymes in a metabolic pathway
Operator	An on/off switch for transcription of operon
Regulatory gene	A gene that codes for a repressor
Inducer	A metabolite that inactivates a repressor
Corepressor	A metabolite that activates a repressor

Regulatory Genes

Thus far, genes that provide coded information for synthesis of proteins (polypeptides) have been discussed. These genes are called **structural genes** because they determine protein structure. Evidence indicates that another class of genes, called regulatory genes, exists. **Regulatory genes** regulate the activity of structural genes. For example, since cells produce only those enzymes that are currently needed for metabolism, regulatory genes probably control the transcription of the genes that code for these enzymes.

Prokaryotic Models

Research with bacteria led French scientists Francois Jacob and Jacques Monod to develop two models that explain how genes can be turned on and off. These regulatory models have the following three components (fig. 10.17; table 10.5).

Operon: A group of structural genes that codes for enzymes active in a particular metabolic pathway, such as the enzymes 1, 2, 3 in this pathway.

$$A \xrightarrow{\ 1\ } B \xrightarrow{\ 2\ } C \xrightarrow{\ 3\ } D$$

Operator: A segment of DNA that acts as an on/off switch to start or stop transcription of the operon.

Regulatory gene: A gene that codes for a protein that either (1) immediately combines with the operator, preventing transcription, or (2) must first join with a metabolite before it combines with the operator. The two models (fig. 10.17) differ as to whether (1) or (2) controls transcription of the operon.

Models

In figure 10.17*a*, the operon is normally inactive because the regulatory gene codes for a protein, called a **repressor,** that combines with the operator, preventing transcription. The operon becomes active when the repressor joins with an inducer molecule and the complex is unable to bind with the operator. The **inducer,** so named because it induces protein synthesis, may be a metabolite in a metabolic pathway. For example, A (above) could be an inducer. In this so-called **inducible operon model,** then, the presence of the first metabolite can indicate the need for particular enzymes.

In figure 10.17*b,* the operon is normally active because the regulatory gene codes for an inactive repressor that must join with a corepressor before the complex combines with the operator. A **corepressor,** so named because it prevents protein synthesis, is a metabolite in a metabolic pathway. For example, D (above) could be a corepressor. In this so-called **repressible operon model,** then, the end product can indicate that particular enzymes are no longer needed.

Notice that the inducible model accounts for some structural genes that are normally inactive and the repressible model accounts for some structural genes that are normally active. Thus some genes could normally be turned on while others could normally be turned off. Evidence suggests that these models, developed to explain regulation of protein synthesis in prokaryotes, may be applicable, at least in part, to regulation of protein synthesis in eukaryotes. There is also evidence that regulation in eukaryotes may involve mechanisms not present in prokaryotes.

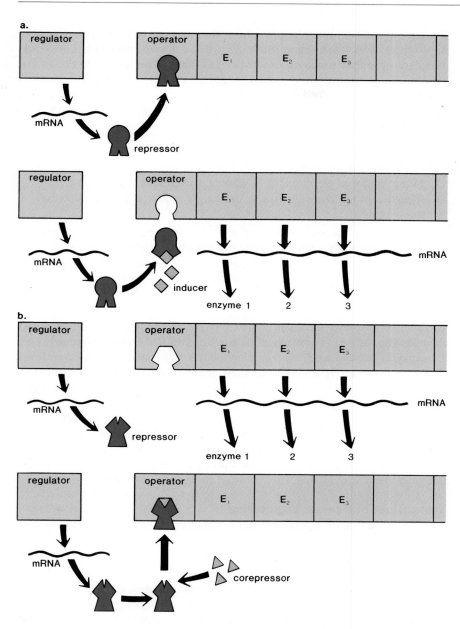

a.

b.

Figure 10.17
Regulation of protein synthesis models. *a.* Inducible model. The regulatory gene codes for a repressor that normally combines with the operator, preventing transcription of an operon. When it combines with an inducer, transcription occurs. This is termed the inducible model because the operon (E_1, E_2, E_3) is normally inactive. *b.* Repressible model. The regulator codes for an inactive repressor that must join with a corepressor before the complex can combine with the operator, preventing transcription of the operon. When the corepressor is not present, transcription occurs. This is termed the repressible model because the operon (E_1, E_2, E_3) is normally active.

Knowledge about regulatory genes is extremely important; genes can best be manipulated when we know how to turn them on and off. Mutations of regulatory genes probably account for some genetic diseases and/or the development of cancer. It's also possible that the mutation of a single regulatory gene could initiate evolution toward speciation since many structural genes would thereby be affected.

Figure 10.18

Gene cloning procedure. Plasmid DNA is removed (*a.*) from *E. coli;* (*b.*) foreign DNA is incorporated or *spliced* into the plasmid, which is then (*c.*) reintroduced into the bacterium. Here it (*d.*) directs protein synthesis and replicates normally.

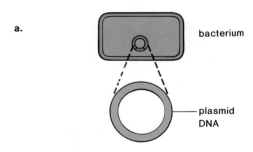

a.

bacterium

plasmid DNA

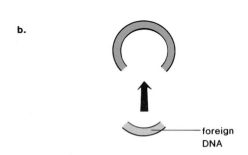

b.

foreign DNA

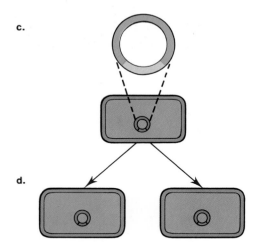

c.

d.

Genetic Engineering

Genetic engineering refers to procedures that manipulate the genes of organisms—from viruses to humans. Both in vitro and in vivo experiments are commonplace in genetic research. *In vitro* means that cell parts and molecules are studied in a test tube, while *in vivo* means that living organisms are used in the experiments. The following procedures are common in experiments involving both prokaryotic DNA and eukaryotic DNA, including human DNA.

1. Isolation of a gene; that is, removal of a particular portion of DNA from a cell. Thereafter it is possible to determine the sequence of nucleotides in the gene.
2. Manufacture of a gene; that is, joining nucleotides together in the sequence of the normal gene, or creating a mutated gene by altering the sequence.
3. Joining the regulatory regions of a viral or bacterial gene to an isolated or machine-made structural gene so that transcription is assured.
4. Placement of the constructed gene in another cell where it undergoes replication and directs protein synthesis.

Gene Cloning

Certain bacteria, such as *E. coli,* have rings of extrachromosomal DNA called **plasmids** (fig. 10.18). These plasmids can be extracted from the bacteria and then enzymatically sliced into fragments. After this is done, DNA taken from another source can be attached to these fragments before the plasmids are reformed. Notice that the plasmids now contain recombined, or **recombinant, DNA** molecules and that such plasmids will be taken up by other bacterial cells (fig. 10.18). Once a plasmid has entered a bacterial cell, the cell is said to be **transformed** because it can now make a protein product it was unable to make before. Further, whenever a bacterium reproduces, the plasmid including the foreign gene is copied. Since exact multiple copies of the plasmid and gene are now available, they are said to have been **cloned.**

Applications of Genetic Engineering

More and more applications have been found for genetic engineering procedures. We will discuss a few of these.

Protein Products as Medicines

All sorts of proteins have been made by *E. coli,* including human insulin and growth hormone. Human insulin went on sale in 1982 and growth hormone is currently undergoing clinical testing. The recombinant DNA procedure can also help us fight disease because *E. coli* can be engineered to make both antibodies, and proteins that can serve as vaccines. A vaccine for foot and mouth disease in animals is now available.

Cloned plasmids can be removed from *E. coli* and introduced into other types of cells. For example, yeast cells have been engineered to make human interferon (p. 451), a chemical useful in immunity and cancer research. Since yeast is used in the production of beer, technology already exists for mass production of these cells. Perhaps, then, interferon will soon be available in bulk.

Protein Engineering

Before a naturally occurring gene is spliced into plasmids, its sequence of bases can be altered. These so-called synthetic mutations cause *E. coli* to produce proteins not normally produced in cells. It is believed that these mutations may eventually tell us more about human genetic diseases. Also, industrial chemists are hopeful that synthetically mutated genes will give us enzymes superior to those now being used as catalysts in the production of important and useful chemicals.

Diagnostic Test for Diseases

Cloned DNA fragments are helping physicians diagnose infectious and genetic diseases. These fragments are called **DNA probes** because they will search out and bind to complementary DNA sequences. For example, a portion of DNA can be removed from an infectious organism and cloned. Blood from a patient is exposed to these fragments and if binding occurs, it can be detected by radioactive or fluorescent techniques. Binding indicates that the infectious organism is present in the patient's blood. Or, the abnormal portion of a gene from a person with a genetic disease can be cloned. If this probe binds to the DNA of a patient's cells, it indicates that the patient has the genetic disease (fig. 10.19).

Alteration of Genetic Inheritance

Genes placed in specially prepared plant cells (fig. 10.20) and in animal eggs remain active as the organism grows and matures. This offers hope that one day it will be possible to engineer plants with genes that promote resistance to pests and/or that fix aerial nitrogen, thus reducing the need for pesticide and fertilizer use. It may also be possible one day to cure human genetic diseases.

Gene Therapy   As discussed in the reading on the next page, researchers have developed the means to attempt to cure human genetic diseases. They have chosen a viral DNA, rather than a plasmid, to serve as the *vector* (carrier) for normal human genes. This particular type of virus has the capability of inserting foreign DNA into host DNA.[1] The virus will be extensively altered. The harmful viral genes will be replaced by the normal human gene needed by the patient. The researchers then plan to infect blood stem cells from the patient's bone marrow. Blood stem cells have been chosen as recipients for the normal genes because they can be removed, treated, and then reinjected into the bone marrow. Hopefully, the active normal genes present in these cells will overcome the detrimental effects of the defective genes in the rest of the patient's body.

[1]The virus is a retrovirus. Cancer-causing retroviruses are discussed on page 213.

Figure 10.19
After studying the chromosomes of hundreds of persons with Huntington's disease, many of whom were members of related families in Venezuela, investigators determined that the gene(s) causing this genetic disease are located on chromosome 4. Using recombinant DNA techniques, a test has been developed that enables investigators to analyze this chromosome in order to predict whether a person will eventually develop the adult-onset disease.

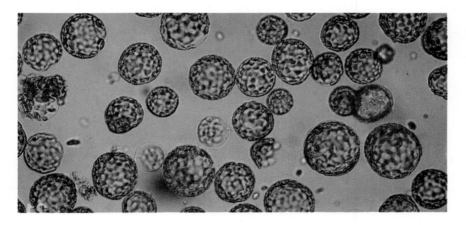

Figure 10.20
Photomicrograph of protoplasts, plant cells that lack cell walls. Whereas plant cells with cell walls cannot take up recombinant plasmids, protoplasts will do so. (These plasmids are from the bacterium *A. tumefaciens*.) Therefore, it is possible to transfer genes for desired characteristics to protoplasts. Given proper treatment, protoplasts will develop into whole plants that display the characteristics. Scientists are hopeful that this technique will allow agricultural plants to be remodeled within the next five or ten years.

Beating Nature's Odds

Nature's lottery is never more tragically played out than in genetic disease. An estimated 25 percent of all hospital beds hold patients suffering from some degree of genetic abnormality. There are approximately 3,000 known genetic diseases. Symptoms can be alleviated in a number of cases: diet therapy in phenylketonuria, blood transfusions in Cooley's anemia, factor VIII replacement in hemophilia, insulin administration in diabetes. But these are just stopgaps. Presently no genetic disease can be cured.

The treatment of disease, genetic and otherwise, may soon shift dramatically. Such a revolution occurred in the 19th century, Lewis Thomas observed (*Science 84*, November), when physicians learned to stop bleeding, purging, and blistering sick patients. Medicine's second revolution, Thomas said, started mid-century when it became possible to cure some infectious diseases with antibiotics. Medicine's third revolution hinges on the applications of molecular biology: the mapping, cloning, and study of thousands of human genes to understand the body's normal functions at the molecular level. The practice of medicine will then become vastly more precise.

Gene therapy is one of the most exciting ramifications of this third revolution. Physicians should be able to treat many disorders by inserting a normal gene into the cells of a patient. If a child is born with a defective gene—for example, a hemoglobin gene that prevents the manufacture of blood, as in Cooley's anemia—a normal gene would be inserted into the appropriate cells so that the disease can be cured. Not just treated with blood transfusions, but cured.

Is this rosy picture really possible in the next 15 years? Yes, probably, for a number of diseases. It is possible now to insert foreign genes into animals and get these genes to function. The first successful gene therapy in a mammal was reported in 1984, when researchers injected a growth hormone gene into a fertilized mouse egg. The mouse would have developed into a dwarf because of a genetic deficiency of growth hormone. The growth hormone deficiency was, in fact, overcorrected in these first experiments: the mouse grew to nearly twice normal size. But the technical ability to cure a genetic disease was demonstrated.

A type of virus known as retrovirus, an RNA tumor virus, can be used to carry genes into an animal's body. The retrovirus is altered so that it functions only as a delivery system: it can no longer cause an infection, but it can carry a functional gene into cells. Researchers remove bone marrow cells from an animal, "infect" the marrow cells with the disabled retrovirus carrying the gene of choice, and then reinject the marrow cells into the animal. In this way they have engineered an active foreign gene into the blood cells of mice.

Before 1986 one or more physicians probably will have submitted clinical protocols requesting permission to carry out this procedure in humans. Because of the revolutionary nature of the treatment, any human gene therapy attempt that receives federal funding must be approved by the National Institutes of Health. In the first cases the open public review conducted by NIH will probably take several months. The Food and Drug Administration may also conduct its own independent evaluation prior to giving its approval. Subsequent approvals presumably will be made more rapidly.

The most promising premier candidate for gene therapy is a disease called adenosine deaminase deficiency. A missing ADA enzyme in immune cells can result in severe combined immune deficiency disease—infants who have little or no immunity and who die from simple childhood infections. They are the "bubble babies" who often cannot survive except in the germ-free environment of a sterile tent or bubble. Inserting a normal ADA gene into the bone marrow cells of these patients should produce normal resistance to infection in them.

Another early candidate will be Lesch-Nyhan disease—a severe neurological disease that results in uncontrollable self-mutilation. Victims bite off their own lips and fingers. The normal version of the defective gene that produces the disease has been isolated. It has been inserted into the bone marrow of mice using a disabled retroviral vector and the human enzyme has been produced in the animals. In addition, the gene has been put into human bone marrow cells growing in culture that have been isolated from patients with Lesch-Nyhan disease. Partial correction of the enzyme deficiency in these cells has been achieved.

Ultimately such genes should be simply packaged and injected into the bloodstream like any other common medication. The packaged gene would have signals directing it to target cells in the patient and, if necessary, into the correct place in the genome of the cell. Thus, the technology should be available for any patient, anywhere in the world, with a genetic disease.

Certainly there will be problems. There is no evidence yet that the technique will even work in human beings. And if it does, there is no assurance that the procedure or the gene itself might not produce other problems. Might the disabled retroviral carrier in some way be reactivated and cause its own disease? Might the insertion of a foreign gene interfere with normal cell function? These questions still must be examined in animal studies before the first attempts to treat humans should be carried out.

Then there is the further question: If we begin changing genes, are we tampering with the essence of our humanness? Should we ever attempt to alter germ line cells so that the patient's offspring would also be corrected? Gene therapy offers enormous hope for the alleviation of human suffering, but we must go forward carefully as we develop greater power to alter the lottery of nature.

From "Beating Nature's Odds," by W. French Anderson, in *Science 85*, pp. 49–50, November 1985. Copyright 1985 by the American Association for the Advancement of Science.

1. Viral gene infects mouse cells.

2. Reverse transcription takes place.

3. Viral DNA is extracted.

human DNA

4. Most of viral genes are removed and human gene is inserted.

5. Transcription of DNA is allowed to occur.

RNA

6. Recombinant RNA is repackaged in viral coat.

retrovirus

7. Virus infects patient's blood cells taken from bone marrow.

8. Reverse transcription occurs.

9. Bone marrow cells carrying corrected human gene are reinjected into the patient.

The virus selected for gene therapy is a retrovirus. Retroviruses have an RNA chromosome. When this chromosome enters a cell, reverse transcription, during which RNA is transcribed to DNA, must take place before reproduction of the virus can begin (fig. 10.2). You can see that this adds steps (numbers 2 and 8) to those described here in order for the virus to serve as a vector for the purpose of gene therapy.

Table 10.6 Mutations

Types of Mutations	Definition
Chromosomal mutation	A rearrangement of chromosome parts, as described in figure 9.12, which may or may not result in a change of the phenotype.
Genetic mutation	A change in the genetic code for a gene or in the expression of the gene. Usually results in a change of the phenotype.
Germinal mutation	A mutation that manifests itself in the gametes so that it is passed on to offspring.
Somatic mutation	A mutation that occurs in the body cells and which very likely is not passed on to offspring.

Table 10.7 Genetic Mutations

Type of Base Change	DNA Code	Effect on Polypeptide
None	TAC'GGC'ATG'TCA	Usual amino acid sequence
Deletion	ACG'GCA'TGT'CA	Amino acid sequence completely altered
Addition	ATA'CGG'CAT'GTC'A	Amino acid sequence completely altered
Substitution	TAG'GGC'ATG'TCA	Change in only one amino acid

Genetic Mutations

Table 10.6 reviews the terminology pertaining to mutations. As you can see, a **genetic mutation** is any alteration in the code of a single gene or any change in its expression.

Types of Genetic Mutations

A change in the sequence of bases can change the DNA code. As mentioned previously, code changes are occasionally introduced when DNA is replicating. Table 10.7 illustrates the profound effect of the deletion or addition of a nucleotide. A single base substitution, on the other hand, may not result in a completely nonfunctional polypeptide. Even so, the results can be devastating; for example, sickel-cell hemoglobin has been traced to a single nucleotide change in DNA.

A surprising finding of late is that cells have built-in mechanisms that produce genetic mutations.

Mechanisms for Genetic Mutations

Transposons

Transposons are specific DNA sequences that have the ability to move within and out of chromosomes. Their movement to a new location sometimes alters neighboring genes, particularly by increasing or decreasing their expression. Although "movable elements" in corn were described by Barbara McClintock (fig. 10.21) forty years ago, their significance was only recently realized. So-called "jumping genes" have now been discovered in bacteria, fruit flies, and humans, and it is likely that all organisms have such elements. Some investigators have suggested that transposons tend to become active during times of environmental stress and in that way increase the likelihood of mutations that can aid survival. In modern genetics laboratories, transposons have been used as vectors to carry selected genes into new hosts (fig. 10.22).

Split Genes

Eukaryotic structural genes (fig. 10.23) are now known to be interrupted by sections of DNA that are not part of the gene. These portions are called *introns* because they are *intra*gene segments. The other portions of the gene are called *exons* because they are ultimately *ex*pressed.

Unlike prokaryotic genes, eukaryotic genes are transcribed in a nucleus. When DNA is transcribed, the mRNA contains bases that are complementary to both exons and introns. But before the mRNA exits from the nucleus, it is *processed*—the nucleotides complementary to the introns are enzymatically removed. Some evidence suggests that the presence of introns facilitates the occurrence of routine mutations. Just as sectional furniture can be arranged differently, it has been shown that exons are like modules that can be selected and arranged to give different sequences of DNA (fig. 10.24). Each rearrangement appears to be a mutation.

Figure 10.21
Barbara McClintock reported forty years ago that her corn experiment results could only be explained by the presence of mobile elements, now called "jumping genes." She was so ahead of her time that the scientific community did not recognize her achievement until she received the Nobel Prize in 1983.

Figure 10.22

Transposons bearing a gene for red eyes were injected into the cells of this brown-eyed *Drosophilia* (*left*) when it was an embryo. Subsequently, its offspring (*right*) had red eyes, indicating that the transposons had successfully incorporated the normal gene into the chromosomes of the parent.

Figure 10.24

What is the function of introns? Perhaps they permit mutations by allowing different sections of DNA to be selected and subsequently pieced together to give various genes.

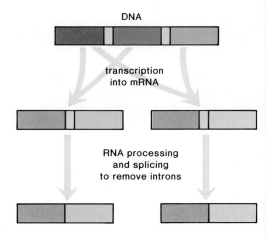

DNA

transcription
into mRNA

RNA processing
and splicing
to remove introns

Figure 10.23

The chromosomes of higher organisms contain segments that do not code for polypeptides. Repetitive DNA is noncoding and consists of sequences of base pairs repeated many times, one after another. Structural genes, themselves, are interrupted by intervening sequences or introns, which do not code for polypeptides. During transcription, the entire gene, including these segments, are copied into mRNA. Before the mRNA leaves the nucleus, these segments are spliced out. Structural genes are flanked at one end by regulatory genes that control whether transcription takes place or not. There is a start code at the beginning and a stop code at the end of a structural gene.

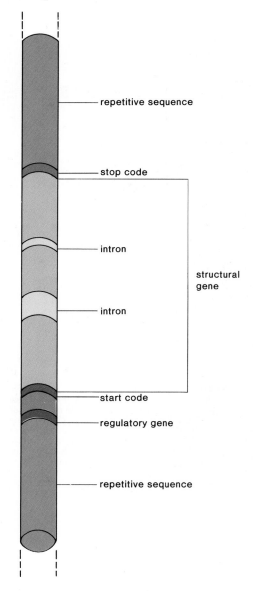

repetitive sequence

stop code

intron

intron

structural gene

start code

regulatory gene

repetitive sequence

Figure 10.23 shows that the eukaryotic chromosome also contains *repetitive DNA*—the same short-to-intermediate length DNA sequence is repeated over and over. The exact function of repetitive DNA has not been determined, but perhaps these sequences are transposons that can function as introns. In other words, movement of transposons sometimes causes a structural gene to be interrupted by portions that are not part of the gene. Another view is that repetitive DNA is "selfish DNA"; DNA that has no function and simply is getting a free ride as it is duplicated and passed from generation to generation.

View of DNA

In the past, most biologists thought of DNA as fixed and static, but the recent findings discussed in the preceding have changed this view. Indeed, DNA has been shown to be labile; it is constantly changing as portions move here or there or are joined in new and different ways prior to transcription.

Figure 10.25
Transformation from normal to cancerous cells. *a.* Normal fibroblasts are flat and extended. *b.* After being infected with Rouse sarcoma virus, the cells become round and cluster together in piles. The virus carries an oncogene.

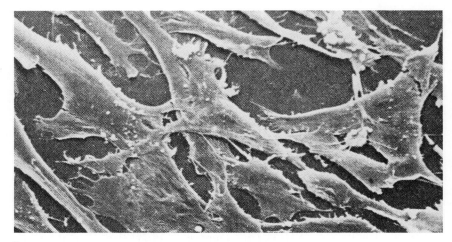

a.

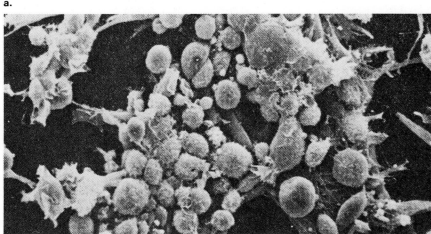

b.

Figure 10.26
Summary of the development of cancer. A virus can pick up an oncogene from one cell and pass it to another cell. Or a normal gene, called a proto-oncogene, can become an oncogene due to a mutation caused by a chemical or by radiation. In this diagram, the oncogene causes the production of growth factor. Now the cell is a cancerous cell that repeatedly divides and never becomes specialized.

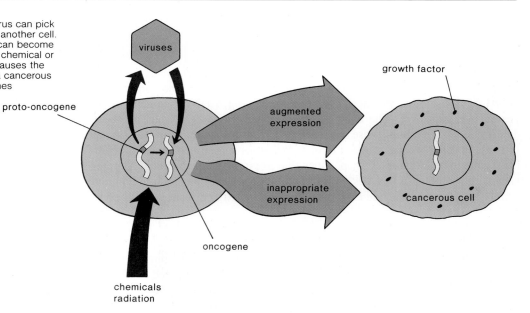

viruses

proto-oncogene

growth factor

augmented expression

inappropriate expression

cancerous cell

oncogene

chemicals radiation

Cancer

Cancer is characterized by irregular (fig. 10.25) and uncontrolled growth of cells that do not stay in the organ where they arose. For a time, the cancer remains at the site of origin, but eventually cancer cells invade underlying tissues, become detached, and are carried by the lymphatic and circulatory systems to other parts of the body where new cancer growth may begin. The process by which cancer spreads to other parts of the body is called **metastasis,** and it is this tendency of malignant cells to metastasize widely that usually results in the death of the patient.

It has long been believed that cancer begins with a change in the DNA, but the exact nature of the change was unknown. Recently, investigators have been able to determine that cells contain genes, called **proto-oncogenes** (*proto* = before; *onco* = tumor), that can become **oncogenes,** cancer-causing genes (fig. 10.26). These genes are not alien to the cell; they are normal, essential genes that have undergone a mutation. By using recombinant DNA techniques, for example, investigators have shown that an oncogene that causes both lung cancer and bladder cancer differs from a normal gene by a change in only one nucleotide. It is now believed that almost any type of mutation can convert a proto-oncogene into an oncogene. For instance, a chromosomal translocation may place a normally dormant structural gene next to an active regulatory gene. If this structural gene is a proto-oncogene, the translocation may cause it to become an oncogene. Or the movement of a transposon might suddenly cause the transformation of a proto-oncogene into an oncogene. Also, environmental **mutagens** (any factor that increases the chances of a mutation), such as chemicals and X rays, can cause cancer when they bring about the conversion of a proto-oncogene to an oncogene. On the other hand, cancer-causing viruses most likely bring active oncogenes into a cell (fig. 10.26).

All cancer-causing viruses discovered so far are retroviruses. **Retroviruses** contain a core of RNA rather than a core of DNA and contain an enzyme that carries out the transcription of RNA to DNA. This is the *reverse* of normal transcription, and contradicts the central dogma discussed on page 197. Retroviruses normally insert their newly transcribed DNA into host DNA, where it replicates along with host DNA. Only after a time does it cause the production of new viruses. The fact that retroviruses insert their DNA in host DNA allows them to plant a foreign oncogene into host DNA.

Sometimes it seems that a cell becomes cancerous when it contains one oncogene, and sometimes several oncogenes are required. Perhaps this difference can be attributed to the influence of the environment on the cell. Oncogene-bearing cells surrounded by normal neighbors remain normal, but if these normal cells are removed, the presence of a single oncogene seems to induce uncontrollable growth.

Function of Oncogenes

Cells transformed (fig. 10.25) by oncogenes often require few or no growth factors in the medium in order to grow and divide without limit. *Growth factors* are proteins that can stimulate a cell to grow and divide once it is received by a particular receptor on the cell's membrane. As figure 10.27 shows, it now appears that any gene that codes for a growth factor and any one that codes for a growth-factor receptor can be considered a proto-oncogene. Growth-factor genes typically become relatively inactive as the cell matures, which accounts for the fact that most cells are capable of only a few divisions during their life span. A mutation that suddenly causes a growth-factor gene to become active or more active than usual also accounts for the gene's conversion from proto-oncogene to oncogene.

Figure 10.27

Function of oncogenes. *a.* Oncogenes sometimes code for a growth factor (colored dots) that exits the cell where it is received by cell membrane receptors. Then the cell begins to grow and divide. *b.* Other times, oncogenes code for a growth factor receptor that due to a malformation is capable of bringing about stimulation even when no growth factor is present.

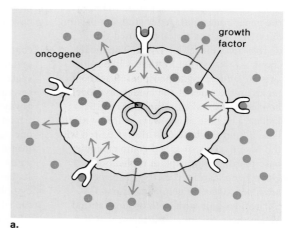

a.

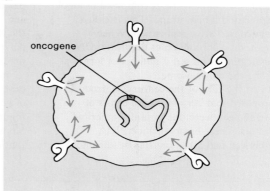

b.

Summary

In eukaryotes, the chromosomes contain both DNA and histone proteins. Hershey and Chase showed that only DNA from the phage T2 (and not the protein coat) entered a bacterium and directed the cell to produce more phages. This indicated that DNA is the genetic material. After studying existing data, Watson and Crick proposed the double-helix model for DNA. The model shows a DNA structure that is like a twisted ladder; sugar-phosphate backbones make up the sides of the ladder; hydrogen-bonded bases make up the rungs of the ladder. The base A is always paired with the base T, and the base C is always paired with the base G. This so-called complementary base pairing helps explain replication. During replication, the DNA becomes unzipped, and new complementary nucleotides pair with each original strand. When these new nucleotides are joined, each new double helix is composed of an old strand and a new strand. This is called semiconservative replication. When eukaryotic chromosomes duplicate, DNA replication begins at a number of locations along the chromosome.

At the turn of the century, Garrod proposed that certain inborn errors of metabolism were due to faulty inherited factors. These errors include PKU, albinism, and alkaptonuria. Beadle and Tatum later proposed the one gene–one enzyme theory after experimenting with *Neurospora*. The one gene–one enzyme hypothesis has been changed to the one gene–one polypeptide theory because some proteins are not enzymes and some proteins have more than one polypeptide chain.

DNA controls the phenotype by controlling protein synthesis. DNA and several forms of RNA participate in protein synthesis. RNA differs from DNA in several respects (table 10.1). DNA, which always stays within the nucleus, contains a triplet code: a series of three bases codes for one particular amino acid. During transcription, messenger RNA is made complementary to one of the DNA strands. It then contains codons and moves to the cytoplasm, where it becomes associated with the ribosomes. During translation, transfer RNA molecules attached to their own particular amino acid travel to the mRNA and, through complementary base pairing, the tRNAs and thus the amino acids in a polypeptide chain become sequenced in a predetermined order.

Bacterial research has permitted the development of two models for the control of gene transcription. In one model, a regulatory gene codes for a repressor that binds with an operator gene in such a way that transcription of an operon is normally impossible. When this repressor joins with an inducer, however, the complex cannot bind with the operator; therefore, operon transcription begins. In the other model, the regulatory gene codes for a repressor that must join with a corepressor before the complex can bind with the operator. In this instance, transcription is normally possible.

Recombinant DNA techniques allow eukaryotic DNA to be placed in the bacterium *E. coli,* where it usually replicates and functions normally. If so, this cloned DNA can be removed and put in viral DNA or into other types of cells. Recombinant DNA techniques can help produce products needed by humans, diagnostic tests for diseases and genetic diseases, and may even some day help produce better crops and cure genetic diseases.

New mechanisms for mutations have been discovered recently. All cells seem to contain transposons, and the eukaryotic chromosome contains introns that allow exons to be variously placed. Mutations are of special interest not only because they can lead to genetic diseases, but also because they may cause cancer to develop. Various oncogenes have been discovered and thus far, they all seem to affect the amount of growth factor available to the cell or the function of a growth-factor receptor.

Objective Questions

1. Chargaff found that the amount of guanine in DNA always equals the amount of _____ .
2. Franklin's X-ray diffraction photograph told researchers that DNA is _____ .
3. The backbone of DNA is made up of _____ and _____ molecules.
4. Replication of DNA is semiconservative, meaning that each new helix is composed of an _____ strand and a _____ strand.
5. In the 1900s, Archibald Garrod referred to biochemical hereditary diseases as _____ .
6. The base _____ in DNA is replaced by the base uracil in RNA.
7. The DNA code is a _____ code, meaning that every three bases stands for an _____ .
8. The three types of RNA that are necessary to protein synthesis are _____ , _____ , and _____ .
9. Which of these types carries amino acids to the ribosomes? _____
10. When a repressor combines with a corepressor we would expect protein synthesis to _____ . (Choose either begin or stop.)
11. Plasmids that are carrying a foreign gene contain _____ DNA.
12. *E. coli,* transformed by recombinant DNA, multiplies and makes many copies of a foreign gene. This gene is said to have been _____ .
13. Another name for transposons is _____ .
14. When mRNA is processed within the eukaryotic nucleus, the portions complementary to _____ in DNA are removed.
15. Cancer-causing genes are termed _____ .

Study Questions

1. Name the three functions of hereditary material.
2. Describe the structure of eukaryotic chromosomes.
3. Describe an experiment with phage T2 that showed that DNA is the genetic material.
4. Describe the structure of DNA, including the double-helix model.
5. Describe how DNA replicates. How do eukaryotic chromosomes duplicate?
6. The idea of "inborn errors of metabolism" was first associated with what metabolic pathway? What were the noted defects in this pathway?
7. Describe Beadle and Tatum's experiments with *Neurospora* by referring to a hypothetical pathway supposedly under investigation.
8. What is the central dogma of modern genetics?
9. Describe the processes of transcription and translation. Include in your description the roles of DNA, mRNA, rRNA, and tRNA.
10. If the code were *AATGCGCAT*, what would the codons be? the anticodons? Using table 10.2, what tripeptide is expected?
11. Describe two models based on prokaryotic research that indicate how transcription might be regulated.
12. Describe the recombinant DNA technique. What are some of the applications of this techinque?
13. Name two types of genetic mutations. Discuss each type.
14. Much of eukaryotic DNA does not code for functional polypeptides. Describe the location and discuss the possible function of this DNA.
15. Discuss the origination of oncogenes and the manner in which they contribute to the development of cancer.

Selected Key Terms

bacteriophage (bak 'tir ē ə ,fäj) 188
purine ('pyür ,ēn) 190
pyrimidine (pi 'rim ə ,dēn) 190
phenylketonuria (,fen l ,kēt n ur ē ə) 193
messenger RNA ('mes n jər) 196
transcription (trans 'krip shən) 197
translation (trans 'lā shən) 197
codon ('kō ,dän) 198

ribosome ('ri bə ,sōm) 200
anticodon ('an tē 'kō ,dän) 202
polysome ('päl i ,sōm) 201
operon ('äp ə ,rän) 204
plasmid ('plaz məd) 206
genetic mutation (jə 'net ik mü 'tā shən) 210
oncogene ('öŋ kō jēn) 213

Suggested Readings for Part 4

Ayala, F., and Kiger, J. A. 1980. *Modern genetics*. Menlo Park, CA: Benjamin/Cummings.

Bishop, J. M. March 1982. Oncogenes. *Scientific American.*

Chambon, P. May 1981. Split genes. *Scientific American.*

Chilton, M. June 1983. A vector for introducing new genes into plants. *Scientific American.*

Cohen, S. N., and Shapiro, J. A. February 1980. Transposable genetic elements. *Scientific American.*

Darnell, J. E. October 1985. RNA. *Scientific American.*

————. October 1983. The processing of RNA. *Scientific American.*

Dickerson, R. E. December 1983. The DNA helix and how it is read. *Scientific American.*

Doolittle, R. F. October 1985. Proteins. *Scientific American.*

Felsenfield, G. October 1985. DNA. *Scientific American.*

Gilbert, W., and Villa-Komaroff, L. April 1980. Useful proteins from recombinant bacteria. *Scientific American.*

Lake, J. A. August 1981. The ribosome. *Scientific American.*

Mader, S. S. 1980. *Human reproductive biology*. Dubuque, IA: Wm. C. Brown Publishers.

Nomura, M. January 1984. The control of ribosome synthesis. *Scientific American.*

Novick, R. P. December 1980. Plasmids. *Scientific American.*

Ptashne, M. November 1982. A genetic switch in a bacterial virus. *Scientific American.*

Shepard, J. F. May 1982. The regeneration of potato plants from leaf-cell protoplasts. *Scientific American.*

Weinberg, R. A. November 1983. A molecular basis of cancer. *Scientific American.*

Evolution

I n his book, *Origin of Species,* Charles Darwin cited evidences for the occurrence of evolution and spelled out a mechanism by which evolution occurs. Since that time, the whole field of genetics has developed and now it is possible to include genetic aspects in an examination of the mechanism of evolution. As Darwin suggested, evolution results in speciation, a process that is recognized as usually requiring geographic isolation, followed by reproductive isolation. Members of different species do not reproduce with one another.

The evidences for evolution have also expanded since the time of Darwin. New findings, whether they pertain to the fossil record, biogeography, or to anatomy and physiology, still support evolution. In addition, there have been actual observances of evolution by means of natural selection.

Bleached lizard from White Sands region, New Mexico. Organisms are adapted to their environment. The white appearance of this lizard causes it to blend into the background, protecting it from would-be predators.

CHAPTER 11

The Evolutionary Process

Concepts

Species, not individuals, undergo evolution.

a. The Hardy-Weinberg law defines evolution as a change in gene frequencies in a population.
b. Evolution involves the production of variations and the transmission of only selected variations in subsequent generations.
c. Both genetic drift and natural selection bring about changes in gene frequencies in a population.
d. An extreme change in gene frequencies in a population sometimes results in the origin of a new species.

The modern theory of organic evolution emphasizes that populations, not individuals, evolve. A **population** is all the members of the same species that live in one locale. A population could be all the green frogs in a frog pond or all the field mice in a barn or all the English daisies on a hill. Each member of a population is assumed to be free to reproduce with any other member, and when reproduction occurs, the genes of one generation are passed on in the manner described by Mendel's laws. Therefore, in this so-called Mendelian population of sexually reproducing individuals, the various alleles of all the gene loci in the individuals make up a **gene pool** for the population. It is customary to describe the gene pool of a population in terms of **gene frequencies.**

For example, suppose it is known that one-fourth of all flies in a *Drosophila* population are homozygous dominant for long wings, one-half are heterozygous, and one-fourth are homozygous recessive for short wings. Using the key given in figure 8.9, we can describe the population in this manner:

$$1/4\ LL + 1/2\ Ll + 1/4\ ll$$

By inspection, it is obvious that the frequency of either allele in the gene pool is 50 percent, or 0.5.

It is now possible to calculate the expected allele frequencies for this gene locus in the next generation. Necessarily, the homozygous dominant individuals will produce one-fourth of all the gametes of the population, and these gametes will all carry the dominant allele, L; the heterozygotes will produce one-half of all the gametes, but one-fourth will be L and one-fourth will be l; the homozygous recessives will produce one-fourth of all the gametes and they will be l. Therefore, in summary, one-half of the gametes will be L and one-half will be l, as is expected if the frequency for each allele in the gene pool is 50 percent.

Assuming that all possible gametes have an equal chance to combine with one another, then as the Punnett square shows, the next generation will have exactly the same ratio of genotypes as the previous generation.

	$1/2\ L$	$1/2\ l$
$1/2\ L$	$1/4\ LL$	$1/4\ Ll$
$1/2\ l$	$1/4\ Ll$	$1/4\ ll$

Results: $1/4\ LL + 1/2\ Ll + 1/4\ ll$

This indicates that (*1*) dominant alleles do not necessarily take the place of recessive alleles and that recessive alleles do not disappear, and (*2*) sexual reproduction in and of itself cannot bring about a change in the allele frequency of the population.

Population Genetics

Hardy-Weinberg Law

Two independent investigators, G. H. Hardy, an Englishman, and W. Weinberg, a German, realized that the binomial equation, expressed here as

$$p^2 + 2pq + q^2$$

could be used to directly calculate the genotype frequencies of a population.

Figure 11.1
Knowing the proportion of the recessive phenotype, exemplified by the albino deer shown here, allows one to calculate the frequency of both the recessive and dominant alleles in a population.

Consider that any present generation of a large population is the result of random matings between members of the previous generation: each type of sperm had an equal chance to fertilize each type of egg. Assign the letter p to one allele and the letter q to the other allele; then:

	p	q
p	p^2	pq
q	pq	q^2

Results: $p^2 + 2pq + q^2$

Compare these results to the results just calculated for the *Drosophila* population and notice that they are comparable. Therefore,

$$p^2 = \text{homozygous dominant individuals}$$
$$q^2 = \text{homozygous recessive individuals}$$
$$2pq = \text{heterozygous individuals}$$

To use this equation to calculate the genotype frequencies in a population, it is necessary to know the frequency of each allele in the population. (Usually, these frequencies are given in decimals rather than fractions because the frequencies of the two alleles are not always easily converted to a fraction.) For example, suppose the dominant allele *(p)* has a frequency of 0.7 and the recessive allele *(q)* has a frequency of 0.3. Then the genotype frequencies of the population would be

p^2 (homozygous dominant)	=	.49
q^2 (homozygous recessive)	=	.09
$2pq$ (heterozygous)	=	.42
		1.00

Notice $p + q$ (frequencies of the two alleles) must equal 1.00 and $p^2 + q^2 + 2pq$ (frequencies of the various genotypes) must also equal 1.00.

When the percentage of individuals in a population that show the recessive phenotype is known, it is possible to calculate the genotype frequencies and the allele frequencies. For example, suppose by inspection we determine that 16 percent of a human population has a continuous hairline. This automatically means that 84 percent of the population has the dominant genotype, a widow's peak. Of these, how many are homozygous dominant? How many are heterozygous?

To answer these questions, first convert 16 percent to a decimal. Then we know that $q^2 = .16$ and, therefore, $q = 0.4$. Since $p + q = 1.0$, we know that $p = 0.6$ and, therefore, p^2 (frequency of the population that is homozygous dominant) $= .36$. To determine the frequency of the heterozygote, we simply realize that thus far we have accounted for only .52 of the population; therefore, $.48 =$ heterozygote. Or, if you prefer, calculate that $2pq = .48$. In summary we have found that

homozygous recessive	$= .16$	$= 16\%$ have a continuous hairline
homozygous dominant	$= .36$	$\left. \right\}= 84\%$ have a widow's peak
heterozygous	$= .48$	

Practice Problems

1. A student places six hundred fruit flies with the genotype Ll and four hundred with the genotype ll in a culture bottle. Assuming evolution does not occur, what will be the genotype frequencies in the next generation and each generation thereafter?
2. Four percent of the members of a population of pea plants are short. What is the frequency of both the recessive allele and the dominant allele? What are the genotype frequencies in this population?
3. Twenty-one percent of a population is homozygous dominant, 50 percent is heterozygous, and 29 percent is recessive. What percentage of the next generation is predicted to be recessive?

*Answers to problems appear on page 234.

Qualifications

The binomial expression stated previously can always be used to calculate the genotype frequencies of a pair of alleles at a gene locus in an established population. It can also be used to predict the genotype frequencies of the next generation as long as there have been no gene pool changes. The Hardy-Weinberg law states that the gene and genotype frequencies in a population will remain the same in each succeeding generation as long as

1. the population is large enough to be unaffected by random gene changes. If the population is large, small changes in gene frequencies is not expected to affect significantly the genotype frequencies of the next generation.
2. mating is random. That is, every individual must have an equal opportunity to reproduce with any other individual in the population.
3. there is no gene flow. Immigration and emigration of individuals must not produce any change in the gene pool of the population.

4. no mutations occur or else there must be mutational equilibrium. Genetic changes in one direction must be balanced by an equal number of mutations in the opposite direction.

5. there is no natural selection. No one phenotype has a greater chance of reproductive success than another.

Evolution

Since all these conditions are rarely, if ever, met, *we do expect gene and genotype frequency changes in each subsequent generation*. That is, we expect evolution to occur. For example, suppose that 64 percent of a bear population is homozygous dominant for a heavy coat of hair, 32 percent is heterozygous, and 4 percent is homozygous recessive. We predict, as indicated in figure 11.2, that the genetic makeup of the next generation will be exactly the same as the parental generation. When we sample the next generation, however, we find a change in the allele frequencies at this gene locus. This change indicates that the population is evolving. *Evolution involves changes in gene frequencies in populations from generation to generation.*

Population genetics lends support to Darwin's theory of organic evolution as discussed in chapter 2. For example, Darwin was unable to explain why members of a species differ from one another. We now recognize that phenotypes vary because genotypes vary. Darwin was unable to explain why evolution occurs gradually, causing the origin of a new species to usually take many thousands of years. Changes in gene frequencies of a population have been shown to take place slowly; this accounts for the usual slowness of the evolutionary process.

Darwin's theory of natural selection states that the more fit or better-adapted organisms produce a greater number of offspring, and as a consequence, the species becomes better adapted to its environments. Population genetics points out that the selection process causes an increase in certain gene frequencies. This, in turn, increases the proportion of adaptive genotypes and phenotypes among members of a population. Nonadaptive changes can also occur by chance, but these changes are not expected to account for the origin of species nor the history of life (table 12.1). Only the process of natural selection and adaptation do this.

Synthetic Theory of Evolution

The explanation of evolution in terms of modern genetic principles is a synthesis; it takes data and hypotheses from all sources and blends them into one whole. According to the synthetic theory, the evolutionary process requires two steps:

1. Production of genotype and therefore phenotype variations
2. Sorting out of these variations through successive populations

Table 11.1 lists the processes involved in both these steps.

Production of Variations

Previously we assumed phenotype and genotype differences between members of a species. Now we are concerned with the source of these variations. Genetic variations in the gene pool of sexually reproducing diploid organisms have three sources: mutations, gene flow, and recombinations.

Figure 11.2
Normally, (*a.*) the gene pool is constant generation after generation, but (*b.*) if certain gametes are selected for reproduction at the expense of others, gene pool frequencies do not remain constant.

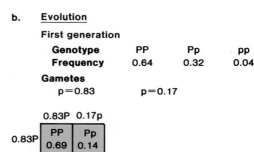

a. **No evolution**

First generation

Genotype	PP	Pp	pp
Frequency	0.64	0.32	0.04

Gametes
P=0.8   p=0.2

0.8P   0.2p

	0.8P	0.2p
0.8P	PP 0.64	Pp 0.16
0.2p	pP 0.16	pp 0.04

Second generation

Genotype	PP	Pp	pp
Frequency	0.64	0.32	0.04

b. **Evolution**

First generation

Genotype	PP	Pp	pp
Frequency	0.64	0.32	0.04

Gametes
p=0.83   p=0.17

0.83P   0.17p

	0.83P	0.17p
0.83P	PP 0.69	Pp 0.14
0.17p	pP 0.14	pp 0.03

Second generation

Genotype	PP	Pp	pp
Frequency	0.69	0.28	0.03

Table 11.1 Mechanism of Evolution

Produce Variation	Reduce Variation
Mutations	Genetic drift
Gene flow	Natural selection
Recombination	

Figure 11.3

A mutation in sheep produced the Ancon breed, represented by the ewe. This breed, which has shorter legs than most breeds, can be maintained only by inbreeding because the unique characteristic is transmitted as a recessive gene.

Figure 11.4

The water flea *Daphnia* normally requires a water temperature of about 20°C to survive, but one mutant requires a temperature between 25°C and 30°C. The head of this multicellular animal is to the right and a brood chamber is to the left. Usually, eggs develop into the young we see here without benefit of fertilization. When environmental conditions are poor, sexual reproduction provides the variation that may be required for the species to adjust to a changing environment.

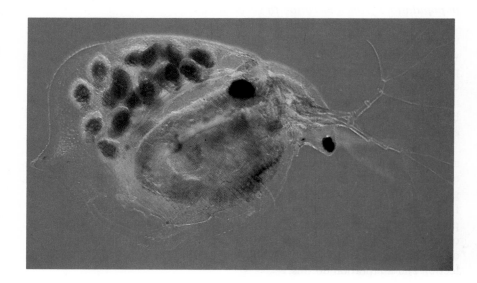

Mutations

There are chromosomal mutations (p. 181) and there are genetic mutations (p. 210) as discussed previously. Mutations are the only original source of allele changes. Even so, mutations do not contribute greatly to the immediate occurrence of evolution. First of all, mutation rates are quite low. A rate of 10^{-5} per gene loci per generation is common in sexually reproducing organisms. Second, a new mutation is likely to be adaptively neutral or harmful (fig. 11.3).

Mutations are expected to eventually contribute to evolution, however. For example, the water flea *Daphnia* (fig. 11.4) normally thrives at temperatures around 20°C and cannot survive at 27°C or more. There is, however, a mutant straint of *Daphnia* that requires temperatures between 25° and 30°C and cannot survive at 20°C. If bodies of water should happen to be only at the higher temperature, only the mutant would survive.

Figure 11.5
The rat snake *Elaphe obsoleta* has such a large
geographical range that subspecies can be distinguished.
The subspecies, or races, interbreed where their individual
ranges meet.

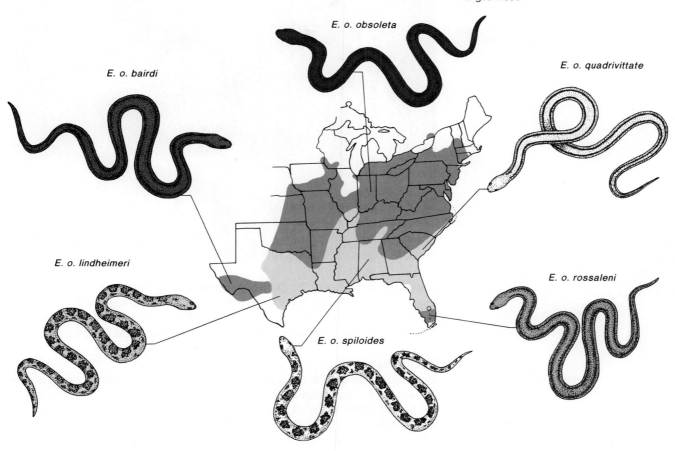

E. o. obsoleta

E. o. bairdi

E. o. quadrivittate

E. o. lindheimeri

E. o. rossaleni

E. o. spiloides

Gene Flow

Gene flow occurs when migrating individuals reproduce with members of a
new population. As a result, new genes may be introduced into the gene pool
of this population. Gene flow usually occurs between populations belonging to
the same species. These populations may be different enough to be called **subspecies** or races. For example, six subspecies of the rat snake *Elaphe obsoleta*
are recognized (fig. 11.5). Notice that each has been given a third name, the
common way to designate subspecies. Since these snakes interbreed wherever
their ranges meet, they all share the same gene pool.

On occasion, gene flow occurs between populations that are separate
species. The resulting genotype is called a **hybrid.** If each species is adapted
to entirely different environments, the hybrids may be at a disadvantage. On
the other hand, if each species is inbred and carries undesirable or even harmful

Figure 11.6

Comparison of an old variety of wheat (in background) with a new highly productive dwarf wheat (foreground). The dwarf wheat was produced, under the direction of Norman Borlaug, by crossing wheat varieties from around the world. Japanese varieties gave these plants their short stature, which prevents them from toppling over even though they produce excessive amounts of grain. Mexican varieties gave them their ability to grow under various conditions.

genes, the hybrid may be more viable and is said to have **hybrid vigor (heterosis)**. Human beings have used heterosis to produce high-yield crops. Hybrid seeds are also used in planting wheat (fig. 11.6), rice, and rye, and they have helped grow more food for the ever-increasing human population.

Recombination

Recombination of alleles occurs during meiosis due to crossing over between homologous chromosomes and independent assortment of chromosomes (p. 140). It also occurs at fertilization when two different gametes join. Recombination gives the offspring a different genotype than has either parent. Mutation is the ultimate source of variability, but the recombined genotype supplied by sexual reproduction is usually of much greater importance. This is because the entire phenotype is subject to natural selection, and individual alleles, particularly if recessive, may not change the phenotype that much.

Among prokaryotes, mutations are more important because recombination does not regularly occur. Also, these organisms are haploid with only one copy of each gene; therefore, mutations can be immediately tested by natural selection. This is not disadvantageous to them because they produce a very large number of offspring within a short period of time. Harmful mutations may temporarily reduce the population size, but will not eliminate the population.

Reduction in Variation

Genetic drift and natural selection both act in such a way that variation in a population is sorted out and reduced (fig. 11.7). But only natural selection is expected to consistently result in adaptation.

Figure 11.7
The genes in the gene pool recombine to give various genotypes that develop into various phenotypes. *a.* A chance sampling of these phenotypes can lead to genetic drift. *b.* Selection of adapted phenotypes can lead to adaptation to the environment. Mutation, gene flow, and recombination are sources of genetic variation in each generation.

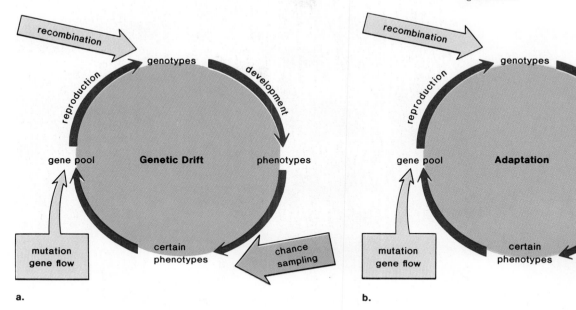

a.

b.

Genetic Drift

Genetic drift is a change in the gene pool that occurs purely as a result of chance. Usually these random changes are insignificant. For example, Mendel reported slight deviations from his predicted 3:1 and 9:3:3:1 ratios.

However, in its most extreme form, genetic drift can lead to the loss of certain alleles, making the homozygous condition the norm (fig. 11.8). This is most apt to happen when only a few offspring from an originally large population mate to produce the next generation. Therefore, as in figure 11.7, drift is placed following a chance sampling of the original population. "Chance sampling" means only a few individuals, among all the phenotypes available, produce offspring. Genetic drift is not expected to result in adaptation to the environment because these phenotypes have not been selected to reproduce; rather chance alone has determined which organisms will reproduce.

For example, as described in the reading on page 220, a study of cheetahs shows that there is little variation among the proteins found in their cells. It is hypothesized that the cheetah population underwent a severe constriction some time past. Unfortunately, the limited gene pool is now causing reproductive problems among these animals.

Figure 11.8
Possible random changes in the frequency of an allele over time. Between generations ten and fifteen, allele *A* is completely eliminated from the population in this example. Therefore, allele *a* is now "fixed" in the population. By "fixing" certain alleles, genetic drift reduces the amount of variation present in the gene pool.

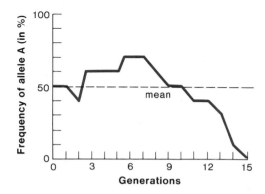

Figure 11.9
The various colonies of the California cypress differ from one another in shape of tree, type of bark, color foliage, and size of cones. These variations in features are believed to be due at least in part to the founder effect.

Founder Effect It has been observed that genetic drift is particularly evident after a new population is started by only a few members of an originally large population. Imagine that, by chance, a small number of individuals, representing a fraction of the gene pool, found a colony. Also by chance, only certain of the founders' alleles are passed on to the next generation. As a result, genetic variation will be severely reduced in the new population. For example, it has been observed that the trees in California cypress groves (colonies) (fig. 11.9) tend to have similar phenotypes. In some colonies, there are longitudinally shaped trees and in others there are pyramidally shaped trees. The bark is rough in some colonies, smooth in others. The leaves are gray to bright green or bluish, and the cones may be small to large in different groves. Each colony is believed to have been started by a single or a few trees. These differences between the colonies do not seem to be due to adaptations to the environment.

Natural Selection

Natural selection (fig. 11.7b) is the process by which populations become adapted to their environments. Individuals having characteristics that make them more suited to environmental conditions are more likely to produce a greater number of offspring than less-suited individuals. Natural selection operates by means of any agent in the environment that influences reproductive ability. Biotic agents include other organisms in the environment, and abiotic agents include physical aspects (fig. 1.7).

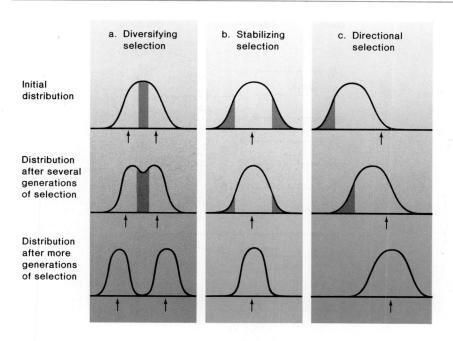

a. Diversifying selection

b. Stabilizing selection

c. Directional selection

Initial distribution

Distribution after several generations of selection

Distribution after more generations of selection

Figure 11.10
One of three types of natural selection can affect the distribution curve for a particular trait among individuals of a population. The curves show that the characteristics for a trait vary continuously throughout the population (fig. 8.16). The arrow(s) point to the characteristic that is favored by selection. The shading indicates those characteristics that are not favored and will disappear from the population. In (a.), diversifying selection, two widely different phenotypes are favored and a separate curve for each is eventually seen. In (b.), stabilizing selection, the curve eventually constricts because the most common phenotype is favored. In (c.), directional selection, an extreme phenotype is favored and the curve eventually moves to the right.

The members of a population have many traits, each one of which is being acted on by natural selection. Three types of selection (fig. 11.10) are possible for any particular trait. In **diversifying selection,** two extreme phenotypes for a given trait are favored over the average phenotype in a population. This would happen if a population previously exposed to a fairly homogeneous environment is exposed to very different conditions in various parts of its range. The population tends to become divided into subpopulations, such as is illustrated in figure 11.5.

Stabilizing selection helps conserve the genotype. It tends to eliminate atypical phenotypes for a given trait and to improve adaptation of the population to those aspects of the environment that remain constant. **Directional selection,** on the other hand, produces changes in gene frequencies due to changing aspects of the environment. These may make an extreme phenotype the better adapted one for a trait and so cause it to be selected. Stabilizing and directional selection can and usually do occur in the same population at the same time—certain traits are affected by one type while other traits are affected by the other type of selection. In order to determine which type of selection is affecting which traits, one must be aware of the population's previous composition. For example, in figure 2.10, the earlier population of giraffes had medium-length necks, while the more recent population has long necks. Having noted this, we know that directional selection is taking place.

Adaptation versus Adaptability

Even though each population becomes adapted to its environment, some variability is always maintained. There are possible genetic reasons for the maintenance of variation (table 11.2). For example, diploidy permits the maintenance of a hidden reserve of recessive genes. Also, as in heterosis, the heterozygote may have an advantage that neither homozygote has. This can cause **balanced polymorphism,** an allele frequency equilibrium in which the two homozygotes are maintained due to the superiority of the heterozygote. An example of balanced polymorphism is given on page 250.

Table 11.2 Maintenance of Variation

Genetic Causes	Ecological Factors
Diploidy	Opposing ecological pressures
Diploidy and multiple alleles	Geographic differences
Heterosis	Periodically changing environment
Polygeny and pleiotropy	

Figure 11.11
Diagram illustrating that one gene locus can affect many characteristics of the individual (pleiotropy), and one characteristic can be controlled by several gene loci (polygeny).

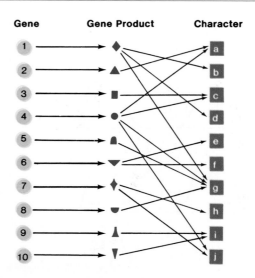

When a trait is controlled by multiple alleles, the continual presence of several different genotypes is expected. In this case too, the population exhibits **polymorphism,** the coexistence in the same population of distinct hereditary types. To take a different example, the same character may be controlled by several gene loci (polygeny), or one gene locus may control more than one character (pleiotropy) (fig. 11.11). Gene interactions such as these cause the genes to become adapted to one another, unifying the genotype. The well-integrated genotype resists selection of only adaptive characteristics and maintains variation.

Ecological factors can also maintain polymorphism. For example, a wide geographic range can cause a single population to exhibit different phenotypes. The British land snail *Cepaea nemoralis* has shells of different colors decorated with as many as five bands that are usually dark in color, but may be pink or white. Song thrushes feed on those snails that are more easily seen. They break open the snails on a rock, eat the soft parts, and leave the shells behind. These rocks are called "thrush anvils." In low-vegetation areas (grass fields and hedgerows) dark unbanded shells tend to be found, and in forest areas, light-colored banded shells tend to be found around these anvils (fig. 11.12). Because the snails have a geographic range including both low vegetation and forest areas, the different shell phenotypes are maintained in the entire population, even though some phenotypes are selected against in the areas mentioned.

The more variation a population contains, the greater its potential to adapt to a change in its environment. Each population seems to strike a balance between loss of variability so as to become better adapted to a particular environment, and retention of variability so as to avoid extinction (fig. 11.13). Populations very closely adapted to certain environments may not be able to adapt to new conditions and will then become extinct.

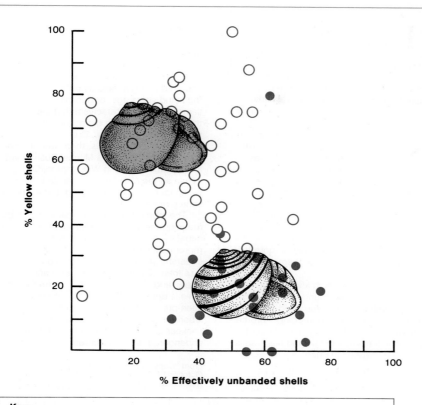

Figure 11.12
Graph showing the relationship between predation and phenotype in the snail *Cepaea*. Thrushes (birds) are more apt to feed on dark, unbanded snails in low vegetation areas (light circles) and light-banded snails in forest areas (dark circles). As a result of these environmental effects, both phenotypes are maintained throughout the range of the snail.

Key:

● forest areas—oakwoods, mixed deciduous woods

○ low vegetation areas—hedgerows, rough herbage

Figure 11.13
Out on an evolutionary limb? Some believe the saber-toothed cat became extinct because its canine teeth became too long to use. But its real problem seems to have been a lack of speed. When its large, slow-moving prey became extinct, it could not adjust to catching smaller, more speedy prey.

Copycat Cheetahs

A South African cheetah at rest with look-alike cubs.

If you have trouble telling one cheetah from another, you're not alone. When researchers at the National Institutes of Health checked blood samples of fifty-five cheetahs from two separate areas of South Africa, they found them to be almost genetically identical.

The scientists took blood from cheetahs born wild in Namibia and the Transvaal Province of South Africa and from offspring bred in a research center in Pretoria. They analyzed forty-seven different enzymes, each of which can come in several different forms. But all the cheetahs carried exactly the same form of every one of the forty-seven enzymes. By contrast, in sample populations of household cats, only 78 percent of the enzymes are identical; in human populations only 68 percent of them match exactly. In another test of more than 150 proteins, 97 percent of them matched in the cheetahs.

Such remarkably high levels of genetic uniformity are usually found only in specially bred laboratory mice. "You need at least twenty generations of inbreeding—brother-sister mating—before you lose all genetic variability, the way the cheetah has," says geneticist Stephen O'Brien.

O'Brien's group theorizes the cheetah population was nearly wiped out generations ago. Perhaps, says O'Brien, they were slaughtered by 19th-century cattle farmers protecting their herds, captured by Egyptians as pets four thousand years ago, or decimated by the same mysterious cataclysm that caused the great mammalian extinction tens of thousands of years ago. During the crisis, "for every 1,000 animals, there were maybe three or four left," says Mitchel Bush, a vet at the National Zoo in Washington, D.C. "Then as the population expanded there was a lot of forced inbreeding and a limited source of new genes."

The inbreeding has now taken its toll by reducing the cheetahs' reproductive capacity. According to David Wildt, a reproductive physiologists at the zoo, the cheetahs show sperm counts averaging less than 10 percent that of lions and tigers, and 70 percent of the sperm they do have is defective. That sort of abnormality, seen previously only in inbred livestock and laboratory mice, would explain why zookeepers have had trouble getting the endangered animal to breed.

Figure 11.14
A geep, a crossbreed between a goat and a sheep. Does such a curiosity contradict the biological definition of a species? No, because this geep was produced not by sexual reproduction, but by the in vitro mingling of cells from a goat embryo and a sheep embryo.

Speciation

A change in gene frequency within a gene pool, either by means of genetic drift or natural selection, can be considered evidence of evolution, but it does not constitute **speciation,** the origin of a species. Speciation requires additional steps. This section will consider these steps that are believed to be necessary.

First, we have to be able to recognize speciation once it has occurred. There are two primary criteria by which a species can be recognized: (1) the members of each species resemble each other and are structurally different in some respects from the members of any other species, and (2) the members of a species are able to interbreed and produce fertile offspring only among themselves (fig. 11.14). It is said that each species is **reproductively isolated.** For example, the red maple and the sugar maple are both found in the eastern half of the United States (fig. 11.15). They remain distinct from one another because they do not interbreed. This criterion for recognizing a species is known as the **biological concept of a species.**

Some groups of organisms (species and/or populations) are prevented from reproducing simply because they are geographically isolated from one another. These groups are said to be **allopatric.** Other groups of organisms do not reproduce with one another even though they are present in the same locale. These groups are said to be **sympatric.**

Table 11.3 lists the mechanisms by which reproductive isolation is maintained.

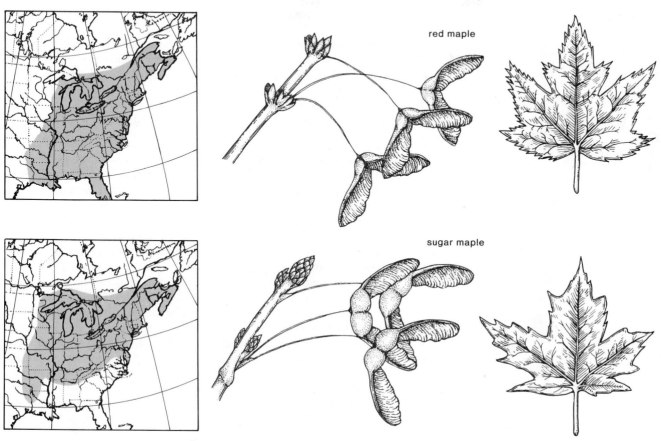

Figure 11.15
The red maple and sugar maple do not interbreed; therefore these two species remain separate in the same locales. The upper illustration shows U.S. distribution of the red maple, and the lower shows U.S. distribution of the sugar maple. The anatomical differences between the two plants are also shown.

red maple

sugar maple

Table 11.3 Reproductive Isolating Mechanisms

Isolating Mechanisms	Example
Premating	
Habitat	Species at same locale occupy different habitats
Temporal	Species reproduce at different seasons or different times of day
Behavioral	In animals, courtship behavior differs or they respond to different songs, calls, pheromones, or other signals
Mechanical	Genitalia unsuitable for one another
Postmating	
Gametic mortality	Sperm cannot reach or fertilize egg
Zygote mortality	Hybrid dies before maturity
Hybrid sterility	Hybrid survives but is sterile and cannot reproduce
F_2 Fitness	Hybrid is fertile but F_2 hybrid has lower fitness

Figure 11.16

Allopatric speciation usually occurs in two stages. In (a.) local populations of a single species are represented by circles; the arrows indicate that crossbreeding may occur when individuals migrate from one population to another. Stage I (b.) begins when two groups of populations become geographically isolated, which makes any further exchange of genes between them impossible. The isolated groups adapt to local conditions and gradually diverge genetically. In Stage II (c.) the populations again make contact, but reproduction, whenever attempted (broken arrows), is unsuccessful. Natural selection then favors the development of premating isolating mechanisms. In (d.) isolation is complete and two separate species coexist.

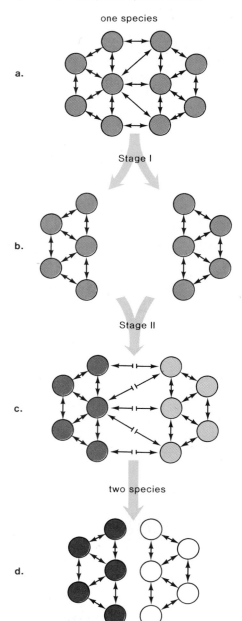

one species

a.

Stage I

b.

Stage II

c.

two species

d.

Premating isolating mechanisms are those that prevent reproduction from being attempted and **postmating isolating mechanisms** are those that prevent hybrid offspring from developing or breeding if reproduction is attempted. Two or more mechanisms tend to operate between any two species. Temporal isolation tends to be more common in plants than in animals, and behavioral isolation is more common in animals than in plants.

Formation of a Species

Speciation is believed to occur in one of two ways. Allopatric speciation requires geographic separation before reproductive isolation can be achieved, but sympatric speciation does not require prior geographic separation.

Allopatric Speciation

Allopatric speciation (fig. 11.16) involves two main stages:

Stage I During stage I (11.16b), *geographic isolation takes place.* Imagine, for example, a grassland that contains several populations of a particular grass. Animals are allowed to overgraze a central portion of the grassland until it becomes a desert. The resultant gap prevents the grasses in one area from reproducing with the grasses in the other area.

Stage II During stage II, *reproductive isolation occurs.* Continuing our example from stage I, we might suppose that surface mining has caused the soil in one area to become contaminated with heavy metals like lead and copper. The grasses living in this area exhibit an increase in the frequency of genes that make them tolerant of these metals. Now the government initiates a reclamation effort that does away with the desert that separated the two groups of grasses. Even though the grasses are no longer separated, the two groups of populations do not interbreed—the grasses tolerant of heavy metals do not reproduce with those that are not tolerant. At first, only *postmating isolating mechanisms,* represented by broken arrows in figure 11.16c, may be evident. For example, cross-fertilization may occur but the seeds are not viable. Later, *premating isolating mechanisms,* shown in figure 11.16d by the absence of arrows, are evident. For example, one type of grass might flower in the spring and the other might flower in the fall. Natural selection favors this difference. Plants exhibiting the premating isolating mechanism have more offspring than those not exhibiting the mechanism. Eventually there are two separate species of grasses whereas before there was only one species.

Sympatric Speciation

Allopatric speciation seems more common than sympatric speciation, which is the occurrence of reproductive isolation without prior geographic isolation. The best evidence for this type of speciation is found among plants where instantaneous speciation can occur by means of polyploidy. A particular plant has inherited, due to nondisjunction, a multiple of the chromosome number; for example, 3N, 4N, or more. This plant is a new species if it can only self-pollinate and can no longer breed with the parental strain.

Sympatric speciation has also been hypothesized in animals. It is noted, for example, that two types of corn borer moths do not mate with one another because each responds to a different pheromone mixture. (Insects locate the opposite sex by means of detecting chemicals called pheromones.) It is suggested that these two species arose sympatrically because of simultaneous genetic mutations in both males and females.

Higher Categories of Classification

Can the speciation processes we have described explain the origin of higher categories of classification, such as genera, families, order, classes, and phyla? They can when we realize the immense amount of time that has elapsed since the beginning of life and the method by which organisms are classified. For example, imagine in figure 11.17 that speciation has occurred and (a) is incapable of reproducing with (b). Perhaps the fossil remains of (a) would be distinguishable enough from (b) for a taxonomist to classify these two types of organisms as separate species. In the diagram, (1) is a descendant of (a), and (2) is a descendent of (b). The descendents vary slightly from the original populations. Now if there were no fossils of (a) and (b), a taxonomist may have no difficulty in classifying the fossil remains of (1) and (2) as separate genera. In other words, the absence of intermediate forms promotes classification of organisms into higher categories.

Figure 11.17
If fossil remains of (a) and (b) were found, these two organisms might be classified as only separate species since their anatomies would indicate that they are closely related. If fossil remains of (1) and (2) were found instead, the two organisms might be classified as separate genera, since their anatomies would indicate that they are more distantly related. This indicates the manner in which the higher categories of classification may come about.

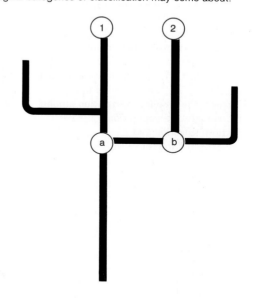

Summary

The Hardy-Weinberg law summarized in the binomial equation $p^2 + 2pq + q^2$ can be used to calculate the gene frequencies within the gene pool of a population. Since sexual reproduction alone cannot significantly alter gene frequencies, gene frequencies in large populations are predicted to remain the same unless mutation, gene flow, or natural selection occurs. Since one or more of these nearly always occurs, there is indeed a change in gene frequencies and evolution does occur. Population genetics defines evolution as changes in gene frequencies within populations over time.

Evolution requires two steps: the production of variations and the sorting out and reduction of these variations. Mutation is the ultimate source of genes, but gene flow moves genes in or out of a given population and recombination produces new genotypes. Genetic drift and natural selection sort and reduce variation, but only natural selection is expected consistently to result in adaptation. Genetic drift is most apt to occur when a few individuals found a new colony. Chance alone determines which genes they bring with them, and most likely there will be a change in gene frequencies. Three types of natural selection have been identified: stabilizing selection occurs as certain characters become better adapted to nonchanging environmental factors; directional selection occurs if the environment changes; and diversifying selection occurs if the environment begins to differ for various segments of the population. Although populations become adapted to current conditions, some variability is maintained because of genetic and ecological effects. This is desirable because different adaptations may be required if the environment changes. Extinction is brought about by the inability to adapt to new conditions.

Speciation involves changes in gene frequencies so great that interbreeding between all former populations of a species is no longer possible. Reproductive isolation is accomplished by premating and postmating mechanisms. Postmating isolation is hypothesized to develop first, followed by premating isolation mechanisms. During allopatric speciation, geographic isolation precedes reproductive isolation. While the populations are geographically isolated, they diverge to the point that interbreeding is no longer possible. Sympatric speciation is well substantiated only in plants.

Objective Questions

1. If the Hardy-Weinberg law holds, evolution will _____ (choose always or not) occur.
2. Twenty-one percent of a population is homozygous recessive. If the Hardy-Weinberg law holds, what percentage is expected to be homozygous recessive in the next generation? _____
3. Can sexual reproduction in and of itself cause evolution to occur? _____
4. Recombination of genes, mutations, and _____ are expected to increase the amount of variation among individuals of a population.
5. Natural selection and _____ are expected to reduce the amount of variation among individuals of a population.
6. The fossil record shows that the brain size of humans has steadily increased since they evolved. This is an example of _____ selection.
7. There are genetic reasons for the maintenance of variation among individuals of a population, but _____ reasons are also important.
8. During the first stage of allopatric speciation, populations become _____ .
9. During the second stage of allopatric speciation, which type of reproductive isolating mechanism precedes the other? _____
10. Two species of butterflies have different courtship behavioral patterns. This is an example of a _____ mating _____ isolating mechanism of the _____ type.

Study Questions

1. Define species, population, variation, and gene frequency.
2. State the Hardy-Weinberg law along with its qualifications.
3. If $q = 0.2$, what is the frequency of the heterozygote genotype?
4. Name the three ways variations are produced, the two ways they are reduced, and discuss each.
5. Give an example of genetic drift, including the term "founder effect."
6. What are the three observed ways by which natural selection operates? What is meant by the phrase "adaptation versus adaptability"?
7. What is the biological definition of a species?
8. Name several pre- and postmating mechanisms by which reproductive isolation is maintained.
9. Describe the two stages of allopatric speciation and give a hypothetical example of this type of speciation.
10. Describe sympatric speciation and give a hypothetical example of this type of speciation.

Answers to Practice Problems
1. $Ll = .6$; $ll = .4$
2. recessive allele $= .2$ and dominant allele $= .8$; homozygous dominant $= .64$; homozygous recessive $= .04$; heterozygous $= .32$
3. 29%

Selected Key Terms

population (ˌpäp yə 'lā shən)   218
gene pool ('jēn 'pül)   218
gene frequency ('jēn 'frē kwən sē)   218
subspecies ('səb ˌspē shēz)   223
hybrid ('hi brəd)   223
hybrid vigor ('hi brəd 'vig ər)   224
heterosis (ˌhet ə 'rō səs)   224

genetic drift (jə 'net ik 'drift)   225
balanced polymorphism ('bal ənst ˌpäl i 'mȯr fiz əm)   227
polymorphism (ˌpäl i 'mȯr fiz əm)   228
speciation (ˌspē shē 'ā shən)   230
allopatric ('a l ə 'pa trik)   230
sympatric (ˌsim 'pa trik)   230

E volution explains the unity and diversity of life. Therefore, it is possible to find evidences for evolution in all fields of biology, from paleontology (study of fossils) to biochemistry.

Evidences of Evolution

Fossil Record

The fossil record tells us that *species are not immutable*. The species we see about us today are not the ones that have always existed. This is to be expected if life forms are continually evolving.

    Fossils (fig. 12.1) are the remains of organisms that lived long ago. Usually when an organism dies it is either consumed by scavengers or it undergoes bacterial decomposition. But occasionally, it is buried quickly and in a way that decomposition is never completed or completed only slowly. In the meantime, the organism may leave an imprint, or better still a cast, that clearly indicates its external and perhaps its internal structure.

Figure 12.1
Fossils. *a.* Impression of fern leaves on a rock. *b.* Cast of a turtle. The shell (now gone) formed a mold that was filled by hardened sediment.

a.

b.

CHAPTER 12

Evidences for Evolution

Concepts

1. Evolution explains both the unity and diversity of life.
   a. Evidences of evolution may be taken from the fossil record
      biogeography
      anatomy and embryology
      comparative biochemistry

2. Evolution occurs by means of natural selection.
   a. Gene frequency in a population with sickle-cell anemia provides an example of stabilizing selection.
   b. Resistance to pesticides and industrial melanism are examples of directional selection.

3. Certain patterns of evolution can be observed.
   a. Macroevolution is exemplified by adaptive radiation and phyletic evolution.
   b. Various patterns of evolution are detectable at the species and higher levels of classification.

Figure 12.2

Sedimentary rock formation. Terrestrial rock is broken down by weathering and transported to the sea by streams and rivers. In the still waters of the ocean, the particles are sorted and compacted according to their weight. They are cemented together by minerals present in the water. The table shows that sedimentary rocks are classified according to the size of the particles of which they are made. Limestone is an exception; it is a precipitate of calcium carbonate. Fossils are often found within limestone.

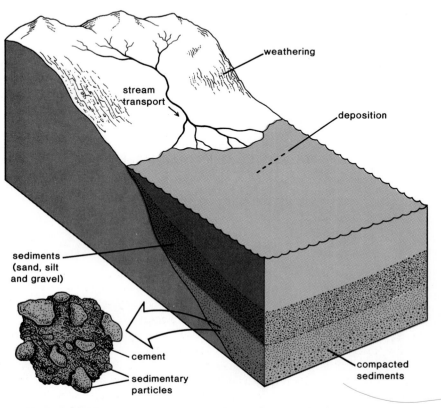

Sedimentary rock	Kind of rock	Size of sedimentary particle
	shale	silt-size particles
	sandstone	sand-size particles
	conglomerate	gravel-size particles
	limestone	precipitated calcium carbonate (often contains fossils)

Figure 12.3
Field researchers examine rocks near North Pole, Australia, for evidence of fossils, remains of previously existing organisms. At this location, researchers have found fossils for the earliest forms of life.

Preserved fossils are most often found in sedimentary rock (fig. 12.2). Weathering produces sediments that are carried by streams and rivers into the oceans and other large bodies of water. There they slowly settle and are converted into sedimentary rock. Later, sedimentary rocks are uplifted from below sea level to form new land. Now researchers are freely able to search for any fossil remains trapped in the rocks (fig. 12.3).

Around 1900 an English civil engineer, William Smith, was among the first to notice that different layers of sedimentary rock had unique fossils. He used this observation as a means to relatively date the layers. Soon it became apparent that simpler organisms are found in earlier layers and more complex organisms are found in later layers. Extinction is apparent when an organism disappears from later layers.

Eventually it was possible to develop the **geological history of life** (table 12.1), which provides a broad outline of how life forms (species) have slowly changed (evolved) from the time of the origination of life. The earth is 4.6 billion years old and the earliest single-celled fossils are found in rocks that are about 3.5 billion years old as determined by present-day radioactive dating techniques. But it was less than a billion years ago (600 million to be exact) when evolution began in earnest. Vertebrates (animals with backbones) made their appearance about 435 million years ago. At that time, too, plant life left the oceans and appeared on land. From these plants, emerged the tree-sized

Table 12.1 The Geological Time Scale—Major Divisions of Geological Time with Some of the Major Evolutionary Events of Each Geological Period

Era	Period	Millions of Years Ago	Plant Life	Animal Life
Cenozoic	Quaternary	2.5	Increase in number of herbaceous plants	Age of human civilization
	Tertiary		Dominance of land by angiosperms	First hominids appear Dominance of land by mammals and insects
		65		
Mesozoic	Cretaceous		Angiosperms spread Gymnosperms decline	Dinosaurs become extinct
		130		
	Jurassic		First angiosperms appear	Age of dinosaurs First mammals and birds appear
		180		
	Triassic		Dominance of land by gymnosperms and ferns	First dinosaurs appear
		230		
Paleozoic	Permian		Land covered by forests of primitive vascular plants	Expansion of reptiles Decline of amphibians
		280		
	Carboniferous		Age of great coal-forming forests, including club mosses, horsetails, and ferns	Age of amphibians First reptiles appear
		350		
	Devonian		Expansion of primitive vascular plants over land	Age of fishes First insects appear First amphibians move onto land
		400		
	Silurian		Primitive vascular plants appear on land	Expansion of fishes
		435		
	Ordovician		Nonvascular marine plants abundant	Invertebrates dominate the seas First fishes (jawless) appear
		500		
	Cambrian		Unicellular marine algae abundant	Age of invertebrates Trilobites abundant
		600		
Precambrian (Proterozoic)			Aquatic life only Origin of invertebrates Origin of complex (eukaryotic) cells Origin of photosynthetic organisms Origin of primitive (prokaryotic) cells	
Archeozoic		4,600	Formation of earth and rest of solar system	

plants that flourished during the Carboniferous period (fig. 12.4) and whose remains provide the coal we still burn today. It was 130 million years ago that the dinosaurs dominated the earth and the first flowering plants appeared. Mammals appeared 65 million years ago, but the fossils we classify as humans are only about 2.5 million years old.

The fossil record not only provides our most concrete evidence of evolution, it also sheds light on abiotic (nonliving) factors that may have affected the rate of the evolutionary process throughout the history of life. The reading on page 246 looks at such evidence from the fossil record.

Figure 12.4
A Paleozoic forest typical of the Carboniferous period. At
upper left are giant club mosses with fine leaves and
conelike strobrilae. At left are ferns, and at right are
horsetails with whorled branches. These same plants
today are of insignificant size and most trees are either
cone-bearing or flowering plants.

Biogeography

Biogeography, the study of the distribution of plants and animals about the
world, tells us that *the very many forms of life are variously distributed.*

The first taxonomist, Carolus Linnaeus, knew of only fifteen thousand
species, but today it is estimated that there may be as many as 30 million
species. These species are not equally distributed. As early as 1857, Philip
Sclater, a British bird specialist, had divided the world into six biogeographical regions. These same regions were recognized by Alfred Wallace in his
book *The Geographic Distribution of Animals* (1876) (fig. 12.5). The land
areas in each region have characteristic plants and animals (fig. 12.6), and
the fossils that have been dug up in a particular region resemble the organisms
found only in that region and not in others. All becomes clear when we notice
in figure 12.5 that these regions are separated by impassable barriers. Therefore it is not surprising that each has a unique evolutionary history and mix
of plants and animals.

There may be similar environmental conditions in two different regions,
but even then the same plants and animals are not necessarily found. Instead,
these organisms only resemble each other because they are similarly adapted.
For example, plants of the cactus family are found in the deserts of southwestern North America, while members of the spurge family are found in
the deserts of Africa. Both types of plants have made similar adaptations to
arid habitats. Figure 12.7 shows that there are North American mammals
and African mammals that are similarly adapted to living in a grassland
environment.

Figure 12.5

Major biogeographical regions of the world. Some natural barriers that separate the regions are also indicated: (*1*) the Sahara and Arabian deserts; (*2*) very high mountain ranges, including the Himalayas and the Nan Ling mountains; (*3*) deep water marine channels among islands of the Malay archipelago (A. R. Wallace recognized and wrote about this barrier, which is called Wallace's line); (*4*) the transition between highlands in southern Mexico and the lowland tropics of Central America.

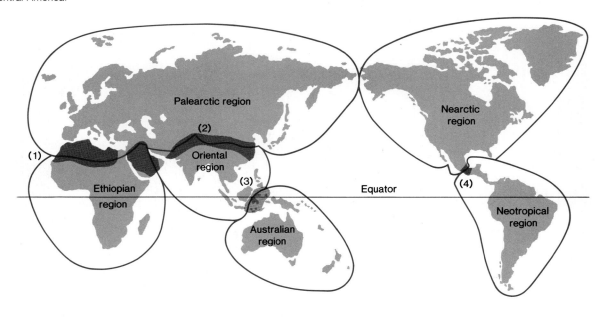

Figure 12.6

This male Count Raggi's bird of paradise is hoping to attract a female by displaying. Birds of paradise are endemic to the oriental biogeographical region.

North America   **Africa**

a. leaping herbivorous mammals

b. burrowing mammals; feed above ground

c. burrowing mammals; feed underground

d. running herbivorous mammals

e. running carnivorous mammals

Figure 12.8
History of the placement of today's continents. *a.* At one time there was only the one continent of Pangaea. *b.* Pangaea broke up into Laurasia and Gondwana during the Paleozoic era. *c.* These latter two continents fragmented into today's continents before the Cenozoic era. The arrows indicate the direction that the continents are drifting yet today.

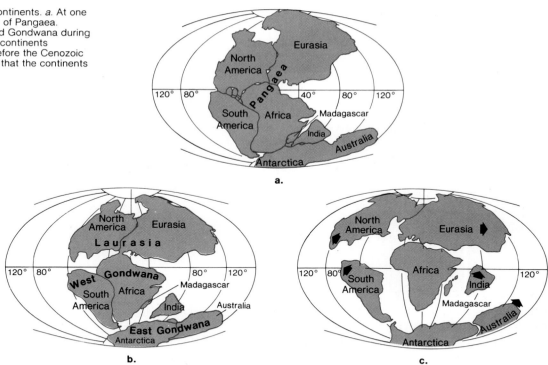

a.

b.

c.

Impassable land barriers separate the biogeographical regions, but they are also separated by oceans. Plentiful evidence now suggests that this was not always the case. Originally there was only one huge continent called Pangaea (fig. 12.8). Pangaea eventually broke up so that during the Paleozoic era and until late in the Mesozoic era, there were two major land masses—Gondwana and Laurasia. In the Cretaceous period, these two were fragmented into the seven present-day continents. Even now, however, these continents are slowly drifting further apart. Dramatic discoveries in the late 1960s provided a mechanism by which drift might be occurring. There are ridges in the oceans where volcanic activity forms new crust. This crust spreads slowly away from the ridges until it is consumed by oceanic trenches (fig. 12.9). The crust is like a conveyor belt that carries along the drifting continents. Today, the combined study of sea floor spreading and continental drift is called **plate tectonics.**

Continental drift further explains the distribution of organisms. Since amphibians and reptiles arose during the Paleozoic era, they should have a wider distribution than mammals, which arose after the final fragmentation. Fossil remains support this conclusion: Fossil amphibians and reptiles have been found on all the continents that were originally a part of Gondwana, while each type of mammal tends to be found on a single present-day continent. Exceptions to this tendency, though, do exist. For example, moose are found in both Europe and North America. Most likely land bridges existed until recent times that allowed migration of animals between certain continents.

Figure 12.9
Diagram illustrating plate tectonics. The continents drift
because they are on plates, or mammoth pieces of crust,
that are formed at the midoceanic ridges and disappear at
oceanic trenches.

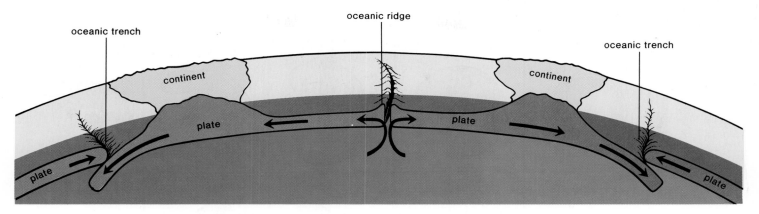

Anatomy and Embryology

A study of the anatomy of members of a taxonomic group or of the development of these members tells us that a *group of related species* has a *unity of plan*. The reproductive organs of all flowering plants are basically similar even though they vary in some details. All vertebrates at some time during development have a supporting dorsal rod, called a notochord, and all later exhibit gill pouches (fig. 12.10). In every vertebrate, the notochord is converted to the vertebral column, but only in fishes and amphibians do the pouches become functional gills.

Homologous Structures

Among adult vertebrates, we can note that their forelimbs are organized similarly and contain the same bones (fig. 12.11), even though the animals themselves are adapted to different ways of life. The explanation for this remarkable unity of plan is that all vertebrates are descended from a common ancestor. The basic body plan originated with this ancestor and was simply varied among the descendent groups as each one continued along its own evolutionary pathway. Structures like these that are similar in structure but not function are termed **homologous structures.** The presence of homologous structures tells us that organisms are related to one another.

Analogous Structures

Sometimes organisms have structures that are constructed differently but appear to be similar because they serve the same function. An insect's wing is a stiffened membrane supported by hard, chitinous veins, while the bird's wing has an internal bony skeleton covered by muscle, skin, and feathers. Even so, both types of wings have the same broad flattened shape (fig. 12.12) because they enable the animal to fly through the air. Structures that have a different structure but serve a similar function are called **analogous structures.** While analogous structures do not indicate relationships, they are still evidences for evolution because they show the degree to which dissimilar structures can still be similarly modified to suit a particular environment.

Figure 12.10
Embryonic comparisons in four animals. Note that development is at first quite similar in all four. For example, in the third frame all the animals have markings that indicate the location of gill pouches.

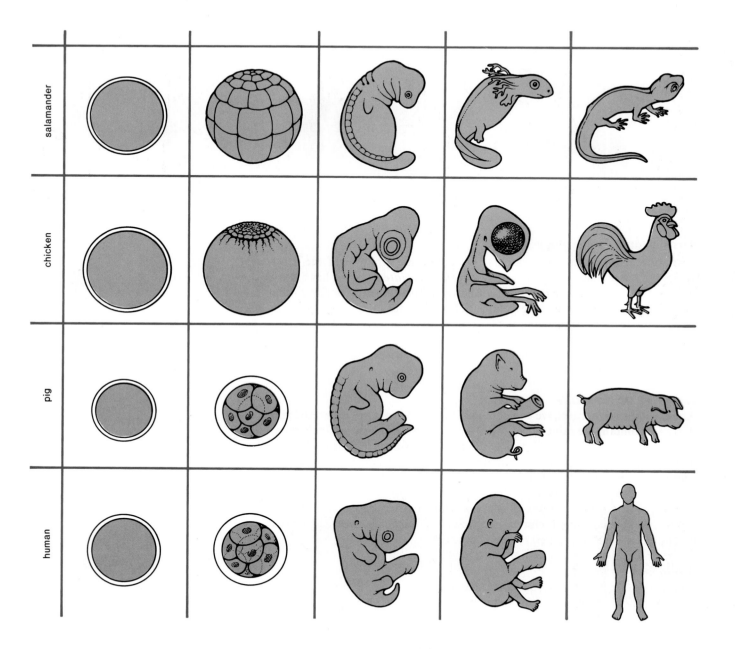

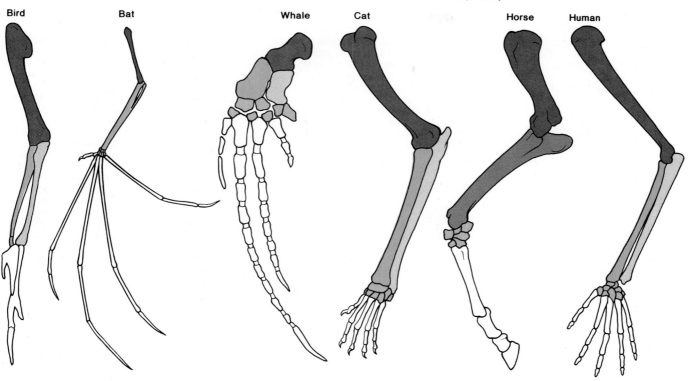

Bird   Bat

Whale   Cat

Horse   Human

Figure 12.11
The unity of plan among vertebrates is exemplified by the bones in the forelimbs. (The bones are color coded.) Although the same bones are present, the limbs are adapted to different functions. All the vertebrates are descended from a common ancestor, but each is evolving separately.

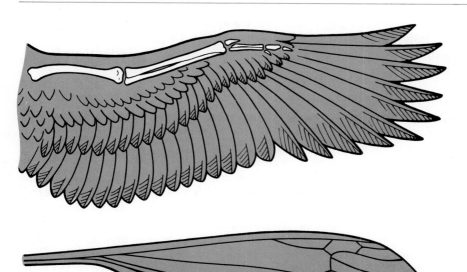

Figure 12.12
A bird's wing (eagle) is analogous to an insect's wing (crane fly). Although the internal structure is completely different, the wings have the same overall shape because they are both adapted to flying in the air.

New Questions from the Fossil Record

There is no doubt that the fossil record supplies data to support the theory of evolution. The record shows that life began with simple organisms and advanced to the complex organisms in existence today. Also, direct ancestors for every living thing can be found in the fossil record.

A few years ago, Stephen J. Gould and Niles Eldredge, two paleontologists, wondered if the fossil record also gives us specific information regarding the tempo (rate) of evolutionary change. Based primarily on experimental genetic data in the laboratory, most biologists believe that a single species changes slowly—over thousands or even millions of years—into other species. This so-called *gradualistic model* of evolutionary change implies that the fossil record contains a plentiful supply of *intermediate forms,* which are fossil remains of life having characteristics of two different groups. Some intermediate forms (a.) have been found, but not an overwhelming number.

In contrast to the gradualistic model, Gould and Eldredge, upon close examination of the fossil record, proposed a *punctuated equilibrium model* of evolutionary change. The record shows that each species tends to remain the same for hundreds of thousands or millions of years. Then, relatively suddenly, new species appear. In other words, an *equilibrium* phase (long periods without change) is *punctuated* by a rapid burst of change.

According to this model, there would be few intermediate forms in the fossil record because such forms existed for only a short period of time. Gould and Eldredge also suggested that, without the presence of any other pertinent factors, such rapid change might require that natural selection select entire species for survival and not just individuals.

Thus far, evolutionists have been unable to supply additional data to support one model over the other. Very recently and unexpectedly, though, they have received some help from geologists and astronomers. For many years, evolutionists and other biologists have tried to explain the mass extinction of the dinosaurs (b.). Dinosaurs dominated the earth for 135 million years, but suddenly they and much of the rest of life on earth vanished. In 1979, a Berkeley geologist, Walter Alvarez, and his colleagues found that Cretaceous clay contained an abnormally high level of iridium. This could have been caused by a worldwide fallout of radioactive material created by an asteroid impact at the time of the dinosaur disappearance. In 1984, paleontologists David Raup and John Sepkoski of the University of Chicago discovered that the dinosaurs are not alone; rather, the fossil record of marine animals shows that mass extinctions have occurred every 26 million years or so.

Surprisingly, astronomers, taking a hint from Alvarez, can give an explanation for these occurrences. Our solar system is in

---

Figure 12.13
Orchids of the genus *Ophrys* have a peculiar-shaped lower petal that is brown and black and hairy in appearance. The plant produces no nectar, but male insects under the impression they have found a female visit it regularly and attempt to mate with the flower. The evolution of this orchid can be explained in this way: Those flowers that began to resemble the female bee were the very ones that reproduced the most.

Vestigial Structures

Rudimentary **vestigial structures** are indicators of relationships because they are fully developed and functional only in related organisms. Humans lack the well-developed tail found in other mammals, but do have vestigial caudal vertebrae and the muscles that move the tail in other mammals (fig. 12.14).

Comparative Biochemistry

Biochemical studies of many different species tell us that the *same type molecules are found in all living things*. Not only is DNA the genetic material in all species, the near universality of its code (p. 197) suggests that all living things have evolved from a single beginning.

Biochemists have the capability to compare the sequencing of amino acids in proteins and the sequencing of bases in DNA between species. The degree of difference can indicate how closely related two species are. In addition, investigators have discovered that there is a fairly steady rate of mutation in all lineages. These rates, called **molecular clocks,** can be used to estimate the date that two species diverged from a common ancestor.

Data from biochemical studies and anatomical data based on the fossil record have been used independently to construct evolutionary trees. As discussed previously (p. 26), an **evolutionary tree** is a diagram that shows the evolutionary relationship between organisms and their common ancestors.

a.

b.

the Milky Way, a starry galaxy that is 100,000 light years[1] in diameter and 1,500 or so light years thick. Our sun moves up and down as it orbits in the Milky Way. Scientists predict that this vertical movement will cause our solar system to approach certain other members of the Milky Way every 26–33 million years, producing an unstable situation that could lead to comet bombardment of the earth. This bombardment can be likened to a worldwide atomic bomb explosion. A cloud of dust will mushroom into the atmosphere and shade out the sun, thus causing plants to freeze and die. Mass extinction of animals will follow. Once the cosmic winter is over, plant seeds will germinate. This new growth will serve as food for the few

remaining animals. Just as Darwin's finches rapidly evolved on the Galápagos Islands because of the lack of competition, these animals will also undergo rapid evolution.

This scenario suggests that the punctuated equilibrium model of evolutionary change does have merit. The equilibrium phase, during which evolutionary change is extremely slow, would occur between times of mass extinction, and the punctuated phase would occur immediately after mass extinctions. It is only necessary to return to the fossil record to see if this hypothesis holds.

[1]One light year, the distance light travels in a year, is about 6 trillion miles.

a. Artist's representation of Archeopteryx, an intermediate form between reptiles and birds. Although fossil remains indicate that it was well-feathered, Archeopteryx had several reptilian features, including jaws with teeth and a long, jointed reptilian tail. b. Artist's representation of Triceratops (right) and Tyrannosaurus (left), two extinct dinosaurs of western North America. The only reptiles to survive the end of the Cretaceous period were those few groups still living today, such as the turtles, the lizards and snakes, and the crocodiles.

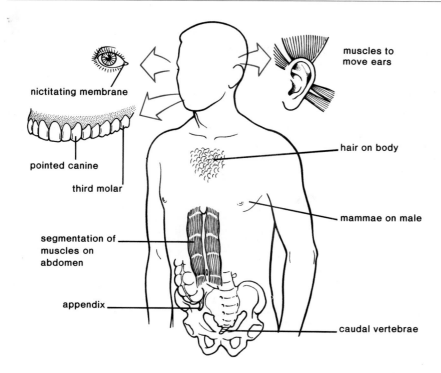

Figure 12.14
Some of the vestigial structures in the human body. These structures are fully functional in other animals and their existence in humans indicates an evolutionary relationship with these animals.

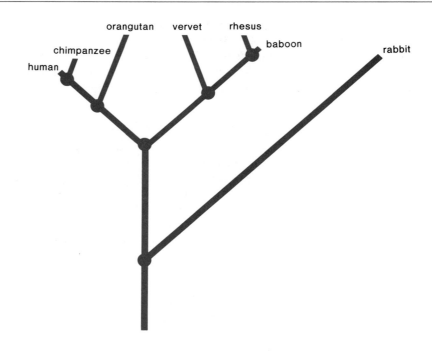

Figure 12.15
Evolutionary tree of various primates and the rabbit, which shows that chimpanzees are most closely related to humans, and rabbits are far more distantly related to humans. This tree is based on differences in the sequence of 116 amino acids in carbonic anhydrase, an enzyme. The dots represent the location of common ancestors and the length of the lines are based on the number of nucleotide substitutions that must have occurred.

Figure 12.16
The quagga was striped on the front end like a zebra, but had a solid, chestnut-colored coat like a horse on the hind end. Although it became extinct a century ago, scientists have been able to clone a portion of its DNA chemically extracted from tissue removed from museum hides. Comparative biochemical studies indicate that the quagga is a zebra. Such a study emphasizes only the physical and chemical aspects of life and does not in the least suggest that biochemists can bring whole organisms back from the dead.

Evolutionary trees based on biochemical data (fig. 12.15) have a similar appearance to those based on comparative anatomy. This is to be expected since diversity is the result of evolution, a process that has a genetic basis.

Whereas biochemists formerly restricted their studies to living organisms, they have recently begun to study extinct organisms and even fossils. For example, investigators have extracted proteins and DNA from a scrap of muscle on the pelt of a 140-year-old museum specimen of a quagga (fig. 12.16), an animal that became extinct a century ago. Cloning provided enough DNA (see page 206) to establish that a quagga is a zebra, not a horse. Protein studies showed that a quagga is most closely related to a plains zebra. Such studies are only possible because all living things share the same types of chemical molecules.

Observation of the Evolutionary Process

Deductive evidences have thus far been presented for evolution, but in recent years, it has been possible to actually observe the occurrence of evolution by natural selection. In the last chapter (fig. 11.10), we identified three types of natural selection, two of which are of primary importance. Our first example here exemplifies *stabilizing selection,* the maintenance in a population of the most adaptive phenotype. The next two that follow are examples of *directional selection,* a change in phenotype toward one characteristic only.

Bacteria and Insects

Indiscriminate use of antibiotics and pesticides has caused bacteria and insects to become resistant to these chemicals. The mutation enabling them to survive the unfavorable environment is already present before exposure; the chemicals are merely acting as selective agents. For example, when bacteria are grown

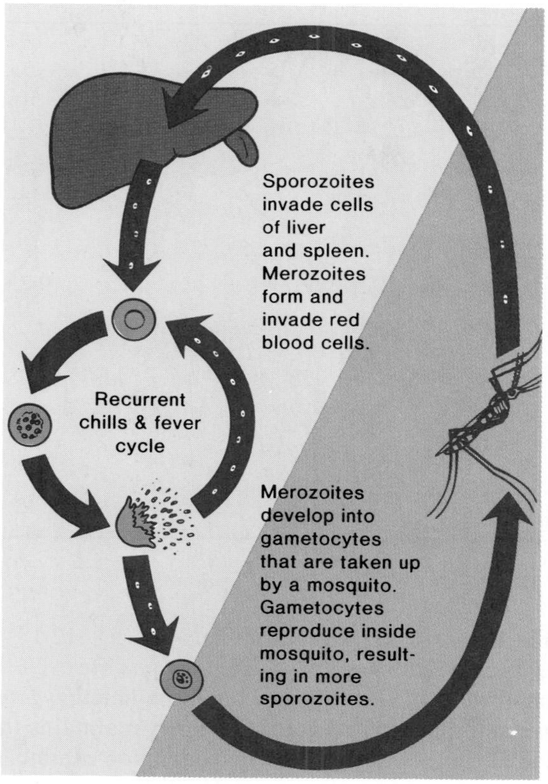

Figure 12.17
Life cycle of *Plasmodium,* the protozoan that causes malaria.

Sporozoites invade cells of liver and spleen. Merozoites form and invade red blood cells.

Recurrent chills & fever cycle

Merozoites develop into gametocytes that are taken up by a mosquito. Gametocytes reproduce inside mosquito, resulting in more sporozoites.

on medium containing streptomycin, a few survive. These few can grow on medium both with and without streptomycin. In any case, however, this new generation of bacteria is now resistant to the antibiotic.

A more recent example is the human struggle against malaria, a disease caused by an infection of the liver and red blood cells (fig. 12.17). The dreaded *Anopheles* mosquito transfers the disease-causing protozoan *Plasmodium* from person to person. In the early 1960s, international health authorities thought that malaria would soon be eradicated. A new drug, chloroquine, was more effective than quinine that had been used previously against *Plasmodium,* and DDT spraying killed the mosquitoes. But in the mid-1960s, *Plasmodium* showed signs of chloroquine resistance and, worse yet, mosquitoes were becoming resistant to DDT. A few drug-resistant parasites and a few DDT-resistant mosquitoes had survived and multiplied, making the fight against malaria more difficult than ever. New tactics have to be devised. Recombinant DNA procedures have enabled researchers to identify proteins that will hopefully be effective as vaccines.

Sickle-cell Anemia

Individuals with **sickle-cell anemia** (p. 156), having the genotype $Hb^S Hb^S$, tend to die of this condition at an early age. Heterozygous $Hb^A Hb^S$ individuals with the sickle-cell trait are better off, but they are still expected to be at a disadvantage compared to individuals who are homozygous normal $Hb^A Hb^A$. However, when geneticists studied populations of black Africans, in whom the disease is most common, they found the recessive S allele has a higher frequency (0.2 to as high as 0.4 in a few areas) than would have been predicted. Further research showed the heterozygote has an adaptive advantage

Figure 12.18

There is a correlation between the distribution of sickle-cell anemia and the distribution of malaria. *a*. Frequency of the sickle-cell anemia allele Hb^s in Africa and adjoining countries. In the homozygous condition this allele causes sickle-cell anemia. *b*. Geographic distribution of malaria in Africa and adjoining countries.

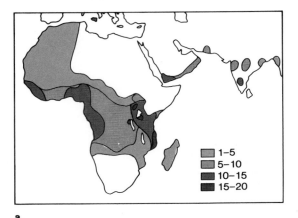

▨	1–5
▨	5–10
▨	10–15
▨	15–20

a.

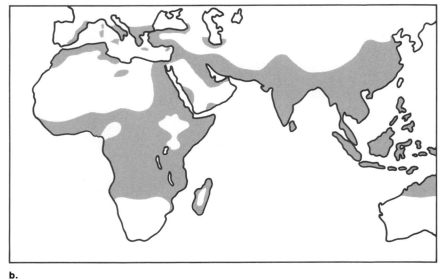

b.

Figure 12.19

Dark and light forms of the peppered moth (*Biston betularia*) *a*. The light form is visible on a soot-blackened oak tree in Birmingham, England. *b*. The dark form is visible on a light, lichen-coated oak tree in an unpolluted region.

a.

b.

over the normal homozygote in malarial regions (fig. 12.18). The malarial parasite enters normal red blood cells to complete its life cycle, but it is unable to make use of sickle cells. For this reason, more heterozygotes survive than do homozygotes and this **heterozygote superiority** maintains the frequency of the sickle-cell allele in the population. The population exhibits *balanced polymorphism* (phenotypes in balance) because all three genotypes persist from generation to generation.

Industrial Melanism

Industrial melanism is also an example of directional selection. Before the industrial revolution in England, collectors of the peppered moth, *Biston betularia,* noted that most moths were light-colored, although occasionally a dark- (melanistic) colored moth was captured. Several decades after the industrial revolution, however, black moths made up 90 percent of the moth population in air-polluted areas.

Moths rest on the trunks of trees during the day (fig. 12.19); if they are seen by predatory birds, they are eaten. As long as the trees in the environment are light in color, the light-colored moths live to reproduce. But if the trees turn black from industrial pollution, the dark-colored moths survive and reproduce to a greater extent than the light-colored moths. The dark-colored phenotype then becomes the more frequent one in the population. If pollution is then reduced and the trunks of the trees regain their normal color, the light-colored moths again increase in number.

To provide experimental evidence that body color influences which moths are captured by birds, H. B. D. Kettlewell of Oxford University released equal numbers of the light- and dark-colored moths into two areas: (1) a nonindustrial, unpolluted area and (2) a highly industrial, polluted area. The released moths were marked so Kettlewell could identify them when he later recaptured his moths. Kettlewell set up blinds in each area from which he observed birds preying on moths. Birds in the unpolluted area preyed on dark-colored moths more than light-colored moths, and birds in polluted areas preyed on light-colored moths more than dark-colored moths. From the unpolluted area,

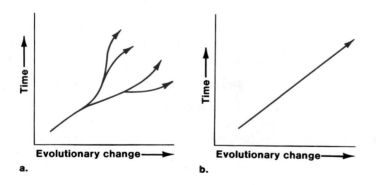

a.

b.

Figure 12.20
Two patterns of evolutionary change. *a*. The population splits into different groups that will eventually achieve the status of recognizable species. Darwin refers to this form of change as "the origin of species"; today it is generally called adaptive radiation. *b*. The population of organisms undergoes change from one state to another. Darwin refers to this type of change as "descent with modifications"; today it is called phyletic evolution.

he recaptured 14.6 percent of the light-colored moths and only 4.7 percent of the melanistic form. From the polluted area, he recaptured 27.5 percent of the melanistic form and only 13 percent of the light-colored moths. This showed that the better adapted moths in each area were more likely to avoid being eaten by predatory birds, which were acting as a selective agent.

Patterns of Evolution

In the previous section, examples of natural selection resulting in microevolution were provided. In this section, we will examine patterns of evolution that are apparent when viewing the results of speciation.

Macroevolution

Macroevolution is evolution of species and even higher categories of classification. Two patterns of macroevolution seem predominant—adaptive radiation and phyletic evolution (fig. 12.20). Adaptive radiation is called *evolution in space* because a single ancestor has given rise to a number of different species. Darwin's finches are a well-known example of adaptive radiation. Phyletic evolution is called *evolution in time* because it results from gradual changes in a single lineage. Evolution of the modern-day horse is a well-known example of phyletic evolution.

Adaptive Radiation

There are fourteen species of finches that live on the Galápagos Islands (fig. 2.5). All these birds are believed to be descended from a type of mainland finch that chanced to invade the islands. As the parent population increased in size, daughter populations were established on the various islands. Each population was then subjected to the process of natural selection as it became adapted to the particular conditions on its island. In time, the various populations became so genotypically different that now when they are found together on the same island they do not interbreed and instead become more phenotypically specialized.

An examination of three species of ground finches supports this scenario. Among the ground finches are three species having bills usually adapted to feeding on small (*G. fuliginosa*), medium (*G. fortis*), or large-sized seeds (*G. magnirostris*) (fig. 12.21). When all three species occur on the same island, they do not mate with each other and their bill sizes (beak depth) are quite distinctive. However, when *G. fortis* and *G. fuliginosa* are on separate islands, their bills tend to be the same intermediate size because specialization in this instance has no particular advantage.

Figure 12.21
Beak depth differences among three species of ground finches. On the large islands of Isabela and Santa Cruz, all three species are present and each has a distinctive beak size (*a*.). Only the species *G. fortis* is on Daphne Major (*b*.), and (*c*.) only the species *G. fuliginosa* is on Los Hermanos. In these instances, both species tend to have the same intermediate beak depth.

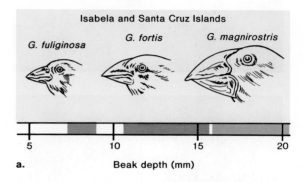

a.

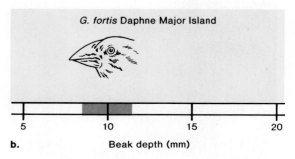

b.

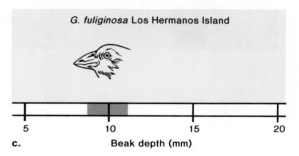

c.

a.

b.

Adaptive radiation has also been observed in plants. On the island of New Caledonia, various species of *Dacrydium* have adapted to various microclimates. For example, *D. araucariodes,* with narrow, incurved leaves, is apparently adapted to a dry environment, while *D. taxoides,* with broad leaves, seems adapted to a moist environment (fig. 12.22).

Phyletic Evolution

The fossil record of the modern horse covers some 60 million years (fig. 12.23). It begins with *Eohippus,* a dog-sized mammal with a small head and small, low-crowned molars. The feet, with four toes on each front foot and three toes on each hind foot were padded. *Equus,* the modern horse, is much larger than *Eohippus.* It has an increased skull size, bigger molars with flattened high crowns, and single-hoofed toe on each foot.

These anatomical changes are attributed to a change in environment. *Eohippus* was adapted to the forestlike environment of the Eocene (table 13.2). For example, the low-crowned teeth, having a relatively simple pattern of enamel were appropriate for browsing on leaves. This small-sized animal could have hidden among the trees for protection. In the Miocene and Pliocene, grassland began to replace the forests. Now the ancestors of the modern horse, *Equus,* were subjected to selective pressure for the development of strength, intelligence, speed, and durable grinding teeth. A large size provided the strength needed for combat; a large skull made room for a larger brain; the elongated legs terminating in hooves gave speed to escape enemies; and the durable grinding teeth enabled horses to feed efficiently on grasses.

By picking and choosing only certain fossils, it is possible to trace the lineage of the modern horse as if all these changes occurred step by step. Actually, however, it would be necessary to forego mentioning a number of other fossils. As *Eohippus* evolved into *Equus,* there were at least three adaptive radiations. These horses became extinct and only *Equus* survived.

As we trace the evolution of any group of organisms for any great length of time, it seems that actually both patterns of evolution occur. Phyletic evolution is followed by adaptive radiation, which is followed by phyletic evolution ad infinitum.

Other Patterns of Evolution

When we view the world of living things, we can pick out certain other patterns of evolution that are the end result of the evolutionary process. Some of these patterns are observed between closely related groups of organisms. **Divergent evolution** occurs at the species level. For example, the various types of maples have diverged from one other: the red maple (*Acer rubrum*) and the sugar maple (*Acer saccharum*) (fig. 11.15) are different enough that they do not reproduce with one another. Even after divergence occurs, closely related groups can evolve similarly. This is called **parallel evolution.** Presumably, for example, the African ostrich and the South American rhea (fig. 2.4) are descended from a common ancestor. Even after these two groups of birds split, they evolved similarly because they were subjected to the same environmental pressures.

Other patterns of evolution are observed between more distantly related groups of organisms. In **convergent evolution,** different groups develop the same adaptations because they exist in the same type of environment. For example, figure 1.7 shows that mammals, reptiles, and fishes have made similar adaptations to the marine environment. In **coevolution,** two unrelated groups act as selective agents on each other. As the bat becomes more skillful in catching the moth, the moth becomes more skillful in escaping the bat (fig. 25.8). One of the most frequently used examples of coevolution is the adaptation that

Figure 12.23
Evolution of the horse. The fossil history of horses dates back to early Cenozoic times, about 60 million years ago. The modern horse evolved from the dog-sized *Eohippus* via a number of intermediate forms. Note the gradual reduction in the number of digits of the hoof. A number of side branches of horse evolution have ended in extinction.

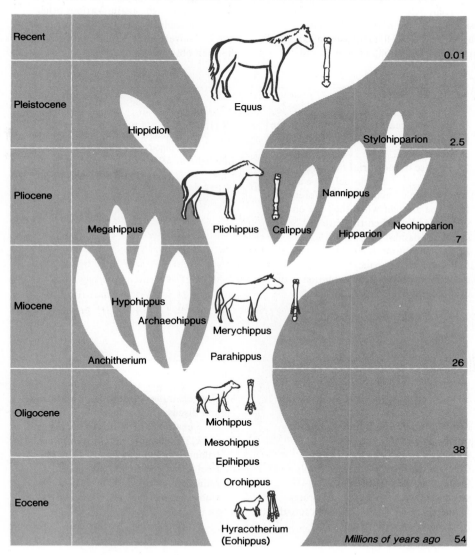

developed between flowers and pollinators. The flower in figure 14.25 produces nectar, has an ultraviolet coloration (fig. 24.8), and a sweet smell that attracts the bee. To make the bee's food-gathering work easier, it has a sucking tongue for collecting nectar, and pollen baskets on the upper segment of the third pair of legs. In this instance, coevolution benefits both the flower and the pollinator. The flower uses the bee to achieve cross-fertilization, and the bee uses the flower as a source of food.

Summary

Since Darwin's time, scientists have accumulated data to support the theory of organic evolution. Paleontologists have discovered a fossil record showing an increase in complexity since organisms first evolved. Those who study biogeography have found that various regions have a unique evolutionary history. Anatomists have found that related animals develop similarly and their organs are structured according to basic plans. The most closely related organisms have homologous organs as opposed to analogous organs. The presence of vestigial structures indicates a relationship with other organisms in which these structures are functional. Finally, comparative biochemists have found molecular homologies. By detecting differences in the amino acid sequence of proteins or of the bases in DNA, it is sometimes possible to construct evolutionary trees that show the degree of relatedness between organisms. These trees generally agree with those based on the fossil record.

In recent years, it has actually been possible to observe natural selection occurring. Sickle-cell anemia is an example of stabilizing selection, while pest and bacterial resistance and industrial melanism are examples of directional selection. Considering a longer length of time, there appear to be two types of macroevolution adaptive radiation and phyletic evolution. Darwin's finches provide an example of adaptive radiation, and the evolution of the horse is an example of phyletic evolution. Actually, in most lineages, we expect these two patterns to be repeated one after the other, over and over again. Various other patterns of evolution, including divergent, parallel, convergent, and coevolution have also been observed.

Objective Questions

Match the phrases in questions 1–4 with those in this key:

   a. biogeography
   b. fossil record
   c. biochemistry
   d. anatomy

1. species are not immutable
2. forms of life are variously distributed
3. a group of related species has a unity of plan
4. same types of molecules are found in all living things
5. $Hb^A Hb^S$ is better adapted than either $Hb^S Hb^S$ or $Hb^A Hb^A$. This is an example of _____ .

6. In polluted areas, the black form of the peppered moth is apt to survive because _____ .
7. In the case of industrial melanism, the _____ are the selective agent.
8. Are Darwin's finches more different from one another when they are on the same island or when they are on separate islands? _____
9. Phyletic evolution is evolution in _____ (choose time or space).
10. An example of phyletic evolution is the evolution of the _____ .

Answers to Objective Questions

1. b 2. a 3. d 4. c 5. heterozygote superiority
6. the trees are blackened by pollutants
7. predatory birds 8. same island 9. time
10. horse

Study Questions

1. List at least five evidences that evolution does occur.
2. What in general does the fossil record tell us about the history of life?
3. Describe an example of convergent evolution.
4. What is continental drift and how might this phenomenon affect the distribution of fossils and present-day organisms?
5. In what way does embryological development give evidence of evolution?
6. Define and contrast homologous structures with analogous structures.
7. Name three vestigial structures found in humans.
8. Give examples of molecular homologies and discuss their importance.
9. Give examples of natural selection at work, including one that exemplifies balanced polymorphism.
10. Evolution proceeds by adaptive radiation and phyletic evolution. Compare and contrast these two, giving an example of each.

Selected Key Terms

fossil ('fäs əl) 235
biogeography (,bī ō jē 'äg rə fē) 239
plate tectonics ('plāt tek tän iks) 242
homologous structure (hō 'mäl ə gəs 'strək chər) 243
analagous structure (ə 'nal ə gəs 'strək chər) 243
vestigial structure (ve 'stij əl 'strək chər) 246
molecular clock (mə 'lek yə lər 'kläk) 246

evolutionary tree (,ev ə 'lü sh ner ē trē) 246
sickle-cell anemia ('sik əl 'sel ə 'nē mē ə) 249
heterozygote superiority (,het ə rō 'zī gōt sù ,pir ē 'òr ət ē) 250
convergent evolution (kən 'vər jənt ,ev ə 'lü shən) 252

Suggested Readings for Part 5

Ayala, R. J. September 1978. The mechanisms of evolution. *Scientific American.*

Ehrlich, P. R., Holm, R. W., and Parnell, D. R. *The process of evolution.* 2d ed. New York: McGraw-Hill.

Grant, P. R. 1981. Speciation and the adaptive radiation of Darwin's finches. *American Scientist* 69:653.

Lewontin, R. C. September 1978. Adaptation. *Scientific American.*

Mayr, E. September 1978. Evolution. *Scientific American.*

Mossman, D. J., and Sarjeant, W. A. S. January 1983. The footprints of extinct animals. *Scientific American.*

Racle, F. A. 1979. *Introduction to evolution.* Englewood Cliffs, NJ: Prentice-Hall.

Rensberger, B. April 1982. Evolution since Darwin. *Science* 82.

Scientific American. 1978. *Evolution.* San Francisco: W. H. Freeman.

Stebbins, G. L. 1977. *Processes of organic evolution.* 3d ed. Englewood Cliffs, NJ: Prentice-Hall.

Stebbins, G. L., and Ayala, F. J. July 1985. The evolution of Darwinism. *Scientific American.*

Volpe, E. P. 1981. *Understanding evolution.* 4th ed. Dubuque, IA: Wm. C. Brown Publishers.

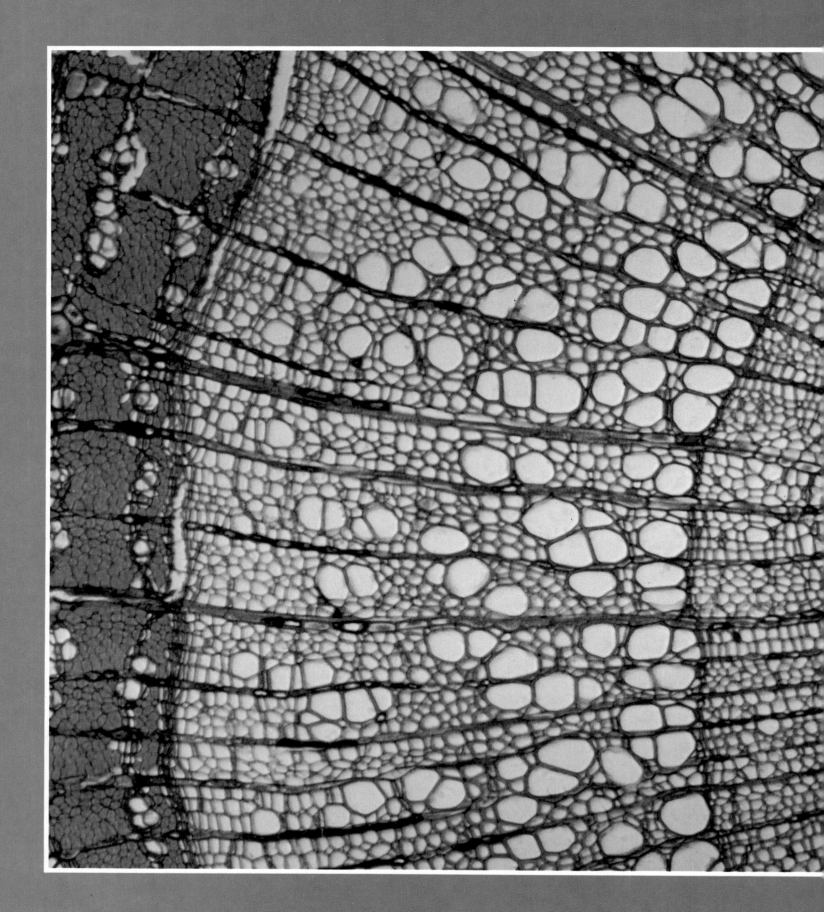

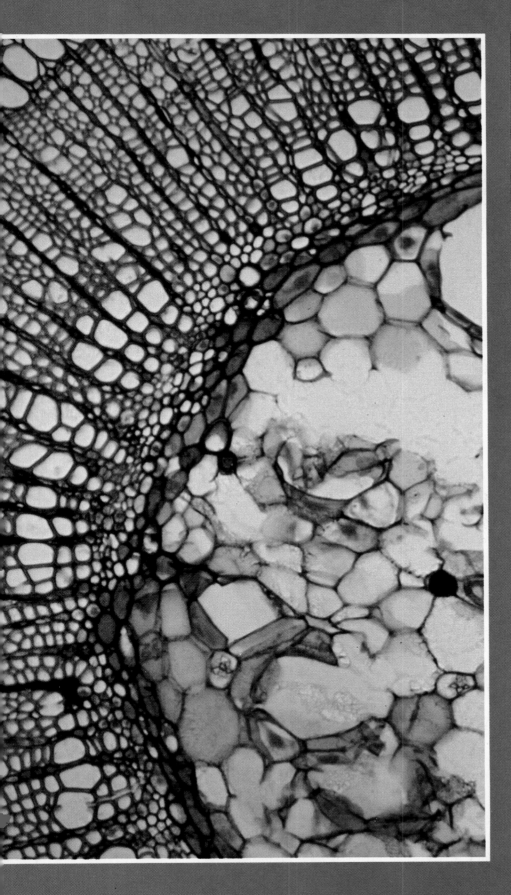

Diversity

There never were in the world two opinions alike, no more than two hairs or two grains. The universal quality is diversity.

—MICHEL DE MONTAIGNE

Dicot cross section of *Tilia* stem.

PART 6

Diversity of Life

Diversity is a striking characteristic of living things. The innumerable types of organisms presently existing are the result of repeated speciation and adaptation to different combinations of environmental factors. There is a type of organism that is specialized to interact with virtually every type of environment.

The fossil record indicates that many species of organisms existing in the past have become extinct. Extinction is often caused by an inability to adapt to a changing environment. As the earth's environment has changed, the existing combination of species has also changed. The history of the present combination of groups of organisms can sometimes be traced by examining the fossil record since related groups have shared a common ancestor in the past. However, since the fossil record is incomplete, comparative anatomy is sometimes helpful in deciding which groups of organisms are most closely related.

The diversity of life is well exemplified by insects, which are adapted to feed on most every type of plant and animal. These potato beetles are found not only in North America, but also in Europe and Russia. The tendency of humans to plant only certain crops can encourage the spread of pests that feed on these crops.

CHAPTER 13

Life's Beginnings

Concepts

1. Evolution includes a chemical evolution that precedes organic evolution.
   A chemical evolution is believed to have produced the first cell(s), after which organic evolution began.

2. Organic evolution proceeds slowly.
   The prokaryotes (kingdom Monera) were the first types of living things, and they alone were present for about two billion years.

3. Some living organisms may have changed minimally since they first evolved.
   The kingdom Protista contains organisms that may resemble the first eukaryotic cells. Although these organisms are unicellular, their internal structure is quite complex.

4. Various characteristics are used to distinguish one group of organisms from another.
   Fungi are placed in a separate kingdom because they are the most complex organisms to rely on saprophytic nutrition.

L ife has a history; it came into existence and then diversified into the many forms that have inhabited the earth. Most of the diversity is the evolutionary product of the last 600 million years, whereas the steps leading up to this stage took nearly 4 billion years (fig. 13.1).

Figure 13.1
Time scale showing that the great diversity of life occurred within the past 600 million years. Some authorities relate this to an increase in atmospheric oxygen at this time.

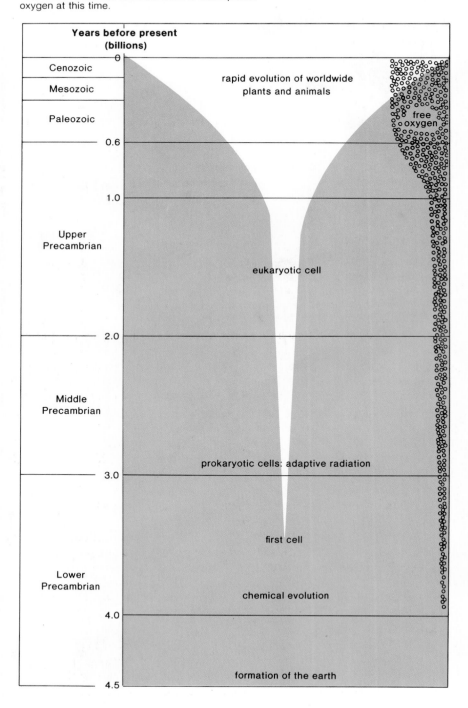

The Beginning

The sun and the planets probably formed from aggregates of dust particles and debris about 4.6 billion years ago. Intense heat produced by gravitational energy and radioactivity of some atoms caused the earth to become stratified into a core, mantle, and crust. Heavier atoms of iron and nickel became the molten liquid core, dense silicate minerals became the semiliquid mantle, and less dense silicate minerals became the crust. Severe convection currents within the mantle caused the thin crust to move continually, so there were no stable land masses during the time life was evolving.

Primitive Atmosphere

The primitive atmosphere was not the same as today's atmosphere. Harold Urey proposed in the 1950s that the earliest atmosphere must have contained a lot of hydrogen (H_2) because it is the most abundant element in our solar system. Later it was suggested that lightweight atoms, including hydrogen, would have been lost as the earth formed because its gravitational field was not strong enough to hold them. It is now thought that the primitive atmosphere was produced after the earth formed by outgassing from the interior, particularly by volcanic action. In that case, the atmosphere consisted of water vapor (H_2O), nitrogen gas (N_2), and carbon dioxide (CO_2), with only small amounts of hydrogen (H_2) and carbon monoxide (CO). The primitive atmosphere lacking free oxygen was a reducing atmosphere as opposed to the oxidizing atmosphere of today.

Chemical Evolution

With further cooling, water vapor condensed to form the oceans, and the gases trapped within the oceans underwent a chemical evolution, an increase in complexity that eventually produced the first life molecules and then cells (table 13.1). As early as the 1920s, Soviet biochemist A. I. Oparin proposed that organic molecules could be produced from the gases of the primitive atmosphere in the presence of such an energy source as lightning or ultraviolet (u.v.) radiation. In 1953 Stanley Miller provided support for Oparin's ideas through an ingenious experiment (fig. 13.2). Miller placed a mixture resembling the primitive atmosphere (methane, ammonia, hydrogen, and water) in a closed system, heated the mixture, and circulated the gases past an electric spark (simulating lightning). After a week's run, Miller discovered that a variety of amino acids and organic acids had been produced. Since that time, others have shown that a less reducing mixture of gases will also result in organic molecules. On the basis of these experiments, the primitive oceans were most likely a thick organic soup.

The next step was the condensation of small organic molecules to produce the macromolecules characteristic of living things: polynucleotides, polypeptides, and polysaccharides. It is possible that macromolecules could have formed in the ocean, but it is more likely that small organic molecules were washed ashore where they adhered to clay particles and were exposed to dry heat that encouraged polymerization. S.W. Fox has heated mixtures of amino acids at 130° to 180° to form amino acid polymers that he calls **proteinoids.** When proteinoids are placed in water, they form **microspheres** (fig. 13.3), structures that have some cell-like properties. They are, so to speak, the beginnings of cells, or **protocells.**

In contrast, Oparin showed that mixtures of macromolecules could join together in water to form **coacervate droplets.** Both microspheres and coacervates can take up molecules from the environment to form a lipid-protein film like a simple cell membrane. When they take up enzymes, they carry on

Table 13.1 Origin of the First Cell
Primitive Earth
outgassing
Gases: No free O_2
energy (e.g., lightning, ultraviolet radiation)
Small molecules: Amino acids, glucose, nucleotides, fatty acids
polymerization
Macromolecules: Protein, nucleic acids
organization
Protocell: Heterotrophic fermenter
organic evolution
Prokaryotic cell: Reproduction possible
Autotrophic cells: Including photosynthesizers
oxygen
Aerobic heterotrophs: Aerobic respiration

Figure 13.2

In Miller's experiment, gases are (a.) admitted to the apparatus, (b.) circulated past an energy source, (c.) and cooled to produce (d.) a liquid that can be withdrawn for testing.

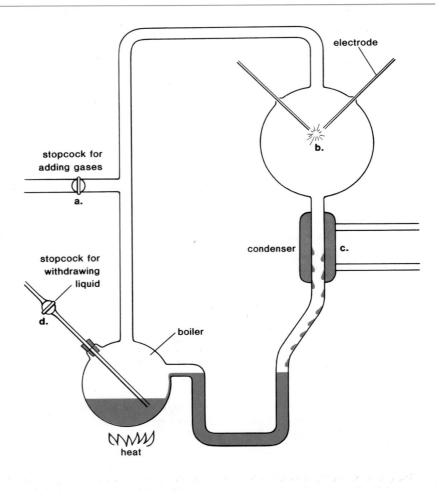

electrode

stopcock for adding gases

a.

b.

condenser

c.

stopcock for withdrawing liquid

d.

boiler

heat

Figure 13.3

a. Scanning electron micrograph of proteinoid microspheres, which may be similar to the structure of protocells. b. Micrograph of the interior of proteinoid microspheres.

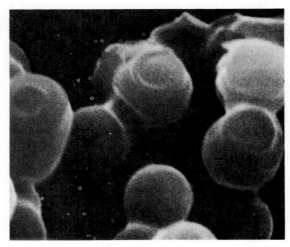

a.

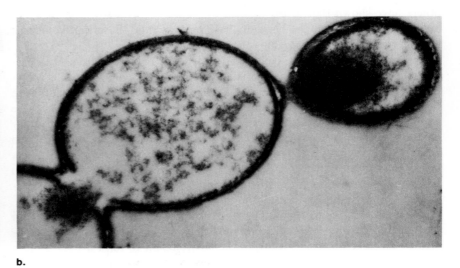

b.

some metabolic functions. Fox believes that after protocell formation, proteins and nucleic acids could have gradually evolved and increased in complexity until a true cell capable of reproduction came into being.

Some biologists, Francis Crick for instance, disagree with this hypothesis. They believe chemical evolution may have proceeded by the initial development of nucleic acid genetic material rather than protein coacervates or microspheres. The most primitive nucleic acids might have reproduced themselves and subsequently acquired the ability to direct the synthesis of peptides. Regardless, the first true cell had to perform two functions: metabolism and reproduction.

Organic Evolution

The first primitive cell(s) possessing rudimentary enzymes, genes, and a selectively permeable membrane must have been a heterotroph living off preformed organic molecules in the primitive ocean, and it must have carried on anaerobic respiration since there was no oxygen (O_2) in the atmosphere. It was then a **heterotrophic fermenter.**

The Earth is quite literally poised between fire and ice. Consider, for example, what would happen if we somehow moved the Earth slightly closer to the sun.

As the oceans grew warmer, more and more water vapor would begin to steam into the atmosphere. Once there, the vapor would begin to act like glass in a greenhouse, preventing heat from radiating back into space. So the Earth would grow warmer still, until oceans began to dry and the carbonate rocks—the limestones and dolomites—began to bake in the heat and release carbon dioxide.

The greenhouse effect caused by this gas is famous: By burning fossil fuels we are already releasing enough carbon dioxide to warm the climate measurably. The carbonate rocks, however, contain billions upon billions of tons of it, enough to trigger a "runaway greenhouse." In the end our planet would become a twin of unfortunate Venus, the next planet inward to the sun: a gaseous, dry, searing hell, its surface covered with clouds, oppressed by a massive atmosphere of carbon dioxide, and hot enough to melt lead.

Suppose, on the other hand, we moved the Earth further out from the sun. As the planet grew colder, glaciers would grind southward over Canada, Europe, and Siberia, while sea ice crept northward from Antarctica. The ice would reflect more sunlight back into space, cooling the planet even more. Step by step, the ice would extend toward the equator. In the end, the Earth would gleam brilliantly—but its oceans would be frozen solid.

Thus, the climate is balanced precariously indeed—so precariously that many geologists now believe that tiny, cyclic variations in the Earth's orbit, known as the Milankovitch cycles, were enough to have triggered the ice ages.

But geologists also have found fossilized marine microbes in rocks more than 3.5 billion years old (see accompanying figure) and they assure us that the oceans of the Earth have remained warm and liquid throughout its 4.6-billion-year history.

Perhaps this is a lucky accident—after all, if the Earth had not formed at just the right distance from the sun to have liquid oceans, we would not be here to worry about it. But the astrophysicists point out that things aren't quite that simple.

The sun, they say, is a quiet and stable star. But like others of its type, it is inexorably getting hotter with age. In fact, it is about 40 percent brighter now than when the Earth was born. So how could the climate possibly stay constant? If the Earth is comfortable now, then billions of years ago, under a colder sun, the oceans must have been frozen solid. But they were not. On the other hand, if the oceans were liquid then, why has the sun not broiled us into a second Venus by now?

One theory, advocated by a number of biologists, is that the early Earth started out with a good deal more carbon dioxide in the air than it has now, which gave enough of a greenhouse effect to keep the planet warm even when the sun was cool. If nothing had changed, the greenhouse eventually would have "run away" as it did on Venus. (There is some evidence that Venus started out with oceans much like ours.) But fortunately for us, about three billion years ago, certain blue-green algae [cyanobacteria] devised a way of taking carbon dioxide out of the air and turning it into organic carbon compounds. We call that process photosynthesis. In the eons since, as the algae and their descendants, the plants, have evolved and multiplied, the decreasing levels of atmospheric carbon dioxide have just about kept pace with the warming sun. Thus, say the biologists, it was life that saved the world for life.

Why Is the Earth neither Too Hot nor Too Cold?

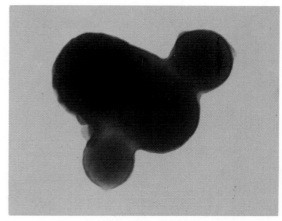

Electron micrograph of organic matter found in 3.5 billion-year-old sedimentary rock. It is possible that these forms are microorganisms in which case they are the oldest ever pictured. The picture was taken by geologist Miryam Glikson.

Once the preformed organic molecules were depleted, organic evolution would have favored any cell capable of making its own food. The first autotrophs probably lacked a light-absorbing pigment and therefore were not photosynthesizers. They could have been chemosynthetic organisms that extracted electrons from inorganic molecules (other than water) and used these electrons to generate ATP by electron transport. Or even if they did possess a pigment able to capture light energy, in order to reduce carbon dioxide, they probably did not at first use water as a hydrogen source and thus did not give off oxygen. The first autotroph to use water would have been selected because water is such a plentiful molecule. Now oxygen would have been given off as a by-product of photosynthesis.

Today's Atmosphere

The presence of free oxygen (O_2) changed the character of the atmosphere; it became an oxidizing atmosphere instead of a reducing atmosphere. Abiotic synthesis of organic molecules was no longer possible because any organic molecules that happened to form would have been broken down by oxidation. As oxygen levels increased, cells capable of aerobic respiration evolved. This is the type of respiration used by the vast majority of organisms today.

The buildup of oxygen in the atmosphere caused the development of an ozone (O_3) layer. This filters out ultraviolet rays, shielding the earth from dangerous radiation. Prior to this time, organisms probably lived only deep in the oceans where they were not exposed to the intense radiation striking the earth's surface. Now life could safely spread to shallower waters and eventually move onto the land. Berkner and Marshall have attributed the great increase in living forms that took place at the beginning of the Paleozoic era (table 13.2) to increases in oxygen concentration (fig. 13.1).

Will have matching series on test

Table 13.2 Geological Time Scale

Era	Period	Epoch	Millions of Years Ago
Cenozoic	Quaternary	Recent	
		Pleistocene *Man*	2.5
	Tertiary	Pliocene	7
		Miocene	26
		Oligocene	38
		Eocene	54
		Paleocene	65
Mesozoic	Cretaceous		130
	Jurassic		180
	Triassic *alligators*		230
Paleozoic	Permian		280
	Carboniferous		350
	Devonian		400
	Silurian		435
	Ordovician		500
	Cambrian		600
Precambrian			4,600

use 70 dominance of mammals & flowering plants

Start of Mesozoic mammals

memorize all our years Eras—Periods—Epoch

Kingdom Monera

Kingdom Monera is one of the five kingdoms recognized in this text (table 13.3). A kingdom is the largest category of classification, a topic that is discussed on page 25. Also, there is a classification of organisms in the appendix that you may wish to refer to from time to time. Kingdom Monera contains the prokaryotes, the first types of cells to evolve. These organisms provide the first fossils—dated 3.5 billion years ago (p. 263)—and the fossil record indicates that prokaryotes reigned supreme for at least two billion years. We see today the results of an adaptive radiation that took place during this time and there are still many modern successful species of prokaryotes.

The prokaryotes include the bacteria and cyanobacteria. The latter were formerly termed blue-green algae. The term "algae" will be used in this text merely to mean an aquatic photosynthesizing organism. As we shall see, different types of algae are classified in the kingdoms Monera, Protista, and Plants.

Anatomy and Physiology

Structure

The prokaryotes are quite small, having an average size between 1 m and 5 m. As discussed in chapter 4 and listed in table 4.5, these cells have no nuclei, mitochondria, endoplasmic reticulum, or other eukaryotic organelles. They have DNA but no nuclear envelope; instead, their DNA is located in a discrete part of the cytoplasm called the *nucleoid*. They have respiratory enzymes, but these are found on the cell membrane, not in the mitochondria. If they have chlorophyll, it is found in flattened vesicles instead of in chloroplasts. They have ribosomes, but these are smaller and lighter than eukaryotic ribosomes.

Microscopic examination of prokaryotes (figs. 13.4 and 13.7) often reveals a cell wall in addition to a cell membrane. This wall, which gives the cell its shape and rigidity, contains long chains of unique amino sugars cross-linked by peptide chains. Its composition is entirely different from the cell wall of any eukaryotic cell. Outside the cell wall some prokaryotes have a shiny or gelatinous capsule composed of polysaccharides or polypeptides. This capsule protects infectious bacteria against phagocytosis by host white cells.

Reproduction

Prokaryotes reproduce by means of binary fission, a process depicted in figure 7.23. The single circular chromosome replicates, then the two copies separate as the cell enlarges. Newly formed cell membrane and cell wall separate it into two cells. Mitosis, which requires the formation of a spindle apparatus, does not occur in prokaryotes.

Table 13.3 Classification

Name of Kingdom	Representative Organisms
Monera	Bacteria and cyanobacteria
Protista	Protozoans, unicellular algae of various types
Fungi	Molds and mushrooms
Plants	Green algae, mosses, ferns, various trees, and flowering plants
Animals	Sponges, worms, insects, fishes, amphibians, reptiles, birds, and mammals

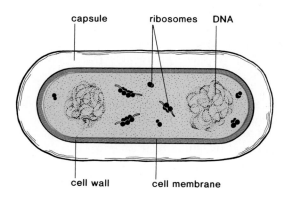

capsule   ribosomes   DNA

cell wall   cell membrane

Figure 13.4
Generalized drawing of a bacterium. Note the lack of membrane-bound organelles.

Figure 13.5

Clostridium botulinum containing an endospore. This organism causes the type of food poisoning known as botulism.

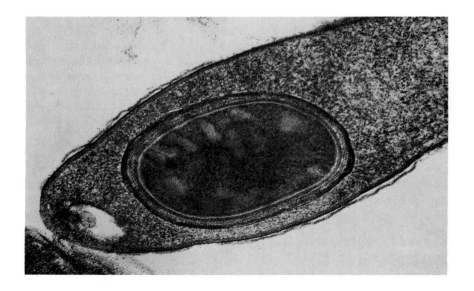

Figure 13.6

Scanning electron micrographs of bacteria. *a.* Spherical-shaped (round) bacteria. *b.* Rod-shaped bacteria. *c.* Spiral-shaped bacteria with flagella used for locomotion. (*a. and c.*) Kessel, R. G., and Shih, C. Y.: *Scanning Electron Microscopy in Biology.* © 1976 Springer-Verlag, Berlin.

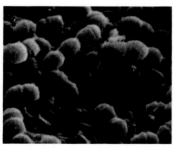

a.

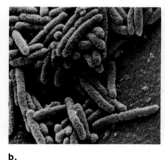

b.

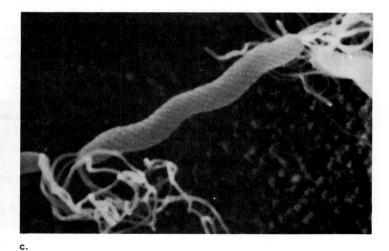

c.

Spore Formation Although not a form of reproduction, some prokaryotes can form spores, bodies that permit survival during unfavorable environmental conditions. Endospores (fig. 13.5), which develop inside some bacteria, have a thick wall of several layers that encloses DNA, ribosomes, and other cytoplasmic constituents. Endospores are very resistant and can survive hours in boiling water, extreme cold, extreme pH, and drying conditions. When exposed to proper stimuli, the spores germinate and produce new vegetative cells.

Bacteria

Bacteria (**phylum Schizophyta**) were discovered by Anton van Leeuwenhoek, who published drawings of them in 1684. There are many different types, but we will consider only the eubacteria, which have three main shapes: round or spherical (coccus), rod (bacillus), and spiral (a curved shape called a spirillum) (fig. 13.6). The rods or cocci sometimes associate, forming clusters or linear filaments.

Table 13.4 Symbiotic Relationships

Name	Definition	Examples
Mutualism	Both organisms benefit	Bacteria in plant nodules Dinoflagellates in coral animals Lichens
Commensalism	One organism gains and the other is unharmed	Bacteria on human skin
Parasitism	One organism gains and the other is harmed	Bacterial, fungal, and protozoan infections

A large variety of bacteria are motile and most of these employ flagella to propel themselves. Bacterial flagella are quite different from eukaryotic flagella; they are built up from several intertwined, helical chains of subunits of a protein called flagellin. Each flagellum is attached to the bacterial cell wall and plasma membrane by a basal structure consisting of a hook and shaft. It is thought that the inner ring portion of the shaft rotates somewhat like an electric motor, and this twists the rigid flagellum, moving the bacterium at speeds up to 80 to 90 μm per second.

Nutrition

The adaptive radiation of prokaryotes is most evident when one considers that every type of nutrition is found among bacteria except holozoism (eating whole food). Most bacteria are heterotrophs and feed on dead organic matter by secreting digestive enzymes and absorbing the products of digestion. These so-called **saprophytic decomposers** can break down such a large variety of molecules that there probably is no natural organic molecule that cannot be digested by at least one bacterial species. Some are very specific and degrade only certain molecules; others are chemical omnivores. *Pseudomonas multivorans,* for example, can use over ninety different organic molecules as sources of carbon and energy. The decomposing bacteria play a critical role in recycling matter and making inorganic molecules available to photosynthesizers, as described in figure 5.1. The metabolic capabilities of bacteria are exploited by human beings, who use them to perform services ranging from digestion of sewage, pectin, and oil to production of such products as alcohol, vitamins, and even antibiotics. By means of gene splicing, bacteria are now used to produce human insulin and growth hormone.

Heterotrophs may be independent or symbiotic (table 13.4), forming mutualistic, commensalistic, and parasitic relationships. **Nitrogen-fixing bacteria** with the capability of reducing aerial nitrogen (N_2) to ammonia (NH_3) are mutualistic because they provide organic nitrogen to their host when they reside in root nodules of soybeans, clover, and alfalfa. Intestinal bacteria in humans release vitamins K and B_{12}, which we can use, and in the stomachs of cows and goats they digest cellulose, enabling these animals to feed on grass. Commensalistic bacteria live on organisms and cause them no harm, but parasitic bacteria are responsible for both plant and animal diseases. Common human infections caused by bacteria are listed in table 13.5.

A few bacteria are autotrophic, being either chemosynthetic or photosynthetic. The chemosynthetic bacteria oxidize compounds like ammonia (NH_3), nitrite (NO_2), and sulfur (S_2) to acquire energy to produce ATP. Purple and green bacteria are photosynthetic, using light as an energy source and carbon dioxide as a carbon source. But their chlorophyll and other pigments are different from those of plants. They do not oxidize water but instead extract electrons from inorganic molecules, such as H_2S. So they are usually found a few meters below the surface of lakes where there is sufficient light

Table 13.5 Infectious Diseases Caused by Bacteria

Respiratory Tract
Strep throat (sometimes causing rheumatic and scarlet fever)

Pneumonia
*Whooping cough
*Diphtheria
*Tuberculosis

Skin Reactions
Staph (pimples and boils)
*Gas gangrene (wound infections)

Nervous System
*Tetanus
Botulism
Meningitis

Digestive Tract
Food poisoning (salmonella, botulism, and staph)
*Typhoid fever
*Cholera

Venereal Diseases
Gonorrhea
Syphilis

*Vaccines are available. Tuberculosis vaccine is not used in this country. Typhoid fever, cholera, and gas gangrene vaccines are given if the situation requires it. Others are routinely given.

Figure 13.7
Generalized drawing of a cyanobacterium. Note the lack of membrane-bound organelles. For example, the photosynthetic lamellae simply lie within the cytoplasm.

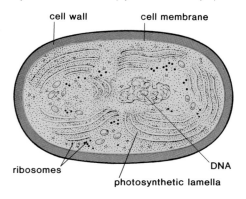

cell wall   cell membrane

ribosomes   DNA

photosynthetic lamella

Figure 13.8
Anabaena is filamentous. *a.* When nitrate (NO₃) is present, *Anabaena* filaments lack heterocysts. *b.* When only N₂ is present as a nitrogen source, *Anabaena* filaments have heterocysts where N₂ is converted to ammonia.

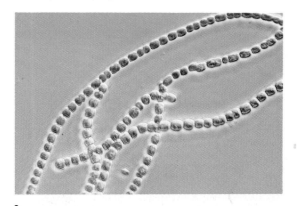

a.
heterocyst

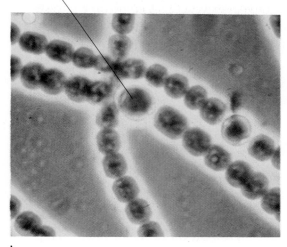

b.

and a supply of H_2S. When water is overloaded with sewage, these bacteria are deprived of light and are unable to function. Then H_2S, which smells like rotten eggs, accumulates and rises to the surface.

Heterotrophic bacteria differ in their need for oxygen. *Obligate anaerobes* cannot live in the presence of air. A few serious human illnesses, such as botulism, gas gangrene, and tetanus, are caused by poisons produced by anaerobic bacteria. Some bacteria are *facultative anaerobes* and survive whether oxygen is present or not. *E. coli,* a well-known facultative bacterium, grows adequately under anaerobic conditions in the human intestine, but flourishes in a laboratory incubator when provided with oxygen. Most bacteria, however, are *aerobic* and, like animals, require a continual supply of oxygen for aerobic respiration.

Cyanobacteria

Cyanobacteria (phylum Cyanobacteria), a group of organisms formerly called **blue-green algae,** is exceptionally diverse, ranging from single cells to colonies and filamentous forms. Some cyanobacteria move by a gliding motion when in contact with a solid surface. The mechanism by which they glide is unknown.

Cyanobacteria (fig. 13.7) carry on photosynthesis in a manner similar to that of plants. They possess chlorophyll *a* and evolve oxygen as a by-product. They also have other pigments that may mask the color of chlorophyll giving them, for example, a red, brown, or black color.

Cyanobacteria are found in almost all environments, including hot springs and deserts. Their nutritional requirements are simple, especially when they possess **heterocysts** (fig. 13.8*b*). Heterocysts are thick wall cells without nuclei where nitrogen fixation occurs. Cyanobacteria are symbiotic (table 13.4) with a number of organisms; they are found within liverworts, ferns, and in invertebrates such as corals. They also associate with fungi, forming **lichens** (fig. 13.20) that can grow on rocks. The cyanobacterium provides organic nutrients to the fungus, while the latter possibly protects and furnishes inorganic nutrients to its partner.[1] It is presumed that cyanobacteria were the first colonizers of land during the course of evolution. Once established, their mass could have provided a physical as well as chemical substrate for the ultimate growth of plants.

Cyanobacteria are ecologically important in still another way. If care is not taken in the disposal of industrial, agricultural, and human wastes, phosphates and nitrates drain into lakes and ponds, resulting in a "bloom" of these organisms. The surface of the water becomes turbid and light cannot penetrate to lower levels. When a portion of the cyanobacteria die off, the decomposing bacteria use up the available oxygen, which causes fish kills.

Cyanobacteria are believed to have evolved from photosynthetic bacteria, although they probably diverged early in the history of prokaryotes. Comparison of their methods of photosynthesis reveals that cyanobacteria are more complex.

Bacteria	**Cyanobacteria**
Photosystem I only	Photosystems I and II
Do not give off O_2	Do give off O_2
Unique type of chlorophyll	Type of chlorophyll found in plants

[1]The relationship has long been thought to be mutualistic, but since the algal cells grow more abundantly when freed of the fungus, there are those who believe the fungus is parasitic on the alga.

Originally, the term virus denoted any infectious agent that caused disease. After a while, it was shown that viruses are very different from cellular disease-causing organisms. At first, size was the primary reason for differentiating viruses from other microorganisms.

In 1892 Dimitri Ivanowsky, a Russian biologist, was performing experiments with tobacco plants infected with tobacco mosaic disease, a condition that takes its name from the wrinkled and mottled appearance of the infected leaves. Ivanowsky, who transmitted the disease to healthy plants by rubbing them with juice extracted from diseased plants, passed the infective extract through a finely meshed porcelain filter. To his surprise, the filtrate was still infective. This meant the disease-causing agent was smaller than any known bacteria. Disease-causing agents that could pass through filters came to be known as filterable viruses and later simply as viruses.

In 1935 W. M. Stanley demonstrated that viruses are remarkably different from bacteria and any other living cell. He was able to crystallize infectious mosaic viruses and show that they were mostly protein in composition.

Regardless of their size and shape (a.–c.), viruses are composed of just two parts: an outer coat of protein and an inner core of nucleic acid (either RNA or DNA). Since they lack a cellular organization and are incapable of independent reproduction, viruses are usually considered to be nonliving. They are called obligate parasites because they only reproduce inside living cells. Reproduction of the T virus, shown on page 189, illustrates that viruses take over the machinery of the cell when they reproduce. Some viruses do not reproduce immediately; instead, they become integrated into the host DNA. These latent viruses only carry on reproduction at certain times.

Some well-studied viral diseases in humans are herpes, flu, mumps, measles, polio, rabies, and infectious hepatitis. New evidence indicates that certain retroviruses may contribute to the development of cancer by bringing oncogenes into the host cell (p. 212). Retroviruses are RNA viruses that have an enzyme capable of transcribing RNA to DNA. These viruses normally insert their newly transcribed DNA into the host DNA.

Most viruses are extremely specific; many even prefer specific tissues. An interesting theory about the origin of viruses suggests that they begin as pieces of DNA or RNA from the tissue they attack. Certainly it seems that viruses may have picked up any oncogene they carry from previous host DNA.

Viruses

Viruses may be (a.) helical like the tobacco virus, (b.) polyhedral like the adenovirus, or (c.) complex like the "T" virus. Note head, tail, and tail fibers.

a.

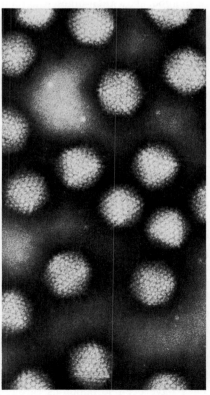

b.

c.

Kingdom Protista

The protists are all eukaryotes. The eukaryotic cell evolved about 1.5 billion years ago in the Precambrian era (fig. 13.1), and today's protists can trace their ancestry to that time. The eukaryotic cell contains the organelles studied in chapter 4 and listed in table 4.3. As discussed earlier, these organelles may have originated when a large prokaryote ingested but did not digest smaller prokaryotes. Another suggestion is that they formed by cell membrane invagination.

In this text we include as protists unicellular (or colonial) organisms whose relationship to either plants, animals, or fungi has not clearly been established. (Unicellular organisms whose relationship is clearer are included as members of these other kingdoms.) While unicellular organisms must cope with the environment without benefit of tissues and organ systems, they are not simple; their complexity resides in the organization of their cells. Their organelles perform the functions we associate with organ systems, and their cell membranes are used for all interactions with the environment. As discussed on page 124, this means that unicellular organisms have no other specialized boundaries for environmental exchange.

Protozoans

Protozoans are animallike heterotrophic unicellular organisms that are characteristically motile. Some are colonial; a **colony** is a loose association of cells, each of which must still take care of its own physiological needs. On occasion, some cells of a colony specialize for reproduction.

In addition to the usual organelles, protozoans have specialized vacuoles. For example, they take in food by simple diffusion or transport or by means of phagocytosis, then digest the food in food vacuoles. Also, freshwater protozoans continually gain water by osmosis and eliminate it by means of "contractile" vacuoles. These vacuoles expand with water drawn from the cytoplasm and then appear to "contract," releasing the water through a temporary opening in the cell membrane.

Although asexual reproduction involving binary fission and mitosis is the rule, many protozoa also reproduce sexually during some part of their life cycle. Sexual exchange is preceded by meiosis.

The protozoans to be studied are placed in four groups according to their type of locomotor organelle (table 13.6).

Table 13.6 Types of Protozoans

Protozoans	Locomotor Organelles	Example
Amoeboid	Pseudopodia	*Amoeba*
Ciliate	Cilia	*Paramecium*
Flagellate	Flagella	*Trypanosoma*
Sporozoa	No locomotor organelles	*Plasmodium*

Amoeboids

Typically, the **amoeboids (phylum Sarcodina)** move about and feed by means of cytoplasmic extensions called pseudopodia. In *Amoeba proteus* (fig. 13.9), a commonly studied member of this group, the pseudopodia form when the cytoplasm streams forward in particular directions. When *Amoeba* feed, they **phagocytize;** the pseudopodia surround and engulf the prey, which may be algae, bacteria, or other protozoans.

Amoeba proteus is found in fresh water, but most amoebas are marine, being either radiolarians or foraminiferans. The radiolarians float near the ocean surface despite an internal skeleton composed of silica or strontium sulfate. Their skeletons are intricate, exquisite, and of almost infinite variety. The foraminiferans have an internal skeleton usually made up of calcium carbonate (fig. 13.10). When they feed, pseudopodia extend through holes called foramina. These protozoa are more often found in the ooze of the ocean floor. They have been so populous through the ages, that following an upheaval, their accumulated shells formed the White Cliffs of Dover along the southern coast of England (fig. 29.14).

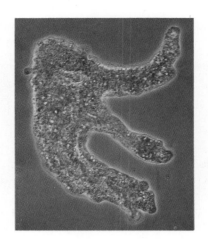

Figure 13.9
a. Photomicrograph of *Amoeba proteus*. Note the presence of pseudopodia by which the animal moves. *b.* Some structural details of *Amoeba proteus*. Note the unique organelles, including the food vacuole and contractile vacuole.

a.

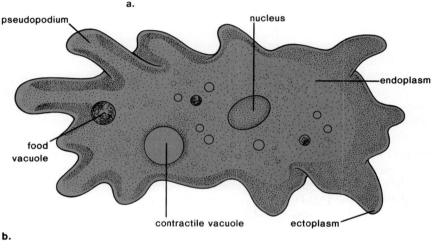

pseudopodium

nucleus

endoplasm

food vacuole

contractile vacuole

ectoplasm

b.

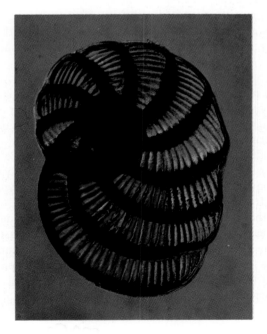

Figure 13.10
Shell of a foraminiferan, an amoebalike protozoan, that has existed for more than 500 million years, time enough for hundreds of lineages to have evolved, diversified, and become extinct. In large shells like this one, each story of the shell is multichambered. Within each chamber lies not only cytoplasm but also symbiotic algae. The algae use the foram's waste products to make food that they share with their host. Without the algae, a foram could not grow so large, live as long, or produce as many offspring.

Figure 13.11
Diagram illustrating the principal structures found in *Paramecium caudatum*. Despite its complexity, a paramecium is a single-celled organism.

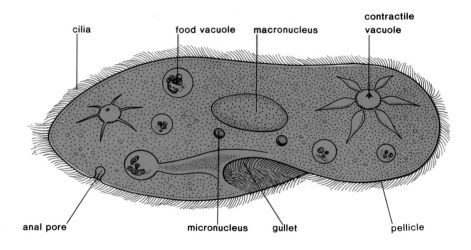

cilia   food vacuole   macronucleus   contractile vacuole

anal pore   micronucleus   gullet   pellicle

Figure 13.12
Two paramecia undergoing conjugation, during which time they exchange nuclear material.
Kessel, R. G., and Shih, C. Y.: *Scanning Electron Microscopy in Biology.* © 1976 Springer-Verlag, Berlin.

Ciliates

The ciliates (**phylum Ciliophera**), such as those in the genus *Paramecium* (fig. 13.11), are the more complex of the protozoans. Hundreds of cilia project through tiny holes in the outer covering, or pellicle. Lying in the ectoplasm just beneath the pellicle are numerous oval capsules that contain **trichocysts.** The contents of the trichocysts may be discharged as long threads, which are used for defense. When a *Paramecium* feeds, food is swept down a gullet to the cell mouth below which food vacuoles form. Following digestion, the soluble nutrients are absorbed by the cytoplasm, and the indigestible residue is eliminated at the anal pore.

The ciliates are quite varied and beautiful. A colony of *Vorticella* looks much like a bouquet of beautiful flowers. *Stentor coeruleus,* a favorite subject for research in regeneration, resembles a giant blue vase decorated with stripes when viewed under the microscope. The barrel-shaped *Didinium* (p. 635) can gobble up a greatly larger *Paramecium* much like a snake swallowing a rabbit. Suctorians have even stranger feeding habits. They are covered with tentacles and rest quietly on stalks until a hapless victim blunders into them. Then they promptly paralyze it and use their tentacles like straws to "suck it dry."

Flagellates

Protozoans that move by means of flagella (**phylum Mastigophora**) are sometimes called zooflagellates to distinguish them from unicellular algae that have flagella. The flagella of these eukaryotic cells have the characteristic 9 + 2 microtubular structure discussed on page 81.

Many zooflagellates enter symbiotic relationships (table 13.4). *Trichonympha collaris* lives in the gut of termites and enzymatically converts wood to soluble carbohydrates easily digested by the insect. The trypanosomes (fig. 13.13) cause African sleeping sickness and are transmitted to the blood of vertebrates by the tsetse fly. The tsetse fly, which becomes infected when it takes a blood meal from a diseased animal, passes on the disease when it feeds on another victim. The white cells in an infected animal accumulate around the blood vessels leading to the brain and cut off the circulation. The lethargy characteristic of the disease is caused by an inadequate supply of oxygen for normal brain alertness.

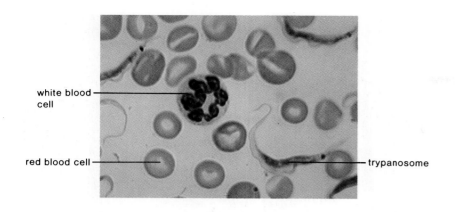

white blood cell

red blood cell — trypanosome

Figure 13.13
Trypanosomes among the blood cells of a patient suffering from African sleeping sickness, an illness that keeps millions of acres in central Africa largely uninhabited because people are fearful of contracting the disease.
Photograph by Carolina Biological Supply Company.

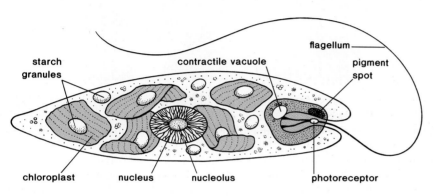

flagellum

starch granules

contractile vacuole

pigment spot

chloroplast   nucleus   nucleolus   photoreceptor

Figure 13.14
Diagram illustrating the principal structures found in *Euglena*. Notice the flagellum (animallike feature) and the chloroplasts (plantlike feature).

Sporozoa

The sporozoans (**phylum Sporozoa**) are also parasitic protozoans whose complicated life cycle almost always involves the formation of infective spores. The most important human parasite among the sporozoa is *Plasmodium vivax*, the causative agent of one type of malaria. When a human being is bitten by an infected *Anopheles* mosquito, the parasite eventually invades the blood cells. The chills and fever of malaria occur when the infected cells burst and release toxic substances into the blood (fig. 12.17).

Malaria is still a major killer of humans despite all efforts to control it. As discussed on page 249 recent resurgence of this disease is caused primarily by the development of insecticide-resistant strains of mosquitoes, and parasite resistance to antimalarial drugs. But recently researchers announced that by using recombinant DNA techniques, they were able to identify a specific protein that may be useful as a vaccine. If so, people could be vaccinated against the disease.

Unicellular Algae

The unicellular algae placed in the kingdom Protista all have chloroplasts and carry out photosynthesis in a manner similar to plants. They do not represent forms that are believed to be direct ancestors of plants, however.

Euglenoids

Euglenoids (**phylum Euglenophyta**) belong to the genus *Euglena* (fig. 13.14) and are freshwater organisms having both animallike and plantlike characteristics. Their animallike characteristics are motility and a flexible body wall.

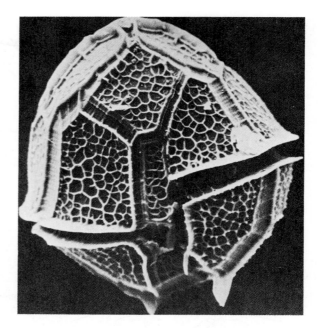

They move by means of two anterior flagella, typically having one much longer than the other. Because euglenoids are bound by a flexible pellicle instead of a rigid cell wall, they can assume different shapes as the underlying cytoplasm undulates or contracts.

Their plantlike characteristics stem from their possession of chloroplasts. A light-sensitive swelling at the base of one flagellum is shaded by a pigment spot, which allows euglenoids to judge the direction of the light. They move toward light so that photosynthesis can take place. Carbohydrates are stored in the pyrenoids, special structures attached to the chloroplasts. It is possible to destroy the chloroplasts by heating them at 35° C for a few generations or by treating them with streptomycin. Then they are permanently bleached and must absorb the nutrients they need since the gullet is not used for food gathering. A "contractile vacuole" located to one side empties its contents into the gullet.

Dinoflagellates

The dinoflagellates (**phylum Pyrrophyta**) are extremely plentiful in the ocean—approximately one thousand species are known. Externally they are typically covered by plates (fig. 13.15) composed of cellulose and have two flagella. One flagellum is located in a transverse groove and encircles the body; the other originates from a posterior groove and extends posteriorly.

Certain species of dinoflagellates multiply rapidly under appropriate conditions and produce a bloom known as the red tide. The toxins they give off at that time cause widespread fish kills and a human condition known as paralytic shellfish poisoning. The toxin accumulates in the tissues of shellfish, and humans take the toxin into their bodies when they eat shellfish that have fed on the dinoflagellates.

Other dinoflagellates have many beneficial effects. They are important sources of food for small animals in the ocean, and they live as symbiotes within the bodies of invertebrates. When they are symbiotic, they are known as zooxanthellae. Corals usually contain large numbers of zooxanthellae, which allows them to grow ten times faster in the light than in the dark.

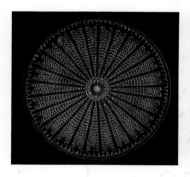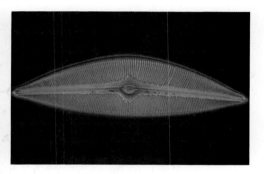

Figure 13.16
Diatoms are single-celled, golden-brown algae with silica-embedded walls. Diatoms are important oxygen and food producers in the ocean.

Diatoms

Diatoms (**phylum Chrysophyta**) are golden brown in color due to an accessory pigment in their chloroplasts that masks the color of chlorophyll. The cell wall of each diatom consists of two silica- (glass) impregnated valves beautifully decorated with delicate, complex designs (fig. 13.16). The valves are of unequal size, with the larger fitting over the smaller like the lid of a box. Following mitotic nuclear division, each daughter cell receives a valve and manufactures a new valve to fit inside it. One daughter cell is always slightly smaller than the parental cell and eventually, after several divisions, gets much smaller than normal. Then the diatoms carry out sexual reproduction, which allows them to return to their original maximal size.

Diatoms are the most numerous of all unicellular organisms in the oceans. As such, they serve as an important source of food for other organisms. Their remains, called diatomaceous earth, accumulate in the ocean floor and are mined and used commercially in making insulation, as an abrasive in toothpaste and silver polish, and in water-filtering systems.

Kingdom Fungi

Fungi are eukaryotic, multicellular organisms that may have evolved during the Precambrian era (table 13.2). They are placed in a separate kingdom because they are the most complex organisms having saprophytic nutrition. Like most bacteria, they usually secrete digestive enzymes that break down organic material in the environment so smaller molecules can be absorbed. Along with the bacteria of decay, fungi help complete the cycling of matter described in figure 5.2. Unfortunately, some fungi are also parasitic, causing serious diseases in plants and animals.

The bodies of all fungi, except unicellular yeasts, are made up of filaments called hyphae. A **hypha** is an elongated cylinder containing a mass of cytoplasm and hundreds of haploid nuclei that may or may not be separated by cross walls. The cell walls are composed largely of chitin, a polymer of an amino sugar also found in bacterial cell walls. The body of a fungus, made up of many hyphae, is called a **mycelium.**

Fungi reproduce both asexually and sexually but always produce spores, small haploid cells that are dispersed by wind currents. Spore formation takes place within specialized hyphae. Following sexual union, a collection of specialized hyphae, called a **fruiting body,** is found in some groups. Classification is based on mode of sexual reproduction.

Figure 13.17

Life cycle of black bread mold *Rhizopus*, including both asexual (lower left) and sexual cycles. Notice that the sexual cycle requires two mating strains.

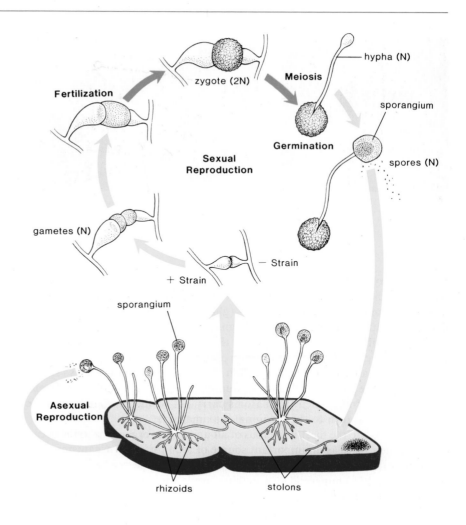

Bread Molds

Black bread molds (**division Zygomycota**), belonging to the genus *Rhizopus,* are often used as an example of the bread molds. Spores are produced in sporangia (sing., sporangium), which turn black when the spores are formed. The life history begins when a spore germinates on bread (fig. 13.17). Laterally extended hyphae, called *stolons,* spread over the surface of the bread, growing with amazing rapidity. Special hyphae called *rhizoids* extend from the stolons into the bread, anchoring themselves and also secreting digestive enzymes so they might absorb digestive products. Other hyphae rise into the air supporting **sporangia.** Mature sporangia eventually burst, releasing asexual spores that are dispersed by air currents.

During sexual reproduction, hyphae of different mating strains (usually referred to as plus and minus) approach each other. Upon contact, swellings of these hyphae are partitioned off by the formation of cross walls. These portions join and the nuclei fuse to form diploid zygote nuclei. The zygote wall then thickens and blackens to give a **zygospore.** After a dormancy of several months, meiosis takes place before germination occurs. One or more haploid sporangia are immediately produced and release spores that germinate after dispersal. The adult fungus is haploid; only the zygospore is diploid.

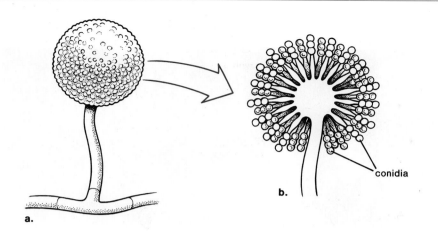

b.

conidia

a.

Figure 13.19
Cup fungus. The mycelium is buried beneath the ground and only the cup-shaped fruiting body is obvious. When microscopic sacs within these cups swell with maturity, they burst and shoot millions of spores into the air.

Sac Fungi

There are many different types of sac fungi (**division Ascomycota**), most of which have hyphae that produce chains of spores called **conidia** (fig. 13.18). During sexual reproduction, sac fungi form spores called ascospores within saclike cells called **asci.** In most species the asci are supported within fruiting bodies. In cup fungi, the largest group of sac fungi, the fruiting body takes the shape of a cup (fig. 13.19).

Yeasts are sac fungi that do not form fruiting bodies. In fact, yeast is different from all other fungi in that it is unicellular and most often reproduces asexually by budding. Yeast, as you know, carries out fermentation as follows:

glucose ⟶ carbon dioxide + alcohol

In the baking industry, the carbon dioxide from this reaction makes bread rise. In the production of wines and beers, it is the alcohol that is desired, and the CO_2 is allowed to escape into the air.

Not all sac fungi are of benefit to humans. A number cause diseases in plants; chestnut tree blight and Dutch elm disease are caused by sac fungi, as are powdery mildews, apple scab, and ergot. Ergot, a disease of cultivated cereals, causes ergotism, a severe illness in domestic cattle and humans who

Figure 13.20
Lichens take different forms including (*a.*) crustose (crusty) lichens and (*b.*) foliose (leafy) lichens.

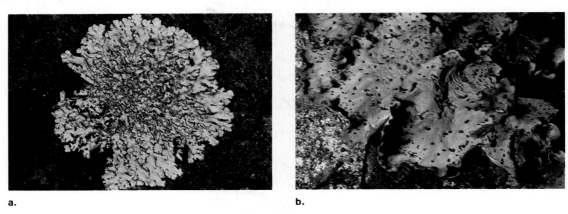

a.　　　　　　　　　　　　　　　　b.

consume infected grain. The condition is accompanied by muscular spasms, paralysis, and convulsions. Ergot is also a source of the psychedelic drug LSD (lysergic acid diethylamide).

Blue-green molds, notably *Penicillium,* are sac fungi. This mold grows on many different organic substances, such as bread, fabrics, leather, and wood. It is purposefully used by humans to provide the characteristic flavor of Camembert and Roquefort cheeses; more importantly, it produces the antibiotic penicillin. However, most penicillin today is synthetically produced. Another mold, the red bread mold *Neurospora,* was used in the experiments that helped decipher the function of genes.

The fungal portion of *lichens* (fig. 13.20) is usually a sac fungus, while the alga may be green or blue-green. Lichens can live on bare rock or poor soil and are able to survive great extremes of heat, cold, and dryness in all regions of the world. Reindeer moss is a lichen that is an important food source for some arctic animals.

Club Fungi

Among the club fungi (**division Basidiomycota**), asexual reproduction may also be accomplished by formation of conidia spores. As a result of sexual reproduction, members of this group form club-shaped structures called basidia, often within fruiting bodies. The mushroom (fig. 13.21), puffball, and shelf fungi (fig. 13.22) are club fungi; the visible portions of these are actually fruiting bodies, and the mycelia lie beneath the surface. On the underside of a mushroom cap, the basidia project from the gills. Within each basidium, a diploid nucleus undergoes meiosis to produce windblown spores called basidiospores.

Many mushrooms are considered gourmet delicacies and have been prized since the time of the Roman Empire. The best-known edible mushroom is *Agaricus campestris,* one of the few gilled mushrooms that can be cultivated commercially. About 65,000 tons of mushrooms are produced annually in this country. Unfortunately, some mushrooms (particularly members of the genus *Amanita*) are extremely poisonous and many people have died after ingesting them.

Rusts and smuts are parasitic club fungi that attack grains, resulting in great economic loss and necessitating expensive control measures. They do not have a conspicuous fruiting body and consist of vegetative hyphae, together with spores of various kinds.

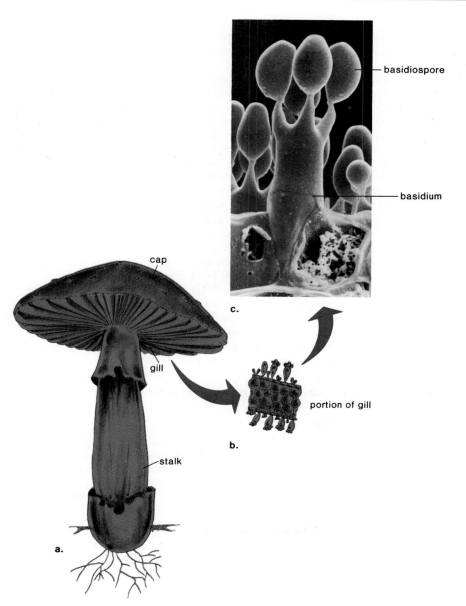

Figure 13.21
A mushroom is a fruiting body. The gills on the underside of the cap are lined with basidia (club-shaped structures). As a result of sexual reproduction, spores called basidiospores are produced here. *a.* Entire mushroom. *b.* Portion of gill showing placement of basidia and basidiospores. *c.* Electron micrograph of basidia and basidiospores.

basidiospore

basidium

c.

cap

gill

portion of gill

b.

stalk

a.

Figure 13.22
Shelf fungi are a common type of club fungus that grow on dead trees, helping to decompose the wood and thereby make inorganic nutrients available to living trees once again.

Summary

The earth was formed about 4.6 billion years ago. Gases escaping from the hot interior of the earth may have formed the first atmosphere, which contained the elements most abundant in living things. A chemical evolution produced small organic molecules that made the primitive ocean a hot organic soup. Polymerization of these produced macromolecules that may have reformed to give protocells. In any case, the first cells were probably anaerobic heterotrophs that lived in the ocean and fed off the preformed small organic molecules. As these were used up, natural selection would have favored development of autotrophic cells that not only made their own food but also gave off oxygen. The presence of oxygen in the atmosphere permitted the organic evolution of aerobic organisms. It also brought about conditions that most likely prevent the evolution of new cells today.

The prokaryotes, represented today by bacteria and cyanobacteria, were alone on the earth from about 1.5 to 3.5 billion years ago. These cells lack a nuclear envelope and other organelles characteristic of eukaryotic cells, and they reproduce asexually by binary fission. Bacteria are either spherical, rod-shaped, or spiral-shaped. They sometimes move by means of flagella made of flagellin. Except for holozoism, every type of nutrition is represented. Cyanobacteria may be single cells, colonies, or filaments. They all possess chlorophyll and photosynthesize in a manner similar to plants. They have such limited nutritional requirements that they are believed to have inhabited land before other types of organisms.

The ancestors of today's protists probably evolved about 1.5 billion years ago. These first eukaryotic cells are represented today by the protozoans and unicellular algae. Protozoans are animallike in that they are heterotrophic and motile.

They are classified according to the type of locomotor organelle employed (table 13.6). The unicellular algae include the euglenoids, the dinoflagellates, and diatoms. They all have chloroplasts and carry on photosynthesis in the same manner as plants. Lichens cannot be neatly classified because they are a symbiotic relationship between a fungus and an alga.

Fungi are saprophytic heterotrophic eukaryotes composed of hyphae that form a mycelium. Along with heterotrophic bacteria, they are organisms of decay. The fungi produce spores during both sexual and asexual reproduction. The major groups of fungi are distinguishable on the basis of their mechanism of sexual reproduction. The black bread molds produce spores in sporangia; the sac fungi produce spores in saclike cells, and the club fungi produce spores in club-shaped structures. The sac and club fungi typically have fruiting bodies.

Objective Questions

1. Preceding organic evolution, there was a _____ evolution that produced the first cell(s).
2. The primitive atmosphere lacked the gas _____ and the protocell carried on what type of respiration? _____
3. The two types of monerans are the _____ and the _____ .
4. Most bacteria are saprophytic decomposers, meaning that they _____ .
5. In contrast, cyanobacteria are _____ , using the energy of the sun to make their own _____ .
6. Amoeba move by means of _____ and ciliates move by means of _____ .
7. Dinoflagellates are classified as what type of protista? _____
8. The body of a fungus is a _____ that contains filamentous _____ .
9. The _____ and the _____ fungi both have fruiting bodies.
10. A lichen is a symbiotic relationship between a _____ and a _____ .

Answers to Objective Questions

1. chemical 2. oxygen, fermentation 3. bacteria and cyanobacteria 4. break down dead organic matter 5. photosynthetic, food 6. pseudopodia, cilia 7. unicellular algae 8. mycelium, hyphae 9. sac, club 10. fungus, algal cell

Study Questions

1. Describe the steps by which the first cell is believed to have evolved. Include in your description the words chemical evolution and organic evolution.
2. Describe the structure and means of reproduction for prokaryotes.
3. Bacteria have what types of nutrition? Which type is most common?
4. Cyanobacteria have what type of nutrition? Compare photosynthesis in cyanobacteria with that of bacteria.
5. Give an example for each group of protozoans in the kingdom Protista.
6. Discuss the significant features of the euglenoids, dinoflagellates, and diatoms.
7. Describe the structure of fungi, their mode of nutrition, and their means of reproduction.
8. Describe the structure of black bread mold and its life cycle.
9. Define a fruiting body and name two groups of fungi that typically have fruiting bodies.
10. Describe the structure of a lichen.

Selected Key Terms

proteinoid ('prō tēn öid) 261
microsphere ('mī krə ,s fir) 261
protocell ('prōt ə 'sel) 261
coacervate droplet (kō 'as ər ,vāt 'dräp lət) 261
saprophytic decomposers (,sap rə 'fit ik ,dē kəm 'pō zer) 267
cyanobacterium (,sī ə nō bak 'tir ē əm) 268

heterocyst ('het ə rō ,sist) 268
lichen ('lī kən) 268
colony ('käl ə nē) 270
amoeboid (ə 'mē 'böid) 270
trichocyst ('trik ə ,sist) 272
hypha ('hī fə) 275
mycelium (mī 'sē lē əm) 275
fruiting body ('frü tiŋ 'bäd ē) 275

Evolution in Plants

Concepts

1. Adaptation to the environment accounts for observed differences among plants.
   a. Some plants are adapted to an aquatic and some are adapted to a terrestrial existence.
   b. The life cycles of plants are also adapted to either an aquatic or terrestrial habitat.
2. Later appearing plants evolved from plants that appeared earlier in the history of life.

   The complex life cycle of seed plants evolved from the life cycles of earlier ancestral plants.

Plants can be defined as living organisms that carry on photosynthesis and cannot move about voluntarily. We will include in this kingdom multicellular photosynthetic forms and those unicellular ones that appear to be very closely related to them.

Plants began to evolve during the Paleozoic era (table 14.1). They first appeared in the seas but eventually became adapted to a land existence. In this chapter survey of present-day plants, we will trace the steps by which adaptation to a land existence may have occurred. In any attempt to trace an evolutionary pathway by studying present-day organisms, it is important to realize that all groups of organisms are continually evolving. In general, the actual organisms alive today are not ancestral to another group. However, certain present-day organisms may closely resemble an ancestor of another group. Therefore, it is permissible to use living groups of organisms to discover how evolutionary adaptations may have taken place.

Algae

During the early Paleozoic era, when the seas were warm and covered much of the land, the algae reigned supreme. The term algae, as explained previously, refers to aquatic photosynthesizing organisms. The term is old but not very exact. In this text the blue-green algae, now called cyanobacteria, are classified in the kingdom Monera; euglenoids, dinoflagellates, and diatoms, being exclusively unicellular, are classified in the kingdom Protista. Green algae, brown algae, and red algae are classified as members of the kingdom Plantae. The algae placed in the plant kingdom all have multicellular representatives.

Table 14.1 The Geological Time Scale: Some Major Evolutionary Events of Plants

Era	Period	Millions of Years Ago	Major Biological Events Involving Plants	
Cenozoic	Quaternary		Increase in number of herbaceous plants	Age of Angiosperms
		2		
	Tertiary		Dominance of land by angiosperms	
		65		
Mesozoic	Cretaceous		Angiosperms spread	
		130	Gymnosperms decline	
	Jurassic		First angiosperms appear	Age of Gymnosperms
		180		
	Triassic		Dominance of land by gymnosperms and ferns	
		230		
Paleozoic	Permian		Land covered by forests of primitive vascular	
		280	plants	
	Carboniferous		Age of great coal-forming forests, including club mosses, horsetails,	Swamp Forests
		350	and ferns	
	Devonian		Expansion of primitive	
		400	vascular plants over land	
	Silurian		Primitive vascular plants	
		435	appear on land	
	Ordovician		Nonvascular marine plants	Age of Algae
		500	abundant	
	Cambrian		Unicellular marine algae	
		600	abundant	

Green Algae

Green algae (**division Chlorophyta**) are believed to be ancestral to the first terrestrial plants because both of these groups possess chlorophylls a and b, store reserve food as starch, and have cell walls that contain cellulose. The green algae are a diverse group that ranges from simple to complex in both structure and sexual reproduction. Most green algae have the life cycle depicted in figure 14.1.

Flagellated Green Algae

The genus *Chlamydomonas* contains unicellular green algae that move by means of a pair of flagella. The single, large, cup-shaped chloroplast contains a conspicuous pyrenoid, which is the site of starch production. The stigma, or "eyespot," is a portion of the chloroplast that aids the organism in moving toward the light. There are two small "contractile vacuoles" at the base of the flagella that rhythmically discharge water.

Chlamydomonas algae reproduce both sexually and asexually (fig. 14.2). During asexual reproduction, an adult cell divides mitotically and produces daughter cells within the parent cell. The parent cell wall then ruptures and releases small flagellated cells called **zoospores.** Each zoospore develops directly into an adult *Chlamydomonas* cell.

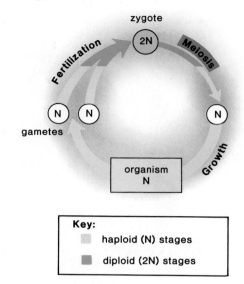

Figure 14.1
Diagram of the sexual life cycle seen in many green algae. The diploid zygote undergoes meiosis; therefore, the only adult stage is haploid. Contrast this life cycle to that diagrammed in figure 14.7a, in which the zygote does not undergo meiosis.

Key:
haploid (N) stages
diploid (2N) stages

Figure 14.2
The structure and life cycle of *Chlamydomonas*, a motile green alga. During asexual reproduction, all structures are haploid; during sexual reproduction, only the zygote is diploid.

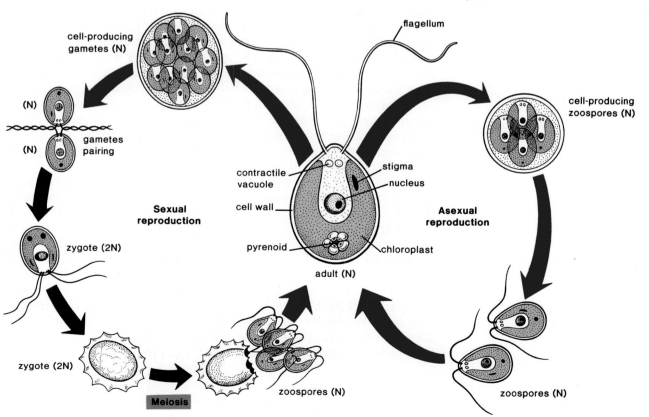

Figure 14.3
The genus *Volvox* contains typical green algal colonies. *a.*
The adult often contains daughter colonies, which are
asexually produced by special cells. *b.* During sexual
reproduction, colonies produce a definite sperm and egg.
Some produce either sperm or egg, and others produce
both sperm and eggs.
Photograph by Carolina Biological Supply Company.

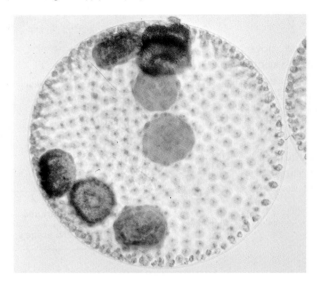

a.

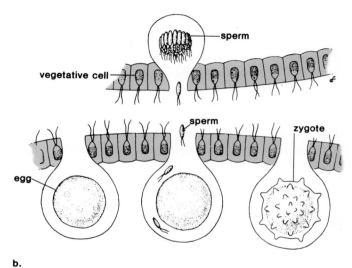

b.

During sexual reproduction, the adult cell divides to produce a number of gametes that have flagella and resemble the adult cell. Gametes of different mating types fuse to form a zygote that secretes a thick wall and becomes dormant. In this state the zygote can withstand adverse conditions, such as a cold winter or dried-up pond. The zygote germinates when conditions improve, and meiosis occurs, producing four haploid flagellated cells. Each of these matures into an ordinary adult, and the sexual cycle is complete.

The zygote is the only diploid stage in the *Chlamydomonas* life cycle. The zoospores, adult cells, and gametes are all haploid. The gametes, identical in appearance, are called **isogametes.** This life cycle, including the nonspecialized gametes, may represent the original pattern of sexual reproduction in plants (fig. 14.1).

A number of **colonial** (loose association of cells) forms occur among the flagellated green algae. *Volvox* is considered the most complex genus among these colonial green algae. A *Volvox* colony is a hollow sphere with thousands of cells arranged in a single layer surrounding a watery interior. The cells of a *Volvox* colony, each one of which resembles a *Chlamydomonas* cell, cooperate in that the flagella beat in a coordinated fashion. Some cells are specialized for reproduction, and each of these can divide asexually to form a new daughter colony (fig. 14.3*a*). This daughter colony resides for a time within the parental colony. A daughter colony leaves the parental colony by releasing an enzyme that dissolves away a portion of the matrix of the parental colony, allowing it to escape. During sexual reproduction there are heterogametes— large nonmotile eggs and small flagellated sperm (fig. 14.3*b*).

Figure 14.4
Members of the genus *Spirogyra* are filaments of cells. *a.*
Anatomy of one cell. *b.* Micrographs depicting
conjugation.

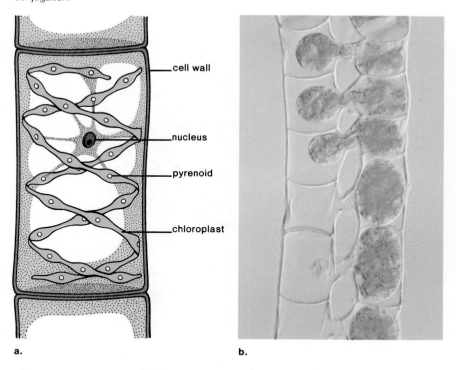

a.
b.

Filamentous Green Algae

Filaments (end-to-end chains of cells) form when cells remain attached after cell division in only one plane.

The genus *Spirogyra* contains common, freshwater, filamentous green algae that take their name from their ribbonlike, *spiral* chloroplasts (fig. 14.4). Asexual reproduction occurs when a filament breaks up and each piece begins to produce new cells. Sexual reproduction, called **conjugation,** occurs when two filaments line up next to one another and conjugation tubes form between their respective cells. The contents of the cells of one filament move into the cells of the other filament, forming diploid zygotes. The zygote can survive the winter, and in the spring it undergoes meiosis to produce a new haploid filament. Just as with *Chlamydomonas,* the zygote is the only diploid cell in the entire life history of the spirogyra.

Among the filamentous algae, there are plants that use heterogametes during sexual reproduction. For example, the genus *Oedogonium* contains filamentous algae in which the cells have cylindrical and netlike chloroplasts. Sexual reproduction (fig. 14.5) occurs when an enlarged specialized cell produces an egg, and other short, disk-like cells each produce two sperm. The sperm, which look like small zoospores, escape and swim to an egg, after which the zygote is released and enters a period of dormancy. Upon germination, the zygote produces four zoospores, each of which may grow into a filament. Also, any vegetative cell may produce a zoospore asexually, and this will develop directly into a filament.

Figure 14.5
Members of the genus *Oedogonium* are filaments that reproduce both asexually and sexually. During sexual reproduction depicted here, the motile sperm swim to the stationary eggs. The zygote is the only diploid portion of the cycle; all the other structures are haploid.

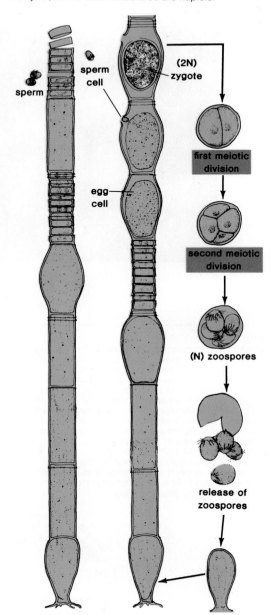

Figure 14.6

a. Ulva. b. Members of the genus *Ulva* undergo alternation of generations in which the sporophyte generation and the gametophyte generation have the same appearance. The sporophyte generation produces spores following meiosis, and the gametophyte generation produces gametes that join to form a zygote.

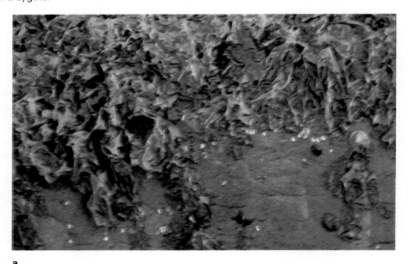

a.

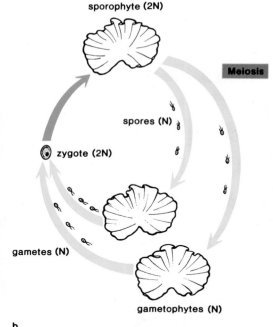

sporophyte (2N)

Meiosis

spores (N)

zygote (2N)

gametes (N)

gametophytes (N)

b.

Figure 14.7

a. Alternation of generations life cycle. In this life cycle, the diploid sporophyte generation produces haploid spores, each of which becomes a gametophyte generation that produces gametes. The haploid gametophyte generation produces gametes. In land plants, the gametes are specialized as the sperm and egg. Following fertilization, the zygote is retained within the body of the gametophyte generation. *b.* Highly magnified photo of *Chara,* a stonewort. The stoneworts have heterogametes and although they have the life cycle depicted in figure 14.1, they do protect the zygote for a while. On the basis of this and recently discovered biochemical evidence, it is suggested that land plants may have evolved from ancestors to the stoneworts sometime during the Silurian period.

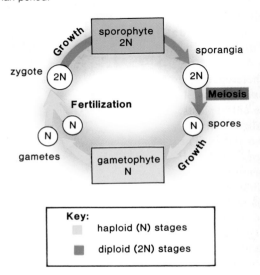

Growth

sporophyte 2N

sporangia

zygote   2N

2N

Meiosis

Fertilization

N   spores

N

N

gametes

gametophyte N

Growth

Key:
haploid (N) stages
diploid (2N) stages

a.

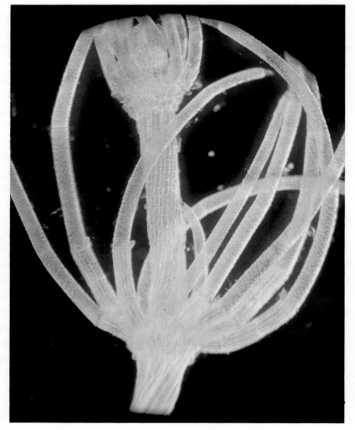

b.

Multicellular Sheets

The genus *Ulva* contains green algae found in the sea close to shore. *Ulva* is commonly called sea lettuce because of its leafy appearance (fig. 14.6). This plant has a two-generation life cycle known as **alternation of generations.**

1. The **diploid sporophyte generation** produces spores by meiosis.
2. The **haploid gametophyte generation** produces gametes.

In *Ulva,* both the gametophyte and sporophyte generations look exactly alike. However, each haploid zoospore produced by the sporophyte generation develops directly into a gametophyte generation, while gametes produced by the gametophyte generation fuse to give a zygote that develops into the diploid sporophyte generation. Notice that the life cycles of both *Chlamydomonas* and *Ulva* contain flagellated zoospores that are capable of swimming. This is an adaptation for reproduction in water.

Origin of Land Plant Life Cycle Like *Ulva,* land plants are multicellular and undergo alternation of generations (fig. 14.7). Therefore, there are those who believe that *Ulva* is ancestral to land plants. Biochemical evidence has now suggested, however, that land plants are most closely related to the green algae known as stoneworts. (Stoneworts have bodies that are encased by calcium carbonate deposits.) Although these plants have the life cycle depicted in figure 14.1, they do have heterogametes and retain the zygote within the haploid parental body for a time. Protection of the zygote is seen in all land plants, and possession of this feature in stoneworts may very well have been the first step toward the type of alternation of generations displayed by land plants.

Seaweeds

Multicellular green algae (such as *Ulva*), red algae, and brown algae are all seaweeds. Their color is dependent on the pigments they contain, which in red and brown algae mask the green color of their chlorophyll. These pigments enable the algae to collect the light they need for photosynthesis. When light strikes seawater, the various wavelengths (p. 99) penetrate to different depths. The accessory pigments in red algae absorb those wavelengths that penetrate deepest, those in brown algae absorb wavelengths that penetrate next deepest, and green algae depend on wavelengths that are quickly filtered out of water. Thus it is possible to predict the depth at which these various algae will be found.

Brown Algae

Brown algae (**division Phaeophyta**) range from small plants with simple branched filaments to large plants between fifty and one hundred meters long. Large brown algae are most often found along the rocky shoreline in the north temperate zone where they are pounded by waves as the tide comes in and exposed to drying at low tide. When the tide is in, these plants are firmly anchored by holdfasts, and their broad, flattened blades are buoyed by air bladders. When the tide is out, they do not dry out because their cell walls contain a mucilaginous, water-retaining material that helps prevent drying.

Both *Laminaria,* a type of kelp, and *Fucus* (fig. 14.8), known as "rockweed," are examples of brown algae that grow along the shoreline attached to rocks. In deeper water, giant kelp, such as *Macrocystis,* often forms spectacular underwater forests. *Sargassum natans* is a free-floating form that makes up much of the dense floating mat of algae in the area known as the Sargasso Sea, a huge eddy in the Atlantic Ocean between the West Indies and the Azores.

Figure 14.8
Representative brown algae drawn smaller than real-life size. *a. Laminaria* (kelp). Notice the blade contains specialized transport tissue as do the bodies of advanced plants. *b. Fucus* (rockweed). The air bladders keep the body afloat in the water; the receptacles contain the sex organs. Fucus has a life cycle more similar to that of animals than land plants because the adult body is always diploid.

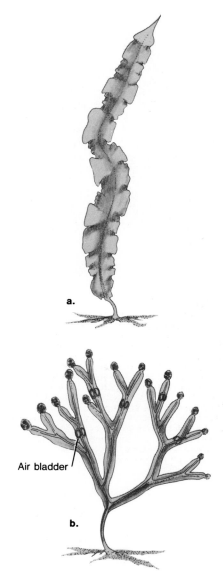

Air bladder

Figure 14.9
Representative red alga (*sebdenia polydactla*). Red algae
are smaller and more delicate than brown algae. Like the
brown algae, though, they too tend to have an advanced
life cycle.

Brown algae provide food and habitat for many marine animals, and
some kelps are used as human food in several parts of the world. Brown algae
have been harvested and processed for fertilizer because they have a high nitrogen content. Algin, a pectinlike material, can be extracted from brown algae.
Algin is added to ice cream, sherbet, cream cheese, and other products to give
them a stable, smooth consistency. It has many other commercial uses.

Red Algae

Red algae (**division Rhodophyta**) (fig. 14.9) are generally found in warm and
deep waters. The water supports and protects their lacy, delicate bodies, which
may be up to a meter long. Coralline algae is a type of red algae that has cell
walls impregnated with calcium carbonate. In some instances, it contributes
as much to the growth of coral reefs as coral animals do.

Red algae are economically important because valuable colloidal substances can be extracted from the outer layer of their cell walls. The best known
of these, **agar,** is used in culture media for bacteria and fungi. It and other
colloids from red algae are used in many foods to produce a smooth consistency and to help retain moistness. Red algae are also used as food in Scotland,
the Orient, and elsewhere.

Adaptation to Land

Prior to the Silurian period (table 14.1), aquatic algae were the only plants
in existence. Since that time, plants (fig. 14.10) have become adapted to living
on the land.

A land existence offers some advantages to plants. One advantage is the
greater availability of light for photosynthesis since even clear water filters
out light. Another advantage is that carbon dioxide and oxygen are present in
higher concentrations and diffuse more readily in air than in water.

Many adaptations (as discussed in detail in chapter 17) are required,
however, for plants to live successfully on land. An internal skeleton is needed
to oppose the force of gravity and to lift the leaves up toward the sun. Plants
must have a way to obtain an adequate supply of water and to transport water
up to and nutrients down from the leaves. Since they are surrounded by air,

Figure 14.10
Representatives of the dominant types of land plants on earth today. *a.* Most plants today are flowering plants. This is a tiger lily. *b.* Mosses are only partially adapted to land. This is haircup moss. *c.* Ferns were much more prominent in days gone by. This is bracken fern. *d.* The gymnosperms are seed plants as are flowering plants. This is a conifer in which the seeds are borne on cones.

a.

b.

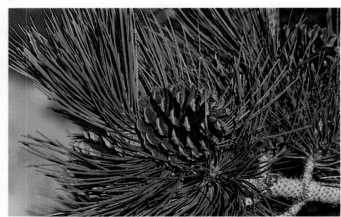

c.

d.

desiccation, or drying out, is a constant threat. Not only the body but also the gametes, zygote, and embryo need to be protected from desiccation. Table 14.2 indicates how the organs of flowering plants are specialized to meet the needs of a land environment.

There are two main groups of plants adapted to living on land. These two groups—the **bryophytes** and **tracheophytes**—can be compared in these ways:

Bryophytes	Tracheophytes
Gametophyte dominant	Sporophyte dominant
No vascular tissue	Vascular tissue (transport)

The dominant generation is the conspicuous generation in the life cycle of a higher plant, the one that lasts longer and is usually considered *the* plant by the layperson.

Table 14.2 Specialization of Flowering Plant Organs

Organ	Function
Roots	Anchor plant and obtain water and minerals from soil.
Stems	Oppose force of gravity by supporting leaves and transport water to and nutrients from leaves.
Leaves	Carry on photosynthesis.
Reproductive structures	Protect gametes, zygote, and embryo from desiccation.

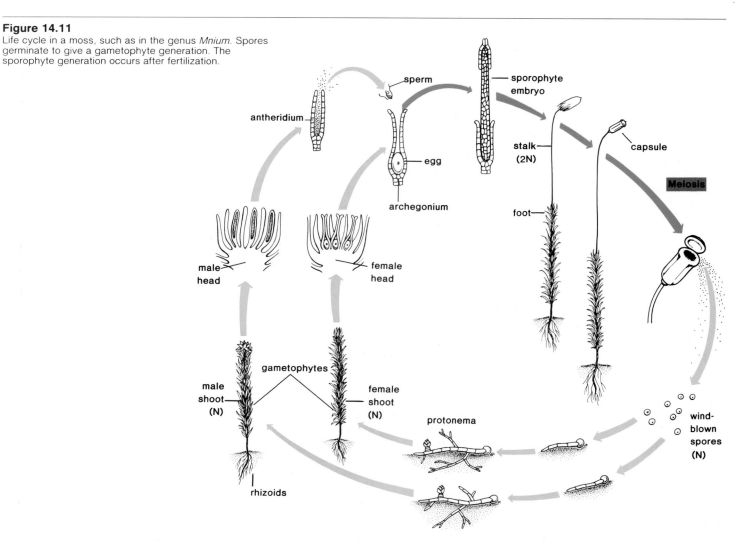

Figure 14.11
Life cycle in a moss, such as in the genus *Mnium*. Spores germinate to give a gametophyte generation. The sporophyte generation occurs after fertilization.

sperm

antheridium

sporophyte embryo

egg

stalk (2N)

capsule

archegonium

Meiosis

foot

male head

female head

gametophytes

male shoot (N)

female shoot (N)

protonema

wind-blown spores (N)

rhizoids

Bryophytes

The bryophytes (**division Bryophyta**) include liverworts and mosses. Some **liverworts** have flattened, lobed bodies, but most species are "leafy" and look like mosses. Examination shows that the body of a liverwort has a distinct top and bottom surface, with numerous rhizoids (rootlike hairs) projecting into the soil. In contrast, a **moss** is a stemlike structure with radially-arranged leaflike structures. Rhizoids anchor the plant and absorb minerals and water from the soil. Since bryophytes do not have vascular tissue, *they lack true roots, stems, and leaves*. Instead, they are said to have rhizoids, stemlike structures, and leaflike structures.

Moss Life Cycle

The moss plant just described is the dominant gametophyte generation that produces the gametes. This is the more permanent and longer lasting generation in bryophytes. In some mosses there are separate male and female gametophytes (fig. 14.11). At the tip of a male gametophyte are **antheridia** in which swimming sperm are produced. After a rain or heavy dew, the sperm

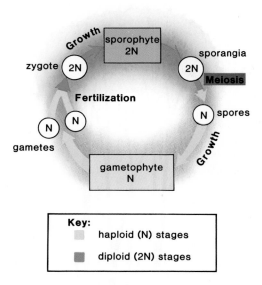

Figure 14.12
Life cycle in bryophytes. There is an alternation of generations in which the gametophyte generation is dominant. Therefore more space is allotted to this generation in the diagram.

swim to the tip of a female gametophyte where eggs have been produced within the **archegonia.** Antheridia and archegonia are both multicellular sex organs, and each has an outer layer of jacket cells that help protect the enclosed gametes from desiccation. After an egg is fertilized, it is retained within the archegonium and begins development as the sporophyte generation. The *sporophyte generation,* which is parasitic on the gametophyte, consists of a *foot* that grows down into the gametophyte tissue, a *stalk,* and an upper capsule or *sporangium,* where meiosis occurs and haploid spores are produced. In some species of mosses, a hoodlike covering is carried along upward by the growing sporophyte. When this covering and a lid to the capsule fall off, the spores are mature and ready to escape. The release of spores is controlled by one or two rings of "teeth" that project inward from the margin. The teeth close the opening when the weather is wet but curl up and free the opening when the weather is dry. This appears to be the mechanism that allows spores to be released at times when they are more likely to be distributed by the wind.

When a spore lands on an appropriate site, it germinates. A single row of cells grows out and then branches. This algalike structure is called a **protonema.** After about three days of growth under favorable conditions, buds appear at intervals along the protonema. Each of these sends down rhizoids and grows upward giving the generation we call a gametophyte. This completes the moss life cycle, which is typical of a bryophyte life cycle (fig. 14.12).

Adaptation

Bryophytes are incompletely adapted to life on land. To prevent water loss, the entire body is covered by a waxy cuticle. The zygote and embryo are also protected from drying out by remaining within the archegonium. The organism is dispersed or distributed to new locations by windblown spores. However, the bryophytes are limited in their adaptations. They lack vascular tissue, and the sperm have to swim in external moisture to reach the egg. It is for these reasons that bryophytes are restricted to moist locations.

Bryophytes and tracheophytes probably share a common ancestor, but whereas the bryophytes never developed vascular tissue, this feature did develop among the tracheophytes.

Figure 14.13
Rhynia major, the simplest and earliest-known vascular plant. The leafless stem was green and carried on photosynthesis. The terminal sporangia apparently released their spores by splitting longitudinally.

Figure 14.14
Generalized life cycle in tracheophytes. There is an alternation of generations in which the sporophyte generation is dominant. Therefore more space is allotted to this generation in the diagram.

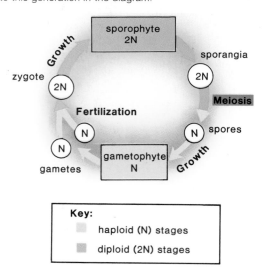

Tracheophytes

There is an extinct plant (*Rhynia major*) in the fossil record that may be ancestral to all tracheophytes (division Tracheophyta) because it first gives evidence of vascular tissue (fig. 14.13). The tracheophytes have two types of vascular tissue. **Xylem** conducts water and minerals up from the soil, and **phloem** transports organic nutrients from one part of the body to another. Because they have vascular tissue, the specialized body parts of tracheophytes can properly be called roots, stems, and leaves.

Among the tracheophytes, the sporophyte, or diploid generation is dominant (fig. 14.14). This is the generation that has vascular tissue. Xylem, with its strong-walled cells, supports the body of the plant against the pull of gravity. The tallest organisms in the world are tracheophytes—the redwood trees of California.

The tracheophytes include those plants that are best adapted to a land existence. One advantage of having the sporophyte dominant is that this is the diploid generation; if a faulty gene is present, it may be masked by a functional gene. In addition, possession of vascular (transport) tissue is a critical adaptation to the land environment. As we shall see, the higher tracheophytes also possess a means of reproduction that is suitable to a dry environment.

a.

b.

c.

Figure 14.15
The Paleozoic swamp forests (fig. 12.4) contained the primitive tracheophytes: whisk ferns, lycopod trees, and treelike giant horsetails. Today the (*a.*) whisk ferns are represented by *Psilotum;* (*b.*) the club mosses are represented by *Lycopodium;* and (*c.*) the horsetails are represented by *Equisetum.* Their giant relatives died out during the Permian period.

Photograph by Carolina Biological Supply Company (a, c).

Primitive Tracheophytes

During the latter half of the Paleozoic era (table 14.1), great forests (fig. 12.4) covered the swampy land. The dominant plants in these forests were the primitive tracheophytes and ferns. The primitive tracheophytes include the whisk ferns (**subdivision Psilopsida**), clubmosses (**subdivision Lycopsida**) and horsetails (**subdivision Spenopsida**). Except for the whisk ferns, these plants were all enormous and treelike. (This is hard to believe because their descendants are insignificant in size today [fig. 14.15].) They died out so quickly that their bodies did not decay; instead they were covered by sedimentary rock. With time the compression was great enough that these plants became the fossil fuels we burn today.

Figure 14.16
Life cycle of a fern. Spores germinate to form a
gametophyte generation. The sporophyte generation
occurs after fertilization.

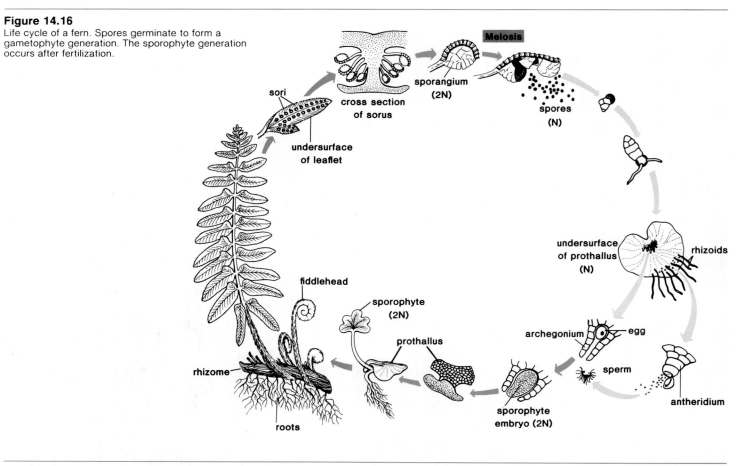

Figure 14.17
Fiddleheads of a fern. Fiddleheads unroll to become the
familiar leafy fronds.

Ferns

Ferns (**subdivision Pteropsida, class Filicineae**) vary in appearance. Figure 14.16 shows a common fern of the temperate zone with a horizontal stem (rhizome) from which hairlike roots project downward and large leaves (fronds) project upward. Young fronds grow in a curled up form called fiddleheads (fig. 14.17) that unroll as they grow. The fronds are often subdivided into a large number of leaflets.

Fronds represent the dominant sporophyte generation in ferns. Sporangia develop in clusters called **sori** (sing., **sorus**) (fig. 14.18a) on the underside of the leaflets. Within the sporangia, meiosis occurs and spores are produced. A band of thickened cells (annulus, fig. 14.18b) snaps in response to moisture changes and flings the windblown spores out.

The gametophyte generation is a small, heart-shaped structure called a **prothallus.** Archegonia develop at the notch, and antheridia are at the tip. Spiral-shaped sperm swim from the antheridia to the archegonia (fig. 14.19), where fertilization occurs. A zygote begins its development inside an archegonium, but the embryo soon outgrows the space available there. The young sporophyte becomes visible as a distinctive first leaf appears above the prothallus and the sporophyte's roots develop below it. Often, gametophyte and sporophyte tissues are distinctly different shades of green. The young sporophyte grows and develops into a mature sporophyte, the familiar fern plant.

Adaptation

Ferns are incompletely adapted to life on land due to the water-dependent gametophyte generation. This generation lacks vascular tissue and is separate from the sporophyte generation. Swimming sperm require an outside source of water in which to swim to the eggs in the archegonia (fig. 14.19). Therefore, ferns are most often found in moist environments where water is available for the gametophyte generation. Some ferns, like the bracken fern *Pteridium aquilinum,* once established, can spread by means of vegetative reproduction into drier areas. As the rhizomes grow horizontally in the soil, the fiddleheads grow up as new fronds.

Figure 14.18
Fern sporophyte anatomy. *a.* A photomicrograph of the underside of a leaflet showing sori. *b.* A scanning electron micrograph of sporangia within a sorus. When the rim contracts, a sporangium breaks open and the spores are released (see right).

a.

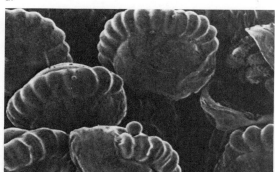

b.

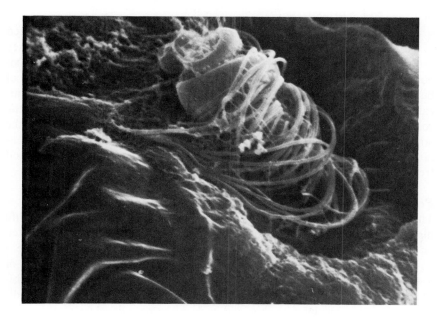

Figure 14.19
A scanning electron micrograph of a sperm, entangled within its flagellum, entering an opening of an archegonium in the bracken fern *Pteridium aquilinum.*
Photograph by Carolina Biological Supply Company.

Figure 14.20

Life cycle in seed plants. The sporophyte generation is dominant and the male and female gametophyte generations are separate.

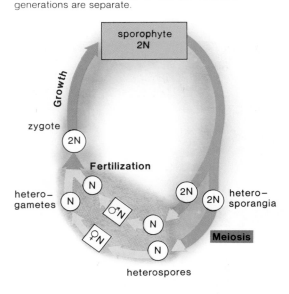

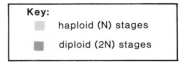

Key:
- ☐ haploid (N) stages
- ■ diploid (2N) stages

Seed Plants

The land was swampy during the late Paleozoic, but it became much drier at the beginning of the Mesozoic (table 14.1). Now the seed plants, which are fully adapted to life on land, came into their own. First the gymnosperms (during the Mesozoic) and then the angiosperms (during the Cenozoic) became the dominant plants on earth. The primary weakness observed in plant life cycles thus far has been the water-dependent gametophyte generation: swimming sperm require outside moisture to swim to the egg. This difficulty has been overcome by the seed plants.

Seed plants produce **heterospores** (fig. 14.20), called microspores and megaspores, instead of homospores, or identical spores. Each microspore develops into a **pollen grain,** an immature male gametophyte, while still retained within a microsporangium. After they are released, pollen grains develop into mature, sperm-bearing male gametophytes. Thus the *pollen grain* in seed plants *replaces the swimming sperm* in nonseed plants. **Pollination,** the transfer of the male gametophyte to the vicinity of the female gametophyte, is dependent on wind or animals, rather than on an external source of water.

Each megasporangium located within an ovule produces only one functional megaspore. Megaspores develop into egg-bearing female gametophytes that are still retained within ovules. After fertilization the zygote becomes an embryonic plant enclosed within a seed. While nonseed plants are dispersed by means of spores, in seed plants the seeds serve to distribute the species. A **seed** contains an embryonic sporophyte generation, plus stored food, enclosed within a protective seed coat. Seeds are resistant to adverse conditions, such as dryness or temperature extremes.

There are two groups of seed plants, the gymnosperms ("naked seeds") and the angiosperms ("enclosed seeds"). Angiosperm seeds are enclosed in fruits, but gymnosperm seeds are not.

Gymnosperms

Gymnosperms (subdivision Pteropsida, class Gymnospermae) include the conifers—pine, cedar, spruce, fir, and redwood trees—and a few other less familiar plants. Many biologists think that these other plants—cycads, ginkgo, and welsitschia—are so dissimilar from the conifers that they should be put in a separate class entirely.

The conifers (fig. 14.21) have evergreen needlelike leaves well adapted to withstand not only hot summers but also cold winters and high winds. They are able to grow on large areas of the earth's surface and are therefore economically important. They supply much of the wood used for construction of buildings and production of paper. They also produce many valuable chemicals, such as those extracted from resin, a waxy substance that protects the conifers from attack by fungi and bark beetles. Perhaps the oldest and the largest trees in the world are conifers. Bristlecone pines in the Nevada mountains are known to be more than 4500 years old, and a number of redwood trees in California are 2000 years old and more than 90 m tall.

Pine Life Cycle

The life cycle illustrated in figure 14.22 is a good example of a conifer's life history. The sporophyte generation is dominant, and its sporangia are located on the scales of the cones. There are two types of cones—male and female.

Typically, the male cones are quite small and develop near the tips of lower branches. Each scale of the male (pollen) cone has two or more microsporangia on the underside. Within these sporangia, each meiotic cell division produces four **microspores,** and each microspore develops into a pollen

Figure 14.21
Conifer trees bear cones, as exemplified here by hemlock. There are actually two types of cones—male and female. The male cones release pollen, the sperm-producing male gametophyte generation. The female cones, such as these, produce seeds, each containing a new sporophyte generation capable of maturing into a full-grown tree after dispersal and germination. The winged seeds are said to be naked since the seeds of gymnosperms are not covered by fruit as are the seeds of angiosperms.

Figure 14.22
Life cycle of pine such as those in the genus *Pinus*. The mature sporophyte (pine tree) has female pine cones, which produce megaspores that develop into female gametophyte generations, and male pine cones, which produce microspores that develop into male gametophyte generations (mature pollen grains). Following fertilization, the immature sporophyte generations are present in seeds located on the female cones.

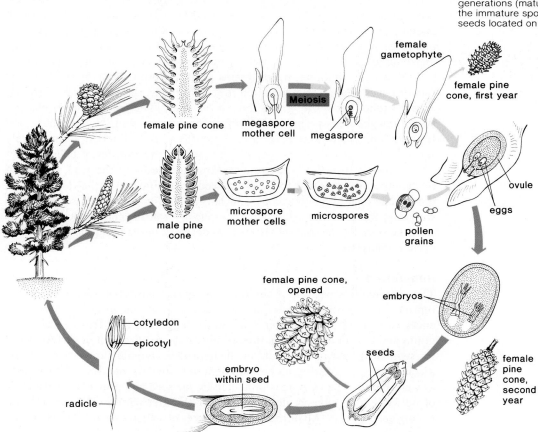

Figure 14.23
Longitudinal section of pine ovule on pine scale.
Pollination has occurred, as evidenced by the presence of
pollen grains (*right*) just outside the ovule (*left*).
Photograph by Carolina Biological Supply Company.

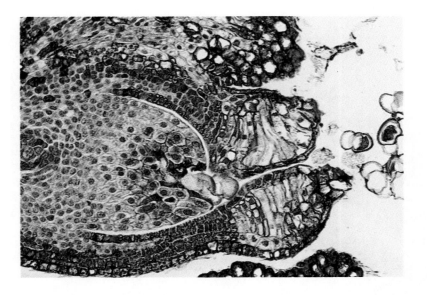

grain, the *male gametophyte generation*. The pollen grain has two lobular wings and is wind blown. Pine trees release so many pollen grains during pollen season that everything in the area may be covered with a dusting of yellow, powdery pine pollen.

The mature female cones are large and located near the top of the tree. Each scale of the female (seed) cone has two ovules that lie on the upper surface. Each ovule (fig. 14.23) is surrounded by a thick, layered coat with an opening at one end. Within the ovule, meiosis produces four megaspores. Only one of these spores develops into a *female gametophyte* with two to six archegonia, each containing a single large egg lying near the ovule opening.

During pollination, pollen grains are transferred from the male to the female cones. Once enclosed within the female cone, the pollen grain develops a pollen tube that slowly grows toward the ovule. The pollen tube discharges two nonflagellated sperm, one of which fertilizes an egg. Notice, then, that fertilization, which takes place fifteen months after pollination, is an entirely separate event from pollination, which is simply the transfer of pollen.

After fertilization, the ovule matures and becomes the seed, composed of the embryo, its stored food, and a seed coat. Finally, in the third season, the female cone, by now woody and hard, opens to release its seeds, whose wings are formed from a thin membranous layer of the cone scale. When the seeds germinate, the sporophyte embryo develops into a new pine tree and the cycle is complete.

Adaptation

The reproductive pattern of conifers has several important advantages over reproduction in other plants that have been considered so far. These differences make the conifers better adapted for terrestrial life. Transfer of pollen grains and growth of the pollen tube eliminate the requirement of surface water for swimming sperm. Enclosure of the dependent female gametophyte inside a cone protects it during its development and shelters the developing zygote as well. Finally, the embryo is protected by the seed and provided with a store of nutrients that support development for the first period of its growth following germination. All of these factors increase chances for reproductive success.

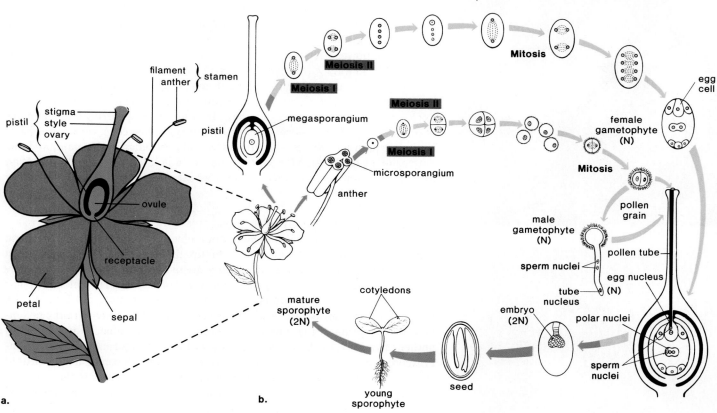

Figure 14.24
Life cycle of flowering plants beginning with the sporophyte generation. *a.* The reproductive parts of a flower are the stamens, whose anthers produce microsporangia, and the pistil(s), each one of which has an ovary within which the ovules contain a megasporangium. *b.* The microsporangia become the pollen grains (male gametophyte generation) that come to rest on the stigma of the pistil following pollination. The megasporangium becomes the female gametophyte generation that produces an egg. After the sperm nuclei travel down the pollen tube, a double fertilization occurs that produces both endosperm and a zygote. Each ovule now matures into a seed, and the ovary develops into a fruit. When the seed germinates and the new sporophyte generation begins to grow, the life cycle is near completion.

Angiosperms

Angiosperms (subdivision Pteropsida, class Angiospermae), or flowering plants (fig. 14.10*a*), are an exceptionally large and successful group of plants. They range in size from tiny pond-surface plants, which are only 0.5 mm in diameter, to very large trees. The oldest fossils definitely recognized as angiosperms come from the Cretaceous period (table 14.1), which began only about 135 million years ago. Yet, during this relatively short evolutionary history, the angiosperms have diversified and increased their numbers tremendously.

In the angiosperms, the micro- and megasporangia are located in the **flower** (fig. 14.24). The flower is advantageous in three ways: it attracts insects and birds that aid in pollination (pollination by wind is also possible); it protects the developing female gametophyte; and it produces seeds enclosed by **fruit**. There are many different types of fruits, some of which are fleshy (e.g., apple) and some of which are dry (e.g., peas in pod). The fleshy fruits are

a.

b.

sometimes eaten by animals, which may transport the seeds to a new location
and then deposit them upon defecation. Fleshy fruits may also provide additional nourishment for the developing embryo. Both fleshy and dry fruits provide protection for the seeds. Many so-called vegetables are actually fruits;
for example, peas enclosed by pods, string beans, and squash. Nuts and berries
and grains of wheat, rice, and oats are also fruits.

Angiosperms have well-developed vascular and supporting tissues that
make them well adapted for terrestrial life. Their xylem tissue is different
from that of other vascular plant groups because they have xylem vessels as
well as tracheids (p. 367). Other vascular plants, including virtually all gymnosperms, have only tracheids in their xylem. Whereas the gymnosperms are
softwood trees, **woody** angiosperm trees have hardwood. Nonwoody angiosperms are called **herbaceous** plants.

Angiosperms are divided into two subclasses: **dicotyledons** (dicots) and
monocotyledons (monocots) (p. 363). The dicots are either woody or herbaceous, with flower parts usually in fours and fives, net veins, vascular bundles
arranged in a circle within the stem, and two cotyledons, or seed leaves. Dicot
families include many familiar plant groups, such as the buttercup, mustard,
maple, cactus, carnation, pea, and rose families. The rose family includes roses,
apples, plums, pears, cherries, peaches, strawberries, raspberries, and a number
of other shrubs. The monocots are almost always herbaceous, with flower parts
in threes, parallel leaf veins, scattered vascular bundles in the stem, and one
cotyledon, or seed leaf. Monocot families include lilies, palms, orchids, irises,
and grasses. The grass family includes wheat, rice, corn (maize), and other
agriculturally important plants.

Flowering Plant Life Cycle

A flower (fig. 14.24) consists of highly modified leaves arranged in concentric
rings that are attached to a modified stem tip, the **receptacle.** The **sepals,** which
forms the outermost ring, are frequently green and quite similar to ordinary
foliage leaves. Next come the **petals,** which are often large and colorful. Their
color often attracts such pollinators as insects and birds (fig. 14.25). Within
the petals, the **stamens** form a whorl around the pistil. Stamens are the pollen-
producing portion of the flower. Each stamen has a slender filament with an
anther at the tip. In the anther, meiosis within microsporangia produces microspores (fig. 14.24). Each microspore divides and becomes a binucleated
pollen grain, the male gametophyte generation. One nucleus is the tube nucleus, and the other is the generative nucleus. The pollen grains are either
blown by the wind or carried by pollinators (fig. 14.25) to the **pistil.** The pistil
of most flowers has three parts—the **stigma, style,** and **ovary.** The ovary contains from one to many **ovules,** depending on the species of plant. Each ovule
contains a megasporangium where meiosis results in one functional megaspore. The latter develops into a multicelled female gametophyte generation.
One of these cells is an egg cell.

When pollen grains are transferred to the stigma, each develops a pollen
tube that penetrates the style and ovary. The generative nucleus enters the
pollen tube and divides to give two sperm nuclei that travel to an opening in
the ovule. One sperm nucleus fertilizes the egg nucleus and the other unites
with two other nuclei (the polar nuclei) of the female gametophyte to form
endosperm, food for the embryo. This so-called double fertilization is unique
to angiosperms. The mature ovule or seed contains an embryo and food enclosed within a protective seed coat. The wall of the ovary and sometimes adjacent parts develop into a fruit that surrounds the seeds. Thus angiosperms
are said to have *covered seeds.*

Adaptation

Like the gymnosperms, angiosperms are well adapted to a land environment. Their vascular tissue is more complex and the seeds are enclosed by fruits. Both of these are selective advantages for the angiosperms.

Extinctions

As the reading below discusses, many flowering plants are now on the verge of extinction. This is caused not by a change in climate or a bombardment by comets, but by the activities of humans. There are many arguments for attempting to preserve all plants, some of which are discussed in the reading.

The California condor, the Maryland darter, the Florida panther and other animals struggling to survive are not the only endangered species. Largely because of man's encroachment, many, perhaps dozens of American plant species are disappearing each year. Indeed, botanists estimate that some 3,000 of the 22,000 species of higher plants native to the U.S. may be facing extinction. Around the world, as many as 40,000 plant species are in trouble.

Now help is on the way, at least for America's vegetation, in the form of the Center for Plant Conservation, which has its headquarters at Harvard University's Arnold Arboretum. With seed money of $500,000, the center has begun an unprecedented program, by far the most comprehensive to date, that aims to preserve every kind of threatened plant in the U.S.

Through a network of 18 affiliated botanical gardens and horticultural research facilities in 14 states, the center this summer coordinated the collection of 92 threatened species, including such exotic plants as the pygmy fringe tree and Gray's lily. It plans to at least double that number next year and hopes to have specimens of most of the nation's endangered plants secured in greenhouses and other protected environments within ten years.

As added insurance, the center will stockpile seeds of some of the species at the Department of Agriculture's Fort Collins, Colo., seed-storage facility. That way, says Frank Thibodeau, the center's scientific director, "despite power losses, hurricanes, fires or any other natural disaster that could befall a greenhouse or garden, we will always have the seeds available for study and propagation."

By growing these rare plants, the center expects eventually to reintroduce some into their natural habitats and to satisfy the needs of both researchers and collectors. The collectors, oddly enough, have contributed to the near extinction of several species. One victim is the Knowlton cactus, the first endangered species cataloged by the center. Says Donald Falk, the center's administrative director: "Collectors will go out and decimate populations, uprooting the cactus to send it back to live on windowsills."

Why spend money and energy to save, say, the frostweed or the small whorled pogonia? Medical benefits alone, says Thibodeau, could justify the center's efforts: "Well over a quarter of all prescription medicines in the U.S. are based on plant products." He points, for example, to antitumor alkaloids found in the Madagascar periwinkle that are now used in the treatment of childhood leukemia and Hodgkin's disease. "The question," says Thibodeau, "is whether you're willing to bet that there isn't another important drug out there among those 3,000 plants or whether you're willing to hold the plants long enough to study them."

Then, too, some of the plants may have as yet undiscovered characteristics important to agriculture: for example, resistance to disease or drought. Using new recombinant DNA techniques, scientists look forward to identifying the genes that confer these traits and transferring them from wild plants to crop plants. By preserving the endangered species, says Falk, "we're building a genetic library." Thibodeau considers the library essential "even if it turned out that these plants have no other identifiable value."

The Living Library of Plants

Gray's lily, an example of a rare plant that is now in the protective custody of the Center for Plant Conservation.

Comparison

We have seen how plants have adapted to living on land. Not all land plants, however, are *completely* adapted to life on land. The later a plant evolved, the more complete its adaptation. Table 14.3 indicates the relative degree of adaptation among bryophyte and tracheophyte plants.

The role of the haploid and diploid stages in the life cycle of various plants may be correlated with their relative adaptation to land (fig. 14.26). Algae, whose life cycle contains haploid structures except for the diploid zygote, are adapted to a water environment. Mosses with a dominant gametophyte and small, dependent sporophyte have a limited distribution on land. Ferns have well-developed sporophyte bodies and vascular tissue, but still require very wet conditions for growth of a small, independent gametophyte and fertilization by swimming sperm.

The gymnosperms and angiosperms are widely distributed on land because the large, dominant sporophyte is well adapted to terrestrial life. Furthermore, the delicate spores, gametophyte, gametes, zygotes, and embryos are enclosed within protective coverings produced by the sporophyte plant.

Table 14.3 Adaptation Summary

Plant	Generations	Reproduction
Nonseed Plants: Windblown spores disperse the species.		
Bryophytes	Both generations lack vascular tissue.	Swimming sperm require a source of outside moisture.
Tracheophytes *Primitive tracheophytes*	Sporophyte generation has vascular tissue; gametophyte generation lacks vascular tissue and is separate and independent of the sporophyte.	Swimming sperm require a source of outside moisture.
Seed Plants: Seeds disperse the species.		
Gymnosperms	Fully adapted to land. Gametophyte generation is retained and protected from desiccation by sporophyte.	Pollen grains replace swimming sperm. Windblown seeds.
Angiosperms (seeds covered)	Fully adapted to land, in the same manner as gymnosperms.	Seeds are further protected by fruit. Seeds often not windblown.

Figure 14.26

The relative importance of the haploid (n) and diploid generation (2n) among plants. In most green algae, with the exception of a few such as those in the genus *Ulva*, only the zygote is diploid. In the rest of the plants depicted, the zygote develops into a diploid sporophyte generation. The haploid generation is then called the gametophyte generation.

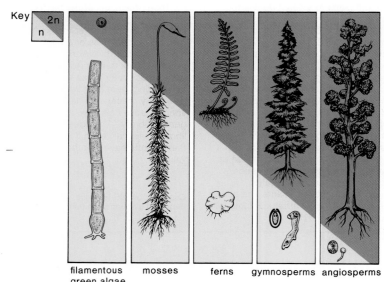

Key
2n
n

filamentous green algae   mosses   ferns   gymnosperms   angiosperms

Summary

The plant kingdom includes multicelluar photosynthetic organisms and closely related unicellular ones. The green algae are a diverse group in which different life cycles are seen. In single cell *Chlamydomonas,* only the zygote is diploid and the gametes are isogametes. In colonial *Volvox* and filamentous *Oedogonium,* there are heterogametes—egg and sperm. In filamentous *Spirogyra,* conjugation occurs. In *Ulva,* there is a diploid sporophyte generation in addition to a haploid generation.

Seaweeds include multicellular green, red, and brown algae. Brown algae are typically found in cool waters along rocky coasts. Red algae are found further from the shore in warm tropical waters.

Until the Silurian period, the algae reigned supreme among plants. Then bryophytes and tracheophytes evolved from green algae and became adapted to a land existence. The bryophytes (e.g., mosses) are not well adapted; the nonvascular gametophyte is dominant and requires external water for swimming sperm. Tracheophytes vary in their degree of adaptation. The vascular sporophyte is dominant in the fern, but there is a separate water-dependent gametophyte and swimming sperm need external water for fertilization to occur. In both mosses and ferns, the species is dispersed by means of spores.

The primitive tracheophytes and ferns grew in great abundance during the second half of the Paleozoic era, but suffered near extinction once the earth became drier during the Mesozoic era. The gymnosperms and then the angiosperms became dominant. Gymnosperms and angiosperms have microspores and megaspores. In the pine, these are located on pine cones. On the male cone, the microspores develop into pollen grains, each a male gametophyte generation. On the female cone, the megaspores are within ovules, where each develops into an egg-producing female gametophyte generation. After pollination, the pollen grain develops a pollen tube within which a sperm travels to an egg. Following fertilization, the ovule becomes a windblown seed.

In flowering plants, the anther portions of the stamens produce pollen grains. The ovules are found in the ovary, a portion of the pistil. After pollination and during fertilization, one sperm nucleus fertilizes an egg and the other initiates development of endosperm. The ovules become seeds, still enclosed by the ovary, which contributes to the development of a fruit. In gymnosperms and angiosperms, seeds disperse the species.

Gymnosperms and angiosperms are well adapted to life on land. The gametophyte generation is small and protected by the vascular, diploid sporophyte generation. The vascular tissue is xylem, which gives trees an internal skeleton that helps them oppose the force of gravity, so that some grow to great height.

Objective Questions

1. In the life cycle of most green algae, the only diploid stage is the _____ .
2. In *Chlamydomonas* the adult is _____ (choose haploid or diploid).
3. *Volvox* has heterogametes, a definite _____ and _____ .
4. In *Spirogyra,* zygotes form following the process of _____ .
5. *Ulva* has the life cycle _____ , as do land plants.
6. Among the seaweeds, *Fucus* is a type of _____ algae.
7. Bryophytes have the _____ generation dominant, while the tracheophytes have the _____ generation dominant.
8. The life cycle of the moss is incompletely adapted to life on land because the sperm must _____ in external water to the egg.
9. In the fern life cycle, the gametophyte generation is independent and _____ from the sporophyte generation.
10. Seed plants have _____ spores and male and female gametophyte generations.
11. Gymnosperms are fully adapted to life on land; the windblown _____ replaces the swimming sperm.
12. Angiosperm have _____ seeds in that they are located inside a _____ .

Answers to Objective Questions

1. zygote 2. haploid 3. sperm and egg 4. conjugation 5. alternation of generations 6. brown 7. gametophyte, sporophyte 8. swim 9. separate 10. heterospores 11. pollen grain 12. covered, fruit.

Study Questions

1. Name all groups of plants studied in this chapter.
2. Describe each of the green algae studied, emphasizing their reproductive patterns, both asexual and sexual.
3. Name several adaptation requirements for a plant living on land. How do the organs of a flowering plant meet these needs?
4. Compare and contrast the bryophytes to the tracheophytes in regard to dominant generation and presence of vascular tissue.
5. Name the primitive tracheophytes. At what time in the history of the earth were they the dominant plants?
6. Compare the life cycles of the moss, fern, pine, and flowering plant, emphasizing adaptation to life on land.
7. Name several plants that are gymnosperms and several that are angiosperms. What do these terms mean?
8. How does the life cycle of seed plants differ from that of nonseed plants?
9. Name several vegetables and grains that would be correctly called fruits.
10. Draw a series of diagrams that illustrate the stage, haploid or diploid, that is dominant in the life cycle of *Chlamydomonas, Ulva,* mosses, ferns, and seed plants. Point out other important differences in these cycles.

Selected Key Terms

zoospore ('zō ə ,spōr) 283
isogamete (,ī sō gə 'met) 284
bryophyte ('brī ə ,fīt) 289
tracheophyte ('trā kē ə ,fīt) 289
xylem ('zī ləm) 292
phloem (,flō em) 292
prothallus ('prō 'thal əs) 295
pollination (,päl ə 'nā shən) 296

gymnosperm ('jim nə ,spərm) 296
angiosperm ('an jē ə 'spərm) 299
dicotyledon ('dī ,kät l 'ēd n) 300
monocotyledon (,män ə ,kät l'ēd n) 300
stamen ('stā mən) 300
pistil ('pis tl) 300
ovary ('ōv ə rē) 300

hile plants are multicellular photosynthetic organisms, animals are *multicellular heterotrophic organisms* that must take in preformed food (fig. 15.1). Unlike the fungi, which rely on external digestion, animals have a central cavity lined with cells that function specifically to obtain and/or absorb nutrients. Animals have a life cycle in which the only multicellular stage is diploid. Meiosis is necessary to production of haploid gametes, which join to give a zygote that develops into the diploid adult. This life cycle was described for a mammal in figure 7.17.

Animals are conveniently divided into **vertebrates**—animals with backbones—and **invertebrates**—animals without backbones. The great majority of animals are invertebrates, and most of these are found in the sea. The evolutionary history of animals (table 16.1) indicates that invertebrates evolved during the Precambrian era and had diversified by the Cambrian period. There are very few fossils for these early animals, presumably because they had soft bodies that rarely left remains.

Figure 15.1
Many animals, including some types of invertebrates, are carnivorous predators. *a.* Spider with dragonfly. *b.* Shrimp feeding on starfish.

a.

b.

CHAPTER 15

Evolution in Animals

Concepts

1. The present-day groups of animals are related.
   It is possible to suggest the manner in which they may have evolved from various common ancestors.

2. Complexity or organization increases from the earliest to the latest evolved.
   A survey of the animal kingdom reveals a gradual increase in complexity.

3. Invertebrates are adapted to various ways of life.
   a. Most groups are adapted to an aquatic existence; only a few major groups are adapted to life on land.
   b. Most are free-living, but some types are parasites of humans and other organisms.

Figure 15.2
Evolutionary tree in animals. Lower invertebrate phyla are
to each side of the lower portion of the tree. In the upper
portion of the tree, the protostomes are found at the left
branch and the deuterostomes are found at the right
branch.

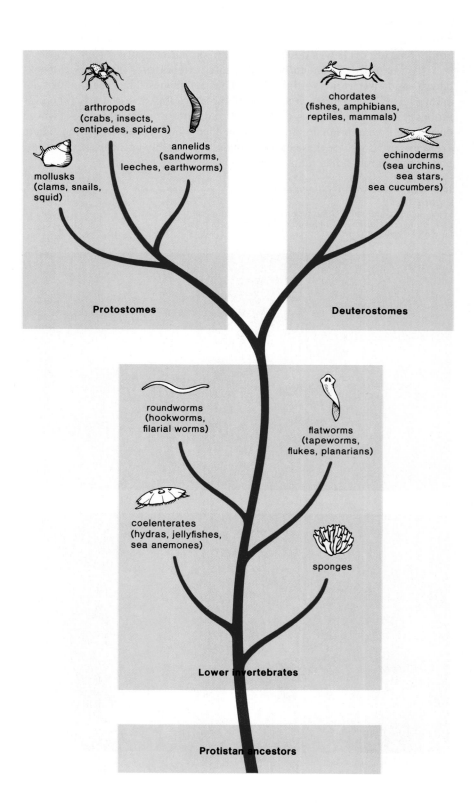

arthropods
(crabs, insects,
centipedes, spiders)

annelids
(sandworms,
leeches, earthworms)

mollusks
(clams, snails,
squid)

chordates
(fishes, amphibians,
reptiles, mammals)

echinoderms
(sea urchins,
sea stars,
sea cucumbers)

Protostomes

Deuterostomes

roundworms
(hookworms,
filarial worms)

flatworms
(tapeworms,
flukes, planarians)

coelenterates
(hydras, jellyfishes,
sea anemones)

sponges

Lower invertebrates

Protistan ancestors

Without an adequate fossil record, biologists have developed possible evolutionary trees such as the one in figure 15.2, based on a study of present-day animal groups. Although classification systems differ,[1] most authorities recognize approximately thirty animal phyla. We will consider only nine, those that most authorities believe are the major ones. These nine phyla have been divided into three groups: the **lower invertebrates,** the **protostomes,** and the **deuterostomes** (fig. 15.2). These groups are distinguished by developmental differences that result in degrees of organizational complexity. These differences will be discussed as we consider each group of animals. The lower invertebrates and protostomes are covered in this chapter, and the deuterostomes are covered in the next chapter, chapter 16.

Sponges

It is possible that sponges (**phylum Porifera**) evolved from a colony of protozoan cells. Like a colony, sponges have a flexible body organization. In 1907, H. V. Wilson cut sponges into small pieces and squeezed the pieces inside bags so that individual cells were separated from each other. These cells, if left undisturbed, moved about and reaggregated to give small but normally functional sponges. Such an ability to reorganize has not been demonstrated in any other type of animal.

Sponges are **sessile filter feeders;** they stay in one spot and filter their food from the water that enters the central cavity through **pores** in the body wall. This body wall is covered by **epidermal cells** (fig. 15.3). The flagella of

[1]A classification of organisms is given in the appendix.

Figure 15.3
Sponge diversity and anatomy. *a.* Yellow tube sponges. *b.* Generalized sponge anatomy. The arrows indicate the flow of water through the animal.

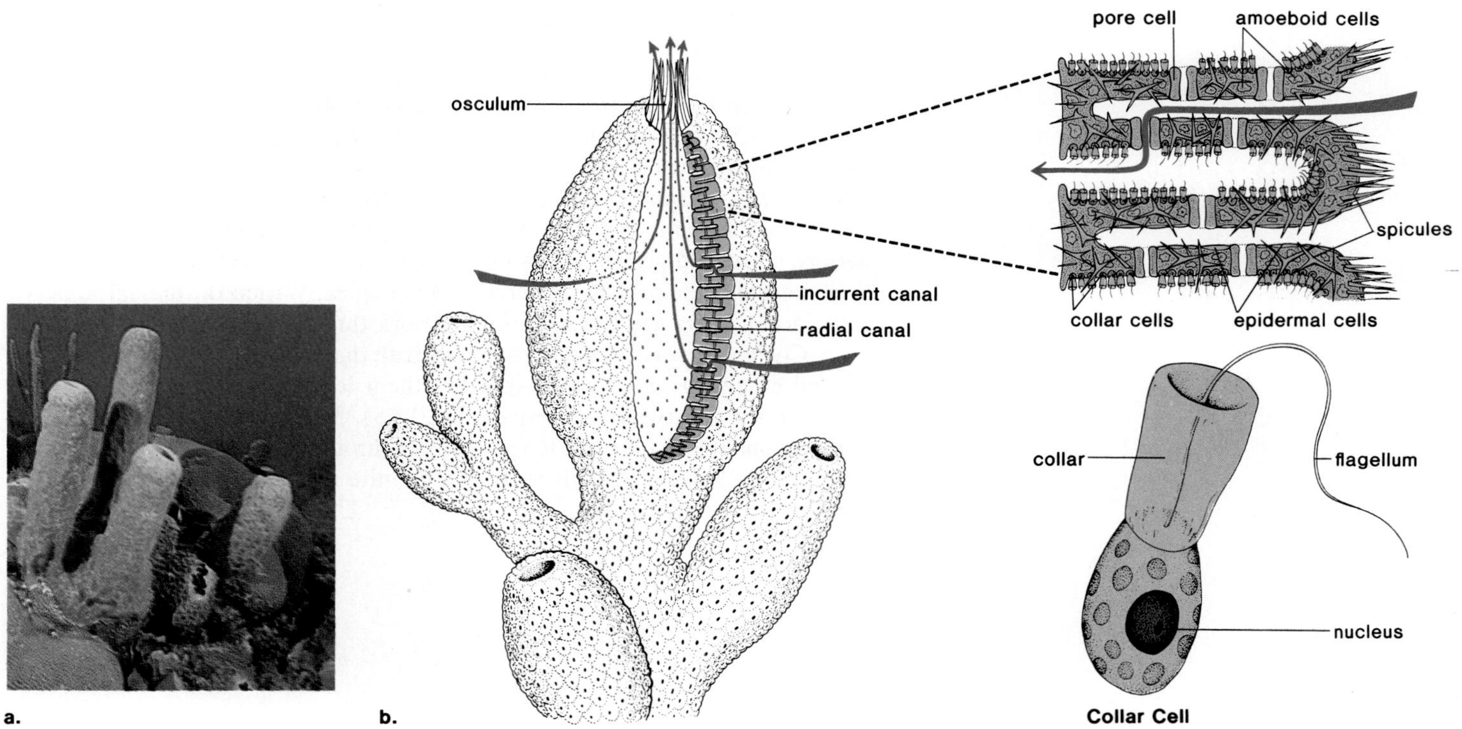

Figure 15.4
Sponge spicules.

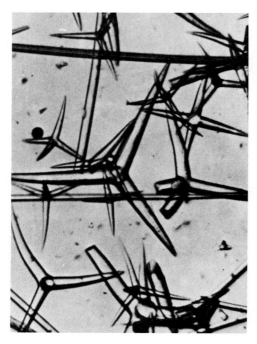

the **collar cells** beat to keep the water moving so that it exits by way of the **osculum,** a single opening. Simple sponges, only 10 cm tall, are estimated to filter as much as 100 l of water each day. The food particles carried by water are engulfed and digested by the collar cells in food vacuoles, or are passed to the **amoeboid cells** for digestion. The amoeboid cells not only act as a circulatory device to transport nutrients from cell to cell, they also produce **spicules** and the gametes. Fertilization results in a zygote that develops into a ciliated **larva** (independent embryonic stage) capable of swimming to a new location. Sponges also reproduce by budding, a process that can produce large colonies of sponges. It is not surprising, then, that sponges can regenerate or regrow an entire organism from a small portion.

Many sponges are simple and small, with pores leading directly from the outside water into the central cavity. Other sponges are more complex and have canals leading to internal pores. Sponges are classified according to the type of spicule. The spicules (fig. 15.4), which act as an *internal skeleton,* are made either of calcium carbonate, silica, or spongin, a fibrous protein. Natural sponges are prepared by beating spongin-containing sponges until all of the living cells are removed and only the skeleton remains. Commercial sponges today, however, are largely synthetic.

Comments

The cellular organization of sponges is different from that of other animals. Also, the main opening of a sponge is used only as an exit, not an entrance; this is unique to sponges. Further, movement is limited to the beating of the flagella, constriction of the osculum, and larval-stage swimming. For these reasons, sponges are believed to be a side branch, and are not believed to have contributed to the evolution of more complex animals. Most likely they and the other truly multicellular animals evolved independently from different protozoan ancestors.

Coelenterates

Coelenterates (**phylum Cnidaria**) and the rest of the animals we will consider are in the mainstream of animal evolution. Coelenterates are a large group of simply organized aquatic animals with a **sac body plan** (fig. 15.10). There is a single opening, a mouth, leading to an internal cavity. This cavity is called the **gastrovascular cavity** because it serves as a location for digestion (gastro) and a means to transport (vascular) nutrients to cells that line the cavity. An outer tissue layer is a protective **epidermis,** while an inner layer, the **gastrodermis,** secretes digestive juices into the gastrovascular cavity. Between these layers is a jellylike packing material, the mesoglea. Within the **mesoglea,** nerve cells form a **nerve net,** a connecting network throughout the body.

Coelenterates are **radially symmetrical;** that is, their bodies are arranged around a central axis so that splitting them lengthwise through the center always results in two equal halves (fig. 15.5). Most other animals are **bilaterally symmetrical;** only one lengthwise cut through the center gives two equal but opposite halves. This gives them a definite right and left half, which the coelenterates do not have.

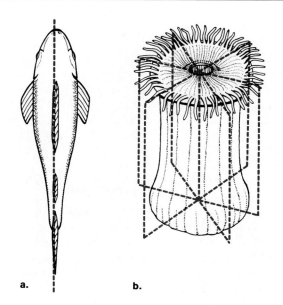

Figure 15.5
Types of animal symmetry. *a.* There is only one plane
along which a bilaterally symmetrical animal can be sliced
to yield two equal but opposite halves. The two equal
halves lend balance to the animal that actively seeks food.
b. Radially symmetrical animals can be sliced on any of
several planes along the main body axis to yield two equal
halves. Sessile inactive animals tend to be radially
symmetrical because this symmetry allows them to reach
out in all directions to capture food.

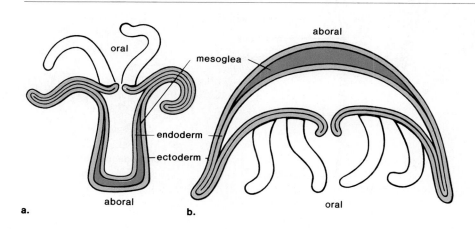

Figure 15.6
The two body forms of coelenterates. The mesoglea is a
packing material between the two germ layers, the
ectoderm and endoderm. *a.* Polyps, with the oral side
uppermost, are usually attached to surfaces. *b.* Medusae,
with the aboral side uppermost, are free-swimming.

The basic coelenterate body plan takes one of two general forms (fig.
15.6). The **medusa** is a bell-shaped jellyfish form that swims about with the
mouth directed downward. In the more slender and usually sessile **polyp,** the
mouth is usually directed upward. Many coelenterates pass through both the
medusa and the polyp forms during their lives.

Coelenterates have special stinging cells. These cells have a fluid-filled
capsule containing a long, spiraled, thread-like fiber called a **nematocyst.** When
the trigger of a stinging cell is touched, the nematocyst is discharged. Some
of these simply entrap the prey; others have spines that penetrate and inject
a paralyzing substance. Once the prey has been subdued, the tentacles about
the coelenterate's mouth maneuver it into the gastrovascular cavity.

Figure 15.7
Hydra anatomy. Note that the nematocyst is contained
within a stinging cell. When a nematocyst is discharged, a
cap opens and the thread shoots out.

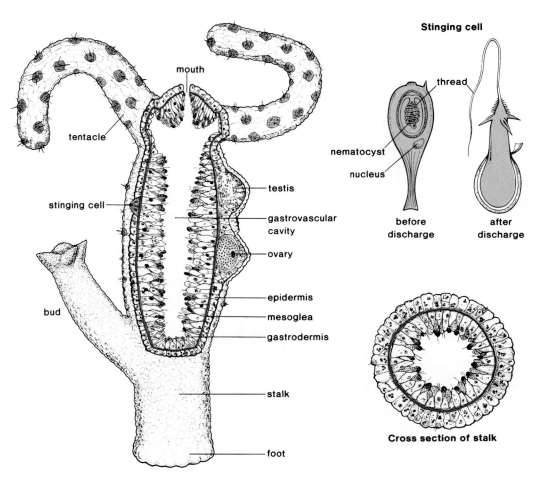

Hydroids

Freshwater hydras (fig. 15.7) are the most familiar hydrozoan coelenterates.
The epidermis of the hydra contains sensory cells, nematocysts, and epitheliomuscular cells. Epitheliomuscular cells contract allowing movement for
feeding and other activities. The gastrodermis also contains contracting cells,
as well as gland cells, that secrete digestive enzymes into the gastrovascular
cavity.

Hydras can reproduce asexually and sexually. They reproduce asexually
by forming **buds,** which are small outgrowths containing the body layers and
an extension of the gastrovascular cavity. When hydras reproduce sexually,
testes develop on the outside of the upper half of the body and an ovary develops lower on the body. The egg, which is produced in the ovary, remains
attached to the adult body; fertilization and early development occur there.
The embryo is surrounded by a hard, protective shell and can stay alive through
unfavorable periods to emerge and produce new polyps.

One of the most unusual hydroids is the Portuguese man-of-war (*Phys-
alia*) (fig. 15.8), which looks as if it might be an odd-shaped medusa. Actually,
it is a colony of polyps. One polyp is specialized as a gas-filled float, which

a.

b.

c.

Figure 15.8
Coelenterate diversity. *a.* Medusae of *Aurelia.* The
coelenterate medusae is a bell-shaped motile structure.
b. Coral polyps. Corals have hard, limy skeletons that
build up layer after layer to form a platform for living
polyps. *c.* Portugese man-of-war. This is actually a colony
of polyps, only one of which is inflated to keep the others
afloat.

provides buoyancy to keep the colony afloat. Other polyps are specialized for
feeding or for reproduction. The Portuguese man-of-war also has stinging
polyps armed with numerous nematocysts. Swimmers who accidentally encounter a Portuguese man-of-war can receive painful, sometimes very serious,
injuries from these stinging polyps.

Other Coelenterates

Other coelenterates (fig. 15.8) include the "true *jellyfishes,*" in which the medusa stage is the dominant phase, and the *sea anemones* and *corals,* which
are polyps having no medusa stage. Sea anemones and corals are more often
found in warm waters, where the anemones are the "flowers of the sea," and
the corals are famous for building reefs. Corals secrete hard, limy skeletons
that remain in place after the polyps die and degenerate. Each coral polyp is
rather small and inconspicuous, but new generations of corals grow on the
skeletal remains of past generations. In this way, massive deposits formed by
corals and the red coralline algae (p. 288) that live with them build up over
the years. These coral formations, frequently found in warmer waters, provide
sheltered habitats within which a great variety of organisms thrive.

Figure 15.9
Germ layers. *a*. In coelenterates, the ectoderm and
endoderm are separated by mesoglea. *b*. In flatworms, a
solidly packed mesoderm layer lies between the ectoderm
and the endoderm.

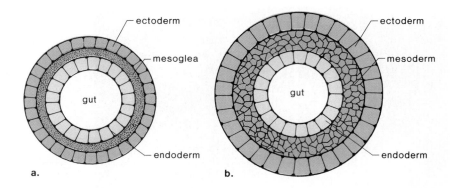

Figure 15.10
Animals have two basic body plans. *a*. The sac plan has
only one opening that must be used both as an entrance
and an exit. *b*. The tube-within-a-tube plan has two
openings; therefore, specialization of parts can develop
along the gut.

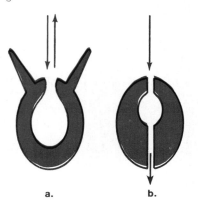

Comments

Coelenterates have *two tissue layers,* the epidermis and the gastrodermis. These
are derived from embryonic **germ layers.** (One of the first events during animal development is the establishment of layers of cells called germ layers.)
Some animals, like the coelenterates, have only two germ layers, called **ectoderm** and **endoderm.** More complex animals have three layers: ectoderm,
endoderm, and **mesoderm,** which lies between the first two (fig. 15.9).

The *radially symmetrical* (fig. 15.5) coelenterates have a *sac body plan;*
the mouth must serve as both an entrance for food and an exit for wastes. More
complex animals are usually bilaterally symmetrical and have a **tube-within-
a-tube body plan** with both a mouth and an anus (fig. 15.10).

Flatworms

Flatworms (**phylum Platyhelminthes**) are *bilaterally symmetrical,* whereas
adult coelenterates are radially symmetrical. It is possible, though, that a bilaterally symmetrical coelenterate larva gave rise to the flatworms. Flatworms
have *three tissue layers;* there is a solidly packed mesodermal layer between
the ectoderm and endoderm (fig. 15.9). This mesoderm contains well-
developed reproductive and excretory organs.

Although flatworms have the organ level of organization they still have
primitive features. The digestive tract has only one opening (fig. 15.10), and
they have no specialized circulatory or respiratory structures. Oxygen and
carbon dioxide diffuse through the body surface or through the lining of the
gastrovascular cavity.

Planarians

Most planarians (**class Turbellaria**) (fig. 15.11) are marine animals, but freshwater ones are commonly studied as representative of the group.

Planarians show **cephalization,** a definite head with a brain and sense
organs. The head often has lateral extensions that function as sense organs
and as chemoreceptors that detect potential food sources and/or enemies. There
are also two pigmented eyespots that are sensitive to light changes. Inside, a
concentration of nerve cells functions as a primitive brain, and a lengthwise
pair of nerve cords (fig. 15.12) are joined by cross-connectives. Planarians are
said to have a **ladder-type nervous system** because the nerve cords plus the
cross-connectives look like a ladder. When planarians move, the action of
ventrally located ciliated cells seems to be aided by rhythmical muscular
contractions.

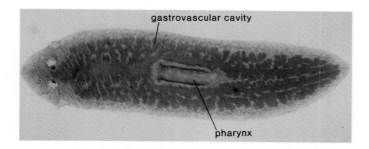

Figure 15.11
Photomicrograph of a planarian. The gastrovascular cavity has a single opening via the pharynx.

Figure 15.12
Planarian anatomy. *a.* Excretory system with flame cell shown in detail. *b.* Nervous system has a ladder appearance. *c.* Reproductive system has both male and female organs, and digestive system has a single opening. The extended pharynx in the worm is shown at upper left of the illustration.

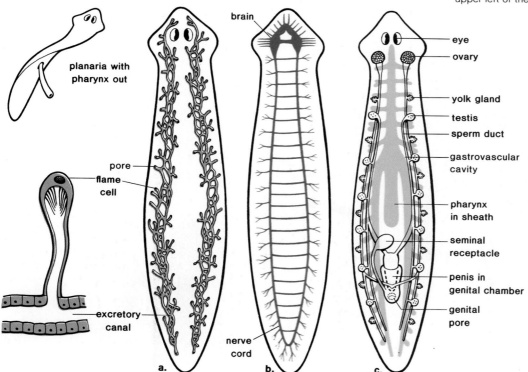

A muscular pharynx is extended through the ventrally placed mouth when a worm is feeding on small dead or living animals. A strong sucking action tears off chunks of food. These chunks are then drawn into the branching gastrovascular cavity, which ramifies enough to make a circulatory system unnecessary. A water-regulating organ consisting of a series of interconnecting canals runs the length of the body on each side. Water drawn into the canals by ciliary action within bulbous cells exits at pores located in the body wall. The beating of the cilia in each bulbous cell looks like the flickering of a flame, so this organ is called a **flame-cell system** (fig. 15.12*a*).

When planarians are cut in half, regeneration occurs: each half develops into a whole worm. Planarians are **hermaphroditic;** that is, each individual has both male and female reproductive organs (fig. 15.12*c*). Zygotes develop directly into small worms, without larval stages.

Figure 15.13

Fluke anatomy. *a.* This drawing illustrates the external anatomy of a fluke. Note the presence of the suckers for attachment to the host. *b.* A scanning electron micrograph of fluke (*Gorgoderina attenuata*) oral sucker. Note the upraised structures, which are believed to be sensory in nature.

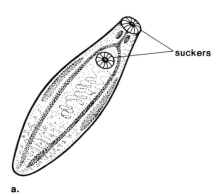

suckers

a.

b.

Figure 15.14

Tapeworm of a *Taenia* species. *a.* Life cycle showing mature proglottid (*right*) and gravid proglottid (*left*) in detail. *b.* A scanning electron micrograph of the scolex.

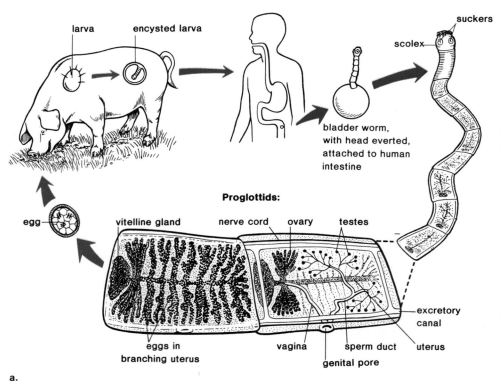

larva   encysted larva

suckers

scolex

bladder worm, with head everted, attached to human intestine

Proglottids:

egg

vitelline gland   nerve cord   ovary   testes

eggs in branching uterus

vagina   sperm duct   uterus

genital pore

excretory canal

a.

b.

Flukes

Flukes (**class Trematoda**) are parasites: they derive nourishment from a living host organism. A few flukes are external parasites, but the majority are **endoparasites** living within the bodies of their hosts. Flukes often have two suckers (fig. 15.13) by which they attach to and feed from host tissues.

Flukes that infect humans are commonly named for the organ they invade. For example, there are liver flukes and blood flukes. It is estimated that nearly half of all the people living in the tropics are infected by blood flukes of the genus *Schistosoma*. The accumulation of vast quantities of fluke eggs within the human body causes the symptoms of **schistosomiasis**—dysentery, anemia, general weakness, and greatly reduced resistance to other infections. Schistosome transmission depends on human feces reaching water, subsequent infection of the appropriate snail that is the intermediate host, and contact between the resulting larvae and human skin in the water. Rice paddy farming, which is the heart of agriculture in many tropical countries, creates a nearly perfect environment for transmission of these flukes because human feces are used for fertilizer and people work in the water tending the rice plants.

Tapeworms

Tapeworms (**class Cestoda**) are very long (6 to 20 m), flattened worms that live as adults in the intestines of vertebrate animals, including humans. They cause problems because they absorb nutrients, excrete toxic wastes, and sometimes interfere with passage of food through the gut.

Instead of a head, a tapeworm has a structure called a **scolex.** The scolex has hooks and suckers that allow the worm to attach to the host's intestinal wall (fig. 15.14). The rest of the body is made of subunits called **proglottids.** When a proglottid is mature, it primarily contains male and female reproductive organs. Thousands (more than 100,000 in some cases) of eggs are stored in the uterus, which expands until it fills the entire proglottid. Such "ripe" proglottids detach and pass out with the host's feces, scattering fertilized eggs on the ground. If pigs or cattle should happen to ingest these, larvae develop that eventually become encysted in muscle that may be eaten as poorly cooked or raw meat by humans. The cyst wall is digested, and a bladder worm develops into a new tapeworm that attaches to the intestinal wall. To prevent tapeworm infections, meat, especially pork, should be cooked thoroughly.

Comments

Although flatworms have a *sac body plan* (fig. 15.10a), they are more complex than coelenterates because they have *three germ layers* (fig. 15.9) and possess true organs. The free-living forms exhibit *cephalization* and *bilateral symmetry.* This is typical of actively moving predaceous animals.

Roundworms

Roundworms (**phylum Nematoda**), as their name implies, have a smooth outside wall, indicating that they are *nonsegmented.* They, like the rest of the animals to be studied, have the *tube-within-a-tube body plan;* that is, the digestive tract is a tube within the rest of the animal, which is the outer tube (fig. 15.10b). Also, roundworms have a body cavity—a **pseudocoelom**—that is incompletely lined with mesoderm (fig. 15.15).

Figure 15.15
A comparison of mesoderm organization. *a.* Flatworms are acoelomate; that is, their mesoderm is packed solidly. *b.* Pseudocoelomate animals have mesodermal tissues, such as muscle, inside their ectoderm, but not adjacent to their gut endoderm. *c.* Coelomate animals also have mesodermal tissue covering their guts. True coeloms are body cavities completely lined by mesodermal tissue. Mesenteries hold organs in place within body cavities.

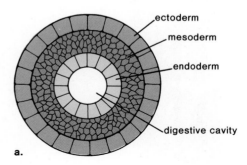

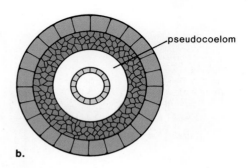

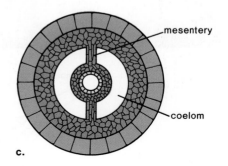

Figure 15.16

Roundworm anatomy. *a.* A photomicrograph of a
roundworm. *b.* Anatomy of female *Ascaris.* The cross
section shows the pseudocoelomate arrangement. There
is mesodermally derived muscle tissue inside the
epidermis, but there is no mesodermal tissue adjacent to
the intestinal wall.

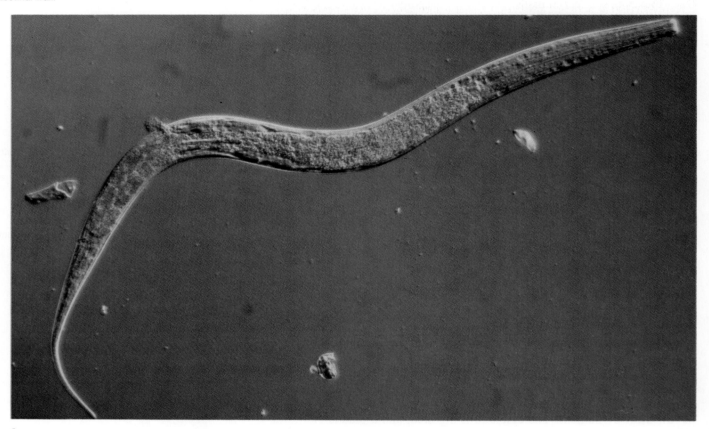

a.

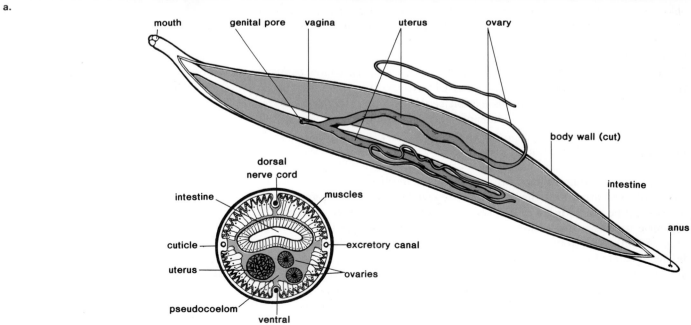

mouth   genital pore   vagina   uterus   ovary

body wall (cut)

intestine

anus

dorsal
nerve cord

intestine   muscles

cuticle   excretory canal

uterus   ovaries

pseudocoelom   ventral
                nerve cord

b.

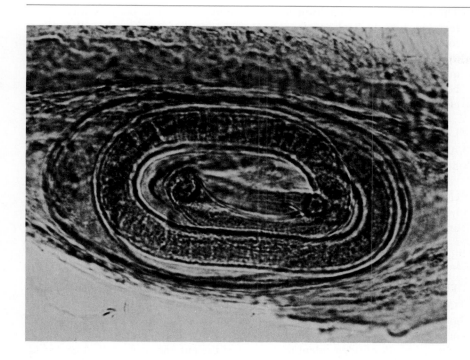

Roundworms are generally colorless and quite small; most are even microscopic. A handful of garden soil can contain hundreds or even thousands of tiny roundworms, some of which do serious damage to plants and cause significant agricultural loss. They can also be found in the muddy bottoms of freshwater lakes and ocean shores, and even thrive in environments provided by humans, such as beer-soaked mats in German taverns and in vats of vinegar.

Ascaris

Ascaris, an intestinal parasite of vertebrates, including humans, is a commonly studied representative of roundworms. Females (20 to 35 cm) tend to be larger than males, but both sexes move by means of a *whiplike motion* because all their muscles run lengthwise. The reproductive system is the best developed of all the organ systems (fig. 15.16). Fertilization is internal and the eggs develop within the uterus before being laid.

After an animal ingests the *Ascaris* eggs, they hatch and the small worms burrow through the intestinal wall, enter blood vessels, and eventually crawl into the lungs. They then work their way up to the throat and from there down into the gut, where they mature. Their tough cuticles protect them from damage by digestive enzymes. A female *Ascaris* can produce as many as 200,000 eggs per day, and a hog lot where infected pigs have lived for several years can contain millions of viable *Ascaris* eggs per square meter. Although uncommon in North America, hundreds of millions of people around the world are infected by *Ascaris*.

Other Roundworm Infections

Intestinal pinworms, which cause more irritation than real damage, infect at least 90 percent of people in every part of the world at some time during their lifetimes. *Trichinella* is a roundworm that encysts in the muscles of pigs and humans (fig. 15.17). When humans eat poorly cooked pork, they may contract an infection called **trichinosis,** in which permanent cysts cause muscle damage.

Figure 15.18
An infection with a filarial worm causes elephantiasis, a
condition in which the individual experiences extreme
swelling in regions where the worms have blocked the
lymph vessels.

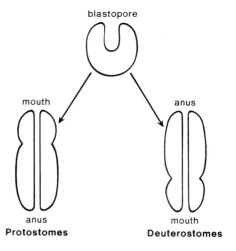

Figure 15.19
Differences in the developmental patterns of two major
groups of animal phyla. In protostome embryos, the
blastopore becomes the mouth. In deuterostome embryos,
it becomes the anus.

The young larvae of **hookworms** usually enter the human body by way of the feet. Once inside, hookworms make their way to the intestine where they cling to the intestinal wall and suck blood. Hookworms cause anemia and susceptibility to other infections. *Filaria* worms are transmitted to humans by mosquitoes in tropical areas and may cause severe swellings in the parts of the body where they block lymphatic vessels (fig. 15.18). This condition is called **elephantiasis.**

Comments

Although the ancestors of roundworms are not believed to have contributed directly to evolution of the higher invertebrates, roundworms possess *two advanced features:* a body cavity and a tube-within-a-tube body plan. The body cavity is a *pseudocoelom* rather than a true coelom (fig. 15.15). The *tube-within-a-tube plan,* in contrast to the sac plan (fig. 15.10), has both a mouth and an anus.

Coelomate Animals

The rest of the animals have a **true coelom** that is completely lined by mesoderm (fig. 15.15*c*). A body cavity seems to have been of selective value to animals with the organ level of organization because it gives room for internal organs to become more complex. Muscles derived from separated layers of mesoderm allow independent movement of the gut and body wall. A coelomic fluid can aid in the movement of molecules, or there may be a circulatory system that serves these functions. Some animals have a **hydrostatic skeleton;** a coelomic fluid supports the body wall and muscle contraction acts against the pressure exerted by the fluid. This is very different from those animals in which the muscles pull against a rigid skeleton when contraction occurs.

Two Branches

Coelomate animals are divided into two separate groups called *protostomes* and *deuterostomes.* During embryonic development, a hollow sphere forms and the indentation that follows produces an opening. In the protostomes this opening becomes the mouth (*proto* = first, *stome* = mouth). In the deuterostomes, this opening becomes the anus (fig. 15.19), and only later does a new opening form the mouth (*deutero* = second, *stome* = mouth). This chapter will consider the protostomes. In the next chapter, we will consider the deuterostomes.

Mollusks

It is possible that all the thousands of types of mollusks (phylum Mollusca) are descended from a bilateral molluskan ancestor (fig. 15.20) that had a definite *head,* and the three common features shared by all mollusks today: a visceral mass, a mantle, and a foot. The **visceral mass** contains the internal organs, including a highly specialized digestive tract, paired kidneys, and reproductive organs. The phylum name comes from the Latin word *mollis,* meaning soft, which refers to this visceral mass. The **mantle** is a covering that lies to either side, but does not completely enclose, the visceral mass. It may secrete a shell and/or contribute to the development of gills or lungs. The space between the folds of the mantle is called the mantle cavity. The **foot** is a muscular organ that may be adapted for either locomotion, attachment, food capture, or a combination of functions. Another feature often present is the

radula, an organ that bears many rows of teeth. As the radula moves back and forth over a cartilaginous tongue, food is torn apart and moved toward the digestive tract.

The ancestral mollusk (fig. 15.20) is presumed to have had a nervous system consisting of a *nerve ring and two longitudinal cords,* one serving the foot and the other serving the visceral mass. In today's mollusks, the *coelom is much reduced* and is largely limited to the region around the heart. In most mollusks, the heart pumps blood through vessels that empty into open spaces, called **sinuses,** before it is collected in vessels and returned to the heart. This arrangement is called an **open circulatory system** because blood is not always enclosed within blood vessels.

Bivalves

Clams, oysters, and scallops are all bivalves (**class Pelecypoda**). The *two shells* of these mollusks are hinged, and are closed by powerful muscles. The so-called *hatchet foot* of bivalves is adapted for locomotion in sandy or muddy soil. Bivalves are *filter feeders*. They glean food from the water that enters and exits the mantle by way of *siphons* located at the posterior end. Food collected on the *gills* that hang down in the mantle cavity is swept toward the mouth by ciliary action. The nervous system consists of *three pairs of ganglia* (nerve centers) located anteriorly, posteriorly, and in the foot. These nerve centers are connected by nerves (fig. 15.21).

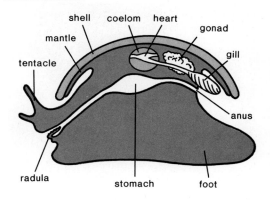

Figure 15.20
A sketch showing the relationship of parts of a hypothetical primitive mollusk. This basic plan has been modified several different ways during the evolution of modern mollusks.

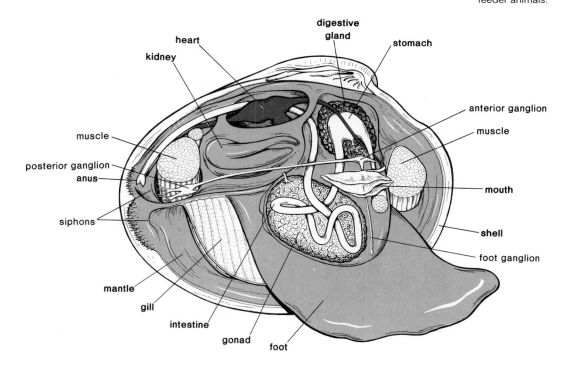

Figure 15.21
Bivalve anatomy is exemplified by the anatomy of a clam. The shell and the mantle on one side have been removed. The heart and dorsal blood vessel are in color. Notice the three-ganglia nervous system and the lack of cephalization, a characteristic that is consistent with filter-feeder animals.

Figure 15.22
Molluskan diversity. *a.* A squid with tentacles outstretched emphasizes its streamlined shape. *b.* A snail moving along a leaf carries its shell behind.

a.

b.

Gastropods

Snails, conches, and other gastropod mollusks (**class Gastropoda**) (fig. 15.22) usually move about by waves of muscular contractions passing along a *ventrally flattened foot.* Although most have coiled shells, the nudibranchs, which are called "sea slugs" and "sea hares," have no shells. Some of the nudibranchs are among the most colorful and beautiful of all animals.

Snails are adapted to life on land. They have three body divisions: a *head* with two pairs of tentacles, one pair of which bears eyes at the tips; the flat, long muscular *foot;* and a visceral mass surrounded by a *coiled shell.* The shell not only offers protection, but also prevents desiccation, or drying out. The mantle is richly supplied with blood vessels and functions as a *lung* when air is moved in and out through respiratory pores. Snails are *hermaphroditic;* in mating, each inserts a penis into the mantle cavity of the other to release sperm for future fertilization of eggs that are later deposited in the ground. Development proceeds directly, without formation of larvae. The presence of a copulatory organ and the absence of a water-dependent larval stage are adaptations to life on land.

Cephalopods

In cephalopods (**class Cephalopoda**), including octopuses, squids, and nautiluses (fig. 15.22), the foot has become *tentacles about the head.* Nautiluses are enclosed in shells, but the shells of squids are reduced and internal. Octopuses lack shells entirely. Both squids and octopuses can squeeze their mantle cavity so that water is forced out, thus propelling them rapidly backward by a sort of *jet propulsion.* They also possess *ink sacs* from which they can squirt out a cloud of brown or black ink. This action often leaves a would-be predator completely confused.

Most cephalopods are fast-moving, predatory animals with well-developed sense organs, including focusing *camera-type eyes* that are very similar to those of vertebrates. Aside from the tentacles that seize the prey, cephalopods have a powerful, *parrotlike beak and a radula* to tear the prey apart. (Experienced divers fear the beak of an octopus much more than its tentacles.) Cephalopods in general and octopuses in particular have well-developed brains and display complex behavior, including impressive learning ability.

Although giant octopuses are a standard part of science fiction, octopuses seldom grow very large. Some giant squids, however, are enormous, being 18 m long and weighing two tons. There are even reports of giant squids that fight back when attacked by sperm whales.

Relationship to Annelids

Present-day mollusks are *nonsegmented,* yet they are believed to be closely related to annelids. Annelids are **segmented** animals, meaning that their bodies are divided into repeating units. In 1952 ten living specimens of a molluskan species, *Neopilina galatheae* (fig. 15.23), previously presumed extinct for about 500 million years, were dredged up from a depth of more than 3,500 m in the ocean near Puerto Rico. A study of the anatomy revealed that the internal organs show evidence of segmentation. Some believe that *Neopilina* may be similar to a common ancestor of both mollusks and annelids. A link between annelids and mollusks is also evidenced by embryologic observations. The zygotes of many annelids and many mollusks develop into top-shaped larvae called **trochophores** (fig. 15.23*b*).

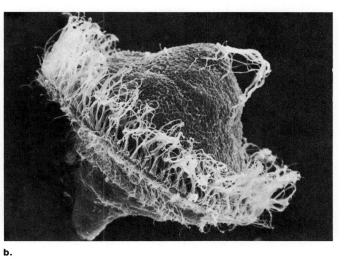

Figure 15.23

The evolutionary relationship of mollusks to annelids. *a.* Dorsal view of *Neopilina* after removal of outer shell. Segmental arrangement of gills, nephridia, and muscles in *Neopilina* suggests that modern mollusks may have evolved from segmented ancestors similar to ancestors of modern annelids. *b.* A trochophore larva. Many mollusks and many annelids have similar early-developmental patterns and trochophore-type larvae.

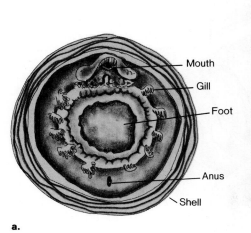

- Mouth
- Gill
- Foot
- Anus
- Shell

a.

b.

Figure 15.24

Annelid diversity. *a.* Marine annelids are represented by a sandworm, which has parapodia on each segment. *b.* The earthworm lacks parapodia, but is still segmented.

a.

b.

Annelids

Annelids (**phylum Annelida**) (fig. 15.24), represented by earthworms and marine worms, are *segmented worms* as is externally evident in the rings that encircle the body. Internally, partitions divide the *well-developed fluid-filled coelom* that acts as a *hydrostatic skeleton*. Annelids have both circular and lengthwise muscles; when lengthwise muscles contract, segments of the body shorten; when circular muscles contract, segments of the body elongate.

Figure 15.25

Earthworm anatomy. The drawing shows the internal anatomy of the anterior part of an earthworm's body. The closed circulatory system has five pairs of hearts. Earthworms are hermaphroditic; each worm has both male and female reproductive structures. The small sketch shows the location of the clitellum.

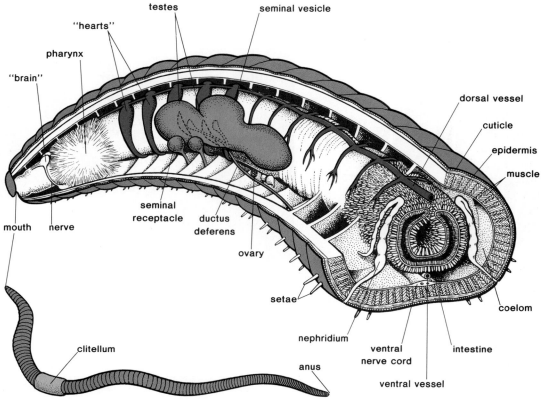

Annelids are *bilaterally symmetrical and have the tube-within-a-tube body plan* (fig. 15.25). The digestive tract shows specialization of parts along its length; for example, there may be a pharynx, stomach, and accessory glands. There is an extensive **closed circulatory system** *in which red blood is* always enclosed within blood vessels that run the length of the body and branch in every segment. The nervous system consists of a *brain connected to a ventral, solid nerve cord with a ganglion in each segment.* The excretory system consists of *paired nephridia,* coiled tubules in each segment that collect waste material from the coelom and excrete it through openings in the body wall.

Marine Annelids

Most annelids are marine and belong to the **class Polychaeta,** named for the presence of many setae. The setae are bristles that anchor the worm or help it locomote. In polychaetes, the setae occur in bundles on **parapodia,** paddle-like appendages found on most segments. Some polychaetes tend to be active, aided by the parapodia, and wander about in search of prey. These worms (fig. 15.14*a*) have a definite head with sense organs. Other polychaetes are sedentary tube worms with tentacles about their head. Water currents created by the action of cilia brings debris within reach of the tentacles, where it adheres. These worms are therefore continuous filter feeders (p. 457). Since the parapodia have a rich supply of blood vessels, they assist the process of gas exchange in all polychaetes.

Figure 15.26
Peripatus has obvious segmentation, but its numerous short legs are not jointed like those of arthropods. There are, however, paired claws on the tips of its feet. A pair of appendages near the mouth has been modified into an organ that can shoot a sticky substance as much as six inches. The animal uses this glue, produced by a modified kidney, for defense and to capture small animals for food.

The polychaetes have breeding seasons and only during these times do the worms have sex organs. Internal sex organs develop anew each year and produce either sperm or eggs. While the gametes sometimes escape by way of the nephridia, in some worms the body literally breaks apart to release them. These events are synchronized, and this circannual rhythm (behavior that occurs yearly) must be controlled by an internal biological clock.

Most polychaete zygotes develop into trochophore-type larvae. Polychaetes are considered to be most similar to the primitive ancestors of all the other modern annelids.

Earthworms

Earthworms (**class Oligochaeta** = few setae) (fig. 15.24*b*), do not have well-developed heads or parapodia. Their setae protrude in clusters directly from the surface of their bodies.

Earthworms are restricted to soils containing adequate moisture because they use a *moist body wall for gas exchange*. Yet, when rainwater fills their underground burrows, they move up to the surface and remain there until the water percolates down and away from their burrows and the surrounding soil.

Earthworms feed by ingesting quantities of soil containing living and decaying organic matter. Much of what they eat passes on through their bodies and is deposited as small piles of dirt (worm casts) on the soil surface. This activity of earthworms is important for soil turnover since, as Charles Darwin calculated, earthworms can carry and deposit as much as 7 to 16 English tons of soil per acre annually.

Earthworms and other oligochaetes are *hermaphroditic* (fig. 15.25), but self-fertilization does not occur. Two worms line up parallel to each other, facing opposite directions. Then the collarlike **clitellum** secretes mucus that protects the sperm as they pass between the worms. Several days later the clitellum secretes a mucus sheath having a hard outer covering. As the worm backs out, eggs and the sperm, received earlier, are released into the sheath where fertilization occurs. The sheath then closes, forming a cocoon in which the zygotes develop into young worms that eventually hatch out of it.

Relationship to Arthropods

Peripatus (fig. 15.26), called the "walking worm" because it has 14 to 400 pairs of short, stumpy legs, possesses organs that either have annelid or arthropod characteristics. Members of this genus resemble fossils from as long ago as the Cambrian period. Apparently, while these "walking worms" have remained relatively unchanged for millions of years, their ancient relatives evolved and diversified into the huge and evolutionarily successful group discussed next: the arthropods.

Figure 15.27
Fossil trilobite, an arthropod that has been extinct for about 225 million years. Note the regular segmentation of this animal named for the three lobes seen here. There are so many trilobites in the fossil record that 3,900 species have been recognized, and even the developmental stages of some species have been studied.

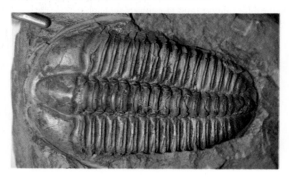

Arthropods

Arthropods (**phylum Arthropoda**) show such diversity (about 900,000 species) and are adapted to so many different habitats that they are often said to be the most successful of all animals. Arthropod means literally "jointed foot." Actually they have **jointed appendages** that are freely movable. Their bodies are segmented, and at one time there was a pair of similar appendages on each body segment (fig. 15.27). In modern arthropods, some segments have fused into regions such as a head, thorax, and abdomen (fig. 15.28), thus reducing the number of appendages. The remaining appendages may be specialized for such specific functions as walking, swimming, reproduction, and eating.

Arthropods have a well-developed nervous system. There is a brain and a *ventral solid nerve cord*. The head bears various types of sense organs, including eyes of two types—compound and simple. The **compound eye** (fig. 15.29) is composed of many complete visual units grouped in one structure. Each visual unit contains a separate lens and a light-sensitive cell.

Figure 15.28
Arthropod diversity. Arthropods are adapted to life in the water (*a.–c.*) and life on land (*d.–f.*). *a.* Cleaner shrimp. *b.* Hawaiian lobster. *c.* Hermit crab. *d.* California sister butterfly. *e.* Grasshopper. *f.* Giant millipede.

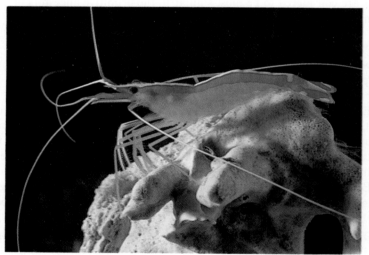

a.

d.

The external skeleton of arthropods is composed primarily of **chitin,** a strong, flexible nitrogenous polysaccharide. Since it is hard and nonexpandable, arthropods must **molt,** or shed, the exoskeleton as they grow larger. Before molting, the body secretes a new, larger exoskeleton, which is soft and wrinkled, underneath the old exoskeleton. After enzymes partially dissolve and greatly weaken the old exoskeleton, the animal breaks it open and wriggles out (fig. 24.14). The new exoskeleton then quickly expands and hardens.

Some arthropods, such as insects and spiders, have a specific number of molts. Others, such as crabs and lobsters, continue to grow and molt throughout their lives. During the times of the year when many lobsters are molting, knowledgeable New Englanders stop buying lobsters because a new exoskeleton takes up a lot of fluid, which artificially adds to the weight. Thus the natives prefer to leave newly molted lobsters for the tourists.

Figure 15.29
A scanning electron micrograph showing that compound eyes are composed of individual units. Each unit has its own lens and retina.

b.

c.

e.

f.

Crustacea

Members of the **class Crustacea** are named for their hard shells. Although most are aquatic, they are very diverse; they differ in structure, habitat, and way of life. The small crustacea represented by the copepods and krill live in the sea where they feed on algae and serve as food for fish and whales. They are so numerous that despite their small size, some believe that we might be able to harvest them and use them for food. Larger crustacea, represented by shrimp, crabs, lobsters, and crayfish, are well known because they are frequently eaten by humans. These all have five pairs of walking legs, the first pair of which are the claws.

The sow bugs and pill bugs are crustacea that live on land even though they have few adaptations for terrestrial life and are restricted to moist places. Barnacles are sessile crustacea and live attached to a substrate. Goose barnacles have a stalk that might seem like a long neck, while acorn barnacles look like acorns attached directly to rocks. Barnacles begin life as free-swimming larvae, but they undergo a metamorphosis that transforms their swimming appendages to cirri, feathery structures that allow them to filter feed. They only extend these when they are submerged so most visitors to the seashore have only seen the calcareous plates that enclose the body of the acorn barnacle.

Insects

Insects (class Insecta) (fig. 15.28*d* and *e*) include more known species than any other group of organisms and are regarded as one of the most successful (possibly the most successful) groups of organisms that has ever lived. Nearly 90 percent of all arthropod species are insects. *Wings* enhance the insects' ability to survive by providing a new way of escaping enemies, finding food, facilitating mating, and dispersing the species.

The grasshopper (fig. 15.30) is often studied as a representative insect. Every system of the grasshopper is adapted to life on land. As part of the exoskeleton there are three pairs of legs, one pair of which is suited to jumping. There are two pairs of wings: the forewings are tough and leathery, and when folded back at rest, protect the broad, thin hindwings. The first segment of the grasshopper bears on its lateral surface a large **tympanum** for the reception of sound waves.

The digestive system is suitable for a grass diet. The food is broken down by the grinding action of mouth parts and by the chemical action of saliva. The food is ground up even finer in the **gizzard.** Finally, chemical digestion is completed by enzymes secreted by the gastric caeca. Excretion is carried out by **Malpighian tubules,** which extend into the **hemocoel,** a network of open spaces where blood flows about the organs. (The hemocoel represents the coelomic cavity in arthropods.) Malpighian tubules form a solid nitrogenous waste that passes out of the animal by way of the digestive tract. This allows the animal to conserve water.

Respiration occurs when air enters small tubules called **tracheae** (fig. 22.4) by way of openings in the exoskeleton called **spiracles.** The trachea branch and rebranch, finally ending in moist areas adjacent to cells where the actual exchange of gases takes place. Breathing by means of trachea may account for the small size of insects since these tubules are so tiny and fragile they would be crushed by any amount of weight.

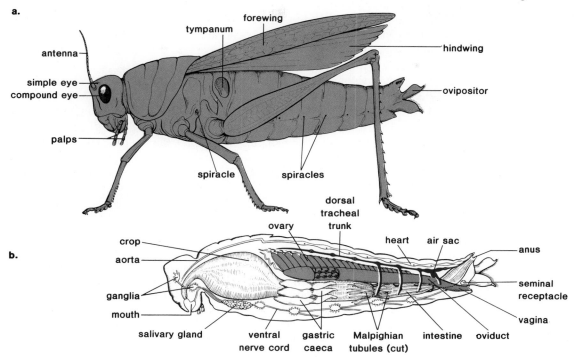

Figure 15.30
Insect anatomy. *a*. External appearance of grasshopper. Note the three pairs of legs attached to the thorax. *b*. Internal anatomy of a female. Sperm are received in the seminal receptacle during mating. The female uses her ovipositor to insert eggs in the ground where embryos remain during winter months.

The grasshopper's reproductive system is adapted to reproduction on land (fig. 15.30). The male passes sperm to the female by way of a penis and fertilization is internal. In this way, both sperm and zygotes are protected from drying out. The female deposits the fertilized eggs in the ground with the aid of her **ovipositor.**

Grasshoppers undergo **incomplete metamorphosis,** a gradual change in form as the animals mature. The immature grasshopper, called a nymph, is recognizable as a grasshopper even though it differs somewhat in shape and form from the adult. Other insects undergo **complete metamorphosis** involving drastic changes in form. At first, the animal is a wormlike larva with chewing mouthparts. Then the larva forms a case or cocoon about itself and becomes a *pupa,* from which the *adult* emerges. This life cycle allows the larvae and adults to make use of different food sources and thus take advantage of seasonal changes in vegetation.

Insects also show remarkable behavioral adaptations well-exemplified by the social systems of bees, ants, termites, and other colonial insects. Insects are so numerous and so diverse that the study of this one group is a major specialty in biology called entomology.

Summary

Animals are multicellular heterotrophic organisms that digest their food. It is possible to arrange the various groups so that an increase in complexity is observed: Table 15.1 contrasts those studied in this chapter.

The sponges should be remembered as exceptional animals because their only large opening is an exit and their locomotion is limited. A sponge seems almost like a colony of cells. Classification is according to the type of spicules, which form an internal skeleton.

The coelenterates are unique in having stinging cells containing nematocysts, which are used to stun prey that is then maneuvered into a gastrovascular cavity by means of tentacles. They may exist as either a polyp or a medusa, or have a life cycle that alternates between the two.

Free-living flatworms, such as *Planaria,* show cephalization. They also have internal organs—digestive, excretory, nervous, and reproductive. Flukes and tapeworms are parasitic flatworms that have a primary and a secondary host.

Roundworms are nonsegmented. They can be quite small so that even a handful of dirt can contain thousands. Although free-living forms are more common, parasites like *Ascaris* are usually studied. Other types of roundworms cause trichinosis, hookworm, and elephantiasis.

Mollusks are a varied group adapted to different ways of life. All mollusks have three parts: a visceral mass, a mantle, and a foot. The foot is modified according to the way of life. For example, the clam has a hatchet foot and the squid's foot has evolved into tentacles. Among the mollusks, snails are adapted to a terrestrial existence.

Neopilina is an animal that has characteristics of both mollusks and annelids. Perhaps mollusks and annelids share a common ancestor, but only the annelids became segmented. Annelids are segmented worms represented by the marine worms and earthworms. Segmentation is obvious externally because of the body rings. Internally the organs repeat; there are nephridia, ganglia, blood vessels, etc., in each segment. A very well-developed coelom is divided by septa. Annelids have a closed circulatory system, and, typical of higher invertebrates, they have a ventral solid nerve cord.

Peripatus has characteristics of both annelids and arthropods. Arthropods are the most numerous and varied of all the animal phyla, especially since insects are arthropods. There are more known different kinds of insects than any other type of animal. Arthropods have an external skeleton and jointed appendages. Although segmented, the segments are fused to form body parts such as a head, thorax, and abdomen. Because arthropods have an external skeleton, they have to molt. Although we know best the larger crustacea such as lobsters, the smaller crustacea are most important in the ocean. Insects are well adapted to life on land. Many have a life cycle that involves metamorphosis, either complete or incomplete.

Table 15.1 Comparison

	Symmetry	Germ Layers	Body Plan	Body Cavity	Organs	Segmentation
Sponges	--	--	--	No	No	No
Coelenterates	Radial	2	Sac	No	No	No
Flatworms	Bilateral	3	Sac	No	Yes	No
Roundworms	Bilateral	3	Tube-within-a-tube	Pseudocoelom	Yes	No
Mollusks	Bilateral	3	Tube-within-a-tube	Coelom (reduced)	Yes	No
Annelids	Bilateral	3	Tube-within-a-tube	Coelom	Yes	Yes
Arthropods	Bilateral	3	Tube-within-a-tube	Coelom (reduced)	Yes	Yes

Objective Questions

1. The lower invertebrates include the _____ , _____ , _____ , and _____ .
2. The function of collar cells in a sponge is _____ .
3. Coelenterates have the _____ body plan and are _____ symmetrical.
4. Planarians have the _____ type of nervous system and a _____ excretory system.
5. The intermediate host for a tapeworm may be _____ or _____ .

6. Pinworm, trichinosis, hookworm, and elephantiasis are all _____ worm infections.
7. In protostomes, the first embryonic opening becomes the _____ .
8. In today's mollusks, the coelom is much _____ and limited to the region around the _____ .
9. A clam is a _____ type of mollusk; a snail is a _____ .
10. Earthworms have external rings signifying that they are _____ animals.

11. Arthropods must _____ because their bodies are covered by an external skeleton.
12. Insects have _____ pairs of legs and _____ pairs of wings.

Study Questions

1. Indicate the supposed evolutionary relationship of sponges, coelenterates, flatworms, roundworms, mollusks, annelids, and arthropods.
2. Name at least three members of each of the groups listed in 1.
3. Substantiate the claim that there is an increase in complexity among the groups of animals studied in this chapter.

4. How are sponges different from all other types of animals?
5. Describe the body plan and life cycle of coelenterates.
6. How is it possible to become infected with a fluke? a tapeworm?
7. Name several roundworm parasites of humans.

8. What three distinguishing characteristics do all mollusks have?
9. Give evidence based on both external and internal anatomy that annelids are segmented animals.
10. List the stages of complete metamorphosis in insects.

Selected Key Terms

vertebrate ('vərt ə brət) 305
invertebrate (,in 'vərt ə brət) 305
protostome ('prōt ō ,stōm) 307
deuterostome ('düt ə rə ,stōm) 307
epidermis (,ep ə 'dər məs) 308
gastrodermis (,gas trō dər məs) 308
mesoglea (,mez ə 'glē ə) 308
medusa (mi 'dü sə) 309

polyp ('päl əp) 309
nematocyst ('nem ət ə,sist) 309
endoderm ('en də ,dərm) 312
mesoderm ('mez ə ,dərm) 312
pseudocoelom (,süd ə 'sē ləm) 315
metamorphosis (,met ə 'mȯr fə səs) 327
ovipositor ('ō və 'päz ət ər) 327

CHAPTER 16

Further Evolution in Animals

Concepts

1. Vertebrates and invertebrates are related. Echinoderms are the invertebrate group most closely related to vertebrates.

2. The classification of animals is based in part on the presence of common characteristics.
   Vertebrates are chordate animals that possess at some time in their life history a notochord, pharyngeal pouches or slits, and a dorsal hollow nerve cord.

3. Adaptation to land includes an appropriate means of reproduction.
   The shelled egg of the reptiles allowed them to be the first vertebrates fully adapted to life on land.

4. Evolution proceeds by means of common ancestors.
   a. Humans shared a common ancestor with apes during the last era of evolutionary history.
   b. It appears that the first member of the genus *Homo* evolved only during the last two million years of evolutionary history.

The *deuterostomes,* the other major division of advanced animals (fig. 15.2), begins with the echinoderms, the invertebrates most closely related to the vertebrates. Most of the evidence linking these two groups is developmental. The digestive system develops in the same way; the second opening for the gut becomes the mouth; and the mesoderm and coelom develop similarly. Even though adult echinoderms are radially symmetrical, echinoderm larvae are bilaterally symmetrical. Therefore, the common ancestor for echinoderms and vertebrates was most likely bilaterally symmetrical or at least had a larval stage that was bilateral. The two groups could have evolved from a larval stage that never matured. Continuance of an immature stage without normal maturation is called **neoteny.**

Echinoderms

Echinoderms (**phylum Echinodermata**) (fig. 16.1) are marine animals that include sea stars (starfishes), sea urchins, sand dollars, sea cucumbers, and sea lilies. The *radial symmetry* of these animals is usually reflected in a five-part body organization. They have an *internal skeleton* (endoskeleton) consisting

Figure 16.1
Echinoderm diversity. *a.* A sea star with one of its arms lifted, showing extended tube feet. The feet attach and contract, pulling the starfish along. When a starfish attacks a clam, it arches its body over the shell and by the concerted action of the tube feet, forces the clam to open. Then it everts a portion of its stomach to digest the contents of the clam. *b.* Sea urchins have fused skeletal plates and long, movable spines that protrude from their skin. Many organisms, ranging from sea stars to mammals, feed on sea urchins. For example, sea otters float on their backs and use stones to crack open urchins held on their chests. Sea urchin eggs are used for human food, especially in Japan where large quantities are imported each year from the west coast of the United States and Canada.

a.

of spine-bearing plates. The spines stick out through the delicate skin (especially in sea urchins), which accounts for their name. Echinoderm comes from the Greek words meaning "spiny skin."

Echinoderms have a *well-developed coelom*. Here coelomic fluid circulates substances, and wandering amoeboid cells clean up particulate wastes. Gas exchange takes place through *gills* that are tiny, fingerlike extensions of the skin. Echinoderms have a central *nerve ring and nerve branches* extending into the multiple body divisions. They are capable of coordinated but slow responses and body movements.

Movement is facilitated by tube feet (fig. 16.2), which are part of a *water vascular system*. Water enters by a *sieve plate* into an arrangement of major canals extending into each portion of the body. Small canals branch off a major one to the **tube feet,** each of which ends in an expanded region called the ampulla. Contraction and expansion of the ampullae control the tube feet. By alternating the attachment and release of the tube feet, the animal moves along.

Sexes are separate in echinoderms. They shed gametes into the water, where fertilization occurs. The zygotes become free-swimming, bilaterally symmetrical larvae that undergo metamorphosis to become radially symmetrical adults.

b.

Figure 16.2
Sea star anatomy. Aside from the water vascular system shown here in color, there are gonads and digestive glands in each arm. The two-part stomach is in the central disc with the anus uppermost and the mouth hidden beneath on the ventral surface.

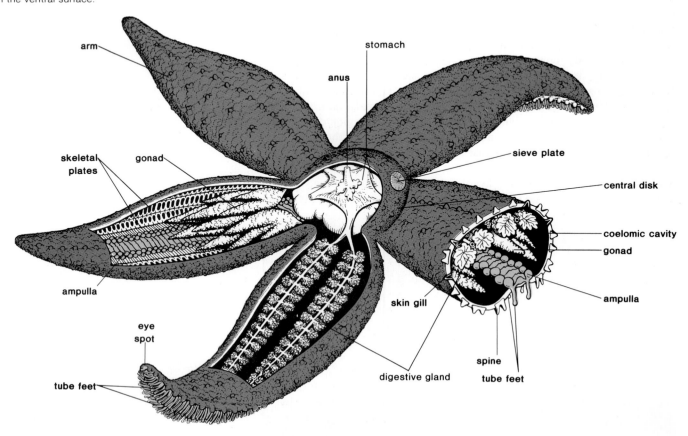

Sea Stars

Sea stars (starfishes) (**class Asteroidea**) are the most familiar of the echinoderms. The body of most species is flattened and has a central disc with five, or a multiple of five, sturdy arms (rays) extending outward from it. Each arm has rows of tube feet in a groove that extends along its ventral surface.

Sea stars commonly feed on clams, oysters, and other bivalve mollusks. To feed, a sea star positions itself over a bivalve and attaches some of its tube feet to the opposite sides of the shell. By working its tube feet in relays, it pulls the shell open. A very small crack is enough for the starfish to push the larger, lower part of its stomach out of its mouth and through the crack so that it contacts the soft parts of the bivalve. The stomach secretes enzymes, and digestion begins even while the bivalve is attempting to close its shell. Later, partly digested food is taken into the sea star's body, where digestion continues in the upper part of the stomach and in the digestive glands found in each arm.

Other Echinoderms

There are several other types of echinoderms. Brittle stars can move around more rapidly than starfishes because they bend their long arms to push themselves along quickly and are not dependent on the slow action of tube feet. When the long arms are injured, they are discarded and new ones grow in their place. Sand dollars have skeletal plates that are fused into a single, flattened unit. The surface has a five-part, flowerlike pattern caused by the presence of pores in the plates. These permit the extension of special tube feet modified for respiration. Sand dollars have furlike spines, not the long, stiff movable kind found in sea urchins (fig. 16.1*b*). Surprisingly, sea urchin spines are used as molds for the production of artificial human blood vessels. To feed, sea urchins extrude teeth from a dental structure called Aristotle's lantern and scrape algae and other food from rocks.

Sea cucumbers, which have reduced skeletons and leathery bodies, differ greatly from other echinoderms. A sea cucumber lies on its side and traps food particles in mucus on the surface of tentacles that are set in a ring around its mouth. Then it puts one tentacle at a time into its mouth and scrapes off the food. If attacked by a predator, a sea cucumber can eject much of its internal organs. Often this gives the sea cucumber an opportunity to move away and begin regeneration of the lost organs, while the startled predator is left to comtemplate a writhing mass of body organs.

Unlike all other echinoderms, sea lilies have mouths directed upward and live attached by a stalk to a substrate. Most sea lilies are suspension feeders, and they have branched arms with small appendages that sweep food out of the water.

Chordates

Among the chordates (**phylum Chordata**) are those animals with which we are most familiar, including human beings. To be considered a chordate, an animal must have the following three basic characteristics at some time in its life history (fig. 16.3).

1. A dorsal supporting rod called a **notochord.** This is replaced by the vertebral column in most adult vertebrates.

2. A **dorsal hollow nerve cord.** By hollow, it is meant that the cord contains a canal that is filled with fluid.

3. **Gill pouches.** These are seen only during embryological development in most vertebrate groups, but break through as gill slits in fishes. Water passing into the mouth and pharynx goes through the *gill slits,* which are supported by *gill arches.*

Figure 16.3
An enlargement of the notochord (*upper*) and a diagram of an idealized chordate (*lower*). All chordates at some time in the life history have a dorsal hollow nerve cord, pharyngeal gill pouches, and a notochord. The notochord is a supporting rod covered by an inner fibrous sheath and an outer elastic sheath.

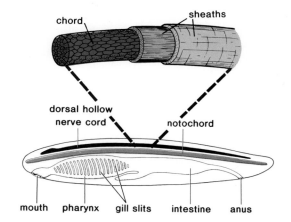

Figure 16.4
Tunicate anatomy. *a.* Photo showing the siphons of a living tunicate. A common name for these animals is sea squirts because water squirts from the excurrent siphon when the animals are irritated. *b.* Drawing of the body of a tunicate. The numerous gill slits in the central cavity are the only chordate feature retained by the adult.

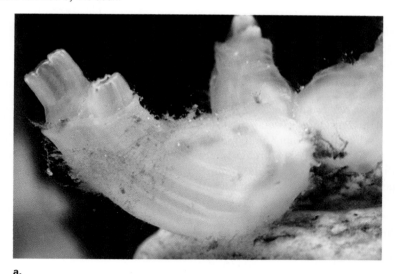

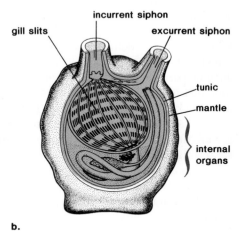

a.

b.

Figure 16.5
The lancelet *Branchiostoma* (amphioxus). *a.* Diagrammatic drawing of anatomy illustrates that the animal possesses the three chordate characteristics as an adult. Lancelets are filter feeders that take food from the water that passes out the numerous gill slits. *b.* Photomicrograph of several individuals in which segmentation of the musculature is obvious. Sometimes lancelets swim freely. Other times, they filter feed while partially buried in the bottom soil.

Photograph by Carolina Biological Supply Company.

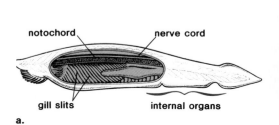

a.

b.

Two groups of animals, **tunicates (subphylum Urochordata)** (fig. 16.4) and **lancelets (subphylum Cephalochordata)** (fig. 16.5), are the protochordates (proto=first). They possess the three typical chordate characteristics in either the larval or adult forms. How these invertebrate chordates are related to the vertebrates has not been exactly determined. However, all three groups must be descended from an ancestor that was in existence sometime during the Cambrian period (table 16.1) when the chordates evolved.

Table 16.1 The Geological Time Scale: Some Major Evolutionary Events of Animals

Era	Period	Millions of Years Ago	Major Biological Events—Animals	
Cenozoic	Quaternary	2	Age of Human Civilization	Age of Mammals
	Tertiary		First hominids appear Dominance of land by mammals and insects	
		65		
Mesozoic	Cretaceous	130	Dinosaurs become extinct	Age of Dinosaurs
	Jurassic		Age of Dinosaurs First mammals and birds appear	
		180		
	Triassic	230	First dinosaurs appear	
Paleozoic	Permian	280	Expansion of reptiles Decline of amphibians	Age of Amphibians
	Carboniferous	350	Age of Amphibians First reptiles appear	
	Devonian		Age of Fishes First insects appear First amphibians move onto land	Age of Fishes
		400		
	Silurian	435	Expansion of fishes	
	Ordovician	500	Invertebrates dominate the seas First fishes (jawless) appear	Age of Invertebrates
	Cambrian	600	Age of Invertebrates Trilobites abundant	

Vertebrates

The vertebrates (**subphylum Vertebrata**), like the annelids and arthropods, have *bilateral symmetry,* the coelomic *tube-within-a-tube body plan,* and are *segmented.* Also, at some time during their life histories, vertebrates have the three chordate characteristics. However, the embryonic notochord is generally replaced by a *vertebral column* composed of individual vertebrae. The *skeleton is internal* and consists not only of a vertebral column, but also of a skull, or cranium, that encloses and protects the brain. In higher vertebrates, other skeletal parts serve for attachment of muscles and for protection of internal organs. All vertebrates have a *closed circulatory system* in which red blood is contained entirely within blood vessels. They show good *cephalization* with sense organs. The eyes develop as outgrowths of the brain. The ears are primarily equilibrium devices in aquatic vertebrates, plus sound wave receivers in land vertebrates. The kidneys are important excretory and water-regulating organs that conserve or rid the body of water as necessary.

In contrast to vertebrates, annelids and arthropods do not have a notochord or gill pouches. They also have a ventral solid nerve cord rather than a dorsal hollow nerve cord. Occasionally, the same types of adaptations are seen

Figure 16.6

Evolutionary tree of the vertebrates. Extinct jawless fishes are the oldest vertebrates in the fossil record. Although they had a well-developed, sometimes armored head, they were most likely bottom feeders. Fishes with jaws that appeared during the late Silurian period evolved into the cartilaginous fishes and the bony fishes, including fishes with lungs and lobed-finned fishes. The ray-finned fishes, which are most common today, are descended from bony fishes with lungs; amphibians came from lobed-finned fishes and were dominant during the Carboniferous period. Stem reptiles also appeared around this time, and from them arose the dinosaurs, who ruled during the Jurassic period, and the first mammals, who did not come into their own until the Tertiary period. Birds may very well be of dinosaurian origin.

birds

mammals

dinosaurs

reptiles

ray-finned fishes

stem reptiles

coelacanths

amphibians

cartilaginous fishes

primitive amphibians

plate-skinned fishes

lobe-finned fishes

bony fishes with lungs

primitive jawed fishes

lampreys and hagfishes

jawless fishes

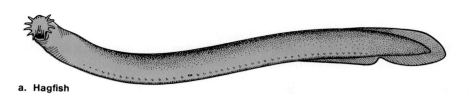

a. Hagfish

b. Lamprey

Figure 16.7
The jawless fishes of today are not at all like their fossil ancestor (fig. 16.6). *a.* The hagfishes are scavengers with sensory tentacles around the mouth and a row of slime glands along each side. *b.* Parasitic lampreys attach themselves to fish with their suckerlike mouth, rasp away the flesh with their horny teeth, and suck out blood.

in both invertebrates and vertebrates; for example, both insects and birds have wings; both fishes and squids have fins. Such structures are analogous rather than homologous, even though they serve the same function, because their basic structural plans are completely different and because they have evolved independently.

Figure 16.6 shows an evolutionary tree of the vertebrates. It is possible to trace vertebrate evolution from fish to amphibians to reptiles to both birds and mammals. Although there is no living representative of primitive ancestors that connect the classes, a number of fossils are intermediate between some of the classes.

Fishes

There are three classes of fishes: the jawless fishes (**class Agnatha**), the cartilaginous fishes (**class Chondrichthyes**), and the bony fishes (**class Osteichthyes**). The earliest vertebrate fossils are jawless fishes of the Ordovician period. They had heavy, bony armor and were probably filter feeders. Living representatives of the jawless fishes are hagfishes and lampreys (fig. 16.7). They are cylindrical, some up to as much as a meter long, with smooth, scaleless skin. The **hagfishes** are scavengers, feeding mainly on dead fish. The **lampreys** may be parasitic, in which case the round muscular mouth serves as a sucker. The lamprey uses this sucker to attach itself to another fish and to tap into its circulatory system. Marine parasitic lampreys that gained entrance to the Great Lakes when a canal was deepened between the St. Lawrence River and the lakes proliferated and caused extensive reduction in the trout population of the lakes in the early 1950s.

The first jawed vertebrates in the fossil record are **placoderms** (fig. 16.8), jawed fishes of the Devonian period (table 16.1). Jaws are a distinct advantage because they allow predation and promote an active life. It is not surprising, then, that these fishes also had paired pectoral fins anteriorly, and paired pelvic fins posteriorly. Paired fins allow a fish to balance and maneuver well in the water. Unknown early jawed fishes were ancestral to the placoderms and to today's jawed fishes.

Figure 16.8
Painting of an extinct placoderm, the first type of jawed vertebrate in the fossil record. The appearance of jaws, which most likely evolved from the first anterior pair of gill arches, was a revolutionary event in the evolutionary history of vertebrates.

Cartilaginous Fishes

Sharks, rays, and **skates** are cartilaginous fishes (fig. 16.9); they have a *skeleton of cartilage* instead of bone. Their bodies are covered with small, toothlike scales called denticles. Thus a shark's skin feels like sandpaper. The menacing teeth of sharks and their relatives are simply larger, specialized versions of these scales.

Rays and skates have horizontally flattened bodies. They swim slowly along the bottom of marine environments and feed on animals that they dredge up. Sharks have beautifully streamlined bodies and are fast-swimming predators in the open sea.

Sharks hold a terrifying fascination for many people, and they have a generally well-deserved reputation as ferocious and somewhat indiscriminate feeders. Tiger sharks, for example, have been known to bite off and swallow chunks of boats. Great white sharks seem to bite first and then decide whether or not to swallow. This may explain why a number of human victims of great white shark attacks seem to have been "spit out." However, some of the largest sharks, basking sharks and whale sharks, are not active predators of large prey, but instead are filter feeders, ingesting tons of small prey, collectively called "krill."

Bony Fishes

Lungfishes   The earliest of the bony fishes are thought to have lived in freshwater habitats. In addition to gills, they possessed simple, *saclike lungs* connected to the anterior end of their digestive tracts. These lungs probably functioned as supplementary gas exchange surfaces that allowed them to breathe air when the stagnant water was oxygen-deficient. There are three living genera of lungfishes (subclass Dipnoi), one each in Australia, Africa, and South America. Today's lungfishes regularly breathe air to supplement gas exchange in their gills.

Ray-Finned Fishes   As figure 16.6 indicates, the primitive lungfishes were a common ancestor to two groups of bony fishes. The **ray-finned fishes** (subclass Actinopterygii), which had fins supported by spinelike rays, evolved to include

Figure 16.9
Cartilaginous fishes. *a.* The southern sting ray has a flattened shape and moves slowly across the bottom dredging up food. *b.* The bull shark is a streamlined predator that moves swiftly through the water to capture its prey.

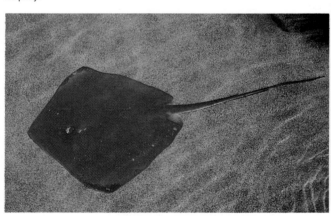

a.

b.

most of today's bony fishes. In these fishes the ancient lung eventually became a swim bladder. By secreting gases into or absorbing gases from this bladder, the fish can change its relative density and thus move up or down in the water. Modern bony fishes have broad, flattened scales covering some or all of their body surfaces. Their gills are located in an enclosed gill chamber that is covered and protected by a hard, bony flap, the **operculum.** Thus they have only one gill opening on each side instead of separate openings for each gill slit, as have sharks.

Ray-finned fishes of today (fig. 16.6) are the most numerous, varied, and familiar of the fishes. They live in almost every kind of aquatic habitat, ranging from seawater to brackish water to fresh water, and from frigid Arctic waters to perpetually warm tropical seas. They range from bright and beautifully colored fish that dart about in coral reefs to jet black fish that lurk in the ocean depths. Their members include snakelike eels having no fins, comical seahorses that hover in the water, angler fishes that dangle a baitlike structure to tempt other fishes within reach, and a host of specializations in still other species.

Lobe-Finned Fishes The **lobe-finned fishes** (subclass Crossopterygii), which had fleshy fins, also evolved from the lungfishes. The lobe-finned fishes could breathe air while they crawled clumsily on stumpy fin lobes from pond to pond during the Devonian period (table 16.1), a time of alternating floods and droughts. Land plants and insects had evolved before land vertebrates, so there was an excellent food supply available on land for any animal that could remain ashore long enough to utilize it. Under these conditions, a selective pressure obviously favored those animals that developed land-dwelling abilities. Animals able to spend more time on land could avoid more vigorous competition in the water. It is probable that the first amphibians evolved from some of the lobe-finned fishes (fig. 16.10).

Figure 16.10
Comparison of limbs of a lobe-finned fish and the limbs of a primitive amphibian. *a.* A lobe-finned fish that probably could move out of the water onto land. *b.* Primitive amphibian. It must have been an awkward walker, but its legs were an improvement for locomotion on land as is made clear by the side drawings. Movement onto land opened up a new phase in the history of vertebrates.

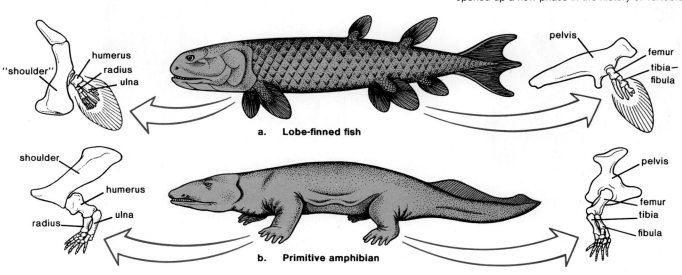

a. Lobe-finned fish

b. Primitive amphibian

The lobe-finned fishes were once believed to have been extinct for 75 million years. However, a living coelacanth was found in marine waters off the coast of South Africa in 1939. Coelacanths are not believed, however, to be closely related to amphibians, which evolved from freshwater lobe-finned fishes.

Amphibians

Improved locomotion on land and other adaptations for terrestrial life led to a diversification of amphibians (**class Amphibia**) during the Carboniferous period (table 16.1), now known as the Age of Amphibians. Many of these became extinct; amphibians are mainly represented today by **newts** and **salamanders,** and by **frogs** (fig. 16.11) and toads. These animals are not fully adapted to life on land. They have small, relatively *inefficient lungs,* and most of them also depend on gas exchange through the skin. For skin to function efficiently as a gas-exchanging surface, it must be quite thin and kept moist. In a dry environment, amphibians are in danger of desiccation and death. However, the most important factor binding amphibians to water is their

Figure 16.11
Frog metamorphosis, during which the animal changes from an aquatic to a terrestrial organism. *a.* Hatching of eggs. *b.* Tadpoles breathe by means of gills. *c.* Development of legs. *d.* Frogs breathe by means of lungs and skin.

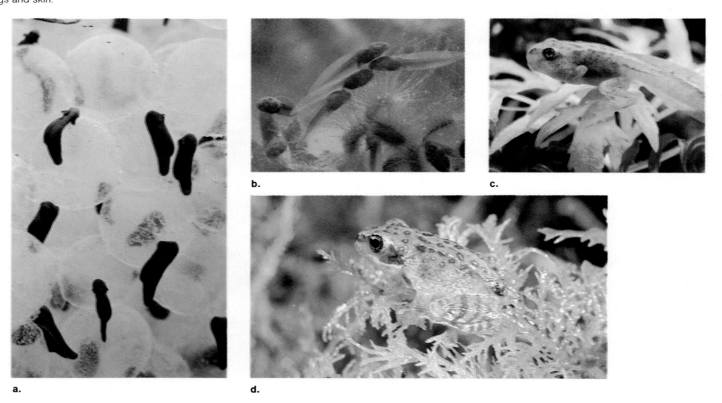

a.

b.

c.

d.

method of reproduction. Amphibians, like most fishes, have *external fertilization;* both eggs and sperm are shed into the water. The eggs are enclosed by a jelly coat that provides no protection against desiccation if the eggs are exposed to the air. Young amphibians hatch from their jelly as aquatic larvae (tadpoles) with gills. Tadpoles feed and grow in the water. Only after *metamorphosis* do amphibians emerge from the water as air-breathing adults (fig. 16.11). Their class name, Amphibia, meaning "double life," accurately describes their life history, which is divided between water and land.

A few amphibians have special reproductive adaptations that enable them to reproduce on land, however. For example, some frogs carry their developing eggs in fluid-filled pouches. Their young proceed to metamorphosis very quickly, becoming tiny adults that are ready to live and grow on land.

Some time before the Permian period, reptiles that were better adapted to life on land evolved from an amphibian related to *Seymouria* (fig. 16.12). They were to become the dominant organisms on earth during the next geological era.

Reptiles

Reptiles (**class Reptilia**) are largely represented today by **turtles, crocodiles and alligators,** and **lizards and snakes** (fig. 16.13). If these animals have limbs, they are designed for land travel. All of them have thick, scaly skin that contains keratin, which makes the skin water impermeable. Since the lungs are efficient, there is no need to use the skin for gas exchange.

Figure 16.12
Drawing of *Seymouria*, a fossil animal related to a possible evolutionary link between amphibians and reptiles.

a.

b.

Figure 16.13
Reptilian diversity. *a.* A snake has a tough, scaly skin that resists drying out. *b.* The slow-moving turtle has an external shell in addition to scaly skin.

Figure 16.14
Reptilian egg, a major advancement because it allowed these animals to reproduce on land. *a.* Baby crocodile hatching out of its shell. Note that the shell is leathery and flexible, not brittle like birds' eggs. *b.* Inside the egg, the embryo is surrounded by membranes. The chorion aids gas exchange; the yolk sac absorbs nutrients; the allantois stores waste, and the amnion encloses a fluid that prevents drying out and provides protection.

a.

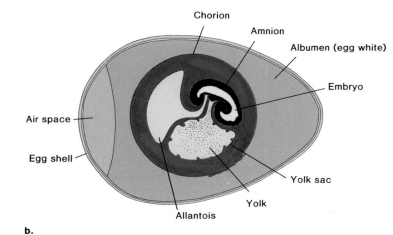

b.

Perhaps the reptiles' most outstanding adaptation to living on land is their means of reproduction. They have internal fertilization and lay eggs that are protected by leathery shells. This shelled egg eliminated the need for a swimming larval stage during development. It contains *extraembryonic membranes* that protect the embryo, remove nitrogen wastes, and provide the embryo with oxygen, food, and water. These membranes (p. 608) are not part of the embryo itself and are disposed of after development is complete. One of the membranes, the amnion, is a sac that fills with fluid and provides a "private pond" within which the embryo develops (fig. 16.14).

The earliest reptiles, called stem reptiles, gave rise to several lines of descent during the Mesozoic era (table 16.1), called the Age of Reptiles. One line produced the **dinosaurs** that underwent adaptive radiation and dominated the earth for more than 100 million years before their relatively sudden extinction, possibly due to asteroid or comet bombardment of the earth as discussed in the reading on page 246. **Birds** may be closely related to dinosaurs. *Archeopteryx* (fig. 16.15), the oldest known fossil bird, had several reptilian

characteristics, including jaws with teeth and a long, jointed reptilian tail. The bird was about the size of a crow and had rather feeble wings for its size, but it was well feathered.

There is another branch from stem reptiles, which, unrelated to the dinosaurs, produced **mammallike reptiles** called **therapsids** (fig. 16.16) early in the Jurassic period. The therapsids had mammallike limbs, skulls, and teeth, and some of them resembled large dogs and were aggressive predators.

Birds

Birds (**class Aves**) are the only modern animals that have feathers, which are actually modified reptilian scales. Scales on the legs and feet of birds are reminders of their reptilian ancestry, as are the claws at the ends of their toes.

Birds' feathers are of two types. *Contour feathers* are attached to the wings in such a way that they overlap to produce a broad, flat surface beneficial to flight. *Down feathers* provide excellent insulation against body heat loss. This is important because birds maintain a constant, relatively high body temperature, which permits them to be continuously active even in cold weather.

There are many orders of birds, including birds that are *flightless* (ostriches), *web-footed* (penguins), *divers* (loons), *fish eaters* (pelicans), *waders* (flamingos), *broad-billed* (ducks), *birds of prey* (hawks), *vegetarians* (fowl), *shore dwellers* (sandpipers), *nocturnal* (owls), *small nectar-feeders* (hummingbirds), and *songbirds,* the most familiar of the birds.

Nearly every anatomical feature of birds can be related to their ability to fly. They have hollow, very light bones with internal air sacs that are a part of the respiratory system (p. 480). A bird's head is relatively small and light because light horny beaks have replaced jaws with teeth. A slender neck connects the head to a rounded, compact torso. The breastbone is enlarged and carries a keel to which the strong flight muscles are attached.

Flight requires well-developed sense organs. Birds have particularly acute vision and excellent visual and muscular reflexes. They can land precisely on small tree branches and swoop from great heights to capture small animals that were spotted while cruising high above the unsuspecting prey.

Figure 16.16
Therapsid, a Jurassic reptile that had many mammalian features including their characteristic means of locomotion and differentiated teeth. Most likely therapsids chewed their food before swallowing, which would have provided quick energy to maintain a warm temperature.

Figure 16.17
Unlike *Archaeopteryx*, modern birds have toothless beaks and an entirely feathered tail instead of a long bony tail. They retained the reptilian egg, except the bird egg is hard-shelled, not leathery. Also, birds are usually devoted parents. Here, a song sparrow feeds its young.

Birds have internal fertilization and lay *hard-shelled eggs*. Many newly hatched birds require parental care (fig. 16.17) before they are able to fly away and seek food for themselves. Complex hormonal regulation and behavioral responses are involved in bird behavior. A remarkable aspect of bird behavior is the annual migration of many species over distances up to thousands of kilometers. Birds navigate successfully by day and night, through sunshine and cloudy weather, apparently using the sun and stars and even variations in the earth's magnetic field to find their way on these impressive journeys.

Mammals

The chief characteristics of **mammals (class Mammalia)** are hair and milk-producing **mammary glands. Hair** provides insulation against heat loss and helps maintain a constant body temperature so mammals can be active in cold weather. Other characteristics include limb positioning so that mammals are able to run faster than most other vertebrates, and well-developed sense organs and an enlarged brain, both important for an active life.

Mammary glands enable females to feed (nurse) their young without deserting them to find food. Nursing also creates a bond between mother and offspring that helps assure parental care when the young are helpless. In most mammals the young are born alive after a period of development in the uterus, a part of the female reproductive tract. Internal development shelters the young and allows the female to move actively about while the young are maturing.

Mammals evolved during the Mesozoic era from therapsid reptiles (fig. 16.16). Some of the first groups, represented today by the *monotremes* and *marsupials,* are not found abundantly today. Mammals remained relatively small until the reptiles declined. Then the marsupials and later the *placental mammals* radiated into many of the habitats previously occupied by the large reptiles, such as dinosaurs.

Figure 16.18
Monotreme versus marsupial. Monotremes, the egg-laying mammals, include the (*a.*) duck-billed platypus, an aquatic organism. Marsupials, the pouched mammals, include (*b.*) the koala bear, a native of Australia.

a.

b.

Monotremes

Monotremes (subclass Prototheria; order Monotremata) are represented by the duck-billed platypus (fig. 16.18) and the spiny anteater, both of which are found in Australia. Mammals are classified according to means of reproduction. The **monotremes** lay their eggs, which resemble reptilian eggs, in burrows in the ground. The female incubates the eggs, as do birds; but after hatching, the young are dependent upon the milk that seeps from modified sweat glands on the abdomen of the female.

Figure 16.19
Placental mammals have three physiological features not
shared by reptiles: maintainence of a constant internal
temperature, development of the higher centers of the
brain, and internal development followed by maternal care
of offspring. Aquatic harbor seals (a.), and forest-dwelling
white-tail deer with fawn (b.) illustrate but two of the many
environments occupied by placental mammals.

a.

b.

Marsupials

The young of **marsupial** (subclass Metatheria; order Marsupialia), or pouched, mammals begin their development inside the female's body, but they are born in a very immature condition. Newborn young crawl up into a pouch on their mother's abdomen. Inside the pouch, they attach to nipples of mammary glands and continue their development.

Today, marsupial mammals (fig. 16.18) are found mainly in Australia; only a few marsupials, such as the American opossum, are found outside that continent. In Australia, marsupials underwent adaptive radiation for several million years without competition from placental mammals, which arrived there only recently. Among herbivores, koala bears are tree-climbing browsers and kangaroos are grazers. The Tasmanian "wolf" or "tiger" is a carnivorous marsupial about the size of a collie dog. There are even marsupial moles in Australia.

Placental Mammals

Developing placental mammals (subclass Eutheria) are dependent on the placenta (p. 611), a region of exchange between maternal blood and fetal blood. Here nutrients are supplied to the growing offspring and wastes are passed to the mother for excretion. While the fetus is clearly parasitic on the female, in exchange she is free to move about as she chooses while the fetus develops.

Placental mammals (fig. 16.19) evolved during the Tertiary period and entered the habitats left vacant when the dinosaurs became extinct. Table 16.2 lists the major orders of placental mammals, including those that are adapted to life in the water, on land, and in the air. In chapter 30 we will see that the various mammals are distributed in a specific manner about the world. For example, the hoofed mammals are likely to be found in grasslands, while the primates are likely to be found in forests. We will now consider primate evolution and in so doing will discuss human evolution.

Table 16.2 Some Major Orders of Placental Mammals

Insectivora (moles, shrews)	Primitive; small, sharp-pointed teeth
Chiroptera (bats)	Digits support membranous wings
Carnivora (dogs, bears, cats, sea lions)	Canine teeth long; teeth pointed
Rodentia (mice, rats, squirrels, beavers, porcupines)	Incisor teeth grow continuously
Perissodactyla (horses, zebras, tapirs, rhinoceroses)	Large, long-legged, one or three toes, each with hoof; grinding teeth
Artiodactyla (pigs, cattle, camels, buffalos, giraffes)	Medium to large; two or four toes, each with hoof; many with antlers or horns
Cetacea (whales, porpoises)	Medium to very large; forelimbs paddle-like; hind limbs absent
Primates (lemurs, monkeys, gibbons, chimpanzees, gorillas, men)	Mostly tree-dwelling; head freely movable on neck; five digits, usually with nails; thumbs and/or large toes usually opposable

Adapted from *Essentials of Biology*, 2d ed., by Willis H. Johnson, et. al. Copyright © 1969 and 1974 by Holt, Rinehart and Winston, Inc. Used by permission of CBS Publishing.

Table 16.3 Classification of Humans and Related Animals

Category	Animals	Category	Animals
Phylum Chordata		**Superfamily Hominoidea**	*Dryopithecus*
Subphylum Vertebrata			*Ramapithecus*
Class Mammalia			Modern Apes
Order Primates	Prosimians		Humans
	Lemurs	**Family Hominidae**	*Australopithecus*
	Tarsiers		*Homo habilis*
	Anthropoids		*Homo erectus*
	Monkeys		*Homo sapiens*
	Apes	**Genus Homo**	*Homo habilis*
	Humans	(Humans)	*Homo erectus*
			Homo sapiens

Figure 16.20

Evolution of primates took place in the Cenozoic era, which includes the Tertiary and Quaternary periods. There are at least four lines of evolution among the primates who arose from mammalian insectivores. Prosimians (before monkeys) include the lemurs and tarsiers; the monkeys include the New World and the Old World monkeys; apes include the gibbons and great apes (orangutan, gorilla, and chimpanzee); and humans.

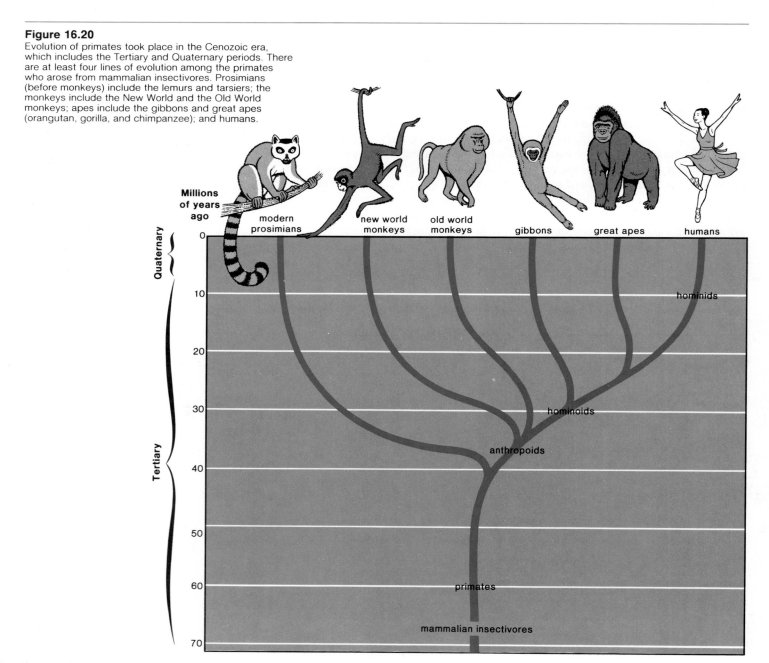

Human Evolution

Primates

Humans belong to that group of mammals designated as primates (table 16.3). **Primates** are believed to have evolved (fig. 16.20) during the early Cenozoic era from small insectivores (insect-eating mammals) that took up an arboreal life, thereby escaping predators on the ground and finding new sources of food in angiosperm trees that were becoming prevalent.

The first primates may have been similar to the living tree shrews of Southeast Asia (fig. 16.21), rat-sized animals with a snout, claws, and sharp front teeth. By 50 million years ago, however, primates had evolved characteristics suitable to moving freely through the trees. Their limbs became adapted to swinging and leaping from branch to branch. While both hands and feet retained five functional digits (fingers and toes), the hands became especially dexterous and mobile. The thumbs were opposable; that is, they closed to meet the fingertips. Thus these animals could easily reach out and bring food, such as fruit, to the mouth. Claws were replaced by nails, which offer protection but still allow a tree limb to be freely grasped and released.

A snout is common to those animals in which a sense of smell is of primary importance. In primates, the sense of sight is more important, so the snout has been considerably shortened, allowing the eyes to move to the front of the head. This resulted in excellent binocular (three-dimensional) vision, permitting primates to make accurate judgments about the distance and position of adjoining tree limbs. The visual portion of the brain, located in the cerebral cortex, also increased in size, as did those centers responsible for hearing and touch, which are also beneficial to survival when living in trees.

As primates increased in size, the care of multiple offspring from a single birth would have been a distinct liability. One live birth at a time became the norm and the period of postnatal maturation was prolonged, giving the immature young an adequate length of time to learn behavioral patterns.

Prosimians

The first primates were **prosimians,** a term which means premonkeys. The living members of the group include lemurs and tarsiers (fig. 16.22). Lemurs, which have a squirrellike appearance, are confined largely to the island of

Figure 16.21
A tree shrew is a small animal with feet adapted for climbing in trees where it feeds on fruit and insects. Notice that the ears are similar to the ears of the primates, being rounded with folded edges. A tree shrew's brain is relatively large, but the olfactory portion is surprisingly small. Tree shrews may resemble the ancient ancestors of modern primates.

Figure 16.22
The ring-tailed lemur, a prosimian. Many lemurs are endangered today, but not the ring-tailed (obviously named for the appearance of its tail). Lemurs have long, bushy tails and long hind limbs.

Figure 16.23
Anthropoids include monkeys and apes. *a.* The spider monkey is a New World monkey. *b.* The orangutan is a great ape that is still arboreal. *c.* Of all the apes, chimpanzees are most closely related to humans.

a.

b.

c.

Table 16.4 Periods and Epochs of the Cenozoic Era

Era	Period	Epoch	Millions of Years before Present*	Biological Events
Cenozoic	Quaternary	Recent	0.01–present	Modern humans
		Pleistocene	2.5	Early humans
	Tertiary	Pliocene	7–2.5	Hominids
		Miocene	25–7	Hominoids
		Oligocene	38	Anthropoids
		Eocene	54	Prosimians
		Paleocene	65	First placental mammals
Mesozoic			230	

*These are approximate dates of the beginnings of these intervals.

Madagascar. Their face is still long and snoutlike; some digits have claws and some have nails. The big toe and thumb are mobile but not opposable. The tarsiers, which are found in the Philippines and East Indies, are curious mouse-sized creatures with enormous eyes suitable for their nocturnal way of life. They have a flattened face and their digits terminate in nails.

Anthropoids

Monkeys, along with apes and humans, are **anthropoids** (fig. 16.23). The anthropoids diverged from prosimians about 38 million years ago during the Tertiary period, which is divided into the epochs noted in table 16.4.

During the Cenozoic era, the climate became cooler and drier, and lush trees were to be found only in southern Pangaea, the original supercontinent of the earth. Here monkeys began to evolve, but the group split when Gondwana separated partially into South America and Africa. Thus, there are *New World monkeys,* which have long prehensile (grasping) tails and flat noses, and *Old World monkeys,* which lack such tails and have protruding noses. Many Old World monkeys have brightly colored buttocks. Some well-known New World monkeys are the spider monkey, the capuchin ("organ grinder's" monkey), and the squirrel monkey. Some of the better-known Old World monkeys are now ground dwellers, such as the colorful mandrill, a familiar baboon, and the macaques, one of which, the rhesus monkey, has been used in medical research.

Hominoids

Humans are more closely related to apes (fig. 16.23) than to monkeys. **Hominoids** include apes and humans. There are four types of apes: *gibbon, orangutan, gorilla,* and *chimpanzee.* The gibbon is the smallest of the apes, with a body weight ranging from 5½ to 11 kg. Gibbons have extremely long arms that are specialized for swinging between tree limbs. The orangutan is large (75 kg), but nevertheless spends a great deal of time in trees. The gorilla, the largest of the apes (180 kg), spends most of its time on the ground. Chimpanzees, which are at home both in trees and on the ground, are the most humanlike of the apes in appearance and are frequently used in psychological experiments.

Hominoid Ancestor

During the Miocene epoch, which began about 25 million years ago (table 16.4), apes became abundant and widely distributed in Africa, Europe, and Asia. Among these Miocene apes, members of the genus *Dryopithecus* are of particular interest because they are thought to be a possible hominoid ancestor.

Figure 16.24
Knuckle-walking as illustrated by this gorilla may have
allowed the hands to remain specialized for grasping.

Figure 16.25
a. The primate hand is capable of grasping objects. b. In
humans this ability is used to grasp tools.

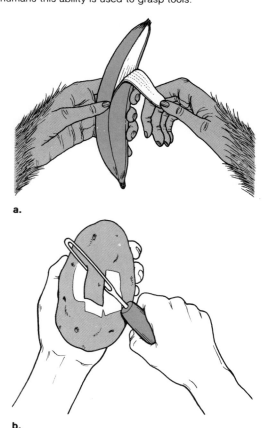

a.

b.

Dryopithecines were forest dwellers that probably spent most of the time in the trees. The bones of their feet, however, indicate that they also may have spent some time on the ground. When they did walk on the ground, however, they walked on all fours virtually all the time, using the knuckles of their hands to support part of their weight. Such "knuckle-walking" is clearly illustrated by the modern gorilla (fig 16.24). Knuckle-walking would have allowed retention of the opposable thumb, a primate characteristic (fig. 16.25).

The skull of *Dryopithecus* had a sloping brow, heavy eyebrow ridges (called supraorbital ridges), and jaws that projected forward. These are features that could have led to those of apes and humans (fig. 16.26). Apes have pronounced supraorbital ridges and the plane of the face projects forward, forming a muzzle. The reason for this pronounced muzzle is the larger number and size of the teeth. Apes' canine teeth are much larger than the adjacent teeth; the two sides of the ape jaw are roughly parallel, giving a rectangular shape to the dental arcade (fig. 16.27). In contrast to apes, modern humans have a high brow and lack the pronounced supraorbital ridges. The plane of the face is flat and the dental arcade assumes a U-shape because the canine teeth are smaller than those of the apes and comparable in size to the other human teeth.

Hominid Ancestor

It is generally agreed that hominids, which include humans and humanlike fossils, shared a common ancestor with apes. Biochemists have compared the structure of proteins and DNA in apes and humans. They believe that the degree of difference between these molecules indicates how long it has been since apes and humans shared a common ancestor. Biochemical evidence suggests that the gibbons began to evolve separately about 18 million years ago and the orangutans split off about 12 million years ago, leaving an apelike creature that led to the evolution of other apes and humans. The close similarity between chimpanzee and human proteins and DNA suggests that these hominids may have had a common ancestor as late as 3 to 4 million years ago (fig. 16.29).

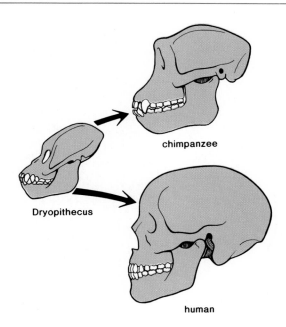

Figure 16.26
Dryopithecus had primitive features, the chimpanzee has apelike features, and humans have features that differ from both of these.

chimpanzee

Dryopithecus

human

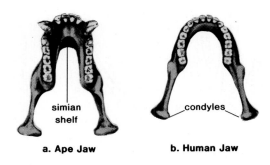

a. Ape Jaw

b. Human Jaw

simian shelf

condyles

Figure 16.27
Dental arcades of an ape and a human. *a.* The two sides of the jaw of an ape are roughly parallel and the canine teeth are much larger than the adjacent teeth; therefore, the dental arcade has a rectangular shape. *b.* In humans, the canine teeth are much reduced and the jaw curves to give a dental arcade that is U-shaped.

Thus far, no one has found fossils that match these dates exactly. A few years ago it was suggested that the fossil remains of *Ramapithecus* may be those of the first hominids. These remains were found in India and Pakistan through the Near East, and in the Balkans to Africa. The remains are dated from as long ago as 17 million to just 8 million years ago. On the basis of only an upper jaw and a handful of molars, it was suggested in the 1960s that the ramapithecines were the first hominids. The molars were large and had a thick enamel covering like human molars and, because the canine teeth were missing, it was believed that the jaw was U-shaped. However, in 1971, when a lower jaw was properly identified, it became obvious that the ramapithecines had apelike features. In the late 1970s and early 1980s, fossilized skulls were found. The close-set eye sockets, eyebrow ridges, protruding jaw, and flaring cheekbones indicate that the ramapithecines were most likely ancestral orangutans. However, because the ramapithecines lived for as long as 10 million years, it cannot be ruled out that they are also ancestral to gorillas, chimps, and humans. One interesting observation is that the ramapithecines, orangutans, and humans all have a thick layer of tooth enamel.

Figure 16.28
Comparison of skeleton of a "knuckle-walking" ape (a gorilla) with the skeleton of a bipedal human. Note differences in proportions of hindlimbs and forelimbs, in the shape of the rib cage and of the pelvis, and in the curvature of the spine and the angle and position of attachment of the vertebral column to the skull. The head of the ape faces forward when knuckle-walking because the vertebral column attaches to the rear of the skull.
From "The Antiquity of Human Walking" by John Napier. Copyright © 1967 by Scientific American, Inc. All rights reserved.

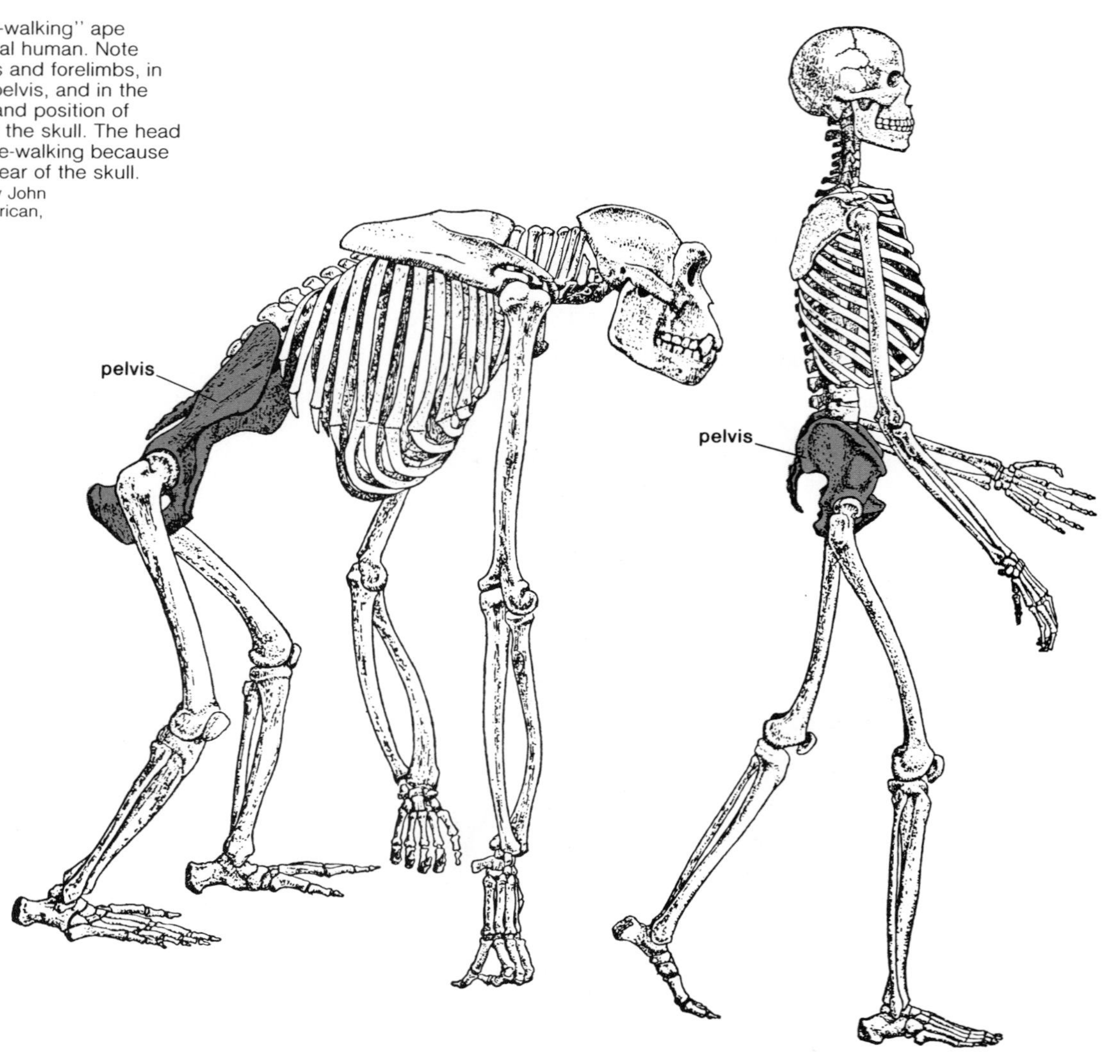

pelvis

pelvis

Large molars having a thick layer of enamel is believed to be significant because it suggests that the diet included not just fruit, but also tough morsels of food such as seeds, grass, stems, and roots. These foods are plentiful on the ground, but not in trees. During the Miocene epoch, the weather was becoming cooler, and even in Africa the tropical and subtropical forests were being replaced by savannas, grasslands with only occasional trees.

The first hominids must have come out of the trees and begun to assume an upright bipedal posture (walking on two legs). This would have enhanced survival on the savanna because it would have allowed hominids to see over tall grass, to spot predators or prey, and would have left the hands free to perform other functions, such as throwing rocks. The transition to erect posture required a number of adaptive changes. Figure 16.28 contrasts the skeleton of a "knuckle-walking" ape with a human skeleton. In the ape, the pelvic region is very long and tilts forward, whereas that in the human is short and upright. The ape shoulder girdle is more massive than that of a human, and ape arms are longer than the legs. In the ape the toe is opposable and the feet have no pronounced arch. The head hangs forward because of the angle of attachment of the vertebral column to the skull. The foramen magnum is a

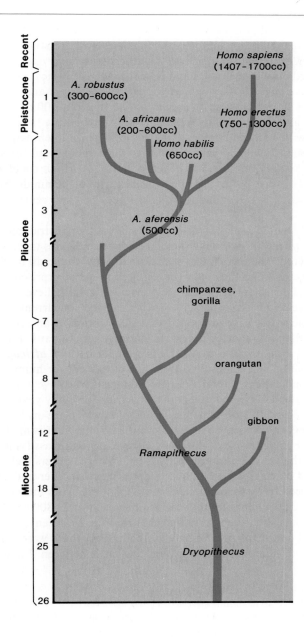

Figure 16.29
Possible evolutionary tree of hominoids (*below*) and hominids (*above*). The hominoid portion of the tree is largely based on biochemical evidence because the fossil record after *Dryopithecus* is virtually blank. Biochemical evidence suggests that modern-day gibbons began to evolve separately from other hominoids about 18 million years ago, followed by the orangutan about 12 million years ago. The hominids and African apes (gorillas and chimpanzees) may have had a common ancestor about 5 to 8 million years ago. The fossil record begins again about 3 to 4 million years ago with the fossil remains of *Australopithecus afarensis* (Lucy). Whether or not this fossil is a common ancestor, as shown here, for all the other hominids is presently in dispute. Also, it has not been determined whether *Homo habilis* is directly ancestral to those fossils generally accepted as humans (*Homo erectus* and *Homo sapiens*). The cubic centimeter (cc) numbers are approximate brain volumes.

hole in the skull through which the spinal cord passes from the vertebral column to join the brain. In apes this opening is well to the rear of the skull; in humans it is almost directly in the bottom center of the skull. Also in humans the spine has an S-curve, allowing a better weight distribution and improved balance when the body is upright.

Hominids

New ideas regarding speciation, particularly the concept of adaptive radiation (p. 251), have been used to interpret the hominid fossil record. Whereas scientists formerly attempted to place each hominid fossil in a straight line from the most primitive to the most advanced, it is now reasoned that several hominid species could have existed at the same time (fig. 16.29).

Figure 16.30

Australopithecus africanus. Some authorities believe this hominid is directly ancestral to humans, while others believe that they share a common ancestor. This picture gives the impression that the australopithecines were hunters, but Pat Shipman of the Johns Hopkins University has recently presented evidence, based on cut marks left by hominids on animal bones, that hominids were most likely scavengers that sought and confiscated the kill of coexisting predators.

Figure 16.31

Homo erectus. Evidence indicates that *Homo erectus* was a successful species found throughout Africa and Eurasia.

Australopithecines

In 1924 a fossil skull, discovered in a limestone quarry in South Africa, was sent to Raymond Dart, an anatomist in Johannesburg, for study. Dart named the fossil *Australopithecus africanus* (southern ape man) and proclaimed it to be intermediate between apes and humans. Many physical characteristics of *A. africanus* (fig. 16.30) are transitional between apes and humans. The brain capacity varies between just over 300 cc to about 600 cc. The brow is more distinct and the face is flatter than that of apes because the dental arcade is U-shaped with reduced canines (fig. 16.27). The foramen magnum is still not quite centered, but the pelvis is shorter than in apes and the spine is S-shaped (fig. 16.28). Therefore, it is believed that *A. africanus* could walk erect, and members of this species are now considered to be hominids. So, too, are the closely related fossils called *A. robustus* because their remains are much larger.

In 1974 Donald Johanson discovered still another australopithecine that he named *A. afarensis.* Although skeletal evidence indicates that this hominid walked fully erect, the rectangular dental arcade and the small brain capacity of only 500 cc are like those of apes. Biologists were surprised to find this combination of characteristics because previously it was believed that walking erect and increased brain capacity probably evolved concurrently.

Some biologists have suggested that *A. afarensis* is ancestral to *A. africanus* and that this species is in turn ancestral to humans. But Johanson thinks that *A. afarensis* is a common ancestor for both *A. africanus* and humans (fig. 16.29). Still another possible line of descent for humans has been suggested. The Leakeys, a family of paleontologists who have hunted hominid fossils at Olduvai Gorge and other sites in Africa for many years, found a 2-million-year-old skull they named *Homo habilis* which means "handy man." They chose this name because they believe that *Homo habilis* made the stone tools that were found nearby. This, they believe, justifies classifying this fossil as a *Homo* although the brain capacity is only 650 cc. Others feel that *Homo habilis* should be classified as an australopithecine. Another find by Richard Leakey has been left unnamed and is simply known by its museum number, "skull 1470." Leakey believes that the brain capacity was 800 cc and that the skull dates from 2.6 million years ago, but this age has been questioned. He feels that this find helps substantiate his belief that there were a number of distinct types of hominids coexisting on the earth between 2 and 3 million years ago.

Humans

The rest of the fossils to be mentioned are in the genus *Homo,* which is the genus for all humans.

Homo erectus

Homo erectus individuals (fig. 16.31) were prevalent throughout Eurasia and Africa during the Pleistocene Age, also called the Ice Age because of the recurrent cold weather that produced the glaciers of this epoch. They had a brain capacity of 750 to 1300 cc compared to 1000 to 1700 cc for modern humans. The face was more projected than in modern humans and the skull bones were thicker. The teeth were large and the lower jaw sloped back so that there was no distinct chin. Even so, the grasp, posture, and locomotion of *Homo erectus* individuals were all similar to those seen in modern humans. We have

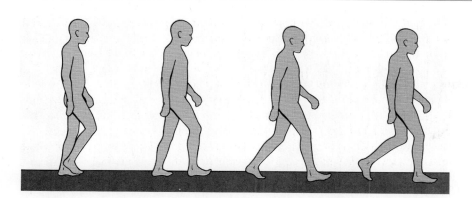

Figure 16.32
The striding gait of modern humans.

a striding gait (fig. 16.32), which means that the legs have alternate phases—the stance phase and the swing phase. When one hindlimb is in the stance phase, the other is in the swing phase. During the swing phase, the knee is first bent and then extended forward. The knee straightens, the heel touches the ground, and, as the foot follows, that limb enters the stance phase.

Homo erectus men made more elaborate tools than *Homo habilis* men and apparently were big-game hunters. In two sites explored near Torralba, Spain, the bones of more than 150 animals have been found, including elephants, deer, horses, wild cattle, and rhinoceroses. Cooperative hunting of big game would have encouraged people to live together. At another site found at Erra Amata in Nice, France, the people apparently lived together in temporary shelters, stopping for a while as they followed the migration of the big-game animals they hunted. Evidence here clearly indicates that *Homo erectus* individuals used fire for cooking.

The size of the brain increased by 50 percent during the years that separate the australopithecines from those classified as *Homo erectus*. This suggests that tool making and increased intelligence may have evolved together during the evolution of humans. There undoubtedly was a selective advantage to an enlarged brain that would facilitate tool use by manipulative hands.

Homo sapiens

About 100,000 years ago, a distinctive group of humans called **Neanderthal** emerged. The name comes from the Neander Valley in Germany where the first fossil of this type of human was unearthed, although actually their fossils have been found in much of southern Europe, Asia, and Africa. There is extensive evidence that the Neanderthals carried out ritualistic practices, which is strongly suggestive of abstract thought, and perhaps they even had a concept of religion and life after death.

Usually the Neanderthals (fig. 16.33) are classified as *Homo sapiens neanderthal*, a variety of modern humans. The distinction is based on primitive skull features such as protruding forehead, no chin, and a flattish skull. But the Neanderthal brain capacity was essentially the same as that of modern humans. Some researchers suggest that the Neanderthals gave rise to modern humans; others believe that Neanderthals were simply a variation of modern humans.

Figure 16.33
Homo sapiens neanderthal. It is now believed that Neanderthals generally had a more modern appearance than this drawing indicates.

Figure 16.34
Cro-Magnon people are the first to be designated *Homo sapiens sapiens*. Their tool-making ability and other cultural attributes, such as their artistic talents, are lengendary.

Fossils generally accepted as modern humans with a cranial capacity of about 1500 cc appear in the fossil record some 40,000 years ago. These have been designated as Cro-Magnon and are classified as *Homo sapiens sapiens,* as are present-day humans. After a period of coexistence, the Neanderthals disappear from the fossil record. Whether the Neanderthals evolved into Cro-Magnon, interbred with them, or were killed off by them is unknown.

Cro-Magnon was also a big-game hunter, and these people have been held responsible by some for the extinction during the Upper Pleistocene of many large mammalian animals, such as the giant sloth, mammoths, saber-toothed tigers, and giant oxen. This predatory life style would have encouraged the evolution of intelligence and the ability to speak. Language would have facilitated their ability to organize to kill prey larger than themselves.

Cooperation, in turn, would have led to socialization and the advancement of culture, customs, and traditions that are passed on from generation to generation. Humans are believed to have lived in small groups, the men going out to hunt by day while the women remained at home with the children.

About 40,000 years ago, a new tool industry spread through the human population. The major characteristic of this industry was the production of the blade, a tool with roughly parallel sides. Cro-Magnon is known not only for his advanced technology but also for his artistic ability. He is believed to have painted the beautiful drawings on cave walls in Spain and France (fig. 16.34), and also to have sculpted many small figurines.

If Cro-Magnon men did cause the extinction of many types of animals, this may account for the transition from a hunting economy to an agricultural economy about 12,000 to 15,000 years ago. This agricultural period extended from that time to about 200 years ago when the Industrial Revolution began, after which many people typically lived in cities, in large part divorced from nature and endowed with the philosophy of exploitation and control of nature. Only recently have we begun to realize that the human population, like all other organisms with which we share an evolutionary history, should work with rather than against nature.

Summary

The deuterostomes in which the second embryonic opening becomes the mouth include the echinoderms and chordates. The echinoderms are radially symmetrical animals with a spiny skin. They move by means of tube feet, which are part of their water vascular system. Their body systems are rather primitive; they have no circulatory system, respiration is by means of skin gills, and the nervous system consists of a central ring with nerve cords in the arms. Sea stars may be the most familiar of the echinoderms, which also include a number of other unusually adapted animals.

All chordates have a dorsal notochord, gill pouches, and dorsal hollow nerve cord at some time in their life history. The invertebrate chordates include tunicates and lancelets. Lancelets are the best example of a chordate that possesses the three chordate characteristics as an adult.

Vertebrates include fishes, amphibians, reptiles, birds, and mammals, all of whom have a vertebral column. The first fishes, represented today by hagfishes and lampreys, were jawless. Sharks, skates, and rays are cartilaginous, while the bony fishes include all other well-known fishes. Bony fishes, represented today largely by ray-finned fishes, originally possessed lungs that remained functional only in lungfishes and in lobe-finned fishes. Amphibians probably evolved from freshwater lobe-finned fishes during the Devonian period.

Amphibians (e.g., frogs and salamanders) are not well adapted to life on land, particularly since they must return to water for reproduction. Reptiles (e.g., turtles, crocodiles, snakes) lay a shelled egg, which allows them to reproduce on land. One main group of ancient reptiles, the stem reptiles, produced a line of descent that evolved into both dinosaurs and birds. Another line of descent from stem reptiles evolved into the mammals.

Birds are feathered, which, since birds are warm blooded, helps them maintain a constant body temperature. Birds are adapted for flight; their bones are hollow, their shape compact, their breastbone is keeled, and they have well-developed sense organs.

Mammals are animals with hair and mammary glands. The former helps them maintain a constant body temperature, and the latter allows them to nurse their young. Monotremes lay eggs; marsupials have a pouch in which the newborn matures; and the placentals retain offspring inside the uterus until birth. Mammals remained small and insignificant while the dinosaurs existed, but when the latter became extinct during the Cretaceous period, mammals became the dominant land organisms.

Humans are primates, most species of which are adapted to an arboreal existence. Prosimians are more primitive than the anthropoids, which include monkeys, apes, and humans. During the Miocene epoch, a mammal appeared that was a common ancestor of humans and apes. Later, a hominid ancestor began to walk erect when the savanna was replacing forests. He could have been a knuckle-walker, thus preserving the opposable thumb that humans use in making and using tools. We are certain that *Australopithecus,* of which there are three known species, was a hominid and walked erect.

Was *Homo habilis,* who made tools during the Pleistocene epoch about 2 million years ago, the first human? *Homo erectus,* who had a brain capacity averaging 1000 cc and who utilized fire, is accepted by most as human. *Homo sapiens neanderthal* is believed to have originated earlier than Cro-Magnon, who is classified along with modern humans as *Homo sapiens sapiens.* Cro-Magnon people created a culture, a trait that sets humans apart from all other animals.

Objective Questions

1. The water vascular system of echinoderms consists of canals and _____ feet.
2. The stomach of a sea star is located in the _____ , while the digestive glands are in the _____ .
3. The three chordate characteristics are a _____ , _____ , and _____ .
4. The _____ and _____ are primitive chordates.
5. The three classes of fishes, _____ , _____ , and _____ , indicate in general their order of evolution.
6. Most of today's bony fishes are _____ fishes.
7. Amphibians evolved from _____ fishes that had primitive lungs.
8. Whereas amphibians must return to the _____ to reproduce, reptiles lay _____ that contain _____ membranes.
9. Both _____ and mammals maintain a constant internal _____ .
10. There are three types of mammals: _____ , _____ , and _____ .
11. Dogs, cats, horses, mice, rabbits, bats, whales, and humans are all _____ mammals.
12. Primates are adapted to an arboreal life in the _____ .
13. Anthropoids include _____ , _____ , _____ .
14. The fossil known as Lucy could probably walk _____ , but had a _____ brain.
15. The two varieties of *Homo sapiens* from the fossil record are _____ and _____ .

Study Questions

1. Describe the anatomy and physiology of an echinoderm such as a sea star.
2. What are the three chordate characteristics, and which animal possesses them as an adult?
3. Name the three types of fishes and give examples of each group. Which type of fish was probably ancestral to amphibians?
4. Evaluate the adaptation of amphibians to a land existence.
5. Indicate ways in which reptiles are better adapted to land than are amphibians.
6. Discuss the adaptations of birds for flight.
7. Name the three types of mammals and give examples of each type.
8. Discuss the adaptations of primates for arboreal life.
9. Name the epochs of the Cenozoic era and relate them to the evolution of humans.
10. Name the common ancestor that humans may have shared with apes and discuss the first possible hominid ancestor.
11. Name and discuss the other hominids, including those that are classified as *Homo*.
12. Name and discuss characteristics by which humans may be anatomically distinguished from apes.

Selected Key Terms

neoteny ('nē ə tē nē) 330
tube feet ('tüb 'fēt) 331
notochord ('nōt ə ,kôrd) 333
tunicate ('tü ni kət) 334
lancelet ('lan slət) 334
placoderm ('plak ō ,dərm) 337
operculum (ō 'pər kyə ləm) 339
therapsid (thə 'rap səd) 343

mammal ('mam əl) 344
monotreme ('män ə ,trēm) 345
marsupial (mär 'sü pē əl) 347
primate ('prī ,māt) 349
prosimian (prō 'sim ē ən) 349
anthropoid ('an thrə ,pȯid) 351
hominoid ('häm ə ,nȯid) 351

Suggested Readings for Part 6

Butler, P. J. G., and Lug, A. November 1978. The assembly of a virus. *Scientific American.*

Cairns-Smith, A. G. June 1985. The first organisms. *Scientific American.*

Day, W. 1979. *Genesis on planet earth.* East Lansing, MI: House of Talos.

Dickerson, R. E. September 1981. Chemical evolution and the origin of life. *Scientific American* 239.

Goreau, T. F., et al. August 1979. Corals and coral reefs. *Scientific American.*

Gosline, J. M., and De Mont, M. E. January 1985. Jet-propelled swimming in squids. *Scientific American.*

Gould, S. J. The five kingdoms. *Natural History* 85(6):30.

Hay, R. L., and Leaky, M. D. February 1982. The fossil footprints of Laetoli. *Scientific American.*

Hickman, C. P., Jr., et al. 1982. *Biology of animals.* 3d ed. St. Louis: C. V. Mosby.

Mader, S. S. 1982. *Inquiry into life.* Dubuque, IA: Wm. C. Brown Publishers.

Margulis, L. August 1971. Symbiosis and evolution. *Scientific American.*

Raven, P. H., et al. 1981. *Biology of plants.* 3d ed. New York: Worth.

Roper, C. F. E., and Boss, K. J. April 1982. The giant squid. *Scientific American.*

Russell, Hunter. 1979. *A life of invertebrates.* New York: MacMillan.

Schopf, J. W. September 1978. The evolution of the earliest cells. *Scientific American.*

Simons, K., et al. February 1982. How an animal virus gets into and out of its host cell. *Scientific American.*

Stanier, R. Y., et al. 1976. *The microbial world.* 4th ed. Englewood Cliffs, NJ: Prentice-Hall.

Stern, K. R. 1982. *Introductory plant biology.* 2d ed. Dubuque, IA: Wm. C. Brown Publishers.

Vertebrate structure and function: Readings from Scientific American. 1974. San Francisco: W. H. Freeman.

Vidal, G. February 1983. The oldest eukaryotic cells. *Scientific American.*

PART 7

Plant Structure and Function

Plants are specialized to carry on photosynthesis. Flowering plants, representative of plants in general, are composed of a root system and a shoot system; the latter includes the stems, leaves, and flowers. The roots and leaves of a plant function as specialized regions that carry on exchanges with the environment. The roots take up mineral ions and water from the soil, and the leaves absorb carbon dioxide from and add oxygen to the air. A transport system takes mineral ions and water to the leaves, where photosynthesis occurs, and takes the products of photosynthesis to the roots.

Plant growth is coordinated so the seed germinates and the plant grows and flowers during those times of the year that are best for the continuance of the species. Woody plants are adapted to survive a cold or a dry season and continually increase in size during each new growing season.

Flowering plants are only one of many types of plants. The flower is composed of modified leaves with coloring that attracts pollinators. Pollination, the transfer of pollen from one plant to the other, is necessary for reproduction to occur.

CHAPTER 17

Plant Structure and Organization

Concepts

1. Flowering plants are adapted to a terrestrial existence as evidenced by their structure and organization.

2. The reproductive process in flowering plants assures that the gametes, zygote, and embryo are protected from drying out.

3. The organs of a plant (roots, stems, leaves, and flowers) have structures that are suited to their functions.
   a. The roots anchor a plant, absorb water and nutrients, and store the products of photosynthesis.
   b. The stem supports the leaves and transports water and nutrients.
   c. The leaves are the main organs for photosynthesis.
   d. The flowers are the sexual reproductive organs.

Table 17.1 Comparison of Water Environment with Land Environment

Water	Land
1. The surrounding water prevents the organism from drying out; that is, it prevents desiccation.	1. To prevent desiccation, the organism must obtain water, provide it to all body parts, and possess a covering that prevents evaporation.
2. The surrounding water buoys up the organism and keeps it afloat.	2. An internal structure is required for a large body to oppose the pull of gravity.
3. The water prevents desiccation and allows easy transport of reproductive units, such as zoospores and swimming sperm.	3. The organism may provide a water environment for swimming reproductive units. Alternately, the reproductive units must be adapted to transport by wind currents or by motile animals.
4. The surrounding water prevents the fertilized egg (zygote) from drying out.	4. The developing zygote must be protected from possible desiccation.
5. The water maintains a relatively constant environment in regard to temperature, pressure, and moisture.	5. The organism must be capable of withstanding extreme external fluctuations in temperature, humidity, and wind.

I n this chapter we will study the structure and organization of flowering plants, which are the most complex, most recently evolved, and most widely distributed plants. Flowering plants evolved on land and are adapted to a terrestrial environment. Compared to water, land is a more difficult environment for organisms, as table 17.1 indicates.

We will begin this chapter with a consideration of the seed and end it with a study of the flower that produces the seed.

The Seed

Seeds in flowering plants are enclosed in either fleshy or dry fruits that help with dispersal of the species. Dry fruits sometimes burst open, propelling the seeds into the air where they may be carried by the wind. In contrast, the one-seeded dandelion fruit has a terminal tuft of long hairs that help it float in the air. Some dry fruits have hooked bristles that cause them to cling to passing animals. Animals often eat fleshy fruits, and in the process, the seeds may be swallowed then deposited unharmed some distance away.

The mature seed contains an embryonic plant composed of several parts. One of the main features is the cotyledon(s), or seed leaf (leaves). **Cotyledons** provide nutrient molecules for growing embryos before photosynthesis begins. Some embryos have one cotyledon; plants that produce such embryos are known as monocotyledonous, or **monocots.** Other embryos have two cotyledons; plants that produce such embryos are known as dicotyledonous, or **dicots.** Adult monocots and dicots have several structural differences, as illustrated in figure 17.1.

Another important difference between the two types of plants is the manner in which nutrient molecules are stored in the seed. In a monocot the cotyledon rarely stores food; rather, it absorbs food molecules from the **endosperm** (p. 300) and passes them to the embryo. On the other hand, during the development of a dicot embryo, the cotyledons (fig. 17.2) do store the nutrient molecules that the embryo uses.

Monocot

One cotyledon in seed

Flower parts in threes
and multiples of three

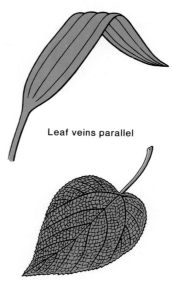

Vascular bundles
scattered in stem

Leaf veins parallel

Dicot

Two cotyledons in seed

Flower parts in fours or
fives and multiples of
four or five

Vascular bundles
in a definite ring

Leaf veins form a net pattern

Figure 17.1
Flowering plants are either monocots or dicots. These four
features are used to distinguish monocots from dicots.

Figure 17.2
Stages in development of a dicot embryo and seed.
a. The single-celled zygote lies beneath the endosperm
nucleus. *b. and c.* The endosperm is a mass of tissue
surrounding the embryo. The embryo itself is actually the
cells above the suspensor. *d.* The embryo becomes heart-
shaped as the cotyledons begin to appear. *e.* There is
progressively less endosperm as the embryo differentiates
and enlarges. As the cotyledons bend around, the embryo
takes on a torpedo shape. *f.* The mature seed has a seed
coat that protects the enclosed embryo. The embryo
consists of the epicotyl, represented here by the shoot
apex; the hypocotyl; and the radicle, which forms the first
seedling root.

triploid
endosperm
nucleus

endosperm

embryo

cotyledons
appearing

bending
cotyledons

shoot apex

cotyledons

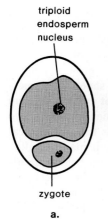

zygote

a.

suspensor

b.

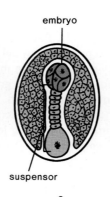

c.

d.

e.

hypocotyl

radicle

seed coat

f.

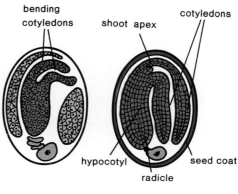

Plant Structure and Organization   363

Figure 17.3
Monocot seed and germination. *a.* Longitudinal section of
mature seed. Notice the presence of both endosperm and
a cotyledon in the monocot seed. *b.* When a seed
germinates, the epicotyl and hypocotyl give rise to those
portions above ground and the radicle becomes the root.

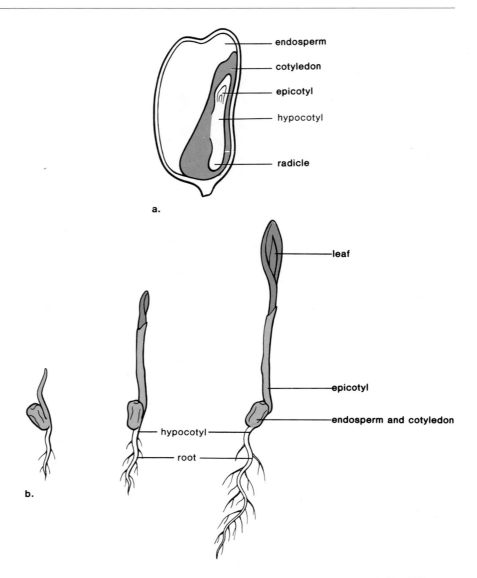

Once the cotyledons develop, it is possible to distinguish the **shoot apex** and the **root apex** (fig. 17.2). Each contains **meristem,** a type of tissue that continuously divides to produce new cells. Plants continue to grow as long as they live because they always contain meristem tissue. The **shoot apical meristem** is responsible for above-ground growth, while the **root apical meristem** is responsible for below-ground growth. The **epicotyl,** a portion of the embryo that lies above the attachment of the cotyledon(s), contains the shoot apical meristem and sometimes bears young leaves, in which case it may be called a *plumule.* The **hypocotyl,** a portion of the embryo that lies below the attachment of the cotyledon(s), becomes a portion of the stem. The lowest part is the **radicle,** which contains the root apical meristem and becomes the first (primary) root of the seedling. The embryo and stored food, either within the cotyledon(s) or within endosperm, are covered by a *seed coat.*

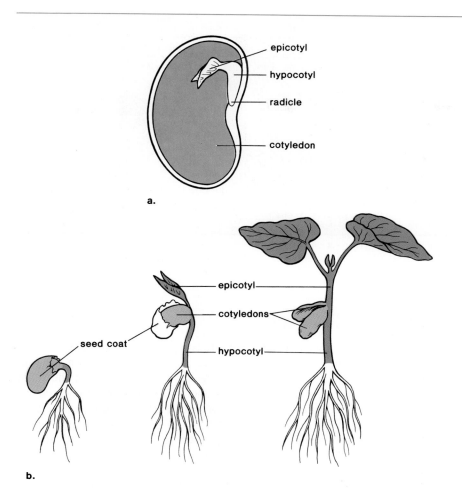

Figure 17.4
Dicot seed and germination. *a.* Longitudinal section of
mature seed after one cotyledon has been removed.
b. When a dicot germinates the cotyledons are present
above ground where they can be seen until they
disappear after all nourishment has been removed from
them by the growing seedling.

Seed Germination

Some seeds will not **germinate** (begin to grow) until they have been dormant
for a while. For seeds, **dormancy** is a period of time during which growth does
not occur even though conditions may be favorable for growth. Seed dormancy
may not occur in the tropics where it has little protective value, but dormancy
in the temperate zone may not be broken until the seeds have been exposed
to a period of cold weather. This requirement helps assure that seeds do not
germinate until the most favorable growing season has arrived. Dormancy is
over and germination may begin once the seed coat bursts. Water, bacterial
action, and even fire act on the seed coat and cause it to break open. Germination now takes place if there is sufficient water, warmth, and oxygen to sustain growth (oxygen is needed for cellular respiration).

Normally, the first event observed in a germinating seed (figs. 17.3 and
17.4) is the emergence of the root. This is followed shortly thereafter by the
appearance and expansion of the shoot. As the dicot seedling emerges from
the soil, the shoot is hook-shaped to protect the delicate plumule. But once
the ground is broken, the stem straightens out and the leaves expand as photosynthesis begins.

Figure 17.5
A land plant is divided into two main portions: the root system below ground and the shoot system containing the stems and leaves above ground. A node is a region of the stem where one or more leaves are attached.

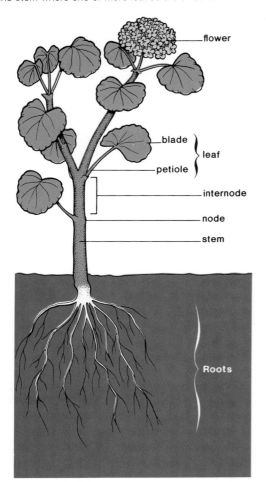

flower

blade
leaf
petiole

internode

node

stem

Roots

Table 17.2 Vegetative Organs and Major Tissues

	Roots	Stems	Leaves
Function	Absorb water and minerals	Transport water and nutrients	Carry on photosynthesis
Tissue Epidermis*	Root hairs absorb water and minerals	Protect inner tissues	Stomata carry on gas exchange
Cortex†	Store products of photosynthesis	Carry on photosynthesis, if green	———
Vascular‡	Transport water and nutrients	Transport water and nutrients	Transport water and nutrients
Pith†	———	Store products of photosynthesis	———
Mesophyll† Spongy layer Palisade layer	———	———	 Gas exchange Photosynthesis

Plant tissues belong to one of three tissue systems.
*Dermal tissue system
†Ground tissue system
‡Vascular tissue system

Organs and Tissues

The body of a flowering plant is divided into two portions: the **root system** and the **shoot system** (fig. 17.5). The root system anchors the plant in the soil, absorbs water and minerals, and stores organic nutrients. Within the shoot system, the stem transports water to the leaves and lifts them up to catch the rays of the sun. The leaves take in carbon dioxide from the air and carry on photosynthesis.

The **vegetative organs** of a plant (roots, stems, and leaves), which are not concerned with reproduction, have a relatively simple anatomy. Some of the specialized tissues found within each organ, along with their primary functions, are listed in table 17.2. These tissues are derived from *meristem,* which continuously divides to produce more meristem. Some of these cells differentiate into the cell types shown in figure 17.6. Parenchyma and sclerenchyma cells are found in most tissues. **Parenchyma** cells are relatively unspecialized and correspond best to a generalized plant cell (fig. 4.12*b*). **Sclerenchyma** cells are hollow, nonliving cells with extremely strong walls that support plant tissues and organs. Two tissue types, epidermis and endodermis, generally contain only epidermal and endodermal cells, respectively. *Epidermis* covers the entire body of a nonwoody plant and of a young woody plant. In stems and leaves, the walls of epidermal cells contain cutin, a waxy substance that forms a continuous layer called a **cuticle** where the cells are exposed to air. Epidermis protects inner body parts and prevents the plant from drying out. In addition, the epidermis has specialized structures and functions that are discussed as each organ is considered. In contrast to epidermis, **endodermis** is found only in roots and stems and contains cells having walls that are modified in a way to be discussed later (p. 371).

Transport (Vascular) Tissue

A tissue called **xylem** transports water and minerals from the roots to the leaves; a tissue called **phloem** transports organic nutrients most often from the leaves to the roots. Xylem contains two types of conducting cells: tracheids and vessel elements (fig. 17.7). Both types of conducting cells are hollow and nonliving,

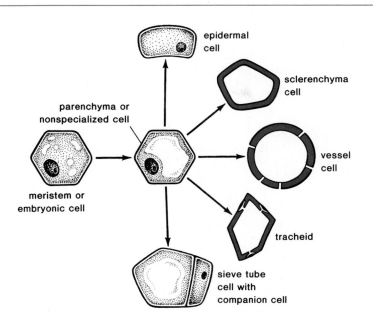

epidermal cell

sclerenchyma cell

parenchyma or nonspecialized cell

vessel cell

meristem or embryonic cell

tracheid

sieve tube cell with companion cell

Figure 17.6
Plant cell types. A meristem cell is embryonic and, by the process of maturation, may become a relatively nonspecialized parenchyma cell or one of the highly specialized cells shown. Nonliving cells that lack cytoplasm are shown in color. Sclerenchyma cells are support cells. Vessel cells and tracheids are in the transport tissue called xylem (fig. 17.7). The sieve tube cell and the companion cell are in the transport tissue called phloem (fig. 17.8). The epidermal cell is found in epidermis, the outermost tissue in all organs of a plant.

Figure 17.7
Xylem structure. General organization of xylem at far left followed by an external view of vessel elements stacked one on top the other and a longitudinal view of several tracheids. Tracheids and vessel elements usually conduct water in the direction shown.

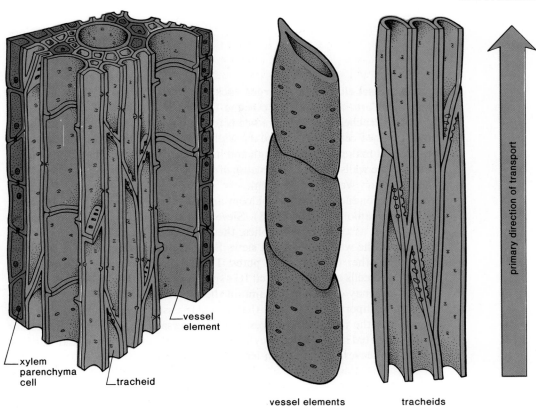

vessel element

xylem parenchyma cell

tracheid

vessel elements

tracheids

primary direction of transport

Figure 17.8

Phloem structure. General organization of phloem at far left followed by external view of two sieve tube cells and their companion cells.

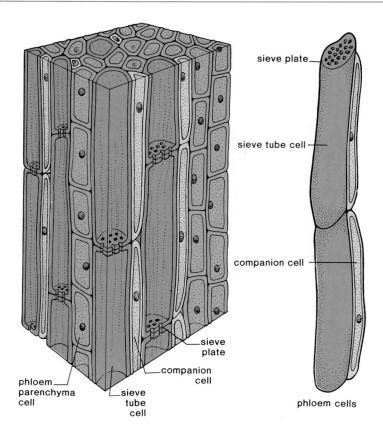

sieve plate

sieve tube cell

companion cell

sieve plate

companion cell

phloem parenchyma cell

sieve tube cell

phloem cells

but the **vessel elements** are larger, lack transverse or end walls, and are arranged to form a continuous pipeline for water and mineral transport. The elongated **tracheids** with tapered ends form a less obvious means of transport. Both types of cells have secondary walls that contain **lignin,** an organic substance that makes the walls tough and hard. Even so, water can move laterally between the walls of the cells because of the presence of pits, depressions where the secondary wall does not form.

The conducting cells of phloem are sieve tube elements, each of which has a companion cell (fig. 17.8). **Sieve tube cells** contain cytoplasm but no nuclei, and, as their name implies, these cells have pores in their end walls that make the walls resemble a sieve. Strands of cytoplasm extend from one cell to the other through these pores. The smaller **companion cells** are more generalized cells and have nuclei. It is speculated that the nucleus of the companion cell may control and maintain the life of both cells.

It is important to realize that vascular tissue (xylem and phloem) extends from the root to the leaves, and vice versa. In the roots, the vascular tissue is located in the vascular cylinder; in the stem, it forms vascular bundles; and in the leaves, it is found in leaf veins.

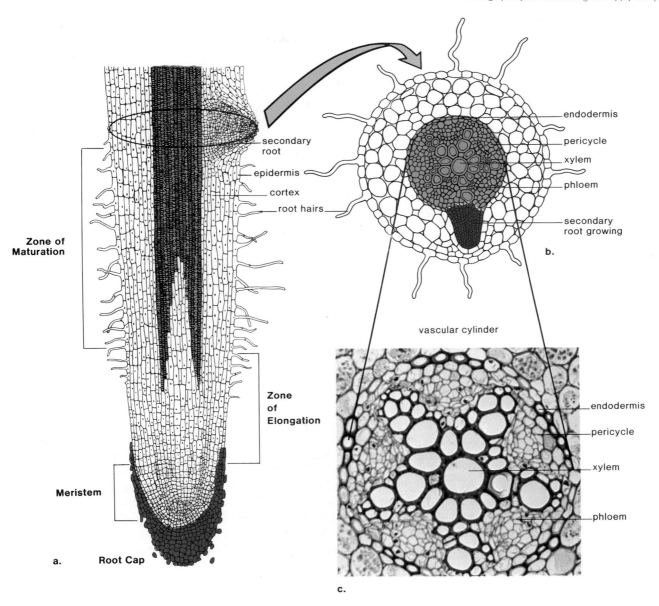

Zone of Maturation

Zone of Elongation

Meristem

secondary root

epidermis

cortex

root hairs

endodermis

pericycle

xylem

phloem

secondary root growing

b.

vascular cylinder

endodermis

pericycle

xylem

phloem

a.

Root Cap

c.

Root System

One of the most striking features of a plant's root system is that it grows continuously as the apical meristem produces new cells. Figure 17.9*a,* a longitudinal section of a root, reveals zones where cells are in various stages of differentiation. Below the **zone of cell division** (apical meristem), cells are continuously being added to the **root cap,** a protective cover for a root tip. Cells in the root cap have to be replaced constantly because they are ground off as

Figure 17.10
Root hair cycle depicted with radish seedling. Because of the cycle, root hairs appear at only one region of a root tip.

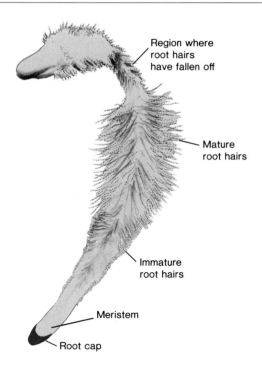

Region where root hairs have fallen off

Mature root hairs

Immature root hairs

Meristem

Root cap

the root pushes through rough soil particles. Above the zone of cell division, cells elongate as they are becoming specialized. Therefore, this region of a root is called the **zone of elongation.** Finally, there is a region in which the cells are now mature and specialized. This region is called the **zone of maturation.** The zone of maturation is recognizable even in a whole root because this is where root hairs are borne by many of the outer epidermal cells. It is even possible to observe evidence of a **root hair cycle.** Root hairs appear, mature and function briefly, and then fall off (fig. 17.10). The root hair cycle causes the root hair zone to be only a short section (1–6 cm) of the length of a root.

Even though each region we have been discussing is short and measures at most several centimeters, the entire root system of a plant is quite extensive. In his study of rye roots in the 1930s, Dittmer estimated that a single four-month-old rye plant had roots totaling more than 600 km in length. There were literally millions of branched and rebranched roots. He estimated that the plant had about 14×10^9 root hairs and that the total surface area of roots and root hairs was almost 640 m². Root hairs add tremendously to the total absorptive surface area of roots and to the absorption functions of roots.

Tissues

Figure 17.9*b* is a cross section of a root cut through the region of maturation to show the specialized tissues found there.

Epidermis

The outer layer of the root, the *epidermis,* consists of only a single layer of cells. The majority of epidermal cells are thin-walled and rectangular, but in the region of maturation, many epidermal cells have root hairs. These project out as far as 5 to 8 mm among the soil particles.

Figure 17.11

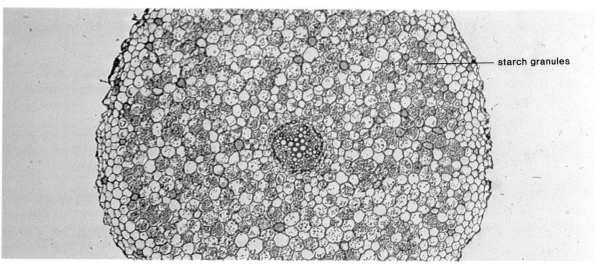

starch granules

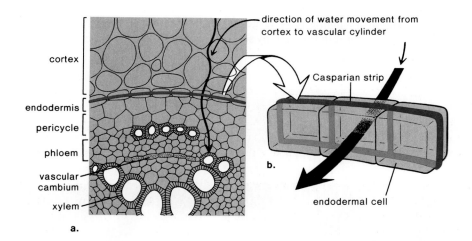

cortex

endodermis

pericycle

phloem

vascular cambium

xylem

a.

direction of water movement from cortex to vascular cylinder

Casparian strip

endodermal cell

b.

Figure 17.12
Movement of solutes from cortex to vascular cylinder by way of endodermal cells. *a.* Diagram illustrating the relationship of root tissues, and the position of the Casparian strip. *b.* Endodermal cells in detail also showing location of the Casparian strip. Because of the Casparian strip, water and solutes (large arrow) must pass through the cytoplasm of endodermal cells. In this way the endodermal cells regulate the passage of materials into the vascular cylinder.

Cortex

Moving inward, next to the epidermis are large, thin-walled parenchyma cells that make up the **cortex** of the root. These irregularly shaped cells are loosely packed to keep free space among them. The cortex functions in food storage; each cell can store starch granules (fig. 17.11).

A single layer of rectangular endodermal cells forms the *endodermis,* a boundary between the cortex and the inner vascular cylinder. The endodermal cells fit snugly together and are bordered on four sides (fig. 17.12) (but not the two sides that contact the cortex or the pericycle) by a strip of waxy material known as the **Casparian strip.** This strip will not permit water and solutes to pass between adjacent cell walls; therefore, the only access to the vascular cylinder is through the endodermal cells themselves, as shown by the arrow in figure 17.12*b.*

Vascular Cylinder

The *pericycle,* the first layer of cells within the vascular cylinder, can start the development of branch or secondary roots (fig. 17.9*b*). The main portion of the vascular cylinder, though, does contain vascular tissue. The xylem appears star-shaped (fig. 17.9*c*) because several arms of tissue radiate from a common center. The phloem is found in separate regions between the arms of the xylem. A single layer of a meristematic tissue, called **vascular cambium** (fig. 17.12*a*), lies between the xylem and phloem in an older root. In *perennial plants* (those that grow for more than one season), vascular cambium produces a new layer of xylem and phloem each year. The xylem builds up to make the roots thicker.

Root Adaptations

In some plants the first or primary root grows straight down and remains the dominant root of the plant. This so-called **taproot** (fig. 17.13*a*) is often fleshy and stores food. Carrots, beets, turnips, and radishes have taproots that we consume as vegetables.

In other plants a number of slender roots develop, and there is no single main root. These slender roots and their lateral branches make up a **fibrous root** system (fig. 17.13*b*). Even the branches of fibrous roots are sometimes fleshy and store food, in which case they are called tuberous roots. Sweet potatoes have tuberous roots, for example.

Sometimes a root system develops from an underground stem or from the base of an above-ground stem. These are **adventitious roots** whose main function may be to help anchor the plant. If so, they are called *prop roots* (fig. 17.13*c*). Mangrove plants have large prop roots that spread away from the plant to help anchor it in the marshy soil where mangroves are typically found.

Figure 17.13
Root systems. *a.* The taproot of a dandelion. *b.* The fibrous root stem of a grass. *c.* Prop roots of a screw pine.

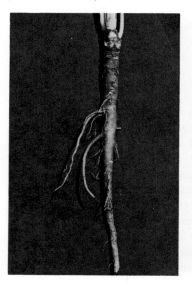

a.   b.   c.

Stems

The *shoot system* of a plant (fig. 17.5) consists of stems and their lateral appendages, the leaves. Although leaves have a shape and arrangement on the stem that is typical of the particular species, leaves are always arranged so each one is exposed to the rays of the sun. The area or region of a stem where a leaf or leaves are attached is called a *node,* and the stem regions between nodes are called *internodes*.

Primary Growth

At the very tip of the stem is a **terminal bud**—the apical meristem protected by newly formed leaves (fig. 17.14). As maturation occurs, each leaf expands and spreads away from the stem. As the internodes lengthen between the leaves, they become spaced out on the stem. At the junction of each leaf and the stem is a group of meristem cells that can become flowers, side branches, and or additional leaves. This group of cells is called a **lateral,** or axillary, **bud.**

Beneath the terminal bud and some of the lateral buds, newly formed cells gradually elongate and differentiate into the various types of tissues characteristic of the mature tissues of a stem (table 17.2). Thus, there is a zone of cell division, a zone of elongation, and a zone of maturation associated with the terminal bud and some of the lateral buds. Formation of these zones constitutes the primary growth of the main stem and branches.

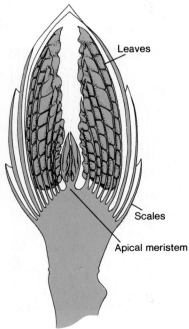

Figure 17.14
A terminal bud and two lateral buds on a shagbark hickory (*Carya ovata*), and a diagram of a longitudinal section of a terminal bud. These buds have emerged from their bud scales, which protected them during the winter.

Leaves

Scales

Apical meristem

Figure 17.15
Herbaceous dicot stem anatomy. *a.* Cross section of
alfalfa (*Medicago*) stem shows that the vascular bundles
occur in a ring. *b.* Drawing of a section of the stem, with
tissues in the bundle and in the stem identified.

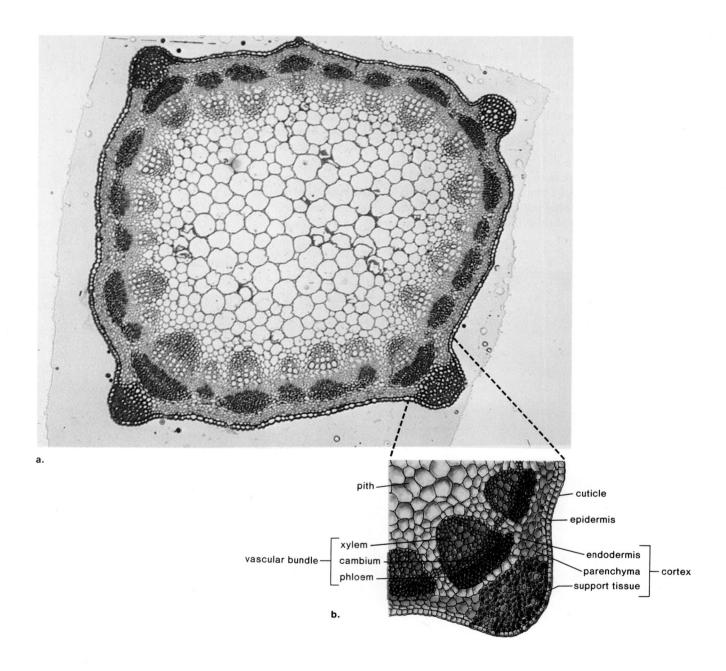

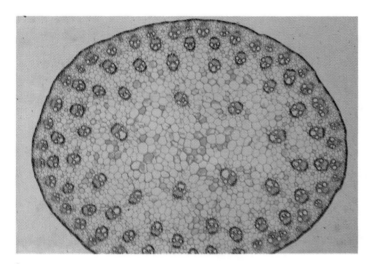

a.

Figure 17.16
Monocot stem anatomy. *a.* Cross section of corn (*Zea mays*), showing that the vascular bundles are scattered. *b.* Enlargement of stem showing vascular bundle in more detail. *c.* Enlargement of one vascular bundle shows the arrangement of tissues in a bundle.
Photograph by Carolina Biological Supply Company.

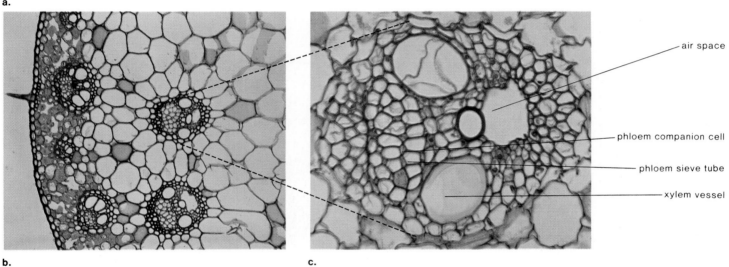

b.

c.

air space

phloem companion cell

phloem sieve tube

xylem vessel

Herbaceous Stems

Herbaceous stems are nonwoody. The outermost tissue of these stems is the epidermis covered by a cuticle to prevent water loss. Herbaceous stems also have **vascular bundles** where xylem and phloem are found. In each bundle, xylem is typically found toward the inside of the stem and phloem is found toward the outside.

In the **dicot herbaceous stem** (fig. 17.15), the bundles are arranged in a *distinct ring* that separates the **cortex** from a central tissue called **pith**. The cortex is sometimes green and carries on photosynthesis, while the pith may function as a storage site. In the monocot stem (fig. 17.16), the vascular bundles are *scattered* throughout the stem, and there is no well-defined cortex or well-defined pith.

Figure 17.17

Drawing of a woody twig as it would appear in winter. The bud contains leaves and meristematic tissue (fig. 17.14) that will begin to grow once the tough protective bud scales open in the spring. Bud scale scars mark the location of the previous terminal bud and thus can be used to determine how much growth occurred during the previous growing season. Leaf scars show where leaves were located. These areas also contain the vascular bundle scars because the conducting tissue of the stem is continuous with that of the leaf.

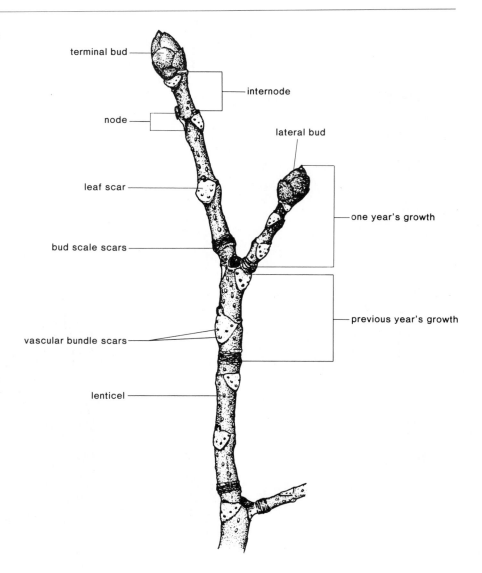

terminal bud

internode

node

lateral bud

leaf scar

one year's growth

bud scale scars

previous year's growth

vascular bundle scars

lenticel

Woody Stems

Some dicots have woody stems (fig. 17.17) that overwinter and grow again the next season. Buds (fig. 17.14) formed at the end of one growing season become active at the beginning of the next growing season. Overwintering is possible because the buds are enclosed within several layers of highly modified leaves, the **bud scales.** These scales may be waxy or hairy, or modified in some other way to protect the delicate embryonic leaves and apical meristem in the bud. In the spring, as the buds absorb water, they burst out of their protective coats. Eventually, the bud scales drop off but leave bud scale scars that show where each season's growth began. Therefore, they can be used as annual time markers to indicate the age of a woody twig (small stem).

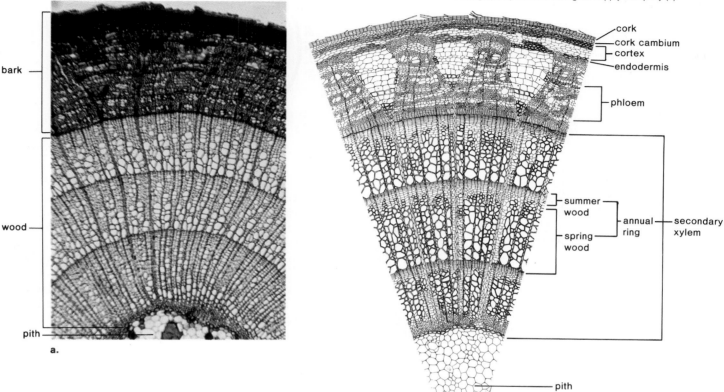

Figure 17.18
Woody dicot stem after three years' growth.
a. Photomicrograph. *b.* Drawing indicating location of cork,
phloem, and xylem that accumulates to give annual rings.
Photograph by Carolina Biological Supply Company (a).

Labels on a.: bark, wood, pith

Labels on b.: cork, cork cambium, cortex, endodermis, phloem, summer wood, spring wood, annual ring, secondary xylem, pith

Secondary Growth

A cross section of woody stem (fig. 17.18) shows an entirely different type of organization than that of a dicot herbaceous stem. There are no vascular bundles. Instead, a young woody stem has three distinct areas: the **bark,** the **wood,** and the **pith.** Between the bark and the wood is a layer of meristem tissue called lateral or **vascular cambium.** Vascular cambium produces new xylem and phloem, called secondary xylem and phloem, each year. The layers of xylem remain and form the **annual rings** that are counted to tell the age of a tree. It is easy to tell where one ring begins and another ends. In the spring, when water is plentiful, the xylem elements are much larger than in the late summer, when water is scarcer. In large trees only the more recently formed layer of xylem, the **sapwood,** functions in water transport. The older inner part, called the heartwood, becomes plugged with deposits, such as resins, gums, and other substances. **Heartwood** may help support a tree, although some trees stand erect and live for many years after the heartwood has rotted away.

Figure 17.19

Secondary growth. Activity of vascular cambium producing secondary tissue in a woody stem. The circumference of the cambium ring increases as the stem grows.

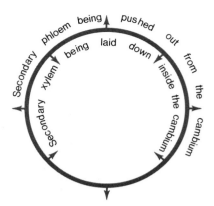

Phloem does not accumulate in rings as does xylem. Phloem is formed on that side of vascular cambium next to the bark (fig. 17.19). The cells are more delicate than those found in xylem, and the outer layers of phloem become crushed as they are pushed against the bark. The newest layer of secondary phloem, however, remains as a functional part of the bark. Since phloem is in the bark, even partial removal can seriously damage a tree.

As the width of a woody stem increases, the epidermis is replaced by **cork** produced by cork cambium, a meristem tissue derived from the cortex. Cork is made up of dead cells impregnated with the same type of waxy material found in the epidermal cell walls of younger stems. **Lenticels** (fig. 17.20) are small pockets of loosely organized tissue where gas exchange can take place.

Types and Uses of Stems

Stem modifications often assist vegetative or asexual reproduction. Above-ground horizontal stems, called **runners** or **stolons**, produce new plants where nodes touch the ground. The strawberry plant (fig. 17.21) is a common example of this type of propagation.

Underground horizontal stems (**rhizomes**) may be long and thin as in sod-forming grasses, or thick and fleshy as in *Iris*. Rhizomes survive the winter and also contribute to asexual reproduction because each node bears a bud. Some rhizomes have enlarged portions called tubers that function in food storage. For example, white potatoes are tubers. The eyes of potatoes, since they are located at nodes, can produce new potato plants.

Corms are bulbous underground stems that lie dormant during the winter just as rhizomes do. They also produce new plants the next growing season. Gladiolus corms are referred to as bulbs by lay persons, but the botanist reserves the term bulb for a structure composed of modified leaves.

Vertical stems, which are of course above ground, are used by humans for various purposes. The stem of the sugarcane plant is our primary source of table sugar. The spice cinnamon and the drug quinine are derived from the barks of two different plants. The softwood of gymnosperms is used for numerous purposes, including lumber for construction and the production of paper and rayon. Hardwood from angiosperms is used for flooring, furniture, and interior finishes.

Figure 17.20
Cross section of a portion of the bark of a woody stem in which the cork is stained a reddish color. A lenticel is a portion of the cork where an indentation among the cells permits gas exchange to occur.
Photograph by Carolina Biological Supply Company.

Figure 17.21
Vegetative reproduction is fairly common in plants. The strawberry plant sends out runners (horizontal stems). New plants appear where the nodes of these runners touch the ground.

Paper Is Secondary Xylem

Over 500 lb., or about 1.4 lb. per day, is a close approximation of the paper used annually by each person in America. Figured on a per capita basis, we use paper products at nearly twice the rate of the next highest consumer, Canada, and at better than 25 times the rate of the Communist bloc nations taken as a whole.

Immediately, of course, we think of paper as a means of communication; historically, that was nearly its exclusive use for over a thousand years. Paper appears to have been invented in China around A.D. 105. Paper remained an exclusively Chinese product for about 500 years until it appeared in Japan. It wasn't until another 500 years had passed that it found its way westward into Egypt, and finally Europe in the twelfth century. Its use for other than communication purposes was virtually nil until the mid-1800s when paper bags and boxes were invented. Around the turn of this century, we find milk cartons, cups, plates, food wrappers, and all kinds of things being made of paper.

What is this wonderful stuff made of? Plants, of course! Actually, the fibers used in manufacturing paper come from rags, straw, grasses, old newspapers, and a number of other sources. But the ultimate source is always plant material, and most of it, better than 90%, is wood. Cellulose, the basic material for paper, is a material in the cell walls of secondary xylem, which is wood. You may have the feeling that paper manufacturing is very complicated. Actually the basic process is quite simple and hasn't changed much in its nearly 2000-year history. There are four steps: (1) make the pulp, (2) make the stock (optional), (3) make wet paper, and (4) dry it out.

Pulp is simply a suspension of pulverized wood cells. In the trade these cells are called fibers, but as plant biologists we know that most of them are really tracheids and vessels. This suspension of fibers is made in either of two ways. The logs with the bark removed may simply be ground against a grinding stone in the presence of water. Paper made from this kind of pulp is relatively inexpensive but weak and generally of poor quality. Newsprint, for example, is primarily of this type. . . . Alternatively, the logs may be fragmented into chips and then pulverized, using hot sulfites, soda, or sulfates, depending on the various processes. These chemicals remove impurities from the chips, leaving only the cellulose; consequently the yield is only about 40–50% of the original wood, but the product is of much finer quality. The pulp is then bleached so that the paper will be white instead of wood colored. It can be used directly to make paper, but usually it passes through an intermediate step during which it is made into stock.

Pulp is refined into stock simply by further grinding. The purpose is to roughen and fray the extracted fiber fragments and to make them more uniform in size. As a result of this treatment, the fibers adhere to one another more readily, and the strength and quality of the paper are improved. Also dyes, strengtheners, and other additives may be introduced at this point.

The third step is to make wet paper out of the pulp or stock. This is accomplished by applying the pulp or stock to a wire-screen filter, draining off the liquid and thereby leaving only the wood fibers which form a thin, matted network on the surface of the screen. In times past this was done

Leaves

Leaves are the organs of photosynthesis in vascular plants. A **leaf** usually consists of a flattened **blade** and a **petiole,** which connects the blade to the stem. The blade may be single or composed of several leaflets. Externally, it is possible to see the pattern of the **leaf veins** that contain vascular tissue. Leaf veins have a net pattern in dicot leaves and a parallel pattern in monocot leaves (fig. 17.1).

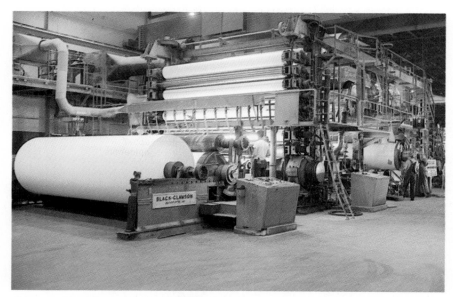

Paper machine take-up spool at Potlatch Corporation, Lewiston, Idaho.

by simply immersing a screen in a vat which contains the stock, and then drawing the screen upward, causing the wood fibers to collect on it. This was done by hand, and one man could produce perhaps 750 sheets per day. Today you'll find manual operation replaced by a revolving wire-screen belt up to 30 feet wide which moves at rates up to 88 feet per second or about 60 miles per hour. This belt picks up the stock and drains off the water rapidly by using a vacuum on the underside. . . . At such rates one machine can turn out in less than 2 seconds the equivalent of one man working all day using the techniques of the early 1800s.

The final step is to press the paper and dry it. This is accomplished by squeezing the paper between heated rotating drums. Finally, the paper is collected in giant rolls.

From *Botany: A Human Concern* 2nd ed., by David L. Rayle and Hale L. Wedberg. Copyright © 1980 by Saunders College/Holt, Rinehart and Winston. Reprinted by permission of Holt, Rinehart and Winston, CBS Publishing.

Figure 17.22a shows the cross section of a typical leaf for a temperate zone plant. At the top and bottom is a layer of epidermal tissue that often bears protective hairs and glands that produce irritating substances. These features may prevent the leaf from being eaten by insects. Regardless of whether these additions are present, the epidermis always has an outer waxy cuticle that keeps the leaf from drying out. Unfortunately, it also prevents gas

Figure 17.22

a. Drawing of a leaf from a C₃ plant that is adapted to
temperate climate. In a C₃ leaf the mesophyll is divided
into a palisade and spongy layer. The bundle sheath cells
lack chloroplasts. *b.* A scanning electron micrograph of
the mesophyll. *c.* Drawing of a leaf vein from a C₄ plant
that is adapted to a hot, dry climate. In a C₄ leaf the
mesophyll is arranged around the bundle sheath, and the
bundle sheath cells have chloroplasts.

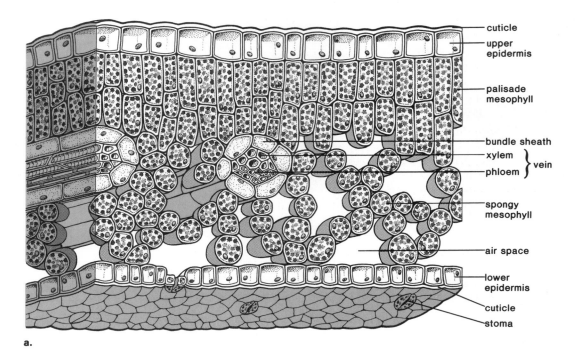

cuticle

upper
epidermis

palisade
mesophyll

bundle sheath

xylem ⎫
        ⎬ vein
phloem ⎭

spongy
mesophyll

air space

lower
epidermis

cuticle

stoma

a.

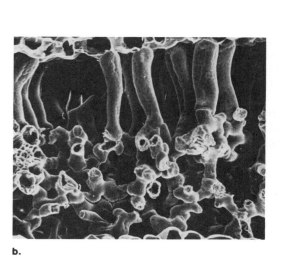

b.

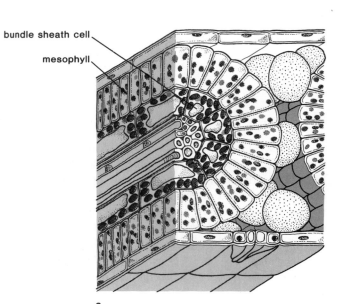

bundle sheath cell

mesophyll

c.

exchange because the cuticle is not gas permeable. However, the lower epidermis, in particular, contains openings called **stomata** (sing., **stoma**) that do allow gases to move into and out of the leaf. Ordinarily, epidermal cells do not contain chloroplasts; however, the **guard cells** surrounding the stomata do contain chloroplasts. These guard cells regulate the opening and closing of the stomata, as discussed on page 394.

The body of a leaf is composed of **mesophyll tissue,** which has two layers of cells: the palisade layer containing elongated cells, and the spongy layer containing irregular cells bounded by air spaces. The parenchyma cells of the **palisade layer** have many chloroplasts and carry on most of the photosynthesis for the plant. The **spongy layer** carries on gas exchange. The loosely packed arrangement of the cells in the spongy layer increases the amount of surface area for gas exchange.

The cells of the spongy layer are protected from drying out by a thin film of water that is constantly evaporating. Loss of water by evaporation at the leaves is called **transpiration.** At least 90 percent of water taken up by the roots is eventually lost by transpiration. This means the total amount of water lost by a plant over a long period of time is surprisingly large. For example, a single *Zea mays* (corn) plant loses somewhere between 135 and 200 l of water through transpiration during a growing season. If this water loss is multiplied by the number of corn plants in a heavily planted cornfield, one can understand why farming requires so much water. The amount of water required by different agricultural crops to produce a given quantity of food can also be compared: millet requires about 225 kg of water for every kilogram of food produced; wheat requires about 500 kg of water per kilogram of food; and potatoes require about 800 kg of water per kilogram of food.

Leaf Veins

A cross section of a leaf shows that leaf veins consist of a strand of xylem and a strand of phloem surrounded by a **bundle sheath** (fig. 17.22a). The bundle sheath differs in C_3 plants as compared to C_4 plants. As discussed on page 108, these terms refer to the number of carbon atoms in the first detected molecule after the start of photosynthesis. The bundle sheath cells in the leaves of C_4 plants characteristically have chloroplasts, and some of the mesophyll cells are often closely packed around the bundle sheath in a radial fashion. These mesophyll cells pass CO_2 to the bundle sheath cells to keep a high concentration of CO_2 in bundle sheath cells. This allows photosynthesis to continue even when the air within the leaf contains a low concentration of CO_2.

Types and Uses of Leaves

Leaves are adapted to environmental conditions. Plants that usually live in the shade tend to have broad, wide leaves, while those that live where it is dry tend to have reduced leaves with sunken stomata. The leaves of a cactus are the spines attached to the succulent stem. Other succulents, however, have leaves adapted to hold moisture.

Leaves can also be specialized for food storage. An onion is a **bulb** with leaves surrounding a short stem. A head of cabbage has a similar construction, except the large leaves overlap one another. The petiole, too, can be thick and fleshy, as in celery and rhubarb.

Figure 17.23
Carnivorous plants have leaves modified for catching and
digesting insects and other small animals. The Venus
flytrap (*a.*) has a bilobed leaf (*b.*) that snaps shut when
triggered by an insect. Once shut, the leaf secretes
digestive juices that break down the soft parts of the
insect's body. The sundew plant (*c.*) has leaves that are
covered by sticky, hairlike structures (*d.*) that trap insects
landing on them. The structures then secrete enzymes
that digest the prey, and the products of digestion are
absorbed.

Photograph by Carolina Biological Supply Company (a, b, d).

a.

b.

c.

d.

Climbing leaves, such as those of peas, are modified into tendrils that can attach to nearby objects. The leaves of a few plants specialize in catching insects (fig. 17.23). The leaves of a *sundew* have sticky epidermal hairs that entrap insects and then secrete digestive enzymes. The *Venus flytrap* has hinged leaves that snap shut and interlock when an insect triggers sensitive hairs. The leaves of a *pitcher plant* resemble a pitcher and have downward-pointing hairs that lead insects into a pool of digestive enzymes. Insectivorous plants commonly grow in marshy regions where the supply of soil nitrogen is severely limited. Feeding on insects provides them with a source of organic nitrogen.

Flowers

A *flower* (fig. 17.24) is composed of modified leaves attached to a portion of the stem called the *receptacle*. The outermost *sepals* are usually green and cover the other flower parts before the bud of the flower opens.

The *petals,* which lie just inside the sepals, are usually brightly colored or white. They often attract insects and birds, which come to the flower to gather *nectar,* a sweet liquid produced by various parts in different flowers.

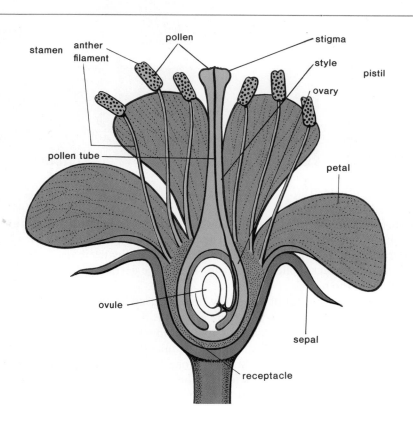

Figure 17.24
Diagram illustrating flower parts. Monocots and dicots differ in the number of flower parts (fig. 17.1). Monocots have flower parts in multiples of three, whereas dicots have flower parts usually in multiples of four and five. This drawing depicts the maturation of the pollen grain. A pollen grain produces two sperm nuclei that travel down a pollen tube to the ovule. Within this structure, one of these nuclei fertilizes an egg nucleus. The resulting zygote becomes the embryonic plant. The other sperm nucleus contributes to the formation of the endosperm (stored food). The ovule will develop into a seed enclosed within a fruit derived from the ovary and, at times, other parts of the flower such as the receptacle.

Figure 17.25
Each type of seed plant has its own type pollen grain, but each one contains two sperm nuclei, one of which will fertilize an egg cell found within an ovule. *a. Thuja Occidentalis* (arbor vitae) pollen. *b. Eucryphia glutinosa* (a small evergreen tree) pollen. *c. Cannabis sativa* (marijuana) pollen.

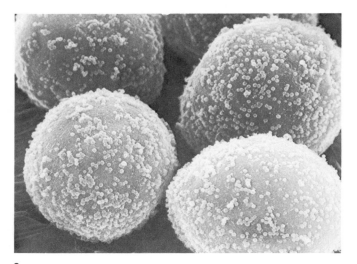

a.

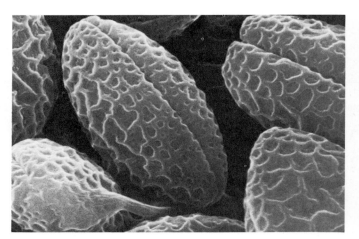

b.

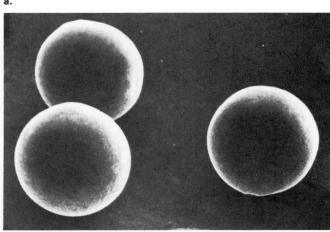

c.

When insects and birds are at the flower, their hair or feathers inadvertently pick up *pollen* produced by the *stamens,* the pollen-producing parts of the flower. Each stamen has a slender filament and terminates in an *anther* that has four pollen sacs. Pollen grains (fig. 17.25) are carried by wind or insects and birds to the pistil.

The *pistil* consists of the ovary, the style, and the stigma. The **ovary** contains the *ovules* within which there are at one particular time several cells, including an egg. The style supports the *stigma* where pollen is received (fig. 17.24). Thereafter, the pollen grain forms a tube through which sperm nuclei travel to the ovule. Within this structure, one of these nuclei fertilizes an egg nucleus. The resulting zygote will become the embryonic plant. The other sperm nucleus contributes to the formation of the endosperm, which is stored food.[1] Thus, when the mature ovule becomes a seed, the *seed* contains the embryonic plant and stored food. The seed coat is derived from the wall of the ovule. Further development of the embryonic plant and germination of a seed is discussed at the beginning of this chapter.

[1]The life cycle of angiosperms is discussed in more detail on page 300.

Angiosperm seeds are enclosed within a *fruit* derived from the wall of the ovary and, at times, other accessory parts of the flower such as the receptacle. As the ovary wall thickens, it becomes divided into three layers that may be fleshy or dry according to the fruit. In a peach, the skin comes from the outer layer; the fleshy portion comes from the middle layer; and the stone comes from the inner layer of the ovary. The seed is enclosed within the stone. A peach is a simple fruit derived from one ovary. A strawberry is an aggregate fruit derived from several ovaries joined together.

Types and Uses of Flowers

Most flowers have both stamens and pistils, but some have one or the other. Maize (corn) has both male and female flowers on the same plant. The male flowers are in the tassel at the top of the stem, and the female flowers are in the ears. After fertilization, each ovary becomes an individual corn kernel or grain. The grains of wheat and rice are one-seeded fruits—the outer layer of the ovary and seed coat are completely fused together.

Some vegetables are derived from flower parts. String beans are fruits, while shelled peas, soybeans, navy beans, and lima beans are the seeds only. Flowers are also the source of various other exotic products. Black and red peppers are from fruits, while coffee beans are seeds, and nutmeg is taken from a seed. The cotton boll is an opened ripe fruit; cotton fibers are hairs attached to the seeds. Marijuana comes from the dried flowers and leaves of the *Cannabis* plant (Indian hemp). The opium poppy fruit (fig. 17.26) is the source of opium, from which both morphine and codeine are extracted.

Adaptation to Land

Flowering plants are adapted to life on land, as is illustrated by a review of some of the main features of their anatomy. The above-ground organs of a flowering plant are protected by a covering that prevents drying out. Leaves and herbaceous stems are covered by a waxy cuticle, and woody stems are surrounded by bark. Both of these features prevent water loss. Also, roots take up water from the soil, and vascular tissue transports water to all parts of a flowering plant. In the next chapter you will study how water is lifted to the top of even the tallest trees.

Flowering plants are able to oppose the force of gravity. In herbaceous plants, turgor pressure helps maintain an upright position. Support cells, such as sclerenchyma cells, are also helpful. Woody plants depend on an internal skeleton composed of wood to give them the support they need to stand tall. As additional xylem is produced, the amount of wood increases with each growing season.

Reproduction in flowering plants is suitable to a land environment. The pollen grain is either windblown or carried by an animal to the pistil where it develops a tube by which sperm nuclei travel to the ovule. The resulting zygote is protected within the ovule, which matures and becomes a seed. A seed contains the zygote and stored food enclosed by a protective seed coat. The seeds themselves are encased in fruit that not only protects them but often helps with the dispersal process. Dispersal is by means of wind or animal transport.

Environmental conditions fluctuate on land to a greater degree than they do in a water medium. Although flowering plants are unable to maintain a constant internal environment, they do become dormant when unfavorable conditions, such as dryness or low temperature, are present.

Thus, it can be seen that flowering plants fulfill the requirements for land adaptation listed in table 17.1.

Figure 17.26
*a.*Opium plant. *b.* Opium oozing from the cuts made in a young fruit capsule of the opium poppy.

a.

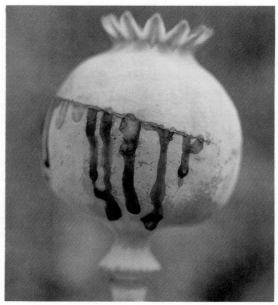

b.

Summary

The seeds of angiosperms are enclosed in a fruit that helps disperse the species. A seed contains an embryonic plant that has three parts. The epicotyl lies above the attachment of the cotyledon(s) and the hypocotyl lies below the attachment of the cotyledon(s). The lowest part is the radicle, which forms the first seedling root. The names and functions of the main tissues of the vegetative organs of a plant (roots, stems, and leaves) are given in table 17.2. Differences between monocot and dicot plants are given in figure 17.1.

Examination of a root tip reveals that the apical meristem produces root cap cells below it and cells that elongate and mature into various specialized tissues above it. There are four regions to a root tip: root cap, zone of cell division, zone of elongation, and zone of maturation. A cross section of the latter shows the outer epidermis with its extended root hairs, the cortex ending in a single layer of endodermal cells, and the vascular cylinder containing xylem and phloem.

Apical meristem tissue found within a bud produces cells that mature into the organs and tissues of the shoot system. Herbaceous dicot stems have a ring of vascular bundles that separate the cortex from the pith. Monocot stems have scattered vascular bundles with no definite demarcation between these two tissues. A woody dicot contains vascular cambium that produces new secondary xylem and phloem each year. Only the xylem builds up to form annual rings that indicate how old the stem is. Cork cambium is also present and produces cork cells that replace the epidermis in woody stems. There are three parts to a woody stem: bark (from the outside to vascular cambium), the inner wood (annual rings), and the central pith. The bark contains functional phloem, and the wood contains sapwood (functional xylem) and heartwood (nonconducting but supportive).

A leaf is protected, top and bottom, by epidermis that contains the stomata surrounded by guard cells. The mesophyll tissue inside the leaf of a C_3 plant has two layers: the palisade layer containing columnar cells and the spongy layer containing irregular cells surrounded by air spaces. Here CO_2 absorption (and oxygen release) occurs along with transpiration. Vascular tissue from the stem terminates in leaf veins within the leaf.

The flower contains the reproductive organs of a flowering plant. The stamens are the pollen-producing parts of a flower. Such pollinators as insects and birds are attracted to flowers by their petals, and when they gather nectar, they inadvertently pick up some pollen grains and carry them to the pistils of other flowers. Here each pollen grain develops a tube, and the sperm nuclei travel down the tube to an ovule within the ovary at the base of the pistil. After fertilization of the egg nucleus within the ovule, it develops into a seed, and the ovary contributes to fruit formation.

Throughout this chapter we have discussed, as summarized in table 17.3, ways in which humans utilize various plant parts. Information given in the chapter makes it apparent that flowering plants are adapted to life on land.

Table 17.3 Food Plants

Plant Part	Foods
Roots	Sweet potato, beets, radish, carrot, turnip, parsnip
Stems	White potato, sugarcane, asparagus
Leaves	Cabbage, kale, spinach, lettuce, tea leaves
Petioles*	Celery, rhubarb
Seeds†	Shelled peas, navy beans, lima beans, nuts, coffee beans
Fruits†	Wheat, rice, corn, oats, rye, string beans, apple, orange, peach, tomato, squash

*Part of a leaf
†Derived from flower parts

Objective Questions

1. Flowering plants are classified as _____ or _____ , according to the number of cotyledons in the seed among other features.
2. The plumule in a seed has tiny _____ , while the radicle forms the _____ .
3. The _____-type plant cell is embryonic and is ever capable of cell division.
4. The conducting cells of phloem are _____ cells, and the larger conducting elements of xylem are _____ cells.
5. On the root, the _____ in the zone of _____ increase the absorptive ability of a root tip.
6. Eventually all materials that enter a root tip must enter the vascular cylinder. Most do so by way of endodermal cells, which are bordered on four sides by the _____ strip.
7. In dicot herbaceous stems, the vascular bundles are arranged in a _____ . In monocot stems the vascular bundles are _____ .
8. Between the bark and the wood in a woody stem, there is a layer of meristem tissue called _____ , which produces new xylem and phloem every year.
9. The body of a C_3 leaf is composed of _____ tissue, which has two layers of cells: the _____ layer and the _____ layer.
10. The sperm within the pollen produced by the _____ travel down a pollen tube, and one of these fertilizes an egg within an _____ located within an ovary.

Answers to Objective Questions

1. monocot, dicot 2. leaves, root 3. meristem 4. sieve tube, vessel 5. root hairs, maturation 6. Casparian 7. ring, scattered 8. vascular cambium 9. mesophyll, palisade, spongy 10. stamens (or anthers), ovule

Study Questions

1. Describe in general the structure and germination of a mature seed.
2. Name and state the function of the main tissues within each plant organ.
3. Name and discuss the zones of a root tip. Trace the path of water and mineral ions from the root hairs to xylem.
4. Discuss the modifications of roots by naming and describing several different types.
5. Describe a monocot, an herbaceous dicot, and a woody stem.
6. Discuss the adaptation of stems by giving several examples.
7. Describe the structure and organization of a leaf.
8. Give several ways in which leaves are specialized.
9. Name the parts of a flower and describe the events leading to fertilization in flowering plants.
10. Name several products derived from the parts of flowers.

Selected Key Terms

endosperm ('en də ‚spərm) 362
meristem ('mer ə ‚stem) 364
parenchyma (pə 'ren kə mə) 366
sclerenchyma (sklə 'ren kə mə) 366
cuticle ('kyüt i kəl) 366
endodermis (‚en də 'dər məs) 366
tracheid ('trā kē əd) 368
vessel element ('ves əl 'el ə mənt) 368

cortex ('kȯr ‚teks) 371
pith ('pith) 377
stoma ('stō mə) 383
guard cell ('gärd 'sel) 383
mesophyll tissue ('mez ə ‚fil 'tish ‚ü) 383
transpiration (trans pə 'rā shən) 383

CHAPTER 18

Nutrition and Transport in Plants

Concepts

1. Organisms require environmental nutrients to sustain themselves. Plants require only inorganic nutrients.

2. Multicellular organisms have specialized regions for nutrient intake. The roots of a flowering plant are specialized to take up water and minerals; the leaves are specialized to take up carbon dioxide.

3. There is need for transport between specialized regions of multicellular organisms. In flowering plants, the xylem transports water and minerals, and the phloem transports organic nutrients.

Table 18.1 Mineral Elements Required by Plants

Element	Some Functions	Deficiency Symptoms
Nitrogen (N)	Part of proteins, nucleic acids, chlorophyll	Relatively uniform loss of color in leaves, occurring first on the oldest ones
Potassium (K)	Activates enzymes; concentrates in meristems	Yellowing of leaves, beginning at margins and continuing toward center; lower leaves mottled and often brown at tip
Calcium (Ca)	Essential part of middle lamella; involved in movement of substances through cell membranes	Terminal bud often dead; young leaves often appearing hooked at tip; tips and margins of leaves withered; roots dead or dying
Phosphorus (P)	Necessary for respiration and cell division; high-energy cell compounds	Plants stunted; leaves darker green than normal; lower leaves often purplish between veins
Magnesium (Mg)	Part of the chlorophyll molecule; activates enzymes	Veins of leaves green but yellow between them with dead spots appearing suddenly; leaf margins curling
Sulphur (S)	Part of some amino acids	Leaves pale green with dead spots; veins lighter in color than the rest of leaf area
Iron (Fe)	Needed to make chlorophyll and in respiration	Larger veins remaining green while rest of leaf yellows
Manganese (Mn)	Activates some enzymes	Dead spots scattered over leaf surface; all veins and veinlets remain green; effects confined to youngest leaves
Boron (B)	Influences utilization of calcium ions, but functions unknown	Petioles and stems brittle; bases of young leaves break down

It should be noted that all the micronutrients are harmful to plants when supplied in excessive quantities. For example, boron has been used in weed killers. Even macronutrients are harmful if present in heavy amounts, although nonessential elements sometimes can counteract their toxicity.

From Stern, Kingsley R., *Introductory Plant Biology*, 3d ed. © 1982, 1985 Wm. C. Brown Publishers, Dubuque, Iowa. All Rights Reserved. Reprinted by permission.

Plants, being autotrophic, require only inorganic nutrients to produce all the organic compounds that make up their bodies. By the eighteenth century it was known that carbon dioxide and water are the primary inorganic nutrients of plants. By the end of the nineteenth century, it was recognized that plants also need mineral elements that, like water, are normally obtained from the soil by the roots. Carbon, hydrogen, and oxygen make up 96 percent of a plant's dry weight so it is evident that these mineral elements (table 18.1) are needed in very small amounts for a plant to remain healthy (fig. 18.1).

390

Figure 18.1
Symptoms of mineral nutrient deficiencies. *a.* Calcium
deficiency symptoms in tomato fruits. "Blossom end rot"
forms on the side away from the stem. *b.* Boron deficiency
symptoms include early death of stem tips. Leaves
become crinkled, and stems and leaf stalks crack. *c.* Iron
deficiency symptoms in tomato leaves. Chlorosis
(yellowing) occurs between veins. There is little sign of
dying tissue.

a.   b.   c.

Mineral Requirements

The essential mineral elements are usually classified as either **macronutrients,**
those needed in greater quantities, or **micronutrients,** those needed in lesser
quantities. The macronutrients are nitrogen, potassium, calcium, phosphorus,
magnesium, sulfur, and iron. The micronutrients that a plant may require include chlorine, copper, manganese, zinc, molybdenum, boron, and sodium. In
addition to these widely required elements, specific plants may require others.
For example, certain algae apparently also require vanadium, silicon, or iodine, while some ferns utilize aluminum.

  Table 18.1 outlines some of the reasons certain elements are required,
and also lists the symptoms that develop when plants lack these elements. As
the table indicates, mineral ions generally are incorporated into metabolic and/
or structural compounds. For example, all plants require nitrogen because it
is found in proteins, nucleic acids, and other important compounds. Similarly,
phosphorus is needed for the metabolic compound ATP and for the phospholipids found within cell membranes. Potassium activates a number of enzymes
and also helps maintain osmotic pressure needed to maintain turgor
pressure (p. 394).

Figure 18.2

Sketch of a hydroponics (liquid culture) experiment. The buckwheat plant on the left was grown in a nutrient solution without potassium. The culture on the right had a complete nutrient solution.

Determination of Mineral Requirements

If you burn a plant, the macronutrient nitrogen is given off as ammonia and other gases, but most other mineral elements remain in the ash. However, the presence of a particular element in the ash does not necessarily mean that the plant normally requires it. For example, plants take up the man-made[239] plutonium and the naturally occurring [90]strontium. These radioactive elements are subsequently found in their ashes. This uptake of radioactive elements by plants is unfortunate because it intensifies the concerns and worries of a nuclear power plant mishap. Complicating the situation still further is the knowledge that if cows feed on plants containing [90]strontium, their milk will contain this radioactive element in even greater concentration. Therefore, humans are likely to receive a dose of radioactivity when they drink the milk.

Since analysis of the ash does not allow an investigator to determine the mineral requirements of a plant, a method for making this determination was developed at the end of the nineteenth century. This method is called water culture, or **hydroponics** (fig. 18.2). Hydroponics allows plants to grow well if they are supplied with all the mineral nutrients they need. The investigator omits a particular mineral and observes the resultant effect on plant growth. If plant growth suffers, it can be concluded that the omitted mineral is a nutrient requirement. This method has been more successful for macronutrients, however, than for micronutrients. For studies involving the latter, the water and mineral salts used must be absolutely pure, and purity is difficult to attain because even instruments and glassware can introduce contaminants. Then, too, the element in question might already be present in the cutting to be used in the experiment. These factors complicate the determination of plant micronutrients by means of hydroponics.

Nonmineral Requirements

Carbon Dioxide

Carbon dioxide is one of the primary nonmineral nutrients of a plant. Figure 18.3 illustrates that the leaf of a flowering plant is specialized for gas exchange, including the uptake of carbon dioxide (CO_2). In many leaves the spongy layer of the mesophyll contains air spaces bounded by cells. Carbon dioxide enters the leaf by diffusing across the membranes of these cells. When photosynthesis is occurring, the cells of a leaf take up carbon dioxide, even though its concentration in air is usually only about .03 percent. Some plants, especially those adapted to hot, dry environments, have structural and metabolic adaptations that allow them to acquire carbon dioxide even when it falls below this level inside the leaf (p. 108).

Leaves are unable to take up carbon dioxide without also losing water. For one thing, the cells within the spongy layer must be kept moist if diffusion is to occur. Also, the movement of water in xylem (p. 400) depends on *transpiration,* the evaporation of water from the leaf. The trade-off between water loss and carbon dioxide uptake in a leaf has to be regulated in some way. The upper and lower epidermis of a leaf is covered by a waxy cuticle as a protection against desiccation. However, the lower epidermis, in particular, contains *stomata* that, when open, allow air to enter the leaf. The *guard cells,* one on either side, regulate the opening and closing of a stoma. When stomata are open, gas exchange and water loss are occurring. When stomata are closed, water loss is prevented but CO_2 uptake is impossible.

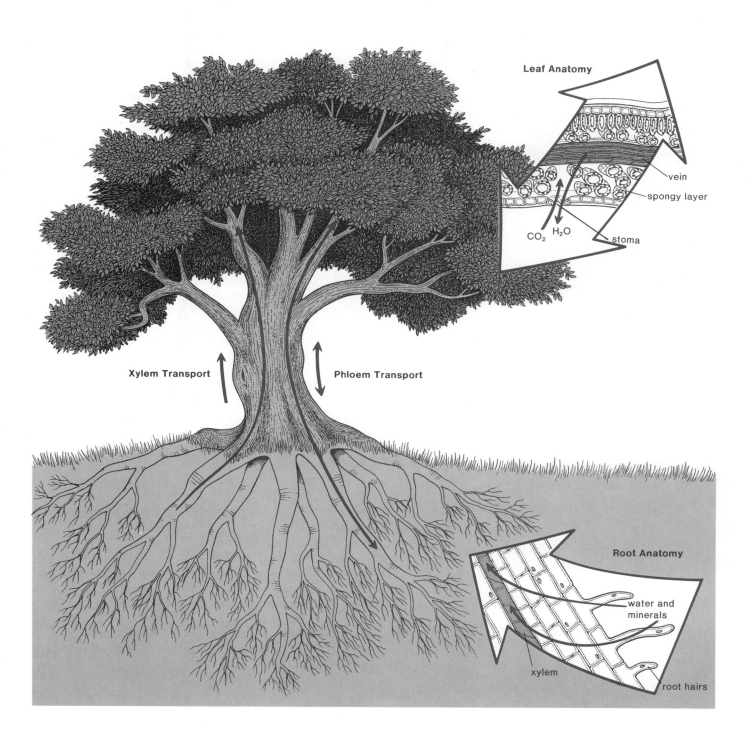

Figure 18.3
Leaves and roots are regions of a flowering plant body that are specialized to interact with the environment. The roots take up water and minerals. The leaves carry on gas exchange, but at the same time they lose water to the environment.

Leaf Anatomy

vein

spongy layer

CO_2 H_2O

stoma

Xylem Transport

Phloem Transport

Root Anatomy

water and minerals

xylem

root hairs

Figure 18.4

a. Opening of stoma is due to osmotic pressure. K$^+$ enters guard cells causing water to enter, followed by opening of stoma. b. Electromicrograph of open stoma.

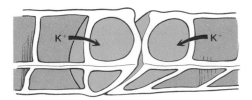

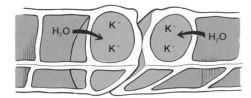

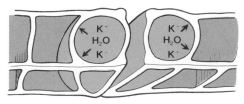

a.

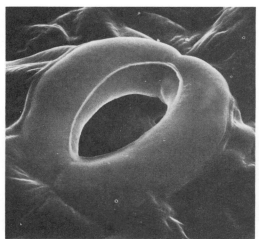

b.

Regulation of Stomata Opening and Closing

Guard cells have thickened inner cell walls and are tightly bound together at each end of a stoma. This arrangement causes guard cells to expand outwardly (fig. 18.4) when they swell, after taking up water. Now the stoma is open. Conversely, when guard cells lose water, the loss of turgor causes them to be as they were. Now the stoma is closed.

This explanation for the opening and closing of stomata has been known since 1856, but botanists are still gathering data to explain what causes the guard cells to either take up or lose water. For many years, it was accepted that the buildup of metabolic products following starch breakdown caused the guard cells to take up water and open. In 1968, however, R. A. Fischer floated small strips of the lower epidermis from a broad bean plant *(Vicia fava)* on various concentrations of potassium (K$^+$) chloride. He found the stomata opened *in the light if the air lacked carbon dioxide* because guard cells took up K$^+$. Subsequently, it has been shown in many plants that water enters guard cells after they take up K$^+$ (fig. 18.4) because of an increase in osmotic pressure. Apparently, K$^+$ is being actively transported into the guard cells when photosynthesis is occurring. It is reasoned that photosynthesis brings about a decreased concentration of CO$_2$ *inside* the leaf; this prompts in some unknown way the uptake of K$^+$ by guard cells, which leads to opening of the stomata. Presumably, the movement of K$^+$ out of guard cells leads to closing of the stomata as guard cells lose turgor.

The opening and closing of stomata is linked not only to photosynthesis, however, but also to the amount of water available to leaf cells. Whenever these cells become water stressed, as when wilting occurs, the stomata close. This is caused by a hormone called abscisic acid (ABA), which is produced by cells in wilting leaves. ABA in some undetermined way causes K$^+$ to move out of guard cells so that they subsequently lose water and close.

Transport of CO$_2$

Terrestrial plants have minute air spaces between the cells of their tissues and rely on diffusion alone to transport gases through these intercellular spaces about the plant. Gases diffuse much faster in air than they do in water, which facilitates this means of distribution.

Water

Water deficiency seems to be a common cause of terrestrial plant death. Normally, water moves from the spaces around soil particles into the *root,* up through the xylem, within vascular tissue, and out into the air spaces of the spongy layer of mesophyll tissue as transpiration occurs (fig. 18.3). Minerals needed by a plant also follow this **transpiration stream.**

As figure 18.5 shows, water can enter the root of a flowering plant simply by moving *between* the cells of the epidermis and cortex via the porous cell walls. Eventually, however, the *Casparian strip* (fig. 17.12) prevents any further progress between the cells, and the water must enter the endodermal cells if it is to reach the xylem. Alternately, water can move directly into the root hair cells and then progress from cell to cell across the cortex and endodermis until it reaches the xylem. The pathway is facilitated by the plant cells being joined one to another by cytoplasm strands.

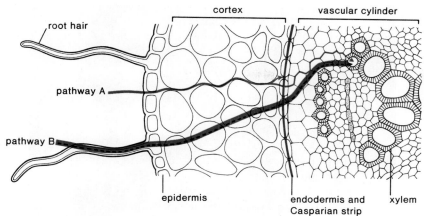

cortex   vascular cylinder

root hair

pathway A

pathway B

epidermis

endodermis and
Casparian strip

xylem

Figure 18.5
Water and minerals can travel one of two pathways across the cortex to the xylem of the vascular cylinder. Pathway A shows that water and minerals can travel via porous cell walls between the cells of the cortex, but must eventually enter endodermal cells because of the Casparian strip (see also fig. 17.12). In this way, the endodermal cells control the passage of materials into the vascular cyclinder and xylem. Pathway B shows that water and minerals can immediately enter a root hair and thereafter move from cell to cell to finally enter the xylem.

Regardless of the route by which water crosses the epidermis and cortex, it must at some time enter root cells. The cytoplasm of root cells usually has a higher concentration of solutes than does soil water; therefore, water can enter root cells passively by means of osmosis. There is evidence that, in some instances, plants exert energy to prevent water from entering root cells; however, they do not exert any energy to have water enter.

Mineral Uptake

Plants absorb minerals in the form of ions, either a group of atoms or a single atom that bears a charge. Like water, mineral ions can move past the epidermis and through the cortex by way of porous cell walls (fig. 18.5); eventually, because of the Casparian strip, they must enter the cytoplasm of the endodermal cells if they are to proceed further and enter the vascular cylinder. Cell membranes are freely permeable to water but not to all mineral ions. Mineral ions that cannot pass through the cell membranes of endodermal cells can go no further. In this way, the endodermis regulates the entrance of minerals into the vascular cylinder.

Often mineral ions are actively absorbed into the cytoplasm at the root hair region of a young root. Only occasionally, when particular mineral ions are highly concentrated in soil water, such as after a fertilizer application, do mineral ions enter root cells by simple diffusion. Figure 18.6 presents data from one experiment showing that active transport is most likely involved in the uptake of minerals by a plant. The rate of respiration and the rate of bromine ion absorption by thin discs of potato increased as more oxygen was made available by bubbling the gas through the culture solution. This indicates that energy was being used to take up the bromine ion.

As the reading on page 397 relates, roots seem to possess an astonishing ability to concentrate ions; that is, to absorb them until they are many times more concentrated than in the surrounding medium. For example, sometimes they absorb all and many times they absorb more than 50 percent of a mineral ion present in a culture solution. Once mineral ions have entered root cells, they diffuse down a concentration gradient and enter vascular tissue.

Figure 18.6
Bromine ion absorption by a potato tuber. As respiration increases, bromine ion absorption also increases. This shows that active transport is involved in nutrient uptake by plants.

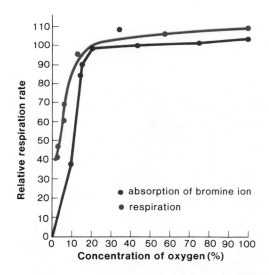

- absorption of bromine ion
- respiration

Relative respiration rate

Concentration of oxygen (%)

Figure 18.7

Symbiosis and mineral nutrition. *a.* Nodules on soybean roots that contain nitrogen-fixing bacteria. Their red color is caused by a pigment produced while active nitrogen fixation is occurring. *b.* Fungus roots in a beech tree (*Fagus sylvatica*). Fungus covers most of the root and its branches. *c.* Loblolly pine root systems. Roots at left were untreated and those at right were inoculated with a fungus. Fungus roots absorb minerals and water more efficiently.

Photograph by Carolina Biological Supply Company.

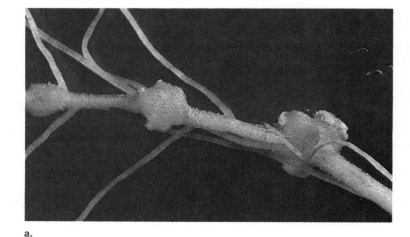

a.

Figure 18.8

Epiphytic bromeliad growing on a tree. An epiphyte is a plant that attaches itself to a larger plant but does not derive nourishment from it. Water and solutes fall into pockets formed by the bases of the leaves.

Adaptations   Two symbiotic relationships are known to assist roots in supplying mineral nutrients to the rest of a plant. Plants are unable to make use of nitrogen (N_2) in the air; in most cases, the roots take up either NO_3^- or NH_4^+ from the soil. But some roots, such as those of the legumes soybean and alfalfa, are infected by bacteria of the genus **Rhizobium** that can incorporate aerial nitrogen into substances a plant can use. This process is called nitrogen fixation. The bacteria live in nodules (fig. 18.7*a*) where they are supplied with carbohydrates by the host plant; the bacteria, in turn, fix nitrogen from the air and furnish their host with nitrogen compounds. This is such a useful service that research is now being directed toward finding ways to make more plants receptive to infection and even, with the use of gene-splicing techniques, to introduce the bacterial genes that control nitrogen fixation (called *nif* genes) into plant chromosomes. Less fertilizer would then be needed to grow crops.

The second symbiotic relationship involves fungi, particularly club fungi. The roots of some vascular plants are invaded by fungal hyphae. The fungus covers portions of the roots and extends out into the soil, making these so-called **fungus roots** (fig. 18.7*b*) shorter, thicker, and more highly branched than ordinary roots. The fungus increases the surface area available for mineral and water uptake and even breaks down organic matter, releasing nutrients that the plant can use. Fungus roots seem particularly important for certain trees, notably pines, that grow poorly if the fungus is not present (fig. 18.7*c*). Sometimes tree seedlings are deliberately infected with fungi before they are transplanted so they will grow better.

As a special adaptation, some plants have poorly developed roots or no roots at all because minerals and water are supplied by other mechanisms. The carnivorous plants discussed on page 384 have poorly developed roots and rely on their leaves to absorb compounds containing minerals. **Epiphytes** are "air plants" that do not grow in soil at all but grow on larger plants that give them support (fig. 18.8). They do not get nutrients from their host, however. Rather, some epiphytes have roots that absorb moisture from the atmosphere, and many catch rain in special hollow leaves.

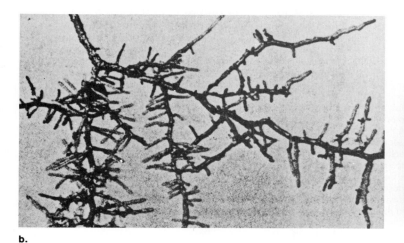

b.

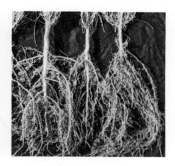

c.

Roots

All living things require many different elements to survive. Sixteen are found in all organisms; we humans require about twenty, including sodium, potassium, phosphorus, calcium, magnesium, and iron. Most of these life-supporting minerals are found in the soil, but their concentrations are very low—typically one part per million or less. Even if we could eat soil, we would have to consume tremendous quantities of it to extract the minimum amounts of minerals needed to sustain ourselves. Fortunately for our digestive tracts, plants extract our minerals for us.

A plant faces two problems in obtaining minerals from the soil. First, it must penetrate the ground in such a way as to gain the greatest possible contact with the soil (or, more precisely, with soil water in which minerals are dissolved). Second, the plant must be able to concentrate the extremely diluted minerals it has taken up from the soil.

The plant solves these problems by means of its root system. As the root system grows, it branches and branches again. The roots, in turn, extend millions of tiny, fingerlike *root hairs* into the soil. Roots and root hairs combined put an astoundingly large surface area in contact with the soil moisture and the minerals dissolved in it. For example, the root system of a single four-month-old rye plant is nearly 11,000 kilometers (7000 miles) long and has a surface area of 630 square meters (7000 square feet)!

Plant roots also have remarkable mineral-concentrating capabilities. Experiments have shown that the concentration of certain minerals within roots is as much as 10,000 times greater than in the surrounding soil. Once taken up, minerals are transported to growing parts of the plant, where they are incorporated into proteins, fats, vitamins, and many other organic compounds. Plants, then, are a concentrated source of essential minerals for consumers in both grazing and detritus food webs.

Plants "mine" staggering quantities of minerals from the soil—as much as 6 billion metric tons every year, according to one estimate. This is more than 6 times the amount of iron, copper, lead, and zinc ore produced by all worldwide mining operations in 1973.

We see, then, that plant roots are vital to our survival and the survival of other terrestrial animals. Without roots we could not obtain sufficient calcium to build bones and teeth, sodium to maintain blood pressure, and iron to help carry oxygen in the blood, as well as the many other minerals our bodies need to function properly.

Fertilizer Use

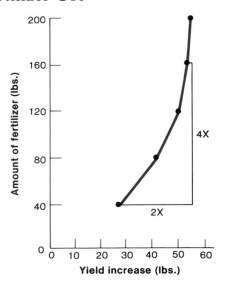

The United States Department of Agriculture data on which this graph is based show that to double an increase in yield, four times as much fertilizer is required. The developed countries can afford to apply excess amounts of fertilizer, but the developing countries cannot.

The plentiful harvests of modern agriculture in the United States depend on generous use of fossil fuel energy, pesticides, irrigation, and fertilizer. Ecological problems are involved with each of these, but just now we will consider only those that arise from the use of fertilizer.

The graph at left indicates that increased crop yield does correlate with increased fertilizer application; however, as more fertilizer is applied, the amount of increase begins to fall off. In other words, fertilizer application is less and less effective as more and more is used. Wealthy industrialized countries of the world, such as the United States, are accustomed to using large amounts of fertilizer, while poorer nonindustrialized countries are accustomed to using very small amounts of fertilizer.

Poorer countries would find it difficult to use more fertilizer for at least two reasons. First of all, they do not have ready access to the raw materials required to make fertilizer. The best supplies of phosphate are in Morocco, the United States, the Soviet Union, and Tunisia; potash (for potassium) is available in Germany, Canada, the Soviet Union, and the United States; and natural gas is plentiful only in Mexico and the Middle East. Natural gas is the source of the hydrogen needed to reduce aerial nitrogen (N_2) to ammonia (NH_3) during fertilizer production. Second, fertilizer production requires a large input of fossil fuel energy, a scarce commodity in poorer countries. In the United States it is estimated that 28 percent of the total amount of energy that agriculture consumes is devoted to fertilizer production. Poorer countries do not have sufficient resources and energy to produce fertilizer in quantity. Nor do they have the funds to buy enough to increase crop yield dramatically.

Countries that do use large amounts of fertilizer face possible environmental degradation. Rainwater washes the fertilizer from the soil (called fertilizer runoff), and in some areas of the United States, 30 to 50 percent of the nitrate in groundwater and surface water has been traced to fertilizer use. Nitrate in groundwater used for drinking can cause illness. If the level exceeds the public health limit of 10 milligrams of nitrate per liter, children under four months of age can become ill and even die. Doctors in the Central Valley of California recommend that only pure bottled water be used to prepare infants' formulas. Excess nitrate in surface water also contributes to cultural eutrophication, a process that in the short run leads to algal bloom and fish kill, and in the long run causes ponds and lakes to fill in and disappear.

Both industrialized countries and poorer countries would benefit if there were another means to supply nitrogen to plants. At present, only legumes (for example, soybeans and alfalfa) form root nodules in which nitrogen-fixing bacteria reduce aerial nitrogen, thereby making nitrogen available for use by the plant host. For years, farmers have used crop rotation in which leguminous plants and cereal grains are alternately planted on a plot of land to increase the amount of nitrate in the soil. Perhaps, however, it will eventually be possible to develop strains of nitrogen-fixing bacteria to live in nodules on the roots of grain plants, which provide most of the world's food. Or even better, there is hope that nitrogen-fixing genes (nif genes) might one day be incorporated into plant cell DNA. It is now a common practice to develop plants from single cells by tissue culture techniques. Perhaps it would be possible to inoculate these cells with nif genes to make the resultant plants capable of direct nitrogen fixation. Both universities and new companies specializing in recombinant DNA products are conducting research to achieve these goals by means of gene splicing.

Transport of Xylem Sap and Phloem Sap

In flowering plants, water and minerals are obtained by roots located underground where light energy is not available for photosynthesis. Photosynthesis occurs in leaves that are usually located well above ground where light is available. Thus, there is a need for transport between these two distant regions: the leaves require water and minerals absorbed from the soil, and the roots require organic nutrients produced in the leaves.

Figure 18.9
a. Cross section of xylem vessels from a cucumber root.
Note the sculptured appearance of the inside of the
vessel elements. *b.* The elements stacked one above the
other form a continuous pipeline that is hollow.

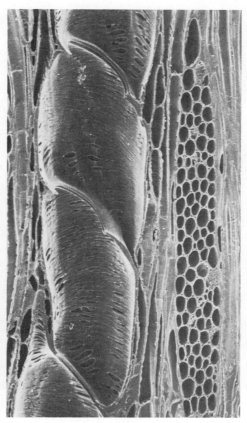

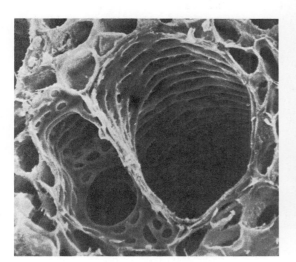

a.

b.

Transport of Xylem Sap

Once water and mineral ions enter the xylem, they are transported to all parts
of a plant. Transport need not be rapid, but the mechanism employed must
be capable of transporting materials over long distances. You will recall from
the previous chapter that xylem contains two types of conducting elements:
vessels and tracheids. The vessels (fig. 18.9) offer the best route, as they form
a continuous, completely hollow pipeline from the roots to the leaves.

When water enters root cells by means of osmosis (p. 68), it creates **root
pressure.** Root pressure tends to push xylem sap (water and minerals) upward.
This can be proved by attaching a glass tube to the cut end of a short stem
and watching the water rise. Water will rise as much as a meter above a cut
tomato stem and several times that height in some kinds of vines. Root pressure is also responsible for **guttation:** when transpiration is not taking place
and soil is well watered, drops of water are forced out of vein endings along

Figure 18.10
Drops of guttation water on the edges of a strawberry
leaf. Guttation, which occurs at night, is thought to be due
to root pressure.

the edges of leaves (fig. 18.10). While root pressure may contribute somewhat to the upward movement of water in xylem vessels, it is not nearly as important as transpirational pull.

Cohesion-Tension Theory of Water Transport

The **cohesion-tension theory** of water transport explains the transport of water to great heights against the direction of gravitational force. First, we must realize that water molecules have a great tendency to cling together because they are polar molecules (p. 42). This *cohesion property* of water means that a column of water can be pulled without its breaking. In fact, it is easier to pull apart the molecules in fine wires made of some common metals than it is to pull apart a column of water in a small-diameter, airtight tube. Secondly, we must also remember that water is continuously lost at the leaves by way of stomata because of transpiration (fig. 18.3). When water evaporates from the mesophyll cells in a leaf, these cells become less turgid. This creates an osmotic gradient that causes water to move out of xylem into these cells. Transpiration therefore creates a *tension* that can pull water upward in the xylem.

Stephen Hales, an English botanist working at the beginning of the eighteenth century, first suggested that transpiration pulls water up through plants. In 1914, Dixon and Joly coupled the idea of transpiration pull with their knowledge of the cohesion of water. Because water molecules tend to cling to one another, transpiration could be the force causing water to rise in the hollow and continuous xylem vessels.

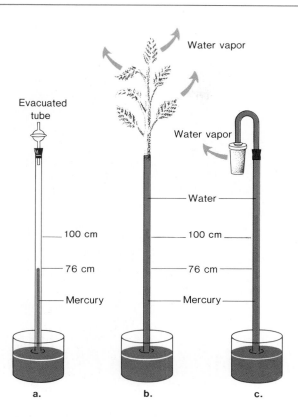

a. b. c.

Figure 18.11
Measurement of the forces involved in transpiration pull.
a. Mercury is raised only 76 cm up an evacuated tube by
atmospheric pressure. *b.* When leaf transpiration pulls
water through the xylem of a living stem, mercury rises
100 cm or more. *c.* Water evaporating from a porous clay
cup also exerts more pulling force than simple suction.

Water vapor

Evacuated
tube

Water vapor

Water vapor

Water

100 cm

100 cm

76 cm

76 cm

Mercury

Mercury

An experiment shows that transpiration aids the movement of water in
xylem (fig. 18.11). A pan filled with mercury is placed under an evacuated
tube. The mercury rises 76 cm from the force of atmospheric pressure. This
shows that atmospheric pressure, even combined with root pressure, could not
raise water to treetops. But when transpiration pull is added, either in the form
of a twig with leaves or a porous clay bulb, the mercury column rises to a
height of 100 cm or more. If this height is expressed in terms of water rather
than mercury, it shows that transpiration must definitely be a factor in pulling
water in xylem vessels to the tops of tall trees. Indeed, other observations have
shown that water in the branches of a tree starts to move earlier in the day
than does water in the trunk (fig. 18.12).

There is an important consequence to the manner in which water is
transported in plants. As mentioned previously (p. 383), transpiration causes
a plant to lose by evaporation at least 90 percent of the water taken in at the
roots. When the ground is dry and the plant is under water stress, the stomata
(fig. 18.4) close. Now the plant will lose little water because the leaves are
protected against water loss by their waxy coated layers of upper and lower
epidermis. However, when the stomata are closed, carbon dioxide cannot enter
the leaves, and plants are unable to continue to photosynthesize. Photosynthesis requires an overabundant supply of water only because of the manner
in which transport of water from the roots to the leaves is achieved.

Transport of Phloem Sap

As long ago as 1679, Marcello Malpighi suggested that bark is involved in
translocating sugars from leaves to roots. He observed the results of removing
a strip of bark from around a tree, called **girdling.** If a tree is girdled below
the level of the majority of leaves, the bark swells just above the cut and sugar
accumulates in the swollen tissue. We know today that when a tree is girdled,

Figure 18.12
Velocity of water movement in a tree. The water
movement in the upper branches begins earlier in the day
than the water movement in the trunk. This is expected, if
water is pulled upward as is believed.

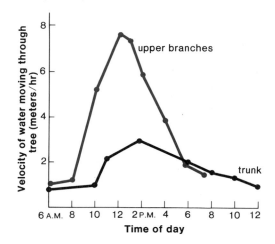

Figure 18.13

Use of radioactive tracers in plants. *a.* Radioactive CO_2 is supplied to the leaves by the method shown. *b.* Later, the leaf is pressed against a sheet of X-ray film so that the radioactivity given off by the leaf is detected. *c.* Thin slices of plant tissue can also be used to prepare microscope slides. The slides are used to prepare a film that indicates which type cells contain radioactivity.

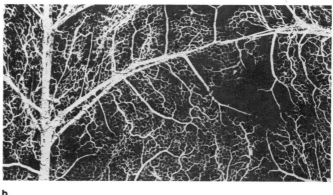
b.

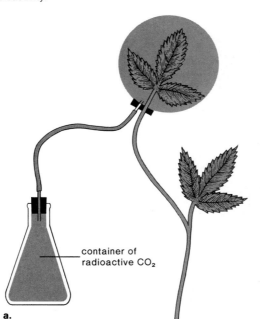

container of
radioactive CO_2

a.

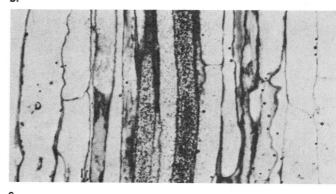
c.

the phloem is removed but the xylem is left intact. Therefore, girdling suggests that phloem is the tissue that transports sugars. Radioactive trace studies (fig. 18.13) have confirmed this. When [14]carbon-labeled CO_2 is supplied to mature leaves, radioactively labeled sugar is soon found moving down the stem into the roots. This labeled sugar is found mainly in the phloem, not in the xylem. Radioactive tracer studies have also confirmed the role of phloem in transporting other types of substances, such as amino acids, hormones, and even mineral ions. Hormones are transported from their production sites to target areas where they exert their regulatory influences. In the autumn, before leaves fall, mineral ions are removed from the leaves and taken to other locations in the plant.

Chemical analysis of phloem sap shows that its main component is sucrose, and the concentration of nutrients is usually 10 to 13 percent by volume. It is difficult to take samples of sap from the phloem without injuring the phloem, but this problem was solved by utilizing aphids (fig. 18.14), small insects that are phloem feeders. The aphid drives its stylet, a sharp mouthpart that functions like a hypodermic needle, between the epidermal cells and withdraws sap from a sieve tube cell. If the aphid is anesthetized by ether, the body may be carefully cut away, leaving the stylet, which exudes phloem that can be collected and analyzed by the investigator.

The conducting cells of phloem are sieve tube cells lined up end to end with their sieve plates abutting (fig. 18.15). Cytoplasm extends through the sieve plates of adjoining cells to form a continuous sieve-tube system that extends from the roots to the leaves, and vice versa.

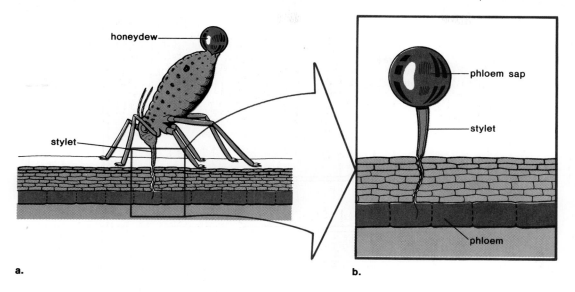

Figure 18.14
Aphids are small insects that remove nutrients from phloem by means of a hypodermiclike mouthpart called a stylet. *a.* Aphid with stylet in place. *b.* When the aphid's body is removed, phloem sap is available to the experimenter.

honeydew

phloem sap

stylet

stylet

phloem

a.

b.

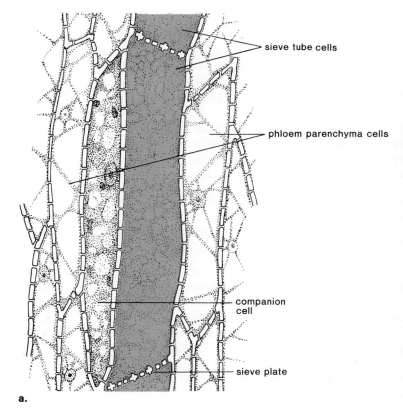

Figure 18.15
Phloem structure. *a.* Longitudinal section shows that sieve tube cells in phloem lie end to end. *b.* Cross section of a sieve plate, by which these elements make contact with one another and thereby form a continuous pipeline.

sieve tube cells

phloem parenchyma cells

companion cell

sieve plate

a.

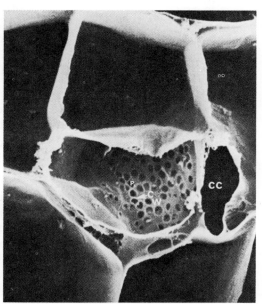

cc

b.

Figure 18.16

Pressure-flow hypothesis. *a.* Model system. Water tends to enter (*1*) because the osmotic gradient across its membrane is so much greater than for the membrane at bulb (*2*). This creates a pressure difference that causes water to flow from (*1*) to (*2*). As water flows, it carries sucrose with it. *b.* Diagram of phloem. At source (*1*) (for example, an actively photosynthesizing leaf), sucrose is actively transported into phloem sieve tubes. At sink (*2*) (for example, in a root), sucrose is actively transported out of phloem sieve tubes. So much water enters the source end of the phloem by osmosis that water is forced out the sink end. Sucrose is transported in the phloem by this flow of water.

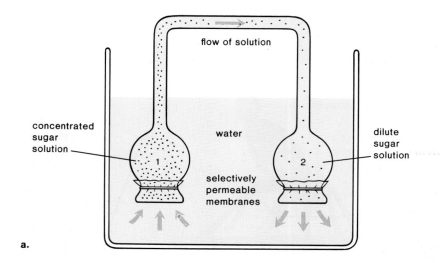

flow of solution

concentrated sugar solution

water

dilute sugar solution

selectively permeable membranes

a.

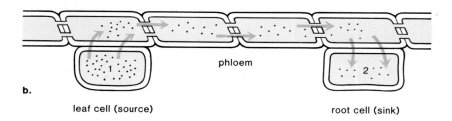

phloem

b.

leaf cell (source)   root cell (sink)

Pressure-Flow Theory

An explanation of phloem transport must account for the movement of fairly large amounts of organic material for long distances in a relatively short period of time. The movement rate of carbon-labeled sugar has been determined by analysis of sap withdrawn through aphid stylets between two areas of the stem (fig. 18.14). Materials appear to move through phloem at a rate of 60 to 100 cm per hour and possibly up to 300 cm per hour.

A **pressure-flow hypothesis** currently offers an explanation of the movement of organic materials in phloem. This hypothesis, first proposed by Ernst Munch in 1927, is demonstrated in figure 18.16. Note the following:

1. There are two bulbs connected by a glass tube. The first bulb contains solute at a higher concentration than the second bulb.
2. Each bulb is bounded by a selectively permeable membrane, and the entire apparatus is submerged in distilled water.

Distilled water will enter the first bulb because it has the higher concentration of solute. This entrance of water creates a *pressure* that causes a *flow* toward the second bulb. The pressure flow will not only drive water toward the second bulb, it will also provide enough drive to force water *out* through the membrane of the second bulb, even though the second bulb contains a higher concentration of solute than does the distilled water.

The artificial system just described will not continue to function after the flow of water brings solutes from the first to the second bulb. Once the second bulb contains as much solute as the first bulb, water will no longer enter the first instead of the second bulb, and the pressure flow will be eliminated. In other words, the concentration gradient within the system must be maintained in order for the system to continue working.

The activities of the living plant cells are able to maintain a concentration difference between parts of the phloem. When a plant is photosynthesizing, the leaves serve as a "source" of sucrose that enters sieve tube cells by means of active transport (fig. 18.16b). *The leaves, like the first bulb, have a higher concentration of solute.* On the other hand, the roots, which are storage areas, are always removing sucrose by means of active transport; they therefore serve as a "sink" for sucrose. *The roots, like the second bulb, have a lower concentration of solute.* In the plant, water enters the sieve tube cells of the phloem at the leaves because of the higher concentration of solute in that portion of phloem. (This water comes from the water in leaf veins.) This creates a pressure flow of water that carries solutes within the phloem usually from the leaves toward the roots.

However, the pressure-flow hypothesis can also account for the observed reversal of flow in the sieve tube system. All that is required is a reversal of the source-sink relationship; for example, whenever the plant is not photosynthesizing, the roots serve as a source of sucrose, and the other parts of the plant serve as sinks.

There are still many questions to be answered about phloem transport. For example, do sieve tubes serve simply as living but passive pipes through the plant? Perhaps not, because if metabolic poisons are applied to a ring around a tree, or if this ring is heated with steam, flow through the phloem stops. This may be simply a response to injury, or it is possible that living, metabolically active tissue is needed for phloem transport to continue. There is much to be learned about phloem transport and its control.

Summary

Plants require mineral nutrients as determined by chemical analysis of plants and, especially, by hydroponics experiments. While macronutrient requirements have been known for some time, micronutrient requirements have been difficult to determine since it is difficult to exclude trace amounts from culture experiments.

Carbon dioxide, one of the main nonmineral nutrients of a plant, enters leaves by way of stomata that open and close depending on the need to acquire carbon dioxide versus prevent water loss. It has been shown that opening and closing of stomata are caused by an increase in osmotic pressure in guard cells after they have taken up K^+.

Water and minerals enter a plant at the roots. While both can enter the spaces between root cells, they must eventually enter endodermal cells because of the Casparian strip. Root cells are freely permeable to water but not to minerals that often are actively absorbed. Two symbiotic relationships assist in nutrient uptake by roots: nitrogen-fixing bacteria live in root nodules of some plants and provide them with a supply of nitrogen compounds, and fungus roots help some plants absorb minerals.

When water enters the root due to osmotic pressure, it is thereby pushed up the xylem of the root. However, root pressure cannot account for movement of water over long distances. Water and mineral transport in xylem is then explained by the cohesion-tension theory. Water is always evaporating at the leaves, and this evaporation, called transpiration, pulls water from the roots because water molecules exhibit cohesion. Nutrient transport in the phloem is explained by the pressure-flow hypothesis. Solute is actively transported into sieve tubes at a source and actively transported out of them at a sink. At the source, water tends to enter sieve tubes by osmosis with such pressure that it is forced out at the sink. This creates a flow of water that is believed to carry solutes with it.

Objective Questions

1. The elements _____ , _____ , and _____ account for about 96 percent of a plant's dry weight.
2. The element copper is a _____ nutrient for most plants.
3. Carbon dioxide enters a plant at the _____ by way of the _____ . Water enters a plant at the _____ and is transported in the _____ .
4. After guard cells take up K⁺, water enters guard cells and they then _____ (choose open or close).
5. The presence of the Casparian strip eventually forces water and minerals to enter the vascular cylinder by way of _____ cells.

6. The close association between certain fungi and roots _____ (choose helps or hinders) the ability of roots to take up water and minerals.
7. Transpiration can pull water up from the roots because water molecules exhibit the property of _____ .
8. At the leaves evaporated water exits by way of the _____ , which close when a plant is under water stress.
9. The _____ hypothesis accounts for the movement of sugars in the phloem.
10. Normally, the sieve tube cells within leaves have a _____ concentration of sugar than the roots. This creates a pressure flow of water that carries solutes from the _____ to the _____ within phloem.

Study Questions

1. Name the elements that make up most of a plant's body.
2. How are mineral requirements categorized and why are they categorized in this manner? Discuss a plant's need for nitrogen, phosphorus, and potassium.
3. Briefly describe two methods used to determine the mineral nutrients of a plant.

4. Describe the manner in which carbon dioxide is taken up by a leaf and explain how this is related to water loss. How are gases, including carbon dioxide, transported about a plant?
5. What events precede the opening and closing of stomata by guard cells?
6. Give two pathways by which water and minerals can cross the epidermis and cortex of a root. What feature allows endodermal cells to regulate the entrance of molecules into the vascular cylinder?

7. Contrast the manner in which water and minerals are absorbed by root cells.
8. Name two symbiotic relationships that assist plants in taking up minerals, and two types of plants that tend not to take up minerals by their roots.
9. Describe and give evidence for the cohesion-tension theory of water transport.
10. What data are available to show that phloem transports organic compounds? Explain the pressure-flow theory of phloem transport, referring to figure 18.16.

Selected Key Terms

macronutrient ('mak ‚rō 'nü trē ənt) 391
micronutrient ('mī ‚krō 'nü trē ənt) 391
hydroponics (‚hī drə 'pän iks) 392
fungus root ('fən gəs 'rüt) 396
epiphyte ('ep ə ‚fīt) 396
root pressure ('rüt 'presh ər) 399

guttation (‚gə 'tā shən) 399
cohesion-tension theory (kō 'hē zhən 'ten chən 'thē ə rē) 400
girdling ('gərd liŋ) 401
pressure-flow hypothesis ('presh ər 'flō hī 'päth ə səs) 404

In previous chapters, we mentioned that plants grow throughout their entire lives (fig. 19.1). It is not surprising, then, that plants respond to internal and external stimuli by changes in their growth patterns. Often, a change in growth pattern is preceded by the production of hormones. A **hormone** is a chemical messenger produced in small amounts by one part of the body that is active in a different part of the body. Generally, plant hormones are produced by the meristematic regions of a plant and responses may

Figure 19.1

Plants grow their entire lives. Meristem tissue is located in the shoot and root apex and in vascular cambium (see chap. 17). This tissue divides repeatedly, providing the cells that differentiate into other types of plant tissues found in, for example, an (a.) oak seedling, (b.) oak sapling, and (c.) oak tree. Seed germination, plant growth, and dormancy are tied to the seasons; therefore, it is important that plants have a mechanism by which they can respond to external stimuli. The production of hormones by meristem tissue provides this mechanism.

a.

b.

c.

CHAPTER 19

Control of Plant Growth and Development

Concepts

1. Growth and development in complex organisms are coordinated.

   Growth and development in plants are coordinated by plant hormones, some of which are stimulatory and some of which are inhibitory.

2. Photoperiodic responses theoretically require a biological clock system. Flowering is an example of such a response, and research is being directed toward finding and determining the specific role of each participant.

Table 19.1 Plant Hormones

Type	Primary Example	Notable Function
Promoters of Growth		
Auxins	Indolacetic acid (IAA)	Cell elongation
Gibberellins	Gibberellic acid (GA)	Stem elongation
Cytokinins	Zeatin	Cell division
Inhibitors of Growth		
Abscisic acid	Abscisic acid (ABA)	Dormancy
Ethylene	Ethylene	Abscission

be observed in most every part of the plant's body. Table 19.1 lists the types of hormones we will be considering. These hormones sometimes interact to control any particular physiological process. Different combinations bring about different effects.

Each naturally occurring hormone has a specific chemical structure. Other chemicals, some of which differ only slightly from the natural hormones, also affect the growth of plants. These and the naturally occurring hormones are sometimes grouped together and called **plant growth regulators.** The reading below discusses the various uses of plant growth regulators.

Plant Growth Regulators

Now that the formulas for many plant hormones are known, it is possible to make them and related chemicals in the laboratory. Collectively these substances are known as plant growth regulators. Many scientists hope plant growth regulators will bring about an increase in crop yield, just as fertilizers, irrigation, and pesticides have done in the past.

Since auxin was first discovered, researchers have found many agricultural and commercial uses for it. Auxins cause the base of stems to form new roots quickly, so that new plants are easily started from cuttings. When sprayed on trees, auxins prevent fruit from dropping too soon. Because auxins inhibit the growth of lateral buds, potatoes sprayed with an auxin will not sprout and thus have a longer storage period.

In high concentrations, auxins are used as herbicides to prevent growth of broad-leaved plants. The synthetic auxins known as 2,4D and 2,4,5T were used as defoliants during the Vietnam war. Even though 2,4D has been known for over thirty-five years, we do not yet know how it works. Apparently, it is structurally different enough from natural auxin that enzymes cannot break it down. The concentration rises until metabolism is disrupted, cellular order is lost, and the cells die.

The other plant hormones studied in this chapter also have agricultural and commercial uses. Gibberellins are used to increase plant size. Treatment of sugarcane with as little as two ounces per acre increases the cane yield by more than five tons. The application of either auxins or gibberellins can cause an ovary and accessory plants to develop into fruit, even though pollination and fertilization have not taken place. In this way, it is sometimes possible to produce seedless fruits and vegetables or bigger, more uniform bunches with larger fruit.

Because cytokinins retard the aging of leaves and other organs, they are sprayed on vegetables to keep them fresh during shipping and storage. The treatment of holly allows it to be harvested many weeks prior to its use as a holiday decoration.

Plant growth regulators are making it possible to grow plants from a few cells in laboratory glassware. It may even be possible to develop new varieties of food plants with particular characteristics, such as tolerance to heat, cold, toxins, and drought. By means of gene-splicing techniques, it might eventually be possible to create plants capable of utilizing aerial nitrogen.

Several synthetic inhibitors are used to oppose the action of the auxins, gibberellins, and cytokinins normally present in plants. Some of these can cause leaf and fruit drop at a time convenient to the farmer. Removing leaves from cotton plants aids harvesting of cotton, and

Hormones That Promote Growth

Notice in table 19.1 that certain plant hormones promote activities associated with growth, such as cell division and cell elongation. Each of these hormones is also associated with specific growth patterns in plants.

Auxins

The most common naturally occurring auxin is **indolacetic acid (IAA).** Auxin has long been thought to be involved in two well-known tropisms (growth responses) observed in plants: phototropism and gravitropism.

Phototropism

Phototropism, the bending of plants toward the light, is easily observed (fig. 19.2). However, this phenomenon was studied for quite some time before its cause was determined. Around 1881, Charles Darwin and his son Francis reported on experiments they had performed with grass and oat seedlings. When

Figure 19.2
Phototropism is easily observed in this field of sunflowers. All the flowers point toward the sun and even move as the sun moves—they track the sun from sunrise to sunset. This is obviously an adaptive trait of plants.

Gibberellic acid (GA₃) on Thompson seedless grapes (Vitis vinifera). (left) Control grapes. (right) GA₃ sprayed at bloom and at fruit set. Almost all grapes sold in stores are now treated with gibberellic acid.

thinning the fruit of young fruit trees produces larger fruit as these trees mature. Retarding the growth of other plants sometimes increases their hardiness; for example, an inhibitor has been used to reduce stem length in wheat plants so the plants do not fall over in heavy winds and rain. Other synthetic inhibitors mimic the action of ethylene and cause ripening of fruit and other crops when they are to be harvested. Fields and orchards are now sprayed with synthetic growth regulators just as they are sprayed with pesticides.

Figure 19.3

a. An oat seedling is at first protected by a hollow sheath called a coleoptile. Later the leaves break through this sheath. *b.* Experiments with oat seedlings show that the tip is necessary and must be exposed to light in order for oat seedlings to bend toward the light. *1.* If the tip of a seedling is cut off, it will not bend toward unidirectional light. *2.* If the tip is covered by a black cap, the seedling will not bend toward the light. *3.* If the seedling is left intact, it will bend toward the light.

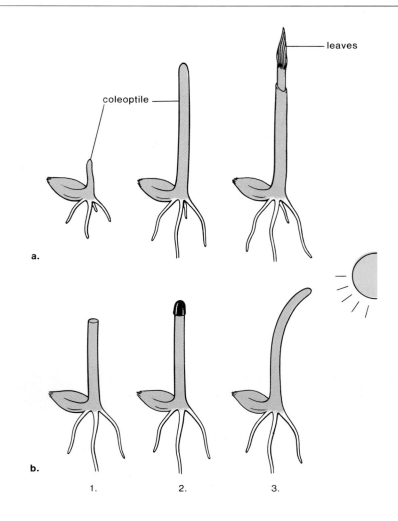

Figure 19.4

This experiment shows that a chemical produced by the coleoptile tip causes the bending of a seedling toward the light. *a.* Coleoptile tips are cut off and placed on agar (a gelatinlike material). *b.* After a time the tips are removed and the agar is cut into small blocks. A block is placed to one side of a decapitated coleoptile. *c.* Bending occurs even in the absence of a light stimulus.

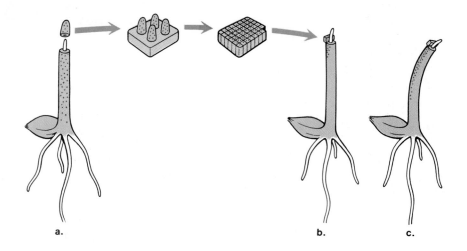

these plants first appear, each seedling is covered by a sheath called the **coleoptile.** Soon the leaves break through this covering of the tiny shoot (fig. 19.3). The Darwins found that if the coleoptile was kept intact, the seedlings would bend toward a unidirectional light source. But if the tip of the seedling was cut off or was covered by a black cap, the seedling would not respond to light. They concluded that some influence is transmitted from the coleoptile tip to the rest of the shoot, which causes bending.

Building on the results of the Darwins and others, Frits W. Went performed still more experiments in 1926 (fig. 19.4). He cut off the tips of coleoptiles and placed them on agar (a gelatinlike material). After about an hour, he removed the tips and cut the agar into small blocks. When an agar block was placed to one side of a decapitated coleoptile, the shoot would bend away from that side. The bending occurred even though the seedlings were not exposed to light.

Went concluded that the agar blocks contained a chemical that had been produced by the coleoptile tips. It was this chemical, he decided, that had caused the shoots to bend. He named the chemical substance **auxin** after the Greek word *auxein,* which means "to increase."

Action of Auxin   It can be shown that when a plant is exposed to unidirectional light, auxin is transported to the shady side. Elongation of the cells on this side brings about the characteristic bending toward the light (fig. 19.5). Auxin causes the affected cell to degrade some of the polysaccharides in the cell wall. The cell thereafter elongates as it is less able to resist the expansion caused by osmotic movement of water into the cell. The direction of cell growth is dependent on where the wall is weakest.

Gravitropism

Not only do plants respond to light, they also respond to gravity, **gravitropism.** A stem displays **negative gravitropism** (geotropism) because it grows opposite to the direction of gravity. Roots display **positive gravitropism** because they grow in the same direction as gravity. These responses are particularly apparent after a plant has been placed in a horizontal position (fig. 19.6).

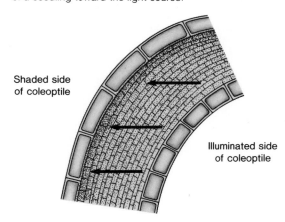

Figure 19.5
Auxin is transported from the illuminated side to the shaded side of a coleoptile sheath as indicated by the arrows. This unequal concentration of auxin, represented by the colored dots, causes the cells on the shaded side to elongate. Elongation of the cells results in the bending of a seedling toward the light source.

Shaded side of coleoptile

Illuminated side of coleoptile

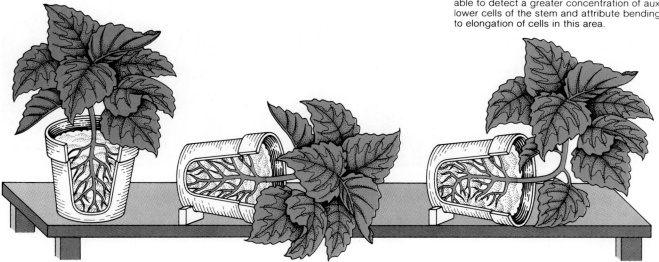

Figure 19.6
When an upright plant is turned on its side, in time the roots bend down and the stem bends up. Roots therefore demonstrate positive gravitropism while the stem demonstrates negative gravitropism. Investigators are able to detect a greater concentration of auxin in the lower cells of the stem and attribute bending of the stem to elongation of cells in this area.

Investigators have found a difference in auxin concentration between the cells located in the upper side of the stem and those located in the lower side. This concentration difference is believed to contribute to the greater elongation of cells located on the lower side. It is this unequal elongation that causes the stem to eventually bend upward. Although plant physiologists have also long attributed the positive gravitropism of roots to auxin distribution, they have thus far been unable to find a difference in auxin concentration between the upper and lower side of roots. Therefore, auxin is no longer believed to be responsible for the response of roots to gravity. Apparently, an inhibitor is produced by the root tip that causes cells on the lower side to elongate less than those on the upper side. (Some believe that abscisic acid is this inhibitor.) In any case, the unequal elongation of cells causes the root to bend downward.

Other Effects

Auxin has been found to affect many other aspects of plant growth. For example, auxin produced by the apex of the main shoot prevents bud development and growth for some distance from the apex. Once this so-called **apical dominance** has worn off, branching does occur (fig. 19.7).

Figure 19.7
The amount of apical dominance present in a plant determines its shape. Apical dominance is due to inhibition of lateral buds by auxin. *a.* A high degree of apical dominance results in a conical-shaped tree. *b.* A low degree of apical dominance results in a bushy appearance.

a.

b.

The application of a weak auxin solution to a woody cutting will cause roots to develop. Auxin production by seeds also promotes the growth of fruit. As long as auxin is concentrated in leaves or fruits, rather than in the stem, leaves and fruits will not fall off. Therefore, trees can be sprayed with auxin to keep mature fruit from falling to the ground. These uses of auxin are discussed in the reading on page 409.

Gibberellins

We know of about seventy gibberellins that chemically differ only slightly. The most common is **GA₃** (the subscript designation distinguishes it from other gibberellins). **Gibberellins** are likely present in newly developing plant organs since they are growth promoters that bring about cell division and enlargement of the resulting cells. When gibberellins are externally applied to plants, the most obvious effect is *stem elongation* (fig. 19.8).

Gibberellins were discovered in 1926, the same year that Went performed his classic experiments with auxin. Kurosawa, a Japanese scientist, was investigating a fungal disease of rice plants called "foolish seedling disease." The plants elongated too quickly, causing the stem to weaken and the plant to collapse. Kurosawa found that the fungus produced an excess of a chemical he called gibberellin.

Many years later, in 1955, British workers reported that minute amounts of externally applied gibberellic acid would cause genetically dwarfed pea plants to grow to a normal size. At first it was thought that dwarfed plants would contain little gibberellin, but surprisingly they contain more gibberellin than do normal plants. For some reason, they are unable to utilize the gibberellin present in their cells. As mentioned previously, it is often the balance of hormones in a plant cell that produces an effect, rather than the presence or absence of a particular hormone only.

Because the application of gibberellin can also cause certain plants to flower, it was once thought that perhaps GA₃ was florigen, the long-sought flowering hormone. But this idea has been abandoned because GA₃ does not cause all plants to flower. It does cause cabbage plants and other biennials to "bolt" (elongate rapidly) and flower. Typically these plants spend their first year as rosettes (low-growing compressed shoots) and only after exposure to cold during the first winter do they bolt and flower.

Gibberellin research provides an example of how a plant hormone might act as a chemical messenger. Barley seeds have a large starchy endosperm that must be broken down into sugars to provide energy for growth. After the embryo produces gibberellin, cells just inside the seed coat synthesize an enzyme that breaks down the starch. It is hypothesized that gibberellin turns on the gene that codes for the necessary enzyme. In this particular instance, then, a plant hormone is functioning as illustrated in figure 23.24*b*.

Cytokinins

The **cytokinins** are a class of plant hormones that promote cell division; *cytokinesis means cell division.* These substances are derivatives of the purine adenine, one of the nitrogenous bases in DNA and RNA (p. 190). A naturally occurring cytokinin, termed **zeatin,** has been extracted from corn kernels. Kinetin also promotes cell division, but it must be classified as a growth regulator because it has not been found to occur naturally.

Figure 19.8
The plant on the right was treated with gibberellin; the plant on the left was untreated. Gibberellins are often used to promote stem elongation in economically important plants, but its exact mode of action still remains unclear.

Figure 19.9
Growth of plant tissues in culture vessels enables
investigators to test the effect of various combinations of
plant hormones on differentiation. Strip of a tobacco plant
(a.) taken from the stem of a flower. Depending on the
ratio of auxin to cytokinin, the acidity of the culture
medium, and the particular mix of oligosaccharins
(chemical fragments from the cell wall) the strip becomes
(b.) undifferentiated callus; (c.) roots; (d.) vegetative
shoots and leaves; (e.) floral shoots.

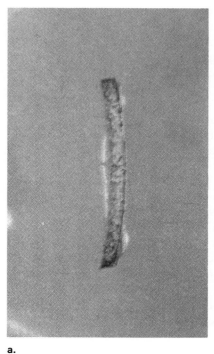

a.

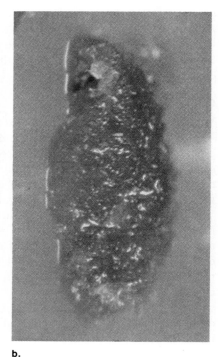

b.

The cytokinins were discovered as a result of attempts to grow plant
tissue and organs in culture vessels (fig. 19.9) in the 1940s. It was found that
cell division occurred when coconut milk (a liquid endosperm) and yeast extract were added to the culture medium. Although the effective agent or agents
could not be isolated, they were collectively called cytokinins. Not until 1967
was the naturally occurring cytokinin, zeatin, isolated from coconut milk.

Plant tissue culturing is now a very common practice and experimenters
are well aware that the ratio of auxin to cytokinin and the acidity of the culture medium determine whether the plant tissue will form an undifferentiated
mass, called a callus, or differentiate to form roots, vegetative shoots, leaves,
or floral shoots. Recently, investigators P. Albersheim and A. G. Darvill reported that chemicals called oligosaccharins (chemical fragments released from
the cell wall) are also effective in directing differentiation (fig. 19.9). They
hypothesize that the function of auxin and cytokinin is actually to activate
enzymes that release these more specific chemical messengers from the cell
wall. They similarly feel all plant hormones probably have such a function,
which explains the pleiotropy (many and various effects) of any specific plant
hormone. If plant hormones do regulate the activity of enzymes, they would
be functioning as shown in figure 23.24a.

Other Effects

When a plant organ such as a leaf loses its natural color, it is most likely
undergoing an aging process called **senescence.** During senescence, large molecules within the leaf are broken down and transported to other parts of the
plant. Senescence need not affect the entire plant at once; for example, as some
plants grow taller, they naturally lose their lower leaves. It has been found
that senescence can be prevented by the application of cytokinins. Not only
can cytokinins prevent death of plant organs, they can also initiate growth.
Lateral buds will begin to grow despite apical dominance when cytokinin is
applied to them.

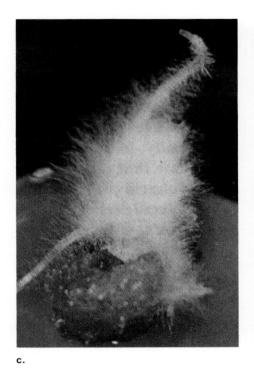

c.

d.

e.

Plant Growth Inhibitors

As is evident in table 19.1, certain plant hormones inhibit growth; that is, they bring about those effects that are associated with a discontinuance of growth. We would expect these hormones to be especially present and active whenever the plant faces conditions that are unfavorable for continued growth, such as lack of water or cold temperatures.

Ethylene

In the 1900s, it was common practice to ready citrus fruits for market by placing them in a room with a kerosene stove. Since heat alone did not have the same effect, it was finally realized that an incomplete combustion product of kerosene, namely **ethylene,** was responsible for the ripening of fruit. Ethylene ripens fruit by increasing the activity of enzymes that soften a fruit. For example, it stimulates the production of *cellulase,* an enzyme that hydrolyzes the cellulose of plant cell walls.

Ethylene, being a gas, moves freely through the air—a barrel of ripening apples can induce ripening of a bunch of bananas even some distance away. Since ethylene is an incomplete combustion product, it is also part of automobile exhaust. Is it possible that increased concentration of this molecule in the atmosphere might be affecting plants? As discussed in the reading on the following page, investigators have been studying this problem.

Abscission

Ethylene is also involved in **abscission,** the falling off of leaves, fruits, and flowers from a plant. As mentioned previously, most likely a decrease of auxin and perhaps gibberellin in these areas of the plant compared to the concentration in the stem initiates abscission (fig. 19.10). But once the process of abscission has begun, ethylene is produced. This stimulates such enzymes as cellulase that cause leaf, fruit, or flower drop.

Figure 19.10
Abscission zone. Before a leaf falls, a special band of cells develops at the base of the petiole, the leaf stem. Here the hormone ethylene promotes the breakdown of plant cell walls so that the leaf finally falls. Here also a layer forms a leaf scar (fig. 17.17) that protects the plant from possible invasion by microorganisms.

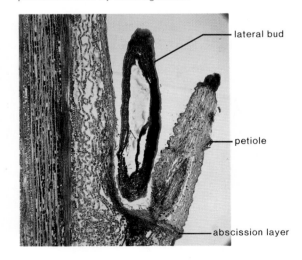

lateral bud

petiole

abscission layer

Control of Plant Growth and Development   415

Plants See Hormone as Toxic Pollutant

Ethylene gas, a by-product of incomplete combustion, is a pollutant associated not only with industrial manufacturing but also with urban automotive exhaust. Ironically, the simple hydrocarbon is also a natural plant hormone. Because ethylene can encourage premature ripening, it is used agriculturally to make an entire crop ripen at once for a single harvesting or to put the color in early-picked fruit.

What George E. Taylor and his colleagues at Oak Ridge (Tenn.) National Laboratory were curious to find out was whether plants respond to chronic low-dose exposures of this chemical as though it were a toxic pollutant—regardless of any subtle, slow-acting hormonal action that might also be occurring. Their experiments with several important crops now suggest that some plants indeed respond immediately and adversely to ethylene gas.

The researchers worked with corn, soybeans, peanuts, tobacco and seedlings of the green ash tree. Except for corn, which for unknown reasons showed no adverse reaction to any concentration of ethylene, the plants responded in a dose-dependent fashion by reducing both their photosynthesis and respiration.

These changes, says Taylor, suggest that the site of ethylene action is the stomata—the little openings on the underside of leaves through which gas exchange occurs. However, since stomata changes were not noticed in all of their adversely affected plants, Taylor says a closing down of stomata apertures may be secondary to changes in a more fundamental mechanism regulating photosynthesis, involving an unhealthy buildup of carbon dioxide concentrations in leaves.

Writing in the *May Environmental Science and Technology,* Taylor and co-workers report that after three- to six-hour exposures to ethylene at concentrations of 3.7 parts per million (ppm) in air, the photosynthetic rate fell in some of the more sensitive species, including soybeans and peanuts, by more than 60 percent. A mere 0.9-ppm exposure cut their photosynthesis by almost a third. And such effects occur very rapidly, Taylor says, sometimes "in less than half an hour."

Although only a handful of published studies have quantified ethylene concentrations in air, Taylor says, some of the reported values do approach levels that affected plants in his study. Moreover, ethylene levels tend to peak when photosynthetic activity is at its highest. If decreases in photosynthesis occurred for prolonged periods, plant growth and crop yields would be stunted, he says.

Figure 19.11
Stomata are openings in the leaf epidermis that are opened and closed by guard cells. Because of transpiration, a plant ordinarily looses a great deal of water by way of the stomata; therefore, it is beneficial for the stomata to be closed when a plant is under water stress. The hormone ABA is believed to bring about the closing of the stomata when water stress occurs.

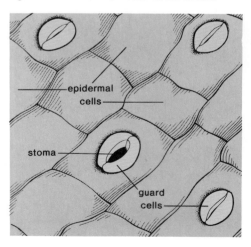

Abscisic Acid

Abscisic acid (ABA) is sometimes called the stress hormone because it initiates and maintains seed and bud dormancy and brings about the closure of stomata. While the external application of absisic acid will promote abscission, this hormone is no longer believed to function naturally in that process.

Dormancy occurs when a plant organ readies itself for adverse conditions by stopping growth (even though conditions at the time are favorable for growth). For example, it is believed that abscisic acid moves from leaves to vegetative buds in the fall, and thereafter these buds are converted to winter buds. A winter bud is covered by thick and hardened scales (fig. 17.17). In the spring, dormancy is broken when seeds germinate and buds begin to send forth new growth. A reduction in the level of abscisic acid and an increase in the level of gibberellins is believed to break seed and bud dormancy.

Abscisic acid is called the stress hormone not only because it promotes dormancy, but also because it brings about stomatal closing (fig. 19.11) when a plant is under water stress. In some unknown way, ABA causes K^+ to leave guard cells. Thereafter, the guard cells lose water and the stomata close. Notice that this is the opposite sequence of events to those that cause stomata to open (fig. 18.4).

Photoperiodism

The term **photoperiod** refers to the length of daylight compared to the length of darkness. The effect of day length on plants is particularly obvious in the temperate zone. In the spring, plants respond to increasing day length by initiating growth; in the fall, they respond to decreasing day length by stopping growth processes (fig. 19.12). Since changes in day length are due to the regular rotation of the earth about the sun, day length is a reliable indication of the season. Other factors such as the degree of temperature or the amount of moisture fluctuate too widely to be as reliable. Therefore, it is adaptive for plants to respond to day length as an indication of the season.

Day length also regulates flowering in some plants. For example, there are those plants, such as violets and tulips, that flower in the spring. Others, such as asters and goldenrods, flower in the fall. Investigators have studied this example of photoperiodism in plants.

Figure 19.13
Day length (night length) effect on two types of plants.
a. Short-day (long-night) plant. *1.* When the day is shorter
(the night longer) than a critical length, this type of plant
flowers; *2.* it does not flower when the day is longer (night
is shorter) than the critical length; *3.* it also does not
flower if the longer-than-critical-length night is interrupted
by a flash of light. *b.* Long-day (short-night) plant. *1.* When
the day is shorter (the night longer) than a critical length,
this type plant does not flower; *2.* it does flower when the
day is longer (the night shorter) than a critical length; *3.* it
does flower if the longer-than-critical-length night is
interrupted by a flash of light.

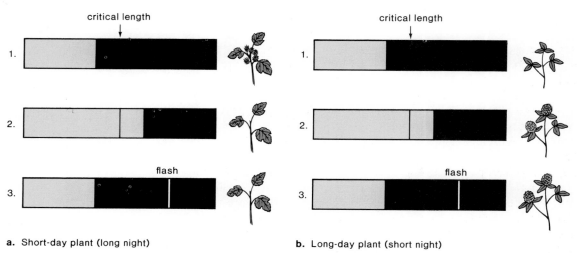

a. Short-day plant (long night) **b.** Long-day plant (short night)

Flowering

In the 1920s, U.S. Department of Agriculture scientists working in Beltsville, Maryland, began to study the control of flowering in plants. They soon realized that it was day length that often initiated flowering. After growing plants in the greenhouse where they could artificially alter the photoperiod, they came to the conclusion that plants can be divided into three groups (fig. 19.13):

Short-day plants—flower when the photoperiod is shorter than a critical length. (Good examples are cocklebur (fig. 19.16) poinsettias, chrysanthemums.)
Long-day plants—flower when the photoperiod is longer than a critical length. (Good examples are wheat, barley, clover, spinach.)
Day-neutral plants—flowering is not dependent on a photoperiod. (Good examples are tomatoes and cucumbers.)

Further, we should note that both a long-day and a short-day plant can have the same critical length: spinach is a long-day plant having a critical length of fourteen hours; ragweed is a short-day plant with the same critical length. Spinach, however, flowers in the summer when the day length increases to fourteen hours or more, and ragweed flowers in the fall when the

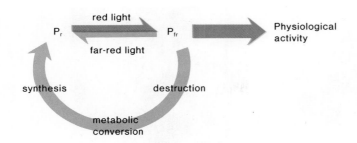

Figure 19.14
The $P_r \rightleftarrows P_{fr}$ conversion cycle. As indicated by the coloring, the inactive form P_r is prevalent during the night. At sunset or in the shade when there is more far-red light, P_{fr} is converted to P_r. Also during the night metabolic processes cause P_{fr} to be replaced by P_r. As indicated by the coloring, P_{fr}, the active form of phytochrome, is prevalent during the day because at that time there is more red light than far-red light.

day length shortens to fourteen hours or less. (Ragweed must mature before it will flower, which is why it does not flower in the spring even though the day length is less than fourteen hours.)

In 1938, K. C. Hammer and J. Bonner began to experiment with artificial lengths of light and dark that did not necessarily correspond to a normal, twenty-four-hour day. Then they discovered that the cocklebur, a short-day plant, will flower as long as the dark period is continuous for eight and one-half hours, regardless of the length of the light period. Further, if this dark period is interrupted by a brief flash of light, the cocklebur will not flower. (Interrupting the light period with darkness had no effect.) Similar results have also been found for long-day plants. They require a dark period that is shorter than a critical length regardless of the length of the light period. However, if a longer-than-critical-length night is interrupted by a brief flash of light, long-day plants will flower. We must conclude, then, that it is the length of the dark period that controls flowering, not the length of the light period. Of course, in nature shorter days always go with longer nights and vice versa.

Phytochrome

If flowering is dependent on day and night length, plants must have some way to detect these periods. Many years of research by other U.S. Department of Agriculture scientists stationed in Beltsville, Maryland, led to the discovery of a plant pigment called phytochrome. **Phytochrome** is a blue-green leaf pigment that alternately exists in two forms. Figure 19.14 indicates that

> P_r (photochrome red) absorbs red light (of 660 nm wavelength) and is converted to P_{fr}.
> P_{fr} (phytochrome far-red) absorbs far-red light (of 730 nm wavelength) and is converted to P_r.

Sunlight contains more red light than far-red light; therefore, P_{fr} is apt to be present in plant leaves during the day. In the shade and at sunset there is more far-red light than red light; therefore P_{fr} is converted to P_r as night approaches. There is also a slow metabolic replacement of P_{fr} by P_r during the night. It was thought for some time that this slow reversion might provide a means for the plant to measure the length of the night. However, it is now assumed that the active form of phytochrome, P_{fr}, signals a biological clock, an unknown internal timekeeper that keeps track of the passage of time. In turn, this timekeeper brings about flowering (fig. 19.15).

Figure 19.15
A biological clock system has three components. *a.* An internal timekeeper mechanism represented here by the face of a clock. *b.* A means of detecting light and dark periods. (Perhaps phytochrome is such a detector.) This portion of the system keeps the timekeeper accurate and helps the system determine seasonal changes. *c.* A means to communicate (for example, by hormone or enzyme level) to the body that it is time for circadian (daily) or circannual (yearly) changes.

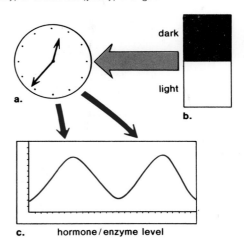

Figure 19.16
a. Cocklebur plant. This plant is often used in flowering experiments because it will respond to a single exposure to appropriate lighting. Other plants require appropriate lighting for several days and nights in succession.
b. Flowering experiment. *1.* A cocklebur plant is kept on an inappropriate long-day exposure except for a single leaf that is covered so it alone receives a short-day exposure. *2.* The cocklebur plant flowers, indicating that a substance has passed from the single leaf to the flower buds.

a.

b.   1.   2.

Figure 19.17
A day-neutral plant (color) is grafted to a long-day plant. When the plants receive (*a.*) long-day lighting, both plants flower, but when the plants receive (*b.*) short-day lighting, neither plant flowers. Because the amount of light has no effect on the flowering of day-neutral plants, it is assumed that inhibiting hormones passed from the long-day plant to the day-neutral plant, overcoming stimulatory hormones for flowering.

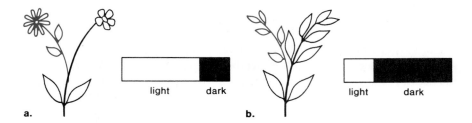

Circadian Rhythms

Plants show other evidences of having a **biological clock mechanism,** as described in figure 19.15. Like other eukaryotic organisms, they display circadian rhythms. **Circadian** (about a day) **rhythms** repeat just about every twenty-four hours. For example, beans and other plants spread their leaves to a horizontal position in the daytime and fold them up or down at night. Remarkably, this periodic change keeps on occurring even when the plant is kept in total darkness. Therefore, plants must have some internal means to tell time, and this internal system automatically initiates the physiological processes that result in circadian rhythms. Notice in figure 19.15 that a biological clock mechanism has three components: the *timekeeper,* a *method of communication,* and a *detector* of the actual change in time according to the sun. The latter is needed to keep the clock's timekeeping accurate and to signal the clock to seasonally related photoperiod changes.

The $P_r \rightleftarrows P_{fr}$ conversion cycle might be the means by which the plant tells the clock about photoperiod changes. Once a particular photoperiod is noted, the clock initiates the hormonal (followed by enzymatic) changes that promote flowering.

Flowering Hormone(s)

What is the possible hormonal control of flowering? For a very long time, researchers have been seeking a flowering hormone that is already called **florigen.** They have been able to extract substances from a plant that will cause a vegetative shoot to produce flowers, but no one has yet been able to chemically identify a specific hormone present in these extracts. Experiments do indicate that the hormone (if it exists) is transported from the leaves to the shoot apex. For example (fig. 19.16), a cocklebur plant is exposed to a photoperiod that is too long to induce flowering. If a single leaf is covered by black paper so that it receives short-day lighting, however, the plant will flower. This is taken as evidence that there is a florigen hormone.

On the other hand, perhaps flowering is controlled by the balance between stimulatory and inhibitory plant hormones, such as those discussed in this chapter. For example, when a day-neutral plant is grafted to a long-day plant, appropriate lighting for the long-day plant promotes flowering in the day-neutral plant, but inappropriate lighting for the long-day plant inhibits flowering in the day-neutral plant (fig. 19.17).

Other Functions of Phytochrome

The $P_r \rightleftarrows P_{fr}$ conversion cycle is now known to control other growth functions in plants aside from flowering. It promotes seed germination and inhibits stem elongation, for example. The presence of P_{fr} apparently indicates to seeds that sunlight is present and that conditions are favorable for germination. Following germination, the presence of P_r indicates that stem elongation may be needed in order to reach sunlight. Seedlings that are grown in the dark *etiolate,* that is, the stem increases in length and the leaves remain small (fig. 19.18). Once the seedling is exposed to sunlight and P_r is converted to P_{fr} the seedling begins to grow normally—the leaves expand and the stem branches.

Figure 19.18
Phytochrome has other functions besides regulating flowering. For example, if far-red light is prevalent as it is in the shade, the stem of a seedling elongates and the leaves remain small (left-hand side of photo). However, if red light is prevalent as it is in bright sunlight, the stem does not elongate and the leaves expand. These effects are due to phytochrome.

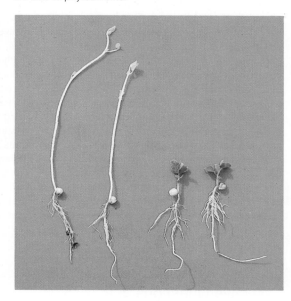

Summary

Both stimulatory and inhibitory hormones control the growth patterns of plants. These patterns account for the plants' response to internal and external stimuli. There are hormones that stimulate growth and hormones that inhibit growth. Among those that stimulate growth, the auxins promote cell elongation and are involved in both phototropism and gravitropism. They also bring about apex dominance, fruit maturation, and the prevention of leaf and fruit drop. Gibberellins promote stem elongation, thus external application of them can cause dwarf plants to reach normal size. They also seem to help break seed and bud dormancy. Cytokinins cause cell division, which is especially obvious when plant tissues are being grown in culture vessels. They can also prevent leaf senescence. Among the plant hormone inhibitors, ethylene causes fruit to ripen and promotes abscission. Abscisic acid is no longer believed to be involved in abscission, but it does promote bud dormancy and causes stomata to close when a plant is under water stress. ABA might also be involved in the negative gravitropism of roots.

Photoperiodism is a plant's cyclical response to day length. For example, short-day plants flower only when the days are shorter than a critical length, and others, called long-day plants, flower only when the days are longer than a critical length. Actually, investigation has shown that it is the length of darkness that is critical; interrupting the dark period with a flash of white light prevents flowering in a short-day plant and induces flowering in a long-day plant.

Phytochrome is a pigment that responds to both red light and far-red light. This ability apparently causes phytochrome to be involved in flowering. Daylight causes phytochrome to exist as P_{fr}, but during the night it is converted back to P_r by metabolic processes. It appears that phytochrome must communicate with a biological clock that measures time. As with other biological clock systems, the timekeeper may be responsible for communicating that it is time to flower. Investigators are still seeking a particular hormone, called florigen, that initiates flowering. However, flowering might be caused by a proportion of stimulatory to inhibitory hormones. In addition to being involved in the flowering process, P_{fr} promotes seed germination, leaf expansion, and stem branching. On the other hand, when P_r predominates, the stem elongates, an adaptive response that assists the plant in seeking sunlight.

Objective Questions

1. The plant hormones that promote growth are _____ , _____ , and _____ .
2. In response to a unidirectional light source, auxin is transported to the _____ side of a stem and causes _____ of cells on this side.
3. Stems exhibit _____ gravitropism, and roots exhibit _____ gravitropism.
4. The shape of a conical tree shows a marked degree of _____ dominance.
5. The application of the hormone _____ would cause a dwarf plant to grow to a normal size.
6. The relative proportions of auxin and _____ affect the differentiation of plant tissues in culture vessels.
7. The hormone _____ causes stomata to close when a plant is under water stress.
8. A short-day plant flowers when darkness _____ (choose is less than or is greater than) a certain critical length.
9. Phytochrome is a pigment that changes from the form designated as _____ to _____ during daylight hours.
10. Phytochrome most likely signals a _____ , which is an internal timekeeping device.

Answers to Objective Questions

1. auxin, gibberellins, cytokinins 2. shady, elongation 3. negative, positive 4. apical 5. gibberellin 6. cytokinin 7. abscisic acid 8. is greater than 9. P_r to P_{fr} 10. biological clock

Study Questions

1. Name three types of hormones that promote growth processes, and give several specific functions for each.
2. Explain the process of phototropism, including the distribution and function of auxin.
3. Discuss gravitropism and the contrasting responses of stem and roots. Also provide possible causes for this contrast.
4. Name two types of hormones that inhibit growth processes, and give several specific functions for each.
5. Define photoperiodism and discuss its relationship to flowering.
6. Explain the phytochrome conversion cycle and the possible functions of phytochrome in a plant.
7. Relate the participants of the flowering process in plants to the components of a biological clock.
8. Discuss two theories regarding hormonal control of flowering, and give evidence for each theory.

Selected Key Terms

hormone ('hȯr ˌmōn) 407
indolacetic acid ('in ˌdōl ə ˌsēt ik 'as əd) 409
phototropism (fō 'tä trə ˌpiz əm) 409
auxin ('ȯk sən) 411
gravitropism (ˌgrav ə 'trō ˌpiz əm) 411
gibberellin ('jib ər 'rel ən) 413
cytokinin (ˌsīt ə 'kī nən) 413
abscisic acid (ˌab ˌsiz ik 'as əd) 416

photoperiod ('fōt ə 'pir ē əd) 417
short-day plant ('shȯrt ˌdā 'plant) 418
long-day plant ('lȯŋ 'dā 'plant) 418
day-neutral plant ('dā 'nü trəl 'plant) 418
phytochrome ('fīt ə ˌkrōm) 419
circadian rhythm (sər 'kād ē ən 'rith əm) 421
florigen ('flōr ə jən) 421

Suggested Readings for Part 7

Albersheim, P., and Darvill, A. G. September 1985. Oligosaccharins. *Scientific American.*

Bold, H. C. 1980. *Morphology of plants and fungi.* 4th ed. New York: Harper & Row, Publishers, Inc.

Brill, W. J. March 1977. Biological nitrogen fixation. *Scientific American.*

Epel, D. November 1977. The program of fertilization. *Scientific American.*

Heslop-Harrison, Y. February 1978. Carnivorous plants. *Scientific American.*

Jansen, W., and Salisbury, F. B. 1971. *Botany: An ecological approach.* Belmont, CA: Wadsworth.

Raven, P. H., et al. 1981. *Biology of plants.* 3d ed. New York: Worth.

Rayle, D., and Wedberg, H. L. 1980. *Botany: A human concern.* Boston: Houghton Mifflin.

Salisbury, F. B., and Ross, C. W. 1978. *Plant physiology.* 2d ed. Belmont, CA: Wadsworth.

Shepard, J. F. May 1982. The regeneration of potato plants from leaf-cell protoplasts. *Scientific American.*

Tippo, O., and Stern, W. L. 1977. *Humanistic botany.* New York: W. W. Norton and Co.

Zimmerman, M. H. March 1963. How sap moves in trees. *Scientific American.*

PART 8

Processing and Transporting in Animals

An animal's body selectively exchanges materials with its external environment, which assists the maintenance of a relatively stable internal environment. In a single-celled organism, the cell membrane is the boundary between the organism and the outside world. A survey of multicellular animals shows an increasing tendency to depend on specific body regions for localized exchanges with the environment. The most recently evolved animals have digestive, respiratory, and excretory systems with specialized surfaces for this purpose. The digestive system changes bulk food to molecules that can be absorbed, the respiratory system carries on gas exchange, and the excretory system rids the body of metabolic wastes. There is also a circulatory system that transports materials throughout the body, including the specialized regions of these organ systems.

Alaskan brown bear catching dinner. Bears are omnivores, feeding on both animal and plant life. Following digestion, the resulting nutrient molecules are transported by the bloodstream to the tissues where they serve as a source of energy or are used for the synthesis of macromolecules.

CHAPTER 20

Circulation and Blood

Concepts

1. Multicellular animals have specialized regions to interact with the environment. A transport system assists communication between these regions and the cells of the body.
    a. Some invertebrates and all vertebrates have closed circulatory systems in which the blood is always contained within the blood vessels.
    b. Every component of blood has a specific function. For example, in humans, red cells transport oxygen, white cells combat disease, and platelets function in blood clotting.

2. The evolutionary history of organisms shows an increase in complexity.
    a. Evolution of the four-chambered heart (two atria and two ventricles) in land vertebrates allows oxygenated blood to remain separate from deoxygenated blood.
    b. The lymphatic system serves as an adjunct to the blood circulatory system. It takes up excess tissue fluid and returns it to the blood circulatory system.

E ach cell requires a supply of oxygen and nutrient molecules and must rid itself of waste molecules. Single-celled organisms (fig. 20.1a) make these exchanges directly with the external environment that surrounds their bodies. Even some small multicellular animals (fig. 20.1b and c) do not require an internal transport system in order to take care of the needs of their cells. Each cell is still able to exchange materials directly with water or with nearby cells that are in contact with water.

The other larger and more complex multicellular animals, however, require a **blood circulatory system** to transport molecules because the body cells are far from specialized regions of exchange with the external environment. Further, in these animals the body cells are surrounded by a fluid called **tissue fluid.** Blood and tissue fluid together create an internal environment for cells. Because the internal environment remains relatively constant, the cells are protected from extreme changes in the external environment. Claude Bernard, the great French physiologist, was the first to recognize the presence and relative constancy (*homeostasis*) of this **internal environment,** which he called *milieu intérieur.* Experimental studies on mammalian circulatory systems during the mid-nineteenth century led him to conclude that these animals could be active in most any type of external environment because the internal environment stayed relatively constant.

Invertebrates

The evolutionary tree of animals (fig. 15.2) shows that invertebrates can be divided into the lower invertebrates and the higher invertebrates. The lower invertebrates do not have a circulatory system, whereas the higher invertebrates do have such a system.

Lower Invertebrates

In coelenterates (fig. 15.7), cells are either a part of an external layer or they line the gastrovascular cavity. In either case, each cell is exposed to water and can independently exchange gases and get rid of wastes. The cells that line the gastrovascular cavity are specialized to carry out digestion. They pass nutrient molecules to other cells by diffusion. In flatworms (fig. 15.12), the tri-lobe gastrovascular cavity ramifies throughout the small and flattened body. No cell is very far from one of the three digestive branches so nutrient molecules can diffuse from cell to cell. Similarly, diffusion meets the respiratory and excretory needs of the cells.

Higher Invertebrates

Some higher invertebrates have an open circulatory system, while others have a closed system.

Open Circulatory System

In an **open circulatory system,** the blood is not always contained within blood vessels (fig. 20.2). A heart pumps blood into vessels, but these vessels empty into body cavities where blood bathes the internal organs or into sinuses located within the organs themselves. In an open circulatory system, the blood ebbs and flows in a sluggish manner.

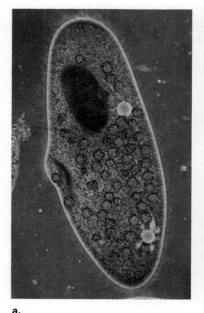

a.

b.

Figure 20.1

These aquatic organisms do not have a circulatory system. *a.* A paramecium is a single-celled organism that carries on gas exchange across its cell surface. Food particles that flow into a specialized region called a gullet are enclosed within food vacuoles (color) where digestion occurs. Molecules leave these vacuoles as they are distributed about the cell by a movement of the cytoplasm. Any nondigestible material is discharged at an anal pore. *b.* A sea anemone is a coelenterate in which there are only two layers of cells separated by a jellylike substance called mesoglea. Digestion takes place inside the gastrovascular cavity, so named because it (like a vascular system) makes digested material available to the cells that line the cavity. These cells can also acquire oxygen from the watery contents and discharge their wastes here. *c.* A planarian is a flatworm whose gastrovascular cavity ramifies throughout the body bringing nutrients to the interior of the worm. Diffusion is sufficient to pass molecules from cell to cell from either the cavity or from the worm's surface.

Photograph by Carolina Biological Supply Company (b).

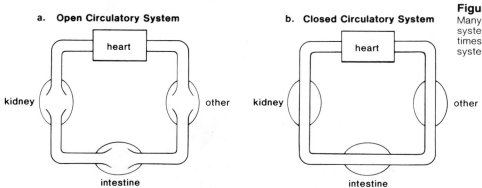

c.

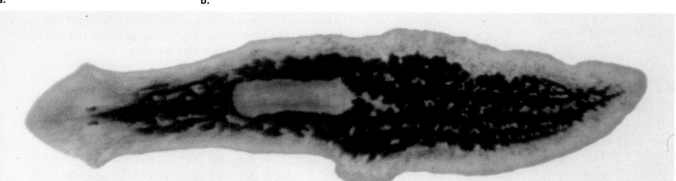

a. Open Circulatory System

kidney   other

intestine

b. Closed Circulatory System

kidney   other

intestine

Figure 20.2

Many animals have either an open or a closed circulatory system. *a.* In an open circulatory system, the blood at times is not contained in vessels. *b.* In a closed circulatory system, the blood is always contained within vessels.

For example, in the grasshopper the dorsal heart pumps the blood into an anterior vessel, which empties into a body cavity termed a *hemocoel* (fig. 20.3a). The hemocoel, appropriately named, is filled with blood. When the heart contracts, openings called **ostia** are closed; when the heart relaxes, the blood is sucked back into the heart by way of the ostia.

The blood of a grasshopper is colorless because it does not contain a respiratory pigment. Oxygen is taken to and carbon dioxide is removed from cells by way of air tubes that are found throughout the body. Flight muscles require an efficient means of receiving oxygen so an open circulatory system would not likely suffice.

Closed Circulatory System

In a closed circulatory system, the blood is always contained within blood vessels. The pumping of a heart keeps the blood moving in this system (fig. 20.2b).

For example, in the segmented earthworm (fig. 20.3b), the five pairs of anterior hearts pump blood into the ventral blood vessel, which has a branch in every segment of the worm's body. Blood moves through these branches into capillaries where exchanges with tissue fluid take place. It then moves into branches of the dorsal blood vessel. This vessel returns the blood to the heart for repumping.

The earthworm has red blood because it contains the respiratory pigment **hemoglobin.** Hemoglobin is dissolved in the blood and is not contained within cells. The earthworm, though, has no specialized boundary for gas exchange with the external environment. Gas exchange takes place across the body wall, which must remain moist for this purpose.

Figure 20.3
a. The grasshopper has an open circulatory system. A hemocoel is a body cavity filled with blood that freely bathes the internal organs. The heart keeps the blood moving, but this open system would not likely supply oxygen rapidly enough to wing muscles. They receive their oxygen from air tubes. b. The earthworm has a closed circulatory system. The dorsal and ventral vessels are joined by five pairs of hearts at the anterior and by branch vessels in the rest of the worm. The blood is red because it contains the respiratory pigment hemoglobin.

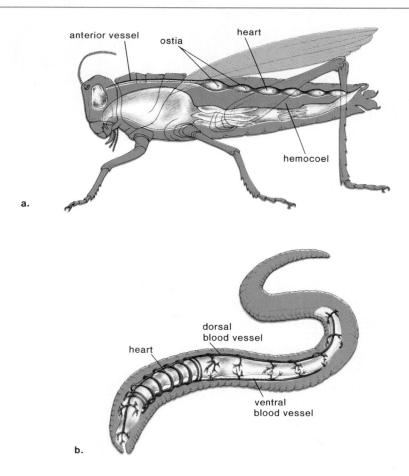

Vertebrates

All vertebrate animals have a closed circulatory system (fig. 20.4). There is a strong, muscular heart in which the **atria** (sing., atrium) primarily receive blood, and the muscular **ventricles** pump blood out through the blood vessels. There are three kinds of blood vessels (fig. 20.5). **Arteries** carry blood away from the heart, **capillaries** exchange materials with tissue fluid, and **veins** return blood to the heart.

Arteries, the strongest of the blood vessels, are resilient: they are able to expand and constrict because their walls have substantial layers of both elastic and muscle fibers. **Arterioles** are small arteries whose constriction can be regulated by the nervous system. Arteriole constriction and dilation affect blood pressure. The more vessels dilated, the lower the blood pressure.

Arterioles branch into *capillaries,* each of which is an extremely narrow, microscopic tube having a wall composed of only one layer of cells (fig. 20.5). Capillary beds (many capillaries interconnected) are present in all regions of the body; consequently, a cut to any body tissue draws blood. Capillaries are the most important part of a closed circulatory system because exchange of nutrient and waste molecules takes place across their thin walls (fig. 20.17).

Not all capillary beds are open at the same time. After a person has eaten, the capillary beds around the digestive tract are usually open; during muscular exercise, the capillary beds of the skeletal muscles are open. The distribution of blood in the various capillary beds is regulated by **sphincters**, circular muscles that open and close tubular structures. In this instance, when the sphincter relaxes, blood can enter a capillary bed. When the sphincter contracts, a capillary is closed and no blood can enter. Contraction and relaxation of sphincters is controlled by the nervous system.

Figure 20.4
All vertebrates have a closed circulatory system containing a heart (pumping device), arteries that take blood away from the heart, and veins that take blood to the heart. Capillaries are the smallest of the vessels and exchanges take place across their thin walls. Blood and the fluid that surrounds the cells (tissue fluid) make up an internal environment that remains relatively constant because the circulatory system transports blood between the body's cells and specialized exchange boundaries with the external environment.

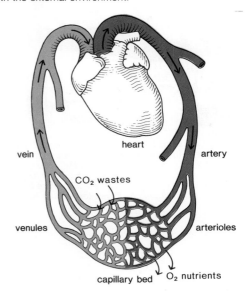

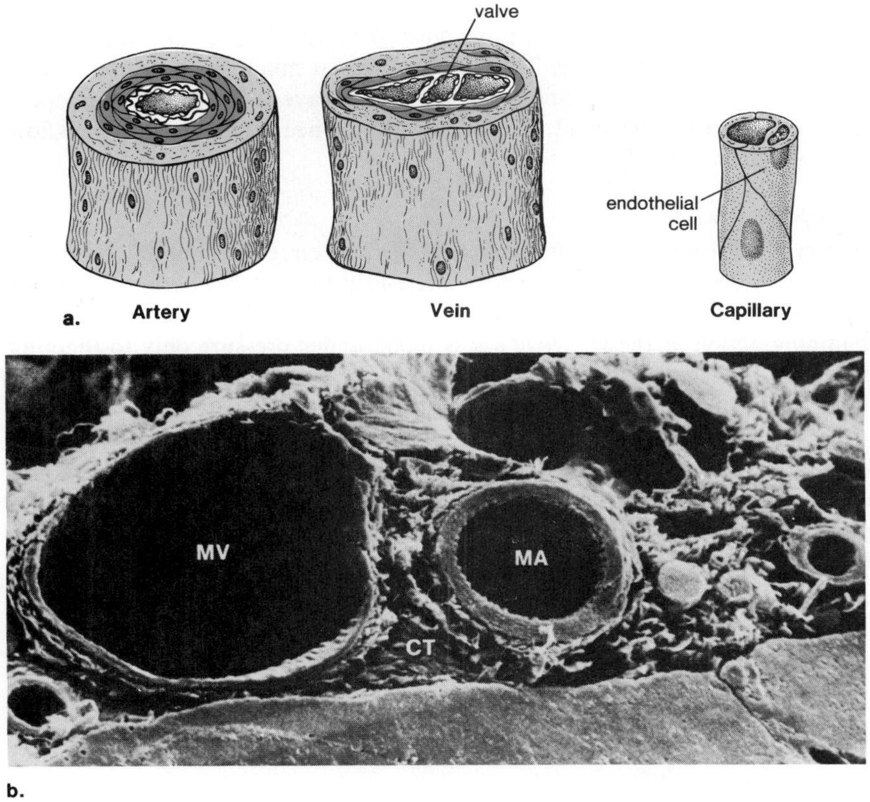

Figure 20.5
The three types of vertebrate blood vessels differ in construction. *a.* The artery has a strong muscular wall. Arteries can constrict or dilate in order to help regulate blood pressure and the distribution of the blood. The veins have weak walls, but valves keep the blood moving toward the heart. A capillary has a thin wall of only one layer of cells, which facilitates the exchange of molecules between blood and tissue fluid, the fluid that surrounds the cells. *b.* A scanning electron micrograph of a major artery (MA) and a major vein (MV), illustrating the difference in wall thickness and vessel size.
Kessel, R. G., and Kardon, R. H.: *Tissues and Organs: A Text-Atlas of Scanning Electron Microscopy.* © 1979 by W. H. Freeman and Co.

William Harvey

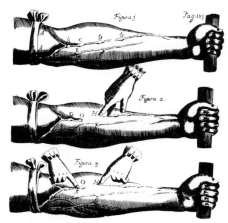

To show that the veins return blood to the heart, Harvey suggested that (Figure 1) *an arm be tied above the elbow to observe the accumulation of blood in the veins;* (Figure 2) *blood can be forced past a valve from H to O, but* (Figure 3) *not in the opposite direction. Therefore it can be deduced that blood ordinarily moves toward the heart in the veins.*

William Harvey was the first to offer proof that the blood circulates in the body of humans and other animals. He was an English scientist of the seventeenth century, a time of renewed interest in the collection of facts, use of the hypothesis, experimentation, and respect for mathematics. The seventeenth century was the time of the scientific revolution.

After many years of research and study, Harvey hypothesized that the heart is a pump for the entire circulatory system, and that blood flows in a circuit. In contrast to former anatomists, Harvey dissected not only dead but live organisms and observed that when the heart beats, it contracts, forcing blood into the aorta. Had this blood come from the right side of the heart? To do away with the complication of the lungs (pulmonary circulation), Harvey turned to fishes and noticed that the heart first received and then pumped the blood forward. He observed that blood in the fetus passes directly away from the right side of the heart through the septum to the left side. He felt confident that in mature, higher organisms, all blood moves from the right to the left side of the heart by way of the lungs.

Harvey then wanted to show an intimate connection between the arteries and veins in the tissues of the body. Again, using live lower organisms, he demonstrated that if an artery is slit, the whole blood system

empties, including arteries and veins. He measured the capacity of the left ventricle and found it to be 2 ounces. Since the heart beats 72 times a minute, in one hour the left ventricle will force into the aorta no less than $72 \times 60 \times 2 = 8640$ ounces $= 540$ pounds, or three times the weight of a heavy man! Could so much blood be created and consumed every hour? The same blood must return again and again to the heart.

Harvey studied the valves in the veins and suggested their true purpose. By the use of ligatures (see the illustration), he demonstrated that a tight ligature on the arm causes the artery to swell on the side of the heart, a slack ligature causes the vein to swell on the opposite side. He said, "This is an obvious indication that the blood passes from the arteries into the veins . . . and there is an anastomosis of the two orders of vessels."

Harvey's methods showed how fruitful research might be done. He established that physical and mechanical evidence would provide data for a theory of circulation. He, himself, erred when he speculated on the function of the heart and lung. He thought the heart heated the blood, and the lung served to cool it or control the degree of heat. His basic method, however, contributed to the scientific revolution and set an example for others to follow.

Venules and veins collect blood from the capillary beds and take it to the heart. First the **venules** drain the blood from the capillaries and then they join together to form a *vein*. The wall of a vein is much thinner than that of an artery because the muscle and elastic fiber layers are poorly developed. Valves within the veins point, or open, toward the heart, preventing a backflow of blood (fig. 20.13).

Comparisons

A survey of vertebrate animals shows that their circulatory pathways differ. In fishes, blood follows a *one-circuit* (single-loop circulatory) pathway through the body. The heart has a single atrium and a single ventricle (fig. 20.6). The pumping action of the ventricle sends blood under pressure only to the gills where it is oxygenated. After passing through gill capillaries, there is little blood pressure left to distribute the oxygenated blood to the tissues.

In contrast, the other vertebrates have a *two-circuit* (double-loop circulatory) pathway. The heart pumps blood to the lungs, called **pulmonary circulation,** and pumps blood to the tissues, called **systemic circulation.** This double pumping action assures adequate blood pressure and flow to both circulatory loops.

Figure 20.6
Comparison of circulatory systems in vertebrates. *a*. In a fish, the blood moves in a single loop. The heart has a single atrium and ventricle and serves to pump the blood into the gill region where gas exchange takes place. Blood pressure created by the pumping of the heart is dissipated after the blood passes through the gill capillaries. This is a disadvantage of this single-loop system. *b*. Amphibians have a double-loop system in which the heart pumps blood to both the lungs and the body itself. The system is not very efficient because oxygenated blood and deoxygenated blood mix in the single ventricle. *c*. The pulmonary and systemic systems are completely separate in birds and mammals since the heart is divided by a septum into a right and left half. The right side pumps blood to the lungs and the left side pumps blood to the body proper.

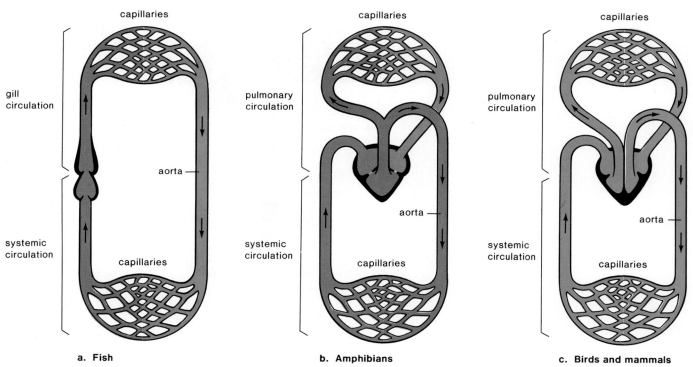

a. Fish   b. Amphibians   c. Birds and mammals

In amphibians the heart has two atria, but there is only a single ventricle. Therefore, the pulmonary system is not separate from the systemic system. Oxygenated blood mixes with deoxygenated blood in the single ventricle. The hearts of other vertebrates are partially (most reptiles) or completely (some reptiles, all birds and mammals) divided into right and left halves. The right ventricle pumps blood to the lungs and the left ventricle pumps blood to the rest of the body. This arrangement increases the likelihood of adequate blood pressure for both the pulmonary and systemic systems.

Figure 20.7

Diagrammatic view of the human circulatory system indicating the path of blood. In order to trace blood from the right to the left side of the heart, you must consider the lung capillaries. In order to trace blood from the gut capillaries to the right atrium, you must consider the hepatic portal system and hepatic vein.

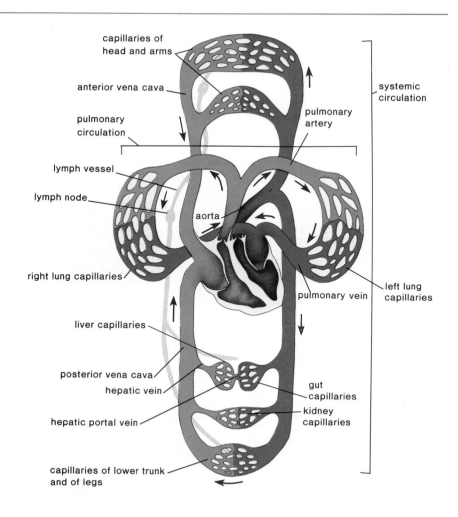

Human Circulation

Humans, like all other vertebrates, have a closed circulatory system. Blood flows away from the heart through arteries and arterioles to capillary networks, where exchanges between blood and tissue fluid take place. Blood leaves the capillaries and returns to the heart through venules and veins. Like other mammals, humans have both pulmonary and systemic circulatory pathways (fig. 20.7). Humans also have a typically mammalian four-chambered heart, with right and left atria and right and left ventricles (fig. 20.8).

Path of Blood

Pulmonary Circulation

All blood entering the right atrium is deoxygenated. Blood passes from the right atrium into the right ventricle, which then pumps it out through the **pulmonary trunk.** The pulmonary trunk branches into two pulmonary arteries

Figure 20.8
Ventral view of the human heart. The atrioventricular valves are supported by strings that allow the valves to shut but keeps them from reverting into the atria once the ventricles fill with blood.

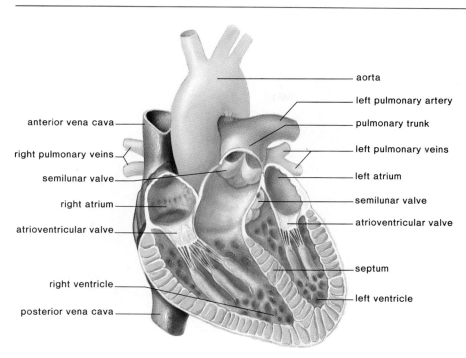

aorta

left pulmonary artery

pulmonary trunk

anterior vena cava

right pulmonary veins

left pulmonary veins

semilunar valve

left atrium

right atrium

semilunar valve

atrioventricular valve

atrioventricular valve

septum

right ventricle

left ventricle

posterior vena cava

that carry blood to arterioles and capillaries in the lungs. After passing through lung capillaries located around the alveoli (see p. 481), blood returns to the left atrium of the heart through **pulmonary venules and veins.**

Systemic Circulation

Blood returning from the pulmonary circulation is oxygenated. This oxygenated blood passes from the left atrium into the left ventricle, which then pumps it out through the **aorta,** the large arterial trunk that supplies the entire systemic circulation.

The aorta sends branches to all parts of the body. The first to branch off the aorta are the **coronary arteries** that are a part of a **coronary circulation** for the heart. Individual heart cells cannot exchange material with the blood being pumped through the chambers; therefore, blood flow through coronary arteries is critical for normal functioning of the heart. Blockage of any coronary artery quickly results in damaged heart muscle and impaired heart function.

Anterior to the heart, the aorta arches to the left and then passes posteriorly through the body. Branches off the arch of the aorta supply blood to the upper parts of the body. As the aorta descends through the abdominal cavity, branches are given off to the digestive organs, the kidneys, the body wall, the legs, and other posterior parts.

Blood returns from the systemic circulation to the right atrium of the heart by way of two large veins, the **anterior** (superior) **vena cava** and the **posterior** (inferior) **vena cava.** The anterior vena cava returns blood from the head, arms, and chest; the posterior vena cava returns blood from the remainder of the systemic circulation.

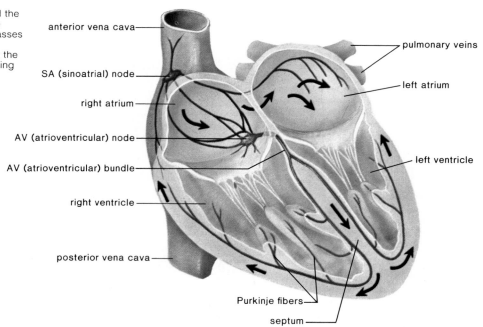

anterior vena cava

SA (sinoatrial) node

right atrium

AV (atrioventricular) node

AV (atrioventricular) bundle

right ventricle

posterior vena cava

pulmonary veins

left atrium

left ventricle

Purkinje fibers

septum

Portal System

A portal system is one that begins and ends in capillaries. The human body has only one such major system, the **hepatic portal system** (fig. 21.18). In this system, the first set of capillaries occurs at the digestive organs and the second occurs in the liver. Blood passes from the capillaries about the digestive organs into venules that join a major vein (the *hepatic portal vein*) that takes the products of digestion to the liver. Here, these products may be processed or stored until they are needed to maintain the constancy of blood composition within the *hepatic vein,* a vessel that leaves the liver to enter the vena cava.

The Human Heart

The human heart (fig. 20.9) is a remarkably efficient and reliable pumping device, and its regular, continual beating is essential to life. Cessation of heartbeat is one of the most obvious indications of death. The heart can never stop its regular beating for even a short time; some organs can survive brief pauses in circulatory supply but others cannot. The brain, for example, is critically sensitive to even short failures of circulation. If circulation to the brain stops for as little as five seconds, oxygen is depleted and consciousness is lost. A four-minute break in circulation to the brain causes death of significant numbers of brain cells, and a nine- to ten-minute circulatory failure causes massive, irreversible brain damage and, usually, death.

During each minute of life, a normal, resting person's heart beats about seventy times and pumps out about five liters of blood, a volume approximately equal to the total blood volume in the body. All this work is done by a relatively small organ—the average human heart weighs only 300 g, less than 0.5 percent of total body weight. Even the heart of a highly trained athlete weighs only about 1 percent of total body weight. This small organ sustains its heavy work load from the time it begins beating early in embryonic development until the time of death, and it never rests continuously for more than a second.

a.

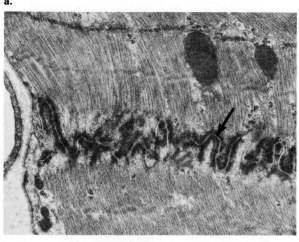

b.

Figure 20.10
Cardiac muscle tissue. Photomicrograph (*a*.) showing its striated appearance. The dark lines are intercalated disks, one of which is shown in detail in (*b*.), an electron micrograph.

Heartbeat

The contraction of the heart is intrinsic to the heart; it does not require outside nervous stimulation. When individual cardiac muscle cells are dissected from the heart, they beat spontaneously and rhythmically. These cells are **striated,** having both light and dark bands, and are branched. When cardiac muscle cells mesh within the heart, they form the **myocardium,** the muscular wall of the heart. Where cardiac muscle cells meet, there is a band known as an **intercalated disk** (fig. 20.10). Here portions of the two cell membranes fuse, and the excitation causing muscular contraction can be transmitted from one cell to another.

The heartbeat is normally regulated by the **pacemaker,** a small node of myocardial cells termed the **SA (sinoatrial) node.** The SA node, embedded in the right atrial wall (fig. 20.9), initiates a wave of excitation every 0.85 seconds. This wave of excitation spreads quickly throughout the atria, which contract while the ventricles are still resting. After the wave of excitation makes contact with the AV (atrioventricular) node, a second node located at the base of the right atrium, the ventricles contract. While the ventricles contract, the atria rest. Following ventricular systole, the entire heart rests. Notice in table 20.1, which gives the time sequences for these events, that contraction of a chamber is called **systole,** while relaxation is called **diastole.**

Table 20.1 The Heart Cycle

Time	Atria	Ventricles
0.15 sec.	Systole	Diastole
0.30 sec.	Diastole	Systole
0.40 sec.	Diastole	Diastole

Figure 20.11

Valves of the heart. *a.* This drawing is a view from above the heart, with the valves exposed. The arrow shows the approximate location in the whole heart. *b.* Photograph of the aortic semilunar valve in the closed position. This valve lies between the left ventricle and the aorta.

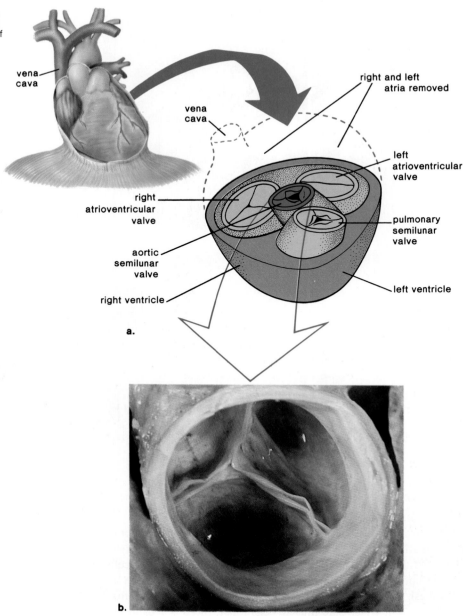

vena cava

vena cava

right and left atria removed

right atrioventricular valve

left atrioventricular valve

pulmonary semilunar valve

aortic semilunar valve

left ventricle

right ventricle

a.

b.

The heartbeat sounds, lubb-dup, are caused by the closing of valves, first those that conduct blood from the atria to the ventricles (atrioventricular valves), then those that conduct blood from the ventricles to their respective arteries (semilunar valves) (fig. 20.11). These valves aid circulation of the blood through the heart because they permit only one-way flow and do not allow a backward flow of blood. **Pulse** is a wave of vibration that passes down the walls of the arterial blood vessels when the aorta expands following ventricle systole. Because there is one arterial pulse per ventricular systole, the arterial pulse rate can be used to determine the heart rate.

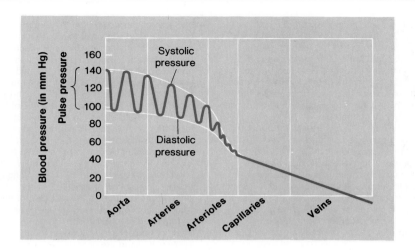

Figure 20.12

Blood pressure in different parts of the human circulatory system. The pulse pressure (includes systolic and diastolic pressures) decreases as blood flows into smaller arteries, and disappears as blood enters capillaries. The slow and even movement of the blood through the capillaries facilitates exchange of molecules here. The blood pressure in the veins is so low that it cannot account for the movement of blood in these vessels. Skeletal muscle contraction pushes the blood along from valve to valve.

Although the beat of the heart is intrinsic, it may still be speeded up or slowed down by nervous stimulation. A heart rate center in the medulla oblongata (p. 516) portion of the brain alters the heartbeat by way of the autonomic nervous system (p. 513). The latter is made up of two divisions: the parasympathetic system promotes normal activities, and the sympathetic system instigates responses associated with times of stress. The parasympathetic system causes the heartbeat to slow down, and the sympathetic system increases the beat. Various factors, such as the relative need for oxygen or the level of blood pressure, determine which of these systems is activated.

Systemic Blood Flow

When the left ventricle contracts, blood is forced into the arteries under pressure. **Systolic pressure** results from blood being forced into the arteries during ventricular systole, and **diastolic pressure** is the pressure in the arteries during ventricular diastole. Human blood pressure is measured with a sphygmomanometer, which has a pressure cuff that permits measurement of the amount of pressure required to stop the flow of blood through an artery. Blood pressures normally are measured on the brachial artery, an artery of the upper arm, and are stated in millimeters of mercury (Hg). A blood pressure reading includes two numbers, for example, 120/80, that represent systolic and diastolic pressures, respectively.

Blood flows from the region of highest pressure, the left ventricle, to the region of lowest pressure in the veins leading to the right atrium. Measurements of blood pressure at various points along the circulatory system show a progressive decrease in blood pressure (fig. 20.12). Pressure is highest in the first part of the aorta, where it is about 140/120. As blood moves through the aorta and then spreads through smaller arteries and arterioles, pressure falls. At the same time, the difference between systolic and diastolic pressure decreases and finally disappears in the capillaries, where there is only a smooth, steady flow of blood.

Hypertension

High blood pressure, or hypertension, is not often listed as the cause of death on death certificates, but it often precedes and accompanies death due to heart attack, stroke, and thromboembolism. Altogether, cardiovascular diseases account for more deaths in the United States than do cancer and accidents combined.

Patients with hypertension frequently have atherosclerosis and vice versa. This common condition involves the formation of plaques within arterial linings. Plaques contain large quantities of cholesterol. Cholesterol is a substance found within the cell membrane of all cells and is modified into hormones by certain glands. Cholesterol is carried in the blood by lipoprotein molecules of two types: LDLs (low-density lipoproteins) and HDLs (high-density lipoproteins).

The cholesterol-LDL molecular combination is the one that triggers the atherosclerotic accumulation. When these molecules are present in the blood, white blood cells called monocytes (fig. 20.16) begin to invade and damage the blood vessel lining. These cells are transformed into macrophages (large phagocytic cells) that absorb the cholesterol and burst, leaving behind fatty streaks. Platelets are also naturally attracted to the damaged area. They release a growth factor that causes the muscle cells of the arterial wall to divide. This accumulating mass constitutes a plaque that can eventually grow so large that it hinders or even stops blood flow through the artery. Eventually calcium collects in the cells of the plaque and harden it.

There has been a great deal of attention given to the role of cholesterol in the development of atherosclerotic plaque. A diet high in saturated fatty acids (abundant in butter and ice cream) and cholesterol (abundant in meats, eggs, and certain cheeses) leads to high blood levels of cholesterol carried by LDL. Recently, investigators discovered that a high intake of omega-3 fatty acids—those in which there is a double bond in the fatty acid chain on the third carbon from the end (omega being the last letter of the Greek alphabet)—reduces the level of LDL in the blood. Omega-3 fatty acids are prevalent in fish oil. Thus a diet high in fish may account for the lower incidence of heart disease among Eskimos.

A significant clinical study showed that lowering the cholesterol level can help prevent heart disease. In 1984, the National Heart, Lung, and Blood Institute in the United States reported the results of a ten-year study involving nearly a half-million middle-aged men who had high blood levels of cholesterol. Half of the men followed a moderate cholesterol-lowering diet and also took a drug that reduces blood cholesterol levels. The other half followed the diet and took a placebo. The group that received the drug had fewer cardiovascular incidences than did the other group. It was concluded that those subjects with a 25 percent lower blood-cholesterol level had their risk of heart attack cut in half.

Heart attack and stroke are common causes of death in individuals with atherosclerosis. The heart muscle requires a large and continuous supply of oxygen. Coronary arteries narrowed by atherosclerosic plaques cannot supply enough oxygen. First, angina (chest pain) and then heart attack (death of heart muscle due to lack of oxygen) may result. Then, too, blood clots often form in vessels damaged by plaque. The clot could develop in the coronary arteries or develop in a larger vessel some distance away and be brought to the coronary arteries by circulating blood. A thrombus is a stationary blood clot, while an embolus is a moving clot; therefore, thromboembolism (blood clots) sometimes causes heart

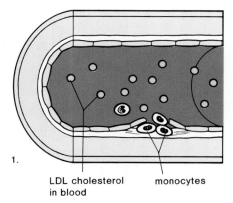
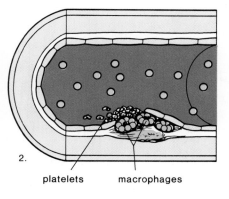
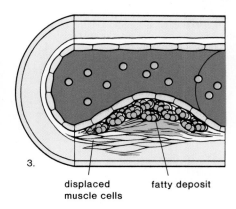

1.

LDL cholesterol monocytes
in blood

2.

platelets macrophages

3.

displaced fatty deposit
muscle cells

Plaque formation. 1. Cholesterol in combination with LDL attracts monocytes that invade the arterial lining. 2. Monocytes become cholesterol-ingesting macrophages that burst, leaving behind fatty streaks. Release of growth factor by platelets causes muscle cells to divide. 3. Plaque that obstructs blood flow consists of all components mentioned.

attacks. Thromboembolism is also a cause of stroke—a blood clot occludes (clogs up) an artery in the brain.

Certainly it would be beneficial for everyone to take all possible steps to prevent cardiovascular disorders. The factors that may contribute to hypertension include the following:

Diet Increased salt intake causes fluid to be retained in the blood vessels. This can cause high blood pressure because there is more fluid to press against arterial walls. In susceptible people, fatty foods and high cholesterol foods may contribute to arteriosclerosis, and therefore to high blood pressure (see table A).

Stress Blood pressure normally rises with excitement or alarm. Perhaps in some persons the blood pressure fails to return to normal once the stressful conditions have passed.

Heredity It appears that people whose parents have high blood pressure are more susceptible to this condition.

Race Blacks are more likely to develop hypertension. This may be due to a genetic defect, but may also be due to diet or exposure to stress.

Obesity All tissues require a supply of blood carried by blood vessels. When weight is gained, the circulatory system also increases in size and the heart must work harder to pump more blood.

Smoking Nicotine constricts blood vessels, cutting off circulation to the extremities and causing blood pressure to rise. The heart must also work harder to pump blood through damaged lungs.

While the effects of heredity cannot be erased, it is clear that good health habits could reduce the incidence of hypertension and cardiovascular disease.

Table A Dietary Reduction of Blood Cholesterol Levels	
Eat Less	**Eat More**
Fatty meats	Fish, poultry, lean cuts of meat
Organ meats (liver, kidney, brain, pancreas), shrimp	
Sausage, bacon, processed meats	Fruit, vegetables, cereals, starches
Whole milk	Skim or low-fat milk
Butter, hard margarine	Soft margarine
High-fat cheese (bleu, cheddar)	Low-fat cheese
Ice cream and cream	Yogurt
Egg yolks	Egg whites
Foods fried in animal fats	Vegetable fats for frying and salad dressing
Commercial baked goods	

Reprinted by permission from the 1986 *Medical and Health Annual*, copyright 1985, Encyclopaedia Britannica, Inc., Chicago, Illinois.

Figure 20.13
When a valve is open, a result of pressure on the veins exerted by skeletal muscle contraction, the blood flows toward the heart in veins. Valves close when external pressure is no longer applied to them. Closure of the valves prevents the blood from flowing in the opposite direction.

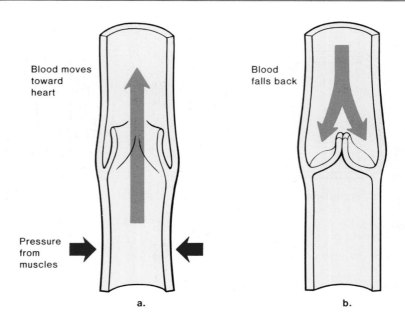

Blood moves toward heart

Blood falls back

Pressure from muscles

a.

b.

Blood pressure in the veins is very low (5 to 10 mm Hg) and finally falls to 2 mm Hg or less in the vena cava. This very low blood pressure is not adequate to move blood back to the heart, especially from lower parts of the body. When body muscles near veins contract, they put pressure on the veins' collapsible walls and on the blood contained in the veins. Because veins have **valves** (fig. 20.13) that prevent backward flow, pressure from muscle contractions effectively helps move blood through veins toward the heart. During periods of inactivity, blood moves very sluggishly through veins in some parts of the body and can accumulate in the veins. For example, blood accumulates in leg veins during long periods of standing.

The Lymphatic System

The lymphatic system is a one-way system rather than a circulatory system and contains lymph veins and capillaries only. The system begins with lymph capillaries that lie near the blood capillaries. They take up any excess *tissue fluid*. Once tissue fluid enters lymph capillaries, it is called **lymph.**

Lymph vessels have a construction similar to that of cardiovascular veins, including valves (fig. 20.14). Lymph in humans moves through the lymph capillaries and vessels as a result of pressure applied by muscle contractions near the vessels. The valves prevent a backward flow of lymph. Lymph moves through smaller vessels that unite to form larger vessels, and finally, two major lymph ducts empty into large blood circulatory veins near the heart.

In addition to returning fluids from the tissues to the blood circulatory system, the lymphatic (fig. 20.15) system has several other important functions. Lymph vessels known as **lacteals** are present within the intestinal villi. The products of fat digestion enter the lacteals and are carried in lymph vessels to eventually enter the circulatory system.

Lymph nodes are small ovoid or round structures composed of lymphoid tissue. These nodes occur at certain strategic points along the medium-sized lymph vessels. **Lymphocytes,** a type of white blood cell, are packed into the spaces of a lymph node, which also filters and traps bacteria and other debris, helping keep the blood purified. When a local infection, such as a sore throat, is present, lymph nodes in that region swell and become painful.

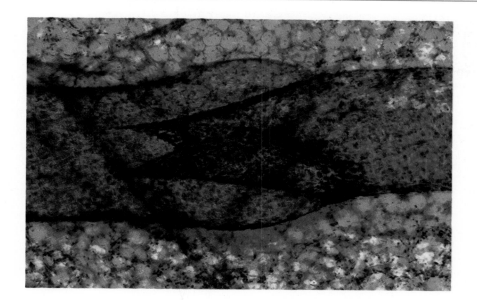

Figure 20.14
Photomicrograph of a valve in a lymph vessel. Lymph is
forced through lymph vessels as contracting muscles
exert pressure on the vessels' walls. Valves maintain one-
way flow through lymph vessels by preventing backward
flow.

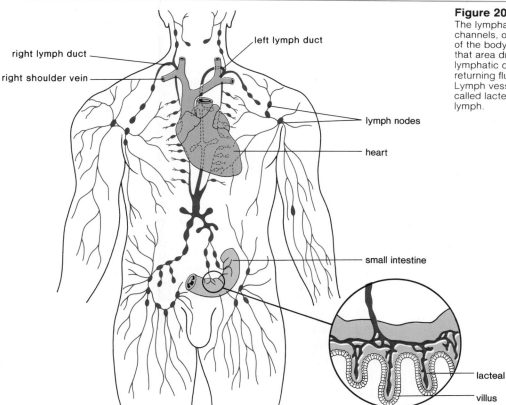

right lymph duct

left lymph duct

right shoulder vein

lymph nodes

heart

small intestine

lacteal

villus

Figure 20.15
The lymphatic system. Lymph vessels flow into two main
channels, one of which drains lymph vessels from all parts
of the body except the upper right portion. Vessels from
that area drain through another duct. These two large
lymphatic ducts drain into the veins of the shoulder, thus
returning fluid from the tissues to the circulatory system.
Lymph vessels in the region of the small intestine are
called lacteals. Nodes filter debris and bacteria out of
lymph.

Lymph nodes along with lymphatic vessels are sometimes removed in
cancer operations because lymph vessels may help spread any remaining cancer
cells. However, this may result in **edema,** a swelling of a portion of the body
caused by an accumulation of excess tissue fluid. An extreme example of edema
is seen in elephantiasis (p. 318) when parasitic roundworms block lymphatic
drainage.

Table 20.2 Components of Blood		
Blood	**Function**	**Source**
1. Cells	Transport oxygen	Bone marrow
Red cells	Clotting	Bone marrow
Platelets	Fight infection	Bone marrow and
White cells		lymphoid tissue
2. Plasma*		
Water	Maintains blood volume and transports molecules	Absorbed from intestine
Plasma proteins	All maintain blood osmotic pressure and pH	
Albumin		Liver
Fibrinogen	Clotting	Liver
Globulins	Fight infection	Lymphocytes
Gases		
Oxygen	Cellular respiration	Lungs
Carbon dioxide	End product of metabolism	Tissues
Nutrients		
Fats, glucose, amino acids, etc.	Food for cells	Absorbed from intestinal villi
Salts	Maintain blood osmotic pressure and pH; aid metabolism	Absorbed from intestinal villi
Wastes		
Urea and ammonia	End products of metabolism	Tissues
Hormones, vitamins, etc.	Aid metabolism	Varied

*Plasma is 90–92 percent water, 7–9 percent plasma proteins, and not quite 1 percent salts. All other components are present in even smaller amounts.

Human Blood

Blood has two main portions: liquid plasma and cells (table 20.2). **Plasma** contains many types of molecules, including nutrients, wastes, salts, and proteins. Salts and proteins buffer the blood, effectively keeping the pH near 7.4. They also maintain blood's osmotic pressure so that water has an automatic tendency to enter blood capillaries.

The cells (fig. 20.16) are of three types: **red blood cells** or erythrocytes, **white blood cells** or leukocytes, and **platelets** or thrombocytes. At this time, we are concerned only with erythrocytes, which are so small that a cubic milliliter of blood contains five million (table 20.3). Each one is so packed with hemoglobin that there is no room for cellular organelles, not even a nucleus. Hemoglobin inside erythrocytes reversibly binds with oxygen to form *oxyhemoglobin* (p. 484). When most of its hemoglobin is in the form of oxyhemoglobin, blood takes on a bright red color. Oxygenated blood, carrying oxyhemoglobin, is found in pulmonary veins and systemic arteries. Hemoglobin that is not carrying oxygen is called *reduced hemoglobin,* and deoxygenated blood, carrying reduced hemoglobin, is a darker, more purplish color. Deoxygenated blood is found in systemic veins and pulmonary arteries. Thus, it is not correct to indicate all arteries by using a red color and all veins by using a bluish color since it is just the reverse in the pulmonary system.

Red blood cells are manufactured in the red bone marrow at the rate of two million per second. They live about four months, after which time they tend to become damaged from squeezing through small capillaries. Damaged erythrocytes are withdrawn from circulation as blood passes through the liver or spleen. Here large cells called **macrophages** derived from monocytes phagocytize and destroy them. When red cells are destroyed, hemoglobin is released. The iron is recovered and returned to the red bone marrow for reuse. The heme portions of the molecules undergo chemical degradation and are excreted by the liver as bile pigments in the bile (p. 464). The bile pigments are primarily responsible for the color of feces.

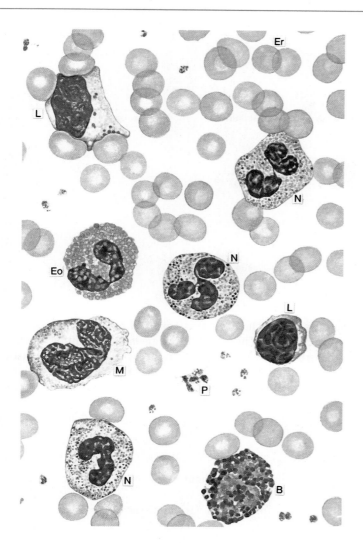

Figure 20.16
A representation of human blood cells. The erythrocytes (Er) are smaller and more numerous than the white cells: eosinophils (Eo), neutrophils (N), and basophils (B), have cytoplasmic granules; monocytes (M) and lymphocytes (L) do not have granules. Notice that the red cells, which are packed full of oxygen-carrying hemoglobin, lack a nucleus. The white cells fight infection in two ways. The lymphocytes produce antibodies; the neutrophils and monocytes engulf bacteria and viruses. Platelets (P) are involved in blood clotting.

Table 20.3 Numbers and Distribution of Cells in Normal Human Blood

Total red cells = 5,000,000 cells per cubic millimeter of blood
Total white cells = 7000 cells per cubic millimeter of blood
Total platelets = 200–400 per cubic millimeter
Percentage of total white cells:
Granular leukocytes
  Neutrophils, 55–65
  Eosinophils, 2–3
  Basophils, 0.5
Agranular leukocytes
  Monocytes, 4–7
  Lymphocytes, 25–33

Figure 20.17

Diagram illustrating the exchanges that take place in a capillary and the forces that aid the process. On the arterial side, blood pressure is stronger than osmotic pressure. On the venous side, osmotic pressure is stronger than blood pressure.

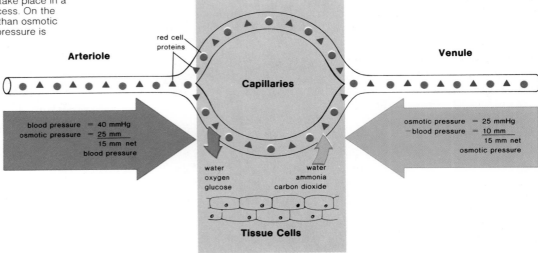

Sometimes the mechanisms that normally balance erythrocyte production and destruction fail. This can result in **anemia,** a condition in which there are fewer than the normal number of erythrocytes (or, in some cases, lower hemoglobin content in the erythrocytes). People suffering from anemia often are fatigued or chilled because their tissues lack adequate oxygen for cellular conversions or heat production. Anemia can be a symptom of a fundamental physiological problem, and proper treatment requires identification and correction of the specific underlying cause.

Capillary Exchange

Capillaries are distributed throughout every tissue of the body, and no cell is far from one or several capillaries. Capillaries form complex networks, and the total number of capillaries in various tissues is so great it defies imagination, much less accurate counting. For example, an area of 1 mm² of guinea pig muscle during maximal exercise may contain more than 3000 open capillaries. The lead in an ordinary pencil has a cross-sectional area of about 3 mm². Thus a piece of muscle of that same size during maximal exercise would contain close to 10,000 open capillaries.

Because the cross-sectional area of capillaries is so large compared to the arterioles, blood flows very slowly through the capillaries. This slow rate allows adequate time for exchange of materials between blood and tissue fluid. Much of this exchange occurs by *diffusion* through thin capillary walls. Lipid-soluble substances pass freely through the cell membranes in the capillary walls, but water-soluble material diffuses mainly through pores located in junctions between cells in the capillary wall. Two forces control movement of fluid through the capillary wall: *osmotic pressure,* which tends to cause water to move from tissue fluid to blood; and *blood pressure,* which tends to cause water to move in the opposite direction. The effective osmotic pressure remains constant at about 15 mm Hg, but blood pressure is about 40 mm Hg at the arteriole end and only 10 mm Hg at the venous end of a capillary.

As figure 20.17 illustrates, blood pressure is higher than osmotic pressure at the arterial end of a capillary. This tends to force fluid out through the pores in the capillary wall. Midway along the capillary, where blood pressure is lower, the two forces essentially cancel one another, and there is no net movement of water. However, solutes now diffuse according to their concentration gradient; nutrients (glucose and oxygen) diffuse out a capillary, and wastes (ammonia and carbon dioxide) diffuse into a capillary. Since proteins are too large to pass out of a capillary, *tissue fluid* tends to contain all components of plasma except proteins. At the venule end of a capillary, where the blood pressure has fallen to 10 mm Hg, osmotic pressure is greater and water tends to move into the capillary. Almost the same amount of fluid that left the capillary returns to it, but not quite. There is always some excess tissue fluid collected by the lymph capillaries.

Blood Clotting

Blood must be a free-flowing liquid if it is to circulate easily through blood vessels, but this liquidity can also cause serious problems. Any injury that breaks a large blood vessel can quickly lead to a serious loss of blood. This is countered by a complex clotting (coagulation) mechanism. Clots form temporary barriers to blood loss until a vessel's walls have healed.

When a blood vessel is injured, blood platelets begin to congregate near the cut or injury, forming a barrier known as the platelet plug. There are 200 to 400 platelets per cubic milliliter of blood. They are fragments of large bone marrow cells called megakaryocytes that disintegrate and discharge platelets into the bloodstream. When platelets come into contact with an injured vascular wall, they swell up, become sticky, and release certain chemicals. Some of these chemicals stimulate the blood vessel to constrict; some increase the tendency of platelets to form a plug; and some help initiate the process of blood clotting.

Blood clotting requires a complex and precise series of reactions that satisfy the requirement for a delicate balance between quick and efficient clot formation as a response to a significant blood vessel break and prevention of accidental formation of clots that interfere with normal circulation.

Since blood clotting is so complex, we will consider only the major events involved:

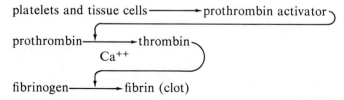

Prothrombin and fibrinogen, two proteins manufactured by the liver, are always present in the plasma. In the presence of calcium ions, **prothrombin** becomes the enzyme thrombin by the action of prothrombin activator. Thrombin is an enzyme that causes **fibrinogen** molecules to become fibrin threads. A clot is a mass of these fibrin threads within which cells and platelets are trapped (fig. 20.18). After a clot has formed, it retracts and its size is reduced as fluid is squeezed out. If clotting has occurred in a test tube (fig. 20.19), this fluid, called **serum,** is easily collected. Serum has the same composition as plasma except that it lacks prothrombin and fibrinogen.

Blood Typing

ABO Grouping

Although there are at least twelve well-known blood grouping systems, the ABO grouping and Rh are most often used to determine blood type. Before the twentieth century, blood transfusions had been attempted although sometimes the results were dire and even caused the death of a recipient. Intrigued

Figure 20.18
A scanning electron micrograph showing a red blood cell caught in the fibrin threads of a clot. Fibrin threads form a mesh that catches many blood cells.

Figure 20.19
When a blood clot contracts, serum is squeezed out. The clot is then a solid plug that prevents further blood loss. Hemophilia is the "bleeder's disease" because the blood will not clot due, most often, to the lack of a blood-clotting factor designated as VIII.

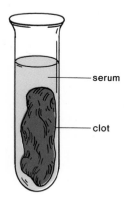

by these occurrences, a newly established physician of Vienna, Karl Lansteiner, began to examine the effect of mixing different samples of blood in the late 1800s. After months of laboratory research, he determined that there are four major blood groups among humans. He and his associates designated them as A, B, AB, and O. The ABO system, of course, still uses these same designations today. It is now known that blood type is based on the type of glycoprotein (p. 67) present on the red cells. Type A blood has type A glycoproteins on the red blood cells; type B blood has B-type glycoproteins; type AB has both of these glycoproteins; and type O has neither of them. The A and B glycoproteins function as antigens; that is, they combine specifically with antibody molecules. When this particular antigen-antibody reaction occurs, the red blood cells clump or **agglutinate.**

Anti-A and Anti-B antibodies are found in human plasma. Table 20.4 shows which type antibody is present in the plasma of which type blood. For a donor to give blood to a recipient, the recipient must not have an antibody that would cause the donor's cells to clump. For this reason, it is important to determine each person's blood type. Figure 20.20 demonstrates a way to use

Table 20.4 Antigens and Antibodies in Human ABO Blood Groups

Blood Type	Antigen Present on Red Blood Cell Membrane	Antibody in Plasma	Incidence of Type in the United States	
			Among Whites	*Among Blacks*
A	A	Anti-B	41%	27%
B	B	Anti-A	10%	20%
AB	A and B	Neither	4%	7%
O	Neither	Anti-A and Anti-B	45%	46%

From Johnson, Leland G., *Biology.* © 1983 Wm. C. Brown Publishers, Dubuque, Iowa. All Rights Reserved. Reprinted by permission.

Figure 20.20
The standard test to determine ABO and Rh blood type consists of putting a drop of Anti-A antibodies, Anti-B antibodies, and Anti-Rh antibodies on a slide. To each of these a drop of the person's blood is added. *a.* If agglutination occurs, as seen at lower left, the person has this antigen on the red cells. *b.* Several possible results.

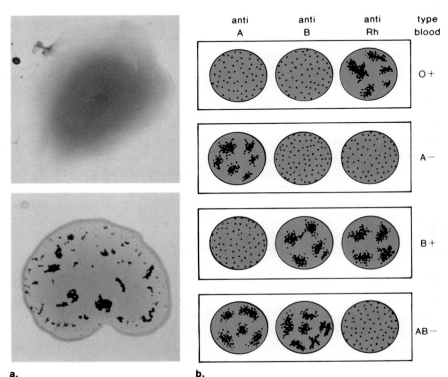

the antibodies derived from plasma to determine the type of blood. If clumping occurs after a sample of blood is exposed to a particular antibody, the person has that type of blood.

Another important antigen in matching blood types is the **Rh factor.** Persons with this particular antigen on the red cells are Rh⁺ (Rh positive); those without it are Rh⁻ (Rh negative). (Only 15 percent of Caucasians are Rh negative.) Rh negative individuals do not normally make antibodies to the Rh factor, but they will make them when exposed to the Rh factor. It is possible to extract these antibodies and use them for blood type testing.

The Rh factor is particularly important during pregnancy. If the mother is Rh negative and the father is Rh positive, the child may be Rh positive. The Rh positive red cells begin leaking across into the mother's circulatory system as placental tissues normally break down before and at birth. This causes the mother to produce Rh antibodies. If the mother becomes pregnant with another Rh positive baby, Rh antibodies (but not Anti-A antibodies and Anti-B antibodies) may cross the placenta and cause destruction of the child's red cells. This problem is solved by giving Rh negative women an Rh immune globulin injection just after the birth of an Rh positive child. This injection contains Rh antibodies that attack the baby's red cells before these cells stimulate the mother to produce her own antibodies.

Defense against Foreign Materials

The outer covering of skin and mucous membranes (fig. 20.21) is the body's first line of defense against foreign substances, including bacteria and viruses. The second line of defense is dependent on the ability of certain white cells to phagocytize foreign materials. The third line of defense is dependent on the ability of the other white cells to produce antibodies against foreign substances.

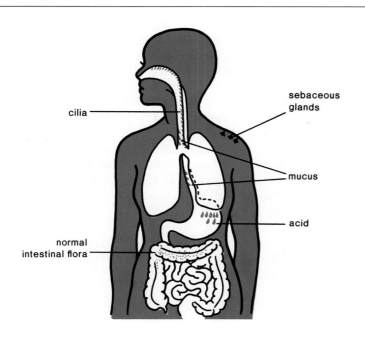

cilia

sebaceous glands

mucus

acid

normal intestinal flora

Figure 20.21
First line of defense against infection are the body's interior and exterior surfaces. The skin is relatively thick and the sebaceous (oil) glands secrete bactericidal acids. The respiratory tract is lined by ciliated and mucus-secreting cells. Impurities entrapped within the mucus are swept by the cilia up into the throat where the mucus can be swallowed. Secretion of mucus along the digestive tract also aids in trapping microorganisms. The acidity of the stomach is low enough to kill off most microorganisms, and any harmful ones that survive have to be able to compete with the normal intestinal flora in order to survive.

Figure 20.22

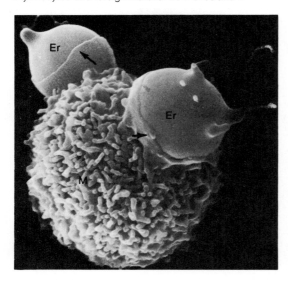

A macrophage (M) ingesting two erythrocytes (ER) (red blood cells). Cytoplasmic extensions (arrows) of the macrophage are in the process of surrounding the erythrocytes. Macrophages continually remove aging erythrocytes and foreign material from circulation.

White Cells (Leukocytes)

White cells (fig. 20.16 and table 20.3) can be distinguished from red cells because they are usually larger, have a nucleus, and, without staining, appear white in color. With staining, white cells appear bluish and may have granules that may bind with certain stains. The latter type of white cell, called **granulocytes,** has a lobed nucleus and is of three types: **neutrophils** have granules that do not take up a dye; **eosinophils** have granules that take up the red dye eosin; and **basophils** have granules that prefer a basic dye that stains them a deep blue color. **Agranulocytes,** white cells that have no granules, have a circular or indented nucleus and are either the larger monocytes or the smaller lymphocytes.

Inflammatory Reaction

When bacteria or viruses enter the body, a response called the inflammatory reaction occurs. Damaged tissue releases **kinins** that cause vasodilation, and **histamines** that cause increased capillary permeability. Blood rushes to the area and the neutrophils, which are amoeboid, squeeze through the capillary wall and enter the tissue fluid where they phagocytize foreign material. The thick yellowish fluid **pus** contains a large proportion of dead neutrophils. Later, monocytes also appear on the scene and are transformed into **macrophages** (fig. 20.22), large phagocytizing cells that often engulf material. The speed with which some macrophages consume bacteria has been timed at less than one hundredth of a second. Macrophages are most apt to appear once an antigen-antibody reaction (discussion following) has occurred.

Immunity

Two types of immunity are recognized. In **active immunity** the body itself is capable of defending its integrity against a foreign substance. In **passive immunity** the defense has been provided for the body.

Active Immunity

Active immunity is dependent on the immune system. The immune system consists of lymphocytes; bone marrow lymphoid tissue, where lymphocytes originate; and the lymph nodes, spleen, and thymus, where lymphocytes are concentrated. The thymus is a bilobed structure located in the upper chest just below the neck. Its size decreases as one grows older.

Lymphocytes are of two types: **T** (thymus dependent) **cells** and **B**[1] (bone marrow dependent) **cells.** While both are originally derived from lymphoid bone marrow stem cells, the stem cells producing T cells have passed through the thymus, while those producing B cells have not (fig. 20.23). T cells are responsible for **cell-mediated immunity,** and B cells are responsible for **antibody-mediated immunity** (table 20.5).

Both T and B cells carry **receptor sites** that react with specific antigens. **Antigens** are portions of substances, most often proteins or polysaccharides, that the body is capable of recognizing as foreign to it. Many times, the antigen is part of a bacterial cell wall, a viral coat, or any cell that is not normally found within the body.

[1]Originally B stood for bursa of Fabricus, a structure found in birds but not in mammals.

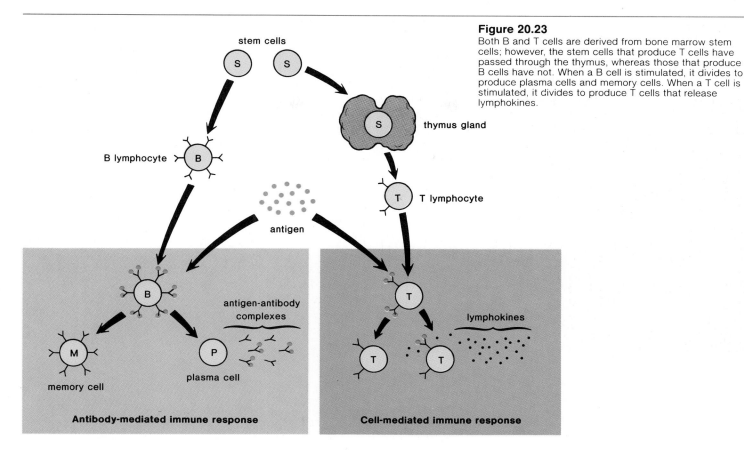

Figure 20.23
Both B and T cells are derived from bone marrow stem cells; however, the stem cells that produce T cells have passed through the thymus, whereas those that produce B cells have not. When a B cell is stimulated, it divides to produce plasma cells and memory cells. When a T cell is stimulated, it divides to produce T cells that release lymphokines.

stem cells

thymus gland

B lymphocyte

T lymphocyte

antigen

antigen-antibody complexes

lymphokines

memory cell

plasma cell

Antibody-mediated immune response

Cell-mediated immune response

Table 20.5 Some Properties of T Cells and B Cells

Property	T Cells (Cell-Mediated Immunity)	B Cells (Antibody-Mediated Immunity)
Antigen-binding receptors on cell surface	Specific receptors	Specific receptors
Response to binding of antigen	Enlarge, multiply, and liberate lymphokines	Enlarge and multiply to produce plasma cells that secrete antibodies
Cytotoxic activity	Antigen-stimulated T cells kill antigen-bearing target cells on contact	None
Function in antibody production	Stimulate antibody production by B cells	Synthesize and liberate antibodies into blood
Effect on macrophages	Stimulate phagocytic activity of macrophages	None

When an antigen combines with a receptor on a T cell, it enlarges, divides, and becomes a "killer cell" exhibiting **cytotoxicity**—the ability to cause the death of the other cell. Activated T cells also release **lymphokines** that stimulate macrophages to become "angry killers," so named because of their increased ability to phagocytize and destroy invading microorganisms. T cells also stimulate B cells to produce antibodies.

Figure 20.24

a. A diagram of antibody structure. Each variable region forms a binding site specific for an antigen. *b.* Since each antibody can bind to two antigens, an antibody-antigen complex often forms.

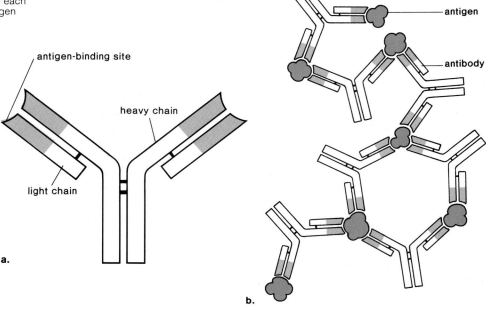

antigen-binding site

heavy chain

light chain

a.

antigen

antibody

b.

Antibodies are specific globulin proteins capable of combining with a specific antigen. All antibodies have the same *Y* shape (fig. 20.24), but each contains two variable regions where their antigen combines in a lock-and-key manner. When an antigen combines with a receptor on a B cell, it is stimulated, particularly in the presence of a T cell, to divide and produce **plasma cells**, which are enlarged, active, antibody-producing lymphocytes. All plasma cells derived from one parent lymphocyte are called a clone. Since one type B cell is chosen by an antigen to clone, this is called **clonal selection.** Some members of the clone do not participate in the current antibody production; instead, they remain in the bloodstream as memory cells, forever capable of producing the antibody specific to a particular antigen. Therefore, the presence of memory cells accounts for active immunity.

Passive Immunity

Some antibodies are able to cross the placenta and/or are present in mother's milk. The former provide passive immunity for one or two months, and the latter provide passive immunity as long as the baby is nursing. *Passive immunity* is short-lived because the person is not actively producing the antibodies needed to combat illness. It is also possible for a person to receive, by injection, antibodies produced by another. This occurs if a person needs protection against a disease-causing microorganism but lacks the time needed to develop active immunity. The necessary antibodies may be obtained from the blood of persons who have already developed immunity against the disease or from a horse that has been inoculated to cause immunity to the disease in question. Occasionally, a patient who receives horse serum becomes ill because the serum contains proteins that the individual's immune system recognizes as foreign. This is called serum sickness.

Monoclonal Antibodies   A unique method of producing human antibodies (fig. 20.25) has recently been devised by Cesar Milstein and Georges Köhler at the Medical Research Council's Laboratory of Molecular Biology in Cambridge, England. Lymphocytes are removed from the body and exposed to a particular antigen. Then they are fused with a cancer cell, after which they divide repeatedly. The fused cells are called hybridomas; *hybrid* because they are a fusion of two different cells and *oma* because one cell is a cancer cell. These antibodies are called **monoclonal antibodies** because they are all the same type and are produced by cells derived from the same parent cell. Monoclonal antibodies can be used to confer passive immunity on recipients and possibly to treat many illnesses, including cancer.

Immunity Side Effects

It is now recognized that overreactivity and underreactivity of the immune system account for various illnesses. **Allergies** are caused by an overactive immune system that forms antibodies to substances not usually recognized as foreign substances. The unpleasant symptoms of allergies are generally caused by the release of histamine following antibody-antigen reactions. Histamine promotes the contraction of smooth muscle and is, for example, responsible for the symptoms of asthma.

Tissue rejection following organ transplants is also due to the immune system. Since a transplanted organ is foreign to the individual, it activates the immune system to react to it. The first step in rejection appears to be cell-mediated, followed by the production of humoral antibodies that cause disintegration of the tissue.

Certain diseases are due to autoimmune reactions. The immune system fails to recognize the body's own tissues. For example, one type of rheumatic fever is caused by an immune system that attacks the heart's tissue. Growing evidence suggests that several other diseases are also due to autoimmune reactions. Among these are certain forms of anemia, multiple sclerosis, and even juvenile-onset diabetes.

Interferon

Interferon is a substance produced by cells in response to a viral infection. It enters noninfected cells and in some way prevents viral replication in these cells. Since AIDS is caused by a viral infection of T cells (p. 594) and cancer sometimes has a viral origin (p. 213) too, interferon has been used in the treatment of these conditions. Interferon is *species specific* (only human interferon is effective in humans) and previously it had to be extracted from human white blood cells. Now it is being mass-produced by the recombinant DNA technique (fig. 10.18).

Figure 20.25
One possible method for producing human monoclonal antibodies. *a.* Blood sample is taken from patient. *b.* Inactive lymphocytes from sample are exposed to antigen. *c.* Activated lymphocytes are fused with cancer cells. *d.* Resulting hybridomas divide repeatedly, giving many cells that produce only one type of antibody.

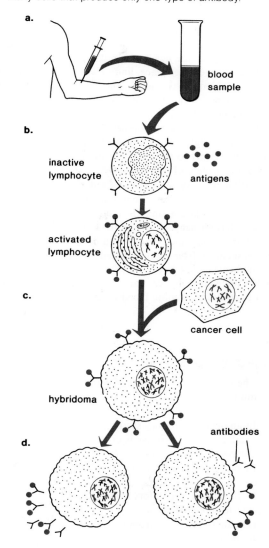

Summary

Some invertebrates do not have a transport system: diffusion alone supplies the needs of cells in coelenterates and flatworms. Other invertebrates do have a transport system: insects have an open system, and earthworms have a closed one. Vertebrates have a closed system in which arteries carry blood away from the heart to capillaries where exchange takes place, and in which veins carry blood to the heart. The circulatory pathway involving respiratory gas exchange differs in vertebrates. Fishes have a single circulatory loop because the heart with a single atrium and ventricle pumps blood only to the gills. The other vertebrates have both a pulmonary circulation and a systemic circulation. Amphibians have two atria but a single ventricle. Birds and mammals, including humans, have a heart with two atria and two ventricles in which oxygenated blood is always separate from deoxygenated blood.

The heartbeat in humans begins when the SA node (pacemaker) causes the two atria to contract, sending blood through the atrioventricular valves to the two ventricles. The AV node causes the two ventricles to contract, sending blood through the semilunar valves to the aorta. Now all chambers rest. The heart sounds, lubb-dup, are caused by the closing of the valves.

Blood pressure created by the beat of the heart accounts for the flow of blood in arteries, but skeletal muscular contraction is largely responsible for flow of blood in the veins, which have valves preventing a backward flow.

Blood has two main parts: plasma and cells. Plasma contains mostly water (92 percent) and proteins (8 percent), but also nutrients and wastes. The red cells contain hemoglobin and function in oxygen transport. The platelets and two plasma proteins, prothrombin and fibrinogen, function in blood clotting, an enzymatic process that results in fibrin threads. When blood reaches a capillary, water moves out at the arterial end due to blood pressure. At the venule end, water moves in due to osmotic pressure. In between, nutrients diffuse out and wastes diffuse in.

The lymphatic system is a one-way system. Lymphatic capillaries collect excess tissue fluid that moves in lymph veins to the blood circulatory veins. Lacteals absorb the products of fat digestion, and lymph nodes help combat infection.

Defense against disease depends on the various types of leukocytes. Neutrophils and monocytes are phagocytic and are especially responsible for the inflammatory reaction. Lymphocytes are responsible for immunity; i.e., B cells for antibody-mediated immunity and T cells for cell-mediated immunity. Plasma cells are B cells actively producing antibodies, and memory cells are B cells remaining in the body after the initial infection is over. Passive immunity has taken on new significance since it is now possible to produce monoclonal antibodies in the laboratory. Immunity also has side effects, such as allergy reactions and autoimmunity. Interferon is a substance produced by infected cells that protects other cells from viral infections.

Objective Questions

1. The internal environment of complex multicellular animals consists of _____ and _____ .
2. In an open circulatory system, the blood _____ (choose is or is not) always contained within vessels.
3. Blood that has a color most likely contains a _____ pigment.
4. In vertebrate circulatory systems, arteries take blood _____ (choose to or away) from the heart.
5. In animals with two separate circulatory circuits, the _____ system delivers blood to the lungs, and the _____ system delivers blood to the body proper.
6. The _____ arteries serve the needs of the heart muscle itself.
7. The SA node, also called the _____ , initiates the heartbeat.
8. Blood pressure can account for movement of blood in the _____ vessels, but not in the _____ .
9. The lymphatic system serves to collect excess _____ and to return it to the blood circulatory system.
10. White blood cells are _____ in size than red blood cells and they have a _____ , a structure that red blood cells lack.
11. The pumping of the heart sends blood to the _____ , where exchange of molecules takes place.
12. When a blood vessel is injured, the blood _____ , but when two incompatible blood types are mixed, the red cells _____ .
13. A blood clot consists of red blood cells trapped within _____ threads.
14. T cells, a type of lymphocyte, are responsible for _____ immunity, while B cells are responsible for _____ immunity.
15. Active immunity means the individual has learned to produce his/her own _____ , while in passive immunity these molecules are injected.

Answers to Objective Questions

1. blood, tissue fluid 2. is not 3. respiratory 4. away 5. pulmonary, systemic 6. coronary 7. pacemaker 8. arterial, veins 9. tissue fluid 10. larger, nucleus 11. capillaries 12. clots, agglutinate 13. fibrin 14. cell-mediated, antibody-mediated 15. antibodies

Study Questions

1. Describe transport in those invertebrates that have no circulatory system; those that have an open circulatory system; and those that have a closed circulatory system.
2. Compare the circulatory systems of a fish, an amphibian, and a mammal.
3. Trace the path of blood in humans from the right ventricle to the left atrium; from the left ventricle to the kidneys and return to the right atrium; from the left ventricle to the small intestine and return to the right atrium.
4. Define these terms: pulmonary circulation, systemic circulation, portal system.
5. Describe the beat of the heart, mentioning all the factors that account for this repetitive process. Describe how the heartbeat affects blood flow; what other factors are involved?
6. Describe the structure and function of the lymphatic system.
7. List the major components of blood and give a function for each component.
8. Describe in detail the exchange of molecules between blood and tissue fluid at a capillary, including the forces that facilitate this exchange.
9. Distinguish between antibody-mediated immunity and cell-mediated immunity in as many ways as possible.

Selected Key Terms

artery ('ärt ə rē) 429
capillary ('kap ə ,ler ē) 429
vein ('vān) 429
arteriole (är 'tir ē ,ōl) 429
sphincter ('sfiŋ tər) 429
ventricle ('ven tri kəl) 429
venule ('vēn ,yül) 430

myocardium (,mī ə 'kärd ē əm) 435
systole ('sis tə ,lē) 435
diastole (dī 'as tə ,lē) 435
lymph ('limf) 440
edema (i 'dē mə) 441
antibody ('ant i ,bäd ē) 450
interferon (,int ər 'fir ,än) 451

CHAPTER 21

Digestion and Nutrition

Concepts

1. All organisms require an outside source of nutrient molecules and energy.

   To acquire these, animals take in preformed food that usually has to be digested.

2. Increased complexity is observed during the evolutionary history of organisms.
   a. Primitive animals tend to have a fairly simple digestive tract, and later-appearing animals have a more complex one.
   b. Complexity of organization in more advanced animals is illustrated by an examination of the human digestive tract.

Figure 21.1
A food chain having at least three links is represented here. The wildebeest (gnu) is a herbivore that feeds on photosynthetic grasses. The lion is a carnivore that feeds on herbivores. Both the herbivore and carnivore need to digest food. Digestion is necessary to break down food and to release the nutrient molecules that serve as building blocks and as a source of energy in these animals.

A nimals are heterotrophic organisms that must take in preformed food. However, the various adaptations that have occurred to secure and digest food are extremely varied. We will necessarily have to limit our discussion to a few examples. Some animals are **omnivores;** they eat both plants and animals. Others are **herbivores;** they feed only on plants. Still others are **carnivores;** they eat only other animals (fig. 21.1).

Animal Digestive Tracts

Generally, animals have some sort of **gut** (digestive tract). The gut may have specialized parts for storing, grinding, enzymatically digesting, absorbing nutrient molecules, and finally eliminating undigestible material.

Incomplete versus Complete Gut

The planarian (fig. 21.2) is carnivorous and feeds largely on smaller aquatic animals. Its digestive system contains only a mouth, pharynx, and intestine. When the worm is feeding, the **pharynx** actually extends beyond the mouth. It wraps its body about the prey and uses its muscular pharynx to suck up minute quantities at a time. Digestive enzymes present in the *gastrovascular cavity* allow some extracellular (outside the cells) digestion to occur. Digestion is completed intracellularly (fig. 21.2) by the cells that line the cavity, which branches throughout the body. No cell in the body is far from the intestine and therefore diffusion alone is sufficient to distribute nutrient molecules.

The digestive system of a planarian is notable for its lack of specialized parts. The digestive system is termed **incomplete** and **saclike** because the pharynx serves not only as an entrance for food, but also as an exit for nondigestible material. Specialization of parts does not occur under these circumstances.

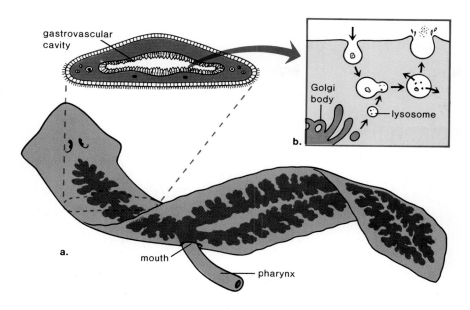

Figure 21.2
A planarian (*a.*) has a very extensive nonspecialized digestive tract, the gastrovascular cavity. Incomplete digestive tracts have little specialization of parts because the single opening serves as both entrance and exit. The insert (*b.*) emphasizes that the planarian relies on intracellular digestion to complete the digestive process. Phagocytosis produces a vacuole that joins with an enzyme-containing lysosome. Notice that the food particles are always surrounded by membrane; therefore, they are never part of the cytoplasm. The digested products pass from the vacuole before any nondigestible material is eliminated at the cell membrane.

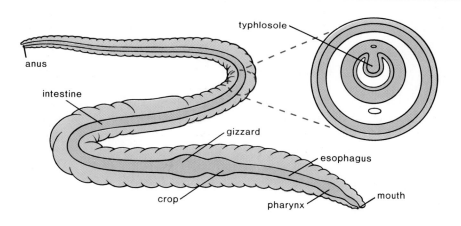

Figure 21.3
The earthworm has a complete digestive tract with both a mouth and an anus and many other specialized parts. The absorptive surface of the intestine is increased by an internal fold called the typhlosole (insert).

Recall that the planarian has some parasitic relatives. The tapeworm (fig. 15.14), for example, has no digestive system at all—it simply absorbs nutrient molecules from the intestinal juices that surround its body. The body wall is highly modified for this purpose: it has millions of microscopic, fingerlike projections that increase the surface area for absorption.

Earthworms (fig. 21.3) feed mainly on decayed organic matter in dirt. The muscular *pharynx* draws in food with a sucking action. The *crop* is a storage area that has expansive thin walls. The *gizzard* has thick muscular walls for churning and grinding the food. Digestion is extracellular—in the *intestine*. The surface area for absorption of nutrient molecules is increased by an intestinal fold called the **typhlosole** (fig. 21.3). Undigested food passes out of the body at the anus.

In contrast to the planarian, the earthworm has a **complete digestive system.** Specialization of parts is obvious because the pharynx, crop, gizzard, and intestine each has a particular function in the digestive process.

Figure 21.4

a. A clam burrows in the sand or mud where it filter feeds while the squid swims freely in open waters. *b.* A flow of water (arrows) brings debris (food particles) into and carries waste out of a clam's mantle cavity. *c.* In contrast, a squid captures other aquatic animals with its tentacles, and bites off pieces with its jaws. A strong contraction of the mantle forces water (arrows), which has entered the mantle cavity, out the funnel resulting in jet propulsion.

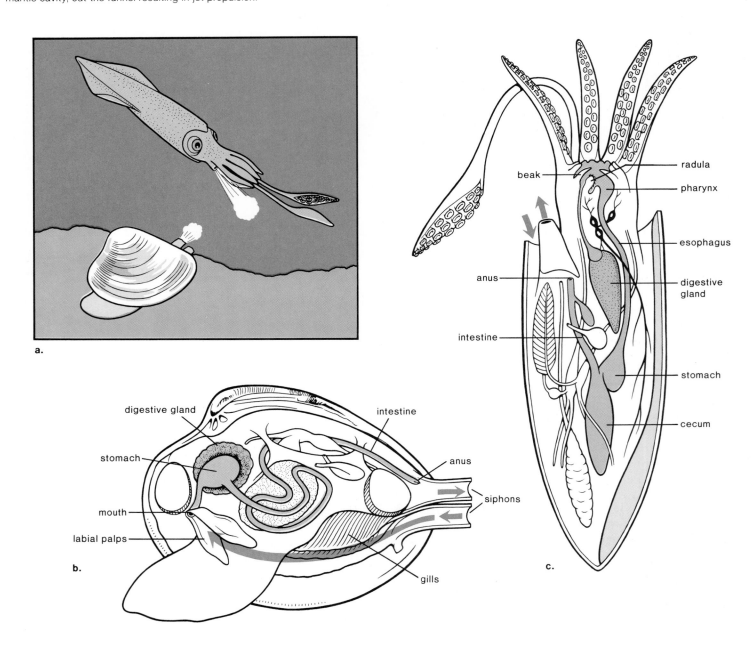

Continuous versus Discontinuous Feeders

In a clam (fig. 21.4), the *incurrent siphon* (slitlike opening) brings water and organic material inside the mantle cavity where it adheres to the gills. Ciliary action moves suitably sized pieces to the *labial palps,* which force these pieces through the mouth into the stomach. Larger particles are immediately eliminated by way of the *intestine* and *anus.* The smaller ones, however, are acted on by digestive enzymes secreted by a large *digestive gland.* Amoeboid cells are present throughout the tract and it is believed that these cells complete the digestive process by means of intracellular digestion.

The clam is a **sessile** (fairly stationary) **filter feeder** that feeds continuously. Water is always moving through the mantle cavity and depositing particles on the gills. The size of the incurrent siphon restricts the entrance of particles to a certain size, and only the smallest of these will adhere to the gills. Further, the stomach rejects any particles that are too large for the digestive system to handle. Therefore, this continual filtering process selects only appropriately small particles for digestion.

There are filter feeders in many other phyla of the animal kingdom. For example, the marine fanworms (fig. 21.5) live in a tube and extend feathery tentacles when they are feeding. Ciliary action not only brings organic material to the tentacles, it also sends appropriately sized particles to the mouth.

Not all filter feeders are sessile. The baleen whale is an **active filter feeder.** The baleen, a curtainlike fringe that hangs from the roof of the mouth, filters krill, a small shrimp, from the water. This type of whale filters up to a ton of krill in a single feeding.

Another mollusk, the squid (fig. 21.4), has a feeding system that is entirely different from the clam. The body of a squid is streamlined and the animal moves rapidly through the water using jet propulsion (forceful expulsion of water from a tubular funnel). It is not surprising, then, that the squid

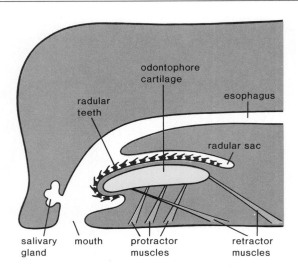

Figure 21.6
Longitudinal section of a radula from a snail. A radula contains rows of teeth that point posteriorly. When a snail is feeding, the radula extends and then scrapes against the surface of an object to bring minute particles of food into the mouth. The squid uses its beak to acquire larger-sized chunks of food and the radula propels them into the mouth.

Figure 21.7
The cheetahs are carnivores that gorge on infrequent feeds. They need teeth that can tear flesh apart, an extendable stomach, but a short intestine because meat is easily assimilated into the system. The zebras are herbivores that feed more often, though they are always ready to take off when necessary. They need teeth that can trim grass and chew it well in order to break up the strong plant cell walls composed of cellulose. Their stomachs contain bacteria that can digest this cellulose.

is a carnivorous predator that feeds on such aquatic organisms as fish, shrimp, and worms. The head of a squid is surrounded by muscular *tentacles,* two of which have teeth bearing suckers at their tips. These teeth seize the prey and bring it to the squid's *beaklike jaws,* which bite off pieces pulled into the mouth by the action of a *radula,* a toothy tongue (fig. 21.6). An **esophagus** leads to a *stomach* and *cecum* (an extension of the stomach), where digestion occurs. The squid also has a *digestive gland* that no doubt secretes digestive enzymes. Absorption takes place in the *intestine,* and undigested material passes out the *anus.*

The squid is a **discontinuous feeder** and requires a storage area for its infrequent meals. The size of the stomach is supplemented by the cecum, which can also hold food. Discontinuous feeders, whether they are carnivores or herbivores, require such a storage area. Note, however, that the gut of a carnivore is short. Meat is easy to digest and assimilate so fewer specializations are required.

Mammalian Herbivores versus Mammalian Carnivores

Figure 21.8 contrasts the dentition (teeth) of the herbivorous horse with the carnivorous lion. The horse has large, strong, sharp front teeth for neatly clipping off blades of grass, its primary food source. The horse's other teeth are large and flat for grinding up the grass so that it can be acted on by the enzymes of the digestive system. Plant material is difficult to digest because it contains a high proportion of cellulose. The stomach of a horse is supplemented by a cecum, and both of these contain a large bacterial population that can digest cellulose. However, a horse must chew its food thoroughly before swallowing. Other mammalian herbivores, like the cow and deer, chew their food at two different times. They graze quickly, sending only partially chewed plant material to a special part of the stomach, called a **rumen,** for bacterial action. This material, now called the **cud,** is regurgitated into the mouth and chewed once more when convenient for these ruminates.

A carnivore bolts its food, swallowing it quickly in large-sized chunks. Its teeth are primarily adapted to tearing the food into pieces small enough to be swallowed. As mentioned before, meat is a richer and more accessible source of nutrients than is plant material. Therefore, the digestive system of carnivores lacks the digestive-tract specializations seen in herbivores.

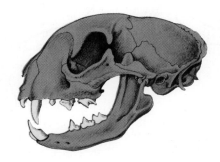

Horse

Lion

Figure 21.8
Dentition of a horse compared to that of a lion. In the horse, the front teeth are incisors that clip the grass, and the back teeth are flat molars that grind the grass well. In the lion, the front teeth include canines that tear at the flesh. The back molars are also pointed.

Human Digestive Tract

The human digestive system is complex, illustrating that the complete digestive tract does permit specialization of parts. Each part has a specific function.

Mouth

Humans are omnivores; they feed on both plant and animal material. Therefore, we would expect human dentition to be nonspecialized, but able to deal adequately with both a vegetable and meat diet. An adult human has thirty-two teeth (fig. 21.9). One-half of each jaw has teeth of four different types: two chisel-shaped **incisors** for biting; one pointed **canine** for tearing; two fairly flat **premolars** for grinding; and three **molars,** well flattened for crushing. The last molar, or wisdom tooth, may fail to erupt, or if it does, it is sometimes crooked and useless.

Three pairs of **salivary glands** secrete saliva into the mouth where it mixes with and moistens the food. Saliva also contains an enzyme, **salivary amylase,** that begins the process of starch digestion by hydrolyzing some of the bonds between the glucose units that make up starch. The disaccharide maltose is a typical end product of salivary amylase digestion (fig. 21.11).

While in the mouth, food is manipulated by a muscular tongue that has touch and pressure receptors similar to those in the skin. The tongue also has **taste buds,** chemical receptors that are stimulated by the chemical composition of food. When food has been chewed and mixed with saliva, the tongue starts the process of swallowing by pushing the food back through the pharynx.

Figure 21.9
Human teeth, in comparison to the horse and lion (fig. 21.8) are nonspecialized. The molars are not as large and flat as the horse's and the canines are not as long and sharp as the lion's.

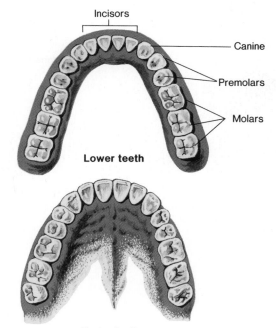

Incisors

Canine

Premolars

Molars

Lower teeth

Upper teeth

Figure 21.10
Human digestive system. Table 21.1 lists the names and functions of the various organs. Perhaps a better name for the small intestine would be the long intestine?

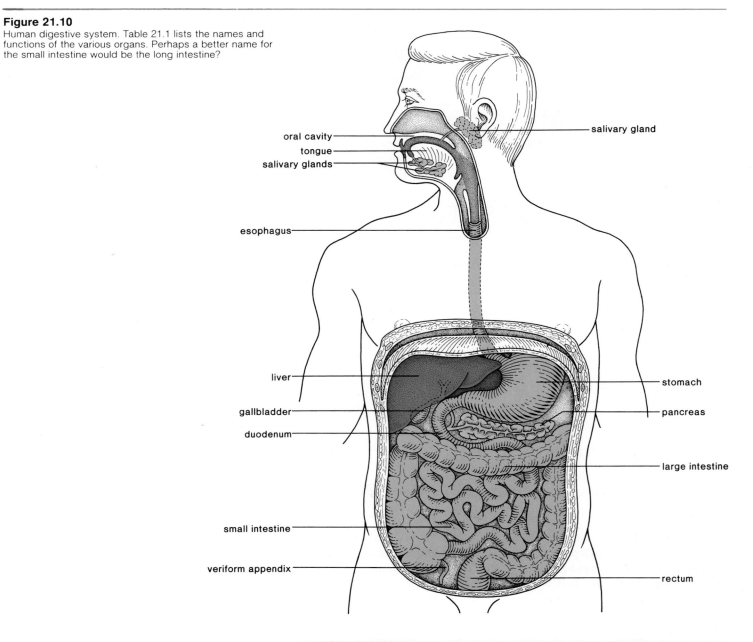

oral cavity
tongue
salivary glands
esophagus
liver
gallbladder
duodenum
small intestine
veriform appendix

salivary gland
stomach
pancreas
large intestine
rectum

Table 21.1 Path of Food

Organ	Special Features	Functions
Mouth	Teeth	Chewing of food; digestion of starch to maltose
Esophagus		Passageway
Stomach	Gastric glands	Digestion of protein to peptides
Small intestine	Intestinal glands Villi	Digestion of all foods and absorption of unit molecules
Large intestine		Absorption of water
Anus		Defecation

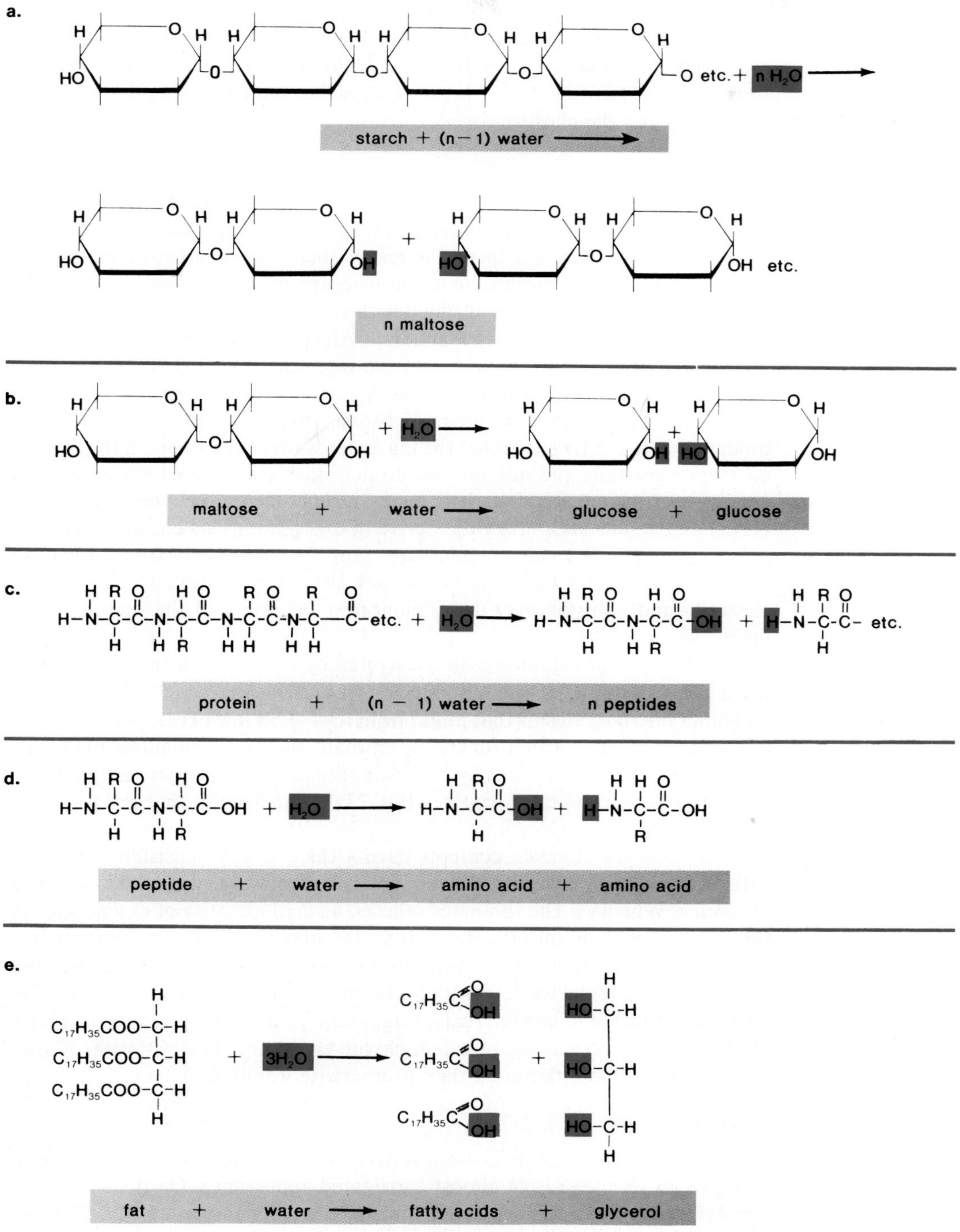

Figure 21.11
If we were to match the enzymatic reactions here to the organs in figure 21.10, the first reaction (a.) occurs in the mouth and small intestine. The second reaction (b.) occurs in the small intestine. The third reaction (c.) occurs in the stomach and small intestine. The fourth reaction (d.) occurs in the small intestine, as does the fifth reaction (e.).

a.

starch + (n − 1) water ⟶

n maltose

b.

maltose + water ⟶ glucose + glucose

c.

protein + (n − 1) water ⟶ n peptides

d.

peptide + water ⟶ amino acid + amino acid

e.

fat + water ⟶ fatty acids + glycerol

Figure 21.12
Swallowing. Respiratory and digestive passages cross in the pharynx. During swallowing, the epiglottis covers the opening into the trachea and prevents food from entering the trachea.

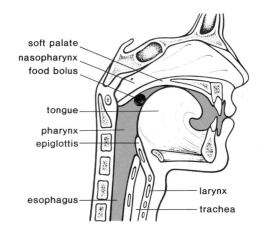

soft palate
nasopharynx
food bolus

tongue

pharynx
epiglottis

esophagus

larynx

trachea

Figure 21.13
A rhythmical contraction, called peristalsis, moves food along the digestive tract.

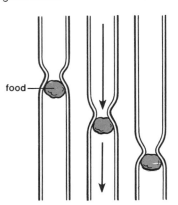

food

Pharynx and Esophagus

The digestive and respiratory passages cross in the pharynx (fig. 21.12). Thus, **swallowing** poses a potential problem because food might enter the trachea (windpipe) and block the path of air to the lungs. Normally, however, swallowing involves a set of reflexes that closes the opening into the trachea. A flap of tissue, the **epiglottis,** covers the opening into the trachea as muscles move the food mass through the pharynx into the **esophagus,** a tubular structure that takes food to the stomach.

When food enters the esophagus, a rhythmical contraction of the wall termed **peristalsis** (fig. 21.13) begins and pushes the food along this tube that passes through the chest cavity.

Stomach

The stomach stores up to two liters of partially digested food, and even more in some cases. Because the stomach stores food, humans can periodically eat relatively large meals and spend the rest of their time at other activities. But the stomach is much more than a mere storage organ, as was discovered by William Beaumont in the mid-nineteenth century. Beaumont was an American doctor who had a French Canadian, Alexis St. Martin, as a patient. St. Martin had been shot in the stomach, and when the wound healed, he was left with a fistula, or opening, that allowed Beaumont to collect **gastric** (stomach) juices and to look inside the stomach to see what was going on there. Beaumont was able to determine that the muscular walls of the stomach (fig. 21.14) contract vigorously and mix food with juices that are secreted whenever food enters the stomach. He found that gastric juice contains **hydrochloric acid** (HCl) and a substance (enzyme) active in digestion. (This enzyme was later identified as **pepsin.**) Gastric juices are produced separately from protective mucous secretions of the stomach, he said. Beaumont's work, which was very carefully and painstakingly done, pioneered the study of the physiology of digestion.

So much hydrochloric acid is secreted by the stomach that it routinely has a pH of about 2.0. Such a high acidity usually is sufficient to kill bacteria and other microorganisms that might be in food. This low pH stops the activity of salivary amylase, which functions optimally at the near-neutral pH of saliva, but it promotes the activity of pepsin. *Pepsin* is a hydrolytic enzyme that acts on protein to produce peptides (fig. 21.11), molecules that are too large to be absorbed.

By now the stomach contents have a thick, soupy consistency and are called **chyme.** At the base of the stomach is a narrow opening controlled by a *sphincter.* Whenever the sphincter relaxes, a small quantity of chyme squirts through the opening into the duodenum, the first part of the small intestine. When chyme enters the duodenum, it sets off a neural reflex that causes the muscles of the sphincter to contract vigorously and to temporarily close the opening. Then the sphincter relaxes again and allows more chyme to squirt through. The slow manner in which chyme enters the small intestine allows digestion to be more thorough than it otherwise would be.

Small Intestine

The human small intestine is not a simple, smoothly lined tube; it has ridges and furrows that give it an almost corrugated appearance. On the surface of these ridges and furrows are small, fingerlike projections called **villi.** Cells on the surfaces of the villi have minute projections called microvilli. A smooth, tubular small intestine would have to be five or six hundred meters long to have a surface area comparable to the human small intestine. Such a huge, long gut would be hard to carry around.

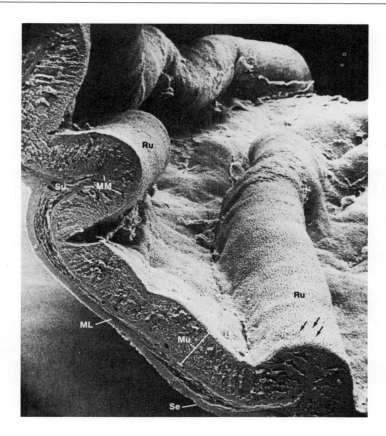

Figure 21.14
This scanning electron micrograph gives you a view of
what Dr. Beaumont saw when he looked through the
opening in St. Martin's stomach. The wall of the stomach
has folds called rugae (Ru) that disappear when the
stomach is full. The arrows indicate the openings to the
gastric glands that produce gastric secretions, including
hydrochloric acid and pepsinogen, a precursor that
becomes pepsin in the stomach. The stomach lining is
called the mucosa (Mu) because its cells secrete mucus
as a protective coating. The other layers to the stomach
wall are MM (muscularis mucosa, a smooth muscle layer),
SU (submucosa, a connective tissue layer), ML
(muscularis layer, another muscular coat), and Se (outer
serosa, a thin connective tissue layer).
Kessel, R. G., and Kardon, R. H.: *Tissues and Organs: A Text-Atlas
of Scanning Electron Microscopy.* © 1979 by W. H. Freeman and
Co.

When chyme enters the duodenum, proteins and carbohydrates have been
only partly digested, and fat digestion needs to be carried out. Considerably
more digestive activity is required before these nutrients can be absorbed
through the intestinal wall. Two important accessory glands, the liver and the
pancreas, send secretions to the duodenum (fig. 21.10). The liver produces **bile**
that is stored in the **gallbladder** and sent to the duodenum by way of a duct.
Bile looks green because it contains pigments that are products of hemoglobin
breakdown. This green color is familiar to anyone who has observed the color
changes of a bruise. Hemoglobin within the bruise is breaking down into the
same types of pigments found in bile. Bile also contains bile salts, which are
emulsifying agents that break up fat into fat droplets.

$$\text{fat} \xrightarrow{\text{bile salts}} \text{fat droplets}$$

When fat is physically broken apart and caused to mix with water, it is emulsified. Emulsified fat is more easily acted on by enzymes.

The pancreas sends **pancreatic juice** into the duodenum, also by way of
a duct. While the pancreas is an endocrine gland when it produces and secretes
insulin into the bloodstream, it is an exocrine gland when it produces and secretes pancreatic juice into the duodenum. Besides digestive enzymes, pancreatic juice contains sodium bicarbonate ($NaHCO_3$), which neutralizes the
chyme and makes the pH of the small intestine slightly basic.

Pancreatic juice contains enzymes that act on every major component
of food. **Pancreatic amylase** digests starch to maltose; **trypsin** digests protein
to peptides; and **lipase** digests fat droplets to glycerol and fatty acids (fig.
21.11*c*).

Figure 21.15
The wall of the small intestine specializes in absorption. Transmission electron micrograph (*a.*) and scanning electron micrograph (*b.*) of microvilli that cover the surface of a (*c.*) villus. Many such fingerlike projections line the intestinal wall. Amino acids and sugars enter the blood capillary while glycerol and fatty acids recombine before entering the lacteals. Mv = microvilli; Mf = microfilaments give support; TW = terminal web that anchors the microfilaments; and ZO = zonula occludens, the juncture between adjacent cells.
a. and b. Kessel, R. G., and Kardon, R. H.: *Tissues and Organs: A Text-Atlas of Scanning Electron Microscopy.* © 1979 by W. H. Freeman and Co.

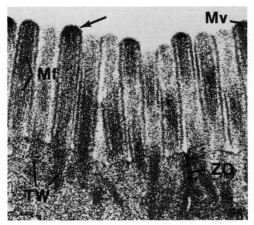

a.

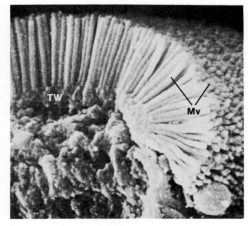

b.

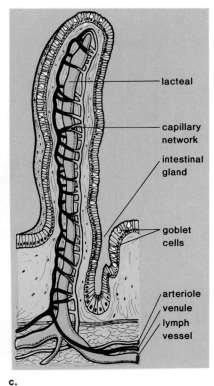

c.

Since there is a digestive gland at the base of every villus (fig. 21.15), there are millions of intestinal digestive glands. These produce **intestinal juice** containing enzymes that complete the digestion of protein and carbohydrates. Peptides are digested to amino acids (fig. 21.11*b*) by **peptidases,** and maltose is digested to glucose (fig. 21.11*a*) by **maltase.** Other disaccharides, each of which is acted upon by a specific enzyme, are digested in the small intestine. Table 21.2 reviews the digestion of food, and table 21.3 reviews the hydrolytic digestive enzymes discussed in this chapter.

Absorption

The small intestine is specialized for absorption. Molecules are absorbed by the projections of the huge number of villi (fig. 21.15). Each villus contains blood vessels and a lymphatic vessel called a **lacteal.** Sugars and amino acids are absorbed into the bloodstream; glycerol and fatty acids enter the lacteals after recombining.

Absorption continues until almost all products of digestion have been absorbed. Thus, absorption involves active transport and requires an expenditure of cellular energy.

Control of Secretions

Under normal circumstances, the digestive secretions are released only when food is present in the digestive tract. Saliva in the mouth and gastric juices in the stomach flow in response to the taste, smell, or sometimes even the thought

Table 21.2 Digestive Enzymes

Reaction	Enzyme	Gland	Site of Occurrence
starch + H₂O → maltose	a. Salivary amylase b. Pancreatic amylase	a. Salivary b. Pancreas	a. Mouth b. Small intestine
maltose + H₂O → glucose*	Maltase	Intestinal	Small intestine
protein + H₂O → peptides	a. Pepsin b. Trypsin	a. Gastric b. Pancreas	a. Stomach b. Small intestine
peptides + H₂O → amino acids*	Peptidases	Intestinal	Small intestine
fat + H₂O → glycerol + fatty acids*	Lipase	Pancreas	Small intestine

*Absorbed by villi.

Food is largely made up of carbohydrates (starch), protein, and fat. These very large macromolecules are broken down by digestive enzymes to small molecules that can be absorbed by intestinal villi. This table indicates the steps needed for carbohydrate digestion (starch and maltose), protein digestion (protein and peptides), and fat digestion (fat), and shows that they are all hydrolytic reactions.

Table 21.3 Comparison of Digestive Enzymes*

Enzyme	Source	Optimum pH	Type of Food Digested	Product
Salivary amylase	Saliva	Neutral	Starch	Maltose
Pepsin	Stomach	Acid	Protein	Peptides
Pancreatic amylase	Pancreas	Basic	Starch	Maltose
Lipase	Pancreas	Basic	Fat	Glycerol; fatty acids
Trypsin	Pancreas	Basic	Protein	Peptides
Nucleases	Pancreas	Basic	RNA, DNA	Nucleotides
Peptidases	Intestine	Basic	Peptides	Amino acids
Maltase	Intestine	Basic	Maltose	Glucose

*All enzymes have a preferred pH which maintains their proper shape to do their job. This table indicates the pH for each of the enzymes in table 21.2.

Figure 21.16
Control of gastric secretions. Especially after eating a protein-rich meal, (a.) gastrin produced by the lower part of the stomach enters the bloodstream, and later, (b.) it stimulates the upper part of the stomach to produce more digestive juices.

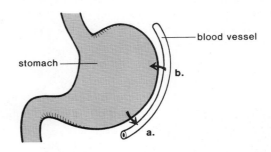

Figure 21.17
Control of intestinal secretions. a. Secretin and CCK produced by the duodenum enter the bloodstream. b. Secretin stimulates the pancreas to release digestive enzymes. c. CCK-PZ stimulates the gallbladder to release bile.

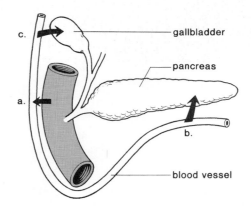

of food. Also, the presence of food in the stomach stimulates sense receptors that signal the brain to stimulate gastric glands in the stomach by way of nerves.

Perhaps the major factor in gastric secretory regulation, however, is the hormone **gastrin,** which is produced by the stomach itself. When proteins contact the stomach mucosa, gastrin-producing cells are stimulated to release gastrin into the bloodstream (fig. 21.16). As soon as gastrin circulates through blood vessels and reaches the acid- and enzyme-secreting cells of the stomach lining, those cells respond by secreting large quantities of HCl and pepsin. As the stomach empties, both the neural reflexes and gastrin release subside, and less HCl and pepsin are secreted.

Similarly, some duodenal cells produce the hormone **secretin** that stimulates the pancreas to release pancreatic juice, especially the sodium bicarbonate component. Other cells release a hormone called **CCK-PZ** (cholecystokinin-pancreazymin) that stimulates the release of bile. CCK-PZ stimulates the gallbladder to empty its contents through a duct into the duodenum. This same hormone also stimulates the pancreas, especially the enzyme secretion of the pancreas, as its full name suggests (fig. 21.17).

Problems with the Digestive Tract

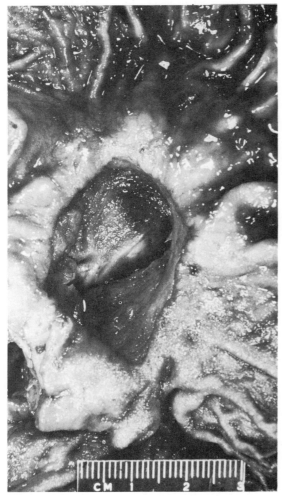

Peptic ulcer in the stomach. Normally, the digestive tract protects itself from digestive juices. An ulcer begins when this protection fails.

The gut tube, for all its twists and expansions, is really just a tunnel through the body. In the strict sense, the contents of the gut are outside the body and remain so unless they pass across cell membranes of gut-lining cells. Indeed, the contents of the gut must not come into direct contact with the blood or other interior tissues of the body because the chyme contains bacteria and other potentially hazardous components. The nature of this risk is graphically illustrated by two rather common digestive ailments—appendicitis and ulcers.

The vermiform appendix, or just appendix for short, is a short blind sac off the beginning of the colon, just past its junction with the small intestine. The human appendix does contain lymphoid tissue and might, therefore, be involved in resistance to infection, but its function is poorly understood. The appendix all too often becomes infected and develops an inflammation called appendicitis. Once appendicitis is diagnosed, standard medical procedure calls for surgical removal of the infected appendix. If untreated, the inflamed appendix can rupture, allowing gut contents to come in contact with the peritoneum, the tissue that covers all the digestive organs and lines the body cavity. The material that escapes from a ruptured appendix causes peritonitis, a life-threatening infection.

Ulcers are breaches of varying depth and position in the wall of the gut. Most are located in the duodenum and are called duodenal ulcers. Somewhat more rarely, gastric ulcers may form in the stomach itself (see the illustration). Ulcers are caused by the combined effect of acid and enzymes eroding the protective lining of the gut. However, what is not well understood is why in some individuals the normal mechanisms that protect the linings of the digestive tract function adequately for a lifetime, and in others they do not. It is widely believed that worry, stress, or frustration can contribute to ulcer development because gastric juice secretion is at least partly controlled by the nervous system. In fact, one of the more drastic treatments for serious ulcers is to cut the vagus nerve, the nerve that stimulates stomach secretion.

Most ulcers are really rather mild, albeit painful, abrasions in the gut lining. Healing can be rapid because the cells lining the gut tube are quickly replaced, but the causative features (often extreme stress) must be removed, and the patient must switch to a bland diet that does not stimulate excess acid secretion. The real dangers come when ulcers begin to bleed or when they perforate (produce a hole directly through the gut wall). Peritonitis can develop following perforation unless prompt medical care is received. Such serious cases require removal of the source of stomach acid. This can be accomplished by cutting the vagus nerve or by removing part of the stomach. An interesting new technique accomplishes the same thing without surgery. The patient swallows a balloon, and then alcohol at a temperature cold enough to freeze and kill stomach lining cells is poured into the balloon. This destroys a large percentage of the acid-secreting cells and has the same effect as surgical treatments. By the time these cells are replaced, the ulcer has had a chance to heal.

Appendicitis and ulcers are not the only digestive ailments. Digestive problems also can occur in the digestive glands, such as the gallbladder and liver. One common condition is gallstones, which result from the precipitation of bile salts in the gallbladder or in the bile duct. Blockage of the bile duct may cause jaundice, a yellowing of the skin due to an abnormal accumulation of bile pigments in the blood. Surgical removal of the gallbladder is the usual solution.

Jaundices also may be a symptom of various types of liver failure, a much more serious condition, given the overall importance of this organ. Inflammation of the liver is known as hepatitis, and it may result from a virus infection (viral hepatitis), poison (toxic hepatitis), or protein deficiency (deficiency hepatitis). The liver does have considerable regenerative ability, but when liver cells are destroyed, they sometimes are replaced by scar tissue. This scarring is called cirrhosis of the liver. All of these liver problems weaken the liver's metabolic capacities and greatly reduce its ability to deal effectively with toxins entering the body. Thus, persons suffering from liver disease, whatever the original cause, are strongly urged not to drink alcohol because alcohol normally is metabolized almost entirely by liver cells.

In addition to being susceptible to appendicitis, ulcers, gallstones, and liver disease, the digestive tract is simply an ideal environment for many disease organisms, from viruses to tapeworms.

Despite all of these potential problems, most human digestive tracts function normally and, except for occasional minor upsets, seldom cause much concern. A normally functioning digestive tract is essential if the body is to receive an uninterrupted supply of vital nutrients.

From Johnson, Leland G., *Biology.* © 1983 Wm. C. Brown Publishers, Dubuque, Iowa. All Rights Reserved. Reprinted by permission.

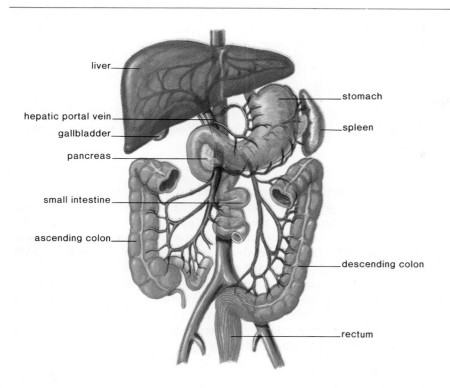

Figure 21.18
Hepatic portal system, a system that begins and ends in capillaries. Blood leaving the digestive tract flows through the hepatic portal vein to the liver, where it enters spaces among the liver cells. Later, it enters the hepatic vein, which joins with the vena cava.

Labels on figure: liver, hepatic portal vein, gallbladder, pancreas, small intestine, ascending colon, stomach, spleen, descending colon, rectum

Liver

Blood vessels from the digestive tract merge to form the hepatic portal vein that leads to the liver (fig. 21.18). The liver has numerous functions, including the following.

1. Detoxifies the blood by removing and metabolizing poisonous substances.
2. Makes the blood proteins.
3. Destroys old red blood cells and converts hemoglobin to the breakdown products in bile (bilirubin and biliverdin).
4. Produces bile that is stored in the gallbladder before entering the small intestine where it emulsifies fats.
5. Stores glucose as glycogen and breaks down glycogen to glucose between meals to maintain a constant glucose concentration in the blood.
6. Produces urea from amino groups and ammonia.

Just now we are interested in the last two functions listed. The liver helps maintain the glucose content of blood at about 0.1 percent by removing excess glucose from the hepatic portal vein and storing it as glycogen. Between meals, glycogen is broken down and glucose is added to the hepatic vein. Glycogen is sometimes called animal starch because both starch and glycogen are made up of glucose molecules joined together (fig. 3.18). If by chance the supply of glycogen and glucose runs short, the liver will convert amino acids to glucose molecules. Amino acids contain nitrogen in the form of amino groups, whereas glucose contains only carbon, oxygen, and hydrogen. Thus, before amino acids can be converted to glucose molecules, **deamination,** or the removal of amino groups from the amino acids, must take place. By an involved metabolic pathway, the liver converts these amino groups to urea.

$$H_2N - \overset{\overset{\textstyle O}{\|}}{C} - NH_2$$

Urea is the common nitrogenous waste product of humans, and after its formation in the liver, it is transported by the bloodstream to the kidneys where it is excreted.

Large Intestine

About 1.5 l of water enter the digestive tract daily as a result of eating and drinking. An additional 8.5 l enter the digestive tract each day carrying the various substances secreted by the digestive glands. About 95 percent of this water is reabsorbed into cells of the large intestine **colon.** This water reabsorption is essential. Failure to reabsorb water can result in **diarrhea,** which can lead to serious dehydration and ion loss, especially in children.

In addition to water, the large intestine absorbs some sodium and other ions from the material passing through it. At the same time, colon cells excrete still other metallic ions into the wastes leaving the body. Thus, the colon functions both in water conservation and ion regulation. Vitamin K, produced by intestinal bacteria, is also absorbed in the colon.

Digestive wastes (feces) eventually leave the body through the rectum and anus. Feces are about 75 percent water and 25 percent solid matter. Almost one-third of this solid matter is made up of intestinal bacteria. The remainder is undigested plant material, fats, waste products (such as bile pigments), inorganic material, and proteins, especially mucus and dead intestinal lining cells.

Nutrition

In this section we will discuss the needs of the human body for the nutrient molecules made available by the digestive process. These molecules are used as building blocks to synthesize the macromolecules found in the body, and as an energy source. In addition to these considerations, the body has to meet its vitamin and mineral needs.

Carbohydrates

The quickest, most readily available source of energy for the body is carbohydrates. Starchy foods, such as bread and potatoes, provide the largest quantity of carbohydrates, but meat and seafood also provide carbohydrates because they contain glycogen.

As mentioned previously, all dietary carbohydrates are digested to glucose, which is stored by the liver in the form of glycogen. In between eating, the liver attempts to maintain the blood glucose level at 0.1 percent, either by breaking down glycogen or by converting amino acids to glucose (p. 467). If necessary, these amino acids are taken from the muscles, even from the heart muscle. This is how people who starve waste away. A constant supply of glucose is necessary because the brain utilizes only glucose as an energy source. Other organs can metabolize fatty acids for energy, but unfortunately this results in **acidosis,** an acid blood pH. In order to avoid this situation, it is suggested that the diet contain at least 100 grams of carbohydrates daily.[1]

[1] A slice of bread contains approximately 14 grams of carbohydrate.

Even so, foods such as candy, ice cream, sugar-coated cereals, soft drinks, and alcohol, which are rich in simple sugars, are labeled "empty calories" by some because they contribute to energy needs and weight gain without supplying any other nutritional requirements. Government agencies charged with advising the public about dietary needs suggest sugar intake be limited (table 21.4).

Fats

Fats are present not only in butter and margarine, but also in meat, eggs, milk, nuts, and a variety of vegetable oils. Animal fats tend to have saturated fatty acids, and plant fats tend to have unsaturated fatty acids. An increase in the amount of fat in the diet can greatly increase the number of calories consumed. The pat of butter or margarine on a potato contains almost as many calories as the potato. This is understandable when you compare the calories derived from a gram of fat to those derived from a gram of carbohydrate (table 21.5).

After being absorbed, the products of fat digestion are transported by the lymph and blood to the tissues. The liver can alter ingested fats to suit the body's needs, except it is unable to produce the fatty acids linolenic and linoleic. Since these are required for phospholipid production, they are considered the **essential fatty acids.** Essential molecules must be present in our food because the body is unable to manufacture them.

Dietary lipids, especially saturated fatty acids and cholesterol, have been found to cause circulatory difficulties such as hypertension and heart attack due to atherosclerosis (p. 438). Not only does the American Heart Association and the governmental agencies mentioned earlier recommend that we limit our fat intake (table 21.4), they also suggest, as described in the reading on page 472, that in doing so we may be protecting ourselves from certain types of cancer.

Proteins

Foods rich in protein include meat, fish, poultry, cheese, nuts, milk, eggs, and cereals. Various legumes such as beans and peas also contain lesser amounts of protein. Following digestion of protein, amino acids enter the blood and are transported to the tissues. Some are incorporated into structural proteins and some are used to synthesize such proteins as hemoglobin, plasma proteins, enzymes, and hormones.

Except for nine amino acids, cells have no difficulty in changing one type of amino acid into another type. These nine amino acids are called the **essential amino acids** because they must be provided by the diet. Some protein sources, which include those available from animal sources, are **complete proteins**—they contain adequate amounts of the amino acids essential to maintaining body tissues and promoting normal growth and development. Other protein sources, such as those from plant sources, are **incomplete proteins** and are unable to maintain body tissue or promote normal growth. Since all of the essential amino acids must be present before protein synthesis is possible, every culture seems to have evolved its own mixture of complementary foods. For example in the Middle East, wheat bread, which lacks adequate levels of the amino acid lysine, is eaten with cheese, which has a high lysine content. Mexicans eat beans and rice, Jamaicans eat rice and peas, and Americans eat breakfast cereals with milk. In less prosperous countries, people usually subsist on diets primarily composed of grains and vegetables.

Table 21.4 Dietary Recommendations

The less-fat recommendations:
1. Choose as protein foods lean meat, poultry, fish, dry beans, and peas. Trim fat off before you eat.
2. Eat eggs and such organ meats as liver in moderation. (Actually, these are high in cholesterol rather than fat.)
3. Broil, boil, or bake, rather than fry.
4. Limit your intake of butter, cream, hydrogenated oils, shortenings, coconut oil.

The less-salt recommendations:
1. Learn to enjoy unsalted food flavors.
2. Add little or no salt to foods at the table and add only small amounts of salt when you cook.
3. Limit your intake of salty prepared foods such as pickles, pretzels, and potato chips.

The less-sugar recommendations:
1. Eat less sweets such as candy, soft drinks, ice cream, and pastry.
2. Eat fruit which is fresh or canned fruit without heavy syrup.
3. Use less sugar—white, brown, raw—and less honey and syrups.

Source: American Dietetic Association based on *Dietary Guidelines for Americans* 1980, U.S. Department of Agriculture and Department of Health, Education, and Welfare.

Table 21.5 Caloric Energy Release

	Calories/Gram
Carbohydrate	4.1
Fat	9.3
Protein	4.1

Figure 21.19
Child with kwashiorkor. The swollen abdomen is caused by edema due to the lack of plasma proteins in the blood. Protein deficiency is the most common form of malnutrition in poorer countries.

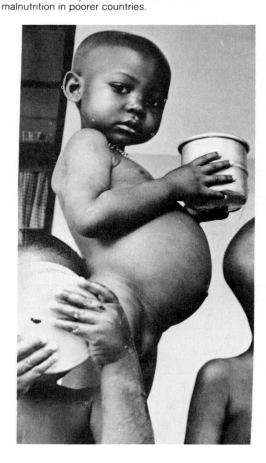

Table 21.6 Vitamins

Vitamin	Disorder	Source
Fat Soluble		
A	Night blindness and skin infections	Carrots, milk, liver, oil, eggs, vegetables
D	Rickets (weak bones)	Sardines, liver, butter
E	None known	Eggs, green vegetables
K	Clotting of blood	Leafy green vegetables
Water Soluble		
Thiamine B_1	Beriberi (damage to nerves and heart)	Yeast, unpolished grains, lean pork
Riboflavin B_2	Inflammation of tongue; damage to eyes	Liver, eggs, cheese, milk, vegetables
Niacin	Pellagra (damage to skin, intestinal lining, and nerves)	Organ meat, yeast, milk
Folic acid	Anemia	Dark-green leafy vegetables, intestinal bacteria
B_{12}	Pernicious anemia	Liver
C (Ascorbic acid)	Scurvy (bleeding gums)	Citrus fruits, tomatoes, peppers

Adapted from *Time* Magazine Special Advertising Section of "Eating Well, Looking Fit, Feeling Better," 1983, by Theodore Berland. Reprinted by permission.

It is often difficult to provide all persons with sufficient dietary protein. In such instances, small children are observed to suffer a protein deficiency, especially after weaning. The malady, known as **kwashiorkor** (fig. 21.19), develops due to a lack of protein even if caloric intake is adequate. Unfortunately, even when a complete protein source is given, recovery is often marked by mental retardation.

The combination of grains and vegetables just mentioned are efficient means of providing dietary protein because, as a rule of thumb, humans store only about 10 percent of the calories available in food. Thus, one hundred calories from grain results in the storage of ten human calories. But if this grain is fed to cattle, it ultimately provides only one human calorie for storage. Nevertheless, in this country the practice of fattening cattle in feedlots where they are fed a rich diet of grain is common because the cattle fatten quickly and the resulting high-cholesterol, fatty beef has been preferred by consumers. This practice may change, however; first, because it requires more fossil fuel energy and, second, because the public is becoming more sensitive to their health needs.

Vitamins and Minerals

Vitamins are organic compounds (other than carbohydrates, lipids, and proteins) that the body is unable to produce and therefore must be present in the diet. Table 21.6 lists some of the more important vitamins and minerals and the best food sources for each one. Various symptoms develop when vitamins are lacking in the diet (fig. 21.20).

Although vitamins are an important part of a balanced diet, they are required only in very small amounts. Many vitamins are portions of coenzymes; for example, niacin is part of the coenzyme NAD, and riboflavin is part of FAD. Coenzymes are needed in only small amounts because each one can be used over and over again. This means that the daily requirement for vitamins is low and a properly balanced diet usually provides the amount needed. However, synthetic vitamins perform the same functions in the body as do those present in food since they are identical chemicals.

The intake of excess vitamins can possibly lead to illness. For example, excess vitamin C is converted to oxalic acid, a product that is toxic to the body. Vitamin A taken in excess over long periods can cause such effects as loss of

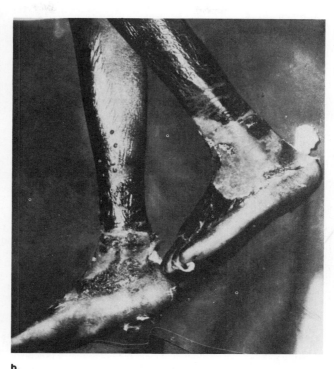

Figure 21.20
Illnesses due to vitamin deficiency. *a.* Bowing of bones
(rickets) due to vitamin D deficiency. *b.* Dermatitis of areas
exposed to light (pellagra) due to niacin deficiency.
c. Bleeding of gums (scurvy) due to vitamin C deficiency.
d. Fissures of lips (cheilosis) due to riboflavin deficiency.

a.   b.

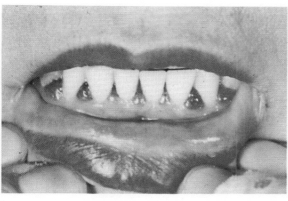

c.   d.

hair, bone and joint pains, and loss of appetite. Excessive vitamin D can cause
an overload of calcium in the blood, which in children leads to loss of appetite
and retarded growth. Megavitamin therapy should always be supervised by a
physician.

In addition to vitamins, various **minerals** are also required by the body.
Minerals are divided into the macrominerals that are needed in gram amounts
per day and the trace elements that are needed in only microgram amounts
per day. The macrominerals sodium, magnesium, phosphorus, chlorine, potassium, and calcium serve as constituents of cells and body fluids and as

Cancer Diet

Health food faddists have been saying it for years: eat right and you may reduce the risk of getting cancer. At a press conference in Washington last week, the National Academy of Sciences not only signaled its agreement with that view, it issued a 500-page report specifying just what is meant by eating right. "Our committee's recommendations should not be regarded as assuring a cancer-free life," said Dr. Clifford Grobstein, University of California biologist and chairman of the NAS panel that released the two-year study, but "by controlling what we eat we may prevent diet-sensitive cancer." Among the recommendations:

Cut fat consumption by 25% or more, both the saturated variety of fat found in meat and whole milk products and unsaturated lipids like those in vegetable oils. Animal tests and human population studies have shown a strong correlation between fat intake and rates of cancer of the breast, colon and prostate.

Eat less smoked, pickled and salt-cured foods, including sausages, smoked fish and bacon. In Japan, China and Iceland, where such foods are frequently consumed, there is a higher incidence of cancers of the stomach and esophagus. These foods also tend to contain nitrosamines and polycyclic aromatic hydrocarbons, chemicals known to cause cancer in animals.

Eat more fruits and vegetables containing vitamin C (oranges, broccoli and tomatoes) and beta-carotene, a precursor of vitamin A found in squash, carrots and other yellow and green vegetables. Both substances inhibit the formation of chemically induced cancers in laboratory tests; both are associated with lower cancer rates in human populations. The committee counseled against high-dose vitamin pills because of insufficient evidence about their health benefits. High doses of vitamin A, it added, can be toxic.

The committee advised Americans to drink only "moderate" amounts of alcohol, although it did not specify how much. Alcohol consumption, particularly when combined with smoking, has been linked to mouth, larynx, liver and lung cancers. Panel members were not, however, able to confirm reports that dietary fiber reduces the risk of bowel cancer. Nor was the evidence sufficient to convince them of the prophylactic benefits of vitamin E or the perils of preservatives, food dyes and other chemical additives.

The report drew criticism from the National Cattlemen's Association, which labeled it "inconclusive and premature," and from the American Meat Institute, which said it was based on "insufficient evidence." Grobstein acknowledged that his panel was "exploring a relationship between two still largely unknowns," but he added: "I don't think we're disseminating unproven theories."

Research has discovered that fruits and vegetables high in vitamins A and C help protect the body against cancer. These particularly include oranges, grapefruit, carrots, tomatoes, broccoli, cauliflower, cabbage, winter squash, and dark green vegetables.

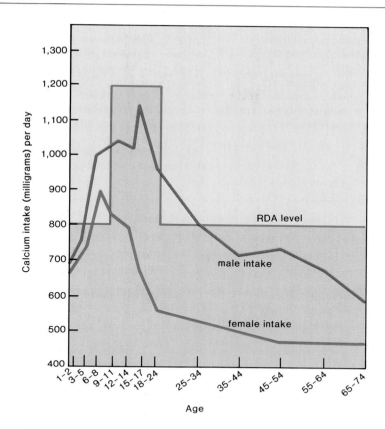

Figure 21.21
This graph shows that the intake of calcium is lower than it should be in adult males and even more so in adult females. This is of considerable concern because older women are susceptible to osteoporosis, gradual bone loss that leads to fractures and a weakened spine with accompanying back pain.

structural components of tissues. For example, calcium is needed for the construction of bones and teeth and also for nerve conduction and muscle contraction (fig. 21.21). The trace elements seem to have very specific functions. For example, iron is needed for the production of hemoglobin, and iodine is used in the production of thyroxin, a hormone produced by the thyroid gland. As research continues, more and more elements have been added to the list of those considered essential. During the past three decades, very small amounts of molybdenum, selenium, chromium, nickel, vanadium, silicon, and even arsenic have been found to be essential to good health.

Summary

Animals are heterotrophic organisms that require preformed food. Often they have a digestive tract that takes care of storing, grinding, enzymatically digesting, absorbing nutrient molecules, and eliminating nondigestible material. Some animals (e.g., planarians) have an incomplete digestive tract that shows little specialization of parts. Other animals (e.g., earthworms) have a complete digestive tract that does show specialization of parts. Some are continuous feeders, like a sessile filter feeder (e.g., clam); others are discontinuous feeders (e.g., squid). Discontinuous feeders need a crop or stomach in which to store food.

There are various adaptations for breaking food up into small pieces for enzymatic digestion: Mammals have teeth. Herbivores need teeth that can clip off plant material and grind it up. Also, the herbivore stomach contains bacteria that can digest cellulose. Carnivores need teeth that can tear and rip meat into pieces. Meat is easier to assimilate so the digestive system of carnivores has less specialization of parts and a shorter intestine.

In the human digestive tract, food is chewed and manipulated in the mouth where salivary glands secrete saliva. Saliva contains salivary amylase, which begins the process of carbohydrate digestion. Food then passes down the esophagus to the stomach. The stomach stores and mixes food with mucus and gastric juice to produce chyme. Pepsin begins protein digestion in the stomach.

Chyme gradually enters the duodenum where bile, pancreatic juice, and intestinal gland secretions are added. These additives then begin to act upon the chyme. Enzymes in the small intestine hydrolyze all of the organic nutrients. Tables 21.2 and 21.3 summarize the enzymes involved in digesting food.

Most nutrient absorption takes place in the small intestine, but some water and mineral ions are absorbed in the colon. Digestive wastes leave the colon by way of the anus.

The liver produces bile, which is stored in the gallbladder. The liver is also involved in the processing of absorbed nutrient molecules and in maintaining the blood concentration of nutrient molecules, such as glucose. The liver converts ammonia to urea and breaks down toxins. The pancreas produces digestive enzymes and hormones involved in the control of carbohydrate metabolism in the body.

Neural and hormonal regulation of digestive-tract secretions ensures that digestive substances, such as enzymes, are released only when food is present.

The body uses the nutrient molecules for the synthesis of macromolecules and as an energy source. Digestion of carbohydrates provides sugars that serve as a quick-energy source, while fat digestion provides a long-term energy source. Unfortunately, certain dietary fats are associated with circulatory disorders. Protein digestion gives us the essential amino acids that are so necessary to continued good health. In addition to these considerations, the body has to meet its vitamin and mineral needs.

Objective Questions

1. Digestion is largely _____ cellular in a planarian, while it is _____ cellular in an earthworm.
2. A clam is a continuous sessile _____ feeder while the predatious squid is a _____ feeder.
3. Mammalian herbivores have _____ (choose flat or pointed) molars, while carnivores have _____ (choose flat or pointed) molars.
4. _____ digestion takes place in the mouth and _____ digestion takes place in the stomach.
5. Digestion is completed in the small intestine, which is specialized for nutrient _____ by the presence of _____ .
6. Before fat digestion can occur, bile _____ fat to fat droplets. These are then acted on by the enzyme _____ .
7. _____ is a hormone that stimulates the gastric glands to secrete _____ and _____ .
8. The liver has numerous functions; among them, the liver maintains the _____ content of the blood by breaking down glycogen.
9. Lack of _____ in the diet can cause the condition kwashiorkor.
10. The mineral _____ is needed for strong bones and to prevent osteoporosis in older women.

Answers to Objective Questions

1. intra, extra 2. filter, discontinuous 3. flat, pointed 4. Carbohydrate, protein 5. absorption, villi 6. emulsifies, lipase 7. Gastrin, HCl, pepsin 8. glucose 9. essential amino acids 10. calcium

Study Questions

1. Contrast the incomplete with the complete gut using the planarian and earthworm as examples.
2. Contrast a continuous with a discontinuous feeder using the clam and squid as examples.
3. Contrast the dentition of the mammalian herbivore with the mammalian carnivore using the horse and lion as examples.
4. Name the parts of the human digestive system and state the function of each, including the function of any enzymes normally present.
5. State the location and discuss the function of both the liver and pancreas.
6. Assume that you have just eaten a ham sandwich. Discuss the digestion of the contents of the sandwich.
7. Define essential amino acids, essential fatty acids, and vitamins. Why is each one required in the diet?

Selected Key Terms

omnivore ('äm ni, vōr) 454
pharynx ('far iŋks) 454
typhlosole ('tif lə ,sōl) 455
esophagus (i 'säf ə gəs) 458
salivary amylase ('sal ə ,ver ē 'am ə ,lās) 459
epiglottis (,ep ə 'glät əs) 462
peristalsis (,per ə 'stòl səs) 462
pepsin ('pep sən) 462
chyme ('kīm) 462
bile ('bīl) 463
trypsin ('trip sən) 463
lipase ('lī ,pās) 463
peptidase ('pep tə ,dās) 464
gastrin ('gəs trən) 465
secretin (si 'krēt ən) 465

Each cell in an animal's body must acquire nutrient molecules and rid itself of metabolic wastes. Some animals are small enough that each cell makes these exchanges directly with the external environment. Other animals are complex and have organ systems that function to maintain an internal environment for the body's cells. This internal environment brings nutrient molecules to and collects the wastes from each cell. Maintenance of a constant internal environment is termed *homeostasis*.

Respiration

A supply of oxygen must be present in the water or air in order for respiration to take place. Without respiration, an animal dies. This was first observed in 1774 by Joseph Priestley, who collected oxygen by heating red mercuric oxide. His contemporary, Frenchman Antione Lavoisier, correctly deduced that both combustion by a lit candle and respiration by an animal remove oxygen from the air. This was one of the first times that physiological processes were explained in the same terms as nonliving mechanisms.

Gas exchange takes place by the physical process of *diffusion*. For diffusion to be effective, the gas-exchange region must be (1) moist, (2) thin, and (3) large in relation to the size of the body. We will compare how these requirements are met in aquatic and terrestrial animals (fig. 22.1).

Aquatic Animals

Water fully saturated with air contains about twenty times less oxygen than does air. This means that aerobic aquatic animals need to have a more efficient means of acquiring oxygen than do terrestrial animals. Even so, some aquatic organisms do not have a specialized region for gas exchange. Their organization is such that they have a large surface area in comparison to their size. Therefore, as in hydras and planarians, it is possible for each cell to carry out gas exchange with the water environment (fig. 22.2).

Many aquatic organisms do have a specialized region for gas exchange. Most respire by means of vascularized (fig. 22.9) *gills,* which can be external or internal. Among invertebrates, gills are typically outgrowths of the body surface. They can be simple, such as those found in sea stars (fig. 16.2), or they can be finely divided, as in the clam (fig. 15.21). Among vertebrates, the gills of fishes are extensions of the digestive tract. Water enters the pharynx and moves through the gill slits past the gills, which may be protected by an outer flap called an **operculum** (fig. 22.3*a*). The gills are composed of *filaments* that are themselves folded into platelike *lamellae* (fig. 22.3*b* and *c*). In the capillaries of each lamella, the blood flows in a direction opposite to the movement of water across the gills. This **countercurrent flow** increases the amount of oxygen that can be taken up; as the blood in each lamella gains oxygen, it encounters water having a higher oxygen content (fig. 22.3*d*). Even though this is a highly effective means of acquiring oxygen, fishes expend up to 25 percent of their energy for breathing, compared to 1 to 2 percent in terrestrial mammals, because water is more dense and much less rich in oxygen than air.

Terrestrial Animals

The earthworm is an invertebrate terrestrial animal that uses its body surface (fig. 15.25) for respiration. Therefore, this surface must remain thin and moist and cannot serve as a protection against drying out. As long as the worm remains buried beneath the moist ground, there is little danger; but an earthworm certainly cannot venture forth on a dry, hot day without dire consequences. Loss of too much water across the body's surface will lead to death.

CHAPTER 22

Respiration and Excretion

Concepts

1. All animals perform gas exchange. They need a continual supply of oxygen for cellular respiration, which gives off carbon dioxide as a by-product.
   a. Small animals tend to exchange gases directly through body surfaces, but larger animals tend to have specialized gas-exchange areas.
   b. Aquatic animals generally exchange gases through gills, while terrestrial animals have tracheal systems or lungs.
   c. Oxygen and carbon dioxide are transported in the blood of larger animals, such as humans.

2. The internal environment of animals is usually regulated to keep its composition within tolerable limits.
   a. Excretion rids the body of nitrogenous wastes and helps regulate ion and water balance.
   b. Excretory organs of animals are diverse, ranging from individual tubules to the vertebrate kidney.

Figure 22.1
Fishes live in water but butterflies and birds live on land.
a. The gills of a fish (fig. 22.3) are kept moist by the water
in which they function. When exposed to air, the gills
collapse and stick together, making them useless. *b.*
Insects have an internal system of rigid air tubes (fig. 22.4)
that take oxygen directly to the cells. *c.* Birds, like humans
(fig. 22.8), breathe air into the lungs, vascularized cavities
where oxygen is picked up by the blood, which then
transports it to the cells.

a.

b.

c.

Figure 22.2
Some aquatic animals do not have gills. Instead, each cell
carries out its own gas exchange. Although they are
multicellular, the organization of certain aquatic animals
permits each cell to carry out its own gas exchange.
Therefore, they do not need a circulatory system to
transport gases to and away from the cells. *a.* In *Hydra,*
the outer layer of cells is in contact with its watery
environment and the inner layer can exchange gases with
the water in the gastrovascular cavity. *b.* In planarians,
their flat bodies permit diffusion of gases to and away
from the external environment.

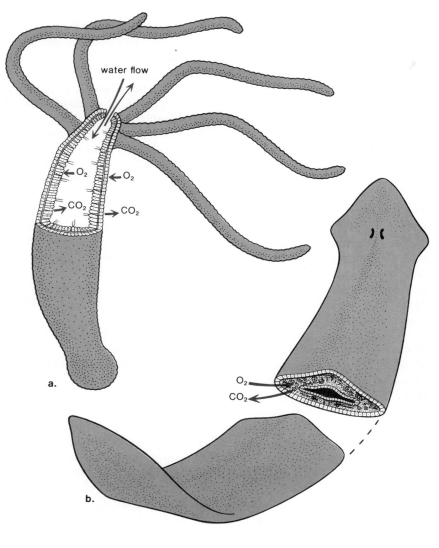

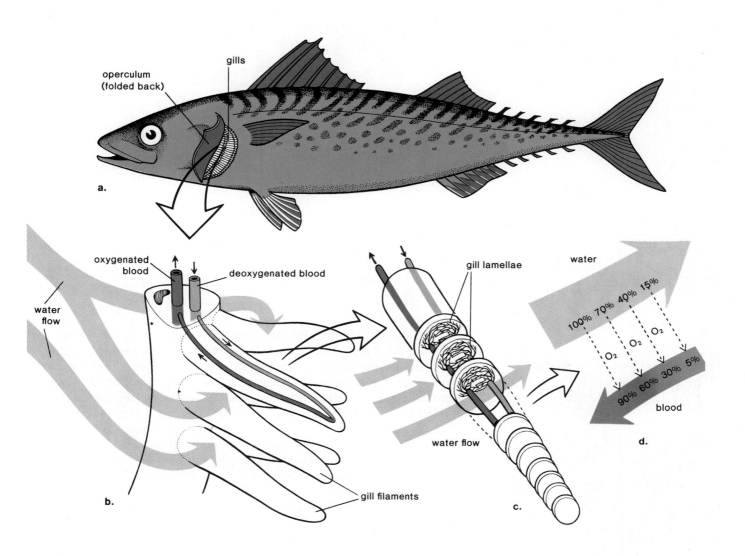

Figure 22.3
Anatomy of gills in detail. *a.* The operculum (folded back) covers and protects several layers of delicate gills. *b.* Each layer of gills has two rows of gill filaments. *c.* Each filament has many thin, platelike lamellae. Gases are exchanged between capillaries inside the lamellae and the water that flows between the lamellae. *d.* The blood in the capillaries flows in the direction opposite to that of the water. This countercurrent flow results in 90 percent oxygen uptake from the water.

Figure 22.4

Respiratory system of insects. *a.* A system of air tubes, the tracheal system, penetrates and travels throughout the body of an insect. The blood has no need to carry oxygen because it is brought to the cells by way of these tubes. *b.* Air enters the trachea at openings called spiracles. From there it moves to the smaller tracheoles that take it to the cells where gas exchange takes place.

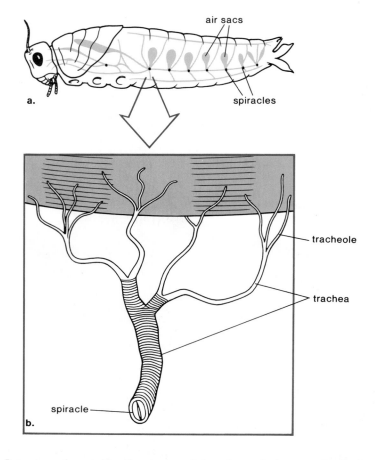

Figure 22.5

In contrast to gills, which are outgrowths (evaginations) of the pharynx, lungs are ingrowths (invaginations) of the same region.

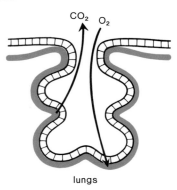

Insects and certain other terrestrial arthropods have a simple but specialized type of respiratory system known as a **tracheal system** (fig. 22.4). Oxygen enters the **spiracles,** valvelike openings on each side of the body, and moves directly to all cells of the body by way of tubes (tracheae) that branch repeatedly into all parts of the body. The trachea end in tiny channels (tracheoles) that are in direct contact with the body cells. Larger insects pump air in and out of the system by body movements, but even so, the efficiency of this system is limited. Notice that utilization of an independent gas-distribution system relieves the blood of this function. The blood of insects is colorless because oxygen is distributed by the respiratory system, not the circulatory system.

Terrestrial vertebrates, in particular, have evolved vascularized internal cavities known as **lungs** (fig. 22.5). The lungs are not highly developed in amphibians (fig. 22.6) because they also make use of the skin for gas exchange. During the courtship season, the male hairy frog (*Astylosternus robustus*) develops dorsal extensions that resemble hair. Extra oxygen is needed because courtship rituals go on for hours. The skin of amphibians, like that of earthworms, must stay thin and moist for gas exchange, so these animals are also restricted to a moist environment.

The inner lining of the lungs is more divided in reptiles than in amphibians (fig. 22.6). The lungs of birds and mammals are even more elaborately subdivided into small passageways and spaces. It has even been estimated that human lungs have a total surface area that is at least fifty times the skin's surface area. This extensive surface area of the lungs helps to assure adequate gas exchange.

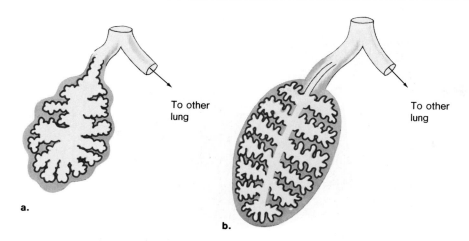

Figure 22.6
The amphibian lung versus the reptilian lung. The frog
lung (a.) has fewer convolutions than does the turtle lung
(b.).

a.

b.

To other
lung

To other
lung

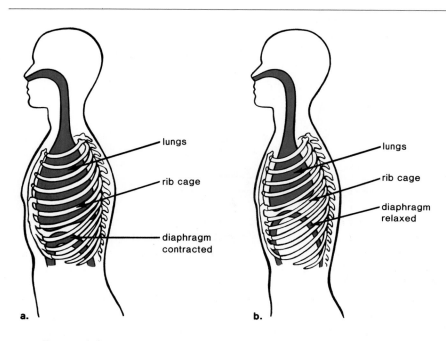

Figure 22.7
Breathing by negative pressure. a. Inhalation occurs as
the rib cage moves up and out and the diaphragm
flattens. As the chest area expands, the lungs are
expanded, and the air comes rushing in. b. Exhalation
occurs as the rib cage moves down and in and the
diaphragm rises. As the chest area is decreased, the
lungs recoil and air is pushed out.

lungs

rib cage

diaphragm
contracted

a.

lungs

rib cage

diaphragm
relaxed

b.

Terrestrial vertebrates ventilate the lungs by moving air in and out of
the respiratory tract. The air is moistened as it moves along the tubes leading
to the lungs, which protects the lungs from the drying effects of air. Frogs use
positive pressure to force air into the respiratory tract. With the nostrils firmly
shut, the floor of the mouth rises and pushes the air into the lungs. Reptiles,
birds, and mammals use **negative pressure** (fig. 22.7). The lungs expand, and
the air comes rushing in. Reptiles have jointed ribs that can be raised to expand the lungs. Mammals have a rib cage that is lifted up and out and a muscular **diaphragm** that is flattened. Both these actions increase the volume of
the lungs, and air is thereby drawn in during **inhalation.** After inhalation, exhalation occurs. During **exhalation,** air is pushed out of the lungs. In reptiles,
lowering the ribs exerts a pressure that forces air out. In mammals, the rib
cage is not only lowered, the diaphragm rises to force the air out of the lungs.

The lungs of amphibians, reptiles, and mammals are not completely
emptied and refilled during each breathing cycle. Instead, the air coming in
mixes with used air still in the lungs. While this helps conserve water, it also

Figure 22.8
The lungs of amphibians, reptiles, and mammals are incompletely ventilated whereas those of birds are completely ventilated. *a.* Inhalation. There is residual "used" air (blue) in the lungs of amphibians, reptiles, and mammals. In birds, the posterior air sacs have been cleared of residual air. *b.* Exhalation. Used air is leaving the lungs of amphibians, reptiles, and mammals, although some will remain. Used air, leaving the lungs of birds, will be completely removed because fresh air (pink) enters by its own pathway as shown in (*a.*).

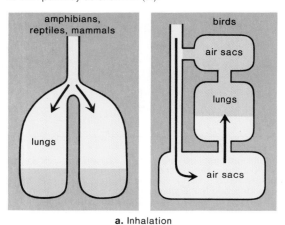

a. Inhalation

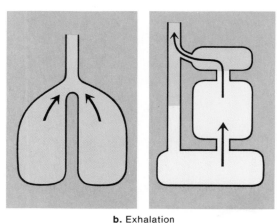

b. Exhalation

Figure 22.9
A circulatory system transports impure blood to the gas-exchange area, where carbon dioxide diffuses out and oxygen diffuses into the blood. It then transports purified blood to the body tissues, where carbon dioxide diffuses into and oxygen diffuses out of the blood.

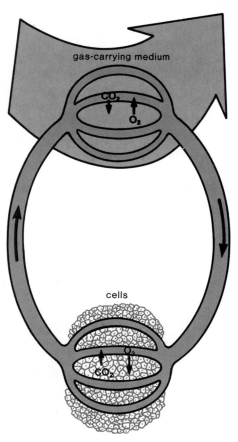

decreases gas-exchange efficiency. The oxygen requirement of birds, whose body size must stay small, cannot be met by this system of *incomplete ventilation*. Therefore, they have developed a system of *complete ventilation* (fig. 22.8). Fresh, oxygen-rich air passes through the lungs in a one-way direction and does not mix with used air. Incoming air is carried past the lungs by a bronchus that takes it to a set of posterior air sacs. The air then passes forward through the lungs into a set of anterior air sacs. From here, it is finally expelled.

In most animals, the exchange area is highly vascularized, being richly supplied with blood capillaries. Oxygen is carried from the exchange surface to the interior of the body by the blood circulatory system (fig. 22.9). The blood of all vertebrates contains the respiratory pigment *hemoglobin,* which helps carry the oxygen. It is hemoglobin that makes the blood red.

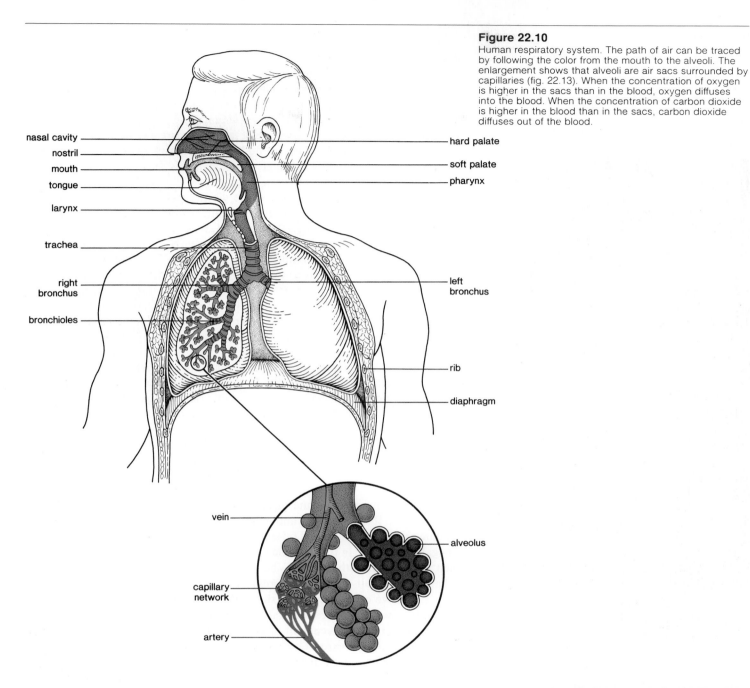

Figure 22.10
Human respiratory system. The path of air can be traced by following the color from the mouth to the alveoli. The enlargement shows that alveoli are air sacs surrounded by capillaries (fig. 22.13). When the concentration of oxygen is higher in the sacs than in the blood, oxygen diffuses into the blood. When the concentration of carbon dioxide is higher in the blood than in the sacs, carbon dioxide diffuses out of the blood.

Human Gas Exchange System

Path of Air

The human gas-exchange (respiratory) system (fig. 22.10) includes all those structures (table 22.1) that conduct air to and from the lungs. The lungs lie deep within the **thoracic** (chest) **cavity** where they are protected from desiccation. As air moves through the **nose, trachea,** and **bronchi,** it is filtered so it is free of debris, warmed, and humidified so it is at body temperature, and saturated with water by the time it reaches the lungs. In the nose, hairs and

Table 22.1 Path of Air

Structure	Function
Nasal cavities	Filters, warms, and moistens
Pharynx (throat)	Connection to surrounding regions
Glottis	Passage of air
Larynx (voice box)	Sound production
Trachea (windpipe)	Passage of air to thoracic cavity
Bronchi	Passage of air to each lung
Bronchioles	Passage of air to each alveolus
Alveoli	Air sacs for gas exchange

Figure 22.11

A scanning electron micrograph showing cilia (Ci) and goblet cells (GC) in the tracheal lining. Goblet cells have microvilli (Mv) on their surfaces. The goblet cells secrete mucus and the cilia sweep debris and mucus up toward the throat.

Kessel, R. G., and Kardon, R. H.: *Tissues and Organs: A Text-Atlas of Scanning Electron Microscopy.* © 1979 by W. H. Freeman and Co.

Figure 22.12

Aortic bodies, located in the aorta, and carotid bodies, located in the carotid arteries, are chemoreceptors that detect changes in CO_2, H^+, and O_2 concentrations in the blood. Sensory nerve fibers carry such information to centers in the brain that control breathing.

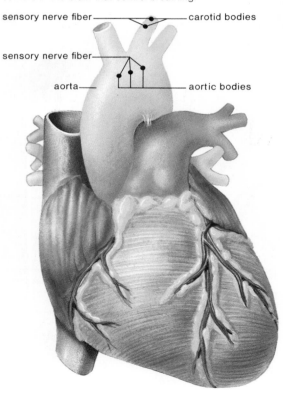

cilia act as a screening device. In the trachea and bronchi, cilia (fig. 22.11) beat upward, carrying mucus, dust, and occasional bits of food that "went down the wrong way" into the throat or pharynx where the accumulation may be swallowed or expectorated.

The hard and soft palates separate the nasal cavities from the mouth, but the air and food passages cross in the pharynx. This may seem inefficient, and there is danger of choking if food accidentally enters the trachea, but this arrangement does have the advantage of letting one breathe through the mouth in case the nose is plugged up. In addition, it permits greater intake of air during heavy exercise when greater gas exchange is required.

Air passes from the pharynx through the **glottis,** an opening into the **larynx** or voice box. At the edges of the glottis, embedded in mucous membrane, are the **vocal cords.** These elastic ligaments vibrate and produce sound when air is expelled past them through the glottis from the larynx.

The larynx and trachea are held permanently open to receive air. The larynx is held open by the complex of cartilages that form the Adam's apple. The trachea is held open by a series of C-shaped, cartilaginous rings that do not completely meet in the rear. When food is being swallowed, the larynx rises and the glottis is closed by a flap of tissue called the **epiglottis.** A backward movement of the soft palate covers the entrance of the nasal passages into the pharynx. The food then enters the esophagus, which lies behind the larynx (fig. 21.12).

The trachea divides into two **bronchi** (sing., **bronchus**) that enter the right and left lungs; each branches into a great number of smaller passages called **bronchioles.** The two bronchi resemble the trachea in structure, but as the bronchial tubes divide and subdivide, their walls become thinner, and rings of cartilage are no longer present. Each bronchiole terminates in an elongated space enclosed by a multitude of air pockets or sacs called **alveoli** (sing., **alveolus**), which make up the lungs.

Breathing

Humans breathe by the same mechanisms used by all other mammals. The volume of the thoracic cavity and lungs is increased by muscular contractions that lower the diaphragm and raise the ribs (fig. 22.7). These movements create a negative pressure in the thoracic cavity and lungs, and atmospheric pressure then forces air into the lungs. When rib and diaphragm muscles relax, air is exhaled as a result of increased pressure in the thoracic cavity and lungs.

Extensive experimentation shows that increased carbon dioxide (CO_2) and hydrogen ion (H^+) concentrations are the primary stimuli that increase breathing rate. The gas content of the blood is monitored by *chemoreceptors* called the **aortic** and **carotid bodies,** specialized structures located in the walls of the aorta and carotid arteries (fig. 22.12). These receptors are very sensitive to changes in CO_2 and H^+ concentrations, but are only minimally sensitive to a lower O_2 concentration. Information from the chemoreceptors goes to the **respiratory center** in the medulla oblongata portion of the brain (p. 515), which then increases the breathing rate. This respiratory center, itself, is also sensitive to the chemical content of the blood reaching the brain.

Based on available statistics, the American Cancer Society informs us that smoking carries a high risk. Among the **risks of smoking** are the following:

Shortened life expectancy A twenty-five-year-old who smokes two packs of cigarettes a day has a life expectancy 8.3 years shorter than a nonsmoker. The greater the number of packs smoked, the shorter the life expectancy.

Lung cancer Smoking cigarettes is the major cause of lung cancer in both men and women. The frequency of lung cancer has risen in women of late because more women are now smoking. Smoker autopsies have revealed the progressive steps by which cancer of the lung develops. The first event appears to be a thickening of the cells that line the bronchi. Then there is a loss of cilia so that it is impossible to prevent dust and dirt from settling in the lungs. Following this, cells with atypical nuclei appear in the thickened lining. A disordered collection of cells with atypical nuclei may be considered to be cancer in situ (at one location). The final step occurs when some cells break loose and penetrate the other tissues, a process called metastasis. This is true cancer (*b.*).

Cancer of the larynx, mouth, esophagus, bladder, and pancreas The chances of developing these cancers are from 2 to 17 times higher in cigarette smokers than in nonsmokers.

Emphysema Cigarette smokers have 4 to 25 times greater risk of developing emphysema. Damage is seen in the lungs of even young smokers. Smoking causes the lining of the bronchioles to thicken. If these become obstructed, the air within the alveoli is trapped. The trapped air very often causes the alveolar walls to rupture and the walls of the small blood vessels in the vicinity to thicken. If a large part of the lungs is involved, the lungs are permanently inflated and the chest balloons out due to this trapped air. The victim is breathless and has a cough. Since the surface area for gas exchange is reduced, not enough oxygen reaches the heart and brain. The heart works furiously to force more blood through the lungs, which may lead to a heart condition. Lack of oxygen for the brain may make the person feel depressed, sluggish, and irritable.

Coronary heart disease Cigarette smoking is the major factor in 120,000 additional U.S. deaths from coronary heart disease each year.

Reproductive effects Smoking mothers have more stillbirths and low-birthweight babies who are more vulnerable to disease and death. Children of smoking mothers are smaller, and underdeveloped physically and socially even seven years after birth.

In the same manner, the American Cancer Society informs smokers of the **benefits of quitting.** These benefits include the following:

Risk of premature death is reduced Do not smoke for 10 to 15 years, and the risk of death due to any one of the cancers mentioned approaches that of the nonsmoker.

Health of respiratory system improves The cough and excess sputum disappear during the first few weeks after quitting. As long as cancer has not yet developed, all the ill effects mentioned can reverse themselves and the lungs can become healthy again. In patients with emphysema, the rate of alveoli destruction is reduced and lung function may improve.

Coronary heart disease risk sharply decreases After only one year the risk factor is greatly reduced, and after 10 years an exsmoker's risk is the same as that of those who never smoked.

The increased risk of having stillborn children and underdeveloped children disappears Even for women who do not stop smoking until the fourth month of pregnancy, such risks to infants is decreased.

Young people who smoke must ask themselves if the benefits of quitting outweigh the risks of smoking.

Risks of Smoking versus Benefits of Quitting

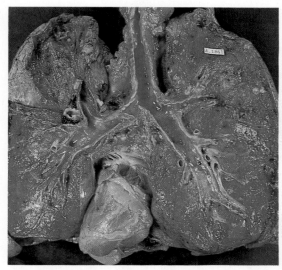

a.

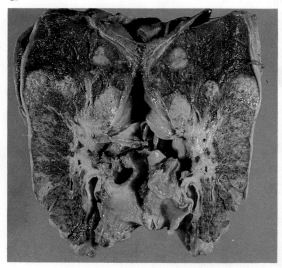

b.

a. *Normal lungs with heart in place. Notice the healthy red color.* b. *Lungs of a heavy smoker. Notice how black the lungs are except where cancerous tumors have formed.*

Figure 22.13

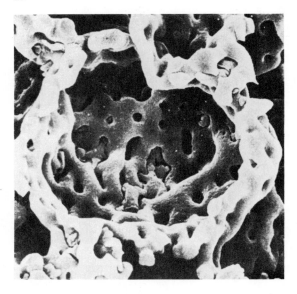

Gas Exchange and Transport

Diffusion primarily accounts for the exchange of gases between the air in the alveoli and the blood (fig. 22.13) in the pulmonary capillaries. Atmospheric air contains little CO_2, but blood flowing into the lung capillaries is almost saturated with the gas. Therefore, CO_2 diffuses out of the blood into the alveoli. The pattern is the reverse for oxygen. Blood coming into the pulmonary capillaries is oxygen poor and the alveolar air is oxygen rich; therefore, O_2 diffuses into the capillaries.

Transport of O_2 and CO_2

Most oxygen entering the blood combines with hemoglobin in red blood cells to form **oxyhemoglobin.**

$$Hb + O_2 \longrightarrow HbO_2$$

Each hemoglobin molecule (fig. 22.14) contains four polypeptide chains, and each chain is folded around an iron-containing structure called **heme.** It is actually the iron that forms a loose association with oxygen. Since there are about 280 million hemoglobin molecules in each red blood cell, each cell is capable of carrying more than one thousand million molecules of oxygen.

Carbon monoxide, present in automobile exhaust, combines with hemoglobin more readily than does oxygen, and it stays combined for several hours regardless of the environmental conditions. Accidental death or suicide from carbon monoxide poisoning occurs because the hemoglobin of the blood is not available for oxygen transport.

Oxygen-binding characteristics of hemoglobin can be studied by examining oxyhemoglobin dissociation curves (fig. 22.15). These curves show the percentage of oxygen-binding sites of hemoglobin that are carrying oxygen at various oxygen partial pressures (PO_2). A partial pressure of a gas is simply the amount of pressure exerted by that gas among all the gases present. At partial pressures of O_2 in the lungs, hemoglobin becomes practically saturated with O_2, but at partial pressures in the tissues, hemoglobin quickly gives up much of its oxygen.

$$HbO_2 \longrightarrow Hb + O_2$$

Figure 22.14

The hemoglobin molecule is a globular protein that contains four polypeptide chains, two of which are alpha chains and two of which are beta chains. Heme groups are shown as planes, and the sphere embedded in each heme group is an atom of iron.

The acid pH and warmer temperature of the tissues also promote this dissociation (breakdown). Hemoglobin that has given up its oxygen is called **reduced hemoglobin.**

In the tissues reduced hemoglobin combines with carbon dioxide to form **carbaminohemoglobin.**

$$Hb + CO_2 \longrightarrow HbCO_2$$

Most of the carbon dioxide, however, is transported as the **bicarbonate ion,** HCO_3^-.

$$CO_2 + H_2O \longrightarrow H_2CO_3 \longrightarrow H^+ + HCO_3^-$$

Carbon dioxide combined with water forms carbonic acid; this dissociates to a hydrogen ion and the bicarbonate ion.

An enzyme in red cells, **carbonic anhydrase,** speeds up this reaction. The released hydrogen ions, which could drastically change the pH of the blood, are absorbed by the globin portions of hemoglobin, and the bicarbonate ions diffuse out of the red cells to be carried in the plasma. Reduced hemoglobin, which combines with a hydrogen ion, can be symbolized as HHb. The latter plays a vital role in maintaining the pH of the blood.

As blood enters the pulmonary capillaries, most of the carbon dioxide is present in plasma as the bicarbonate ion. The little free carbon dioxide remaining begins to diffuse out, and the following reaction is driven to the right.

$$H^+ + HCO_3^- \longrightarrow H_2CO_3 \longrightarrow H_2O + CO_2$$

Carbonic anhydrase also speeds up this reaction, during which hemoglobin gives up the hydrogen ions it has been carrying, HHb becoming Hb.

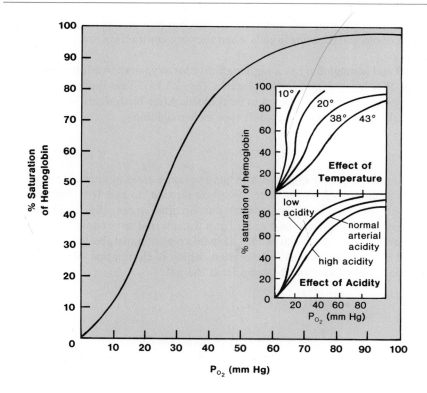

Figure 22.15
The oxyhemoglobin dissociation curve. The large curve shows the percentage of saturation of hemoglobin at 38° and normal arterial blood acidity. As the partial pressure of oxygen (PO_2) decreases, hemoglobin gives up its oxygen. This effect is also promoted by the higher temperature and acidity of the tissues.

Figure 22.16
Fetal mammals, such as (*a.*) a human fetus, have hemoglobin that has a higher affinity for oxygen than adult hemoglobin. This permits fetal hemoglobin to combine with oxygen at a lower PO$_2$ value than (*b.*) the mother's hemoglobin. Therefore, oxygen tends to leave the blood of the mother and move into the blood of the fetus.

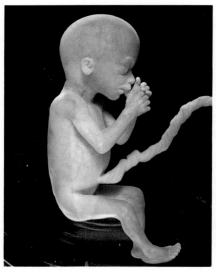

a.

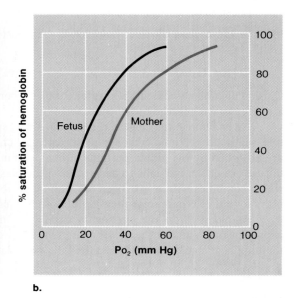

b.

Figure 22.17
The internal environment of cells (blood and tissue fluid) stays relatively constant because the blood is continually refreshed by certain organs of the body. These organs are involved in excretion—ridding the body of metabolic wastes. In particular, the gut excretes heavy metals; the lungs excrete carbon dioxide; and the kidneys and skin excrete nitrogenous wastes. The liver removes toxic substances from the blood and converts them to molecules that are excreted by the kidneys.

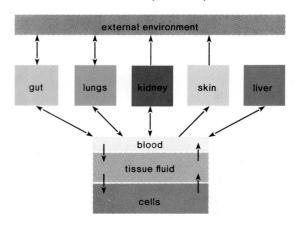

Types of Hemoglobin

Muscle cells contain an oxygen-binding pigment called **myoglobin,** which has a higher affinity for oxygen than does hemoglobin, and which tends to hold oxygen until the PO$_2$ falls very low. Myoglobin provides an excellent reserve source of oxygen for muscle cells when they are contracting and metabolizing rapidly.

Fetal hemoglobin has a higher affinity for oxygen than adult hemoglobin at the PO$_2$ levels found in the placenta (fig. 22.16). This facilitates transfer of oxygen from mother's blood to fetal blood. After birth, fetal hemoglobin is gradually replaced with the adult type of hemoglobin.

Excretion

We have seen that the digestive and gas-exchange areas of animals allow molecules to enter and, in the case of carbon dioxide, to exit from the internal environment. Now let us consider how certain other organs (fig. 22.17) remove nitrogenous wastes, ions, and water from the internal environment. **Excretion** is the elimination of molecules that have taken part in metabolic reactions and should not be confused with **defecation,** which is elimination of nondigested materials, dead cells, and bacteria from the gut.

Nitrogenous Wastes

The breakdown of various molecules, including nucleic acids and amino acids, results in nitrogenous wastes. For simplicity's sake, however, we will limit our discussion to amino acid metabolism. Amino acids, derived from protein in food, can be used by cells for synthesis of new body protein or other nitrogen-containing molecules. The amino acids not used for synthesis are oxidized to generate energy or are converted to fats or carbohydrates that can be stored. In either case, the **amino groups** ($-NH_2$) must be removed (fig. 22.18) because they are not needed for any of these purposes. Once the amino groups have been removed from amino acids, they may be excreted from the body in the form of ammonia, urea, or uric acid, depending on the species (table 22.2). Removal of amino groups from amino acids requires a fairly constant amount of energy. However, the energy requirement for the conversion of amino groups to either ammonia, urea, or uric acid differs as indicated in figure 22.18.

Ammonia

Amino groups removed from amino acids immediately become ammonia by addition of a third hydrogen. Therefore, little or no energy is required to convert an amino group to ammonia. The gas ammonia is quite toxic and can only be used as a nitrogenous excretory product if a good deal of water is available to wash it from the body. The high solubility of ammonia permits this means of excretion in bony fishes, aquatic invertebrates, and amphibians whose gills and skin surfaces are in direct contact with the water of the environment.

Urea

Terrestrial amphibians and mammals usually excrete urea as their main nitrogenous waste. Urea is much less toxic than ammonia and can be excreted in a moderately concentrated solution. This allows body water to be conserved, an important advantage for terrestrial animals with limited access to water.

However, production of urea requires expenditure of energy. Urea is produced in the liver by a set of enzymatic reactions known as the **urea cycle.** In the cycle, some of whose reactions require ATP, carrier molecules take up carbon dioxide and two molecules of ammonia to finally release urea.

$$H_2N - \overset{\overset{\textstyle O}{\|}}{C} - NH_2$$

Uric Acid

Uric acid is excreted by insects, reptiles, birds, and some dogs (for example, dalmatians). Uric acid is not very toxic and is poorly soluble in water. Poor solubility is an advantage for water conservation because uric acid can be concentrated even more readily than can urea. In reptiles and birds, a dilute solution of uric acid passes from the kidneys to the **cloaca,** a common reservoir for the products of the digestive, urinary, and reproductive systems. After water is absorbed by the cloaca, the uric acid passes out with the feces.

Figure 22.18
Amino acid metabolism. Amino acids can join together to form a protein, or they can be broken down. The carbon skeleton can be used as an energy source, but the amino group must be excreted as either ammonia, urea, or uric acid. The energy and water requirements for these vary as indicated.

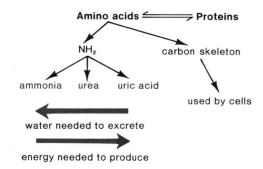

Table 22.2 Nitrogenous Waste Excretion in Relation to Habitat

Product	Habitat	Animals
Ammonia	Water	Aquatic invertebrates Bony fishes Amphibian larvae
Urea	Land	Adult amphibians Mammals
Uric acid	Land	Insects Birds Reptiles

Figure 22.19
Hard-shelled egg. All nutrients and water required during development must be enclosed by the eggshell because only oxygen, carbon dioxide, and water vapor diffuse freely through the shell. Nitrogenous wastes must also be collected and stored within a sac (allantois) until hatching time. At hatching time, the sac is detached from the body and remains inside the broken shell.

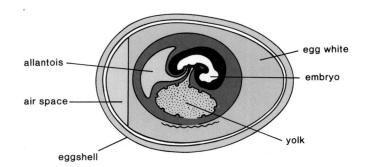

Excretion of uric acid by reptiles and birds is correlated with the necessity to conserve water during development. Embryos of reptiles and birds develop inside completely enclosed eggs (fig. 22.19). All nutrients and water for metabolism and growth of the embryo must be inside the egg before embryonic development begins. Also, all nitrogenous wastes must be stored until hatching occurs; therefore, the production of insoluble, relatively nontoxic uric acid is advantageous for these embryos. Uric acid is stored in a highly concentrated form inside a sac attached to the body of the embryo during development. Upon hatching, the embryo breaks the connection with the sac and leaves it and the uric acid inside the broken eggshell.

Uric acid is synthesized by a long, complex series of enzymatic reactions that require expenditure of even more ATP than does urea synthesis. Here again, there seems to be a trade-off between the advantage of water conservation and the disadvantage of energy expenditure for synthesis of an excretory molecule.

Osmotic Regulation

Excretory organs have the important function of regulating the water and salt balance of the body. Figure 22.20 shows that among animals, only marine invertebrates and cartilaginous fishes have body fluids that are nearly isotonic to seawater. These organisms have little difficulty maintaining their normal salt and water balance. Surprising, though, is the observation that while *they are isotonic,* the body fluids of cartilaginous fishes do not contain the same amount of salt as does seawater. The answer to this paradox is that their blood contains a concentration of urea high enough to match the tonicity of the sea! For some unknown reason, this amount of urea is not toxic to them.

The body fluids of all bony fishes have only a moderate amount of salt concentration. Apparently, their common ancestor evolved in fresh water, and only later did some groups invade the sea. Marine bony fishes (fig. 22.21a) are therefore prone to water loss and become dehydrated. To counteract this, they drink water almost constantly. On the average, marine bony fishes swallow an amount of water estimated to be equal to 1 percent of their body weight every hour. This is equivalent to a human drinking about 700 ml of water every hour around the clock. While they get water by drinking, this habit also causes these fishes to acquire salt. Instead of forming a hypertonic urine, however, they actively transport sodium (Na^+) and chloride (Cl^-) ions into the surrounding seawater at the gills.

It is easy to see that freshwater bony fishes (fig. 22.21b) have the exact opposite osmotic problems as marine bony fishes. Their body fluids are hypertonic to fresh water, and they are prone to gain water. These fishes never

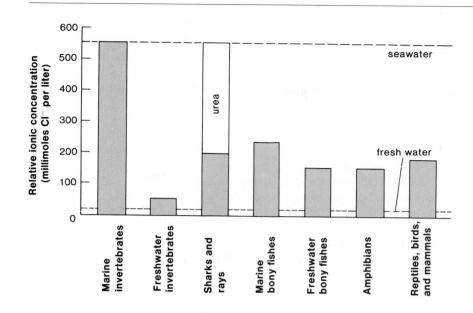

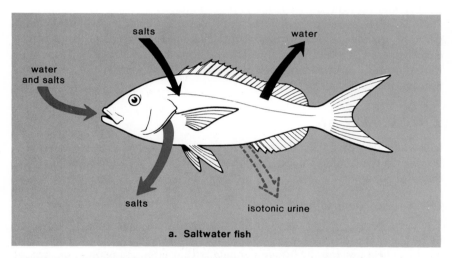

Figure 22.20
Comparison of relative ionic (Cl⁻) concentration of animal body fluids and the ionic concentration in seawater (dotted line above) and fresh water (dotted line below). For example, marine invertebrates are the only animals to have fluids with the ionic concentration of seawater, and freshwater invertebrates are the only animals to approach the ionic concentration of fresh water.

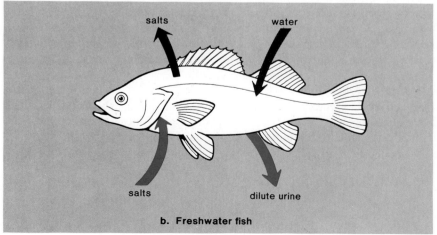

Figure 22.21
Water and salt balances in bony fishes. *a.* Marine bony fish. *b.* Freshwater bony fish. The black arrows represent passive transport from the environment, and the colored arrows represent active transport by the fishes to counteract environmental pressures.

drink water but, instead, eliminate excess water through production of large quantities of dilute urine. They discharge a quantity of urine equal to one-third their body weight each day. Because they tend to lose salts, they actively transport salts across the membranes of their gills.

The difference in adaptation between marine and freshwater bony fishes makes it remarkable that some fishes actually move between the two environments during the life cycle. Salmon, for example, begin their lives in freshwater streams and rivers, move to the ocean for a period of time, and finally return to fresh water to breed. These fish alter their behavior and gill and kidney functions in response to the osmotic changes they encounter when moving from one environment to the other.

Freshwater invertebrates also have to rid the body of excess water. Planarians have a network of tubular excretory canals (fig. 15.12) that open to the outside of the body through pores. Located along the canals are bulblike flame cells, each of which has a cluster of beating cilia that under the microscope look like a flickering flame. The beating of flame-cell cilia propels fluid through the excretory canals and out of the body.

Like marine bony fishes, some animals that evolved on land are also able to drink seawater despite its high toxicity. Birds and reptiles that live at sea have a nasal salt gland that can excrete large volumes of concentrated salt solution. Mammals that live at sea, like whales, porpoises, and seals, most likely can concentrate their urine enough to drink salt water. Humans cannot do this, and they die if they drink only seawater.

Most terrestrial animals need to drink water occasionally, but the kangaroo rat manages to get along without drinking water at all. It forms a very concentrated urine, and its fecal material is almost completely dry. These abilities allow it to survive using metabolic water derived from the breakdown of nutrient molecules alone.

Excretory Organs

Most animals have a special organ that excretes nitrogenous wastes and is also involved in body fluid homeostasis. We will consider several of these, including the human kidney.

Earthworm Nephridia

The earthworm's body is divided into segments, and nearly every body segment has a pair of excretory structures called **nephridia** (sing., **nephridium**). Each nephridium (fig. 22.22) is a tubule with a ciliated opening and an excretory pore. As fluid from the body cavity is propelled through the tubule by beating cilia, certain substances are reabsorbed and carried away by a network of capillaries surrounding the tubule. This process results in the formation of a urine that contains only unwanted metabolic wastes and ions.

Malpighian Tubules

Insects have a unique excretory system consisting of long, thin tubules, called **Malpighian tubules** (fig. 22.23), attached to the gut. Since insects have an open circulatory system, the tubules are surrounded by blood. Water and uric acid

Figure 22.22
The earthworm nephridium. The nephridium has a ciliated opening that leads to a coiled tubule surrounded by a capillary network. Urine can be temporarily stored before being released to the outside via a pore termed a nephridiopore. Most segments contain a pair of nephridia.

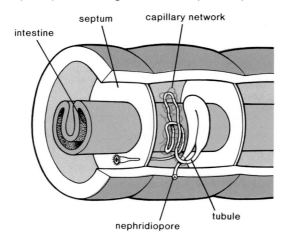

Figure 22.23
Location of Malpighian tubules in the grasshopper. The Malpighian tubules are attached to the gut and have an excretory function. The Malpighian tubules are named for the Italian microscopist Marcello Malpighi (1628–1694).

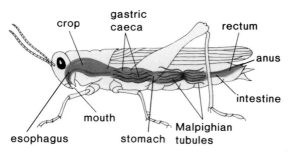

simply flow from the surrounding blood into the tubules before moving into the gut. The water may be reabsorbed, but the uric acid eventually passes out of the gut. Insects living in water and insects eating large quantities of moist food reabsorb little water. But insects in dry environments reabsorb most of the water and excrete a dry, semisolid mass of precipitated uric acid. Mealworms, the larvae of a beetle that lives in dry flour, produce an excretory product that is so dry it actually absorbs water from very humid air.

Human Kidney

The human kidneys are a pair of bean-shaped organs about 11 or 12 cm long that lie at the back of the abdominal cavity. They are a part of the human urinary system (fig. 22.24), which is composed not only of the kidneys that make urine, but also of those organs that conduct urine out of the body. Each kidney is connected to a **ureter,** a duct that carries urine from the kidney to the **urinary bladder,** where it is stored until it is voided from the body through the single **urethra.**

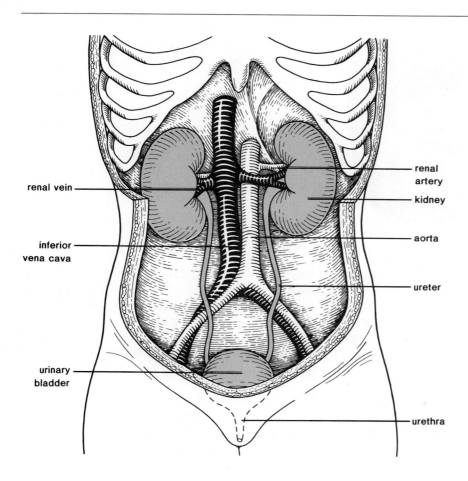

Figure 22.24
Human urinary system. Urine is excreted by the kidneys and passes to the bladder by way of the ureters. After storage in the bladder, it exits when convenient by way of the urethra. These are the only organs in the body to ever contain urine.

renal vein

inferior vena cava

urinary bladder

renal artery

kidney

aorta

ureter

urethra

Figure 22.25
Human kidney structure. *a.* A diagram of a longitudinal section of the human kidney, with an enlargement of one pyramid. *b.* A diagram showing some internal details of the renal blood supply and the location of a nephron.

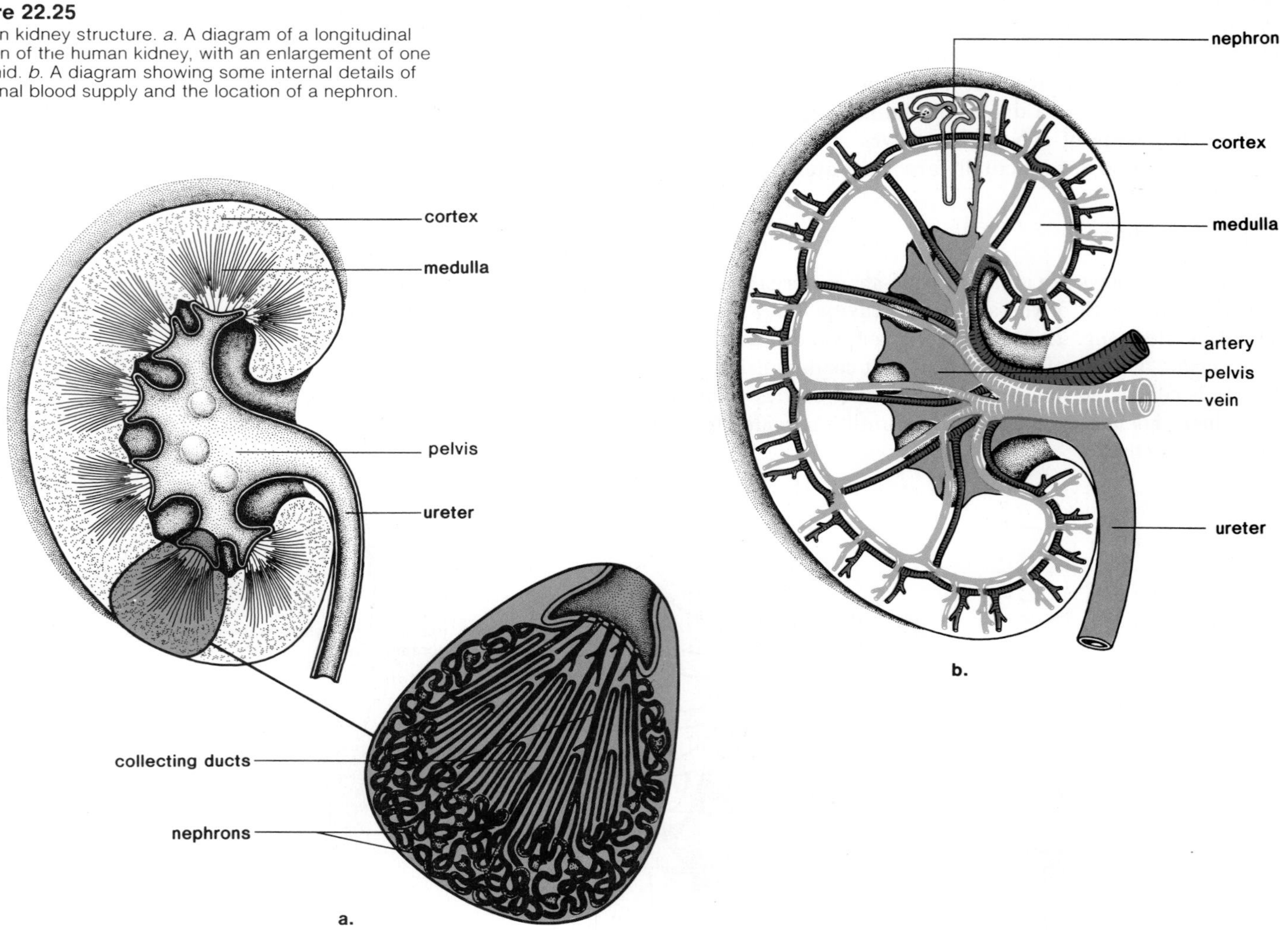

Structure

If a kidney is sectioned longitudinally (fig. 22.25), three major parts can be distinguished. The outer region is the **cortex,** which has a somewhat granular appearance. The **medulla** lies on the inner side of the cortex and is arranged in a group of pyramid-shaped regions, each of which has a striped appearance. The innermost part of the kidney is a hollow chamber called the **pelvis.** Urine formed in the kidney collects in the pelvis before entering the ureter.

Nephrons

Microscopically, each kidney is composed of about one million tiny tubules called **nephrons.** Some nephrons are located primarily in the cortex, but others dip down into the medulla. Each nephron (fig. 22.26) is made of several parts.

Figure 22.26
The human nephron. Each kidney contains over one
million nephrons. The term *proximal* means nearer; *distal*
means farther. In this case, the proximal convoluted tubule
is proximal (closer) to Bowman's capsule and the distal
convoluted tubule is farther from Bowman's capsule.

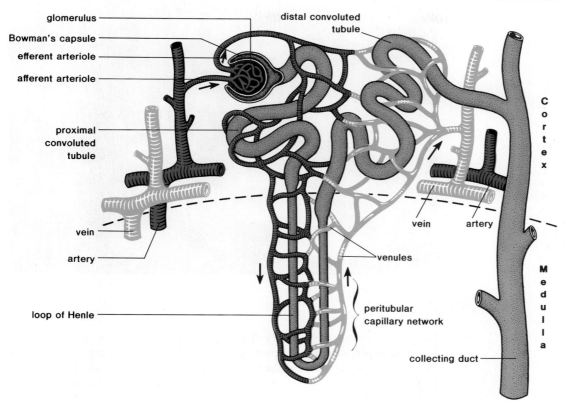

The blind end of the tubule is pushed in on itself to form a cuplike structure
called **Bowman's capsule.** Next, there is a region known as the **proximal** (near
Bowman's capsule) **convoluted tubule** that leads to a narrow U-turn known as
the **loop of Henle.** This is followed by the **distal** (far from Bowman's capsule)
convoluted tubule. Several distal convoluted tubules enter **one collecting duct.**
The collecting duct transports urine down through the medulla and delivers
it to the pelvis. The loop of Henle and the collecting duct give the pyramids
of the medulla their striped appearance (fig. 22.25*a*).

    Each tubule has its own blood supply (fig. 22.26). After the renal artery
leaves the aorta to enter the kidney, it branches into numerous small arteries
(fig. 22.25*b*). These small arteries pass to all parts of the kidney and give off
tiny arterioles, one for each nephron. Each arteriole, called an **afferent arteriole,** divides to form a capillary tuft, the **glomerulus,** which is surrounded by

Figure 22.27

A scanning electron micrograph of a section of kidney cortex, showing a glomerulus (the outer layer of Bowman's capsule has been removed). The holes surrounding the glomerulus are cross sections of tubules.

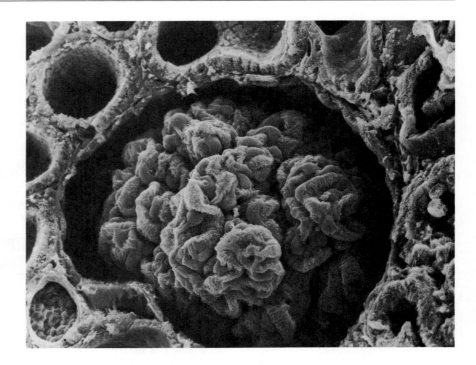

Bowman's capsule (fig. 22.27). The glomerular capillaries drain into an **efferent arteriole,** which subsequently branches into a second capillary network around the tubular parts of the nephron. These capillaries, called **peritubular capillaries,** lead to venules that join the renal vein, a vessel that enters the vena cava.

Urine Formation

Human nephrons function somewhat like earthworm nephridia (p. 490) in that they exchange molecules with the blood. Urine production requires three distinct processes (fig. 22.28):

1. Pressure filtration at Bowman's capsule
2. Reabsorption, including selective reabsorption, at the proximal convoluted tubule in particular
3. Tubular secretion at the distal convoluted tubule in particular

Pressure Filtration

When blood enters the glomerulus, blood pressure is sufficient to cause small molecules, such as nutrients, water, salts, and wastes, to move from the glomerulus to the inside of Bowman's capsule, especially since the glomerular walls are 100 times more permeable than the walls of most capillaries elsewhere in the body. The molecules that leave the blood and enter Bowman's capsule are called the **glomerular filtrate.** Blood proteins and blood cells are too large to be part of this filtrate, and they remain in the blood as it flows into the efferent arteriole. Glomerular filtrate has the same composition as tissue fluid, and if this composition were unaltered in other parts of the nephron, death from loss of water (dehydration), loss of nutrients (starvation), and lowered blood pressure would quickly follow. However, selective reabsorption prevents this from happening.

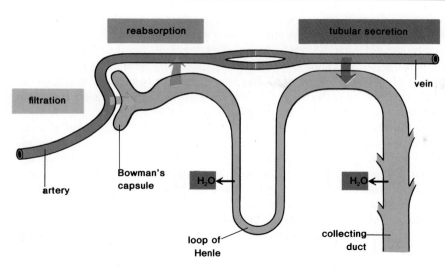

Figure 22.28
Diagram of nephron, showing steps in urine formation: filtration, reabsorption, and tubular secretion. Note also that water enters the tissues at the loop of Henle and the collecting duct.

Reabsorption

Both passive and active reabsorption from the nephron to the blood takes place through walls of the proximal convoluted tubule. Nutrients, water, and even some waste molecules diffuse passively back into the peritubular capillary network. Osmotic pressure further influences the movement of water. The nonfilterable proteins remain in the blood and exert an osmotic pressure. Also, after sodium (Na^+) is actively reabsorbed, chlorine (Cl^-) follows along, and these two ions together increase the osmotic pressure of the blood.

Selective reabsorption of nutrient molecules is brought about by active transport. The cells of the proximal convoluted tubule have numerous microvilli that increase the surface area and numerous mitochondria that supply the energy needed for reabsorption. Reabsorption is selective since only molecules recognized by carrier molecules are actively transported from the tubule to the blood. Usually, almost 100 percent of nutrient molecules are reabsorbed.

The kidney has a threshold level for every substance that is reabsorbed. The threshold level of a substance is its normal level in the blood, and reabsorption occurs until this level is obtained. Thereafter, the substance will not be reabsorbed and will appear in the urine. For example, the threshold level for glucose is .18 g glucose per 100 ml of blood. After this amount is reabsorbed, any excess present in the filtrate will appear in the urine. In contrast to the high threshold level of glucose, urea has a very low threshold level that is quickly reached, and nearly all urea remains in the urine.

Tubular Secretion

Tubular secretion is the means by which other nonfilterable wastes can be added to the fluid as it passes through the tubules. Toxic substances, such as foreign acids and bases that have been absorbed in the gut, are eliminated by tubular secretion. Penicillin and a number of other substances also are excreted in this way. But the process of tubular secretion is not so important to urine formation as are the first two steps studied.

Hypertonic Urine

Reptiles and birds rely primarily on the gut to reabsorb water, but mammals rely on the kidneys. The arrangement of the loop of Henle in relation to the collecting duct enables mammals to excrete a hypertonic urine. A concentration gradient is established in the inner medulla (fig. 22.29) that promotes the reabsorption of water from the descending limb of the loop of Henle and the collecting duct. It now appears that this gradient is due to the extrusion of salt, Na^+Cl^-, by the upper portion of the ascending limb (of the loop of Henle) and by the diffusion of urea from the collecting duct. Because the descending limb of the loop of Henle looses water the ascending limb has a high concentration of salt. You might think this would cause it to take up water, but as figure 22.29 shows, the ascending limb is impermeable to water. Instead it loses salt first by diffusion and then by active extrusion.

The descending limb automatically loses water, but the amount of water that leaves the collecting duct is regulated by hormonal action. The hormone **ADH (antidiuretic hormone),** released by the posterior lobe of the pituitary, increases the permeability of the collecting duct so more water will leave it and be reabsorbed into the blood. If the osmotic pressure of the blood increases, the posterior lobe of the pituitary releases ADH, more water is reabsorbed, and consequently there is less urine. On the other hand, if the osmotic pressure of the blood decreases, the posterior lobe of the pituitary does not release ADH. The resulting impermeability of the collecting duct causes more water to be excreted and more urine to be formed. Drinking alcohol causes diuresis (increased urine flow) because it inhibits the secretion of ADH. Beer drinking also causes diuresis mainly because of increased fluid intake. Drugs called diuretics are often prescribed for high blood pressure. These drugs cause increased urinary excretion and thus reduce blood volume and blood pressure.

Adjustment of pH

The kidneys help maintain the constant pH of the blood within a narrow range, and the whole nephron takes part in this process. The excretion of hydrogen ions and ammonia, together with the reabsorption of sodium and bicarbonate ions, is adjusted to keep the pH within normal bounds. If the blood is acidic, hydrogen ions are excreted in combination with ammonia, while sodium and bicarbonate ions are reabsorbed. This will restore the pH because sodium promotes the formation of hydroxyl ions

$$Na^+ + HOH \longrightarrow Na^+OH^- + H^+$$

while bicarbonate takes up hydrogen ions when carbonic acid is formed.

$$HCO_3^- + H^+ \longrightarrow H_2CO_3$$

If the blood is basic, fewer hydrogen ions are excreted, and fewer sodium and bicarbonate ions are reabsorbed.

Reabsorption and/or excretion of ions (salts) by the kidneys illustrates their homeostatic ability: they maintain not only the pH of the blood but also its osmolarity.

Kidney Failure

The urinary tract is subject to attack by a number of different bacteria. If the infection is localized in the urethra, it is called **urethritis.** If it invades the bladder, it is called **cystitis.** And finally, if the kidneys are affected, it is called **nephritis.** Often nephritis occurs after a strep infection in some other part of the body. Antigen-antibody (p. 450) complexes reach the glomeruli from the arterioles and are trapped there, causing irritation that leads to glomerular inflammation and damage. The glomerular membrane may then become more permeable than usual. Therefore, albumin, white cells, or even red cells may appear in the urine. Also, glomerular damage sometimes leads to blockage of the glomeruli so no fluid moves into the tubules.

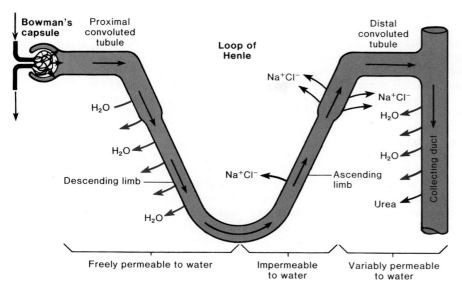

Figure 22.29
Loop of Henle. The ascending limb of the loop of Henle is impermeable to water but not to salt, Na^+Cl^-, which is actively extruded by the upper portion of this limb. (Some salt also diffuses from the lower portion of the limb.) This helps establish a concentration gradient that draws water from the descending limb of the loop of Henle and the collecting duct. Authorities also suggest that the diffusion of urea from the collecting duct into the inner medulla contributes to the concentration gradient needed for mammals to produce a hypertonic urine.

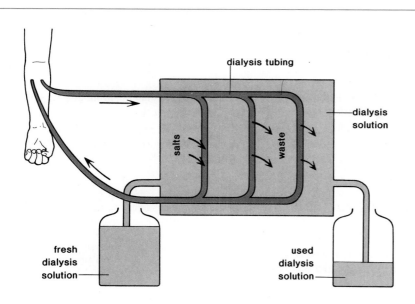

Figure 22.30
Diagram of an artificial kidney. As the patient's blood circulates through dialysis tubing, it is exposed to a solution. Salts enter the blood from the solution and wastes exit from the blood into the solution due to a preestablished concentration gradient.

Artificial Kidney

Persons suffering from either temporary or permanent renal failure can be treated with an artificial kidney machine (fig. 22.30). Blood circulates from an artery in the patient's arm to the kidney machine and returns to a vein. In the process, the patient's blood is passed through a semipermeable, membranous tube that is in contact with a balanced salt solution, or *dialysis fluid*. Substances more concentrated in the blood diffuse into the dialysis fluid. Conversely, substances more concentrated in the dialysis fluid diffuse into the blood. Hence, the artificial kidney can be used either to extract substances from the blood, including waste products or toxic chemicals and drugs, or to add substances, such as ions, to the blood. In the course of a six-hour dialysis, from 50 to 250 g of urea can be removed from a patient, which greatly exceeds the urea clearance of normal kidneys. Therefore, a patient need undergo treatment only about twice a week.

Summary

Some small animals use their entire body surface for gas exchange, but most animals have a localized, special gas-exchange area. Most aquatic animals pass water over gills. On land, insects utilize tracheal systems and vertebrates have lungs. Because lungs are recessed, water loss is minimized, but there is need for ventilation. Some vertebrates use positive pressure but most inhale, using muscular contraction to produce a negative pressure that causes air to rush into the lungs. When the breathing muscles relax, air is exhaled. Birds have a series of air sacs that allow a one-way flow of air over the gas-exchange area.

Table 22.1 lists the structures found in the human respiratory system. Gas exchange takes place between air in lung alveoli and blood in alveolar capillaries. Most oxygen is then bound to hemoglobin, which becomes reduced when oxygen enters the tissues. Carbon dioxide is carried in the blood as the bicarbonate ion. Carbon dioxide and hydrogen ions have been found to promote increased breathing rate.

Animals excrete nitrogenous wastes that differ in the amount of water required to excrete them. Aquatic animals usually excrete ammonia, and land animals excrete either urea or uric acid. Most animals have to adjust their water and ion intake with excretion to maintain the normal concentration in the body fluids.

Animals often have an excretory organ. Earthworm nephridia exchange molecules with the blood in a manner similar to vertebrate kidneys. Malpighian tubules in insects take wastes and water from the hemocoel to the gut, where only water is reabsorbed. Kidneys are a part of the human urinary system (fig. 22.24). Microscopically, each kidney is made up of nephrons, each of which has several parts and its own blood supply (fig. 22.26).

Urine formation requires three steps: pressure filtration at Bowman's capsule, where nutrients, water, and wastes enter the tubule; selective reabsorption when nutrients and some water are reabsorbed; and tubular secretion when additional wastes are added to the tubule. Humans excrete a concentrated urine. The ascending limb of the loop of Henle actively extrudes sodium so the medulla is hypertonic to the contents of the descending limb and the collecting duct. Therefore, water has a tendency to diffuse out of these. ADH, which makes the collecting duct more permeable, is secreted by the posterior pituitary in response to a change in the osmotic pressure of the blood. The kidneys adjust the pH of the blood by excreting or conserving H^+, HCO_3, and Na^+, as appropriate.

Objective Questions

1. In the capillaries of the lamellae found within the filaments making up the gills, the blood flows _____ (choose in the same direction or in the opposite direction) to the flow of water across the gills.

2. In insects, oxygen is carried to the cells by the _____ system, not by the blood.

3. Frogs inhale by a exerting a _____ pressure. Humans inhale when there is a _____ pressure in the lungs.

4. The aortic and carotid bodies constantly monitor the _____ concentration of the blood.

5. Oxygen is carried in the blood by _____ . Carbon dioxide is carried primarily as the _____ ion.

6. You would expect _____ (choose aquatic or terrestrial) animals to excrete uric acid.

7. A saltwater fish excretes _____ across the gills while a freshwater fish excretes a _____ (choose dilute or concentrated) urine.

8. Insects excrete the nitrogenous end product _____ by way of _____ tubules.

9. The _____ take urine from the kidneys to the bladder.

10. In the human nephron, glucose is present in the _____ but absent in the urine because it has been _____ at the proximal convoluted tubule.

Study Questions

1. Compare the respiratory organs of aquatic animals to those of terrestrial animals.
2. Compare inhalation and exhalation among the various vertebrates.
3. Name the parts of the human respiratory system and list a function for each part.
4. Describe in detail the role that hemoglobin plays in the transport of oxygen and carbon dioxide.
5. Relate the types of nitrogenous wastes to the habitats of animals.
6. Contrast the osmotic regulation of a marine bony fish with that of a freshwater fish.
7. Name two invertebrate excretory organs and discuss the function of each.
8. Name the parts of the human kidney and trace the path of urine in the human body.
9. Tell how each part of a nephron contributes to urine formation in humans. Include in your discussion how ADH helps regulate diuresis and blood volume.

Selected Key Terms

diaphragm ('dī ə ‚fram) 479
trachea ('trā kē ə) 481
bronchus ('brän kəs) 481
glottis ('glät əs) 482
larynx ('lar iŋks) 482
bronchiole ('brän kē ‚ōl) 482

alveolus (al 'vē ə ləs) 482
oxyhemoglobin (‚äk si 'hē mə ‚glō bən) 484
nephridium (ni 'frid ē əm) 490
ureter ('yůr ət ər) 491
urethra (yů 'rē thrə) 491

nephron ('nef ‚rän) 492
glomerular filtrate (glə 'mer yə lər fil 'trāt) 494
selective reabsorption (sə 'lek tiv ‚rē əb 'sȯrp shən) 495

Suggested Readings for Part 8

Buisseret, P. D. August 1982. Allergy. *Scientific American.*

Doolittle, R. F. December 1981. Fibrinogen and fibrin. *Scientific American.*

Eckert, R., and Randall, D. 1978. *Animal physiology.* San Francisco: W. H. Freeman.

Feder, M. E., and Burggren, W. W. November 1985. Skin breathing in vertebrates. *Scientific American.*

Guyton, A. C. 1979. *Physiology of the human body.* 5th ed. Philadelphia: W. B. Saunders.

Hickman, C. P., et al. 1982. *Biology of animals.* 3d ed. St. Louis: C. V. Mosby.

Hill, R. W. 1976. *Comparative physiology of animals: An environmental approach.* New York: Harper & Row, Publishers, Inc.

Hole, J. W. 1983. *Human anatomy and physiology.* 3d ed. Dubuque, IA: Wm. C. Brown Publishers.

Human nutrition: Readings from Scientific American. 1978. San Francisco: W. H. Freeman.

Jarvik, R. K. January 1981. The total artificial heart. *Scientific American.*

Kessel, R. G., and Kardon, R. H. 1979. *Tissues and organs: A text-atlas of scanning electron microscopy.* San Francisco: W. H. Freeman.

Koehl, M. A. R. December 1982. The interaction of moving water and sessile organisms. *Scientific American.*

Leder, P. May 1982. The genetics of antibody diversity. *Scientific American.*

Moog, F. November 1981. The lining of the small intestine. *Scientific American.*

Perutz, M. F. December 1978. Hemoglobin structure and respiratory transport. *Scientific American.*

Rose, N. R. February 1981. Autoimmune diseases. *Scientific American.*

Schmidt-Nielson, K. May 1981. Countercurrent systems in animals. *Scientific American.*

Schmidt-Nielson, K. December 1971. How birds breathe. *Scientific American.*

Scrimshaw, N. S., and Young, V. R. September 1976. The requirements of human nutrition. *Scientific American.*

Tonegawa, S. October 1985. The molecules of the immune system. *Scientific American.*

Vander, A. J. 1985. *Human physiology: The mechanisms of body function.* 4th ed. New York: McGraw-Hill.

Zucker, M. B. June 1980. The functioning of blood platelets. *Scientific American.*

PART 9

Integration and Coordination in Animals

T he systems examined in part 8 help maintain a relatively stable internal environment, but for the body to function as an integrated whole, there must be coordination between body parts. Coordination requires communication among body parts that are, in many cases, widely separated from one another. In animals chemical control by hormones provides relatively slower but more sustained regulation of body parts, while nervous mechanisms allow relatively faster but shorter-lived regulation.

A survey of animals shows an increase in complexity of the hormonal and nervous systems when they are arranged from the earliest to the latest evolved. Many animals have specialized receptors that are sensitive to external stimuli and muscle effectors that allow the animal to respond in a noticeable manner. The latter results in the animal having definite behavioral patterns. Just as patterns of anatomy and physiology are inherited and subject to evolutionary change, so are behavioral patterns.

White pelicans. Though awkward looking, pelicans swim, fly, and fish extremely well. They swoop down and scoop up fish to be deposited in distensible pouches beneath the bill.

CHAPTER 23

Nervous and Endocrine Systems

Concepts

1. The activity of the specialized areas within multicellular organisms is coordinated by the nervous and endocrine systems.
   a. Nerve cells, called neurons, conduct the nerve impulse, which is an electrochemical phenomenon involving the movement of ions across the cell membrane.
   b. Transmission from neuron to neuron requires the secretion of substances called neurotransmitters.
   c. The endocrine glands release secretions into the blood, which carries them to target organs.
   d. Although hormones have diverse effects, the manner in which they influence cellular metabolism is similar.
2. The evolutionary history of organisms shows an increase in complexity of the nervous system with time.
   a. The human nervous system is the result of this evolutionary trend.
   b. The human brain has several parts, each of which has specific functions.
   c. The hypothalamus of the brain regulates the blood level of several important hormones.

The ability to respond to stimuli is a characteristic of all living things. In complex animals, such as vertebrates, this ability depends on the nervous, endocrine (hormonal), sensory, and musculoskeletal systems. Working together, the nervous and endocrine systems coordinate all systems of the body to produce effective behavior and to keep the internal environment within safe limits. Because hormones are transported in the blood, it may require seconds, minutes, hours, or even longer for these chemical messengers to produce their effects. The nervous system, on the other hand, communicates rapidly, requiring only thousandths of a second. Both these systems, however, are equally important for continued existence.

Evolution of Nervous Systems

A comparative study of animal nervous systems (fig. 23.1) indicates what steps may have led to the complex nervous system of vertebrates. Coelenterates, such as *Hydra,* have a very simple nervous system that looks like a net of threads extending throughout the body. The net is actually composed of neurons (nerve cells), each one of which has processes that reach out to the other neurons. Impulses spread in all directions along the net that seems primarily to enable *Hydra* to give a local response to a stimulus. External receptor cells that respond to stimuli, like chemicals or pressure, communicate with the nerve net that then causes cells to contract within the two layers making up the body of *Hydra* (fig. 15.7).

Planarians have a more complicated nervous system. They have a head with sense organs and a brain, a concentration of neurons that apparently acts as a relay station between sense organs and muscles. Two longitudinal nerve cords with cross-links between them connect the brain to muscles and other parts of the body.

Segmented worms, such as the earthworm, have an even more elaborate nervous system. The brain and a large **ventral nerve cord** running along the midline of the body form a **central nervous system** that is responsible for integration and coordination. The nerve cord gives off nerves in each segment, and these nerves comprise a so-called **peripheral nervous system.**

When invertebrate nervous systems are compared to vertebrate systems, it is apparent that further evolutionary changes involved vast increases in the number of neurons. For example, an insect's entire nervous system may contain a total of about one million neurons, while a vertebrate nervous system may contain many thousand to several billion times that number.

Human Nervous System

The human nervous system (fig. 23.2), like that of the earthworm, is divided into the central and peripheral nervous systems. The *central nervous system* includes the brain and spinal cord (dorsal nerve cord), which lie in the midline of the body where the skull protects the brain and the vertebrae protect the spinal cord. The *peripheral nervous system* contains both cranial nerves, which originate from the brain, and spinal nerves, which project from either side of

Figure 23.1
Simple nervous system. *a.* The nerve net of *Hydra,* a coelenterate. *b.* The paired nerve cords with cross connections of a planarian, a flat worm, has a ladder appearance. *c.* The earthworm, a segmented worm, has a central nervous system consisting of the brain and a ventral solid nerve cord. It also has a peripheral nervous system consisting of nerves. *d.* A rabbit, like other vertebrates, has a dorsal hollow nerve cord in its central nervous system.

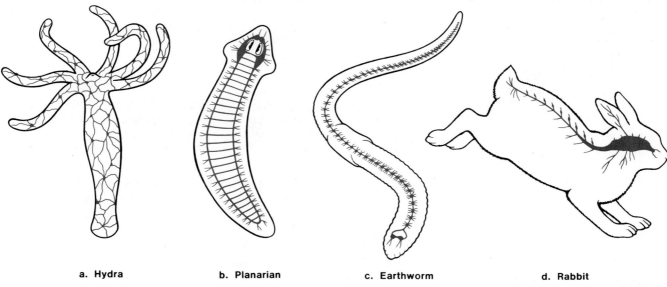

a. Hydra   b. Planarian   c. Earthworm   d. Rabbit

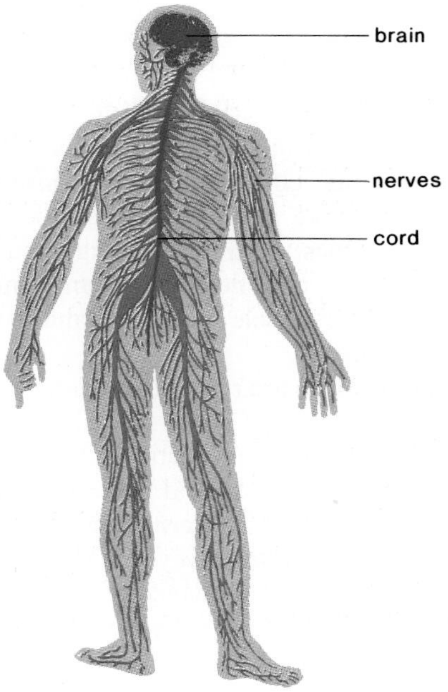

Figure 23.2
The central nervous system (brain and spinal cord) and the nerves of the peripheral nervous system in the human nervous system.

— brain

— nerves

— cord

Figure 23.3
The human nervous system has two main divisions, the central nervous system and the peripheral nervous system. The central nervous system contains the brain and spinal cord. The peripheral nervous system contains nerves, some of which belong to the somatic nervous system and some of which belong to the autonomic nervous system. The autonomic nervous system has two portions, the sympathetic and the parasympathetic systems.

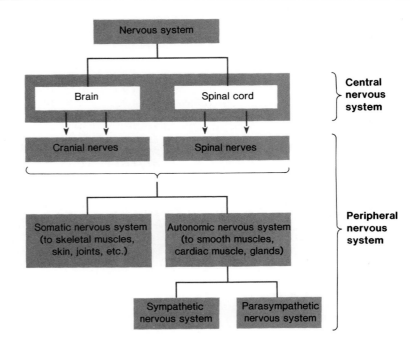

the spinal cord. The peripheral nervous system is further divided into the **somatic division** and the **autonomic (or visceral) division** (fig. 23.3). The somatic division contains nerves that control skeletal muscles, skin, and joints. The autonomic division contains nerves that control internal organs. Before we continue our discussion of the nervous system, we must first examine the anatomy and functions of nerve cells, or neurons.

Neurons

All neurons (fig. 23.4) have three parts: dendrite(s), cell body, and axon. A **dendrite** conducts nerve impulses toward the cell body, and an **axon** conducts nerve impulses away from the cell body. **Sensory neurons,** each with one long dendrite and a short axon, take messages from sense organs to the central nervous system. **Motor neurons,** each with a long axon and short dendrites, take messages away from the central nervous system to muscle fibers or glands. Because motor neurons cause muscle fibers or glands to react, they are said to innervate these structures.

Sometimes sensory neurons are referred to as afferent neurons, and motor neurons are called efferent neurons. These words, derived from Latin, mean running to and running away from, respectively. Obviously, they refer to the relationship of these neurons to the central nervous system.

A third type of neuron, called an **interneuron,** is always found completely within the central nervous system. It conveys messages between various parts, such as from one side of the brain or spinal cord to the other or from the brain to the cord, and vice versa. An interneuron has short dendrites and either a long or a short axon.

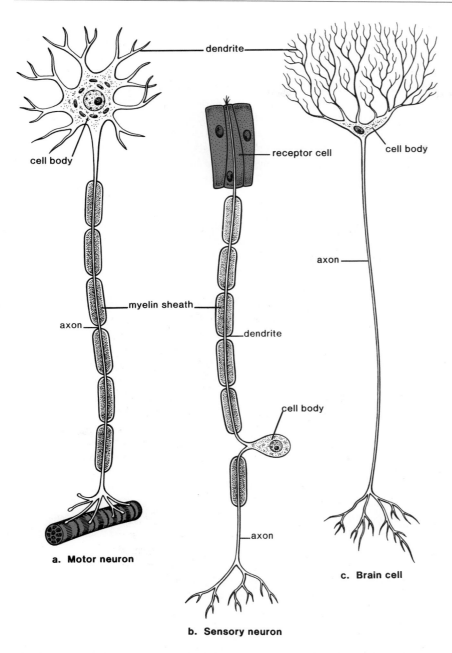

dendrite

cell body

receptor cell

cell body

axon

myelin sheath

dendrite

axon

cell body

axon

a. Motor neuron

b. Sensory neuron

c. Brain cell

The dendrites and axons of neurons are sometimes called fibers or processes. Most long fibers, whether dendrite or axon, are covered by a white **myelin sheath** (fig. 23.4) formed from the membranes of tightly spiraled Schwann cells surrounding these fibers. There are intervals between the Schwann cells, called **nodes of Ranvier** (fig. 23.8), that greatly speed up the conduction of nerve impulses.

Figure 23.5
The original nerve impulse studies utilized giant squid axons and a voltage recording device known as an oscilloscope (fig. 23.6). This drawing shows the location of the axons in the squid. A microelectrode can be inserted in these oversized fibers.

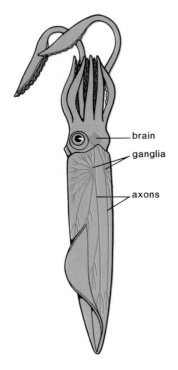

brain

ganglia

axons

Figure 23.6

Scientist working at an oscilloscope, the electrical recording device that measures changes in voltage whenever an electrode is placed on or inserted in a neuron.

Nerve Impulse

An Italian investigator, Luigi Galvani, discovered in 1786 that a nerve could be stimulated by an electric current. But it was realized later that the speed of the nerve impulse is too slow to be simply a flow of electrons or current within a nerve fiber. In the early 1900s Julius Bernstein at the University of Halle, Germany, suggested that the nerve impulse is an electrochemical phenomenon involving the movement of unequally distributed ions on either side of a nerve cell membrane. It was not until 1939 that investigators developed a technique that enabled them to substantiate this suggestion. A. L. Hodgkin and A. F. Huxley, English neurophysiologists, received the Nobel Prize in 1963 for their published results. They and a group of researchers headed by K. S. Cole and J. J. Curtis at Woods Hole, Massachusetts, managed to insert a very tiny electrode into the giant axon of the squid *Loligo* (fig. 23.5). This internal electrode was then connected to a type of voltmeter called an **oscilloscope** (fig. 23.6), which shows a trace or pattern indicating a change in voltage with time. **Voltage** is a measure of the electrical potential difference between two points, which in this case is the difference between two electrodes, one placed inside and another placed outside the axon. When the axon is not conducting an impulse, the oscilloscope records a potential difference across the membrane equal to about -60 mV (millivolts) (fig. 23.7). This is the **resting potential** because the axon is not conducting an impulse.

The existence of a resting potential can be correlated with a difference in ion distribution on either side of the axomembrane. As figure 23.7*b* shows, there is a concentration of sodium ions (Na^+) outside the axon and a concentration of potassium ions (K^+) inside the axon. There are also large organic negative ions in the axoplasm (cytoplasm within the axon) that cause the resting fiber to be negative inside compared to outside the axon. The selectively permeable nature of the axomembrane holds these organic ions inside. The unequal distribution of sodium and potassium is maintained by a form of energy-requiring active transport called the **sodium/potassium pump** (fig. 23.9).

If the axon is stimulated to conduct a nerve impulse by either an electric shock, a rapid change in pH, or a pinch, there is a rapid change in the polarity recorded as a trace on the oscilloscope screen. This change, called the action potential, has an upswing and a downswing (fig. 23.7*c*). As the trace goes from -60 mV to $+40$ mV (upswing), experiments indicate that sodium ions are rapidly moving to the inside of the axon. It is now known that stimulation causes the membrane to suddenly become permeable to sodium (fig. 23.7*d*) because certain pores called "sodium gates" have opened. As the trace goes from $+40$ mV back to -60 mV (downswing), experiments indicate that potassium ions are moving to the outside of the axon, meaning that the membrane has suddenly become permeable to potassium or that the "potassium gates" have opened (fig. 23.7*d*).

Notice that while a nerve impulse is an exchange of ions, the oscilloscope records this exchange as a change in polarization. **Depolarization** occurs when the inside of the fiber goes from negative to positive, as sodium ions enter, and **repolarization** occurs when the inside returns to negative again, as potassium ions exit.

The oscilloscope records from only one location of a fiber, but actually the nerve impulse travels along the length of a fiber—jumping from one node of Ranvier to another if the fiber is myelinated (fig. 23.8). This can account for a nerve impulse speed of over 250 mi/hr in large myelinated fibers.

Figure 23.7
The resting and action potentials. The oscilloscope (a.) reads a resting potential of −60 mV due to (b.) the presence of large negative organic ions in neuroplasm. The action potential (c.) is a change in polarity that may be explained in terms of (d.); first, the movement of Na⁺ to the inside, and second, the movement of K⁺ to the outside of the axon.

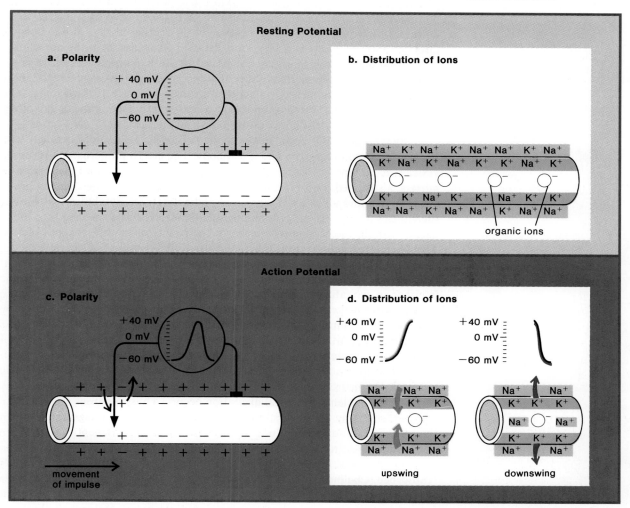

Figure 23.8
The speed of the nerve impulse in myelinated fibers can be accounted for by the fact that it jumps from one node of Ranvier to the next.

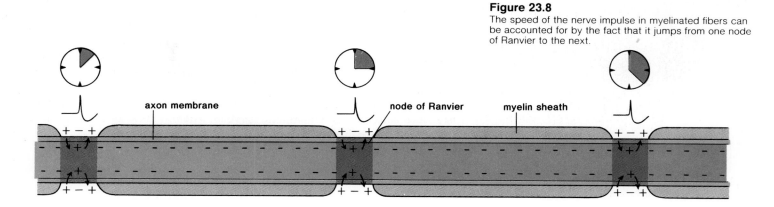

Figure 23.9

Sodium/potassium pump moves sodium (Na^+) to outside the cell and potassium (K^+) to inside the cell by means of the same type protein carrier. Presumably, after Na^+ is taken up by the carrier (a.), a phosphate group (P) (b.) attaches to the carrier as ATP splits. This changes the shape of the carrier so that Na^+ is released and K^+ is taken up (c.). After the phosphate group (P) is released (d.), so is the K^+ (e.).

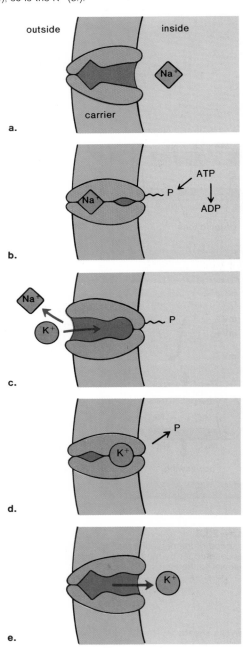

A fiber can conduct a nerve impulse several hundred times because only a very small number of ions are exchanged with each impulse; nevertheless, whenever the fiber rests, the sodium/potassium pump returns the sodium to the outside and the potassium to the inside (fig. 23.9).

Transmission across a Synapse

In 1897 Sir Charles Sherrington of Oxford and other scientists noted two important aspects of nerve impulse transmission between neurons. First, an impulse passing from one vertebrate nerve cell to another always moves in one direction. Second, there is a very short delay in transmission of the nerve impulse from one neuron to another. This led to the hypothesis that there is a minute space between neurons. Sherrington called the region where the impulse moves from one neuron to another a **synapse** (fig. 23.10), meaning "to clasp." A synapse has three components: a presynaptic membrane, a gap now called the **synaptic cleft,** and a postsynaptic membrane.

Synapses usually occur between the end of an axon and a dendrite or between the end of an axon and a cell body. The electron microscope reveals vesicles at the ends of axons (fig. 23.10b), now known to store **neurotransmitters,** also called transmitter substances. When nerve impulses reach a presynaptic membrane, calcium ions enter and activate microtubules that draw the vesicles up against the presynaptic membrane. The vesicles then fuse (fig. 23.11) with the membrane and discharge their contents into the synaptic cleft. The discharged neurotransmitter molecules diffuse across the synaptic cleft and attach in a lock-and-key manner to the postsynaptic membrane at *receptor sites*. This reception alters the potential of the postsynaptic membrane in either an excitatory or inhibitory direction (fig. 23.10c). An excitatory neurotransmitter makes the interior potential less negative, whereas an inhibitory neurotransmitter makes the interior potential more negative. Whether or not a neuron fires (initiates a nerve impulse) depends on the summary effect of all neurotransmitters received. If the amount of excitatory neurotransmitter received is sufficient to overcome the amount of inhibitory neurotransmitter received, the neuron fires. If the amount of excitatory neurotransmitter received is insufficient, only local excitation occurs. Thus, synapses are regions where a "summing up" occurs; therefore, they are regions of integration where the nervous system can fine tune its response to the environment.

Once neurotransmitters have been received, they are either immediately reabsorbed or first quickly broken down by enzymatic action before they are reabsorbed. **Cholinesterase** is an enzyme produced by the dendrite side of the synapse that destroys a neurotransmitter called acetylcholine after it has performed its function. A single molecule of cholinesterase catalyzes the breakdown of 25,000 molecules of acetylcholine per second. The breakdown products are then taken up into the presynaptic neuron and used in the synthesis of new acetylcholine molecules.

Neurotransmitters

Acetylcholine and **norepinephrine** are both excitatory neurotransmitters in the peripheral nervous system. Acetylcholine is the neurotransmitter for both the somatic nervous system and the parasympathetic system, while norepinephrine is also used by the sympathetic nervous system. (These divisions of the nervous system are discussed on p. 513).

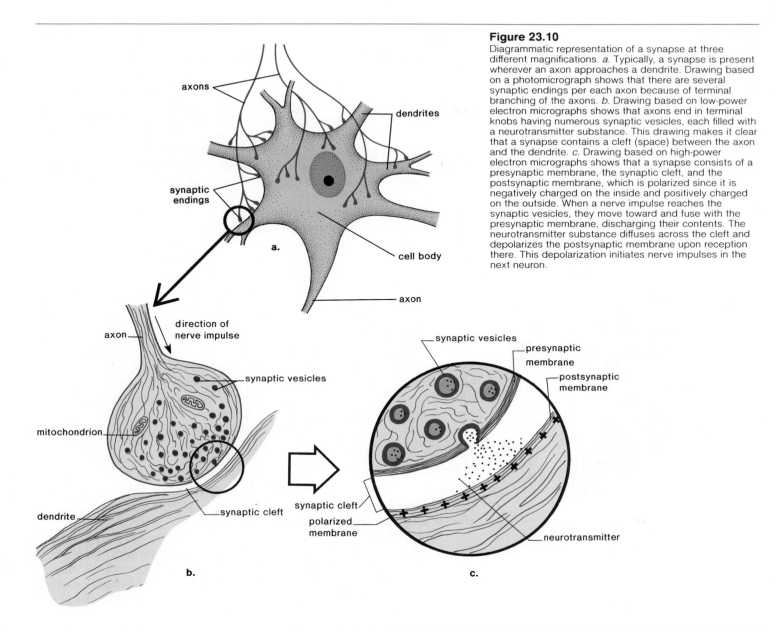

Figure 23.10
Diagrammatic representation of a synapse at three different magnifications. *a.* Typically, a synapse is present wherever an axon approaches a dendrite. Drawing based on a photomicrograph shows that there are several synaptic endings per each axon because of terminal branching of the axons. *b.* Drawing based on low-power electron micrographs shows that axons end in terminal knobs having numerous synaptic vesicles, each filled with a neurotransmitter substance. This drawing makes it clear that a synapse contains a cleft (space) between the axon and the dendrite. *c.* Drawing based on high-power electron micrographs shows that a synapse consists of a presynaptic membrane, the synaptic cleft, and the postsynaptic membrane, which is polarized since it is negatively charged on the inside and positively charged on the outside. When a nerve impulse reaches the synaptic vesicles, they move toward and fuse with the presynaptic membrane, discharging their contents. The neurotransmitter substance diffuses across the cleft and depolarizes the postsynaptic membrane upon reception there. This depolarization initiates nerve impulses in the next neuron.

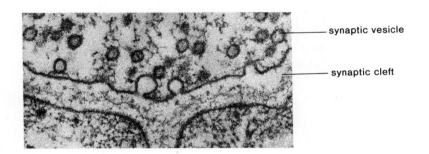

Figure 23.11
Electron micrograph showing synaptic vesicles apparently opening into the synaptic cleft.

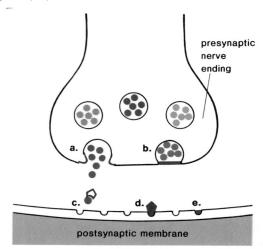

presynaptic nerve ending

a.

b.

c.   d.   e.

postsynaptic membrane

Within the central nervous system, the situation is a bit more complicated. **GABA** (gamma aminobutyric acid) appears to be the most abundant neurotransmitter, but dopamine, serotonin, norepinephrine, and acetylcholine are also present. (Probably many compounds, some as yet unknown, are neurotransmitters.) Various illnesses have been associated with the lack of a particular transmitter. For example, Parkinson's disease, characterized by slow movements, lack of control over muscle actions, and tremors, is known to be caused by a lack of dopamine. Investigators have also found that victims of Alzheimer's disease, characterized by loss of mental capabilities, did not produce as much acetylcholine as other patients. The more serious the outward manifestations of Alzheimer's disease, the less acetylcholine produced.

Effects of Drugs at Synapses

Research on neurotransmitters has revealed how psychoactive drugs affect the nervous system. For example, certain drugs may alter the synthesis of a specific neurotransmitter, its packaging into vesicles, its release, its reception, or its breakdown and recycling at the synaptic cleft (fig. 23.12).

Tranquilizers, such as valium, interact with GABA receptors. Stimulants, such as cocaine and amphetamines, enhance the action of norepinephrine, perhaps by promoting its release or by preventing its resorption.

Hallucinogens, such as LSD and mescaline, are believed to interact with the receptors for the neurotransmitter serotonin. The list could be longer, but the point is that understanding the functions of the synapse enables biologists to better understand the actions of many drugs. Some drugs that act on synapses may prove to be powerful tools for treating neurological disorders and perhaps even various types of mental illnesses.

Nerves

Nerves are a part of the peripheral nervous system (PNS). They consist of long dendrites and/or long axons. There are no cell bodies in nerves because these are found only in the central nervous system (CNS) or in ganglia. **Ganglia** are collections of cell bodies within the PNS.

Humans have twelve pairs of **cranial nerves** and thirty-one pairs of **spinal nerves.** Cranial nerves are either *sensory nerves* (having long dendrites of sensory neurons only), *motor nerves* (having long axons of motor neurons only), or *mixed nerves* (having both long dendrites and long axons). All cranial nerves, except the vagus, control the head, neck, and face. The **vagus nerve** innervates the internal organs.

Spinal nerves are all mixed nerves. Their arrangement shows that humans are segmented animals: there is a pair of spinal nerves for each segment. Spinal nerves project from the spinal cord (fig. 23.13). The spinal cord extends longitudinally along the back where it is protected by vertebrae. The cord contains a tiny **central canal** filled with cerebrospinal fluid; gray matter consisting of cell bodies and short fibers; and white matter consisting of myelinated long fibers.

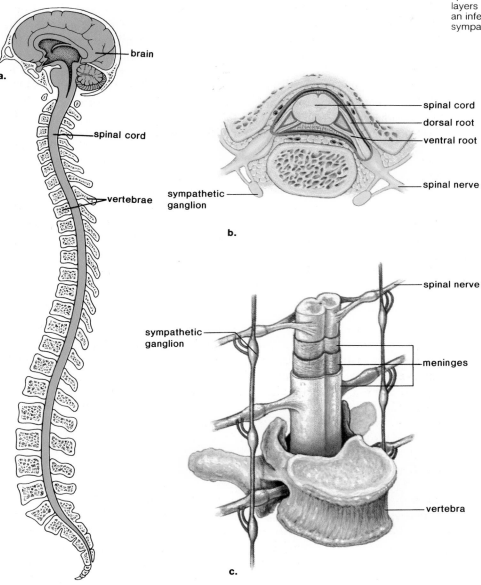

Figure 23.13
Diagrammatic representation of the spinal cord. *a.* This longitudinal section shows that the central nervous system consists of the brain and the spinal cord. The brain is protected by the skull and the spinal cord is protected by the vertebrae. *b.* This cross section of the spinal cord shows that the spinal cord gives off a dorsal root and a ventral root that join to form a spinal nerve on each side of the cord. The human body has a total of thirty-one pairs of spinal nerves. *c.* This segment of the spinal cord shows that the cord is protected by three layers of tissue called the meninges. Spinal meningitis is an infection of these layers. Note the placement of sympathetic ganglia, also shown in figure 23.16.

brain

a.

spinal cord

vertebrae

spinal cord

dorsal root

ventral root

sympathetic ganglion

spinal nerve

b.

spinal nerve

sympathetic ganglion

meninges

vertebra

c.

Simple Reflex

Figure 23.14 diagrams a spinal nerve within the *somatic division* of the nervous system. It can be used to trace the path of a **simple reflex,** a quick reaction to an environmental stimulus that does not require the direct involvement of the brain. Simple reflexes, such as withdrawing the hand from a hot stove, are unlearned and do not require consciousness. First, receptors in the skin generate nerve impulses that move along dendrites of sensory neurons to cell bodies within a ganglion located just outside the dorsal (toward the back) portion of the cord. From here the impulses travel into the cord where they pass to many interneurons, some of which stimulate the dendrites of motor neurons. The axons of these neurons leave the cord ventrally (toward the front). The impulses travel along these axons to muscle fibers that then contract. Now the individual withdraws his hand from the stove. Other responses may also be observed. For example, the person may look in the direction of the stimulus, stamp a foot, or utter appropriate exclamations. This whole series of responses is possible because the stimulated interneurons take impulses to all parts of the CNS, including the brain that makes the individual conscious of the stimulus and his reaction to it.

Figure 23.14

Diagram of a reflex arc shows the detailed composition of a spinal nerve. When the receptors in the skin are stimulated, nerve impulses move along a sensory neuron to the cord. (Note that the cell body of a sensory neuron is in a ganglion outside the cord.) The nerve impulses are picked up by an interneuron, which lies completely within the cord, and passed to the dendrites and cell body of a motor neuron that lies ventrally within the cord. The nerve impulses then move along the axon of the motor neuron to an effector, such as a muscle fiber that contracts. The brain receives information concerning sensory stimuli by way of other interneurons, with long fibers in tracts that run up and down the cord within the white matter.

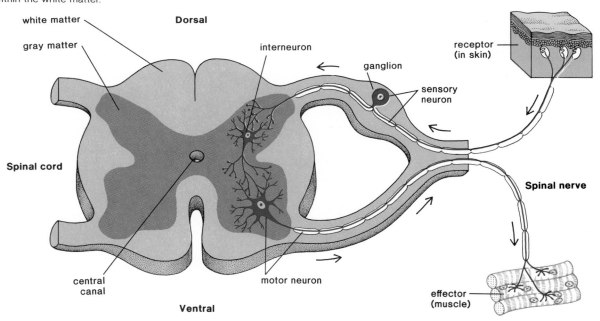

Autonomic Nervous System

The autonomic nervous system is a part of the PNS. It is made up of motor neurons that control the internal organs automatically and usually without the need for conscious intervention. There are two divisions to the autonomic nervous system (table 23.1): the **sympathetic system** and the **parasympathetic system.** Both of these (1) function automatically and usually subconsciously in an involuntary manner; (2) innervate all internal organs; and (3) use two motor neurons for each impulse. Notice in figure 23.15 that the cell body of the first motor neuron is located in the CNS, and the cell body of the second motor neuron is located in a ganglion. The axon that occurs before the ganglion is called the preganglionic fiber, and the axon that occurs after the ganglion is called the postganglionic fiber.

Table 23.1 Sympathetic versus Parasympathetic System

Sympathetic	Parasympathetic
Fight or flight	Normal activity
Norepinephrine is neurotransmitter	Acetylcholine is neurotransmitter
Postganglionic fiber is longer than preganglionic	Preganglionic fiber is longer than postganglionic
Preganglionic fiber arises from middle portion of cord	Preganglionic fiber arises from brain and lower portion of cord

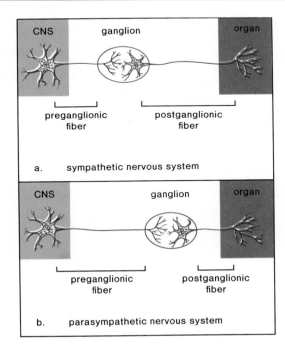

a.   sympathetic nervous system

b.   parasympathetic nervous system

Figure 23.15
Diagram illustrating that both the sympathetic (*a.*) and parasympathetic (*b.*) divisions of the autonomic nervous system use two motor neurons to carry impulses from the CNS to any particular internal organ. Note that the sympathetic and parasympathetic systems differ in the length of the preganglionic and postganglionic fibers.

Preganglionic fibers of the sympathetic nervous system arise from the middle, or thoracic-lumbar, portion of the cord and almost immediately terminate in ganglia that lie near the cord (fig. 23.16). The sympathetic nervous system is especially important during emergency situations and may be associated with "fight or flight." For example, it inhibits the digestive tract but dilates the pupil of the eye, accelerates the heartbeat, and increases the breathing rate. It is not surprising, then, that the neurotransmitter released by the postganglionic axon is norepinephrine, a chemical close in structure to epinephrine (adrenalin), a well-known heart stimulant.

Figure 23.16
The sympathetic and parasympathetic divisions of the autonomic nervous system and the organs they innervate. The sympathetic division contains nerves that arise from the middle, or thoracic-lumbar, portion of the cord, and the parasympathetic division contains nerves that arise from the brain and sacral portion of the cord. The preganglionic fibers of the sympathetic nerves synapse in ganglia that lie along the cord, but the preganglionic fibers of the parasympathetic nerves synapse in ganglia that lie near or within the organ they innervate. Usually an organ is stimulated by both divisions. In general, the sympathetic division produces those reactions associated with stressful times. The parasympathetic system produces reactions associated with rest and relaxation.

Sympathetic Division

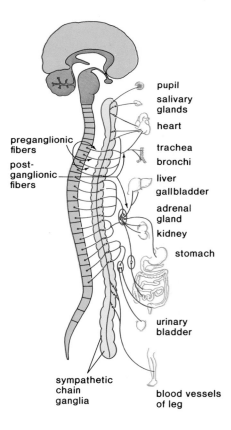

Parasympathetic Division

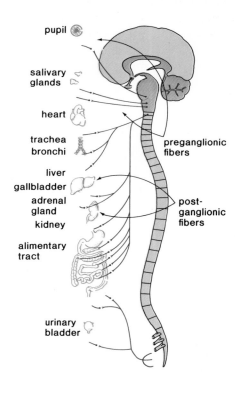

In the parasympathetic nervous system the preganglionic fibers arise from the brain and the bottom (sacral) portion of the cord system. This system is therefore referred to as the craniosacral portion of the autonomic nervous system. The preganglionic fibers terminate in ganglia that lie near or within the organ (fig. 23.16). The parasympathetic system, sometimes called the "housekeeper system," promotes all the internal responses we associate with a relaxed state; for example, it causes the pupil of the eye to contract, promotes digestion of food, and retards the heartbeat. The neurotransmitter utilized by the parasympathetic system is acetylcholine.

The Brain

The brain (fig. 23.17) is protected by the skull and wrapped in three protective membranes known as **meninges** that also surround the spinal cord; spinal meningitis is a well-known infection of these coverings. The interior of the brain contains spaces known as ventricles. Like the central canal of the spinal cord, these are filled with cerebrospinal fluid.

In the same manner as the spinal cord, the brain contains gray matter and white matter. The **gray matter** is made up of cell bodies and short fibers. The **white matter** consists of the long myelinated fibers of interneurons that connect all parts of the brain and spinal cord. These long fibers of interneurons are grouped together in bundles called **tracts.** Longitudinal tracts passing between the brain and spinal cord cross so that the left side of the brain controls skeletal muscles on the right side of the body, and vice versa.

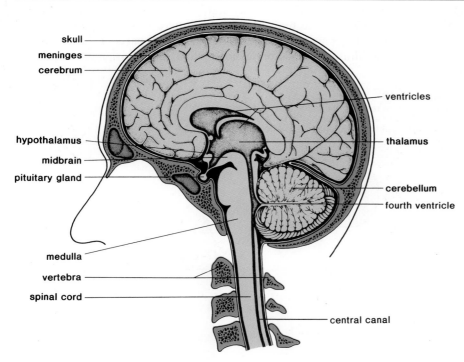

Figure 23.17
The human brain encased in the bony cranium and the spinal cord encased in the vertebral column. Note the central canal in the spinal cord, and the ventricles, which are cavities inside the brain. The spinal cord's central canal and the brain's ventricles are filled with cerebrospinal fluid.

Figure 23.18

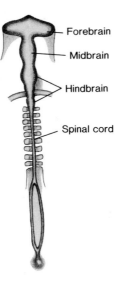

Figure 23.18
The major brain regions—forebrain, midbrain, and
hindbrain—as they appear early in the development of a
vertebrate embryo.

- Forebrain
- Midbrain
- Hindbrain
- Spinal cord

Parts of the Brain

The human brain, like that of all vertebrates, can be divided into three major sections: the hindbrain, the midbrain, and the forebrain (fig. 23.18). The midbrain is largely a relay station that lost its importance in humans as the forebrain increased markedly in size and importance.

Hindbrain

The medulla oblongata and the cerebellum are located in the hindbrain. The **medulla oblongata** (fig. 23.17) contains centers for heartbeat, breathing, and vasoconstriction (blood pressure), and reflex centers for vomiting, coughing, sneezing, hiccoughing, and swallowing. It also functions as a pathway for sensory and motor impulses that are traveling between higher brain centers and the spinal cord.

The **cerebellum** is a bilobed structure that, in general appearance, resembles a butterfly. The cerebellum coordinates muscles by integrating impulses received from the cerebrum. It ensures that skeletal muscles produce smooth and graceful motions. It also maintains normal muscle tone and transmits impulses that maintain posture and balance after receiving information from the inner ear about the position of the body.

Forebrain

The forebrain includes the hypothalamus, thalamus, and cerebrum (fig. 23.17). The **hypothalamus** helps regulate homeostasis and contains centers for hunger, sleep, thirst, body temperature, water balance, and blood pressure. It also contains centers for pleasure, reproductive behavior, hostility, and pain. As discussed on page 524, the hypothalamus controls the pituitary gland.

The **thalamus** is a primary relay station in the brain because sensory pathways traveling up the brain stem synapse here before passing into the cerebrum. The thalamus has connections with various parts of the cerebrum by way of the diffuse thalamic projection system, an extension of the **ARAS**

Figure 23.19
The ARAS (the reticular formation and its extensions) has connections to various parts of the brain and directs the cerebral cortex to pay attention to particular stimuli.

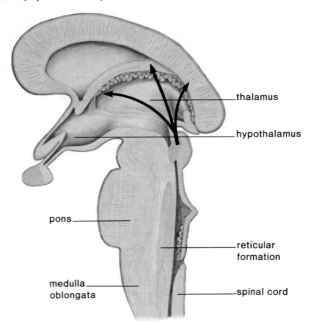

- thalamus
- hypothalamus
- pons
- reticular formation
- medulla oblongata
- spinal cord

(ascending reticular activating system). This system is a complex network of cell bodies and fibers that extends from the medulla to the thalamus and beyond (fig. 23.19). The ARAS, along with the thalamic projection system, is believed to alert higher centers to be prepared to receive information. When one sleeps, the ARAS does not function actively, and brain waves characteristic of sleep appear. The thalamus is sometimes called the gatekeeper to the cerebrum because it monitors the sensory data to be sent on to the cerebrum. For this reason, it is possible to ignore extraneous sensory information unless it becomes important for us to pay attention to it.

The **cerebrum,** which is divided into the right and left **cerebral hemispheres,** is the most well-developed area of the brain. A comparative study of vertebrates (fig. 23.20) indicates a progressive increase in the size of the cerebrum from fishes to humans. In fishes and amphibians, the cerebrum largely

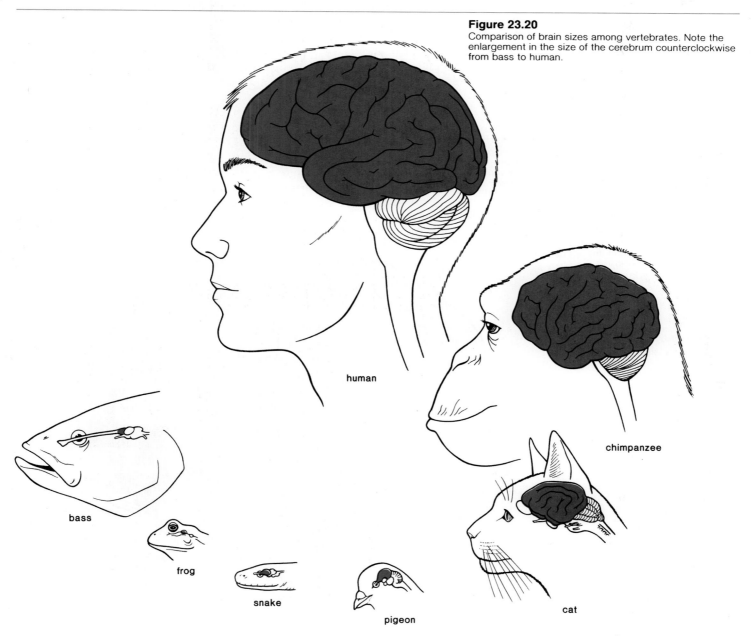

Figure 23.20
Comparison of brain sizes among vertebrates. Note the enlargement in the size of the cerebrum counterclockwise from bass to human.

human

chimpanzee

bass

frog

snake

pigeon

cat

Figure 23.21
The convoluted cortex is divided into the lobes noted here. Each lobe has a particular function. The frontal lobes (*right and left*) initiate motor control over muscles and also are centers for higher modes of thought. The other lobes are involved in receiving sensory information. The limbic system (color) includes portions of the frontal lobes, temporal lobes, the thalamus, and the hypothalamus. Among other functions, the limbic system is thought to produce endorphins, internal opiates that can cause a feeling of euphoria and reduce the threshold for pain. (See also fig. 23.27.)

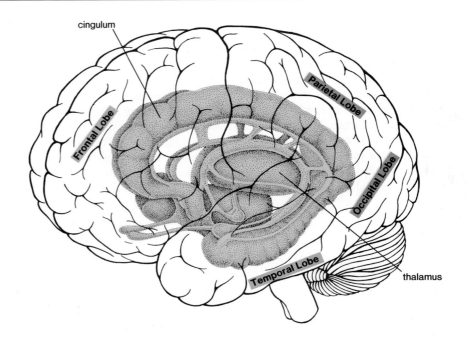

Figure 23.22
The cerebrum can be mapped because each area is associated with a particular function. Over and beyond that, there are those who think that the left and right cerebrum themselves have certain strengths. It is said that the left half of the cerebrum is better at verbal, logical, quantitative, and analytical thinking, and the right half is more visually and spatially oriented, allowing more artistic, musical, emotional, and creative expression. Supposedly, one half is stronger than the other in some individuals. Data to support this difference in function and dominance are not conclusive.

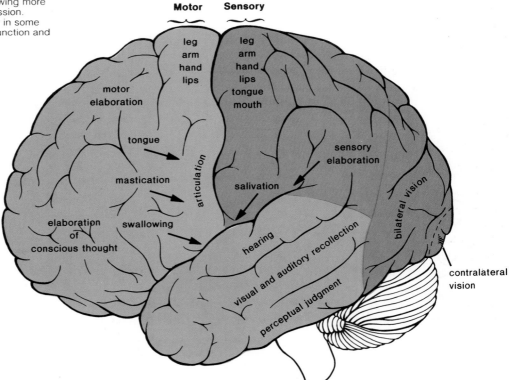

has an olfactory function, but in reptiles, a portion of the cerebrum receives information from other parts of the brain and coordinates sensory data and motor functions. It is this outer portion of the cerebrum, called the **cortex,** that is highly convoluted in mammals, especially humans. The cortex has four lobes: frontal, parietal, temporal, and occipital (fig. 23.21). Different functions are associated with each lobe. For example, the **frontal lobe** controls motor functions and permits us to consciously control our muscles. The **parietal lobe** receives information from receptors located in the skin, such as those for touch, pressure, and pain. The **occipital lobe** interprets visual inputs. The **temporal lobe** possesses sensory areas for hearing and smelling.

Certain areas of the cerebrum have been "mapped" in great detail (fig. 23.22). We know which portions of the frontal lobe control various parts of the body and which portions of the parietal lobe receive sensory information from these same parts. Each of the four lobes of the cerebral cortex contains an *association area* that receives information from the other lobes and associates it into higher, more complex levels of consciousness. These areas are concerned with intellect, artistic and creative ability, learning, and memory processes, which are also believed to require the limbic system (fig. 23.21).

Limbic System

The limbic system is an interconnected group of brain parts that includes portions of the frontal lobes, temporal lobes, thalamus, and hypothalamus, as well as pathways that connect all the parts. Stimulation of different areas of the limbic system causes the subject to experience rage, pain, pleasure, or sorrow. The pleasure centers are sometimes called self-stimulation centers because subjects taking part in an experiment will repeatedly stimulate these centers if given the opportunity. The actual feelings that go along with emotional states may be generated by the frontal lobes, while the manifested behavior may be determined by the balance of excitatory and inhibitory messages arriving at a limbic integrating center in the hypothalamus.

Endocrine System

Hormones have traditionally been defined as substances produced by one part of the body that have an effect on specific target cells located in a different part of the body. In complex animals, hormones are secreted by organs called **endocrine glands.**[1] To distinguish the endocrine system from the nervous system, it is said that the endocrine system works slowly by means of hormones that are distributed by the bloodstream, and the nervous system is said to react quickly by means of nerve impulses that pass from cell to cell.

This distinction is becoming less and less clear-cut as investigators discover that various tissues in the body produce chemical messengers that possibly are for use within the organ itself. We have known for a long time that nerve cells produce neurotransmitter substances that stimulate nearby cells. Now it has been found that brain cells have receptors for the hormone insulin, previously thought to be produced by the pancreas only. Since pancreatic insulin normally has little access to the brain, it would seem that the insulin found there is being made by brain cells. This insulin is acting as a chemical messenger between the brain cells. Similarly, it seems that cells in other organs are also producing chemical messengers that are sent to neighboring cells

---

[1]The term endocrine gland (Greek *endo* meaning "within") indicates that these glands secrete their products directly into the bloodstream, while exocrine (from the Greek *exo,* "outside") glands secrete their products outside the body through ducts. Sweat glands and salivary glands are exocrine glands.

Figure 23.23

a. The substances we call hormones today may have at one time been chemical messengers between like cells. Chemical messengers are sent between like tissue cells (*1*), including nerve cells (*2*). The chemical messengers that are sent between nerve cells include neurotransmitter substances. *b.* In vertebrates, hormones are produced by specialized glands and are distributed to target organs by way of the circulatory system. Tissue cells (*1*), including nerve cells (*2*), produce hormones. For example, the hypothalamus produces hormones that control the anterior pituitary.

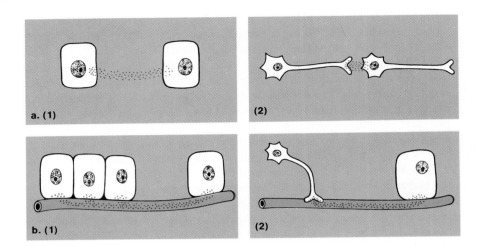

only (fig. 23.23*a*). Perhaps the chemicals we term hormones today were simply chemical messengers between like cells in the first multicellular organisms to evolve. With the evolution of complex animals, specialized glands began to produce hormones and to secrete them into the bloodstream (fig. 23.23*b*).

Even so, there are many other tissues besides these specialized glands that produce hormones. For example, we have already mentioned the production and secretion of gastrin, secretin, and CCK-PZ by the intestinal organs. To facilitate matters, however, we will limit our discussion in this chapter to some major human endocrine glands (fig. 23.25 and table 23.2). Aside from humans, hormones in invertebrates, particularly in insects, have also been studied in great detail. The reading on page 606 discusses the role of hormones in insect development.

Mechanism of Action

In humans, hormones are organic substances that fall into two basic categories: (1) amino acids, polypeptides, or proteins, and (2) steroid hormones, which are complex rings of carbon and hydrogen atoms (fig. 3.25). When hormones of the first types are received by a cell, they bind to specific receptor sites (fig. 23.24*a*) in the membrane. This **hormone-receptor complex** activates an enzyme that produces **cAMP** (cyclic adenosine monophosphate), a compound made from ATP that contains only one phosphate group attached to adenine at two locations. The cAMP then activates the enzymes of the cell to carry out their normal functions.

Unlike peptide hormones, steroid hormones pass through the cell membrane with no difficulty because they are relatively small and lipid soluble. There are receptor molecules present in the cytoplasm (fig. 23.24*b*). After the hormone has combined with the receptor, the hormone-receptor complex moves into the nucleus where it binds with chromatin at a location that promotes activation of a particular gene. Protein synthesis follows. In this manner, steroid hormones lead to protein synthesis.

Figure 23.24
Cellular activity of hormones. *a.* Peptide hormones
combine with receptors located in the cell membrane.
This promotes the production of cAMP, which in turn leads
to activation of a particular enzyme. *b.* Steroid hormones
pass through the membrane to combine with receptors,
and the complex is believed to activate certain genes,
leading to protein synthesis. Some athletes take steroids
to increase the size of their muscles. Unfortunately, the
side effects can be disasterous: Males are feminized,
women are masculinized, and both can suffer from
cardiovascular and liver disease.

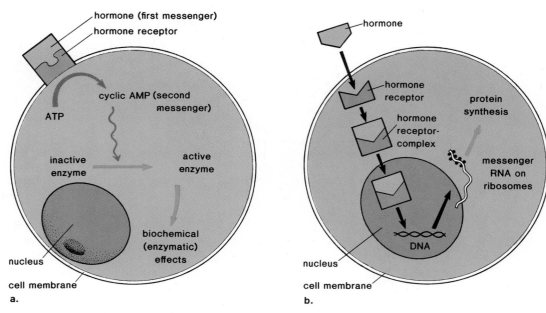

Human Glands and Hormones

Figure 23.25 illustrates the location of the major endocrine glands in the body. Table 23.2 lists these glands and the hormones produced by each.

Pituitary Gland

The pituitary gland, which has two portions called the **anterior pituitary** and the **posterior pituitary**, is a small gland about the size of a garden pea that lies at the base of the brain. Both portions of the pituitary have an anatomical connection to the hypothalamus (fig. 23.26).

Posterior Pituitary

The hormones secreted by the posterior pituitary are products of neurosecretory cells in the hypothalamus; they travel down axons to the posterior lobe, where they are stored until their release. One hormone **ADH** (antidiuretic hormone) was discussed in the previous chapter. It promotes water retention by the kidneys. Another hormone, **oxytocin,** causes contraction of uterine muscles and ejection of milk from mammary glands in females. Although oxytocin does not seem to be necessary for normal childbirth, it can be used to induce labor and childbirth if a pregnancy has continued too long past normal term.

Figure 23.25
Anatomical location of major endocrine glands in the body.
The hypothalamus controls the pituitary, which in turn
controls the hormonal secretions of the thyroid, adrenal
cortex, and sex organs. Both sets of sex organs are
shown; ordinarily an individual has only one set of these.

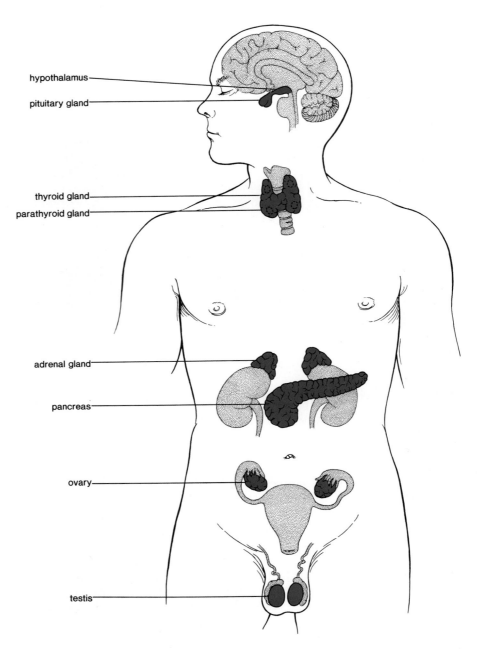

hypothalamus

pituitary gland

thyroid gland

parathyroid gland

adrenal gland

pancreas

ovary

testis

Table 23.2 Some Major Mammalian Hormones

Source	Hormone	Function
Hypothalamus	Releasing hormones	Stimulate or inhibit pituitary
Posterior pituitary	Antidiuretic (ADH)	Water retention by kidneys
	Oxytocin	Uterine contraction
Anterior pituitary	Thyroid stimulating (TSH, thyrotropin)	Stimulate thyroid
	Adrenocorticotropin (ACTH)	Stimulate adrenal cortex
	Gonadotropins*	Stimulate gonads
	Follicle stimulating (FSH)	Stimulate ovarian follicle and ovum formation
	Luteinizing (LH)	Stimulate follicle after ovulation
	Prolactin	Milk production
	Growth hormone (GH, somatotropin)*	Stimulate growth of bones and other tissues
Thyroid	Thyroxin	Increases metabolic rate
	Calcitonin	Calcium blood level
Parathyroid	Parathyroid hormone	Calcium and potassium blood levels
Adrenal cortex	Glucocorticoids (cortisol)	Glucose metabolism
	Mineralocorticoids (aldosterone)	Sodium, potassium blood levels
Adrenal medulla	Epinephrine and norepinephrine	Prepares body for stress
Pancreas	Insulin	Lowers blood sugar
	Glucagon	Raises blood sugar
Testes	Testosterone	Secondary male characteristics
Ovary	Estrogens and progesterones	Secondary female characteristics

*Also called adrenocorticotropic gonadotropic, and somatotropic hormone respectively.

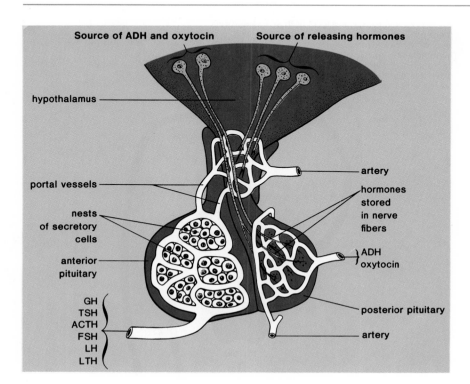

Figure 23.26
The anterior pituitary is connected to the hypothalamus by a portal system; the posterior pituitary is connected by nerves. Consult table 23.2 for the full names of the hormones listed here by initials.

Figure 23.27

The body is now known to produce endorphins, internal opiates that, like morphine, can produce a feeling of euphoria and a higher threshold to pain. The central nervous system, including the limbic system (fig. 23.21), produces one type of endorphin and the pituitary produces another type. It is possible that physical activity causes the release of endorphin and can elevate the mood.

Oxytocin secretion also occurs during suckling and causes milk to be "let down" or released. This is a good example of a neuroendocrine reflex—sensory information received by the nervous system due to sucking by the infant leads to the secretion of the hormone by the pituitary gland.

Anterior Pituitary

The anterior pituitary used to be called the **master gland** because it has an effect on other glands of the body (fig. 23.28). Now we know that the hypothalamus controls the anterior pituitary by liberating substances that travel through the portal system (fig. 23.26) between the hypothalamus and the anterior pituitary. These substances were at first called releasing factors but now they are called **releasing hormones**.

The anterior pituitary produces **GH** (growth hormone) and **prolactin.** If the secretion of growth hormone is insufficient during childhood, the individual becomes a pituitary dwarf, while excess secretion during the same period produces a pituitary giant. While prolactin is produced in both female and male mammals, only its function in females is known. Prolactin acts on the mammary glands, causing them to produce milk after they have been prepared to do so by the female sex hormones.

Thyroid Gland

The thyroid gland (fig. 23.25) can be used to illustrate the relationship between the hypothalamus, pituitary, and other endocrine glands (fig. 23.29). The hypothalamus produces TRH (thyroid releasing hormone) that travels by way of the portal system to the anterior pituitary. This causes the anterior

Figure 23.28

The anterior pituitary gland produces hormones that affect other glands: TSH stimulates the thyroid to produce and release thyroxin; ACTH stimulates the adrenal cortex to produce and release cortisol; FSH and LH stimulate the ovaries to produce and release estrogen and progesterone in females; and ICSH stimulates the testes to produce and release testosterone in males. These hormones control their own level in the bloodstream by a feedback control mechanism shown in figure 23.29.

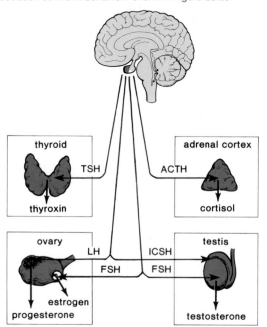

Figure 23.29

The level of thyroxin in the body is controlled in three ways: (a.) the level of TSH exerts feedback control over the hypothalamus; (b.) the level of thyroxin exerts feedback control over the anterior pituitary; and (c.) the level of thyroxin exerts feedback control over the hypothalamus. In this way, thyroxin controls its own secretion. Substitutes of the appropriate terms would also allow this diagram to illustrate control of cortisol and sex hormone levels.

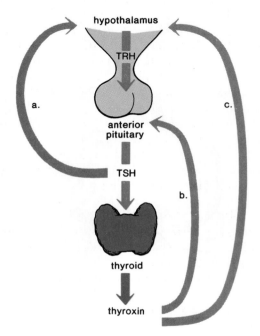

pituitary to secrete TSH (thyroid stimulating hormone). The latter travels by way of the blood to the thyroid gland, which then increases in size and produces **thyroxin.** When the blood concentration of thyroxin increases, TSH secretion temporarily slows down. When a substance slows down its own production, it is called regulation by negative feedback. Many such feedback mechanisms control the secretion of various hormones.

Thyroxin does not act specifically on a target organ; instead, it stimulates most of the body's cells to metabolize at a faster rate. If individuals produce too much thyroxin, they are hyperactive; if they do not produce enough thyroxin, they are lethargic. If thyroxin is not being produced, the pituitary keeps stimulating the thyroid to the point that it enlarges and becomes a goiter, externally visible as a swelling at the base of the neck (fig. 23.30). Iodine must be present for an individual to produce thyroxin, and in most instances goiter is caused by a lack of iodine in the diet. The custom of adding iodine to salt (iodized salt) developed as a way to prevent goiter.

Adrenal Glands

The adrenal glands, as their name implies (ad = near; renal = kidneys), lie atop the kidneys (fig. 23.25). Each consists of an outer portion called the cortex and an inner portion called the medulla.

Adrenal Medulla

This gland secretes the hormones **epinephrine** and **norepinephrine** (also called adrenalin and noradrenalin). These hormones cause reactions associated with stress: enlarged pupils; extreme alertness; increased breathing, heartbeat, and muscular strength. These effects are similar to those brought about by the sympathetic nervous system, not surprising since the tissue from which the adrenal medulla develops originates from the embryonic nervous system.

Adrenal Cortex

While the adrenal medulla may be removed with no ill effects, the adrenal cortex is absolutely necessary to life. Stress is believed to stimulate the hypothalamus to secrete the releasing hormone CRH (corticotropin releasing hormone) that travels by way of the portal system to the anterior pituitary, which in turn secretes **ACTH** (adrenocorticotropic hormone). ACTH is carried by the blood to the adrenal cortex, which secretes glucocorticoids. The **glucocorticoids** (cortisol, hydrocortisone, corticosterone, and cortisone) raise the level of amino acids in the blood by removing them from muscle. The liver then converts these amino acids to glucose. In other words, the adrenal cortex makes sugar available by a means not usually employed.

The adrenal cortex also produces mineralocorticoids, but their secretion is not controlled by ACTH. **Mineralocorticoids** (for example, aldosterone) regulate the level of sodium (Na^+) and potassium (K^+) in the blood by causing the kidneys to reabsorb sodium and excrete potassium. These are important functions; the amount of sodium in the blood affects blood pressure, and the amount of potassium affects cardiac function.

Parathyroid Glands

There are four parathyroid glands embedded in the thyroid (fig. 23.31), although they are functionally separate from the thyroid. Under the influence of **parathyroid hormone,** the calcium (Ca^{++}) level in the blood increases and the phosphate (PO_4^{-3}) level decreases. Parathyroid hormone is counteracted by the hormone **calcitonin,** which is produced by a special group of cells in the thyroid. When necessary, parathyroid hormone causes calcium to be removed from bone to maintain the blood level of this ion.

Figure 23.30
A woman suffering from goiter, an enlargement of the thyroid gland that often accompanies hypothyroidism. Continued TSH stimulation causes massive overgrowth of thyroid tissue.

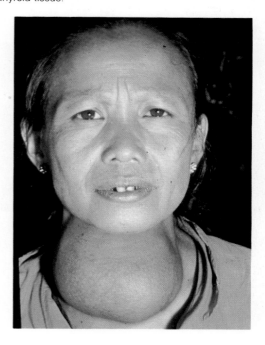

Figure 23.31
The parathyroid glands are embedded in the posterior surface of the thyroid gland.

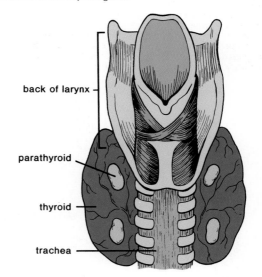

back of larynx

parathyroid

thyroid

trachea

An islet of Langerhans in a section of pancreas. The islets are separated from one another by tissue that secretes the pancreatic digestive enzymes. In each islet, the so-called alpha cells secrete glucagon and the beta cells secrete insulin.

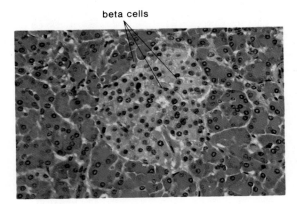

beta cells

Pancreas

The pancreas lies in the abdomen near the small intestine (fig. 23.25). It is composed of two types of tissues (fig. 23.32); one that produces and secretes the digestive juices that go by way of the pancreatic duct to the small intestine, and another, called the **islets of Langerhans,** that produces and secretes the hormones **insulin** and **glucagon** directly into the blood. In humans, the islets make up only 2 percent of pancreatic tissue, yet the pancreas contains more than a million islets. The islets respond quickly to blood glucose levels; when glucose level rises, insulin is secreted; when glucose level falls, glucagon is secreted.

It was known in the early 1900s that the islets of Langerhans underwent degenerative changes in people with diabetes mellitus. But it was not until 1921 that Federick Banting, a young Canadian orthopedic surgeon, and Charles Best, a medical student, succeeded in isolating a physiologically active chemical from the islets. They called this chemical insulin. Almost overnight, the prospects of the diabetic patient changed from certain death to a normal life. It is now known that insulin causes all body cells to take up glucose from tissue fluid. It seems to affect cell membranes; when insulin is present, glucose enters cells about twenty-five times faster than usual. It also promotes glycolysis with concomitant breakdown of glucose. Insulin promotes glycogen synthesis in the liver and muscle cells. The net effect of all these actions is lowered blood glucose levels.

Diabetes Mellitus

A significantly large number of people suffer from diabetes mellitus. As much as 5 percent of the U.S. population may have diabetes and the incidence is increasing yearly. The primary symptom of diabetes is sugar in the urine caused by excess sugar in the blood. One type of diabetes develops slowly and is usually seen in individuals after the age of forty. In this type of diabetes, termed **maturity-onset diabetes,** physiological symptoms are often mild and can usually be controlled by diet alone. Curiously, there is enough insulin in the blood; therefore, this type of diabetes is probably due to a defect in the molecular machinery of tissue cells. There is evidence that the cells have a decreased number of insulin receptors, molecules in the cell membrane that bind to glucose.

The second type of diabetes develops quickly and is usually seen before the age of twenty. In this type of diabetes, termed **juvenile-onset diabetes,** there is a deficiency of insulin in the blood. This deficiency accelerates the breakdown of the body's reserve of fat, resulting in the production of ketones and organic acids. These metabolites lower the pH of the blood, producing a condition known as **diabetic ketoacidosis** that can result in death. These patients usually require daily injections of insulin.

The cause of juvenile-onset diabetes is still being investigated. There is evidence that certain individuals are susceptible to viral infections of the islets of Langerhans. Because of the infections, the immune system begins to destroy the cells that produce insulin. Thus, the disease may involve genetic, environmental, and immunity aspects.

Summary

The nervous and endocrine systems are regulatory systems that coordinate body functions. A comparative study of the invertebrates shows a gradual increase in the complexity of the nervous system.

Neurons, cells that conduct nerve impulses, have three parts: dendrite(s), cell body, and axon. The nerve impulse, which is recognized when the resting potential becomes an action potential, is an electrochemical phenomenon involving the movement of ions across the cell membrane (table 23.3). Transmission across a synapse usually requires neurotransmitters because there is a small space, the synaptic cleft, that separates neuron from neuron. Psychoactive drugs affect in various ways the concentration of specific neurotransmitters at synapses.

The human peripheral nervous system includes cranial and spinal nerves. Spinal nerves extend from the spinal cord and contain both sensory and motor fibers. It is possible to use them to trace the path of a reflex from a receptor to an effector, as table 23.4 illustrates. The autonomic nervous system controls internal organs and includes the sympathetic and parasympathetic nervous systems, which are contrasted in table 23.1.

The human central nervous system includes the spinal cord and brain. The brain has various parts. In the hindbrain, the medulla oblongata contains centers that regulate internal organs, and the cerebellum functions in muscle coordination. In the forebrain, the hypothalamus regulates homeostasis, and the thalamus sends sensory input to the cerebrum, which is responsible for integration and coordination at the highest level. A survey of vertebrates shows a continual evolutionary increase in the size of the cerebrum, with its highest development in humans. The highly convoluted cerebral cortex is divided into lobes, each of which has specific functions.

Most endocrine glands (table 23.2) regulate specific target cells by secreting hormones directly into the blood. The pituitary gland has a posterior portion that secretes two hormones actually produced in the hypothalamus. The anterior pituitary is also under the control of the hypothalamus, which sends so-called releasing hormones by way of a portal system to the anterior pituitary. Aside from growth hormone and prolactin, this gland secretes hormones that control other endocrine glands—thyroid, adrenal cortex, and sex glands. The relationship of the hypothalamus, anterior pituitary, and thyroid can be used to illustrate negative feedback, by which the level of hormone in the blood regulates its own production.

Table 23.3 Summary of Nerve Impulse

Resting potential	Sodium/potassium pump at work, inside negative (-60 mV)
Action potential (nerve impulse)	a. Sodium moves to the inside, making the inside positive compared to the outside ($+40$ mV)
	b. Potassium moves to the outside, making the inside negative again compared to the outside (-60 mV)

Table 23.4 Path of a Simple Reflex

1. Sensory neuron	Receptor generates nerve impulses that travel in the dendrite to the cell body and then to the axon inside cord
2. Interneurons	Impulses picked up by dendrites are passed to axons
3. Motor neuron	Impulses picked up by dendrites inside the cord are passed to the axon outside cord. Axon innervates effector

Objective Questions

1. Even in the earthworm, it is possible to divide the nervous system into the _____ and _____ nervous systems.
2. In a neuron, the axon conducts nerve impulses _____ from the cell body.
3. The animal used in early experiments on nerve impulses was the _____ .
4. Inside the neuron there are large organic ions that carry a _____ charge.
5. As a nerve fiber depolarizes, the ion _____ is entering the fiber.
6. In the somatic nervous system, synaptic vesicles release the _____ substance, called _____ , which diffuses across the synapse to initiate a nerve impulse in the next neuron.
7. In a spinal reflex the _____ neuron takes impulses to the cord.
8. One would expect the _____ nervous division of the autonomic system to speed up the heartbeat.
9. The portion of the brain called the _____ coordinates muscle actions.
10. When one surveys the vertebrates, it becomes evident that there has been a gradual increase in the size of the _____ , the part of the brain responsible for consciousness.
11. Steroids are believed to turn on genes so that certain _____ are produced by the cell.
12. The _____ is an endocrine gland that is often called the master gland, although it is actually controlled by the hypothalamus.
13. The hormone ACTH stimulates the _____ glands to secrete glucocorticoids.
14. The amount of any particular hormone in the body is usually controlled by a _____ feedback mechanism.
15. Diabetes mellitus occurs when there is not enough _____ produced by the _____ .

Answers to Objective Questions

1. central, peripheral 2. away 3. squid 4. negative 5. sodium 6. neurotransmitter, acetylcholine 7. sensory 8. sympathetic 9. cerebellum 10. cerebrum 11. proteins 12. anterior pituitary 13. adrenal 14. negative 15. insulin, pancreas

Study Questions

1. Trace the evolution of the nervous system by contrasting the organization of the nervous system in *Hydra,* the planarian, the earthworm, and humans.
2. Describe the anatomy of sensory and motor neurons.
3. What are the two major events of an action potential and the ion changes that are associated with each event?
4. Describe the action of a neurotransmitter, including how it is stored and how it is destroyed.
5. Trace the path of a reflex action.
6. Contrast the sympathetic and parasympathetic divisions of the autonomic nervous system.
7. Name the major parts of the human brain and give the principal function of each.
8. Name the major human endocrine glands, the hormones each produces, and give the function of each hormone.
9. What mechanisms of action have been suggested to explain how hormones work?
10. Describe the regulation of the hormonal levels of the blood by means of negative feedback. Give a concrete example.

Selected Key Terms

central nervous system ('sen trəl 'nər vəs 'sis təm) 502

dendrite ('den ,drīt) 504

axon ('ak ,sän) 504

synapse ('sin ,aps) 508

neurotransmitter ('nur ō trans 'mit ər) 508

cholinesterase (,kō lə 'nes tə ,rās) 508

acetylcholine (ə ,set əl 'kō ,lēn) 508

norepinephrine ('nȯr ,ep ə 'nef rən) 508

meninges ('men ən 'gēz) 515

tract ('trakt) 515

medulla oblongata (mə 'del ə 'äb ,lȯŋ gä tä) 516

cerebellum (,ser ə 'bel əm) 516

thalamus (thal ə məs) 516

cerebrum (sə 'rē brəm) 517

endocrine gland ('en də krən 'gland) 519

U pon receiving stimuli, receptors (sense organs) initiate nerve impulses that are transmitted to the central nervous system. After integrating and processing these, the central nervous system sends impulses to effectors, signaling them to react to the stimuli. The previous chapter considered the functions of the nervous and endocrine systems; this chapter will consider sense organs and muscle effectors, both of which are needed to enable animals to respond to stimuli.

Sense Organs

Each **receptor** is designed to respond to a particular stimulus. For example, eyes respond only to light and ears only to sound waves (table 24.1). Receptors do not interpret stimuli, they act merely as transducers that transform the energy of stimuli into nerve impulses. Interpretation is the function of the brain, which has a region that interprets for each of the senses. Impulses arriving at a particular sensory area of the brain can be interpreted in only one way; those arriving at the visual area result in sight sensation, and those arriving at the olfactory area result in smell sensation, for example. Usually the brain is able to discriminate the type of stimulus, the intensity of the stimulus, and the origin of the stimulus. On occasion the brain can be fooled, as when a blow to the eyes causes one to "see stars."

Interoceptors located within the body monitor such conditions as blood pressure, expansion of lungs and bladder, and movement of limbs. **Exteroceptors** are located near the surface of the animal and respond to outer stimuli. Both types of receptors send information to the brain and both are needed to maintain homeostasis and to promote appropriate behavioral actions (fig. 24.1). Interoceptors are discussed in connection with the internal organs; just now we are limiting our coverage to exteroceptors.

Chemoreceptors

The receptors responsible for taste and smell are termed **chemoreceptors** because they are sensitive to certain chemical substances in food, liquids, and air. Activation of chemoreceptors influences the behavior, particularly of lower animals, as discussed in the reading on page 532.

CHAPTER 24

Sense Organs and Muscle Effectors

Concepts

Organisms respond to environmental stimuli. Response in multicellular organisms often depends on the presence of sense organs that receive the stimulus and effectors that bring about the response.

a. Each sense organ (receptor) responds to one type of stimulus, but they all initiate nerve impulses.

b. In higher organisms, such as humans, the sensation realized is determined by the region of the cerebrum receiving nerve impulses.

c. Muscles (effectors) are often attached to a skeleton that gives support, protects internal organs, and assists movement.

d. Contraction of a muscle fiber depends on actin and myosin filaments, a ready supply of ATP, and certain ions.

e. Motor axons stimulate muscle fibers to contract. The sequence of events leading to contraction has been studied in detail.

Table 24.1 Sense Organs

Sense Organs	Sense	Stimulus
General		
Temperature	Hot/Cold	Heat flow*
Touch	Touch	Mechanical displacement of tissue†
Pressure	Pressure	Mechanical displacement of tissue†
Pain	Pain	Tissue damage‡
Proprioceptors	Limb placement	Mechanical displacement†
Special		
Eye	Sight	Light*
Ear	Hearing	Sound waves†
	Balance	Mechanical displacement†
Taste buds	Taste	Chemicals‡
Olfactory cells	Smell	Chemicals‡

* Radioreceptors
† Mechanoreceptors
‡ Chemoreceptors

Figure 24.1

Response to an environmental change requires both receptors and effectors. Receptors are of two types: interoceptors are sensitive to internal events, and exteroceptors are sensitive to external changes. Receptors initiate nerve impulses, which travel to the central nervous system (CNS). Thereafter the CNS sends nerve impulses to effectors (glands and muscles) that bring about the response. The response is appropriate for the limits of the organism; for example, birds fly but horses run to escape danger.

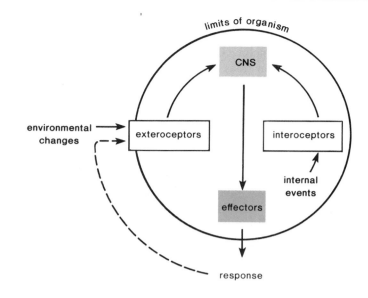

Figure 24.2

Taste bud location and anatomy. *a.* Taste buds on the tongue are located within nipplelike elevations called papillae. *b.* Longitudinal section of two taste buds. A taste bud contains taste cells with microvilli that project through an opening, the taste pore. The microvilli are stimulated by chemicals in the food.

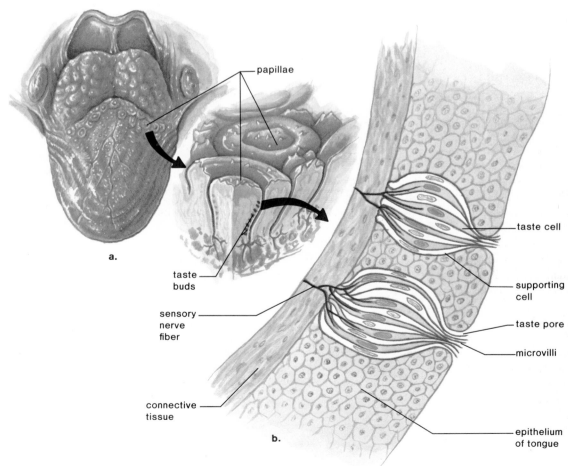

Figure 24.3
Olfactory cell location and anatomy. (*right*) The olfactory area in humans is located high in the nasal cavity. (*left*) The olfactory receptor cells are modified neurons located between supporting cells. The axons of these cells synapse with sensory nerve fibers making up the olfactory nerve. The dendrites of these cells have special structures known as olfactory cilia that project into the nasal cavity where they are stimulated by chemicals in the air.

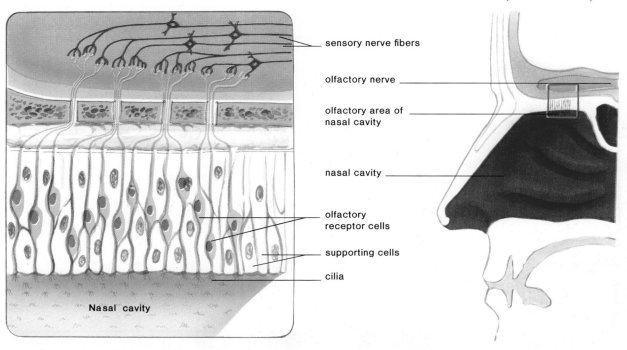

sensory nerve fibers

olfactory nerve

olfactory area of nasal cavity

nasal cavity

olfactory receptor cells

supporting cells

cilia

Nasal cavity

Taste

The receptors of the sense of taste in mammals are the **taste buds.** These are located on the tongue (fig. 24.2), but are also present on the surface of the soft palate, pharynx, and epiglottis. Each taste bud contains a number of elongated cells. These cells have *microvilli* that project through the taste pore, an opening in the taste bud. It is the microvilli that are stimulated by various chemicals in the environment. Humans are believed to have four types of taste buds, each type stimulated by chemicals that result in a bitter, a sour, a salty, or a sweet sensation.

Variations in the sense of taste are genetically inherited so foods taste differently to different people. This can account for the fact that some people dislike a food preferred by others. A taste preference also can be acquired by experience, however.

Smell

Receptors for the sense of smell are the olfactory cells located high in the roof of the nasal cavity in humans (fig. 24.3). The olfactory receptor cells are actually modified neurons that synapse with nerve fibers making up the olfactory nerve. Each olfactory receptor cell has six to eight cilia that are stimulated by many chemicals in the air. Research, resulting in the stereochemical theory of smell, has shown that different smells may be related to the various shapes of molecules rather than to the atoms that make up the molecules. These shapes fit specific olfactory sites on the olfactory cells' cilia.

Pheromones

Most organisms secrete chemical compounds that diffuse into the environment. Some of these, called pheromones, apparently evolved as communication mechanisms within, and sometimes between, species. Pheromones may evoke behavioral, developmental, or reproductive responses.

A number of pheromones that affect behavior patterns have been intensively studied. Well-known examples include trail-marking behavior in ants and territory marking by dogs, deer, and antelope through pheromones secreted in urine or from scent glands. In these cases the pheromone proclaims to other males of the species that the territory belongs to a particular individual.

An even more intriguing example of pheromones are the sex attractants, which have been most extensively studied in insects. The first sex attractant identified and synthesized was bombykol, which comes from females of the silkworm moth *Bombyx mori*. Bombykol can attract male silkworm moths several kilometers away. Sex-attractant pheromones have been discovered in other moths and in many other insect species as well.

For example, the queen bee releases a pheromone during her nuptial flight that entices the drones to follow her and mate.

Sex attractants have a very compelling effect on behavior.

Sex attractants have also been studied in mammals. In a number of mammalian species females emit a special odor during their period of fertility and sexual receptiveness, which is called estrus. Normally docile, well-behaved male dogs will scratch doors, break leashes, or jump fences and roam great distances to reach a female in estrus once they detect her chemical signal.

Recently, a sex attractant was isolated from female rhesus monkeys. This focuses new attention on primate pheromones in general and human pheromones in particular. Whether or not sex attractants and other pheromones exist in humans remains a matter of some controversy. At any rate, modern men and women seem to prefer to wash off natural body chemicals and replace them with laboratory-compounded scents. Who knows what would happen if manufacturers of these scents should succeed in isolating, identifying, synthesizing, and marketing real human pheromones?

From Johnson, Leland G., *Biology*. © 1983 Wm. C. Brown Publishers, Dubuque, Iowa. All Rights Reserved. Reprinted by permission.

The head of a male Polyphemus *moth with its long feathery antennae. The antennae bear receptors with which the moth can detect minute quantities of the sex attractant (released by the female of the species).*

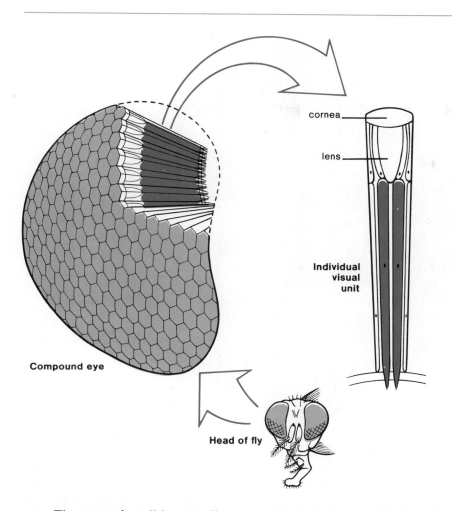

Figure 24.4
The compound eye of insects and other arthropods is composed of many individual units, each of which has its own lens. The light-sensitive portion is present in each lens.

cornea

lens

Individual visual unit

Compound eye

Head of fly

The sense of smell is generally more acute than the sense of taste. The human nose, for example, can detect 1/25 millionth of 1 mg of mercaptan, the odoriferous chemical given off by a skunk. This averages out to approximately one molecule per sensory ending. Yet, humans have a weak sense of smell compared to other vertebrates, such as dogs.

The sense of taste and the sense of smell supplement each other, creating a combined effect when interpreted by the cerebral cortex. For example, when one has a cold, food seems to lose its taste, but actually the ability to sense its smell is temporarily absent. This may work in reverse also. When we smell something, some of the molecules move from the nose down into the mouth and stimulate certain taste buds. Thus, part of what we refer to as smell is actually taste.

Photoreceptors

Many animals have light-sensitive receptors called **photoreceptors.** In its simplest form, a light receptor indicates only the presence of light and its intensity. The "eye spots" of planarians also allow the animal to determine the direction of light. More complex eyes have lenses that focus light on certain receptors. Arthropods have **compound eyes** composed of many independent visual units, each of which has its own lens (fig. 24.4) and views a separate portion of the

Figure 24.5
Detailed anatomy of the human eye. The eye has three coats: the sclera forms the cornea at the front; the choroid is rich in blood vessels; and the retina contains the rods and cones, the latter concentrated in the fovea centralis. Focusing is done by the lens and the humors.

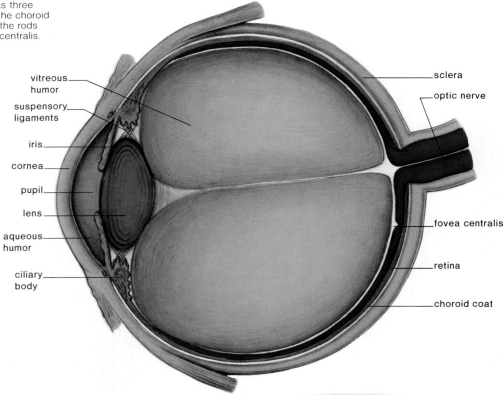

object. How well the brain combines this information to see the entire object is not known, but compound eyes seem especially well suited to detecting motion, as anyone knows who has tried to catch an insect.

Human beings and certain mollusks, like the squid, have a **camera type of eye.** There is a single lens that focuses an image of the visual field on the photoreceptors, which are closely packed together. All of the photoreceptors taken together can be compared to a piece of film in a camera. However, the situation is more complex, as we shall see.

Anatomy of the Human Eye
The human eye (fig. 24.5), which is an elongated sphere about one inch in diameter, has three layers or coats. The outer **sclera** is a white fibrous layer except for the transparent **cornea,** the window of the eye. The middle thin, dark-brown layer, the **choroid,** contains many blood vessels and pigment that absorbs stray light rays. Toward the front, the choroid thickens and forms a ring-shaped muscle, the **ciliary muscle,** and finally becomes a thin, circular, muscular diaphragm, the **iris,** that regulates the size of an opening called the **pupil.** The **lens,** which is attached to the ciliary muscle by ligaments, divides the cavity of the eye into two chambers. A basic, watery solution called **aqueous humor** fills the chamber between the cornea and the lens. A viscous, gelatinous material, the **vitreous humor,** fills the large cavity behind the lens.

Figure 24.6
Receptors for sight. *a*. Drawing of rod and cone. The
photosensitive pigment is located in the lamellae of the
outer segment, which apparently is a modified cilium. A
cilium-like stalk with nine sets of microtubules connects
the outer segment to the inner segment. *b*. Scanning
electron micrograph of rods illustrates their abundance in
the retina.

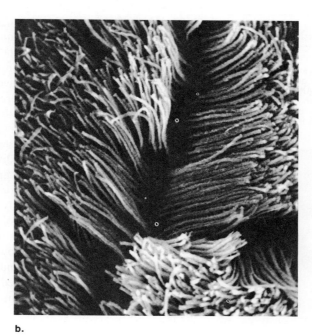

lamellae

connecting cilia

mitochondria

outer segment

nuclei

inner segment

cell bodies and nuclei

fibers

synaptic endings

Rod Cell   **Cone Cell**

a.

b.

Retina   The inner layer of the eye, the **retina,** contains the receptors for sight:
the **rods** and **cones** (fig. 24.6). Nerve impulses initiated by the rods and cones
are passed to neurons, whose fibers pass in front of the retina forming the **optic
nerve.** The optic nerve penetrates the layers of the eye. There are no rods and
cones where it passes through the retina; therefore, this site is a **blind spot,**
where vision is impossible.

The retina contains a special region called the **fovea centralis,** an oval
yellowish area with a depression where there are only cone cells. In the fovea
centralis, vision is most acute in daylight; at night, it is scarcely sensitive. At
this time, the more peripheral rods are active.

Table 24.2 Parts of the Eye	
Part	**Function**
Lens	Refraction and focusing
Iris	Regulate light entrance
Pupil	Admit light
Choroid	Absorb stray light
Sclera	Protection
Cornea	Refraction of light
Humors	Refraction of light
Ciliary body	Hold lens in place
Retina	Contain receptors
Rods	Black-and-white vision
Cones	Color vision
Optic nerve	Transmit impulse
Fovea	Region of cones in retina
Ciliary muscle	Accommodation

Figure 24.7
a. The high concentration of rhodopsin in the alligator's dark-adapted eye heightens sensitivity and gives its interior a rose-colored appearance. *b.* After moments in the light, the concentration of rhodopsin drops and the inner eye appears pale gold. This cycle of bleaching in light and regenerating in darkness was the first clue to the chemistry of vision.

a.

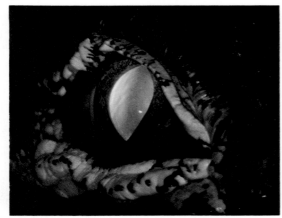

b.

Physiology

Table 24.2 lists the parts of the eye and their functions. Light rays from objects are bent (refracted) and brought to a focus on the retina as they pass through the cornea, lens, and humors. The lens is relatively flat when viewing far objects, but rounds up for close objects because light rays must be bent to a greater degree when viewing a close object. These changes are called **accommodation.** With normal aging, the lens loses its ability to accommodate for close objects; therefore, persons frequently need reading glasses once they reach middle age.

The image on the retina is rotated 180°, but it is believed that this image is righted in the brain. In one experiment scientists wore glasses that inverted and reversed the field. At first, they had difficulty adjusting to the placement of objects, but they soon became accustomed to their inverted world. Experiments such as this suggest that if we see the world "upside down," the brain has learned to see it right side up.

The rods contain **rhodopsin,** a pigment containing a molecule called **visual purple.** Only dim light is required to stimulate rods; therefore, they are responsible for *night vision.* The plentiful rods are especially good at detecting motion, but are unable to allow distinct and/or color vision. This causes objects to appear blurred and to have a shade of gray in dim light. When light strikes rhodopsin (fig. 24.7), it breaks down to a protein, **opsin,** and the pigment portion of the molecule, **retinal.** As breakdown occurs, a nerve impulse is generated. The more rhodopsin that is present in the rods, the more sensitive are our eyes to dim light. During the time required for adjustment to dim light, rhodopsin is being formed in the liver and transported to the rods. Formation of retinal requires *vitamin A.* Vitamin A is abundant in carrots; thus, the notion that we should eat carrots for good vision is not without foundation.

The cones, located primarily in the fovea and activated by bright light, detect the fine detail and color of an object. *Color vision* depends on three kinds of cones that contain either blue, green, or red receptors. Each pigment is made up of retinal and opsin, but there is a slight difference in the opsin of each, which accounts for their individual absorption patterns. Various combinations of receptors are believed to be stimulated by in-between shades of color, and the combined nerve impulses are interpreted in the brain as a particular color.

Fishes, reptiles, and most birds are also believed to have color vision, but among mammals only humans and other primates have color vision. It would seem, then, that this trait must have been adaptive for life in trees, accounting for its retention in just these few mammals. Insects have color vision, but they make use of a slightly shorter range of the electromagnetic spectrum (fig. 4.9) compared to humans. They can see the longest of the ultraviolet rays, and it is speculated that this may enable them to be especially sensitive to the reproductive parts of flowers (fig. 24.8) that reflect particular ultraviolet patterns.

Mechanoreceptors

Mechanoreceptors are sensitive to mechanical stimuli, such as pressure, sound waves, and gravity. Human skin (fig. 24.9) contains various types of mechanoreceptors, called **touch receptors,** that respond to pressure. The most thoroughly studied of these is the Pacinian corpuscle, a receptor shaped like an onion that consists of a series of concentric layers of connective tissue wrapped around the end (dendrite) of a sensory neuron. In contrast, pain receptors are only the unmyelinated ("naked") ends (dendrites) of the fibers of sensory neurons.

Figure 24.8
Marsh marigold as seen by humans (*left*) and insects (*right*). Humans see no markings, but insects see distinct blotches because their eyes respond to ultraviolet rays whereas our eyes do not. These types of markings often highlight the reproductive parts of flowers where insects feed on nectar and pick up pollen at the same time.

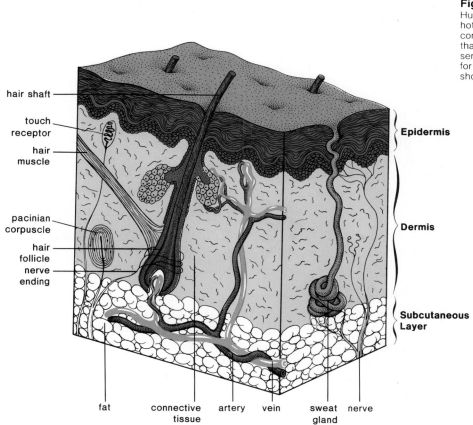

Figure 24.9
Human skin has various receptors for the sensations of hot, cold, touch, pressure, and pain. While there is controversy about the hot/cold receptor(s), it is known that the Pacinian corpuscle is responsible for pressure sensations; free nerve endings (in yellow) are responsible for pain; and that the touch receptors are located where shown.

hair shaft

touch receptor

hair muscle

pacinian corpuscle

hair follicle

nerve ending

Epidermis

Dermis

Subcutaneous Layer

fat connective tissue artery vein sweat gland nerve

Figure 24.10
Lateral line system of fishes. (*upper*) Location of the
system is shown. (*lower*) Longitudinal section of the
system. A main canal has openings to the exterior. Lining
the canal are hair cells (embedded in cupulae) that act as
sense receptors for pressure.

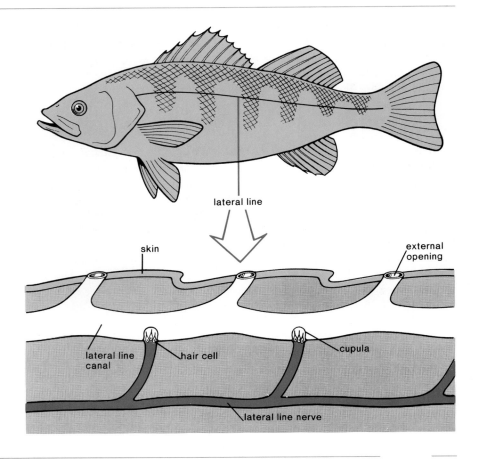

lateral line

skin

external
opening

lateral line
canal

hair cell

cupula

lateral line nerve

Figure 24.11
The anatomy of the human ear. Sense of balance is
controlled by the semicircular canals and vestibule. Sense
of hearing requires the auditory canal, the ossicles
(hammer, anvil, and stirrup), and the cochlea. The cochlea
contains the organ of Corti, which is shown in detail in
figure 24.13.

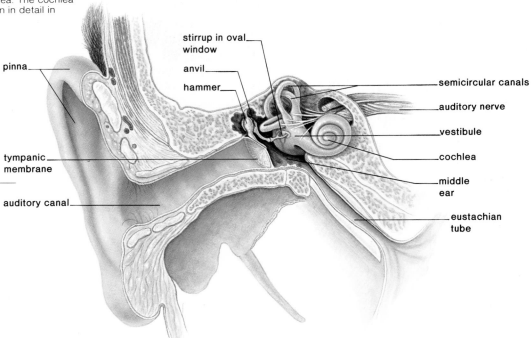

pinna

stirrup in oval
window

anvil

hammer

semicircular canals

auditory nerve

vestibule

cochlea

middle
ear

tympanic
membrane

auditory canal

eustachian
tube

"Hair cells," which are named for the cilia they bear, are often mechanoreceptors in various specialized vertebrate sense organs. The **lateral line** of fishes and amphibians is a series of such receptors that detect water currents and pressure waves from nearby objects. Each receptor is a collection of hair cells with cilia embedded in a gelatinous, wedge-shaped mass known as a **cupula** (fig. 24.10). This mass bends when disturbed, and the hair cells initiate nerve impulses.

Human Ear

The inner human ear contains the receptors for the senses of balance and hearing. The inner ear probably evolved from the lateral line of fishes and amphibians. As figure 24.11 shows, the human ear has an outer and a middle portion, as well as an inner ear. The **outer ear** consists of the pinna (external flap) and auditory canal. The latter is lined with fine hairs that filter the air. In the upper wall are modified sweat glands; these secrete earwax that helps guard the ear against entrance of foreign materials.

The **middle ear** begins at the **tympanic membrane** (eardrum) and ends at a bony wall that has small openings covered by membranes, the **oval window** and the **round window.** Three small bones are located between the tympanic membrane and the oval window. Collectively called **ossicles,** individually they are the **hammer** (malleus), **anvil** (incus), and **stirrup** (stapes), named for their structural resemblance to these objects. The **Eustachian tube** extends from the middle ear to the pharynx and permits an equalization of air pressure. Chewing gum, yawning, and swallowing in elevators and airplanes help move air through the Eustachian tubes during ascent and descent.

Whereas the outer ear and middle ear contain air, the inner ear is filled with fluid. The inner ear has three anatomic areas: the first two, called the **vestibule** and **semicircular canals,** are concerned with balance; the third, the **cochlea,** is concerned with hearing. Table 24.3 reviews the parts of the ear and their functions.

Balance

The vestibule is a chamber between the semicircular canals and the cochlea. It contains two small sacs, the **utricle** and **saccule.** Within each of these are groups of little hair cells whose cilia are embedded in a gelatinous layer containing calcium carbonate granules known as **otoliths.** When the head tilts, the otoliths press on cilia in a certain direction, which initiates nerve impulses that inform the brain about the position of the head. Similar types of organs are also found in coelenterates, mollusks, and crustacea; these are known as statocysts (fig. 24.12). These organs do not result in a sensation of movement and are therefore *static equilibrium* organs.

The semicircular canals (fig. 24.11) are *dynamic equilibrium* organs because they initiate a sensation of movement. At the base of each is a cupula containing hair cells. As the body moves about, there is a slight lag before the fluid within the canals moves; this causes the cupula to move and the cilia to bend. The three semicircular canals are orientated to detect movement in any direction. The brain integrates the different impulses it receives from each of the three canals and thus can determine the direction and rate of movement. Very rapid or prolonged movements of the head may cause uncomfortable side effects, such as the dizziness and nausea that accompany seasickness.

Table 24.3 Parts of the Ear

Part	Function
Outer Ear	
Pinna	Collect sound waves
Middle Ear	
Tympanic membrane and ossicles	Amplify sound waves
Inner Ear	
Oval window	Initiate pressure waves
Canals	Transmit pressure waves
Organ of Corti	Receptor for hearing
Utricle and saccule	Static equilibrium
Semicircular canals	Dynamic equilibrium

Figure 24.12
Generalized statocysts as found in mollusks and crustaceans. A small particle, the statolith, moves in response to a change in the animal's position. Wherever the statolith is resting, it stimulates the underlying cilia of hair cells. These cilia transmit impulses that indicate the position of the body.

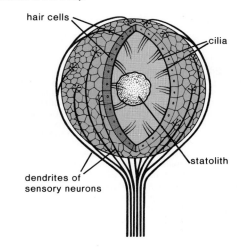

Hearing The cochlea within the inner ear resembles the shell of a snail because it spirals (fig. 24.11). A cross section of the cochlea (fig. 24.13) shows that it contains three canals: the **vestibular, cochlear,** and **tympanic canals.** Along the length of the **basilar membrane,** which forms the floor of the cochlear canal, there are at least 24,000 little hair cells. Just above them is another membrane called the **tectorial membrane.** The hair cells plus the tectorial membrane are the **organ of Corti,** the sense organ for hearing.

Sound waves reach the organ of Corti by way of the outer and middle ear. Ordinarily, sound waves do not carry much energy, but when a large number of waves strike the eardrum, it moves back and forth (vibrates) ever so slightly. The hammer transfers the pressure from the inner surface of the eardrum to the anvil, thence to the stirrup. The pressure is multiplied about twenty times as it moves from the eardrum to the stirrup. The stirrup vibrates the oval window, which transmits pressure waves to the fluid in the inner ear. These waves cause the basilar membrane to move up and down and the cilia of the hair cells to rub against the tectorial membrane. Bending of the cilia initiates nerve impulses that pass by way of the auditory nerve to the brain, where the impulses are interpreted as a sound.

Figure 24.13
Location and anatomy of the organ of Corti. *a.* The organ of Corti is located in the cochlear canal of the cochlea (see fig. 24.11). *b.* The organ of Corti consists of pressure receptor cells with cilia that approach the tectorial membrane. Pressure waves cause the basilar membrane beneath the ciliated cells to vibrate. When the cilia touch the tectorial membrane, nerve impulses are initiated and taken up by sensory nerve fibers within the auditory nerve. The intensity of the sound is determined by how many cells are stimulated; the quality of the sound is determined by which particular cells along the entire organ of Corti are stimulated.

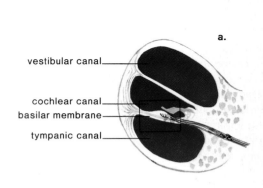

vestibular canal

cochlear canal
basilar membrane
tympanic canal

a.

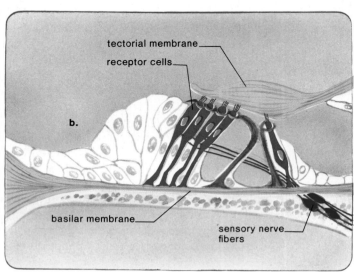

tectorial membrane
receptor cells
b.
basilar membrane
sensory nerve fibers

The organ of Corti is narrow at its base but widens as it approaches the tip of the cochlear canal. Various cells along it are sensitive to different wave frequencies, or **pitch.** Near the apex, the organ of Corti responds to low pitches, such as a bass drum, and near the base it responds to higher pitches, such as a bell or whistle. The neurons from each region along the length of the cochlea lead to slightly different areas in the brain. The pitch sensation we experience depends on which brain area is stimulated. **Volume** is a function of the amplitude of sound waves. Loud noises (measured in decibels, db) cause the fluid of the cochlea to oscillate to a greater degree, and this, in turn, causes the basilar membrane to move up and down to a greater extent. The resulting increased stimulation is interpreted by the brain as loudness. It is believed that **tone** is interpreted by the brain according to the distribution of hair cells that are stimulated.

Picturing the Effects of Noise

We have an idea of what noise does to the ear," David Lipscomb [of the University of Tennessee Noise Laboratory] says. "There's a pretty clear cause-effect relationship." And these photomicrographs of the cochlea's tiny structures graphically document noise trauma to the inner ear.

Hair cells transmit the mechanical energy of sound waves into those neural impulses that the brain interprets as sound. Loud noise can damage or destroy hair cells as these scanning electron micrographs illustrate.

Hair cells come in two varieties: a single row of inner cells and a triple row of outer ones. "Outer cells degenerate before inner cells," notes Clifton Springs, N.Y.-otolaryngologist Stephen Falk. The most subtle change wrought by noise is a development of vesicles, or blister-like

protrusions along the walls of the hair cells' stereocilia. Continued assault by noise will lead to a rupturing of the vesicles and damage. In addition, the "cuticular plate"—base tissue supporting the stereocilia—may soften, followed by a swelling and ultimate degeneration of hair cells.

But sensory hair cells are not the only structures at risk. Adjacent inner-ear cells . . . may undergo vacuolation—development of degenerative empty spaces in cells. Even nerve fibers synapsing at the hair cells' roots may die. In the final phase of noise-induced cochlear damage, the organ of Corti—of which hair cells and supporting cells are a part—is completely denuded and covered by a layer of scar tissue.

Reprinted with permission from *Science News,* the weekly newsmagazine of science, copyright 1982 by Science Service, Inc.

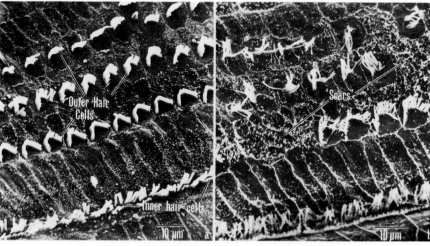

a.   b.

Damage to organ of Corti due to loud noise. a. *Normal organ of Corti.* b. *Organ of Corti after twenty-four-hour exposure to noise level typical of rock music. Note scars where cilia have worn away.*

Figure 24.14
Arthropods have an external skeleton. In order to grow, they must periodically shed this exoskeleton, a process called molting. In this photo, a cicada is just emerging from its discarded skeleton. It will be more vulnerable until its new exoskeleton has hardened.

Musculoskeletal System

Skeleton

The skeleton is the framework of the body. It helps protect the internal organs, and assists in movement. To cause body movements, the force of muscular contractions must be specifically directed against other parts of the body. In some animals, such as segmented worms, pressure of muscular contractions is applied to fluid-filled body compartments. Then the animal is said to have a **hydrostatic skeleton** (p. 321). Animals with hydrostatic skeletons are not as capable of complex movements as are animals with a rigid, supporting exo- or endoskeleton. The protective **exoskeleton** of arthropods is largely composed of chitin that must be periodically shed by molting as the animal grows (fig. 24.14). The **endoskeleton** of vertebrates is composed of cartilage and bone that grows with the animal. These tissues are the storage areas for calcium and phosphorus; they are also the site of blood cell production in adults.

Cartilage and Bone

Both cartilage and bone have cells separated by a matrix. The matrix of cartilage, composed of protein and proteinaceous fibers, is more flexible than that of bone, which contains a protein framework in which salts of calcium and phosphorus are embedded. Cartilage (fig. 24.15*a*) can grow more rapidly than bone, and vertebrate embryos have skeletons that are initially composed of cartilage. Some vertebrates—notably sharks, skates, and rays—have cartilaginous skeletons throughout life. But in most other vertebrates, much of the cartilage of the embryonic skeleton is gradually replaced by bone during subsequent development. In adult humans, for example, cartilage remains only in certain regions, such as the pinna of the ear and the tip of the nose, and covers the ends of bones at movable joints.

Bone, the more rigid supporting tissue, occurs in two forms—spongy bone and compact bone (fig. 24.16). Spongy bone is composed of a latticelike network of strands. The irregular spaces among the bony strands, as well as the hollow cavities inside many bones, contain bone marrow where blood cells are produced in adult vertebrates. Compact bone, which is more solid than spongy bone, consists of structural units called **Haversian systems** (fig. 24.15*b*). Each system has bone cells within cavities arranged in concentric circles around tiny tubes called Haversian canals. These canals contain blood vessels and nerve fibers. Nutrients, gases, and wastes diffuse through tiny passages (processes) that connect the cavities with the central canal.

Figure 24.15

a. Cartilaginous tissue. Cartilage cells are located in cavities called lacunae, which are scattered in a matrix produced by the cells. *b.* Compact bone. Cells are located in lacunae that form concentric circles around Haversian canals.

Figure 24.16

Typical long bone of the human body. Cartilage, compact bone, and spongy bone are located where shown. Spongy bone contains irregularly shaped bony strands and marrow spaces. Compact bone contains Haversian systems, one of which is shown in detail in figure 24.15*b*.

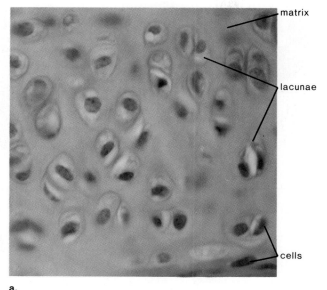

matrix
lacunae
cells

a.

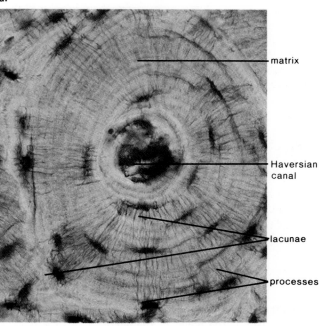

matrix
Haversian canal
lacunae
processes

b.

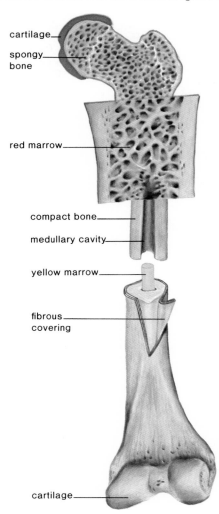

cartilage
spongy bone
red marrow
compact bone
medullary cavity
yellow marrow
fibrous covering
cartilage

Human Skeleton

The human skeleton (fig. 24.17), like all vertebrate skeletons, can be divided into two parts for purposes of description. The **axial skeleton** supports the main body axis—the head, neck, and trunk. It includes the skull, which surrounds and protects the brain; the vertebrae, a series of bones making up the vertebral column or backbone and which protect the spinal cord; and the ribs, curved bones attached to vertebrae that enclose and protect vital organs, such as the heart and lungs.

Figure 24.17
The major bones and muscles of the body. Which bones
are in the axial skeleton and which are in the appendicular
skeleton?

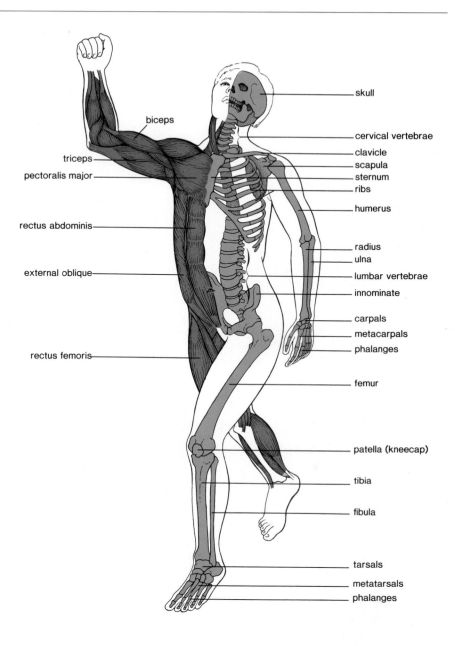

biceps

triceps

pectoralis major

rectus abdominis

external oblique

rectus femoris

skull

cervical vertebrae

clavicle

scapula

sternum

ribs

humerus

radius

ulna

lumbar vertebrae

innominate

carpals

metacarpals

phalanges

femur

patella (kneecap)

tibia

fibula

tarsals

metatarsals

phalanges

The **appendicular skeleton** includes the bones of the paired appendages (arms and legs) as well as the bones of the **pectoral** and **pelvic girdles.** These girdles support the arms and legs, respectively, and connect them to the axial skeleton.

Some bones in the adult, such as those of the skull, are connected to one another by immovable joints that keep them permanently fixed in position relative to one another. But many other bones meet in movable joints that are held together by ligaments. **Ligaments** are tough, but flexible, connecting straps that hold two bones together. Movable joints are flexible enough to permit body movements in response to contractions of skeletal muscles.

Figure 24.18
Vertebrate muscle fibers. *a.* Smooth muscle fibers. Smooth muscle fibers have a single nucleus and lack striations. *b.* Cardiac muscle fibers. Note the branching of the fibers, the central position of the nuclei, and the presence of intercalated discs, which are the complex junctions between adjacent individual cells. *c.* Skeletal muscle fibers. Note striations and the peripheral location of nuclei in the multinucleate fibers.

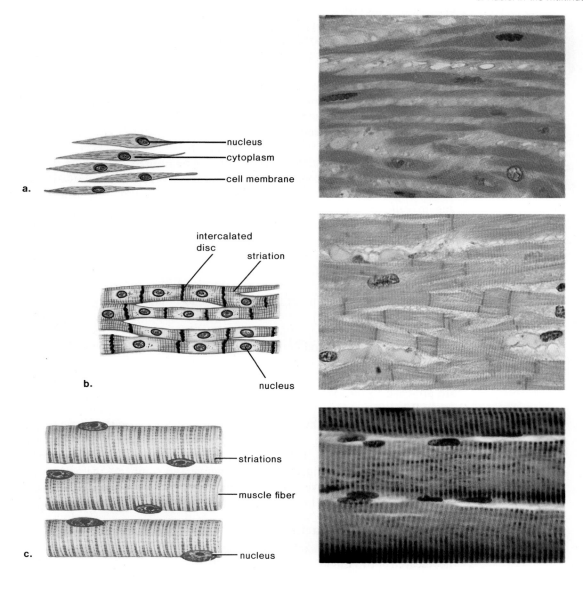

Vertebrate Muscles

Vertebrate animals have three types of muscle (fig. 24.18): smooth, cardiac, and skeletal. The contractile cells of these tissues are called muscle fibers. A **muscle fiber** is a cell that specializes in contraction.

Smooth muscle and cardiac muscle are called *involuntary* because their contractions are not consciously controlled. However, some individuals learn to modify the contraction of these muscles that are normally under the control

Figure 24.19

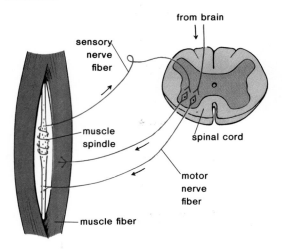

Diagram illustrating how spindle fibers aid in the coordination of muscular contraction by sending sensory impulses to the central nervous system either in the absence of prior motor stimulation or following motor stimulation.

Labels: from brain; sensory nerve fiber; muscle spindle; spinal cord; motor nerve fiber; muscle fiber

of the autonomic nervous system. Biofeedback, which permits the individual to monitor his own physiological states, has helped some people learn to control heartbeat, blood vessel dilation, and intestinal muscle activity.

Smooth muscle fibers are spindle-shaped cells, each with a single nucleus. Most often the cells are arranged in parallel, forming sheets (fig. 24.18a). The striations seen in cardiac and skeletal muscle are not observed in this tissue. Smooth muscle, in contrast to striated muscle, can maintain good tone even without nervous stimulation. A muscle has **tone** when a number of its fibers are contracted at any time. While smooth muscle is slower to contract, it can sustain prolonged contractions and does not fatigue easily.

Cardiac muscle fibers are uninucleated, striated, and branched (fig. 24.18b). This branching allows the fibers to interlock for greater strength. There are **intercalated discs** (fig. 20.10) where the fibers fuse. These permit contractions to spread throughout the atria and ventricles. The heart does not fatigue because cardiac fibers relax completely between contractions. In this way, overexertion is controlled. Also, the many mitochondria present in heart muscle provide a source of continual energy.

Skeletal muscle is called *voluntary* muscle because its contraction is consciously stimulated and controlled by the nervous system. Skeletal muscle fibers are multinucleated and striated (fig. 24.18c). Skeletal muscle striation is due to cellular contractile elements called myofibrils, whose microscopic anatomy and physiology are discussed on the following page. Skeletal muscle tone depends on the presence of sense receptors, called muscle spindles or sometimes **stretch receptors.** Sensory nerve fibers attach to these near the middle of their lengths (fig. 24.19). Whenever a spindle is stretched, sensory stimuli travel to the central nervous system, which then directs muscle fibers to contract either to maintain tone or to produce voluntary movements. Because tone requires repetitive nervous stimulation of some of the fibers, skeletal muscles are subject to paralysis when they fail to receive stimulation. The muscles become flaccid and flabby and degenerate when all motor stimulation by way of nerves is lost as when, for example, paralysis follows poliomyelitis.

Skeletal muscles, which make up over 40 percent of the body's weight, are attached to the skeleton (fig. 24.20) by **tendons,** tough fibrous bands or cords. When muscles contract, they shorten. Thus, muscles can only pull; they cannot push. Because of this, skeletal muscles must work in **antagonistic pairs.** For example, one muscle of an antagonistic pair bends the joint and brings the limb toward the body. The other muscle straightens the joint and extends the limb. Figure 24.20 illustrates this principle and compares the actions of muscles in the limb of a vertebrate, where they are attached to an endoskeleton, with those of an arthropod, where they are attached to an exoskeleton.

Skeletal Muscle Contraction

Present understanding of skeletal muscle contraction has resulted in part from the research of two English scientists. In the 1960s, A. F. Huxley studied muscle contraction mainly with the light microscope. H. E. Huxley studied the fine structure of skeletal muscle with the electron microscope. Each skeletal muscle fiber is striated (fig. 24.18c); it has a pattern of alternate light and dark bands when observed microscopically. The striation of whole fibers arises from the alternating light and dark bands of the many smaller, tubular **myofibrils** contained in each muscle fiber (fig. 24.21). Electron microscopy and biochemical analyses have shown that these bands are due to the placement of the muscle proteins **actin** and **myosin** within myofibrils. Myosin makes up relatively thick

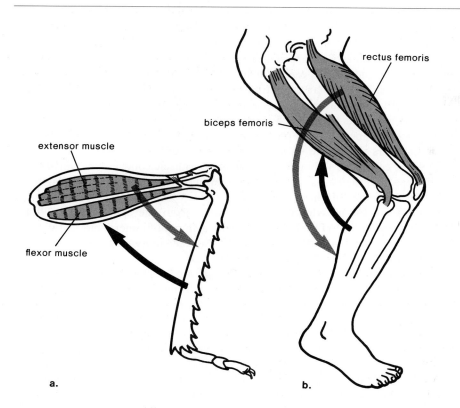

extensor muscle

biceps femoris

rectus femoris

flexor muscle

a.

b.

Figure 24.20
Antagonistic muscle pairs. Muscles can exert force only by shortening. Movable joints are supplied with double sets of muscles that work in opposite directions. *a.* Muscles in an insect's leg as an example of antagonistic muscles attached to the inside of an arthropod exoskeleton. *b.* Muscles in the human leg as an example of antagonistic muscles attached to a vertebrate endoskeleton. As indicated by the arrows, flexor muscles move the lower limb toward the body and extensor muscles move the lower limb away from the body.

Figure 24.21
A whole muscle (*a.*) is made up of (*b.*) muscle cells, or fibers. Within the muscle fibers are (*c.*) myofibrils. A photomicrograph of a longitudinal section of muscle fibers (*d.*). An electron micrograph (*e.*) of a longitudinal section of myofibrils showing one complete sarcomere.

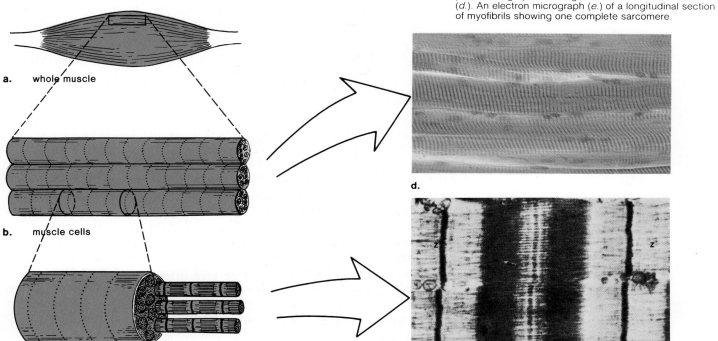

a. whole muscle

b. muscle cells

c. muscle cell showing myofibrils

d.

e.

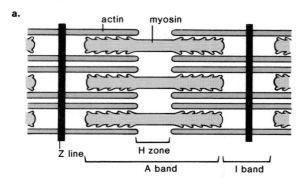

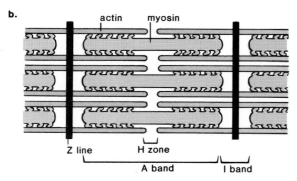

filaments, and actin forms thin filaments. As figure 24.22 shows, the lightest areas of a myofibril (I band) contain only thin filaments, while the darkest regions contain both thick and thin filaments.

The customary unit for describing muscle contraction is the **sarcomere** (figs. 24.21*e* and 24.22), each of which extends from a so-called Z line to another Z line. Notice that the actin filaments are attached to these dark lines, while the myosin filaments are not attached. When a muscle fiber contracts, the actin filaments within each sarcomere *slide* past the myosin filaments and approach one another. This process shortens the myofibrils and, therefore, the muscle fibers that make up a muscle. The movement of actin in relation to myosin is called the **sliding filament theory** of muscle contraction. During this sliding process the muscle fiber shortens, even though the filaments themselves remain the same length.

Energy for muscle contraction comes from the splitting of ATP by myosin, a powerful ATPase. Myosin is not a smooth filament, but instead has a number of globular heads. These heads attach to actin at specific binding sites, forming so-called **cross-bridges** (fig. 24.24). Once cross-bridges have formed, they bend, thus exerting a pulling force on the thin filaments. Hydrolysis of ATP after the cross-bridges have formed provides energy for this bending. The overall formula for muscle contraction

$$\text{ATP} \longrightarrow \text{ADP} + \textcircled{P}$$
$$\text{actin} + \text{myosin} \longrightarrow \text{actomyosin}$$

was worked out by Albert Szent-Gyorgyi in the 1940s. He extracted and purified actin and myosin and then precipitated them as actomyosin threads. He found that these fibers shortened if the solution contained appropriate ions and ATP.

The ATP needed for muscle contraction is made available in three ways.

1. Muscle cells are generously supplied with mitochondria within which, by means of aerobic cellular respiration, ATP is formed.

2. Muscle cells contain *creatine phosphate,* which is used to store a supply of high-energy phosphate. Creatine phosphate does not participate directly in muscle contraction. Instead, it is used to regenerate ATP by the following reaction.

$$\text{creatine} \sim \textcircled{P} + \text{ADP} \longrightarrow \text{ATP} + \text{creatine}$$

3. When all the creatine phosphate has been depleted, a muscle cell can still produce ATP by means of anaerobic respiration (p. 121).

Anaerobic respiration occurs, of course, when the cells are not being supplied with oxygen quickly enough to make aerobic respiration possible. This would occur during times of strenuous exercise. In practice, anaerobic respiration can supply ATP for only a short time because **lactic acid buildup** produces muscular aching and fatigue. At this time, we are also in **oxygen debt** and continue to breathe deeply and pant even while resting. This continued intake of oxygen is required to complete the metabolism of lactic acid that has accumulated during exercise. The lactic acid is transported to the liver where one-fifth of it is completely broken down to carbon dioxide and water by means of the Krebs cycle and respiratory chain. The ATP gained by this respiration is then used to reconvert four-fifths of the lactic acid to glucose.

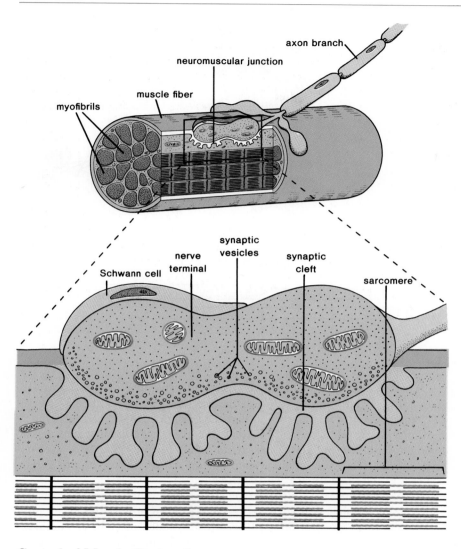

Figure 24.23
Diagrammatic drawing of a neuromuscular junction. (*above*) The longitudinal section shows that a terminal branch of an axon ends in a knob that approaches, but does not touch, a muscle fiber. (*below*) The space between the axon knob and the muscle fiber is a synaptic cleft. When nerve impulses move along an axon, they cause synaptic vesicles to fuse with the membrane discharging their contents. Reception of this neurotransmitter substance sets off a muscle action potential that results in contraction of sarcomeres.

Control of Muscle Contraction

Nerve impulses (action potentials) conducted by the axons of motor neurons stimulate muscle fibers to contract. As in a synapse, a space separates the specialized end (often called the motor end plate) of the motor axon from the membrane of the muscle fiber, called the **sarcolemma.** This synapselike junction is called a **neuromuscular junction** (fig. 24.23).

When an impulse passes down to the tip of a motor axon, acetylcholine is released from vesicles that open at the surface of the motor end plate. Acetylcholine released from the motor axon crosses the gap between the axon and the sarcolemma and binds with acetylcholine receptors in the sarcolemma. The sarcolemma, like the nerve cell membrane, normally is polarized; there is a resting potential across the membrane. *Acetylcholine* binding changes the permeability of the sarcolemma and initiates a muscle action potential that sweeps along the sarcolemma. So far, these events closely parallel those seen in synapses between nerve cells. But how does an action potential transversing the sarcolemma cause filament sliding in all parts of all myofibrils?

Figure 24.24

Control of sarcomere contraction. Note the globular heads of myosin and that a protein twists about actin covering up the cross-bridge binding sites. After calcium (Ca⁺⁺) is released from its storage sacs, it combines with a regulatory protein. Now the cross-bridge binding sites are exposed. The myosin filament extends its globular heads, forming cross-bridges that bind to these sites. The breakdown of ATP by myosin causes the cross-bridges to detach and reattach further along the actin. In this way, the actin filaments are pulled along past the myosin filaments.

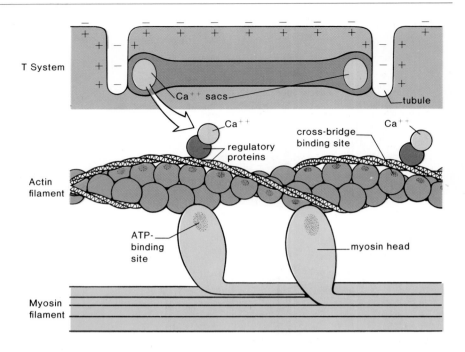

In the late 1940s investigators injected a number of substances into muscle fibers and found that, of all those tested, only calcium salts caused fiber contraction. It was suggested that the action potential changed the permeability of the sarcolemma to let calcium (Ca⁺⁺) diffuse into the muscle fiber and set off contraction of the sarcomeres. But it was quickly pointed out that simple diffusion could not account for the delivery of calcium to all sarcomeres at the same time.

Now it is known that the sarcolemma forms tubules, called **T tubules** (for transverse tubules), that penetrate or dip down into a muscle cell until they come in contact, but do not fuse, with expanded portions of the sarcoplasmic reticulum (endoplasmic reticulum). Calcium is stored in these regions, called **calcium storage sacs.**

Each action potential that passes along the sarcolemma causes a rapidly sequential depolarization of the membranes of all the T tubules. When an action potential reaches the calcium storage sacs, calcium ions are released. Because T tubules penetrate to all parts of the muscle fiber, calcium release is nearly simultaneous in all parts of the fiber. Calcium then binds with regulatory proteins present in thin filaments (fig. 24.24) to expose the myosin-binding sites on actin molecules. This permits cross-bridge formation and filament sliding.

Relaxation follows contraction. At this time an active transport system pumps calcium back into the calcium storage sacs.

Summary

Sense organs are transducers: they transform the energy of a stimulus to the energy of nerve impulses. It is the brain, not the sense organ, that interprets the stimulus.

Human taste buds and olfactory cells are chemoreceptors sensitive to chemicals in water and air. The eye is a photoreceptor. The compound eye of arthropods is made up of many individual units, whereas the human eye is a camera-type eye with a single lens. Table 24.2 lists the parts of the eye and the function of each part. When we see, light has stimulated the rods and cones in the retina.

Mechanoreceptors include the pressure (touch) receptors in human skin. Many mechanoreceptors are hair cells with cilia, such as those found in the lateral line of fishes as well as in the inner ear of humans. When we hear, the organ of Corti has been stimulated by pressure waves. The inner ear also contains the sense organs for balance. Just like the statocysts of invertebrates,

portions of the human inner ear contain calcium carbonate granules resting on hair cells. The movement of these granules gives us a sense of static equilibrium. Movement of fluid past hair cells in the semicircular canals gives us a sense of dynamic equilibrium.

The exoskeleton of invertebrates and the endoskeleton of vertebrates give rigidity to the body, help protect internal organs, and assist movement. The human skeleton is divided into the axial portion, which supports and protects the head, neck, and trunk of the body, and the appendicular portion, which includes the bones in the limbs and the associated girdles.

Vertebrate animals have three types of muscle tissue. Smooth muscle is located in the internal organs. Cardiac muscle is found in the heart. Skeletal muscles work in antagonistic pairs because they only pull on bones; they never push them. When viewed microscopically, skeletal muscle cells have

striations caused by the placement of actin and myosin filaments within sarcomeres. The myosin (thick) filaments attach to and detach from the actin (thin) filaments, which causes them to slide and make a sarcomere shorter. Myosin is an ATPase, and ATP provides the energy for muscle contraction. There are three sources of high-energy phosphate in muscles: aerobic respiration, creatine phosphate, and anaerobic respiration. The latter results in oxygen debt that must be repaid after exercise.

Muscle contraction is stimulated by motor axons at neuromuscular junctions. Here nerve impulses cause synaptic vesicles to release acetylcholine, which diffuses across a synaptic cleft to the sarcolemma of a muscle fiber. The muscle action potential travels down tubules to end near calcium storage sacs. Release of calcium results in filament sliding. Calcium is then returned to the storage sacs during relaxation.

Objective Questions

1. The two human chemoreceptors are the _____ and the _____ cells.
2. The rods and cones are present in the _____ , the innermost layer of the eye.
3. The rods function in _____ light, and the cones function in _____ light.
4. The _____ , held in place by muscle ciliary body, particularly helps with focusing.
5. The Pacinian corpuscle present in human skin is a _____ receptor.
6. The _____ line of fishes contains receptors that detect pressure waves from nearby objects.
7. The ossicles in the middle ear of humans are the _____ , _____ , and _____ .
8. The sense of dynamic equilibrium should be associated with the _____ , present in the inner ear.
9. The auditory nerve carries nerve impulses whenever the cilia of hair cells present on the basilar membrane rub against the _____ membrane.
10. Compact bone contains _____ systems in which bones are arranged in concentric circles.
11. Both _____ and _____ muscle are striated.
12. When a sarcomere contracts, the _____ filaments slide past the _____ filaments.
13. The molecule _____ provides the energy needed for muscle contraction.
14. The junction between a motor nerve fiber and a muscle fiber is called a _____ junction.
15. A muscle action potential causes _____ ions to be released from storage sacs.

Answers to Objective Questions

1. taste, olfactory 2. retina 3. dim, bright
4. lens 5. pressure 6. lateral 7. hammer, anvil,
stirrup 8. semicircular canals 9. tectorial
10. Haversian 11. cardiac, skeletal 12. actin
(thin), myosin (thick) 13. ATP
14. neuromuscular 15. calcium

Study Questions

1. What are two ways to classify receptors?
2. Discuss the structure and function of human chemoreceptors.
3. Name the parts of the eye and give a function for each part.
4. Contrast the location and function of rods to that of cones.
5. Describe the anatomy of the ear and how we hear.
6. Describe the role of the utricle, saccule, and semicircular canals in balance.
7. Distinguish between the axial and appendicular skeletons.
8. Discuss the microscopic anatomy of a muscle fiber and the structure of a sarcomere. What is the sliding filament theory?
9. Cite three ways in which ATP is made available for muscle contraction. What is oxygen debt?
10. What causes a muscle action potential? How does the muscle action potential initiate muscle contraction?

Selected Key Terms

receptor (ri 'sep tər) 529
cornea ('kȯr nē ə) 534
lens ('lenz) 534
retina ('ret ən ə) 535
fovea centralis ('fō vē ə sen 'tral əs) 535

rhodopsin (rō 'däp sən) 536
semicircular canals (,sem i 'sər kyə lar kə 'nalz) 539
cochlea ('kō klē ə) 539
organ of Corti ('ȯr gən uv 'kȯr tī) 540
axial skeleton ('ak sē əl 'skəl ət ən) 543

appendicular skeleton (,ap ən 'dik yə lər 'skəl ət ən) 544
myofibril (mī ō 'fīb rəl) 546
actin ('ak tən) 546
myosin ('mī ə sən) 546
sarcomere ('sär kə ,mir) 548

Figure 25.1

This Australian bower bird has decorated his bower (area between the two tufts of grass) with colored objects and is waiting for a female to choose him as a mate. No doubt this behavior is a reproductive isolating mechanism (p. 231) because only a member of his own species will be attracted by his efforts.

Concepts

1. Animal characteristics are inherited and undergo evolution.
   a. Behavioral patterns are also inherited and so subject to natural selection.
   b. Altruistic behavior by members of a society can be explained on the basis of evolutionary theory.

2. Animal characteristics are dependent on their anatomy and physiology.
   a. Behavioral patterns may be associated with a continuum that ranges from primarily innate at one end to primarily learned at the other.
   b. The internal state of the animal affects the degree to which the animal performs a certain behavior.
   c. Members of a society are able to communicate by chemical, visual, and/or auditory means. They also have ritualistic ways to express aggression.

A t the start of the breeding season, the male bower bird (fig. 25.1) decorates a bower (parlor) with various types of colored objects. Thereafter, he spends most of his time near his bower, renewing his decorations while waiting for females and guarding his work against possible raids by other males. After inspecting many bowers and their owners, a female enters one and mates with the owner.

An **ethologist** (a scientist who studies behavior) might seek answers to several questions about the behavior of the male bower bird. For one thing, it would be of interest to know how this behavioral pattern evolved, and for another, to what degree the pattern is instinctive (innate) or learned. Figure 25.2 diagrammatically presents these two basic aspects of behavioral patterns. It is assumed that an animal's behavior is adaptive; that is, it increases the possibility that the individual's genes will be passed on to the next generation. In other words, behavioral patterns are subject to the process of natural selection just as other aspects of the phenotype are also subject to this process. Those individuals with the most adaptive behavioral patterns will be those that will contribute most to the gene pool of the next population of organisms. This is the manner in which the behavior of the species will evolve. The left-hand side of figure 25.2 concerns the evolution of behavior. The right-hand side of the figure concerns the execution of the behavioral pattern. The phenotype of the organism (a typical one for the species) includes a repertoire of behavioral patterns, and the particular stimulus determines which of these patterns will be executed. The behavioral pattern can range from being primarily innate to being primarily learned. Even if the pattern can be modified by learning, it should be noted that the capacity to learn is still determined by the phenotype and therefore by the genotype.

Figure 25.2
A diagram that emphasizes current interests in behavior. On the left-hand side of the diagram, note that selected phenotypes contribute genes to the gene pool of the population. In this way, behavior can evolve. On the right-hand side of the diagram, note that environmental stimuli cause organisms to display behavioral patterns that differ in the degree that they are innate or learned. In the latter case, the phenotype determines the capacity to learn.

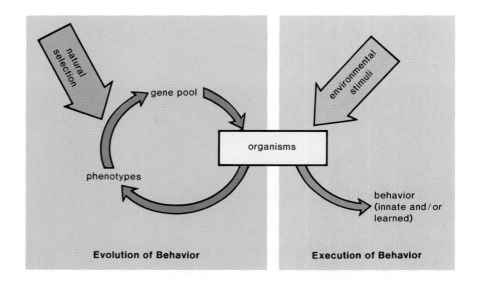

natural selection

environmental stimuli

gene pool

organisms

phenotypes

behavior (innate and / or learned)

Evolution of Behavior   **Execution of Behavior**

Evolution of Behavior

Most pet owners have observed that cats tend to be loners and dogs tend to prefer companionship. The reading on page 556 tells us that today's cats were originally adapted to living in a forest where there was no advantage in hunting or living in groups, while dogs were originally adapted to living in a grassland where they learned to hunt cooperatively. The evolutionary history of these two groups of animals explains their present behavior.

While it is interesting to consider in general the evolution of an animal's behavior, today's ethologists often prefer to study one particular behavioral pattern of an animal. For example, Nikolaas Tinbergen, who wrote *Herring Gull's World* (fig. 25.3) in 1960 and who emphasized the importance of making field observations of animals under natural conditions, observed that black-headed gulls always remove broken eggshells from their nests after the young have hatched. In a series of experiments, he determined that nests in which broken shells remained were subject to more predator attacks by crows than those from which the shells had been removed. It can be reasoned that the gulls that remove broken eggshells from their nests will successfully raise more young than the gulls that do not do so. This behavior pattern has been selected, then, through the evolutionary process because it improves reproductive success.

Genes and Behavior

Another example of behavioral evolution is the nest-building habits of the parrot *Agapornis* (fig. 25.4). Three species of this genus have different methods of building nests. Females of the most primitive species use their sharp bills

Figure 25.3
Nikolaas Tinbergen, one of the fathers of modern ethology, studied gulls in the field during the 1950s. He found that gull behavior was adaptive and therefore subject to the evolutionary process.

Figure 25.4
An *Agapornis roseicollis* parrot cuts strips of nesting paper with its beak and then tucks them in its rump feathers. Using the methods of classical genetics, W. C. Dilger managed to breed birds of this species with those of another species that simply carry the strips in their beaks. The resulting hybrids tried to tuck the strips, but were unable to perform the behavior. Later, they learned to carry the strips in their beaks.

Why Cats Are Cats and Dogs Are Dogs

Pet owners know that their cats are loners while their dogs prefer company.

Large or small, most cats, or felids, tend to be loners while the more sociable dogs, or canids, tend to prefer pairs or family groups.

Why are they so different? The answer, which goes back into evolutionary time, can help owners of cats and dogs better understand their pets. The social systems of wild dogs and cats have been molded over millions of years by where they live, how they hunt and what they eat, and today's domestic animals reflect most of the same distinctive characteristics as their wild ancestors.

Both the canids and felids belong to a group known as the carnivores (meat-eaters), and to begin to understand why their social systems are so different, one must go back 40 million years to a time when the Miacids, the ancestral carnivores, were on the scene. The Miacids were small, forest-dwelling animals that probably looked somewhat like today's civet. The fossil record shows that their paws were adapted for climbing, so they probably spent much of their time in the trees. Some of their teeth were specialized for meat-eating so they were already catching and feeding on insects, small mammals and birds. They were almost certainly solitary in their habits, as most of today's small forest carnivores are, and females probably raised their young alone, without any help from the male.

Some 25 million years ago, a change in the world's weather indirectly provided the Miacids with a new opportunity. As the climate became drier, grasslands appeared, and with them a great array of grass-eating creatures, including small horses, rabbits and hares, and many new species of rodents. The scene was now set for the ancestral carnivores to take advantage of this wealth of prey. The Miacids began to diversify. Bears, dogs, raccoons, weasels, civets, hyenas and cats appeared, each specializing to deal with different foods and different habitats.

Some of the original carnivores remained in the shadows of the forest and became highly specialized as meat-eaters. Their canine teeth became larger and more daggerlike, and they lost the flat, unspecialized crushing molars of their ancestors. They did not become runners but kept their climbing ability and grappling forelimbs, which they also used to ambush prey. They were the group that became the cats as we know them today.

Others moved out to exploit the new food sources in the open areas, and this posed an entirely new set of problems. Prey species that lived in these developing grasslands gradually acquired all kinds of different ways to avoid being eaten: they ran fast, they formed groups or they burrowed out of reach. The carnivore group that took to the plains had to be flexible. These animals needed to be able to run and dig and survive on other foods during the lean times. They became the dogs we know today.

The first dogs were small, and their fossilized skeletons suggest they looked like a cross between a fox and a genet. They kept the dual-purpose teeth of their ancestors, along with an omnivorous diet. These early dogs spent more time on the ground in open areas and became runners and chasers. Over the course of countless generations, they lost the ability to use their front paws for capturing and relied more on running down small animals which they killed with a pounce and a quick shake. . . .

for cutting bits of bark, which they carry in their feathers to a site where they make a padlike nest. Females of a more advanced species cut long, regular strips of material, which are placed for transport only in the rump feathers. These strips are used to make elaborate nests with a special section for the eggs. Females of the most advanced species carry stronger materials, such as sticks, in their beaks. They construct roofed nests with two chambers and a passageway. The steady progression in this genus from primitive to advanced nest building suggests that an evolution of behavioral patterns has taken place.

Cats, on the other hand, are adapted for making great leaps and striking down prey with their forelimbs. Their large, sharp, retractile claws and powerful, mobile forelimbs give them a distinct advantage when it comes to climbing trees or killing large animals. Cats have excellent vision and hunt mostly by sight, though hearing is also important. In keeping with their dappled, spotted or striped coats, almost all cats hunt from cover, using the stalk-and-ambush technique to capture prey. . . .

Why then do most cats lead such solitary lives? The answer is that there is no advantage for them to hunt or live in groups. They are such specialized meat-eaters that their food—be it mice or deer—is usually best hunted alone. Fewer stalking feet make less noise. Hunting in a group might be a real disadvantage in that situation. In addition, forest prey are usually scattered or in small groups. Only on the open plains do large herds exist. . . .

Lions, the only cats with a complex society, have simply extended the mother-young relationship. A pride is composed of a female and her mature daughters. Under normal conditions, all the females in a pride are closely related. The abundance of prey available in large herds means that females can "afford" to tolerate each other. In fact, cooperative hunting becomes an advantage. . . .

Wolf and African wild dog packs, on the other hand, evolved from an extension of the pair bond that exists between canid males and females in the breeding season. Unlike felids, most members of the dog family are monogamous; males mate with only one female and stay with her to help feed and defend the pups. . . . Larger pack-hunting canids have an extended pair bond that continues year-round. The young remain with the parents, and together with their grown pups, they hunt cooperatively. The packs consist of closely related individuals working together to hunt big prey.

Though the dogs' all-purpose canid teeth are a good compromise for a varied diet, they are not very effective when it comes to killing a large deer. However, several dogs working together as a team can chase an animal until it tires and then pull it to the ground with their combined strength. It is not as swift a kill as those made by cats, but the technique works, and it's the only way dogs can kill large prey. . . .

Domestic cats and dogs share many behavioral and character similarities with their wild cousins, but their social behavior has undergone substantial changes during the long process of domestication. Though many dogs live together in pairs or small groups, they rarely show the long-term pair bond of wild canids. Males do not bring food to females and pups, and when dog packs form, they are much more loosely organized than those of wolves or African hunting dogs. Domestic cats have become much more tolerant—often five or six can live together in the same house quite peacefully. But life is still easier if a group consists of a female and her grown kittens. Like lions, they are brought together by an abundant and steady food supply, but at heart, . . . they still prefer to walk by themselves.

From "Why Cats are Cats and Dogs are Dogs," by Fiona Sunquist. Copyright 1985 by the National Wildlife Federation. Reprinted from the September/October 1985 *International Wildlife*.

The fact that genes are controlling these traits can be supported by the following experiment. The species that carries material in its rump feathers has been mated to the species that carries material in the beak. The resulting hybrid birds cut strips and try to tuck them in their rump feathers, but are unsuccessful in their attempts. It is as if the genes of one parent require the offspring to tuck the strips, but the genes of the other species prevents them from succeeding. With time, however, these hybrid birds learn to carry the cut strips in their bills and only briefly turn their heads toward the rump before flying off.

Recently, investigators reported even more in-depth work with the snail *Aplysia* (fig. 25.5). This snail lays long strings of more than a million eggs. It winds the string into an irregular mass and attaches it to a solid object like a rock. Several years ago, a hormone that caused the snail to lay eggs, even though it had not mated, was isolated and analyzed. This egg-laying hormone (ELH) was found to be a small protein of thirty-six amino acids. The ELH acts locally as an excitatory transmitter, augmenting the firing of a particular neuron in the snail's body. It also diffuses into the circulatory system and excites the smooth muscle cells of the reproductive duct, causing them to contract and expel the egg string. Using recombinant DNA techniques (p. 206), the investigators isolated the ELH gene. Surprisingly, the gene's product turned out to be a protein with 271 amino acids. The protein could possibly be cleaved into as many as eleven products, and ELH is one of these that is known to be active. The investigators speculate, therefore, that all the components of the egg-laying behavior could be mediated by this one gene. This work is quite remarkable because it definitely shows that genes control behavior. It further shows the manner in which they do so in a simple organism.

Figure 25.5
The snail *Aplysia* lays long strings of eggs that are wound into an irregular mass for deposition on a nearby object. Using modern recombinant DNA techniques, investigators have isolated a gene whose product contains 271 amino acids, thirty-six of which make up the ELH (egg-laying hormone) peptide. This peptide is partially responsible for the egg-laying behavior. Analysis of the gene shows that there are enough signals for cleavage (vertical lines) to permit the product to include ten other peptides, each of which may have some role in the animal's behavior.

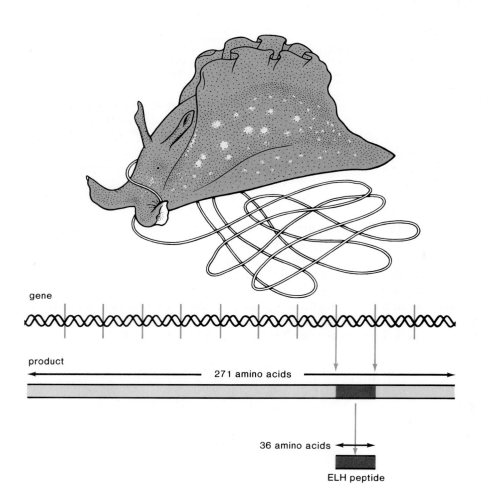

gene

product

271 amino acids

36 amino acids

ELH peptide

Execution of Behavior

Figure 25.2 indicates that behavior occurs as a response to a stimulus. Therefore, it is dependent on the sense receptors, nervous system, and the musculoskeletal system of the organism. Often, however, the animal must be motivated to respond, and motivation is most likely dependent on the physiological, or internal, state of the organism.

The complexity of behavior increases with the complexity of the nervous system. Animals having simple nervous systems tend to respond automatically to stimuli in a programmed way, whereas animals having complex nervous systems are apt to choose behavior that suits the particular circumstance. The first type of behavior, which is inherited, is called **innate** or **instinctive,** while the second type, which requires modification of behavior, is said to involve **learning.** All animals, including humans, have some instinctive behavior, but as higher animals with complex nervous systems evolved, the capacity to learn behavioral patterns developed (fig. 25.6).

Innate Behavior

In innate behavior, the stimulus appears to trigger a fixed response that does not vary according to the circumstances.

Taxes

Orientation of the body toward or away from a stimulus is termed **taxis** in animals. Animals exhibit a number of different types of taxes; **phototaxis** is movement in relation to light, and **chemotaxis** is movement in relation to a chemical. Some insects, for example moths and flies, exhibit phototaxis; they will fly directly toward a light. Often they orient themselves by shifting the body until the light falls equally on both eyes. If one eye is blind, the animal will move in a spiral, forever trying to find the direction in which the light will be balanced between the two eyes.

Chemotaxes are quite common. Insects are attracted to minute quantities of chemicals, called **pheromones,** given off by members of their species. Army ants are so set on following a pheromone trail that if, by chance, it leads them into a circle, they will continue to circle until they die of exhaustion (fig. 25.7). Vertebrates, too, are sometimes highly responsive to chemicals. A few whiffs of a piece of clothing, and bloodhounds are capable of tracking down a single individual.

Figure 25.6
Comparative frequency of modes of adaptive behavior among invertebrates and vertebrates. While innate types of behavior (taxes, reflexes, and instincts) are more frequently found among invertebrates, learned types of behavior are more frequently found among vertebrates. From V. G. Dethier and Eliot Stellar, *Animal Behavior,* Third Edition, © 1970, p. 91. Adapted by permission of Prentice-Hall, Inc., Englewood Cliffs, New Jersey.

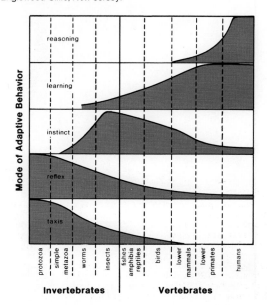

Figure 25.7
These ants are following a pheromone trail in a circle. This is an example of chemotaxis, orientation of the body in relation to a chemical in the environment.

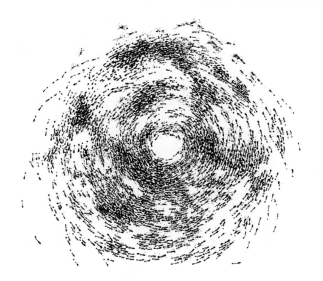

Only a few animals are able to orient themselves by means of *sonar* or *echolocation*. Bats send out series of sound pulses and listen for the echoes of these pulses. The time it takes for an echo to reach the bat indicates the location of both inanimate and animate objects. This enables a bat to find its way through dark caves and to locate food at night (fig. 25.8). Some moths have evolved the ability to hear the sounds of a bat, and they begin evasive tactics when they sense that a bat is near.

Migration and Homing

Migration, which often occurs seasonally, and **homing,** the ability to return home after being transported some distance away, have been studied in birds. Evidence indicates that they use the sun in the day (fig. 25.9) and the stars at night as compasses to determine direction (north, south, east, or west). Like bees, they even allow for the east-to-west movement of the sun during the course of a day. Bees offered a food source can return to their hive no matter where the sun is located in the sky. Therefore, it is believed that they have an internal or *biological clock* (fig. 19.15) that tells them the location of the sun according to the time of day.

Although birds use the sun and stars as compasses, they cannot rely on these to tell them that home is in a particular direction. In other words, suppose you were blindfolded and then transported away from home. Upon being set free, you were given a compass to tell direction. How would you know which direction to select? Some investigators now believe that birds are sensitive to magnetic lines of force, which are dependent on the earth's magnetic field, and this allows them to determine the direction of home.

Salmon use chemotaxis to find their way home. Salmon are born in a tributary of a river but grow to maturity in the open sea. At spawning time, mature salmon travel back up the river to the same spot at which they were born. Experiments have shown that the fish appear to return to the spawning ground by following the chemical scent of their first home.

Reflexes

Reflexes are simple automatic responses to a stimulus over which the individual appears to have little or no control. When a human knee is hit by a mallet, the lower leg jerks in a characteristic manner. Since reflexes are clearly innate, some investigators believe that it might be possible to explain complex behavior of some lower animals, such as army ants, by a series of reflexes, each one of which acts as a stimulus for the one following.

Fixed-Action Pattern   The term **fixed-action pattern** (stereotyped behavior) has been given to complex behavior that occurs automatically, as if it were a composite of reflex actions. A stimulus that initiates a fixed action behavior is called a **sign stimulus.** An animal must possess neural mechanisms (called releasing mechanisms) that are sensitive to the sign stimulus in order to respond in the stereotyped manner. As an example, consider the fact that male robins attack a red tuft of feathers in preference to an exact replica of a male robin without the red breast (fig. 25.10). The color red is a sign stimulus that provokes the releasing mechanism, which controls the attacking behavior. This behavior is therefore a fixed-action pattern. Animals performing fixed-action patterns may seem to be acting in a purposeful manner but, just as with the robin in the previous example, experimentation proves that this is not the case. For example, certain solitary digger wasps dig a hole and then seek out a caterpillar, paralyzing it with a series of stings along the undersurface. The wasp carries the prey to the nest and pulls it in (fig. 25.11). After laying her egg

Figure 25.9

Migration of Canada geese. Many different types of organisms migrate long distances to breeding grounds where they mate and reproduce. Scientists have long wondered how they find their way. Recently, it was shown that these organisms are capable of using the earth's magnetic field to navigate. A strongly magnetic iron-oxide compound called magnetite has been found in their bodies. It is possible that this compound is involved in the process.

Figure 25.10

A male robin is pecking at a red tuft rather than at an exact replica of a robin, only without a red breast. This shows that the color red is a sign stimulus that initiates the release of the hostile behavioral pattern, an innate fixed-action pattern consisting of a series of reflexes carried out by the nervous system without need for prior learning. In some instances (fig. 25.13) fixed-action patterns can be affected by learning, however.

Figure 25.11

A digger wasp is about to pull this paralyzed caterpillar into a hole. The prey will serve as food for an offspring that will hatch from an egg she lays on the caterpillar. This is an example of a fixed-action pattern because if, by chance, an experimenter removes its contents, the wasp will still close the hole.

Figure 25.12

A gull is trying to brood an artificial, oversized egg provided by an experimenter even though the proper-sized egg is in view. Experimenters have found that fixed-action behavioral patterns are more likely to be prompted by supernormal stimuli than by normal stimuli.

on the side of the caterpillar, she begins to close the burrow. If, at this point, the experimenter removes the caterpillar and puts it on the ground nearby, the wasp will continue to cover the hole even though the caterpillar is in full view.

Gulls, too, have fixed-action patterns. When a gull is incubating eggs, it will retrieve any egg that rolls out of the nest. The more speckled the egg, the more strongly the gull is stimulated to roll it into the nest and incubate it. One curious result of experiments like these has been the discovery of **supernormal stimuli.** For example, parent gulls, if given a choice in size, will prefer to retrieve eggs much larger than normal even if they are then unable to brood them (fig. 25.12).

Fixed-Action Pattern and Learning   In order to tell if a specific behavior is wholly innate, it is customary to determine if it is (1) performed by all members of the species in the same manner, and (2) performed by animals that have been raised in isolation and/or have been prevented from practicing it. Experimenters used this technique to determine if the song of a European chaffinch was innate or learned. To their surprise they found that it was a little of both. A male European chaffinch raised in isolation does sing a song, but one that is less complicated than normal (fig. 25.13). If the bird is permitted to hear another adult chaffinch sing the song, however, it quickly learns to sing the normal song. Chaffinches never learn to sing the song of other species. Therefore, as suggested by figure 25.2, the phenotype (and ultimately the genotype) has set a limit on the degree to which behavior can be modified.

Learned Behavior

Learning is a change in behavior as a result of experience. The capacity to learn is inherited and allows an organism to change its behavior to suit the environment. The organism alters its behavior to respond to a specific stimuli in ways that promote its own survival or that of its offspring and/or near relatives.

Kin Recognition

Currently, there is a great deal of interest in detecting how organisms learn to recognize close relatives. For instance, tadpoles of American toads can apparently discriminate between siblings and nonsiblings, and Belding's ground squirrels even know the difference between full siblings and half siblings.

Figure 25.13

Sound spectrograms of chaffinch songs. (*top right*) The adult male usually sings a complicated song. (*bottom right*) A bird raised in isolation sings a less complicated song. This same bird can learn to sing the normal chaffinch song, however. This shows that innate behavior can be modified by learning.

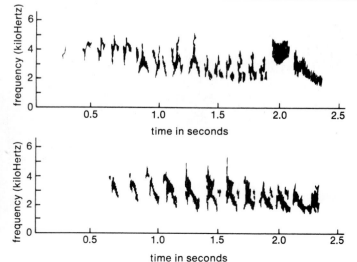

Kin recognition is beneficial in at least four ways. Parents who can recognize their offspring do not waste time rearing unrelated young; young that recognize their parents avoid the risk of being harmed by a nonparent; giving aid to a sibling rather than a nonsibling increases an animal's fitness (the likelihood of passing on one's genes); and mating with an individual that is to a degree unrelated increases the likelihood of fertile offspring.

Thus far, research in kin recognition has been most extensive in the area of imprinting.

Imprinting Konrad Lorenz, another famous ethologist, observed that birds become attached to and follow the first moving object to which they are exposed. He termed this behavior **imprinting** and suggested that it was a means by which organisms learn to recognize their own species. Ordinarily, the object followed is the mother; however, Lorenz caused chicks to imprint on him (fig. 25.14), and then showed that they would choose him over their own mother when given the opportunity.

A variety of animals, in addition to birds, are susceptible to imprinting during a certain critical period of time following birth. Therefore, this term may be used in a larger context to refer to any period of time during which a type of learning is apt to be more successful. For example, puppies take to dog training better than older dogs; and adult humans are better at sports they learned as a child.

Conditioned Learning

In a type of conditioned learning called **associative learning,** an animal learns to respond to an irrelevant stimulus. This was first observed by Ivan Pavlov, the famous Russian scientist who, at the turn of the century, experimented with dogs. Pavlov's dogs expected food and began to salivate when a bell rang because they had learned to associate the ringing of the bell with food (fig. 25.15). Associative learning can explain behavior that seems out of place as in this example; the ringing of a bell in and of itself does not have anything to do with food. At times, though, associative learning can be useful; for example, it has been suggested that mothers can instill love of learning by holding their children in their laps while reading to them.

Figure 25.14
Baby ducks are following Konrad Lorenz, who, like Niko Tinbergen, performed experiments in the field. Lorenz found that ducks learn to follow the first moving object they see, which, of course, is normally their mother. Here they have become "imprinted" on an improper object. Normally, the process of imprinting leads to evolutionary success and therefore has been selected for by the evolutionary process.

Figure 25.15
Conditioned reflex. A conditioned reflex shows that the subject has learned to associate two previously unrelated events. For example, (*a.*) a dog salivates when presented with food. If (*b.*) a bell is rung when the food is presented, the dog (*c.*) then salivates when the bell is rung even though no food is presented.

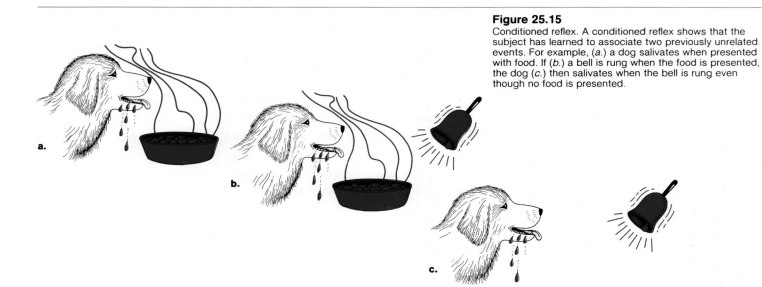

a.

b.

c.

Figure 25.16

Apes are capable of insight learning—reasoning out a solution to a problem without trial and error. On the far left, a chimpanzee is unable to reach a banana. In the middle sketch, the chimp seems to suddenly realize a solution to the problem. On the far right, it stands on boxes to reach the banana. If an animal can reason, does it have a cognitive awareness of itself and its surroundings? Investigators are now wrestling with this question.

Operant Conditioning   Operant conditioning is **trial-and-error learning**. An animal faced with several alternatives is *rewarded* (positively reinforced) for making the proper choice and thereafter learns to make this response without hesitation. This type of learning, employed by animal trainers, can be successfully used to make animals learn all sorts of tricks. Using **positive reinforcement**, B. F. Skinner, a well-known Harvard University behaviorist, has even taught pigeons to play a form of Ping-Pong. Skinner believes that humans, too, are primarily controlled by positive reinforcement mechanisms although many disagree with this supposition.

Insight Learning

Insight, or **reasoning,** is the ability to solve a problem using previous experiences to think through to a solution. To the observer, it seems as if the animal employing insight needs no practice to successfully reach a goal. For example, apes can devise the means to get to bananas placed out of their arms' reach. They will pile up boxes or use a pole in order to reach the food (fig. 25.16).

Perhaps the ultimate in reasoning ability is learning and using a language. When adult humans use language they are able to put words together in various ways to convey new and creative thoughts. As discussed in the reading on page 566, for the past several years investigators have been trying to determine the degree to which other animals can learn to use language in this way. Most of this work has been done with chimpanzees. Investigators have overcome the biological inability of chimpanzees to articulate many sounds by having them learn a sign language. Chimpanzees have learned as many as four hundred signs of an artificial visual language; however, thus far it has not been possible to demonstrate unequivocally that these animals are capable of putting the signs together to create new sentences and meanings. Some have suggested that the animals are merely copying their trainers and have no true understanding of language use.

Of course, it can be questioned whether cognition, or the ability to think, requires the use of language at all. The reading gives several examples that illustrate that even lower animals, like bees, at times seem capable of analyzing a situation. Perhaps animals only differ in the degree to which they are aware or in the manner in which they are aware of themselves and their environments.

Motivation

As mentioned earlier, certain types of behavior recur periodically in animals. These types of behavior seem to require an internal readiness before the animal shows the behavior. Thus it is said that the animal is **motivated** to perform this behavior. Motivated behavior seems to include three stages:

1. An appetitive stage during which the animal searches for the goal
2. A consummatory stage or a series of responses directed at the goal
3. A quiescent stage when the animal no longer seeks the goal

A good example of motivated behavior is the need for food. A hungry animal goes out to look for food; once food is found, it eats the food; and then, being satisfied, it no longer seeks food.

Figure 25.17
Ringdove mating behavior. Successful reproductive behavior requires these steps: (a.) male performs courtship behavior as he bows and coos; (b.) suitable nesting materials and nest site leads to building of the nest; (c.) incubation of eggs; (d.) baby chicks are fed "crop milk"; (e.) cycle begins again. It can be shown that internal hormonal levels motivate the birds to perform this behavioral pattern. Similarly, behavioral patterns in all animals are more likely to occur when the animal has been physiologically primed to perform them.

Motivated behavior requires the supposition that the animal is made ready to perform the behavior because of some internal state. A study of reproduction (fig. 25.17) in ringdoves has indeed shown this to be the case. Ringdoves reproduce in the spring, but when male and female ringdoves are separated one from the other, neither shows any tendency toward reproductive behavior. In contrast, when a pair are put together in a cage, the male begins courting by repeatedly bowing and cooing. Since castrated males do not do this, it can be reasoned that the hormone testosterone readies the male for this behavior. The sight of the male courting causes the pituitary gland in the female to release FSH and LH; these in turn cause her ovaries to produce eggs and release estrogen into the bloodstream. Now both male and female are ready to construct a nest, during which time copulation takes place. The hormone progesterone is believed to cause the birds to incubate the eggs, and while they are incubating the eggs, the hormone prolactin causes crop growth so that both parents are capable of feeding their young crop milk.

Reproductive behavior in ringdoves can be explained on the basis of both *external and internal stimuli*. The external stimuli are processed by the central nervous system, which directs the secretion of hormones. Thus, the nervous and endocrine systems work together to produce physiological changes that lead to appropriate behavioral patterns.

Do Animals Think?

Cognitive psychologists believe their newfound acceptance by other scientists is based on their strict adherence to the tenets of experimental observation. And what they have observed has much to say about intelligence in animals. For example:

Working with a pair of young female bottlenose dolphins, psychologist Louis Herman of the University of Hawaii has been able to achieve what many other scientists have not: proof of an animal's understanding not only of vocabulary, but also of communicative grammar and syntax.

Beginning in 1979, the dolphins—which were kept in two 50–foot pools—were taught two artificial languages, each made up of about 35 words. The languages have their own syntactical rules for word order, from which sentences up to five words long can be constructed. Depending on the order of the words, about 1,000 different sentences can be expressed. In one language, the words are made up of electronically generated whistles, created by a computer. In the second language, the words are formed by hand- and-arm gestures performed by the trainer. ("Ball," for instance, is signaled by a human at poolside, raising his hands quickly above his head and then lowering them.) One dolphin specialized in each language.

The evidence that the dolphins appreciate syntax was found in their ability to differentiate correctly between phrases such as "hoop fetch Frisbee" (in other words, "get the hoop and take it to the Frisbee)" and "Frisbee fetch hoop" ("get the Frisbee and take it to the hoop"), or, "net in basket" ("put the fishing net in the plastic laundry basket") and "basket in net" (the reverse). For research purposes, Herman named the two study dolphins Phoenix (the animal that responds to the trainer's hand gestures) and *Akea Kamai,* which means "Lover of Wisdom" in Hawaiian. "It is a name," remarks Herman, "chosen with hope."

Psychologists R. J. Herrnstein of Harvard, Mark Rilling of Michigan State University and Anthony Wright of the University of Texas have tested pigeons for their ability to make mental discriminations. Some of their most interesting work goes beyond the birds' ability to differentiate between objects; it tests their capacity to form concepts or categories—a higher order of intellectual accomplishment.

To date, the experiments indicate that pigeons can form generalized concepts of such test categories as "human being," "tree," "fish" and "oak leaf." In a typical test, a bird was trained to peck at a small plastic paddle when shown a slide of a human being. It was rewarded with food for a correct answer.

After the bird learned to peck when the picture of a person appeared and not to peck when photos lacking people were shown, the experiment moved to the test phase. In this stage, the bird was shown a large number of photos—1,200 in all—which it had never seen. Some of the new test transparencies were pictures of human beings. Without food reinforcement, the pigeon correctly pecked when photos of people were shown. It even did so when the new pictures portrayed groups of people. The scientists' conclusion: the pigeons must be able to understand the concept of "people." Similar experiments have also been successfully conducted using trees as a photographic subject. And Wright has achieved the same kind of results using different tests with rhesus monkeys as his subjects instead of pigeons.

Robert Seyfarth and Dorothy Cheney, ethologists from the University of California at Los Angeles, spent more than a year in Kenya studying the vocalizations of vervet monkeys. Their research shows that the animals have at least three distinct alarm cries: a raspy bark for the sighting of a leopard, after which the troop retreats to nearby trees; a short grunt for eagles and hawks, which signals a quick glance skyward and a rush for thick vegetation; and a high chutter for a snake, causing the troop members to rise up on their haunches and survey any tall grass in the vicinity. This sort of *specificity* of communication, "culturally transmitted" from adults to young, is clearly within the province once thought to be reserved only for humans.

The *Schwanzeltanz,* or "waggle dance," with which honey bees tell their sisters back at the hive the precise location of a good nectar source, was first decoded by Karl von Frisch in the 1940s (he was finally awarded the Nobel Prize for the discovery in 1973). More recently, researchers such as James Gould of Princeton University have further investigated the complexity of bee communication. "It is," says Gould, "the second most complex language we know of." (The first is the one you are employing to read this article.) For example, it appears that bee language even has dialects—and that a waggle which in effect means five yards to an Egyptian bee means 50 yards to an Austrian bee.

During the course of his work, however, Gould has chanced upon other evidence of seemingly intelligent behavior. For example, the time-honored method of training bees to fly from a hive to a desired location is to begin by placing a small container of sugar water very close to the hive and then progressively move it farther away. That allows the bees to collect the sugar and return to the hive before flying out again. The standard practice is to relocate the sugar source 25 percent farther away with each move. If it was 20 yards from the hive, it was moved to 25 yards. If it was 200 yards from the hive, it was moved to 250 yards. Using this method, Gould and his associates have successfully trained bees to fly to sources six miles from the hive.

However, during the course of the training, a curious thing began to happen. "The bees seemed to have figured out that we were moving the source in a regular way," recalls Gould. "When we moved to a new, more distant area, they were already there waiting for us—as if they had been able to figure out the rule we were using to move the sugar water. And it's hard to explain this behavior as innate, because there is nothing I know of in the natural history of flowers that has allowed them to move."

Lastly, consider the prodigious feats of memory accomplished by the Clark's nutcracker, a food-hoarding relative of the crow that lives in the southwestern United States. Biologist Russell Balda of Northern Arizona University found that the small bird spends the late summer collecting seeds from the piñon pine and burying as many as 33,000 of them in caches of four or five seeds each. According to Balda's research, the nutcracker finds its caches months after their creation by an elaborate system in which the creature memorizes nearby landmarks . . . .

. . . "Now that the legitimacy of animal cognition as a field of inquiry has been demonstrated," says Columbia's Herbert

Herbert Terrace, a psychologist, points to a puppet's mouth and a chimp responds by pointing to its own.

Terrace, a leading cognitive psychologist, "I see little gained by approaching the subject as one would approach human cognition." Maryland's Hodos is even more emphatic: "Attempts to equate animal and human intelligence may be doomed to failure," he argues. "Intelligence is a human concept. Our ideas about it are still evolving. The same behavior may be viewed as intelligence in one situation and not in another. Therefore, it represents a value judgment as much as a biological property. And what we as humans value may not be what an animal values."

The recent move to study animal intelligence on its own terms, in the words of University of Massachusetts psychologist Alan Kamil, "to refocus attention on animals as animals, with less concern for immediate human relevance," is certainly a healthy sign. "It's high time" observes Kamil, "that we stopped looking at animals as proto-humans."

As the search to understand the nature of animal intelligence continues, the quarry is now being viewed in a new light. "I believe," observes Anthony Wright, "that the role of science should be to discover the underlying processes, the foundations, the basic laws."

Most of the researchers are well aware of the scientific dangers. "If you look through the history of this field," says Mark Rilling, "it's littered with corpses rotting in blind alleys." Rutgers' Beer sounds a note of caution: "Most of the work today is quite sober, but you still have to separate the romance from the hard thinking." To which psychologist Roger Thomas of the University of Georgia adds: "We never really know who is more clever, the animal or the experimenter."

Figure 25.18
Example of chemical communication between members of the same species. Male moth antennae are capable of detecting female pheromone from miles away.

Societies

A **society** is a group of individuals belonging to the same species that are organized in a cooperative manner. In order to accomplish cooperation, evidence suggests that members of a society need a means of reciprocal communication and a means of overcoming aggression.

Communication

Communication by chemical, visual, auditory, or tactile (touch) stimuli often includes a social releaser, a sign stimulus used between members of the same species that causes the receiver to respond in a certain way.

Chemical Communication

The term *pheromone* is used to designate chemical signals that are passed between members of the same species. Pheromones can have either releaser effects or primer effects. A pheromone with a **releaser effect** evokes an immediate behavioral response, while a pheromone with a **primer effect** alters the physiology of the recipient, leading to a change in behavior.

Sex attractants are good examples of pheromones with releaser effects. For example, female moths secrete chemicals from special abdominal glands. These chemicals are detected downwind by receptors on male antennae (fig. 25.18). This signaling method is extremely efficient since it has been estimated that only forty out of forty thousand receptors on the male antennae need to be activated in order for the male to respond.

It is well known that male dogs and cats are attracted to the opposite sex by means of scent. Experimentation with apes and humans has also suggested that some individuals may be attracted by body odors. Following puberty, human males are reported to secrete about twice as much exaltolide, a chemical, in their urine as compared to females. Females are also reported to produce a vaginal chemical called copulin that possibly attracts certain males.

The members of ant colonies are controlled by numerous pheromones, each one inducing a particular response. Some are used as alarm signals, causing the ants to move about rapidly and to attack all foreign objects. Others are used to mark trails to food or new nesting sites. Still others cause the adults to take care of and nurture the larvae.

Ants and honeybees provide an example of a pheromone with a **primer effect.** The queen produces a substance that is passed from worker to worker by regurgitation. This substance prevents the workers from raising other queens and it also prevents the ovaries of the workers from maturing. Primer effects have also been seen in mice. Male mice produce a substance that can alter the reproductive cycle of females. When a new male and female are placed together, this substance can cause the female to abort her present pregnancy so that she can then be impregnated by her new mate. Similarly, crowding of female mice causes disturbances or even blockage of their estrous cycles. In both instances, removal of the olfactory lobes prevents these occurrences.

Visual Communication

The communication of honeybees is remarkable because the so-called language of the bees uses not only visual stimuli but other stimuli as well to impart information about the environment and not about the bee itself. Karl von Frisch[1] of the University of Munich, Germany, carried out many detailed bee experiments in the 1940s and was able to determine that when a foraging bee

[1]Karl von Frisch, Konrad Lorenz, and Nikolaas Tinbergen, sometimes called the fathers of modern ethology, jointly received a Nobel Prize in 1973 for their enterprising studies in animal behavior.

Figure 25.19
Visual communication among bees relates information
about the environment, not the bee. Honeybees do a
waggle dance to indicate the direction of food. *a.* If the
dance is done outside the hive on a horizontal surface,
the straight run of the dance will point to the food source.
b. If the dance is done inside the hive on a vertical
surface, the angle of the straight run to that of the
direction of gravity is the same as the angle of the food
source to the sun.

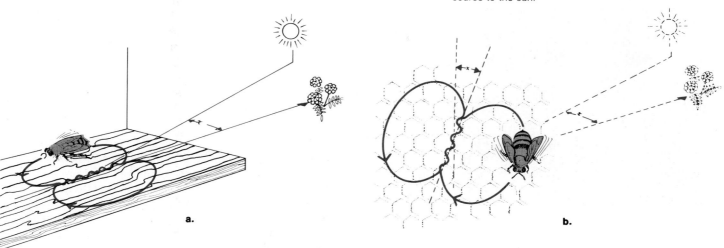

a.   b.

returns to the hive, it performs a **waggle dance** (fig. 25.19). The dance, which
indicates the distance and direction of a food source, has a figure-eight pattern. As the bee moves between the two loops of the figure eight, it buzzes
noisily and shakes its entire body in so-called waggles. *Distance* to the food
source is believed to be indicated by the number of waggles and/or the amount
of time taken to complete the straight run. The straight run also indicates the
location of the food; when the dance is performed outside the hive, the
straightaway indicates the exact direction of the food, but when it is done
inside the hive, the angle of the straightaway to that of the direction of gravity
is the same as the angle of the food to the sun. In other words, a 40° angle to
the left of vertical means that food is 40° to the left of the sun. The bees can
use the sun as a compass to locate food because their biological clocks (fig.
19.15) allow them to compensate for the movement of the sun in the sky.

Visual communication includes many **social releasers.** Male birds and
fish sometimes undergo a color change that indicates they are ready to mate.
Female baboons show that they are in estrus by a reddening of the sex flesh
on the buttocks. On the other hand, visual communication also includes defense and courtship patterns (fig. 25.24) that comprise a type of body language. These patterns are **ritualized,** which means that the behavior, which is
stereotyped, exaggerated, and rigid, is always performed in the same way so
that its social significance is clear. Ritualized behavior is believed to be derived
from body movements associated with such activities as locomotion, feeding,
and caring for the young. Facial expressions of human beings seem to be universal and it has been suggested that they, too, are ritualizations of movements
that were first used for other biological processes.

Auditory Communication
Because auditory (sound) signals are able to reach a larger audience and can
be sent even in the dark, they are sometimes favored even by animals with
good vision.

Figure 25.20
A chimpanzee with an experimenter. Chimpanzees are unable to speak but can learn to use a visual language consisting of symbols. Some researchers believe that chimps are capable of creating their own sentences, but others believe that they only mimic their teachers and never understand the cognitive use of a language. Here the experimenter shows Nim the sign for "drink." Nim copies.

Male crickets have calls and male birds have songs for a number of different occasions. For example, birds may have one song for distress, another for courting, and still another for marking territories. Sound stimuli have been shown to be more important than visual stimuli for birds that live in dense woods where vision is obstructed. In experiments, a male wood thrush attacked models of an unrelated species as long as they were silent. But if the unrelated species' song was played via a loudspeaker, the wood thrush paid the model no heed. Then, too, the wood thrush attacked a model of its own species with vehemency directly proportional to the loudness with which his own species' song was played.

One advantage of auditory communication is that the message can be modified by the sound's intensity, duration, and repetition. In an experiment with rats, an experimenter discovered that an intruder could avoid attack by increasing the frequency with which it made an appeasement sound. Rats isolated from birth could make the sound but had not learned to increase the frequency.

The fact that organisms with good vision often rely on sound communication has been demonstrated in chickens. Hens will react vigorously when they hear a chick peep even if they cannot see the chick. However, they will ignore a chick that is peeping within a soundproof glass container.

Language is the ultimate auditory communication, but only humans have the biological ability to produce a large number of different sounds and to put them together in many different ways. Nonhuman primates have at most only forty different vocalizations, each one having a definite meaning, such as the one meaning "baby on the ground" that is uttered by a baboon when a baby baboon falls out of a tree. As discussed on page 564, although chimpanzees can be taught to use an artificial visual language (fig. 25.20), they never progress beyond the capability level of a two-year-old child. It has also been difficult to prove that chimps understand the concept of grammar or can use their language to reason. It still seems as if humans possess a communication ability unparalleled by other animals.

Competition

Members of the same population compete with one another for resources, including food and mates. **Aggression** is belligerent behavior that helps an animal compete. Thus aggression helps animals establish territories, obtain mates, train and/or defend their young, and maintain status in a group.

Territoriality and Dominance

Territoriality means that a male defends a certain area (fig. 25.21) thus preventing other males of the same species from utilizing it. Territoriality spaces animals and thereby reduces aggression. It also avoids overcrowding, ensuring that the young will have enough to eat. Since animals without a territory do not mate, it also has the effect of regulating population density to some extent.

A **dominance hierarchy** exists when animals within a society form a sequence in which a higher ranking animal receives food and a chance to mate before a lower ranking animal. Dominant males lead the group and maintain order; therefore, a dominance hierarchy assures that the stronger males are in this position.

When the male defends his territory or engages another in contest to determine dominance, blood is rarely shed. Rather, the animals have a repertoire of sign signals that comprise a threat *display* or *ritual*. As in figure 25.22, the display often includes postures that make the body appear larger

Figure 25.21
Two male elephant seals battle over a piece of beachfront
property. This will be the winner's territory. Here he will
collect a harem, females that will willingly mate with him.
Territoriality partitions resources according to the
supposed fitness of the animal.

Figure 25.22
Ritualization of aggression in a great horned owl. Notice
how the feathers are fluffed and the wings are arched to
make the bird appear larger than it is.

and color changes that make the animal more conspicuous. Many of the sign signals in the display are derived from the normal activity of the animal, but now they are used to convey a social message.

When the animals are facing one another, two opposing responses—*approach response* and *avoidance response*—are simultaneously present in each individual. The animals are often in *conflict* as to whether to fight or to escape and this conflict causes them to threaten one another rather than to fight outright. Natural selection favors this situation because a contest decided by threat rather than by fighting is more apt to preserve each animal for the purpose of reproducing. The contest result is decided when one animal backs down and flees or submits.

Appeasement, or submission, occurs when an animal actually exposes itself to the attack of another, a gesture that prevents further attack. For example, a subordinate baboon of either sex turns away from an aggressor and crouches in the sexual presentation posture. Tinbersgen found that gulls have a whole range of postures from actual fighting to appeasement. In gulls, food-begging behavior in adults is appeasement. Appeasement behavior is believed to cause even greater conflict in the aggressor so that inactivity results.

When an aggressor is in conflict, *redirection* of aggression or *displacement* of aggression may occur. As an example of the first of these, a bird might peck at the ground and a human might bang on a table. Displacement of aggression is recognized by the fact that the animal performs an irrelevant activity. A bird might preen its feathers (fig. 25.23), while a human might pull on his chin.

Courtship is a time during which aggression must be at least temporarily suspended in order for mating to take place. At this time, conflict within the male may cause him to vacillate between aggression and nonaggression. For example, in the spring the male stickleback stakes out a territory and builds a nest; at this time, his body becomes highly colored, including a red belly. Any male attempting to enter the territory is attacked as the owner repeatedly darts toward and nips the intruder. (Experiments have shown that the red belly of the male acts as a sign signal.) On the other hand, the owner entices a female to enter the territory by first darting toward her and then away in a so-called zigzag dance (fig. 25.24). Finally, he leads her to the nest, where she deposits her eggs. Investigators have pointed out that the zigzag dance of the male actually contains the same aggressive movements used when the male darts toward and attacks a trespassing male.

Sociobiology

It can be shown that the degree to which members of a species cooperate with one another is dependent on their own phenotype and by the nature of the environment. For example, an investigator reports that in the Kibale Forest in eastern Africa, chimpanzees usually forage in groups of three or four. The group spreads out and when one spots a tree in fruit, it calls to the others. Since they are able to travel on the ground, it is advantageous for them to forage in a group. On the other hand, orangutans are solitary feeders. They literally have four hands that are adapted to gripping tree limbs. Therefore, orangutans do not travel well on the ground. Under these circumstances, it is best for each one to hoard for itself any tree that is in fruit. Or we might take another example from reproductive behavior. Song birds (fig. 16.17) are often monogamous while mammals tend to be polygamous. In the case of birds,

Figure 25.24
Reproductive behavior of stickleback fish. *a.* A red belly on a crude model stimulates females and males more than an exact replica of a fish without a red belly. *b.* The male entices the female to the nest, where he lies flat on his side; the female swims into the nest and lays her eggs when prodded by the male, and the male will later enter the nest to fertilize the eggs. Finally, the male beats his pectoral fins and creates a current of water that flows through the nest. This action supplies oxygen to the developing eggs.

a.

b.

incubating the eggs and/or gathering enough food requires the effort of both parents. If they did not cooperate, neither would have any offspring. In the case of mammals, the female, not the male, is anatomically and physiologically adapted to care for the offspring. Under these circumstances, it is more adaptive for the male to impregnate as many females as possible.

The basic tenet of **sociobiology,** the biological study of social behavior, is that cooperative behavior among members of a society only exists because such behavior increases the fitness of the individual. *Fitness,* of course, is judged by the frequency that an animal's genes are passed on to the next generation.

Altruism

At times, it may seem as if an animal is acting **altruistically;** that is, for the good of the group rather than for self-interest. For example, in ants and bees, only the queen lays eggs and most of the hive consists of female workers who do not have any offspring at all. How can such a society increase the fitness of the workers? The answer lies in the fact that the male parent is haploid; therefore, siblings have three-fourths of their genes in common (fig. 25.25). Since offspring have only one-half of their genes in common with their parents, it actually increases the fitness of the worker ants to raise siblings rather than offspring.

Altruism, exemplified by the willingness of daughters to help the queen rather than to have their own offspring, can be explained by noting that **kin selection** increases the animal's **inclusive fitness.** In other words, survival of close relatives (*kin*) increases the frequency of an animal's genes in the next generation. Therefore the animal's overall (*inclusive*) fitness is also increased.

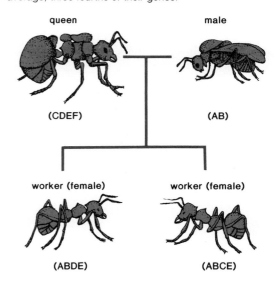

Figure 25.25
Sociobiology and the behavior of ants. It is more genetically advantageous for female ants and bees to assist in caring for siblings rather than their own offspring. Offspring possess one-half of the queen's genes, but since the male parent is haploid, siblings share, on the average, three-fourths of their genes.

queen

male

(CDEF)

(AB)

worker (female)

worker (female)

(ABDE)

(ABCE)

Figure 25.26
Acorn woodpeckers, which live and mate in a group, defend their store of acorns against birds of another group. It can be shown that group living increases their individual fitness.

Communal Groups The concept of kin selection can possibly apply to other instances, as, for example, in the case of acorn woodpeckers (fig. 25.26). These birds received this name because they store acorns in holes drilled in trees. In California and New Mexico, acorn woodpeckers usually live in a social group consisting of about fifteen birds. Each group defends its storage tree from birds that are not a member of the group.

Investigators have noted that only certain members of a group are *breeders* who share mates. Mate sharing, which is rarely seen in birds, does, in this instance, increase the inclusive fitness of the individual bird. First of all, groups have more offspring per male than do single pairs. For a group of three, investigators found there were 1.16 young per adult male on the average; for pairs there were only .92 young per male. Secondly, among the sharers, the males are usually brothers and the females are usually sisters.

In a large group, there are several *nonbreeders,* birds who only help the breeders raise their young. Since nonbreeders are offspring of the breeders, it is possible that they are also increasing their inclusive fitness by their behavior. However, the investigators also note that the adjoining territories are already occupied and the birds have a very difficult time establishing themselves as breeders. Under these circumstances, it may be that it is simply more advantageous for the nonbreeders to remain in a group where they can feed on the stored acorns and wait until there is a vacancy among the breeding birds.

Alarm Callers Sociobiologists suggest that all animal behavioral patterns should be interpreted as selfish acts. For example, some animals that move in groups give an alarm call when a predator approaches. Careful analysis is expected to reveal that the alarm callers are not putting themselves in danger and may, instead, be protecting *themselves* by this action. Sociobiologists maintain that there is always a biological explanation for social behavior that is consistent with and supports the concept of organic evolution by natural selection of the fittest.

Humans Sociobiologists interpret human behavior according to these same principles. For example, parental love is clearly selfish in that it promotes the likelihood that an individual's genes will be present in the next generation's gene pool. People tend to keep their aggression under control with those they recognize as blood relatives. Also, we would expect small towns to have less crime than large cities because small towns have a greater percentage of related residents.

Human reproductive behavior can be similarly examined. Human infants are born helpless and have a much better chance of developing properly if both parents contribute to the effort. Perhaps this explains why the human female has evolved to be continuously amenable to sexual intercourse. (Other mammalian females are receptive only during estrus, a physiological state that recurs periodically; sometimes only once a year.) The fact that sex is continuously available may help assure that the human male will remain and help the female raise the young.

Human beings are usually monogamous (one mate at a time); however, there are exceptions that appear to be adaptations to the environment. Among African tribes, for example, one man may have several wives. This is reproductively advantageous to the male, but is also advantageous to the woman. By this arrangement she has fewer children who are thereby assured a more nutritious diet. In Africa, sources of protein are scarce and early weaning poses a threat to the health of the child. In contrast, among the Bihari hillpeople of

India, brothers have the same wife. Here the environment is hostile and it takes two men to provide the everyday necessities for one family. Since the men are brothers, they are actually helping each other look after common genes.

It should be stated that not all scientists and laypeople are in favor of applying sociobiological principles to human behavior. Many are concerned that it may lead to the perception that human reproductive behavior is fixed and cannot be changed. Particularly, today, when people are exploring new avenues of behavior, some do not wish to unduly stress any potential obstacles to bringing about a new social order.

Summary

Behavioral patterns are inherited and evolve in the same manner as other animal traits. Breeding experiments with the parrot *Agapornis* and recombinant DNA experimentation with the snail *Aplysia* support this contention. Also, the present behavior of animals can often be explained by an analysis of their evolutionary pasts.

Behavior occurs as a response to a stimulus and is often divided into two categories: innate (instinctive) and learned. Innate behavior includes taxes and reflexes. Taxes, together with the possession of an internal clock, can in some cases offer an explanation for the ability of animals to return to certain locations. Fixed-action patterns, which are a series of reflexes, occur as a response to a sign stimulus. These are largely inherited, although they often increase in efficiency with practice.

Learned behavior includes kin recognition, conditioned learning, and insight learning. Much investigative work is being done to determine how aware animals—other than humans—are of themselves and their surroundings, and if these animals can be taught to use a language. An animal's motivation often determines whether or not it performs a behavior.

Members of a society communicate and compete with one another. This communication can be chemical, visual, and/or auditory. If the group is to survive, the members must overcome a certain amount of aggression; enough that cooperation results. Territoriality and dominance are two mechanisms by which aggression is controlled. During courtship, aggression is minimized, although aggressive actions are still detectable.

Sociobiology is the study of social behavior according to the tenets of evolutionary theory. While it may seem as if members of a group are altruistic, it can be shown that their motives are most likely selfish and their behavior serves to increase their own fitness. Sometimes it is necessary to utilize the concept of inclusive fitness when interpreting an animal's behavior on this basis.

Objective Questions

1. Behavioral patterns are _____ and suited to the environment.
2. Breeding experiments with *Agapornis* parrots and biochemical studies with *Aplysia* snails show that behavior is controlled by the _____ .
3. A taxis is the _____ of the body in relation to an environmental stimulus.
4. A stimulus that initiates a fixed-action behavior is called a _____ stimulus.
5. Experiments have shown that the ability of chaffinches to sing the normal adult song is both _____ and _____ .
6. Birds and other animals become _____ to the first moving object they see.
7. Most animal trainers use _____ to teach animals to do tricks.
8. Pheromones are an example of _____ communication.
9. _____ communication includes various sign stimuli.
10. Both territoriality and dominance hierarchies help reduce outright _____ between animals.
11. Sociobiology says that each member of a society is only cooperating with the group because to do so is in its own _____ best interest.
12. Acorn woodpecker offspring help their parents raise young because it actually increases their own _____ to do so.

Study Questions

1. Draw a diagram showing that behavior evolves and occurs as a result of an environmental stimulus.
2. Describe the breeding experiments with the parrot *Agapornis,* and the results of recombinant DNA work with the snail *Aplysia.*
3. Give several examples of taxes in animals. Include an explanation for migration and homing.
4. Define a fixed-action pattern. Give an example of a fixed-action pattern that is modified by learning.
5. Name three types of learning and give an example of each type.
6. Describe the ringdove reproductive experiment and its significance in regard to motivation.
7. Give examples of three means of communication between members of a society.
8. Name and discuss mechanisms that help reduce overt aggression among members of a society.
9. State the basic tenet of sociobiology and give examples to show that members of a society only cooperate to increase their own fitness.

Selected Key Terms

ethologist (ē 'thäl ə jest) 553
taxis ('tak səs) 559
phototaxis (,fōt ō 'tak səs) 559
chemotaxis (,kē mō 'tak səs) 559
pheromone ('fer ə ,mōn) 559
migration (mī 'grā shən) 560
homing ('hōm iŋ) 560
reflex ('rē ,fleks) 560

imprinting ('im ,print iŋ) 563
operant conditioning ('äp ə rənt kən 'dish ə niŋ) 564
insight ('in ,sīt) 564
waggle dance ('wag əl 'dans) 569
social releaser ('sō shəl ri 'lē sər) 569
territoriality (,ter ə ,tōr ē 'al ət ē) 570
sociobiology (,sō sē ō bī 'äl ə jē) 573

Suggested Readings for Part 9

Alcock, J. 1979. *Animal behavior: An evolutionary approach.* Sunderland, MA: Sinauer.

Alkon, D. L. July 1983. Learning in a marine snail. *Scientific American.*

Blaustein, A. R., and O'Hara, R. K. January 1986. Kin recognition in tadpoles. *Scientific American.*

Bloom, F. E. October 1981. Neuropeptides. *Scientific American.*

Bonner, J. T. April 1983. Chemical signals of social amoeba. *Scientific American.*

Cantin, M., and Genest, J. February 1986. The heart as an endocrine gland. *Scientific American.*

Fenton, M. D., and Fullard, J. H. 1981. Moth hearing and feeding strategies of bats. *American Scientist* 69:266.

Fine, A. August 1986. Transplantation in the central nervous system. *Scientific American.*

Ghiglieri, M. June 1985. The social ecology of chimpanzees. *Scientific American.*

Gwinner, E. April 1986. Internal rhythms in bird migration. *Scientific American.*

Hadley, N. F. July 1986. The arthropod cuticle. *Scientific American.*

Horridge, G. A. July 1977. The compound eye of insects. *Scientific American.*

Hudspeth, A. J. January 1983. The hair cells of the inner ear. *Scientific American.*

Israël, M., and Dunant, Y. April 1985. The release of acetylcholine. *Scientific American.*

Jones, D. G. 1981. Ultrastructural approaches to the organization of central synapses. *American Scientist* 69:200.

Keynes, R. D. March 1979. Ion channels in the nerve cell membrane. *Scientific American.*

Levine, J. S., and MacNichol, E. F. February 1982. Color vision in fishes. *Scientific American.*

Ligon, J. D., and Ligon, S. H. July 1982. The cooperative breeding behavior of the green woodhoopoe. *Scientific American.*

Llinas, R. R. October 1982. Calcium in synaptic transmission. *Scientific American.*

Lorenz, K. 1981. *The foundations of ethology.* New York: Springer-Verlag.

Manning, A. 1979. *An introduction to animal behavior.* 3d ed. New York: Addison-Wesley.

Nauta, W. J. H., and Feirtag, M. September 1979. The organization of the brain. *Scientific American.*

Newman, E. A., and Hartline, P. H. March 1982. The infrared "vision" of snakes. *Scientific American.*

Schwartz, J. H. April 1980. The transport of substances in nerve cells. *Scientific American.*

Silver, R. 1978. The parental behavior of ring doves. *American Scientist* 66:209.

Smith, J. M. September 1978. The evolution of behavior. *Scientific American.*

Snyder, S. H. October 1986. The molecular basis of communication between cells. *Scientific American.*

Stevens, C. F. September 1979. The neuron. *Scientific American.*

Wilson, E. O. 1980. *Sociobiology: The abridged edition.* Cambridge, MA: Belknap.

Wurtman, R. J. April 1982. Nutrients that modify brain function. *Scientific American.*

Zwislocki, J. J. 1981. Sound analysis in the ear: A history of discoveries. *American Scientist* 69:184.

PART 10

Animal Reproduction and Development

Members of the same species compete with one another to reproduce because reproduction ensures the passage of an individual's genes. For the species, reproduction is the means by which a population size can be maintained or increased. It may also produce individuals that are better able to exploit a new environment.

Reproductive processes involve specific activities by individual cells. Even the reproductive strategies of multicellular animals include a return to the single-cell level since they produce specialized individual reproductive cells, such as eggs and sperm. Via such specialized cells, a male and female parent each contributes one-half of the genes to a newly formed zygote. The zygote then undergoes a series of developmental stages during which differentiation occurs, even though the same set of genes is received by all cells. It is possible to view and describe the various stages of development, but the manner by which differentiation is controlled is one of the great unsolved problems of biology.

Ring-tailed lemur with young. Placental mammals practice internal fertilization and development. Females have mammary glands that produce milk for the offspring shown nursing here. Usually primates have only one offspring at a time. It is difficult to care for more than one offspring while climbing about in a tree.

CHAPTER 26

Animal Reproduction

Concepts

1. Reproductive strategies are adapted to the environment.
   a. Animals reproduce by asexual and sexual means, both of which are advantageous depending on the environment.
   b. Human beings practice sexual reproduction in a manner that is an adaptation to life on land.
2. In higher animals, male and female are often adapted to the reproductive role they play.
   Hormones control the reproductive process and the sex characteristics of the individual.
3. Unlike other animals, humans are capable of planning reproduction.
   a. Birth control measures are available to help prevent pregnancy.
   b. Measures are also available to help seemingly infertile couples.
   c. Sexually transmitted diseases are a current concern.

Animals expend a considerable amount of energy in reproduction. This is most likely because reproduction ensures that the animals' genes are passed on to the next generation of organisms. The life cycle of any particular animal comes to an end, but its genes can be perpetuated as long as reproduction has taken place. In evolutionary terms, the most fit organisms are the ones that have generated the most offspring. They are the best adapted to the environment, and it is beneficial to the species that they reproduce more than their cohorts.

Patterns of Reproduction

There are two fundamental patterns of reproduction: asexual and sexual. We will examine the methodology and the particular advantage of each pattern.

Asexual Reproduction

In asexual reproduction, there is only one parent (fig. 26.1), and the offspring tends to have the same genotype and phenotype as that parent. This lack of variability is not a disadvantage as long as the environment stays the same. Asexual reproduction often has the tremendous advantage of being able to produce a large number of offspring within a limited amount of time.

Figure 26.1
Asexual reproductive pattern versus sexual reproductive pattern. (*left*) In asexual reproduction, there is only one parent and the offspring tend to be identical to each other and to the parent. Asexual reproduction occurs in various ways: by regeneration, by budding, or by parthenogenesis. (*right*) In sexual reproduction, there are most often two parents. Each parent contributes one-half of the genes to the offspring either by way of the sperm or the egg. Therefore, the offspring tend to be genetically and phenotypically different from either parent.

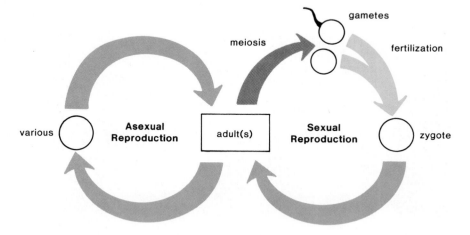

Only certain methods of asexual reproduction are found among animals. Many flatworms can constrict into two halves, each of which become a new individual. This is a form of **regeneration** that is also seen among sponges and echinoderms. Chopping up a sea star does not kill it at all. Instead, each fragment grows into a complete animal, and the total number of sea stars actually increases in the process. Some coelenterates, such as *Hydra,* reproduce by **budding** (fig. 15.7), during which the new individual arises as an outgrowth (bud) of the parent. Insects have the ability to reproduce parthenogenetically (fig. 26.2). **Parthenogenesis** is a modification of sexual reproduction in which an unfertilized egg develops into a complete individual.

Sexual Reproduction

Sexual reproduction (fig. 26.1) involves the utilization of **gametes** (sex cells). The gametes may be specialized into **eggs** or **sperm,** which are produced by the same or separate individuals. Earthworms (fig. 15.25) practice cross fertilization even though they are *hermaphroditic* and have both male and female sex organs. Among vertebrates, the sexes tend to be separate, and it is often easy to tell whether an animal is an egg-producing female or a sperm-producing male (fig. 26.3). When animals reproduce sexually, the offspring inherits half its genes from one parent and the other half of its genes from the other parent. Therefore, the offspring often has a different combination of genes than either parent. In this way, variability may be introduced and maintained. Such variability is an advantage to the species if the environment is changing because an offspring might be better adapted to the new environment than is either parent.

Figure 26.2
Stem covered with aphids. In the summer, many generations of aphid females produce up to one hundred young without prior fertilization. In the autumn, however, aphids reproduce sexually and the zygotes survive the winter.

Figure 26.3
Male and female vertebrates of the same species are biologically distinguishable: The male sex has testes that produce sperm, and the female sex has ovaries that produce eggs. In addition, the sex hormones often bring about the secondary sex characteristics that cause the sexes to differ in appearance. Note that the female lion in the foreground lacks the mane of the male lion in the background.

Figure 26.4
Generalized vertebrate brain shows the location of the pineal gland. The pineal gland is believed to contain a biological clock system (see fig. 19.15). It is sensitive to day length and produces the hormone melatonin. Melatonin is believed to communicate with the reproductive system so that breeding is synchronized with the seasons.

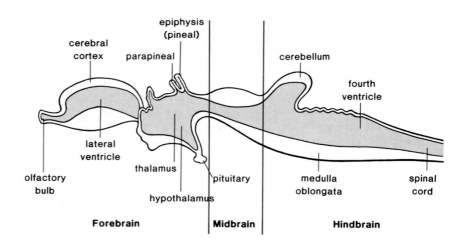

Reproductive Timing

The majority of sexually reproducing animals have a single breeding period each year. Reproduction at this time of year is most favorable to survival of offspring. For example, breeding in wild mammals is timed so that the end of the period of internal development (gestation) occurs when conditions are favorable for the growth of the newborn animals. Thus, many temperate-zone mammals breed in the fall and deliver young in the spring.

Just as we observed with plants, the reproductive cycles of a very large percentage of animals, both vertebrates and invertebrates, are timed by responses to day-length changes. Day-length changes are the most reliable indicators that it is the proper time for reproductive behaviors, including migration to far-distant places. Researchers have found that in many species there is a daily rhythm in the production of the hormone *melatonin* by the *pineal gland* (fig. 26.4). Production of melatonin is stimulated by darkness and inhibited by light. Apparently, the level of melatonin at certain times of the day provides a photoperiodic signal that controls the activity of the reproductive system in seasonally breeding mammals.

External versus Internal Fertilization

Many aquatic animals practice **external fertilization.** They shed their gametes into the water at exactly the same time and same place. The probability of fertilization is increased by the production and release of huge quantities of gametes (a female oyster may release sixty million eggs in one season). External fertilization is possible because environmental water protects the zygote and embryo from drying out. Often the embryo is a swimming larva that is capable of acquiring its own food (fig. 26.5).

Internal fertilization is a mechanism that improves the probability that the sperm will meet the egg. Even some aquatic animals practice internal fertilization. In certain sharks, skates, and rays, the pelvic fins are specialized to pass sperm to the female, and in many of these animals the young develop

a.

b.

c.

d.

Figure 26.5
Life cycle of the frog requires a water environment.
a. Fertilization is external and development occurs in the
water. *b.* The swimming larva is capable of feeding itself.
c. Metamorphosis is occurring; the larva is taking on the
characteristics of the adult frog. *d.* Adult frog.

internally and are born alive. On land, internal fertilization is a necessity because sperm and egg dry quickly when exposed to air. Often the male has a copulatory organ by which the sperm are transferred to the female. Following fertilization, the egg may be encased in a protective covering (fig. 16.14) and laid in a prepared location. In placental mammals, the fetus develops within the uterus where it receives nourishment by way of the placenta. Humans can be used as an example of reproduction in placental mammals.

Figure 26.6
Side view of the male reproductive system. Notice that the urethra in males carries either urine from the bladder or semen from the testes.

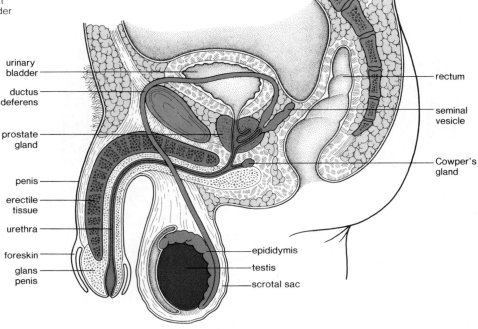

urinary bladder

ductus deferens

prostate gland

penis

erectile tissue

urethra

foreskin

glans penis

rectum

seminal vesicle

Cowper's gland

epididymis

testis

scrotal sac

Table 26.1	Male Reproductive System
Organ	**Function**
Testes	Produce sperm and sex hormones
Epididymis	Maturation and some storage of sperm
Ductus deferens	Conducts and stores sperm
Seminal vesicles	Contribute to seminal fluid
Prostate gland	Contributes to seminal fluid
Urethra	Conducts sperm
Cowper's glands	Contribute to seminal fluid
Penis	Organ of copulation

Human Reproduction

The manner in which humans reproduce is similar to that of other mammals and represents an adaptation to the land environment. When the male penis passes sperm to the female, desiccation not only of the male and female gametes but also of the subsequent embryo is prevented.

Male Reproductive System

Figure 26.6 shows the reproductive system of the male, and table 26.1 lists the organs and their functions. The male gonads are paired **testes** suspended in a saclike structure, the **scrotum.** The testes begin their development inside the abdominal cavity but descend into the scrotum as development proceeds. If the testes do not descend and the male is not treated by administration of hormones or operated on to place the testes in the scrotum, **sterility,** the inability to produce offspring, results. The reason for sterility in this case is that normal sperm production does not occur at body temperature; a cooler temperature is required.

A longitudinal cut (fig. 26.7) shows that internally each testis is composed of compartments called **lobules,** each of which contains one to three tightly coiled **seminiferous tubules.** Altogether, these tubules have a combined length of approximately 250 m. A microscopic cross section of a tubule (fig. 26.7b) shows that it is packed with cells undergoing *spermatogenesis*. These cells are derived from undifferentiated germ cells, called **spermatogonia** (singular, spermatogonium), that lie just inside the outer wall and divide mitotically, always producing new spermatogonia. (Germ cell is a general term for

582   Animal Reproduction and Development

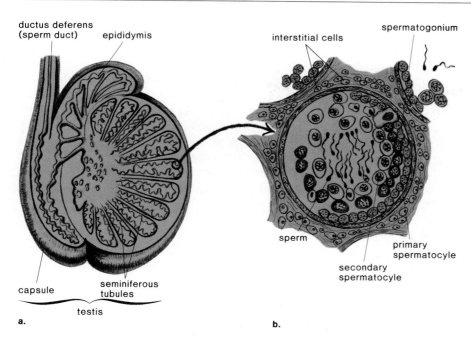

ductus deferens
(sperm duct)   epididymis

capsule   seminiferous
           tubules

testis

a.

interstitial cells   spermatogonium

sperm

secondary
spermatocyle

primary
spermatocyle

b.

Figure 26.7
Sections through a human testis. *a.* Longitudinal section showing lobules containing seminiferous tubules. *b.* Cross section of a tubule showing germ cells in various stages of spermatogenesis.

those cells in males and females that give rise to the sex cells.) Some spermatogonia move away from the outer wall to increase in size and become primary **spermatocytes** that undergo meiosis, a type of cell division described in chapter 7. While these cells have forty-six chromosomes, they divide to give secondary spermatocytes, each with twenty-three chromosomes. Secondary spermatocytes divide to give spermatids that also have twenty-three chromosomes, but that are single-stranded. Spermatids then differentiate into spermatozoa, or mature sperm.

Spermatozoa (fig. 26.18) have three distinct parts: a head, a middle piece, and a tail. The middle piece and tail contain microtubules in the characteristic $9+2$ pattern of cilia and flagella. Wrapped around these and assisting motility in the middle piece are mitochondria and in the tail, nine protein fibers. The head contains a nucleus covered by a cap called the **acrosome** that is believed to store enzymes needed for fertilization. The human egg is surrounded by several layers of cells and a thick membrane. These acrosome enzymes play a role in allowing the sperm to reach the surface of the egg so that one sperm can enter.

Each acrosome may contain so little enzyme that it requires many sperm for one to actually enter the egg. This could explain why so many sperm are required for the process of fertilization. The normal human male usually produces several hundred million sperm per day, assuring an adequate number for fertilization to take place.

Male sex hormones (chiefly **testosterone**) are produced by the **interstitial cells** scattered in the spaces between the seminiferous tubules (fig. 26.7). Sex hormones bring about maturation of the sex organs and are also largely responsible for the secondary sex characteristics that appear as the male undergoes puberty. In humans the secondary male characteristics include a beard, deep voice, and increased muscle strength.

Figure 26.8
The hypothalamus-pituitary-testis control relationship. Testosterone acts on various body tissues and also regulates the amount of hypothalamic GnRH being sent to the pituitary. GnRH affects gonadotropic hormones production by the pituitary. LH regulates the amount of testosterone produced, and FSH controls spermatogenesis. Some investigators suggest that the seminiferous tubules release a substance that is also involved in feedback inhibition.

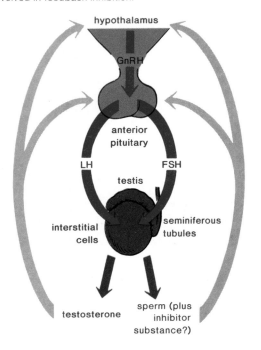

Hormone Regulation

The anterior pituitary produces two gonadotropic hormones, **LH** and **FSH,** in both males and females. LH in males is sometimes given the name interstitial cell stimulating hormone (**ICSH**) because it controls the production of **testosterone** by the interstitial cells. FSH promotes spermatogenesis in the seminiferous tubules. The hypothalamus (p. 524) has ultimate control of the testes' sexual functions, however, because it secretes a releasing hormone (GnRH = gonadotropin releasing hormone) that stimulates the anterior pituitary to produce the gonadotropic hormones. All of these hormones are involved in a feedback relationship that maintains the fairly constant production of sperm and testosterone (fig. 26.8). This functional connection between hypothalamus and pituitary also accounts for the ability of psychological factors, such as visual sexual stimuli, to influence blood testosterone levels and thereby change the general level of sexual responsiveness.

Path of Sperm

Sperm are produced within the seminiferous tubules of the testes, but mature within the **epididymides** (sing., **epididymis**), which are tightly coiled tubules lying just outside the testes. Maturation seems to be required for the sperm to swim and thus to fertilize the egg. Once the sperm have matured, they are propelled into the **ductus** (vas) **deferentia** (sing., **ductus deferens**) by muscular contractions. Sperm are stored in part in both the epididymides and the ductus deferentia. When a male becomes sexually aroused, sperm enter the **urethra,** which is located within the **penis.**

The penis is a cylindrical organ that usually hangs in front of the scrotum (fig. 26.9). Spongy erectile tissue containing distensible blood spaces extends through the shaft of the penis. During sexual arousal, nervous reflexes cause an increase in arterial blood flow to the penis. This increased blood flow fills the blood spaces in the erectile tissue, and the penis, which is normally limp (flaccid), stiffens and increases in size. These changes are called **erection** (fig. 26.9). If the penis fails to become erect, the condition is called **impotency,** which may be due to medical or psychological problems. In many male mammals a bone in the penis gives it a permanent partial rigidity, but the copulatory function of the human penis depends entirely on erection by filling of the blood spaces in the erectile tissue.

Figure 26.9
Erect penis. The position of the flaccid penis and the relaxed scrotum is also represented.

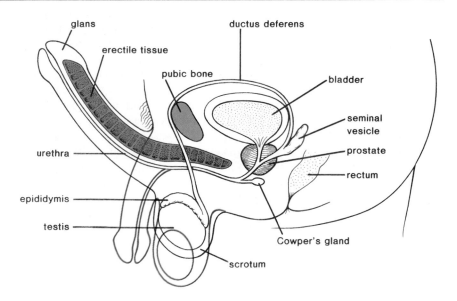

Orgasm in Male

Orgasm in both sexes is characterized by a release of neuromuscular tension, particularly of the genital area but also of the body as a whole. In males, orgasm is obvious because rhythmical muscle contractions compress the urethra and expel semen. This is termed **ejaculation,** after which a male typically experiences a **refractory period** when stimulation does not bring about an erection.

Semen

Semen, a thick whitish fluid, contains sperm and seminal fluid. As table 26.1 indicates, three types of glands contribute to the production of seminal fluid. The first of these are the **seminal vesicles,** which lie between the bladder and the rectum and each of which joins with a ductus deferens to form an ejaculatory duct that enters the urethra. As sperm pass from the ductus deferentia, these vesicles secrete a thick, viscous fluid containing nutrients for possible use by the sperm. Just below the bladder is the **prostate gland,** which secretes a milky alkaline fluid believed to activate or increase the motility of the sperm. In older men the prostate gland frequently becomes enlarged, thus constricting the urethra and making urination difficult. Slightly below the prostate gland on either side of the urethra is a pair of small glands called **Cowper's glands,** which have mucous secretions that have a lubricating effect. Notice from figure 26.6 that the urethra carries, at different times, urine from the bladder and semen from the ductus deferentia.

Female Reproductive System

Table 26.2 lists the female reproductive organs and their functions, and figure 26.10 shows the female reproductive system. The **ovaries,** the gonads in the female, lie in shallow depressions on the lateral wall of the pelvis attached to

Table 26.2 Female Reproductive System

Organ	Function
Ovaries	Produce egg and sex hormones
Oviducts (Fallopian tubes)	Conduct egg
Uterus (womb)	Location of developing offspring
Vagina	Copulatory organ
Clitoris	Sexual response

Figure 26.10
Side view of the female reproductive system. The egg awaits the sperm in the oviduct where fertilization occurs.

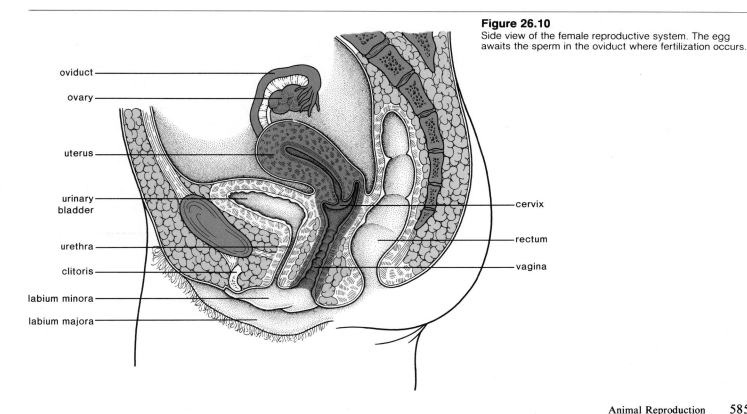

oviduct

ovary

uterus

urinary bladder

urethra

clitoris

labium minora

labium majora

cervix

rectum

vagina

Animal Reproduction   585

Figure 26.11
A diagrammatic longitudinal section of the ovary and the
oviduct. The diagram shows a follicle in all stages of
development, ovulation, and fertilization. A single follicle
goes through all stages in one place within the ovary.

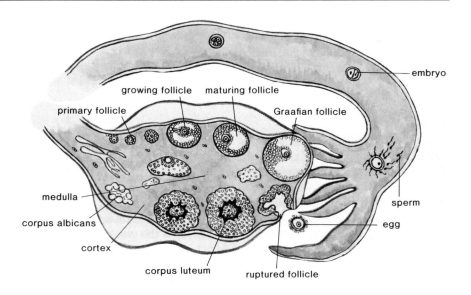

Figure 26.12
Human ovulation. The surface of the ovary and follicle
break open. The oocyte emerges, surrounded by fluid
from the follicle.

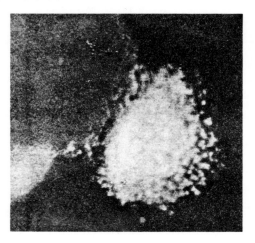

the uterus and pelvic wall by ligaments. A longitudinal section of an ovary
(fig. 26.11) shows that it is made up of an outer **cortex** and inner **medulla.** The
cortex contains **follicles** at various stages of maturation. A female is born with
a large number of immature follicles (400,000 in both ovaries), each containing a germ cell (oocyte) surrounded by a layer of nongerminal cells. Only
a small number of these follicles (about four-hundred) ever mature because
a female produces only one egg a month during her reproductive years. Since
follicles are present at birth, they age as the woman ages. This has been given
as a reason why an older woman is more apt to produce a child with a genetic
defect.

As a follicle undergoes maturation, it develops from a primary to a secondary to a Graafian follicle. In a *primary follicle* an **oocyte** divides meioti-
cally into two cells, each having twenty-three chromosomes. One of these cells,
termed the **secondary oocyte,** receives almost all the cytoplasm, nutrients, and
enzymes. The other is a *polar body* that disintegrates. A *secondary follicle*
contains the secondary oocyte pushed to one side of a fluid-filled cavity. In a
Graafian follicle, the fluid-filled cavity increases to the point that the follicle
wall balloons out on the surface of the ovary and bursts (fig. 26.12), releasing
the secondary oocyte surrounded by the zona pellucida and a few cells. This
is referred to as **ovulation,** and for the sake of convenience, the released germ
cell is called an **ovum,** or egg (fig. 26.18). Actually, the second meiotic division
does not take place unless fertilization occurs.

Besides producing eggs, the ovary also secretes the female sex hormones,
estrogen and **progesterone.** Prior to ovulation a follicle primarily secretes estrogen, but after ovulation a follicle becomes the **corpus luteum** that also secretes increasing amounts of progesterone. If fertilization does not take place,
the corpus luteum degenerates into the **corpus albicans,** a whitish scar. The
series of stages during which the follicle matures and then degenerates is called
the **ovarian cycle.**

Figure 26.13
Scanning electron micrographs of stages of fertilization. Only one sperm enters the egg because after the first sperm begins entry, changes occur that make the egg unresponsive to other sperm. *a.* Sperm have located an egg. *b.* Once the first sperm makes contact with the cell membrane, it develops projections (microvilli) that rise up around the sperm so that the sperm enters the egg. After this, the egg is activated to begin development.

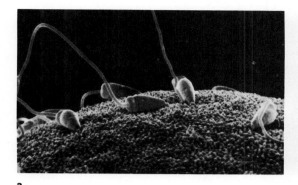

a.

b.

Accessory Organs

The oviducts, uterus, and vagina are the other internal organs of the female reproductive system. The **oviducts,** also called the uterine (or Fallopian) tubes, are about four inches long and extend from the uterus to the ovaries. They end in fingerlike projections called **fimbria** that sweep over the ovary at the time of ovulation. When the egg bursts from the ovary during ovulation, it is usually sucked up into an oviduct where it may be fertilized if sperm are present (fig. 26.13). Whether or not fertilization occurs, cilia lining the tubes and tubular contractions propel the egg toward the uterus. The **uterus** is about the size and shape of an inverted pear and lies between the bladder and the rectum. It is a muscular organ having three parts: the fundus, the body, and the **cervix.** The oviducts join the uterus just below the fundus, and the cervix enters into the vagina at nearly a right angle. The opening of the cervix, called the os or cervical canal, leads to the **vagina,** a small muscular tube that makes a 45° angle with the small of the back. The lining of the vagina lies in folds capable of extension as the muscular wall stretches. This capacity to extend is especially important when the vagina serves as the birth canal, and it may facilitate intercourse when the vagina receives the penis during copulation.

Orgasm in Females

Orgasm in the female is a release from neuromuscular tension within the muscles of the genital area, vagina, and uterus. Prior to orgasm, the vagina is lubricated by mucous secretion and expands for reception of the penis. The **clitoris,** which has a limited amount of erectile tissue within a small shaft and a pealike head or glans, also expands. It is a highly sensitive organ that is located anterior to the opening of the urethra. The urethral and vaginal external openings in the female lie between two folds of skin called **labia minor** and **labia major** (fig. 26.14). At birth, the vaginal opening is partially or wholly occluded by a membrane called the hymen. All types of physical activities, sexual intercourse in particular, can disrupt the hymen.

Figure 26.14
External genitalia of female.

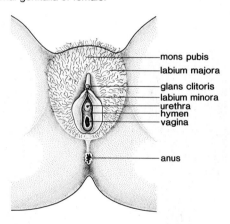

mons pubis
labium majora
glans clitoris
labium minora
urethra
hymen
vagina

anus

Figure 26.15

Plasma hormone levels associated with the ovarian and menstrual cycles. During the follicular phase, FSH produced by the anterior pituitary promotes the maturation of a follicle in the ovary. This structure produces increasing levels of estrogen, which causes the endometrium lining of the uterus to thicken. After ovulation, during the luteal phase, LH promotes the development of the corpus luteum. This structure produces increasing levels of progesterone, which causes the endometrium lining to become secretory. Menstruation begins when LH production declines to a low level.

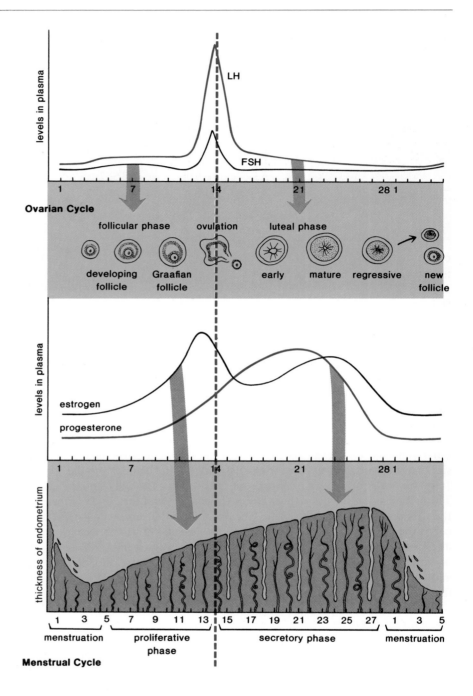

Menstrual Cycle

The term menstrual cycle is derived from menstruation, the periodic shedding of tissue and blood from the **endometrium,** the inner lining of the uterus. Because this shed material flows out through the vagina, it is the most obvious external sign of cyclical changes. But actually, the uterine cycle runs concurrently with the ovarian cycle (fig. 26.15) described on page 586. This coordination allows the uterus to be prepared to receive the developing egg that implants itself in the uterine lining.

Table 26.3 Ovarian and Menstrual Cycles (Simplified)

Ovarian Cycle Phases	Events	Menstrual Cycle Phases	Events
Follicular—Days 1–13	FSH	Menstruation—Days 1–5	Endometrium breaks down
	Follicle maturation Estrogen	Proliferation—Days 6–13	Endometrium rebuilds
O V U L A T I O N—Day 14*			
Luteal—Days 15–28	LH Corpus luteum Progesterone	Secretory—Days 15–28	Endometrium thickens and glands secrete glycogen

*Assuming a twenty-eight-day cycle

The day on which menstruation begins is commonly called "day one" of the menstrual cycle. During the early days of the cycle, the hypothalamus is secreting moderate amounts of GnRH, and the anterior pituitary is secreting *FSH* (*follicle stimulating hormone*). This hormone causes a follicle to begin to mature and produce an egg. The growing ovarian follicle secretes increasing amounts of *estrogen,* which stimulates repair of the endometrium. This so-called proliferation phase (table 26.3) lasts for eight to ten days following the end of menstruation.

Presumably the high level of estrogen in the blood causes the hypothalamus to suddenly secrete a large amount of GnRH at about the midpoint of a twenty-eight-day menstrual cycle. This causes a surge of LH (*luteinizing hormone*) production by the pituitary. This LH surge brings about ovulation and eventually influences the follicle to develop into the corpus luteum that continues hormone production. The maturing corpus luteum secretes both estrogen and increasing quantities of *progesterone.* Progesterone acts on the uterus by stimulating vascularization, glandular development, and glycogen accumulation in the endometrium. All these progesterone-induced changes, which occur during the so-called **secretory phase,** allow an embryo to implant itself easily in the endometrial lining.

Nonpregnant Female   In the usual menstrual cycle, when a pregnancy has not begun, the hypothalamus and then the pituitary respond to the high levels of sex hormones put into circulation by the active corpus luteum. This feedback response causes a decrease in LH production, and the resultant fall in LH levels cause the corpus luteum to begin degenerating about the twenty-second or twenty-third day of a twenty-eight-day cycle. As the corpus luteum is degenerating, its hormone output decreases.

Maintenance of the fully developed uterine endometrium depends on the relatively high progesterone levels present while the corpus luteum is most active. Thus, when circulating progesterone levels decrease as a result of corpus luteum regression, the uterine lining starts to degenerate and slough off, and menstruation begins. Before long, however, the hypothalamus and pituitary respond to decreased circulating estrogen: FSH production rises again, follicle growth and maturation are stimulated, and the whole complex cycle starts over.

Menstruation marks the end of the time when the uterine endometrium could accept an implanting embryo. It is a clear signal that pregnancy has not begun during that particular menstrual cycle.

Around age 45 to 50, the ovary gradually ceases to respond to the anterior pituitary hormones, and no more follicles are eventually produced. Following this occurrence, called the **menopause,** menstruation ceases entirely.

The Reasons for Orgasm

Why does orgasm exist? A glance at the reproduction of primitive forms of animal life shows that such a fancy thing far exceeds what's necessary to pass genes from one generation to another, which even in mammals can be attained by a no-frills male ejaculation and mere passive receptivity on the part of the female. Yet it's unlikely that nature, as economical as it is, would create something so baroque without good reason.

The simpler the creature, the more difficult it is to determine whether its sexual relations are pleasurable, much less orgasmic. However, all mammals show a marked interest in sex, and will even work for it—a good indication that it's rewarding. Scientists prefer low-key terms like "ejaculatory reflex" and "estrus behavior" when discussing animal sexuality; they're reluctant to apply the term "orgasm," because it can't be verified in animals. The best gauge of orgasm is uniquely human: Did you or didn't you?

Male sexual responses, particularly erection and ejaculation, have been better studied than female processes. Even the humble rat, that scrupulously observed mammal, exhibits the basic criteria associated with male orgasm: ejaculation followed by refraction, accompanied by characteristic movements. "Males of most mammalian species have what at least looks like a precursor to human orgasm, demonstrated by skeletal and facial patterns," says Benjamin Sachs, a reproductive behavior researcher at the University of Connecticut. "You may as well call it orgasm. I'd be more cautious about females." Moreover, in addition to appearances, males have an excellent motive for ejaculation: progeny.

Alas for researchers, human females don't always show characteristic muscular movements during orgasm. Even more vexing, they need not have orgasms to conceive—or even to seek and enjoy mating. The best evidence for female orgasm is subjective and anecdotal—the yea or nay of the woman in question. Until there's a way to acquire such information from animals, subhuman female orgasm can only be inferred. Does this mean that for female animals mating is simply a selfless matter of lie still and think of England?* Ronald Nadler, a primatologist at the Yerkes Regional Primate Research Center in Atlanta, believes it's unlikely that

---

Figure 26.16
Urinary excretion of estrogen, progesterone, and chorionic gonadotropin during pregnancy. Urinary excretion rates are an indication of blood concentration of these hormones.

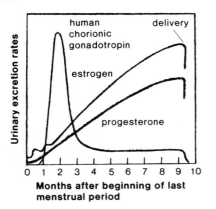

Pregnant Female If fertilization has occurred, an embryo begins development (fig. 26.11) even as it travels down the oviduct to the uterus. The endometrium is now prepared to receive the developing embryo, which becomes embedded in the lining several days following fertilization. This process, called **implantation,** causes the female to become pregnant. During implantation, an outer layer of cells surrounding the embryo produces a gonadotropic hormone (**HCG,** or human chorionic gonadotropin) that prevents degeneration of the corpus luteum and instead causes it to secrete even larger quantities of progesterone (fig. 26.16). The corpus luteum may be maintained for as long as six months, even after the placenta is fully developed.

The **placenta** (fig. 27.13) originates from both maternal and fetal tissue. It is the region of exchange of molecules between fetal and maternal blood, although there is rarely any mixing of the two types of blood. After its formation, the placenta continues production of **HCG** and begins production of progesterone and estrogen. The latter hormones have two effects: they shut down the anterior pituitary so no new follicles mature, and they maintain the lining of the uterus so the corpus luteum is not needed. There is no menstruation during the length of pregnancy.

HCG is produced in such large quantities that pregnant women excrete considerable amounts of it in their urine. The chemical tests for pregnancy are based on the detection of this hormone in the urine. HCG is so readily available that it is routinely used in many teaching and research laboratories. For example, it is even used to induce ovulation in female frogs to obtain eggs for embryological studies.

Occasionally, corpus luteum activity may decrease due to illness of the mother or a failure of the chorion to continue adequate HCG production. The resultant drop in progesterone level initiates menstruationlike breakdown of the endometrium and loss (**miscarriage**) of the implanted embryo. Miscarriage

women are alone among female primates in having orgasms, but the real controversy is not whether, but why, females have orgasms at all.

There are two schools of thought on this. The first maintains that the female has orgasms because the male does, though hers serve no essential reproductive function (an analogy is the male nipple). The female genitals are poorly designed for easy stimulation to orgasm, particularly during intercourse, and if orgasm were really important to procreation, anatomy would have evolved differently to facilitate it. This idea of female orgasm as vestigial is supported by the fact that the inability to have orgasms is common in women yet rare in men, and that orgasm sometimes requires considerable education and practice for women to achieve.

The second school says that female orgasm is designed for a reason or reasons. Sarah Blaffer Hrdy, a primatologist at the University of California at Davis, notes that a female could accumulate sufficient stimulation for orgasm as well as increase her chances of conception via the repeated matings typical of primates during estrus. Furthermore, if the motions of the vagina and uterus during orgasm can be proved to enhance the mobility of sperm, as some researchers speculate, orgasm might be a way to enable a female to be somewhat selective about when and even by whom she becomes pregnant. Although neither camp can prove its thesis, a majority holds that female orgasm has a purpose—without knowing what the purpose is.

While scholars debate, couples unwittingly contend with evolutionary traits that have more practical application. As sex therapists are wont to put it, "Men heat up like light bulbs and women like irons." Trying as this difference in timing can be, it makes sociobiological sense: a female motivated to seek repeated sexual encounters increases her opportunities for pregnancy and assures male benignity to a greater extent than if she were quickly aroused and satisfied. On the other hand, a male, always on the defensive, can't afford lengthy dalliances with his back to potential enemies.

Though many theories about the evolution of orgasm are specific to males or females, the response also makes sense in social, as well as reproductive, terms for both. While the sex lives of other mammals are regulated by hormones—totally so in non-primates like rats, partially so in nonhuman primates—humans appear to be freed from such controls, and even castrated men and women can enjoy sex and have orgasms. There are reliable reports of children seven years old having orgasms, and there's no limit at the other end of the age spectrum—further indications that there's more to sex than reproduction. Roger Short, a reproductive biologist at Monash University in Melbourne, Australia, thinks that because human females, unlike those of other species, will accept intercourse at any time, whether or not they are in a fertile period, most couplings are performed for social purposes such as strengthening pair-bonding and releasing sexual tension, so that the more mundane business of life can be attended to.

*Traditional advice to Victorian brides on their wedding nights.
Winifred Gallagher, © *Discover Magazine* 2/86.

during the first month or so of pregnancy may result in a blood flow somewhat heavier and longer than normal menstruation, but the embryo is still so small that the entire brief and abruptly terminated pregnancy simply may be mistaken for a somewhat delayed menstrual period.

Female Sex Hormones

The female sex hormones, *estrogen* and *progesterone,* have many other effects on the body in addition to their effect on the uterus. Estrogen is largely responsible for the secondary sex characteristics, including female body hair and fat distribution. In general, females have a more rounded appearance than males because of a greater accumulation of fat beneath the skin. Also, the pelvic girdle enlarges in females, and the pelvic cavity has a larger relative size compared to males. This means that females have wider hips. Both estrogen and progesterone are also required for breast development.

Breasts A female breast contains fifteen to twenty-five lobules (fig. 26.17), each with its own milk duct that begins at the nipple and divides into numerous other ducts that end in blind sacs called **alveoli.** In a nonlactating breast the ducts far outnumber the alveoli because alveoli are made up of cells that can produce milk.

Milk is not produced during pregnancy. Prolactin is needed for lactation (milk production) to begin, and production of this hormone is suppressed by the feedback inhibition estrogen and progesterone have on the pituitary during pregnancy. It takes a couple of days after delivery of the baby for milk production to begin; in the meantime, the breasts produce a watery, yellowish-white fluid called **colostrum,** which has a similar composition to milk but contains more protein and less fat.

Figure 26.17
The human female breast contains lobules consisting of ducts and alveoli. The alveoli are lined by milk-producing cells in the lactating breast.

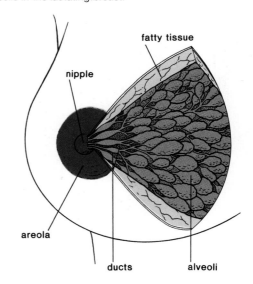

fatty tissue

nipple

areola

ducts

alveoli

Table 26.4 Birth Control Methods

Name	Procedure	Methodology	Effectiveness*	Action Needed	Risk
Vasectomy	Ductus deferentia are cut and tied	No sperm in semen	Almost 100%	Sexual freedom	Irreversible sterility
Tubal ligation	Oviducts are cut and tied	No eggs in oviduct	Almost 100%	Sexual freedom	Irreversible sterility
Pill	Must take medication daily	Shuts down pituitary	Almost 100%	Sexual freedom	Thrombo-embolism
IUD	Must be inserted into uterus by physician	Prevents implantation	More than 90%	Sexual freedom	Infection
Diaphragm	Plastic cup inserted into vagina to cover cervix	Blocks entrance of sperm into uterus	With jelly about 90%	Must be inserted each time before intercourse	—
Condom	Sheath that fits over erect penis	Traps sperm	About 85%	Must be placed on penis at time of intercourse	—
Coitus interruptus (Withdrawal)	Male withdraws penis before ejaculation	Prevents sperm from entering vagina	About 80%	Intercourse must be interrupted before ejaculation	—
Jellies, creams, foams	Contain spermicidal chemicals	Kill a large number of sperm	About 75%	Must be inserted before intercourse	—
Rhythm method	Determine day of ovulation by record keeping; testing by various methods	Avoid day of ovulation	About 70%	Limits sexual activity	—
Douche	Cleanses vagina and uterus after intercourse	Washes out sperm	Less than 70%	Must be done *immediately* after intercourse	—

*Effectiveness is the average percentage of women who did not become pregnant in a population of 100 sexually active women using the technique for one year.

Control of Reproduction

Birth Control

The prevention of pregnancy is called **contraception** (literally, "against conception"). Contraceptive techniques are evaluated on the basis of several criteria. Are they effective in reliably preventing pregnancy? Are they reversible; when couples do wish to have children, can they stop using the technique and successfully initiate a pregnancy? Are they safe and relatively free from physiological side effects? Are they acceptable on personal and social grounds; that is, do they interfere with sexual enjoyment for one or both partners, or do they place an unfair burden of responsibility for contraception on one member of the pair? The birth control techniques currently in use all fall short of these criteria for one reason or another. Thus, the available techniques must be judged on their relative merits weighed against the chances of unwanted pregnancy.

Birth control techniques fall into the following categories: (1) abstinence from intercourse, especially during the portion of the menstrual cycle when conception might occur; (2) suppression of egg or sperm production or release; (3) prevention of contact between egg and sperm by use of physical barriers; and (4) prevention of implantation. Techniques from all these categories are used to control human fertility, but their effectiveness and risk vary considerably (table 26.4).

Infertility

Sometimes couples do not need to prevent pregnancy; conception or fertilization does not occur despite frequent intercourse. The American Medical Association estimates that 15 percent of all couples in this country are unable to have any children and are therefore properly termed sterile; another 10 percent have fewer children than they wish and are therefore termed infertile.

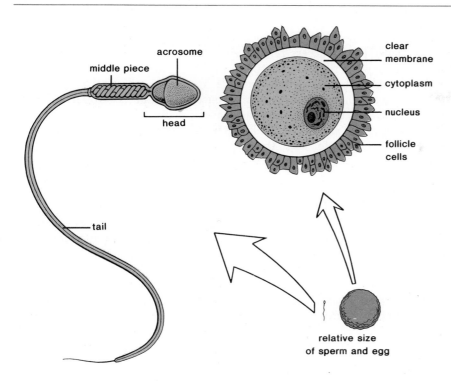

acrosome

middle piece

head

tail

clear membrane

cytoplasm

nucleus

follicle cells

relative size of sperm and egg

Figure 26.18
Anatomy of sperm and egg. A physician can remove an egg from a mature follicle that has moved to the surface of the ovary and is about to burst. This egg can then be placed in laboratory glassware where it is exposed to sperm. In vitro fertilization will occur, and the embryo can then be placed in the womb where it will develop normally if it has successfully implanted itself. In vitro fertilization is often said to produce a "test-tube baby."

Infertility can be due to a number of factors. There may be a congenital malformation of the reproductive tract, or the venereal disease gonorrhea (p. 596) may have caused an obstruction of the oviduct or ductus deferens. Sometimes these physical defects can be corrected surgically.

A hormonal imbalance, which can prevent ovulation in females and cause a low sperm count in males, can be treated medically. As a last resort, it is possible to give females a substance rich in FSH and LH that is extracted from the urine of postmenopausal women. This treatment causes multiple ovulations and sometimes multiple pregnancies, however. If the sperm count in the male is low, it is possible to concentrate the sperm and use this to artificially inseminate the female. However, the emphasis of late has been on the quality of the sperm rather than the total number of sperm since men with very low sperm counts are known to father children naturally.

One area of concern is that radiation, chemical mutagens, and the use of psychedelic drugs can contribute to sterility, possibly by causing chromosomal defects.

In Vitro Fertilization

Assuming that the sperm and egg are normal (fig. 26.18), the new method of external fertilization followed by implantation of the zygote is a possibility for some couples. After appropriate hormonal treatment of the women, eggs are removed from Graafian follicles that have ballooned out of the ovary. Sperm from the males are placed in a solution that approximates the conditions of the female genital tract. When the eggs are introduced, fertilization occurs. **In vitro fertilization** means fertilization in laboratory glassware. The resultant zygotes begin development; after about two to four days they are inserted into the uterus of the woman, who is now in the secretory phase of her menstrual cycle. If an implantation is successful, development continues in the usual manner. Since the first successful birth by in vitro fertilization in 1978, in vitro fertilization clinics have opened in several countries, including the United States. Thus far, the number of birth defects has been no greater using this method than for normally conceived pregnancies. Despite the cost of about $3,000 and the low success rate of only about 20 percent per attempt, the demand is so great that at least 100 to 200 more clinics are expected to open in the United States during the next year or so.

There are those who are morally concerned about in vitro fertilization. Though scientists tend to dismiss the concept, many people feel that it is not right to overcome "natural" infertility. Also, during the early experiments that laid the groundwork for these treatments, human zygotes were cultured and discarded. Some people question whether this can ever be morally justified.

Sexually Transmitted Diseases

There are about a dozen **sexually transmitted diseases,** sometimes called **venereal diseases.** Until recently, bacterial infections, such as gonorrhea and syphilis, were of primary concern; but today, there are two epidemics—the herpes virus epidemic and the AIDS (acquired immune deficiency syndrome) epidemic—of which all people should be aware. Gonorrhea and syphilis are usually curable because they most often respond to antibiotic treatment. But there are no currently available drugs to cure a herpes virus infection or to cure AIDS. A herpes infection is a chronic condition that is not lethal, but 70 percent of those who contract AIDS usually die within two years.

AIDS

Table A AIDS: The Rising Toll

	New Cases Diagnosed in United States	Number Who Have Since Died
1979	11	9
1980	46	42
1981	258	218
1982	987	728
1983	2,691	1,867
1984	5,139	2,546
1985	8,850	3,929
1986*	3,083	656
Total†	21,065	9,995

*As of May 26
†Table totals include eight cases diagnosed prior to 1979; of these eight cases, four are known to have died.
Source: Centers for Disease Control

Since the recognition of the disease AIDS (acquired immune deficiency syndrome) in 1981, the number of diagnosed cases has steadily increased (table A). Approximately 21,000 cases of AIDS is expected to have been reported by mid-1986. AIDS is characterized by the destruction of the immune system, which leads to various infectious diseases and a virulent type of skin cancer (Kaposi's sarcoma). Many victims also show early brain impairment with ever-increasing neuromuscular and psychological consequences.

Prior to having AIDS, an individual will most likely show the symptoms of pre-AIDS or AIDS-related complex (ARC). These symptoms include swollen lymph nodes, weight loss, fatigue, fever, diarrhea, and thrush (a fungal infection of the mouth). About two million people in the United States are thought to have pre-AIDS. Only a portion of these will eventually come down with full-blown AIDS.

There are certain high-risk groups in the United States for contracting AIDS.[1] Of the cases reported:

73 percent are homosexual or bisexual men
17 percent are intravenous drug abusers
3 percent are hemophiliacs and others requiring blood products
1 percent are women who are sexual partners of infected men, and the children of these women

The designation of this last group is recent and may increase in size. In Africa, where the disease is believed to have originated, as many women as men have the disease. This may be due to the use of unsterilized needles for medical procedures and for tatooing the skin. It may also indicate that heterosexual transmission, including both men and women and vice versa, is possible.

AIDS is transmitted in blood as can be deduced by observing that two of the high-risk groups (intravenous drug users and hemophiliacs, etc.) have developed the disease by way of infected blood. Still,

Herpes

Two types of herpes viruses are involved in the current epidemic. *Herpes simplex virus 1* usually causes cold sores and fever blisters on the lips. *Herpes simplex virus 2* usually causes genital lesions. The Center for Disease Control in Atlanta has estimated that 20.5 million Americans now have genital herpes, and an estimated half-million additional cases are expected each year. Some have even estimated that one out of every six sexually active individuals may be capable of spreading the disease. There are no symptoms immediately following infection, but once the virus invades, it begins to multiply rapidly (fig. 26.19).

The individual may experience a tingling or itching sensation before **blisters** appear at the infected site, usually within two to twenty days. Once the blisters rupture, they leave painful **ulcers** that may take as long as three weeks or as little as five days to heal. After the ulcers heal, the disease is only dormant; the blisters can develop repeatedly at variable intervals.

While the virus is dormant, it resides in nerve cells. Type 1 resides in a group of nerve cells located near the brain. Type 2 resides in nerve cells that lie near the spinal cord. Type 1 occasionally travels via a nerve fiber to the eye and causes an eye infection that can lead to blindness. Type 2 has been known to cause a form of meningitis and has also been associated with a form of cervical cancer.

Sunlight, sex, menstruation, and stress seem to cause the symptoms of genital herpes to recur. The ointment form of acyclovir relieves only initial symptoms, but the oral form, only recently developed, prevented outbreaks in about 70 percent of patients in a trial study.

Figure 26.19
Cells infected with herpes virus. Since viruses take over the metabolism of the cell, it is very difficult to kill viruses and not cells with medication. It is possible, however, to develop a vaccine that will activate the immune system to destroy a virus. Researchers are currently working on such a vaccine for sexually transmitted herpes.

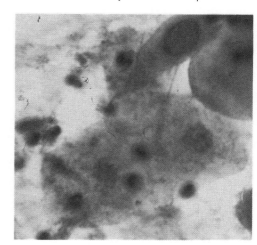

AIDS is a venereal disease and so is also sexually transmitted. Infected semen (and possibly saliva) passes the disease to a sexual partner when it comes in contact with the partner's bloodstream. This can happen if a woman is menstruating, if the partner has an open abrasion, or if the sexual act itself damages the mucosal lining. The wall of the vagina is protected by several layers of flattened squamous cells and therefore heterosexual intercourse does not usually cause any blood vessels to rupture. In contrast, the rectum lacks protective cells and blood vessels lie near the surface; therefore, anal intercourse, which is frequent among homosexuals, is more apt to tear the wall and allow semen to enter the bloodstream.

Scientists in both the United States and France reported in 1984 that a virus causes AIDS. One name given to the virus is HTLV-III (human T cell lymphotropic virus), and its life cycle is known. HTLV-III is a retrovirus that infects lymphocytes known as T-4 cells. The genetic material of a retrovirus is RNA, which must be transcribed by the enzyme reverse transcriptase to give viral DNA. This viral DNA can direct the reproduction of new viruses. First, however, the viral DNA incorporates itself into the cell's own DNA where it can remain dormant for an indefinite period of time. When a person develops AIDS, HTLV-III genes have become activated. There is some evidence that a co-infection with another virus (either hepatitis B or Epstein-Barr virus) is needed to trigger the production of HTLV-III viruses. Once reproduction has started, many T-4 cells die and the immune system is unable to function properly. Death is usually due to the secondary infections that follow.

At first the body is able to mount an offensive against the AIDS virus, as evidenced by the presence of antibodies in the bloodstream. Researchers have developed a test to detect these antibodies, which is used by blood banks to prevent tainted blood from being distributed. It is also used to identify individuals who have been exposed to the AIDS virus. Unfortunately, many people are not able to develop a strong immunity against the disease and so there may never be an effective vaccine to immunize the public against AIDS, although researchers are attempting to produce one. A number of drugs are also being tested, some of which seem to be active against the virus, but there is no generally accepted treatment as yet. An added problem is that the drug will have to be able to pass from the blood into the brain.[2] Drug treatment directed against the virus will have to be accompanied by a therapeutic means to rebuild the immune system. Several possibilities for this aspect of treatment are being explored.

While there is much to be learned about AIDS, we do at least know its means of transmission. There is little danger of contracting AIDS by casual encounters between individuals, such as shaking hands, touching the same utensils, breathing the same air, etc. However, the public is advised to avoid promiscuity. Individuals can best protect themselves by practicing monogamy (always the same sexual partner) with someone who is free of the disease.

[1]Haitians were originally considered a high-risk group but it was later concluded that the infected individuals actually belonged to one of other groups.
[2]There is a so-called blood-brain barrier because the capillaries in the brain are especially constructed to be generally impermeable.

In addition to dealing with the infection, people who have herpes often undergo psychological distress. Those who do not tell potential sexual partners risk passing it on, particularly if the disease is active; and those who do tell potential sexual partners risk being rejected. Since the disease is most common among the promiscuous, some young people have changed their life styles to prevent infection. Although it is possible to learn to cope with an infection, those who already have herpes often feel that their entire lives will be affected.

Gonorrhea

Gonorrhea is the best-known bacterial disease that is transmitted sexually. In men, gonorrhea is usually detected readily because pus is discharged from the penis, and burning sensations during urination develop within a few days after infection. But in women, the infection concentrates in the cervical canal and produces very mild, if any, symptoms. Women may unknowingly develop extended infections of the reproductive tract that result in infection and inflammation of the oviducts. This can lead to partial or complete blockage of the ducts, and thus sterility. Physicians have routinely treated gonorrhea with large doses of penicillin, but penicillin-resistant strains are becoming more common and so treatment is more complex than it once was.

Syphilis

Syphilis is a bacterial disease having an incubation period of about three weeks. After this amount of time, the first symptom, a hard, painful ulcer called a **chancre,** develops at the site of the infection. Even if the disease is left untreated, the chancre disappears after a short time, and the individual sees no more sign of disease for two to four months. At this time, the disease enters its **secondary stage** as a generalized skin rash, and infections of various organs sometimes develop. The disease then goes into a **latent period** that can last throughout life. But in some people, the disease can enter a phase that produces severe nervous system or circulatory system damage and even death. Syphilis, too, usually yields to penicillin treatment, but as in the case of gonorrhea, resistance to common antibiotics is a growing problem.

Immunization

Immunization for these sexually transmitted diseases is not yet possible. Resistance to antibiotic treatment has spurred interest in developing a vaccine for gonorrhea and syphilis. Several promising vaccines for herpes have been prepared and are presently being tested. Of three hundred people who received a herpes vaccine in England, only two developed the infection despite the fact that their sexual partners had the disease. A vaccine for herpes will possibly be available to the general public in the near future.

Summary

Some animals can reproduce asexually by various means, but most reproduce sexually by means of gametes. Asexual reproduction tends to keep the genotype constant, while sexual reproduction tends to introduce variability.

In sexual reproduction a female parent produces eggs in ovaries, and a male parent produces sperm in testes. Most animals have a breeding period at which time both sexes are reproductively active. Aquatic animals often practice external fertilization; land animals practice internal fertilization.

Reproduction in mammals is represented by human reproduction. Within the testes of males, spermatogenesis occurs within seminiferous tubules, and testosterone production occurs within interstitial cells. Sperm mature in each epididymis and are stored in each ductus deferens before entering the urethra just prior to ejaculation. Accessory glands (seminal vesicles, prostate gland, and Cowper's gland) produce seminal fluid. If sexual stimulation is sufficient, the penis becomes erect, and ejaculation of semen occurs. Semen contains sperm and seminal fluid. As many as 500 million sperm may be ejaculated, although only one sperm fertilizes an egg.

Hormonal regulation in the male maintains testosterone at a fairly constant level. The hypothalamus produces GnRH that stimulates the anterior pituitary to produce FSH and LH. In males FSH promotes spermatogenesis, and LH promotes testosterone production. Testosterone stimulates development of the secondary male sex characteristics—those features that are associated with the male body aside from the genitals, the primary sex characteristic.

The female reproductive system consists of the ovaries, oviducts, uterus, and vagina. Oogenesis occurs within the ovaries where one follicle reaches maturity each month. Upon ovulation, the egg is taken up by one of the two oviducts that lead to the uterus. Next, the vagina opens to the outside in a region bounded by the labia minor, which come together at the clitoris, a highly sensitive organ.

The female menstrual cycle parallels the ovarian cycle (table 26.3). During the first days of this cycle, the anterior pituitary is secreting FSH, a hormone that stimulates an ovarian follicle to grow and produce an egg. The growing follicle secretes increasing amounts of estrogens that promote repair and growth of the endometrium of the uterus. About the fourteenth day of a twenty-eight-day cycle, an LH surge occurs that brings about ovulation. Now, the follicle becomes the corpus luteum that secretes progesterone in increasing amounts. Progesterone causes the endometrium to become secretory so it is ready to receive an egg. If implantation occurs, the corpus luteum is maintained because a layer of

cells surrounding an embryo secretes HCG. Regardless, the female sex hormones maintain the secondary sex characteristics of females, including breast development.

The chances of pregnancy can be reduced in sexually active females by appropriate birth control measures. Some females do not get pregnant because of infertility problems that may involve themselves or their partners. In vitro fertilization followed by implantation of the zygote is available to a few infertile couples.

Sexually transmitted diseases include gonorrhea and syphilis. These respond to antibiotic treatment. Herpes and AIDS are two relatively new sexually transmitted diseases that are more difficult to treat.

Objective Questions

1. Asexual reproduction does not involve the use of _____ although sexual reproduction does involve these types of cells.
2. If one were tracing the path of sperm, the structure that follows the epididymis is the _____ .
3. The prostate gland, Cowper's glands, and the _____ all contribute to seminal fluid.
4. The primary male sex hormone is _____ .

5. In the female reproductive system, the uterus lies between the oviducts and the _____ .
6. In the ovarian cycle, once each month a _____ produces an egg. In the menstrual cycle, the _____ lining of the uterus is prepared to receive the zygote.
7. The female sex hormones are _____ and _____ .
8. Pregnancy in the female is detected by the presence of _____ in the blood or urine.

9. Herpes is characterized by the presence of _____ that become _____ in the genital area.
10. Antibiotics are effective against the bacterial venereal diseases _____ and _____ , but not against the viral venereal diseases _____ and _____ .

Answers to Objective Questions
1. gametes 2. ductus deferens 3. seminal vesicles 4. testosterone 5. vagina 6. follicle, endometrial 7. estrogen, progesterone 8. HCG 9. blisters, ulcers 10. gonorrhea, syphilis; herpes, AIDS

Study Questions

1. Give examples of asexual and sexual reproduction among animals.
2. Discuss the human reproductive system as an adaptation to life on land.
3. Discuss the anatomy and physiology of the testes. Describe the structure of sperm.
4. Give the path of sperm. What glands produce seminal fluid?

5. Give the names of the endocrine glands involved in maintaining the sex characteristics of males and the hormones produced by each.
6. Discuss the anatomy and physiology of the ovaries. Describe ovulation.
7. Give the path of the egg. Where do fertilization and implantation occur? Name two functions of the vagina.

8. Discuss hormonal regulation in the female by giving the events of the menstrual cycle and relating these to the ovarian cycle. In what way is menstruation prevented if pregnancy occurs?
9. State the various means of birth control and stress their relative effectiveness.
10. Describe at least three common sexually transmitted venereal diseases.

Selected Key Terms

regeneration (ri ,jen ə 'rā shən) 579
budding ('bəd niŋ) 579
parthenogenesis (,pär thə nō 'jen ə səs) 579
gamete (gə 'mēt) 579
seminiferous tubule (,sem ə 'nif ə rəs 'tü ,byül) 582

testosterone (te 'stäs tə ,rōn) 583
epididymis (,ep ə 'did ə məs) 584
ductus deferens ('dək təs 'def ə ,renz) 584
seminal vesicle ('sem ən əl 'ves i kəl) 585

prostate gland ('präs ,tāt 'gland) 585
Cowper's gland (,kaů pərz 'gland) 585
oviduct ('ō və ,dəkt) 587
uterus ('yüt ə rəs) 587
vagina (və 'jī nə) 587

CHAPTER 27

Animal Development

Concepts

1. Embryology gives evidence that all animals are related.
   All animals go through the same developmental processes and early stages that lead to the establishment of embryonic germ layers.
2. Closely related animals have similar embryonic stages.
   Reptiles, birds, and mammals have similar stages and make use of extraembryonic membranes to carry out life support functions.
3. Development is an orderly process.
   a. Induction, or the ability of one tissue to influence the development of another, contributes to orderly development and partially accounts for differentiation, or specialization of parts.
   b. It is possible to describe the steps in human development from fertilization to the birth of the child.

I n the early days of **embryology** (the study of development), there was a fierce debate as to whether development consists of **preformation** or **epigenesis.** Those who believed in *preformation* thought that a sperm (or egg) already contained the animal, and development simply allowed this creature to unfold. It is not surprising, then, that some early microscopists maintained that they saw tiny men, animalcules, curled up inside each sperm (fig. 27.1). In contrast, those who believed in *epigenesis* thought that neither the sperm nor the egg contained the animal. Instead, the animal came to be as development progressed. If so, it was reasoned, it would be possible to divide an egg and have each half develop into a whole being. Hans Driesch and Wilhelm Roux performed such experiments on animal eggs in the late 1800s. Sometimes a divided frog's egg produced two embryos; sometimes only one embryo. The result depended on how the egg was divided (fig. 27.2). Such experiments contributed to the conclusion that development is largely epigenetic—the animal gradually comes to be—but environmental factors, including original gradients present in the egg's cytoplasm, influence development.

Development

Certain processes occur during development. These processes require a series of stages.

Processes of Development

The following processes are involved in animal development.

Cleavage

Following fertilization (fig. 26.13), the zygote begins to divide; at first there are two, then four, eight, sixteen cells, and so forth. Since increase in size does not accompany this **cleavage,** the embryo is no larger than the zygote was. Cell division during this process is mitotic, and each cell receives a full complement of chromosomes and genes.

Growth

After cleavage, cell division is accompanied by an increase in size of the daughter cells. These two events, occurring simultaneously, constitute growth.

Morphogenesis

Morphogenesis refers to the progressive development of pattern and form as the embryo takes shape. Morphogenesis is first evident when cells are seen to move, or migrate, in relation to other cells. The embryo begins to form various organs and structures by these movements.

Differentiation

When cells take on a specific structure and function, **differentiation** occurs. The first system to become visibly differentiated is the nervous system.

Early Developmental Stages

All chordate embryos go through the same early stages of development, as listed in table 27.1. Each stage can be identified by the events or the result of that stage. However, the presence of **yolk,** dense nutrient material, affects the manner in which embryonic cells complete each stage and hence the appearance of different embryos at the end of each stage. Table 27.2 indicates the amount of yolk in the four embryos discussed in this chapter and relates the

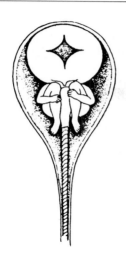

Figure 27.1
Early microscopists of the seventeenth century imagined that they saw little men inside sperm. This supported their belief in preformation, as well as their belief that the male sperm was the "seed" and the female, the "soil" into which new life was planted. It was not until 1827 that Karl von Baer discovered that female mammals produce eggs.

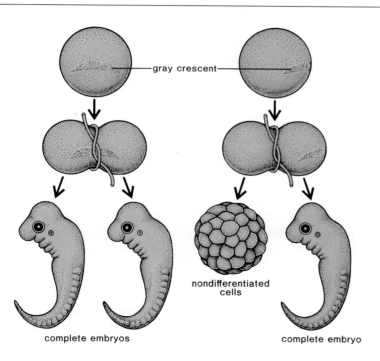

gray crescent

nondifferentiated cells

complete embryos   complete embryo

Figure 27.2
Experiment performed by Wilhelm Roux in 1888. (*left*) If a single frog's egg is divided so that each half receives a portion of the gray crescent (a special part of the cytoplasm), both halves develop into embryos. (*right*) If a single egg is divided so that only one half receives the gray crescent, then only this half develops normally. This shows that the gray crescent contains substances that direct development.

Table 27.1 Early Developmental Stages

Stage	Process	Result
Cleavage	Cell division without growth	Many cells
Blastula	Morphogenesis and growth	Cavity present
Gastrula	Morphogenesis and growth	An embryo with three germ layers
Neurula	Differentiation by induction	Nervous system development

Table 27.2 Amount of Yolk

Animal	Yolk	Location of Development
Lancelet	Little	External in water
Frog	Some	External in water
Chick	Much	Within hard shell
Human	Little	Inside mother

yolk content to the environment in which the animal develops. Thus, the two animals (lancelet and frog) that develop in water have less yolk than the chick because development in these two animals proceeds quickly to a swimming larva that can feed itself. But the chick is representative of animals that have solved the problem of reproduction on land, in part, by providing a great deal of yolk within a hard shell where development continues to a more mature individual capable of land existence.

Early stages of human development resemble those of the chick embryo, yet this resemblance cannot be related to the amount of yolk because the human egg contains little yolk. But the evolutionary history of these two animals can provide an answer for this resemblance. Both birds (for example, chicks) and mammals (for example, humans) are related to the reptiles, and this evolutionary relationship manifests itself in the manner in which development proceeds.

The developmental stages we will discuss are contrasted in figure 27.3 for lancelets, frogs, and chicks.

Cleavage

This first stage is cell division without growth. In a lancelet, the division is equal, and the cells are of uniform size; in a frog the upper cells are smaller than the lower cells. This difference in size occurs because the upper cells at the animal pole contain little yolk, while the lower cells at the vegetal pole do contain yolk. Cells containing yolk are slower to cleave than those not containing yolk. The presence of extensive yolk in the chick causes cleavage to be incomplete, and only those cells lying on top of the yolk cleave. This means that while cleavage in a lancelet and a frog results in a ball of cells called the **morula,** no such ball is seen in a chick; rather, the cells spread out on a portion of the yolk.

Blastula

The **blastula** stage begins when the cells of the solid morula move or position themselves to create a space (fig. 27.3). In a lancelet, a completely hollow ball results, and the space within the ball is called the **blastocoel.** In a frog, the blastocoel is formed within the animal hemisphere only. The heavily laden yolk cells of the vegetal hemisphere do not participate in this step. In a chick a blastocoel is created when the cells lift up from the yolk and leave a space between the cells and the yolk.

Gastrula

The **gastrula** stage is evident in a lancelet when certain cells begin to push, or **invaginate,** into the blastocoel (fig. 27.3). This creates a double layer of cells. The outer layer is called the **ectoderm,** and the inner layer is called the **endoderm.** The space created by invagination will become the gut and is called either the primitive gut or the **archenteron.** The pore or hole created by invagination is called the blastopore, and in a lancelet, as well as in the other animals discussed here, it becomes the anus.

In the frog, the cells containing yolk do not participate in gastrulation and thus do not invaginate. Instead, a slitlike **blastopore** is formed when the animal pole cells begin to invaginate from above. Following this, other animal

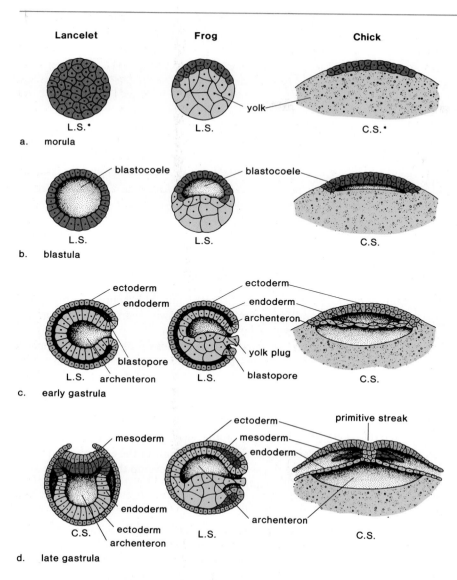

Lancelet	Frog	Chick

a. morula

L.S.*   L.S.   C.S.*

yolk

b. blastula

blastocoele   blastocoele

L.S.   L.S.   C.S.

c. early gastrula

ectoderm
endoderm
blastopore
L.S.   archenteron

ectoderm
endoderm
archenteron
yolk plug
L.S.   blastopore

C.S.

d. late gastrula

mesoderm
endoderm
C.S.   ectoderm
archenteron

ectoderm
mesoderm
endoderm
L.S.
archenteron

primitive streak
C.S.

*L.S. = longitudinal section; C.S. = cross section

Figure 27.3
a. Comparative morula stages for a lancelet, a frog, and a chick. b. Comparative blastula stages. c. Comparative gastrula stages. d. Comparative late gastrula stages.

pole cells move down over the yolk, and the blastopore becomes rounded when these cells also invaginate. Some yolk cells are left in the region of the pore and are called the **yolk plug.**

In the chick, the embryonic cells simply sort themselves out to form two layers of cells lying above the yolk. The upper layer of cells is the ectoderm, and the lower layer is the endoderm.

Table 27.3 Organs Developed from the Three Primary Germ Layers

Ectoderm	Mesoderm	Endoderm
Skin epidermis, including hair, nails, and sweat glands	All muscles	Lining of digestive tract, trachea, bronchi, lungs, gallbladder, and urethra
Nervous system, including brain, spinal cord, ganglia, nerves, and sense receptors	Dermis of skin	Liver
	All connective tissue, including bone, cartilage, and blood	Pancreas
Lens and cornea of eye	Blood vessels	Thyroid, parathyroid, and thymus glands
Lining of nose, mouth, and anus	Kidneys	Urinary bladder
Teeth enamel	Reproductive organs	

Mesoderm Formation

Gastrulation is not complete until three layers of cells have formed (fig. 27.3). The third, or middle, layer of cells is called the **mesoderm.** In a lancelet, this layer begins as outpocketings from the primitive gut. These outpocketings grow in size until they meet and fuse. In effect, then, two layers of mesoderm are formed, and the space between them is the *coelom.*

In a frog, the mesoderm cannot form as outpocketings because the presence of the inert yolk cells prevents this process. Instead, mesoderm arises when certain cells begin to segregate between the ectoderm and endoderm layers. At first, this is a solid layer of cells, but later it splits to form the coelom.

In a chick, the mesoderm forms by involution of cells between the outer ectoderm and inner endoderm along the length of the **primitive streak,** a midline region that extends the length of the embryo. Mesoderm later forms blocklike portions called **somites** in the posterior half of the embryo, and these become muscle tissue. Above the somites, vessels develop that go into the yolk to collect nutrients for the developing embryo.

Germ Layers

Ectoderm, mesoderm, and endoderm are called the **primary germ layers** of the embryo, and no matter how gastrulation takes place, the end result is the same: three germ layers are formed. It is possible to relate the development of future organs to these germ layers, as is done in table 27.3. Von Baer, the nineteenth-century embryologist, first related later development to the germ layers.

Differentiation

Numerous experiments have shown that a parceling out of genes does not account for differentiation. The same quantity and type of genes are found in all the cells that make up the body of an animal. For example, Karl Illmensee of the University of Geneva, and Peter Hoppe of the Jackson Laboratory in Bar Harbor, Maine, performed the following experiment (fig. 27.4). They removed an embryo from the womb of a gray mouse and delicately took the nuclei from the cells making up the embryo. Each nucleus was placed in an egg cell taken from a black mouse. (These egg cells had just been fertilized, so they were prepared by removing both the egg nucleus and sperm nucleus.) The resulting embryo was placed in the womb of a white mouse, which gave birth to a gray mouse as expected. This experiment shows that the nucleus in a differentiated cell contains all the genes necessary for normal development. Genes are not parceled out during development, and all cells in the embryo and adult contain a full complement of genes.

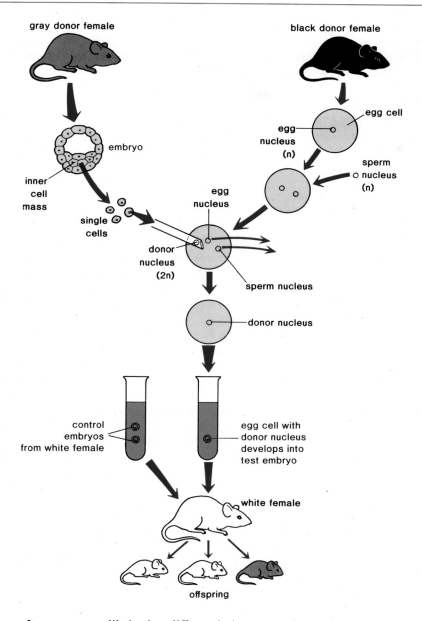

gray donor female

black donor female

egg cell

egg nucleus (n)

embryo

inner cell mass

sperm nucleus (n)

single cells

egg nucleus

donor nucleus (2n)

sperm nucleus

donor nucleus

control embryos from white female

egg cell with donor nucleus develops into test embryo

white female

offspring

Figure 27.4
A cloning (producing exact copies) experiment that was performed in recent years. At the upper left, nuclei are removed from an embryo taken from a gray mouse. At upper right, an egg cell from a black mouse is prepared to receive the donated nucleus. At lower half, development proceeds normally and a gray mouse is born to a white mouse. This experiment shows that the nuclei taken from the differentiated cells of a gray mouse embryo contained a complete set of genes. This is a cloning experiment because all the gray mice have exactly the same genotype and phenotype.

It seems more likely that differentiation occurs because some genes are turned on and some are turned off during development. Evidence for such an idea comes from the Driesch-Roux experiment mentioned earlier (fig. 27.2). A frog's egg contains a special section of cytoplasm called the gray crescent. If an experimenter ties the fertilized egg so that each half has a portion of the gray crescent, both halves develop successfully into a complete embryo. If the experimenter ties the egg so that only one-half has the gray crescent, only that half develops successfully. The other half becomes a mass of nondifferentiated cells. This experiment suggests that substances within the area of the gray crescent direct development.

Early investigators referred to the gray crescent as the primary organizer for the embryo. Following cleavage, an **organizer** is a group of cells that can influence the development of other cells. For example, mesoderm tissue located at the upper lip of the blastopore brings about, or **induces,** the formation of the nervous system in animals.

Figure 27.5

a. Neural plate is an ectoderm layer that will give rise to the neural tube. *b.* As microfilaments contract, cells begin to invaginate. *c.* Continued constriction causes the neural tube to pinch off from the outer ectoderm layer.

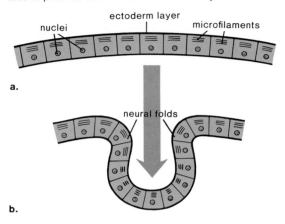

a.

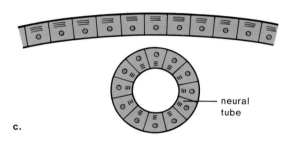

b.

c.

Induction

In vertebrate animals, such as a frog, a central portion of the mesoderm is the future notochord that will later be replaced by the vertebral column. The nervous system develops from an ectoderm layer (the **neural plate**) located just above the *notochord* in the region that will be the dorsal midline of the body. Cells in the plate divide and change shape due to contractions of microfilaments strategically located inside them (fig. 27.5). The collective action of these cell-shape changes causes the neural plate to roll up, first into a pair of **neural folds** and then into a **neural tube** (fig. 27.6). It would seem that the notochord is influencing the ectoderm cells above it to become a neural tube. It is said that the notochord **induces** the formation of the neural tube.

Mechanism of Action

Various experiments indicate that cells will differentiate according to their location in the embryo. As early as 1924, Hans Spemann, a student of Wilhem Roux, performed one such experiment in amphibians (fig. 27.7). He cut out presumptive (potential) neural tube cells and transplanted them to another region of the embryo. He noted that these transplanted cells did not become nerve cells. On the other hand, if he placed presumptive notochord cells beneath what would be belly ectoderm, this ectoderm differentiated into neural tissue.

Taking such an experiment one step further, James Weston of the University of Oregon found that presumptive neural tube cells grown in tissue culture do not become nerve cells unless the chemical fibronectin is added to the culture medium. It is likely that the induction is caused by the release of inducer molecules into the environment. The cells that receive these molecules begin to differentiate in a certain way. The inducer molecules probably bind

Figure 27.6

Development of neural tube and coelom in a frog embryo. *a.* Ectoderm cells that lie above the future notochord thicken to form a neural plate. The cells beneath the archenteron (primitive gut) are laden with yolk and do not participate in the early development of organs. *b.* The neural groove is noticeable as the neural tube begins to form. *c.* A splitting of the mesoderm produces a coelom that is completely lined by mesoderm. *d.* Neural tube and coelom have now developed.

Development of the coelom and nervous system in a frog embryo.

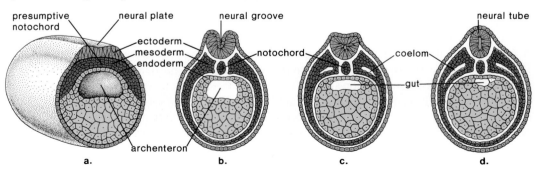

with receptor molecules on the surface of responding cells in much the same way that hormones bind with membrane-bound receptors of target cells (p. 521). This complex of inducer molecule and receptor molecule probably then enters the cell and causes genetic activation leading to differentiation.

Although the nature of most inducer molecules is not known, it is known that various hormones can dramatically affect animal development. For example, hormones control the changes in body form during metamorphosis. The familiar conversion of tadpole to frog is caused by the thyroid hormone thyroxin. If thyroid secretion is inhibited in tadpoles, metamorphosis does not occur. Insect metamorphosis also is hormonally controlled. The egg-larva-pupa-adult sequence in insects is precisely regulated by a group of interacting hormones, as described in the reading on the following page.

Figure 27.7
Experiment performed by Hans Spemann in 1924. In experiment A, the presumptive nervous system does not develop into the neural plate if moved from the normal location. In experiment B, the presumptive notochord can cause the belly ectoderm to develop into the neural plate. This shows that the notochord influences the cells around it, most likely by sending out chemical messengers.

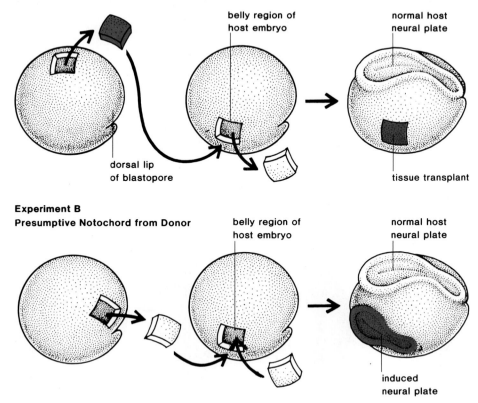

Experiment A
Presumptive Nervous System from Donor

belly region of host embryo

normal host neural plate

dorsal lip of blastopore

tissue transplant

Experiment B
Presumptive Notochord from Donor

belly region of host embryo

normal host neural plate

induced neural plate

Hormones and Insect Development

This electron micrograph shows a starving three-headed armyworm that has been fed a molting-process inhibitor in a pest-control experiment. The heads—each about the size of a pinhead—bury the insect's functional mouthparts.

During their development, insects molt because their rigid outer body covering must be shed when they grow larger. Molting is followed by the expansion and hardening of a new, soft, larger exoskeleton that has developed underneath the old exoskeleton.

Quite often as insects develop, they go through a series of larval stages, each one terminated by molting. (Finally, as discussed on page 327, pupation occurs during which metamorphosis produces the adult insect.) There are two powerful hormones that affect the number of larval molts and also the rate with which the insect hastens toward pupation. Ecdysone is sometimes called the molt and maturation hormone because it promotes not only molting but also maturation. During larval stages, however, response to ecdysone is modified by the action of juvenile hormone. This hormone, as its name implies, causes an insect to remain immature and to molt more often than usual.

Research is being directed toward using these hormones as insecticides. The possibility exists that spraying larval insects or their food with one or the other hormone might cause them to develop abnormally and die or fail to reach an appropriate stage in their life cycle to survive the winter.

Since the hormones are effective at very low concentrations, occur naturally, and are biodegradable, they would not be toxic to humans. Because the hormones are species specific, it would be possible to affect one particular insect life cycle without affecting other insects as well. This is especially important to growers concerned about the well being of beneficial insects, such as honeybees. It seems unlikely that insects would develop resistance to these hormones because their own development depends on their being responsive to very small amounts.

Some plants produce compounds that are similar or identical to ecdysone or juvenile hormone. These compounds apparently protect the plants by disrupting the development of insects that feed upon them. It is possible that these compounds can be extracted from plants and used as insecticides. For example, an East African medicinal plant, *Ajuga remota,* produces antiecdysones that inhibit molting. If the fall armyworm caterpillar is fed these compounds, it does not molt properly and instead develops three heads, burying its functional mouthparts (see the figure).

Figure 27.8
During development, the forebrain induces formation of a lens of the eye. *a.* Ectodermal cells above the optic vesicle thicken. *b. and c.* Lens vesicle forms as invagination occurs. *d.* Lens is now located within the developing eye.

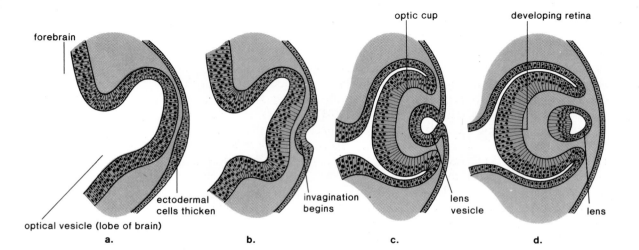

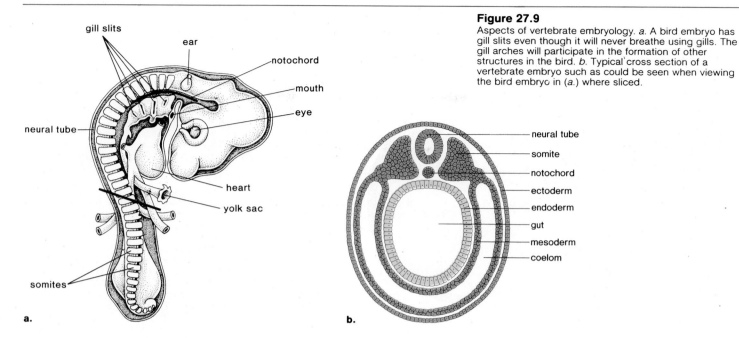

Figure 27.9
Aspects of vertebrate embryology. *a.* A bird embryo has gill slits even though it will never breathe using gills. The gill arches will participate in the formation of other structures in the bird. *b.* Typical cross section of a vertebrate embryo such as could be seen when viewing the bird embryo in (*a.*) where sliced.

Labels for part a: gill slits, ear, notochord, mouth, eye, neural tube, heart, yolk sac, somites

Labels for part b: neural tube, somite, notochord, ectoderm, endoderm, gut, mesoderm, coelom

Orderly Development of the Embryo

The process of **induction** can help explain the orderly development of the embryo. One tissue is induced by another, and this in turn induces another tissue, and so on, until development is complete. Induction can in large measure account for the stepwise and timely progression of development. For example, once closure of the neural tube is complete, the presumptive forebrain induces formation of the lens of the eye. First the sides of the brain bulge out and widen just beneath overlying ectoderm. This seems to trigger cell division of these ectoderm cells. Then the bulge dips in to form a cuplike structure that will be the future eyeball, while the overlying thickened ectoderm grows into a ball of cells to form the lens of the eye (fig. 27.8).

Vertebrate Cross Section

A diagram of a chordate embryo (fig. 27.9*b*) illustrates development to the point of the nervous system. It is a generalized cross section because *all* chordates at sometime show these same features. Correlation of figure 27.9 with table 27.3 helps relate the formation of vertebrate structures and organs to the three embryonic layers of cells: ectoderm, mesoderm, and endoderm. The skin and nervous systems develop from the ectoderm; muscles, skeleton, kidneys, circulatory system, and gonads develop from the mesoderm; and the digestive tract, lungs, liver, and pancreas develop from the endoderm.

The diagram illustrates that embryonic chordates have a notochord and a dorsal hollow nerve cord. Another characteristic of all vertebrate embryos is the presence of gill pouches or slits (fig. 27.9). Only the lower vertebrates (fishes and amphibian larvae) use the gill slits as functioning structures. The fact that higher forms go through this embryonic stage indicates that the higher forms are related to the lower forms. The phrase "*ontogeny* (development) *recapitulates* (repeats) *phylogeny* (evolutionary history)" was coined some years ago as a dramatic way to suggest that all animals share the same embryonic stages. This theory has been modified since embryos proceed only

Figure 27.10

Comparison of extraembryonic membranes in (*a.*) bird embryo and (*b.*) human embryo. Notice that two of the membranes participate in the formation of the umbilical cord in humans.

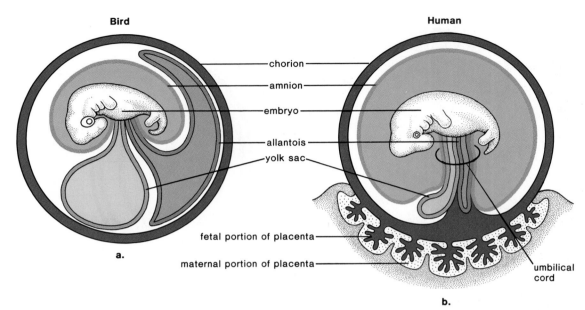

Bird

Human

chorion

amnion

embryo

allantois

yolk sac

fetal portion of placenta

maternal portion of placenta

umbilical cord

a.

b.

through those stages that are necessary to their later development. For example, in higher vertebrates the actual slits never form; instead, the first gill pouch becomes the cavity of the middle ear and Eustachian tube. The second pouch becomes the tonsils; the third and fourth pouches become the thymus and parathyroids, respectively. Only the fifth pouch disappears. Thus, gill pouches develop because they are necessary to later development.

Extraembryonic Membranes

In a chick, the three germ layers extend beyond the body of the embryo and form the **extraembryonic membranes.** Evolution of these membranes in the reptile made development on land possible. If an embryo develops in water, the water supplies oxygen for the embryo and takes away waste products. The surrounding water prevents desiccation and provides a protective cushion.

On land all these functions are performed by the extraembryonic membranes. Figure 27.10*a* shows the chick within its hard shell surrounded by these membranes. The **chorion** lies next to the shell and carries on gas exchange. The **yolk sac** surrounds the remaining yolk. The **allantois** collects nitrogenous waste, and the **amnion** contains the amniotic fluid that bathes the developing embryo.

As figure 27.10*b* indicates, humans also have these extraembryonic membranes. The chorion develops into the fetal half of the placenta, a region of exchange of molecules between fetal and maternal blood; the yolk sac is present but largely nonfunctional; the allantoic blood vessels become the umbilical blood vessels; and the amnion contains fluid to cushion and protect the fetus. The function of the membranes has been modified to suit internal development, but their very presence indicates the human relationship to birds and reptiles. It is interesting to note that all animals develop in water, either directly or within amniotic fluid.

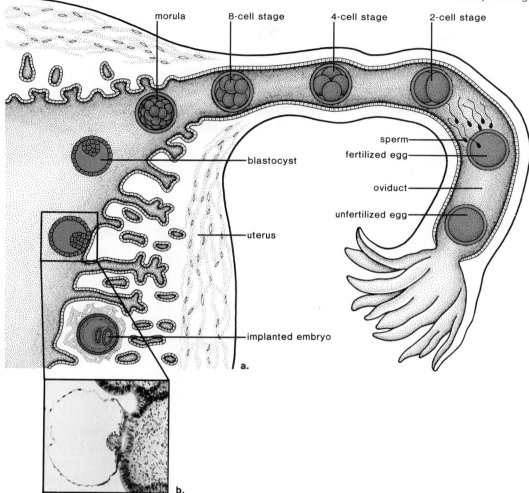

Figure 27.11
a. Diagram illustrating fertilization and movement of the embryo down the oviduct to the uterus, where implantation occurs. *b.* Photomicrograph of a monkey embryo undergoing implantation.

morula   8-cell stage   4-cell stage   2-cell stage

blastocyst

sperm

fertilized egg

oviduct

uterus

unfertilized egg

implanted embryo

a.

b.

Human Development

Human development is often divided into embryonic development (first two months) and fetal development (three to nine months). The embryonic period consists of early formation of the major organs, and fetal development is a refinement of these structures.

Early human development is most easily understood in the following way. The human egg requires little yolk because the mother supplies the needs of the developing embryo by way of the extraembryonic membranes; thus, it is not surprising that the extraembryonic membranes develop early in humans. This is the major change from the development of a chick, discussed earlier.

Embryonic Development

First Month

Cleavage   Following fertilization within the oviduct, the human zygote begins to undergo cleavage as it travels down the oviduct to the uterus (fig. 27.11). The first cleavage division is completed within about a day after fertilization, and subsequent divisions occur at intervals of eight to ten hours. Passage of

Figure 27.12

Stages showing the early appearance of the extraembryonic membranes and the formation of the umbilical cord in the human embryo. *a.* The blastocyst is surrounded by the trophoblast. *b.* As gastrulation occurs, the amniotic cavity appears. *c.* The chorion and yolk sac are now apparent. *d.* With completion of gastrulation, the body stalk is apparent. *e. to g.* Embryo becomes more differentiated as the umbilical cord forms.

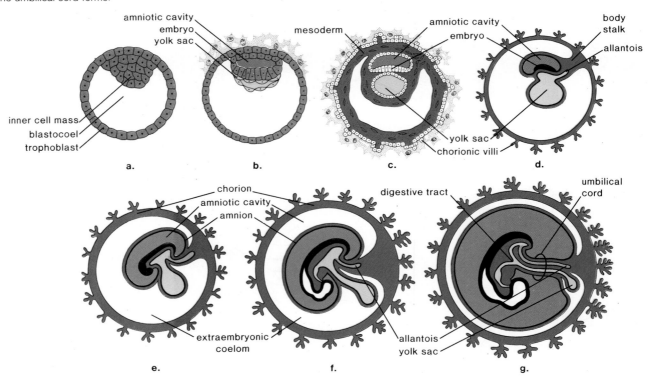

the embryo through the oviduct takes from three to three and a half days. While the embryo lies free in the uterine cavity, the morula (p. 600) becomes the **blastocyst,** consisting of the **inner cell mass** with a single layer of surrounding cells known as the **trophoblast.** The trophoblast signals the formation of the first extraembryonic membrane since it will be the outer layer of cells of the chorion (fig. 27.12).

At the end of the first week, the embryo begins the process of implanting itself in the wall of the uterus. During this process, the trophoblast secretes enzymes to digest away some of the tissue and blood vessels of the uterine wall (fig. 27.11*b*). After the blastocyst has entered the endometrium, uterine tissue grows over the surface and heals the implantation site.

Gastrulation   As implantation occurs, the embryo undergoes gastrulation. During the first stage of gastrulation, the inner cell mass becomes the **embryonic disc,** consisting of a layer of endodermal cells beneath an ectodermal layer of cells (fig. 27.12*c*). An *amnionic cavity* is now visible above the embryo, and a *yolk sac* cavity is visible below. Now two more extraembryonic membranes have made their appearance. As gastrulation continues, mesoderm formation occurs. Mechanics of mesoderm formation in the human embryo are virtually identical to those described for bird embryos. Some cells leave the embryonic disc and invaginate along a primitive streak to assume a new position between ectoderm and endoderm. This new, third layer is the mesoderm of the human embryo. The three primary body layers (germ layers) of the human embryo

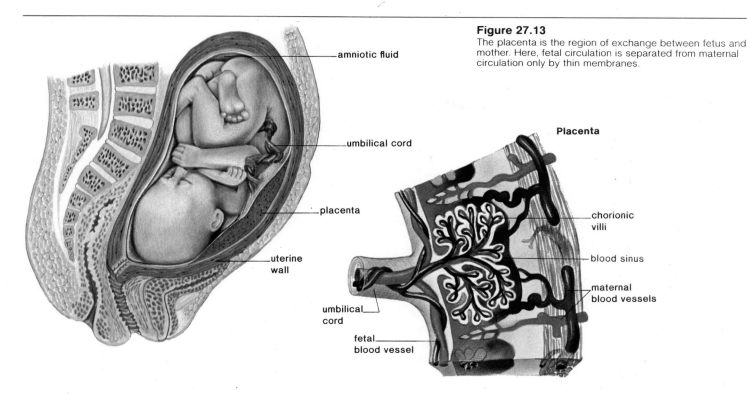

amniotic fluid

umbilical cord

placenta

uterine wall

Placenta

chorionic villi

blood sinus

maternal blood vessels

umbilical cord

fetal blood vessel

can now be identified. From top (amnion side) to bottom (yolk-sac side) they are ectoderm, mesoderm, and endoderm. The edges of the embryonic disc later fold under, and the flat disc is converted into a tubular, three-layered body. This body folding brings the germ layers into their permanent relationship with one another (fig. 27.9*b*): ectoderm on the outside, endoderm on the inside, and mesoderm between the two.

Placenta Formation

As implantation occurs, a placental relationship is developing between the embryo and its mother. A series of fingerlike outgrowths, **villi** (sing., villus), develop over the *chorion*'s surface (fig. 27.12*d*). Only villi on one side are destined to grow; they branch elaborately and participate in the development of the placenta. The remainder regress and later disappear. Within the body stalk, blood vessels of the vestigial *allantois* contribute to the development of the umbilical blood vessels, which take blood to and from the placenta within the umbilical cord.

The **placenta** (fig. 27.13) has a fetal side (the chorion) and a maternal side that lie in close contact so wastes and carbon dioxide can move from the fetal to the maternal side of the placenta, and nutrients and oxygen can move from the maternal to the fetal side. Notice in figure 27.13 how the chorionic villi are surrounded by maternal blood sinuses. Yet, the blood of mother and fetus do not mix; instead, exchange takes place across cell membranes.

After its formation, the placenta produces *HCG* (human chorionic gonadotropin hormone), which prevents degeneration of the corpus luteum in the ovary for as long as six months. In the meantime, the placenta also begins production of progesterone and estrogen to maintain the supply of these hormones after the corpus luteum degenerates. The high level of progesterone and estrogen in the pregnant woman's body assures, through the feedback mechanism discussed in the previous chapter, that no new follicles will mature, and it also maintains the uterine lining so embryonic growth can continue.

Figure 27.14
Human embryo at twenty-one days. The neural folds still need to close at the anterior and posterior of the embryo. The pericardial area contains the primitive heart, and the somites are the precursors of the muscles.

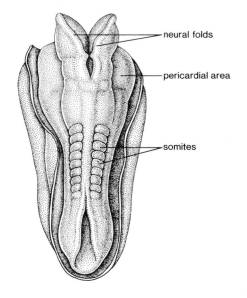

neural folds

pericardial area

somites

Figure 27.15
At six weeks, the head in the human embryo is disproportionately large and seems to rest on the bulging heart. The limb buds are beginning to form into arms and legs. In this photo, you can see the amnion, an extraembryonic membrane that forms the fluid-filled amniotic cavity.

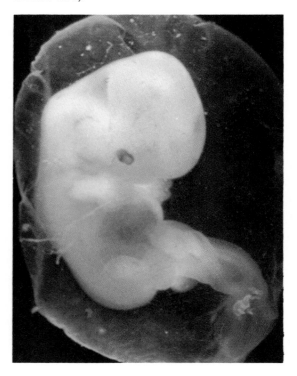

Organ Formation   By the middle of the third week, the mesoderm cells of the central axis have formed a notochord. The central nervous system begins development above this. Neural folds arise from the ectoderm and fuse to give a hollow tube that swells anteriorly to form the brain (fig. 27.14). Chunks of mesoderm, called *somites,* appear on both sides of the neural tube. They will become the dermis of the skin, the muscle tissue, and the vertebrae and other bones.

The heart and blood vessels also arise from the mesoderm. At first the heart is two tiny tubes; these fuse to give a single chamber as the third week ends. This primitive heart bulges and is easily observed as it begins to pump blood through simple arteries and veins. These take blood to the developing esophagus, stomach, liver, pancreas, and intestine, all emerging from the endoderm.

During the fourth week, eyes, ears, and nose appear as pits on an oversized head that is flexed and appears to rest on the heart bulging beneath it. The embryo is still quite tiny, being only 5 to 6 mm. Only now is the mother beginning to suspect she is pregnant, even though the embryo has many organs already forming.

Second Month

During the second month, the arms and then legs begin as **limb buds.** At first the buds look like paddles. As they grow, hands, forearms, upper arms, and shoulders appear. The newly formed arms bend at the elbow, and the legs bend at the knee to allow these appendages to reach across the body, fingers and toes nearly touching.

At first the fetus's skeleton is cartilage, but it begins turning to bone during the sixth week. Just as the growth of the arms is faster than the growth of the legs, so the growth of the head outstrips the growth of the trunk. At the end of the second month the embryo measures roughly 2.5 cm from crown to rump; about half of this is head. The eyes, nose, mouth, and ears are all recognizable, and the embryo now has a distinctly human appearance.

During these two months, the embryo is especially susceptible to harmful agents that disrupt normal development. Precautions can be taken to reduce this susceptibility, as suggested in the following reading.

Fetal Development

Fetal development is marked by an extreme increase in size. Weight multiplies six hundred times, going from less than an ounce to seven pounds. In this time, too, the fetus grows to about twenty inches in length. The genitalia finally make their appearance, and it is possible to tell if the fetus is male or female. With this accelerated growth, proportions change. The head is now only a half to a fourth of the body's total length.

Soon, hair, eyebrows, and eyelashes add finishing touches to the face and head. In the same way, fingernails and toenails complete the hands and feet. A fine downy hair (lanugo) covers the limbs and trunk, only to later disappear. The fetus looks like an old man because the skin is growing so fast it wrinkles. A waxy, almost cheeselike, substance (vernix caseasa) protects the wrinkly skin from the watery amniotic fluid.

The fetus at first only flexes its limbs and nods it head, but later it can vigorously move its limbs to avoid discomfort. The mother will feel these movements from about the fourth month on. The other systems of the body also begin to function. The fetus begins to suck its thumb, swallow amniotic fluid, and urinate for example.

At least one in sixteen newborns has a birth defect, either minor or serious. It is estimated that only 20 percent of all birth defects are due to heredity. These birth defects can be detected by subjecting embryonic and fetal cells to various tests. A new method called chorionic villi sampling allows physicians to collect embryonic cells as early as the fifth week. The doctor inserts a long, thin tube through the vagina into the uterus. With the help of ultrasound, which gives a picture of the uterine contents, the tube is placed between the lining of the uterus and the chorion. Then, suction is used to remove a sampling of the chorionic villi cells. Chromosomal analysis and biochemical tests for about two hundred different genetic defects can be done immediately on these cells.

Previous to the development of chorionic villi sampling, physicians had to wait until about the sixteenth week of pregnancy to perform amniocentesis or fetoscopy. In amniocentesis, a long needle is used to withdraw a small amount of amniotic fluid along with fetal cells. Since the fluid contains only a few cells, testing must be delayed until cell culture has caused the cells to grow and multiply. Thus, another two to four weeks may pass before the prospective parents are told whether their child has a genetic defect.

Fetoscopy is a procedure more complicated than amniocentesis. In fetoscopy, a tiny periscope with a cold light source is inserted through a small incision in the uterus. This allows the physician to see small portions of the fetus and to take skin and blood samples for testing purposes. This is the only current method for acquiring samples of blood.

Treatment of the fetus in the womb is a rapidly developing area of medical expertise. Biochemical defects have been treated by giving the mother appropriate medicines. For example, a healthy baby was born to a woman who took massive doses of vitamin B_{12} because her baby was unable to produce this vitamin. Similarly, a fetus was treated for a biotin (another type of vitamin) disorder by giving the mother large amounts of biotin. Structural defects have also been corrected by surgery. Using a needle to guide placement, a physician inserted a tube into the brain of a fetus that had water on the brain. Similarly, a tube was inserted into the bladder of a fetus that was unable to pass urine. These tubes provided a means for proper drainage in both cases. Physicians are also developing techniques that will allow them to correct common structural defects, including neural tube defects, on babies that have been lifted from the womb just long enough for the surgery to be done. They have found that if pregnancy continues to term, it is too late to perform these same operations.

Many babies are born with defects not caused by heredity. These congenital defects can often be avoided. It is recommended that all girls be immunized for German measles before their child-bearing years. German measle viruses can cause defects such as deafness. Drugs, like aspirin, caffeine (present in coffee, tea, and cola), and alcohol, should be severely limited during pregnancy. The mother most likely should not take mood-altering drugs at all. It is not unusual for babies of drug addicts and alcoholics to display withdrawal symptoms and to be mentally retarded. Most people are also aware that women taking the tranquilizer thalidomide produced children with deformed arms and legs. If pregnancy is suspected, a woman should discontinue taking birth control pills and similar hormone medications. Sex hormones can possibly cause abnormal fetal development including abnormalities of the sex organs.

A pregnant woman should avoid so-called fetotoxic chemicals, such as pesticides and many organic industrial chemicals. Cigarette smoke includes some of these very same chemicals; as a result, babies born to smokers are often underweight and subject to convulsions. The mother also should avoid X-ray diagnostic therapy during pregnancy because X rays are mutagenic to a developing embryo or fetus. Children born to women who have received X-ray treatment are apt to have birth defects and/or develop leukemia later on.

Upon giving birth, an Rh negative woman who has had a child by an Rh positive man should receive a RhoGam injection to prevent the production of Rh antibodies. During a subsequent pregnancy these antibodies can cause birth defects, including nervous system and heart defects.

Now that physicians and lay people are aware of the various ways in which birth defects can be prevented, it is hoped that all birth defects, both genetic and congenital, will decrease dramatically.

After twenty weeks, the fetal heartbeat is heard through a stethoscope. A fetus born at twenty-four weeks does have a chance of survival, although the lungs are still in an immature state and often cannot capture the oxygen the fetus needs. Weight gain during the last couple of months increases the likelihood of survival.

Figure 27.16
Stages in human development. *a.* Four-to-five-month fetus.
b. Six-to-seven-month fetus. *c.* Eight-to-nine-month fetus.

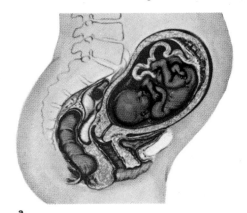

a.

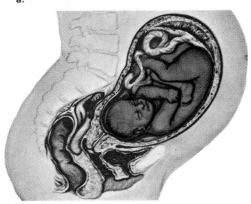

b.

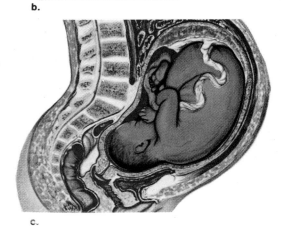

c.

Birth

The uterus contracts gently and periodically throughout pregnancy. During the ninth month, however, the contractions gradually increase in frequency, culminating in **labor,** the strong contractions that will expel the fetus from the womb.

Why labor begins is not completely understood, but most likely it involves hormonal changes in the woman's body. As pregnancy progresses, the level of estrogen in the bloodstream exceeds the level of progesterone (fig. 26.16). Progesterone inhibits uterine contractions, and estrogen promotes uterine irritability.

It is also possible that the fetus and placenta initiate labor by releasing chemical messengers. If the hypothalamus or pituitary is destroyed in fetal sheep, labor can be postponed indefinitely. Similarly, labor is delayed in humans whenever the fetus's pituitary gland is poorly developed. The inference, then, is that the fetal pituitary secretes a chemical message that brings on labor. And although the mother does not show an increased level of oxytocin in the blood, this hormone, too, is a powerful inducer of contractions. It is often administered to induce labor or to increase the strength of uterine contractions. Also, prostaglandins are known to prepare the cervix for dilation.

Parturition, which includes labor and expulsion of the fetus, is usually considered to have three stages. During the first stage, the cervical canal dilates; during the second, the baby is born; and during the third, the afterbirth is expelled. Prior to the start of parturition, the baby's head is directed anteriorly, but the baby usually turns and the head is directed posteriorly. If this does not occur, the birth, called a breech birth, is usually far more difficult.

Stage One

Prior to or concomitant with the first stage of parturition, there may be a "bloody show" caused by expulsion of a mucus plug from the cervical canal. This plug prevents bacteria and sperm from entering the uterus during pregnancy.

Uterine contractions during the first stage of labor occur in such a way that the cervical canal slowly disappears (fig. 27.17*b* and *c*) as the lower part of the uterus is pulled upward toward the baby's head. With further contractions, the baby's head acts as a wedge to assist cervical dilation. The baby's head usually has a diameter of about 10 cm; therefore, the cervical canal has to dilate to this diameter to allow the head to pass through. If it has not already occurred, the amniotic membrane is apt to rupture now, releasing the amniotic fluid, which escapes out the vagina. The first stage of labor ends once the cervix is completely dilated.

Stage Two

During the second stage, the uterine contractions occur every one to two minutes and last about one minute each. They are accompanied by a desire to push or bear down. As the baby's head gradually descends into the vagina, the desire to push becomes greater. When the baby's head reaches the exterior, it turns so the back of the head is uppermost (fig. 27.17*d*). Since the vagina

Figure 27.17
Birth occurs in several stages. *a.* Dilation of cervix.
b. Amnion bursts. *c.* Fetus descends into pelvis. *d.* Head
appears. *e.* Rotation. *f.* Delivery of shoulders. *g. and h.*
Expulsion of afterbirth.

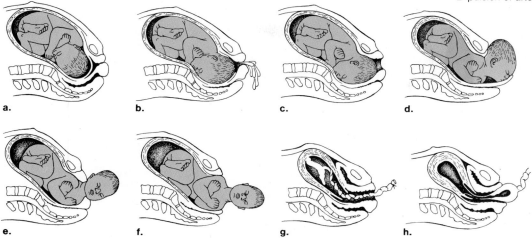

a.   b.   c.   d.

e.   f.   g.   h.

may not expand enough to allow passage of the head without tearing, an
episiotomy is often performed. This incision is stitched later and heals more
perfectly than a tear would. As soon as the head is delivered, the baby's shoulders rotate and the baby faces either to the right or left. The physician may
at this time hold the head and guide it downward while one shoulder and then
the other emerges. The rest of the baby follows easily.

Newborn infants must make rapid respiratory and circulatory adjust-
ments during the transition from complete dependence on the placenta to independent life outside the uterus. Circulation to the placenta normally shuts
down after delivery. This deprives the newborn of placental gas exchange and
leads to a falling oxygen concentration and a rising CO_2 concentration in the
blood. Brain respiratory centers respond to this increasing level of CO_2 by
sending impulses to chest and abdominal muscles, thereby stimulating the
contractions needed to initiate breathing.

At birth, also, the oval window, an opening in the septum of the fetal
heart, closes. Expansion of the lungs causes blood to enter the lungs in quantity, after which return of this blood to the left side of the heart causes a flap
to cover the opening. Once the baby is breathing normally, the umbilical cord
is cut and tied (fig. 27.18), severing the child from the placenta. The stump
of the cord shrivels and leaves a scar, which is the navel.

Stage Three

The placenta, or **afterbirth,** is delivered during the third stage of labor (fig.
27.17*g* and *h*). About fifteen minutes after delivery of the baby, uterine muscular contractions shrink the uterus and dislodge the placenta. The placenta
is then expelled into the vagina. As soon as the placenta and its membranes
are delivered, the third stage of labor is complete.

Figure 27.18
A newborn human infant just after delivery. The umbilical
cord will be cut, severing once and for all the connection
between the offspring and the placenta.

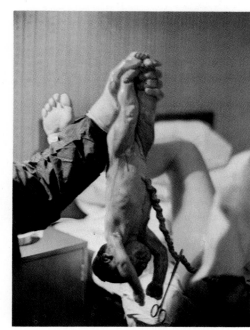

Summary

In this chapter we have seen that all animals, including human beings, develop similarly. The processes of development are cleavage, growth, morphogenesis, and differentiation. Cleavage results in many cells that become first a morula, then a blastula, and finally, a gastrula. Gastrulation is affected by the amount of yolk, a dense nutrient material, present in the embryo. Yolk or yolk-laden cells do not undergo invagination; nevertheless, the result of gastrulation is the establishment of three germ layers: ectoderm, endoderm, and mesoderm. Later development can be related to these germ layers.

The control of differentiation, whereby tissues take on specific structures and functions, may not be accounted for by a parceling out of genes. Rather, it appears that the cytoplasm may at first control which genes are turned on or off. Also, during development certain cells affect other cells. For example, the notochord induces formation of the nervous system, and the forebrain induces formation of the lens.

All vertebrates at some time in their development portray a similar cross section that displays typical vertebrate embryonic characteristics: dorsal hollow nerve cord, notochord, and coelom completely surrounded by mesoderm. The presence of extraembryonic membranes in reptiles and birds makes development on land possible because these membranes carry out necessary functions for an embryo developing out of water. The chorion serves for gas exchange; the yolk sac provides nourishment; the allantois collects nitrogenous waste; and the amnion surrounds the embryo with fluid for protection. Humans also have these membranes, but their function has been modified for internal development: the chorion becomes the fetal part of the placenta; the yolk sac and allantois are largely nonfunctional; and the amnion again surrounds the embryo with fluid. The human embryo lies within the amniotic cavity and is connected to its lifeline, the placenta, by way of the umbilical cord. During the first two months, all major organs are formed, and the embryo takes on a human appearance. After this, we refer to the developing new life as the fetus. Various features of the fetus are refined during the next several months, while the last few months serve largely to increase the size of the soon-to-be newborn.

The process of birth requires three stages: first, the cervix increases in size to permit the passage of the baby; second, the baby is forced out into the world by strong contractions of the uterus and abdominal wall; third, after expulsion of the fetus, the membranes and placenta follow as the afterbirth.

Objective Questions

1. When cells take on a specific structure and function, _____ occurs.
2. The presence of _____ determines whether cleavage is complete or incomplete.
3. Gastrulation is complete once the germ layer _____ forms.
4. Every cell in an animal's body retains a full complement of _____ , and thus can theoretically be used in a cloning experiment.
5. The notochord _____ the formation of the nervous system.
6. The coelom is a body cavity _____ lined by mesoderm.
7. In humans, the _____ embryonic membranes develop early. This facilitates the timely formation of the _____ from the chorion and the uterine wall.
8. During the first month of human embryonic development, chunks of mesoderm called _____ appear on both sides of the neural tube. From these _____ tissues will develop.
9. Fetal development in humans encompasses the months _____ to _____ . This is largely a time of _____ since most organs have already formed.
10. If delivery of the child is normal, the _____ appears before the rest of the body.

Study Questions

1. List and define the processes of development. List and describe the stages of animal development.
2. What is the germ layer theory? What structures are associated with each germ layer?
3. Describe an experiment that indicates genes are not "parceled out" during development.
4. Give examples of induction during the development of vertebrates. What is the significance of this process?
5. Draw a cross section of the typical vertebrate embryo, and label its parts.
6. What are the extraembryonic membranes for both chicks and humans? Compare and contrast the function of the membranes in these animals.
7. Describe in detail development in humans during the first month. Describe in general, listing significant happenings, the other months.
8. Outline briefly the events during each stage of birth.

Selected Key Terms

preformation (ˌprē fȯr 'mā shən) 598
cleavage ('klē vij) 598
morphogenesis (ˌmȯr fə 'jen ə səs) 598
differentiation (ˌdif ə ˌren chē 'ā shən) 598
morula ('mȯr ə lə) 600
blastula ('blas chə lə) 600
blastocoel ('blas tə sēl) 600

gastrula ('gas trə lə) 600
archenteron (är 'kent ə ˌrän) 600
blastopore ('blas tə ˌpōr) 600
induction (in 'dək shən) 607
chorion ('kōr ē 'än) 608
allantois (ə 'lant ə wəs) 608
amnion ('am nē ˌän) 608
parturition (ˌpärt ə 'rish ən) 614

Suggested Readings for Part 10

Baconsfield, P.; Birdwood, G.; and Baconsfield, R. August 1980. The placenta. *Scientific American.*

DeRobertis, E. M., and Gurdon, J. B. December 1979. Gene transplantation and the analysis of development. *Scientific American.*

Gehring, W. J. October 1985. The molecular basis of development. *Scientific American.*

Lein, A. 1979. *The cycling female: Her menstrual rhythm.* San Francisco: W. H. Freeman.

Mader, S. S. 1980. *Human reproductive biology.* Dubuque, IA: Wm. C. Brown Publishers.

Rugh, R., et al. 1971. *From conception to birth: The drama of life's beginnings.* New York: Harper & Row, Publishers, Inc.

Slotkin, T. A., and Lagercrantz, H. April 1986. The "stress" of being born. *Scientific American.*

Vander, A. J., et al. 1980. *Human physiology.* 3d ed. New York: McGraw-Hill.

Volpe, E. P., 1983. *Biology and human concerns,* 3d ed. Dubuque, IA: Wm. C. Brown Publishers.

Environment

In the working of silver or drilling in turquoise, the Indians had exhaustless patience. . . . The land and all that it bore they treated with consideration; not attempting to improve it, they never desecrated it.

—WILLA CATHER
from *Death Comes for the Archbishop*

The moose is a very large long-legged deer of cold northern climates. It feeds in summer in lakes and rivers.

PART 11

Biology of Populations

In mature natural ecosystems the population sizes tend to remain within the carrying capacity of the environment. This is due particularly to biotic interactions, such as competition and predation. Because the population sizes remain constant, a natural ecosystem tends to require the same amount of energy and chemicals each year.

Humans have created their own ecosystem, consisting of the country and city, which differs from natural ecosystems in that the human population often increases in size and/or ever-greater amounts of energy and raw materials are used each year. Since energy is not used efficiently and raw materials are not recycled, the human ecosystem adds excess heat and wastes to the biosphere. Pollution results when the human ecosystem overwhelms the capacity of the biosphere's chemical cycles to absorb these wastes.

The biosphere contains aquatic and terrestrial ecosystems. The aquatic ecosystems along the coast are the most productive and the most endangered by pollution. Among the terrestrial ecosystems, the tropical rain forests, which in some instances are still relatively natural, are now being increasingly utilized for human purposes. Many feel that it would better serve the human race to protect the tropical rain forests and all that remains of the other natural ecosystems as well.

Leaf collar ant with leaf. This ant is gathering food for the colony's fungal garden, which the ants feed upon. Their relationship is an example of social mutualism in which both the ants and the fungi benefit.

CHAPTER 28

Population Ecology

Concepts

Populations of organisms are adapted to both the abiotic and biotic aspects of their environment.

a. Each population has a characteristic growth curve, survivorship curve, and age pyramid; these reflect the overall strategy for continued existence.

b. Population size is held in check by density-independent and-dependent effects. The latter are population interactions, such as competition, predation, and parasitism.

c. Populations competing for the same resources tend to evolve to occupy different niches.

d. Population interactions include not only those that tend to reduce population size, but also cooperative interactions that tend to increase population size.

Figure 28.1
Temperate deciduous forest and members of various populations: (*a.*) general appearance; (*b.*) chipmunk whose cheeks are bulging with stored nuts; (*c.*) round-lobed hepatica.

a.

b.

c.

German zoologist Ernst Haeckel coined the word "ecology" from two Greek roots: *oikos,* meaning "household," and *ology,* meaning "study of." He recognized that organisms do not exist alone; rather, they are part of a household, a functional unit that includes other types of organisms and the physical surroundings. Haeckel considered **ecology** to mean the study of the relationship of an organism to both its biotic (living) environment and its abiotic (physical) environment.

Today we envision the **biosphere,** a thin region surrounding the earth where organisms may be found, as divided into functional units called ecosystems. For example, a portion of a temperate forest (fig. 28.1) is an ecosystem. The organisms that live here—trees, shrubs, flowers, mice, birds, ground squirrels, rabbits, foxes, and so forth—belong to populations. A *population* is all the members of a species living in a particular locale. The various populations make up a **community,** and within the community populations interact with one another and the physical environment. The community (the biotic environment) and the physical environment (the abiotic environment) function together to form a system, termed an **ecosystem.**

This chapter considers the biotic interactions and how they serve to regulate the size of each population. We will begin by describing, in general terms, population growth and maintenance of normal size.

Table 28.1	Exponential Growth (r = 2.0%)				
Unit Time	N	I	Unit Time	N	I
0	10,000.00	200.00	7	11,486.86	229.74
1	10,200.00	204.00	8	11,716.60	234.33
2	10,404.00	208.08	9	11,950.93	239.02
3	10,612.08	212.24	10	12,189.95	243.80
4	10,824.32	216.49	11	12,433.75	248.68
5	11,040.81	220.82	12	12,682.43	253.65
6	11,261.63	225.23			

Figure 28.2
The exponential growth curve is J shaped, indicating that the population is experiencing a steady rise in population size.

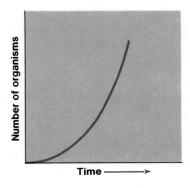

Population Growth

Populations increase in size whenever the number of births exceeds the number of deaths. Suppose, for example, a population has 1,000 individuals (N = 1,000), a **birthrate** (b) of 30 per year, and a **death rate** (d) of 10 per year. The annual **rate of natural increase** (r) would be

$$r = \frac{b - d}{N} = \frac{30 - 10}{1000} = .02 = 2\% \text{ per year.}$$

The rate of natural increase does not include any change in population size from either immigration or emigration, which for the purpose of our discussion can be assumed to be equivalent.

Populations with a positive rate of natural increase grow larger each year. The expected increase (I) can be calculated by multiplying the rate of natural increase (r) by the current population size (N)

$$I = rN$$

This formula indicates that population growth is **exponential:** at the end of each year, N is larger, and therefore, I will also be larger, as demonstrated in table 28.1. This means population size increases by an ever larger amount each year. This is apparent when population size is plotted against time. When an arithmetic scale is used for population size, the resulting growth curve resembles that in figure 28.2. It starts off slowly and then increases dramatically, giving a **J-shaped curve.**

Biotic Potential

The **biotic potential** of a population is the maximum rate of natural increase (r_{max}) that can possibly occur under ideal circumstances. These include plenty of room for each member of the population, unlimited resources, and no hindrances, such as predators or parasites. Since these conditions rarely occur, it is very difficult to determine the r_{max} for various species. A few approximations, however, are given in table 28.2.

Whether the biotic potential is high or low depends on such factors as the following:

1. Usual number of offspring per reproduction
2. Chances of survival until age of reproduction
3. How often each individual reproduces
4. Age at which reproduction begins

Maximum growth rate is not experienced by any population for any length of time. This is readily apparent when considering, for example, that if a single female pig had her first litter at nine months and then produced two litters a year, each of which contained an average of four females (which,

Table 28.2	Biotic Potential
Kind of Organism	Approximate Biotic Potential r_{max} (per year)
Large mammals	0.02–0.5
Birds	0.05–1.5
Small mammals	0.3–8
Larger invertebrates	10–30
Insects	4–50
Small invertebrates (including large protozoans)	30–800
Protozoa and unicellular algae	600–2000
Bacteria	3000–20,000

From *Principles of Ecology*, by Richard Brewer. Copyright © 1979 by W. B. Saunders Company. Reprinted by permission of Holt, Rinehart and Winston, CBS College Publishing.

Figure 28.3

The number of fruit flies in a colony was counted every other day. When these numbers were plotted, the result was a sigmoidal (S-shaped) growth curve.

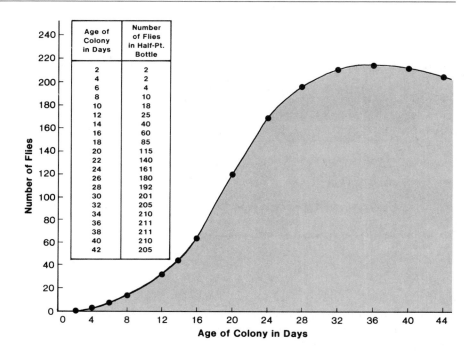

Age of Colony in Days	Number of Flies in Half-Pt. Bottle
2	2
4	2
6	4
8	10
10	18
12	25
14	40
16	60
18	85
20	115
22	140
24	161
26	180
28	192
30	201
32	205
34	210
36	211
38	211
40	210
42	205

Figure 28.4

The logistic or S-shaped growth curve. The population initially grows exponentially but begins to slow down as resources become limited. Population size remains stable when the carrying capacity of the environment is reached.

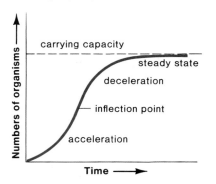

Table 28.3	Environmental Resistance

Density-Independent Effects
  Climate and weather
  Natural disasters
  Requirements for growth
Density-Dependent Effects
  Competition
  Emigration
  Predation
  Parasitism

in turn, reproduced at the same rate), there would be 2,220 pigs by the end of three years. Due to exponential growth, even species with a low biotic potential would soon cover the face of the earth with their own kind. For example, as Charles Darwin calculated, a single pair of elephants would have over 19 million live descendants after 750 years.

Environmental Resistance and Carrying Capacity

Populations do not undergo exponential growth for very long. Even in the laboratory setting, where crowding may be the only limiting factor, growth of populations eventually levels off. Figure 28.3 gives the actual growth curve for a fruit fly population reared in a culture bottle. This type of growth curve is often called a **logistic curve,** or an S-shaped (sigmoidal-shaped) curve. During an acceleration phase, when births exceed deaths, r is positive; during a deceleration phase, when births and deaths are more nearly equal, r is declining. Finally, when births exactly equal deaths, r is zero, and the population maintains a steady state (fig. 28.4).

Population size is believed to level off at the **carrying capacity** (K) of the environment. In other words, the environment is capable of sustaining a population of a limited size. As the population increases in size, there will be more and more competition for available space and food, for example. Eventually, this will effect population size. It is said that the environment increasingly resists the biotic potential of the population. Therefore, **environmental resistance** depends in part on the number of organisms in the population; that is, it is, in part, density dependent (table 28.3). The expression

$$\frac{K - N}{K}$$

describes this type of resistance. It also tells us what proportion of the carrying capacity still remains. If N is small, the expression approximates 1, and most of the carrying capacity still remains. If N is large, the expression approaches 0, which indicates that most of the carrying capacity is being utilized.

If our argument to this point is sound, then insertion of the expression

$$\frac{(K - N)}{K}$$

into the equation for exponential growth should result in an S-shaped curve.

$$I = r \frac{(K - N)}{(K)} N$$

At low population densities, when the expression approximates 1, growth would be nearly equal to rN; therefore, growth would be exponential. But as the density increases, the expression approaches 0, and growth levels off. Therefore, it seems that the concept of carrying capacity explains the logistic growth curve. Further, we can predict that if a population happens to overshoot the carrying capacity, it would have to decline once again, since the expression

$$\frac{(K - N)}{K}$$

would then have a negative value.

Fluctuations are sometimes seen in population sizes. In figure 28.5, this fluctuation is due to an abrupt change in carrying capacity from season to season. During the summer, when resources are more than adequate, the bobwhite population increases in size. In winter, when the environment is less favorable, the bobwhite population declines in size. As a result, population size can fluctuate from year to year. The mean of the series of ups and downs over time represents the average carrying capacity of the environment.

Among some species, notably insects and certain plants, populations grow rapidly for a limited period of time and then quickly die out. The growth curve for these populations is at first J-shaped, but the exponential growth is followed by a sharp decline. Rapid population decline can occur because of a change in the environment. For example, a frost will kill all the tomato plants regardless of how many plants there are. In other instances, population size sharply declines because of the density of organisms. For example, a reindeer herd introduced on St. Matthew Island, Alaska, underwent a rapid expansion and then declined just as quickly when food resources were depleted (fig. 28.6) by the overbrowsing of the herd that had grown too large.

Practical Applications

The concepts of biotic potential and environmental resistance can be used to devise the means of regulating population growth. When a population is a human food source, it should be harvested only to the inflection point (fig. 28.4), where the greatest amount of population growth occurs. This amount of cropping results in a **maximum sustained yield** as long as enough mature individuals capable of reproduction remain. Unfortunately, many marine populations, as with whales and commercial fisheries, have been overexploited to the point where it will be many years before a maximum sustained yield will again be possible.

For pests, such as rats and mosquitoes, it is far better to reduce breeding sites and food supply or to increase predators than it is to try to kill off the animals. Killing the animals does not restrict their growth potential, which may come into full force as soon as the population size is reduced. On the other hand, when trying to preserve an endangered species, it is best to place them in a protective environment, such as a reserve, which provides for all their needs. In this case, reduction in environmental resistance will permit population growth to occur.

Figure 28.5
Population fluctuations in a bobwhite quail population. Open bars represent spring population sizes, and colored bars represent fall population sizes. Note that quail populations increase from spring to fall and decrease from fall to spring.

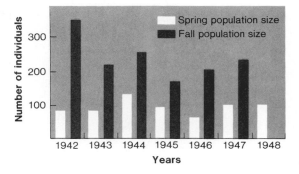

Figure 28.6
The St. Matthew Island reindeer herd grew exponentially for several decades and then underwent a sharp decline as a result of overgrazing the available range.

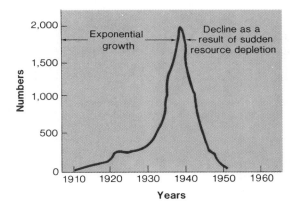

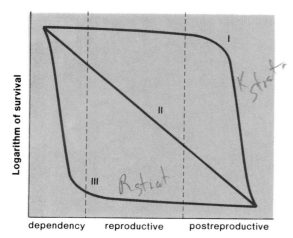

Figure 28.7
The three basic types of survivorship curves. Curve I occurs when the members of a species usually live the full physiological life span. Curve II occurs when the rate of mortality is fairly constant at all age levels. Curve III occurs when there is a high mortality early in life.

Mortality and Survivorship

Rates of increase in population size are a function of additions by birth and losses by death. The rate of increase sometimes becomes greater, not because of a growing birthrate, but because of a decline in the death rate. This decline in deaths is one reason for the explosive growth of the human population in certain countries of the world.

Individuals that make up a population have a limited life span, and death is the means by which most organisms are lost to a population. (The other means is emigration.) The number of individuals in a population that are dying per unit of time per unit of population is the *death rate*. The death rate is usually determined by dividing the number that have died during a given time span by the number alive at the beginning of the time period.

Because living members of the population are more important to the population than are dead ones, another way of looking at mortality is **survivorship,** the number of individuals in a given population alive at the beginning of each age interval. A survivorship curve can be drawn by plotting the number of survivors against time.

In theory, there are three types of survivorship curves (fig. 28.7). A type-I curve shows that most individuals within a designated group live out their allotted life span and then die of old age. Under very favorable environmental conditions, certain mammals, such as humans, and annual plants that live one season tend to approach this curve. The type-II curve results when organisms die at a constant rate through time. Such curves are typical of birds, rodents, and some perennial plants that live more than one season. Type-III curves indicate that many organisms tend to die early in life. This curve is typical of fishes, many invertebrates, and many perennial plants.

Age Structure

Populations have three age groups: *prereproductive, reproductive,* and *postreproductive.* One characteristic of some populations is the number of individuals in each age group. This is best visualized when the proportion of individuals in each group is plotted on a bar graph, thus producing an **age structure diagram.** If the birthrate is high and the population is undergoing exponential growth, a *pyramid shape* (fig. 28.8*a*) results because each successive generation is larger than the previous one. When the birthrate equals the death rate, a *bell-shaped* (fig. 28.8*b*) age diagram is expected because the prereproductive and reproductive age groups become more nearly equal. The postreproductive group is still the smallest, however, because of mortality. If the birthrate falls below the death rate, the prereproductive group will become smaller than the reproductive group. The age structure diagram will be *urn-shaped* (fig. 28.8*c*) because the postreproductive group is now the largest.

In a study of a honeybee hive, these changes in the age structure diagram were observed in a single season. At the beginning of the season, when the population was expanding, the age structure diagram was pyramid shaped; in midseason, when the population size was stationary, the age structure diagram was bell shaped; and finally, as the season came to a close and the population was dying off, the age structure diagram was urn shaped.

Strategies for Existence

Populations do not tend to increase in size year after year because environmental resistance, including both **density-independent** and **density-dependent** effects (table 28.3), regulates number. Some populations are regulated primarily by density-independent effects. These tend to have J-shaped growth curves until some environmental change causes them to decline within a short

Figure 28.8
Age structure diagrams for three different populations.
a. Age structure diagram is a pyramid, indicating that the
population is undergoing exponential growth. *b.* Age
structure diagram is tending to be bell shaped, indicating
that population size is fairly stationary. *c.* Age structure
diagram is tending toward an urn shape, indicating that
the population is on the decline.
From "The Human Population," by Ronald Freedman and Bernard
Berelson. Copyright © 1974 by Scientific American, Inc. All rights
reserved.

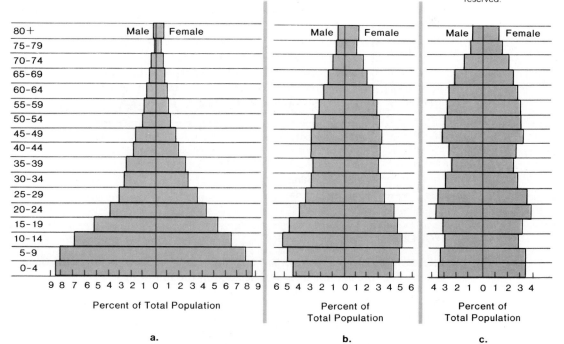

time. Most likely, these populations are **opportunistic species** that move in and
occupy new environments. Ragweeds in gardens and along roadsides, and tent
caterpillars in cherry and apple trees are opportunistic species. During a short
period of time, they produce many offspring that require little care. Therefore,
these populations usually have a survivorship curve similar to type III in figure
28.7. But chances are some of these offspring will quickly disperse to colonize
new habitats, as opportunistic species usually have efficient dispersal mechanisms. From an evolutionary point of view, such species have undergone selection to maximize their rates of natural increase and, for this reason, are
said to be **r-selected** or to be **r-strategists.**

Some populations are regulated primarily by density-dependent effects.
These tend to have sigmoidal-shaped (S-shaped) growth curves. Such populations are termed **equilibrium species.** For example, oak trees and whales are
relatively large, long-lived, and slow to mature. They produce a small number
of well-developed offspring and expend considerable energy in producing each
individual because resources are limited and competitive superiority is a necessity for survival. These organisms tend to have a survivorship curve like
type I in figure 28.7. Equilibrium species are strong competitors and, once
established, can dominate or exclude opportunistic species. They are specialists rather than colonizers and tend to become extinct when their normal way
of life is destroyed. When undisturbed, the population size remains stable at
or near the carrying capacity of the environment. Equilibrium species are said
to be **K-selected** or to be **K-strategists.**

Figure 28.9
Data from a scramble competition experiment. Because food was limited to larvae, scramble competition resulted in few adults whenever there were many larvae.

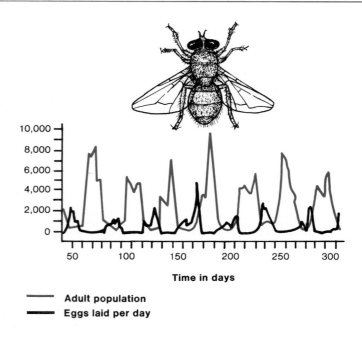

— Adult population
— Eggs laid per day

Population Interactions

Interactions between members of a population or between populations themselves affect population sizes as discussed in the following.

Competition

Such resources as food, water, light, and space determine the carrying capacity of the environment and set a limit on population sizes. Competition for these resources may occur between members of the same population (**intraspecific competition**) or between populations (**interspecific competition**).

Intraspecific Competition

When members of the same population compete, it is possible to distinguish two types of competition: scramble competition and contest competition.

Scramble competition In **scramble competition,** each member of the population tries to acquire as much of a resource as possible. A. H. Nicholson, an English animal ecologist, showed that this type of competition could lead to population size fluctuations. He chose blowflies, which have the usual insect life cycle consisting of egg, larva, pupa, and adult, as his experimental material. Because the larvae live on decaying meat and the adults feed on sugar-rich foods, he was able to regulate the amount of the flies' food resource independently. In one experiment, he rationed food to the larvae but provided an unlimited amount to the adults. When larval density was high, only a few grew large enough to pupate and the few adults they produced were small. These small adults laid a small number of eggs, which, in turn, resulted in few larvae. But more members of the small new larval population now had access to sufficient food, and in the end, there was a larger adult population than before. Thus, the adult population size fluctuated (fig. 28.9) from one generation to the next, its size dependent on the degree of competition for food among the larvae.

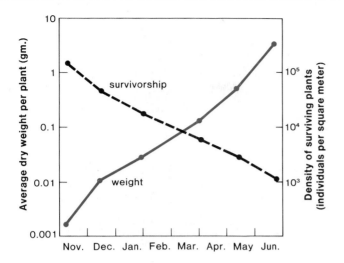

Figure 28.10
A self-thinning curve for an experimental planting. Horseweed was sown at a density of 100,000 seeds per square meter. As the density (number) of the surviving plants decreases, the weight per plant increases. Multiplying these values gives the weight (biomass) that the environment can support—the environment's carrying capacity.
From Moran, et. al., *Introduction to Environmental Science.* Copyright W. H. Freeman and Company, New York. Reprinted by permission of W. H. Freeman and Company, and the British Ecological Society.

Scramble competition is also seen among plants. When horseweeds were sown at low densities, each plant tended to grow vigorously. When they were sown at high densities, many did not survive, and those that did were smaller than usual. Data from a number of plants produced the self-thinning curve shown in figure 28.10. This curve indicates the carrying capacity of the environment since combinations of density and size (weight) lying outside the curve did not occur.

Scramble competition is apt to cause animals to **emigrate**—to move away from centers of great density. Emigration must not be confused with migration, which is an annual movement of animals to more receptive feeding or reproductive areas, as when birds go south for the winter. Some animals, notably rodents, undergo emigrations about every four years due to an increase in population size. Bees or grasshoppers swarm, or emigrate, when the colony becomes crowded.

In laboratory experiments with mice, adequate food and water are provided, but emigration is prevented. When the density increases to the point that crowding interferes with normal reproductive behavior, the mice undergo endocrine changes that prevent fertilization, pregnancy, and normal birth. Delayed spermatogenesis occurs in males, and a prolonged estrous cycle and inadequate lactation occur in females. These, and observations in other animals, suggest that crowding can actually cause biological effects that act as density-dependent environmental resistance factors.

Contest Competition In contrast to scramble competition, **contest competition** tends to keep the animal population within the capacity of the environment without extreme oscillations in population size. As discussed on page 570, some animals live in groups with *dominance hierarchies* that determine the social position and importance of each animal. Dominance is determined by a contest—either actual fighting or simply a ritual contest of who can frighten whom (fig. 28.11). The dominant males eat first and have a chance to mate before lower ranking animals. The other males either assist the dominant males or leave the group to live alone or with bachelor males. In either case, they do not produce offspring.

Figure 28.11
A male baboon in full threat displaying some of the characteristics that may frighten competitors. Such threats enable him to establish his dominance without actually fighting.

Figure 28.12

Competition between two species of paramecium. When grown alone in pure culture, *Paramecium caudatum* and *Paramecium aurelia* exhibit sigmoidal growth. When the two species are grown together in mixed culture, *P. aurelia* is the better competitor and its population increases. *P. caudatum* cannot grow in the presence of a growing population of *P. aurelia*, so it dies out.

From G. F. Gause, *The Struggle for Existence*. Copyright © 1934 by Williams and Wilkins Company.

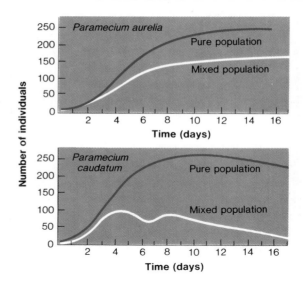

Figure 28.13

Competition prevents two species of barnacles from occupying as much of the intertidal zone as possible. The farthest right area indicates the area of competition between *Chthamalus* and *Balanus*.

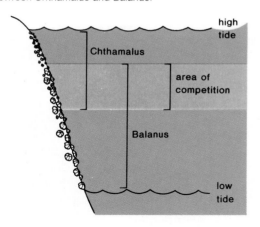

With the same end result, some species exhibit *territoriality* in which successful male competitors establish territories that they defend against other males (fig. 25.21). The male animal mates and the offspring are born within the territory. The best defenders often have the best territories—those with the best food resources and, among birds, the best nesting sites. The less successful defenders must make do with less desirable territories, and some have no territories at all. Therefore, these males produce fewer offspring than dominant males, or they produce no offspring at all.

Interspecific Competition

When populations of different species compete with each other, it is possible that one population could be so successful that the other would die out. This was demonstrated in the laboratory by G. F. Gause, who grew two species of paramecium in one vial containing a fixed amount of bacterial food. Whereas each population survived if grown separately, only one survived when they were grown together (fig. 28.12). The successful paramecium population had a higher rate of natural increase than the unsuccessful paramecium population.

Instead of extinction in natural ecosystems, there may be restriction in the range of one of the species, as observed by Joseph Connell when he studied the distribution of barnacles on the Scottish coast. *Chthamalus stellatus* lives on the high part of the intertidal zone, while *Balanus balanoides* lives on the lower part (fig. 28.13). Free-swimming larvae of both species attach themselves to rocks at any point in the intertidal zone, and after doing so, they develop into the sessile adult form. In the lower zone, faster-growing *Balanus* individuals either force *Chthamalus* individually off the rocks or grow over them. *Balanus,* however, is not as resistant to drying out as is *Chthamalus* and does not survive well in the upper intertidal zone, thus allowing *Chthamalus* to grow there.

The effects of competition have also been observed after humans have inadvertently introduced a species into a new location. The carp, a fish imported to the United States from the Orient, is better able to tolerate polluted

Table 28.4 Aspects of Niche	
Plants	**Animals**
Season of year for growth and reproduction	Time of day for feeding and season of year for reproduction
Sunlight, water, soil requirements	Habitat and food requirements
Relationships with other organisms	Relationships with other organisms
Effect on abiotic environment	Effect on abiotic environment

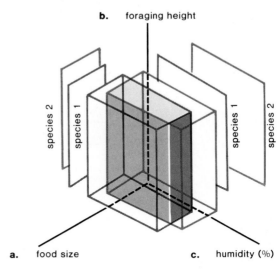

Figure 28.14
As shown here, a niche may be represented by a hypervolume. (In this illustration, the hypervolume is given a rectangular shape.) There is competition if the hypervolumes of two species overlap. The amount of overlap for (a.) food size, (b.) foraging height, and (c.) humidity is shown in color.

water than are native fishes. Therefore, this fish is often more prevalent than native species. An ornamental tree, the melaleuca, was introduced into Florida and has since invaded the Everglades, where it is drying up the cyprus swamps and reducing the survival rate of natural vegetation. And in Australia, the prickly pear cactus introduced from the United States overran 60 million acres before it was brought under control.

Exclusion Principle

Competition between species does not always lead to expansion of one population and restriction of another. Gause showed in a laboratory experiment that when two different species of paramecium occupied the same tube, both survived because one species fed on bacteria on the bottom, while the other fed on bacteria suspended in solution. This experiment illustrated that two similar species can live in concert when they occupy different niches. Only if the two species are attempting to occupy the same niche will one tend to die out. The fact that no two species can occupy the same niche is called the **competitive exclusion principle.**

Niche is used to refer to the role a species plays in a community of organisms. To describe a species' niche, it is necessary to state all requirements and activities of that species. Table 28.4 lists factors to be included when describing the niche of a particular plant or animal species. One of these factors is **habitat,** the exact location of the population's existence in the ecosystem. For example, in a forest ecosystem, a particular population might live in the trees, on the ground, or even in the ground.

Some investigators have suggested that it is possible to represent a niche as a many-sided figure or abstract multidimensional hypervolume. In figure 28.14, a three-dimensional box is created by describing environmental variables important to a particular bird species. For example, an important environmental variable for a species might be food size. Therefore, this is plotted as a line in one dimension (fig. 28.14a). A second variable might be foraging height, plotted as a line in a second dimension (fig. 28.14b). These lines then enclose a rectangle that represents two dimensions of a niche. Yet a third dimension can be added for another variable—humidity. The three lines now describe three dimensions of a niche (fig. 28.14c). This process could theoretically be continued indefinitely, resulting in a multidimensional hypervolume.

Diversity

Figure 28.14 also indicates that it is possible for the niches of various populations (species) to overlap. For example, with Darwin's finches (p. 251), birds that prefer medium-sized seed could most likely feed on some seed also eaten

Figure 28.15
Resource partitioning among five species of coexisting warblers. The diagrams that appear beside the birds represent spruce trees. The time each species spent in various portions of the trees was determined, and it was discovered that each species spent more than half its time in the regions indicated.

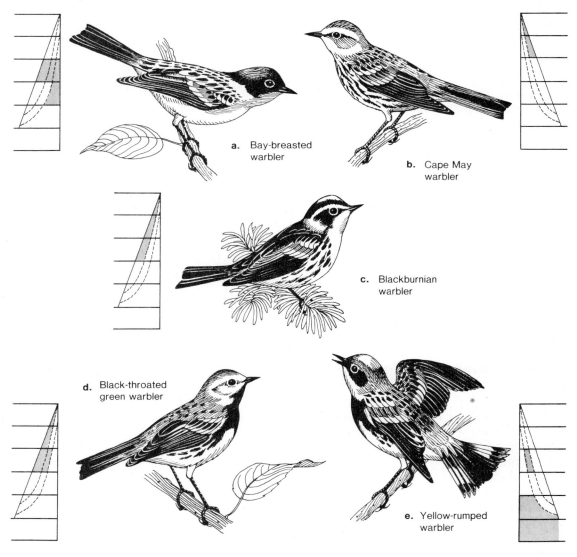

a. Bay-breasted warbler

b. Cape May warbler

c. Blackburnian warbler

d. Black-throated green warbler

e. Yellow-rumped warbler

by birds that prefer small-sized seed. The greater the degree of niche overlap, the greater the degree of competition between populations. Such competition causes species to evolve to fill different niches, and therefore leads to diversity.

While it may seem as if several species living in the same area are occupying the exact same niche, it is usually possible to find slight differences. For example, Robert MacArthur demonstrated how five species of North American warblers partition resources in their northern spruce forest habitat (fig. 28.15). He divided spruce trees into zones, recorded the length of time each species spent in the different zones, and determined where each species did most of its feeding. He discovered that different species used different parts of the tree canopy, although some overlap did occur. This partitioning of resources permitted all five species to feed in the same forest without seriously competing with one another.

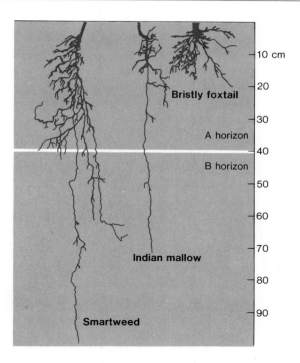

Figure 28.16
Resource partitioning by three species of annual plants in an abandoned field one year after cultivation. The A horizon is the upper layer of soil that contains more organic material and variable amounts of moisture. The B horizon below contains inorganic material and a more constant amount of moisture.

Resource partitioning is also observed in plants. Three species of annual plants, found in an abandoned field one year after cultivation ceased, exploited a different part of the soil. Bristly foxtail has shallow fibrous roots that take water only from the top part of the soil; Indian mallow has a sparsely branched taproot that extends to an intermediate depth; and smartweed has a taproot that is moderately branched in the upper soil layer, but develops primarily below the rooting zone of the other species (fig. 28.16).

Predation

Predation can be defined simply as one organism feeding on another. Examples include deer feeding on plants, a robin feeding on an earthworm, or a cougar feeding on a deer. Through the evolutionary process, predators become specialized to capture their prey. The anatomy of browsers enables them to reach high into trees—giraffes have long necks and elephants have long snouts. Birds of prey have talons and sharp beaks that enable them to capture, hold, and kill their prey. Carnivores also need keen eyesight. Hawks have a resolving power that is eight times better than that of the human eye due to the concentration of cones in two foveas instead of one.

The efficiency of predation is related to the amount of prey available, but the correlation is not as simple as might be predicted. In an experiment with European titmice, which feed on insect larvae, predation was less than expected when the prey species was rare. Once the prey increased in number, however, predation was greater than expected. Finally, when the prey species was most numerous, predation was again less than expected (fig. 28.17). Because of these results, it has been suggested that predation requires a **search image,** a mental picture of the prey. When the prey is rare, predation is low because it is not economical for the animal to form a search image; but when the prey starts to increase in number, the animal develops a search image and begins to actively seek this particular food. When the predator has achieved its maximum potential for capturing the prey because its belly is full, there is no further increase in predation regardless of the amount of prey.

Figure 28.17
Size of prey population versus risk of predation. The risk of bird predation (dotted line) was found to be smaller than expected when an insect's population was low, and greater than expected when the population was increasing. This indicates that perhaps as a particular prey becomes more abundant, birds develop a search image that helps them find that prey.

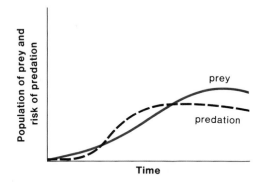

Figure 28.18

Changes in abundance of lynx and hare as determined from the number of pelts received by the Hudson's Bay Company.

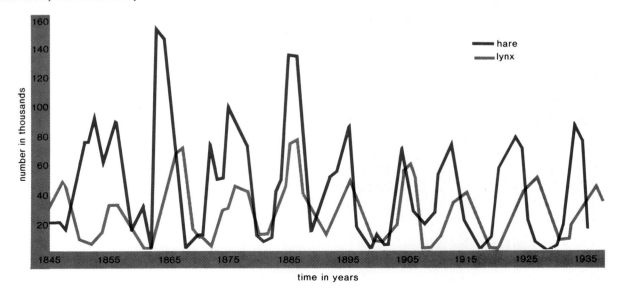

Other data suggest that predators might overkill their prey population, eventually leading to a reduction in the size of the predator population. For example, the population size of the lynx in Canada decreases approximately every ten years (fig. 28.18). Lynx population size fluctuations closely follow cyclic changes in the population size of their principal prey, the snowshoe hare. Ecologists are not convinced, however, that these oscillations are due to the predator-prey relationship. Possibly these changes are due to a periodic decline in the availability of food for the hares; as their population size declines, the lynx population declines in turn. This seems reasonable because the size of the hare population fluctuated even in the absence of lynxes.

One other consideration in the predator-prey relationship was illustrated by other experiments performed by G. F. Gause. In an experiment involving two protozoans, the didinia actually captured all the paramecia and then died out (fig. 28.19). However, if Gause provided refuge for the prey, in this case sediment on the bottom of the laboratory glassware, part of the prey population survived. This suggests that predator-prey relationships can stabilize when predator pressure eases as the prey population decreases in size. In nature, this can be a result of the prey being able to hide, or of the predators switching to a more abundant and hence more economical prey species.

This experiment tells us that rather than an oscillation in predator and prey populations, there could be a steady relationship. For example, on **Isle Royale,** an island in Lake Superior, a population of about one thousand moose has coexisted with a population of about twenty-four wolves year after year since 1948. Ecologists who have studied this association tell us that the wolves are able to capture only the old, sick, or very young animals. The inability of the wolves to capture more moose keeps the wolf population in check.

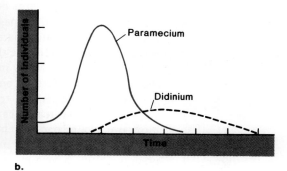

a.

b.

The regulatory effects of predation on the prey population is apparent when it is known that in 1930, before there were wolves on Isle Royale, the moose population grew to the point where it exceeded the carrying capacity of the island and suffered a crash. Once the vegetation recovered, the moose population again grew to the point of overexploitation and suffered another crash in the 1940s. The chance introduction of wolves has kept the moose population within the carrying capacity of the island since this time.

Sometimes humans have neglected to take into consideration that predators can keep prey populations in check. Coyotes have been killed off in the West without regard to the fact that they tend to keep the prairie dog population under control. Similarly, when the dingo, a wild dog in Australia, was killed off because it attacked sheep, the rabbit and wallaby populations increased. Some farmers now recognize the benefits of maintaining populations of owls because they capture rodents. In other instances, too, predators have been used to control pest populations. The prickly pear cactus in Australia was finally brought under control by the introduction of moth borers (a, p. 639), and a weevil has been used to reduce the amount of water hyacinth in southern Florida. The cottony-cushion scale, a small, sap-sucking insect named for the large white egg mass deposited by the female on tree limbs, threatened the California citrus industry. It was brought under control by importing its natural predator, a ladybug beetle called the vedalia. Such examples indicate that agricultural pests could be biologically rather than chemically controlled. Control of agricultural pests is discussed in the reading on page 638.

Table 28.5 Animal Antipredator Adaptations

Name	General Behavior
Sensory methods	Cryptic coloration and behavior
	Startle display
	Distraction display
	Death display
Vigilant method	Detection of predator
Escape behavior	Evasion
Repellent behavior	Having a chemical or mechanical defense
	Resembling animals that have a chemical or mechanical defense

Figure 28.20
Concealment as an antipredator defense. Young green heron birds strike a pose that makes them resemble tree branches.

Prey Defenses

While predators have evolved strategies to secure the maximum amount of food with minimal expenditure of energy, prey organisms have evolved strategies to escape predation. Plants are not an exception to this statement. *Acacia* trees and shrubs occur in tropical and subtropical regions. In Africa and tropical America, where browsers are indigenous, the acacia species are protected by thorns that are often well developed. In Australia, where there are no indigenous browsers, most species of acacia lack thorns entirely. Similarly, the sharp spines of the cactus; the pointed leaves of the holly; and the tough, leathery leaves of oak trees all discourage predation by insects. Plants even produce poisonous chemicals, some of which interfere with the normal metabolism of the adult insect, and others which act as hormone analogues that interfere with the development of insect larvae (p. 606). Humans often make use of these chemicals; certain ones are the active ingredients of spices, medicines (digitalis, morphine, quinine), stimulants (caffeine), and other drugs (nicotine, marijuana, opium).

Animal **antipredator defenses** are also quite varied (table 28.5). Some animals take the path of passive resistance and attempt to blend in with the background. Stick caterpillars look like twigs; katydids look like sprouting green leaves; and moths resemble the barks of trees. Often there is a backup defense in case of discovery. Many moths have eyelike spots on their underwings that they can flash to startle a bird long enough to allow an escape.

Larger animals, too, try to conceal themselves by looking like their environment. Decorator crabs cover themselves with debris, and green heron birds, even young ones, attempt to look like straight tree branches (fig. 28.20).

Other types of prey immediately try to frighten or attack a would-be predator. Some have a means of making themselves appear much larger than they are, such as the frilled lizard of Australia that can open up folds of skin around its neck (fig. 28.21). The porcupine sends out arrowlike quills with barbs that dig into the enemy's flesh and penetrate even deeper as the enemy struggles after impact.

Poisonous animals tend to be brightly colored as a warning to those that may want to prey upon them. The spotted arrow-poison frog found in tropical regions is so poisonous that South American Indians have long used it, as its name implies, to poison their arrow tips. When the arrow strikes, it paralyzes the victim instantly.

Flocks of birds, schools of fish, and herds of mammals stick together as protection against predators. Grazing herbivores are constantly on the alert; if one begins to dart away, they all run. Baboons who detect predators visually and antelope who detect predators by smell sometimes forage together, giving double protection against stealthy predators. The gazellelike springboks of southern Africa jump stiff-legged straight up and down, 8 to 12 feet, a number of times as a warning. Such a jumble of shapes and motions might confuse an attacking lion, and the herd can escape.

Mimicry

Mimicry occurs when one species resembles another that possesses an overt antipredator defense. For example, the brightly colored monarch butterfly (fig. 28.22*a*) causes predaceous birds to become sick and sometimes to vomit. Monarch butterflies have this capability because as larvae they feed on milkweed plants. These plants contain digitalislike compounds that activate nerve centers controlling vomiting. (Larvae incorporate a modified form of the poison

Figure 28.21
Scare tactics as an antipredator defense. The frilled lizard of Australia gives the appearance of a much larger beast by opening its mouth wide and unfurling folds of skin around its neck.

Figure 28.21
Scare tactics as an antipredator defense. The frilled lizard of Australia gives the appearance of a much larger beast by opening its mouth wide and unfurling folds of skin around its neck.

Figure 28.22
The monarch butterfly (a.) and its mimic, the viceroy butterfly (b.).

a.

b.

into their own tissues, retaining it even as they develop into adults.) Birds that have eaten a poisonous monarch butterfly avoid all monarch butterflies in the future.

The viceroy butterfly (fig. 28.22b) mimics the monarch butterfly's coloration, but is not toxic. Birds eagerly eat viceroy butterflies unless they have had previous experience with a poisonous monarch. Then, because the butterflies closely resemble each other, birds avoid both types of butterflies. The queen butterfly also mimics the monarch, although unlike other viceroys, the queen is actually poisonous.

A mimic that lacks the defense of the organism it resembles is called a **Batesian mimic;** the viceroy butterfly is a Batesian mimic of the monarch. A mimic that also possesses the same defense is called a **Mullerian mimic;** the queen butterfly is a Mullerian mimic of the monarch butterfly.

Biological Pest Management

The use of chemical pesticides to kill agricultural pests has often created more problems than it has solved. Indeed, percentage crop losses to insects are now over three times what they were in the 1940s, before the modern pesticides became available. Pesticides invariably reduce the numbers of predators and parasites that naturally control the pests, more than they reduce the numbers of pests themselves. Then, after the pests become resistant to the pesticide, there are no natural enemies to hold the pests in check. The population size then increases manyfold as does that of other plant-eating insects whose natural enemies were also destroyed by the pesticide.

Pesticides also find their way into the water supply of humans, and/or they concentrate in food chains causing humans to receive larger doses than were applied in the first place. In Hawaii, cows were fed pineapple leaves that had been sprayed, and the cows' milk then passed the pesticide on to human infants.

These problems with pesticide use have caused many to become interested in integrated pest management, which utilizes resistant plants, natural enemies of the pest, and environmental modification to attempt to reduce a pest population. This system seeks to control the pest rather than to eradicate it.

The use of natural predators, parasites, and pathogens can control agricultural pests. The vedalia ladybug was imported from Australia in 1888 to save citrus trees that were literally covered with cottony-cushion scale. The ladybug proved to be such a good predator that the pest was nearly exterminated within two years of the initial release of just twenty-eight beetles. The praying mantis and lacewing are predators that feed on many types of pests. When lacewings were released in cotton fields, they reduced the boll weevil population by 96 percent and increased cotton yield threefold. A predatory moth was used to reclaim 60 million acres in Australia overrun by the prickly pear cactus (see *a.* and *b.*), and another type of moth is now used to control alligator weed in the southern United States. Similarly, in this country the Chrysolina beetle controls Klamath weed, except in shady places where a root-boring beetle is more effective.

Parasites are also sometimes helpful. Many parasitic wasps lay their eggs in caterpillars, which are then used as food for the wasp larvae. Cereal leaf beetle, which attacks grain crops, and olive scale, which attacks olive trees, can both be controlled by specific parasitic wasps. Bacteria and viruses can also successfully control certain pest populations. For example, milky spore disease caused by a bacterium kills the larvae of Japanese beetles, and a virus kills the cabbage looper, which damages the leaves of cotton, cabbage, and several other crops.

The use of pheromones is also a type of biological control. Traps are baited with synthetic pheromones to collect male and/or female insects. Outbreaks of gypsy moths have been curtailed in this way. Recently, scientists in Sweden and Norway synthesized the sex pheromone of the Ips bark beetle, which attacks spruce trees. Almost 100,000 baited traps collected females that are normally attracted to

Symbiosis

Symbiotic relationships (table 28.6) are close relationships between two different species. As with predation and prey, coevolution occurs, and the species are closely adapted to one another. In parasitism, the parasite benefits, but the host is harmed; in commensalism, one species benefits, but the other is unaffected; and in mutualism, both species benefit.

Parasitism

Parasitism is similar to predation in that the parasite derives nourishment from the host. Usually, however, the host is larger than the parasite, and the parasite does not immediately kill the host. While viruses are the only group of organisms that are obligate parasites (p. 269), there are also parasites among bacteria, protista, plants, and animals. The smaller parasites tend to be endoparasites that live within the bodies of the hosts. The larger parasites tend to be exoparasites that remain attached to the exterior of the host by means of a specialized organ.

Table 28.6 Symbiosis	Species 1	Species 2
Parasitism	+	−
Commensalism	+	0
Mutualism	+	+

Key: + = benefits
 − = harmed
 0 = no effect

males emitting this pheromone. The use of compounds that interfere with insect development has an even more direct effect than do pheromones. Two such compounds isolated from *Ajuga remota*, an East African plant, cause the fall armyworm caterpillar, a common cotton field pest, to develop three heads, burying its functional mouthparts (p. 606).

Sterile males have also been used to reduce pest populations. The screwworm fly parasitizes cattle in the United States. Flies were raised in the laboratory and made sterile by exposure to radiation. The entire southeast was freed from this parasite when female flies mated with sterile mates and then laid eggs that did not hatch.

Various farming techniques can modify the environment to control pests. When farmers use strip farming and crop rotation, pests do not have a continuous food source. Maintaining hedgerows, weedy patches, and certain trees can provide habitats and food for predators and parasites that help control pests. For example, cottonwood and willow trees are normally inhabited by three species of flies that will kill the cotton bollworm, and wild blackberry bushes are the alternate host for wasps that parasitize grape leafhoppers.

Biological control, which requires in-depth knowledge of pests and their life cycles, is a more sophisticated method than chemical control of pests. Its great advantage is that it avoids the problems associated with the use of pesticides. However, it does not have an immediate effect on pests; its effect is invisible and thus tends to go unnoticed by the farmer or gardener who benefits from it.

a.

b.

Example of biological control. a. *Prickly pear cactus infestation in Australia made land unfit for human use.* b. *The same area after the introduction of Argentine moths* (Castoblastis cactorum).

Just as predators can dramatically reduce the size of a prey population that lacks a suitable defense, so parasites can reduce the size of a host population that has no defense. Thousands of elm trees have died in this country from the inadvertent introduction of **Dutch elm disease,** which is caused by a parasitic fungus. A new method of treating Dutch elm disease utilizes bacteria that produce a fungus-killing antibiotic when injected into the tree. Another means of curbing this fungal infection of trees is to control the bark beetle that spreads the disease.

Many other parasites use a secondary host for dispersal. For example, *tapeworms* utilize cattle or pigs, and *flukes* employ snails and sometimes fish to complete their life cycle (p. 315). The **hookworm,** judged to be the most troublesome parasitic worm in the United States, does not require a secondary host. New World hookworm males are from 5 to 9 mm in length, and females are usually about 1 cm long. The head of the worm is sharply bent in relation to the rest of the body, accounting for its characteristic hooklike appearance. Adult hookworms attach themselves to the intestinal wall (fig. 28.23) of their host, and pass eggs out with the feces. When deposited on moist, sandy soil,

Figure 28.23
Longitudinal section through a hookworm attached to the intestinal wall.

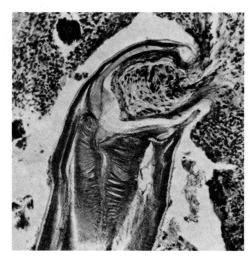

Figure 28.24
Slave-making Amazon ants invade the colony of a slave species and carry off cocoons. When ants emerge from the cocoons, they serve as slaves only to increase the size of the capture population. This is a form of social parasitism in which one species benefits and the other is harmed.

the larvae develop and hatch within twenty-four to forty-eight hours. After a period of growth and development, the worms extend their bodies into the air and wave about in this position until they come into contact with the skin of a suitable host, such as a human. Penetration usually occurs through the feet. Once in the blood vessels, the worms are passively carried to the lungs, where they invade the alveoli. The larvae migrate from the lungs to the trachea to be swallowed and passed along to the small intestine where they mature. They attach to the intestinal wall by means of their stout mouth parts and suck blood and tissue juices from the host. Symptoms of hookworm infection include abdominal pains, nausea, diarrhea, and finally iron deficiency anemia.

Social Parasitism

Social parasitism occurs when one species exploits another species. The cuckoo lays eggs in nests of songbirds, and the newly hatched cuckoo ejects its nestmates so the songbird parents attend only to it. Slave-making amazon ants of the species *Polyergus rufescens* raid the ant colonies of a slave species *Formica fusca* (fig. 28.24). They destroy any resisting defenders with their mandibles, which are shaped like miniature sabers. *Polyergus* ants are so specialized they can only groom themselves. To eat, they must beg slave workers for food. The slave workers not only provide food for the slave-making ants but also care for the eggs, larvae, and pupae of their captors.

Commensalism

In a **commensal** relationship only one species benefits, but the other is neither benefited nor harmed. Often the host species provides a home and/or transportation for the benefited species. Barnacles, which attach themselves to the backs of whales and the shells of horseshoe crabs, are provided with both a home and transportation. **Remoras** are fish that attach themselves to the bellies of sharks by means of a modified dorsal fin that acts as a suction cup. The remoras obtain a free ride and also feed on the remains of the shark's prey. **Epiphytes** (fig. 18.8) grow in the branches of trees where they receive light, but they take no nourishment from the trees; instead, their roots obtain nutrients and water from the air. **Clownfish** (fig. 28.25) live within the tentacles and gut of sea anemones and thereby are protected from predators. Perhaps this relationship borders on mutualism, since the clownfish may attract other fish on which the anemone can live. The sea anemone's tentacles quickly paralyze and seize other fish as prey.

Figure 28.26

Organisms are adapted to their way of life, such as the hummingbird. The bird's long bill allows it to collect the nectar of tubular flowers. This is a form of mutualism, in which the flower species is benefited when the hummingbird carries pollen to another flower, thus allowing cross-fertilization to occur. The flower species, in return, provides food for the hummingbirds.

Mutualism

Mutualism is a symbiotic relationship in which both members of the association benefit. Mutualistic relationships often allow organisms to obtain food or avoid predation.

Bacteria that reside in the human intestinal tract are provided with food, but they also provide humans with vitamins, molecules we are unable to synthesize for ourselves. Termites would not even be able to digest wood if it were not for the protozoans that inhabit their intestinal tract. These organisms digest cellulose, which termites cannot digest. Algae and fungi live together as lichens. Lichens, but not algae, can grow on rocks, probably because the fungi provide dissolved minerals from the rock. The algae carry on photosynthesis and return carbohydrates to the fungi. **Mycorrhizae** (p. 396) are symbiotic associations between the roots of plants and fungal hyphae. Mycorrhizal hyphae improve the uptake of nutrients for the plant, protect the plant's roots against pathogens, and produce plant growth hormones. In return, the fungus obtains carbohydrates from the plant.

Flowers and their pollinators (fig. 14.25) have coevolved ever since flowering plants first appeared. For this reason, flowers that attract bees or birds are brightly colored, and flowers that attract bees, beetles, or bats have strong scents. Sometimes the plant nectaries, which produce a sugar solution, are located at the base of a tubular corolla. Here they are accessible only to the moth or bird that has evolved a long sucking tongue, such as the hummingbird (fig. 28.26). Flowers that attract bees have a landing platform and a structure that requires the bee to brush up against the anther and stigma as it moves toward the nectaries. The coevolution of flowers and their pollinators is of mutual benefit; the pollinator receives food, and the flower achieves cross-pollination.

A number of mutualistic relationships involve ants and other organisms. **Leaf cutter ants** keep fungal gardens. These ants were so named because they gather flowers and leaves, cut them into pieces, and transport them to underground nests. After preparing the leaves, they implant them with fungal mycelia, which then grow in profusion. This relationship gives the fungi a competitive edge over other fungal species, but also provides the ants with a source of food.

In tropical America the bullhorn acacia (fig. 28.27) is adapted so as to provide a home for ants of the species *Pseudomyrmex ferruginea*. Unlike other acacias, this species has swollen thorns with a soft, pithy interior where ant larvae can grow and develop. In addition to housing the ants, the acacias provide them with food. The ants feed from nectaries at the base of the leaves and eat fat- and protein-containing nodules, called Beltian bodies, that are found at the tips of some of the leaves. The ants constantly protect the plant from herbivorous insects because, unlike other ants, they are active twenty-four hours a day. Because the ants are so diligent, the acacias have leaves throughout the year, while related acacia species lose their leaves during the dry season.

Cleaning symbiosis (fig. 28.28) is a phenomenon believed to be quite common among marine organisms. There are species of small fish and shrimp that specialize in removing parasites from larger fish. The large fish line up at the "cleaning station" and wait their turn, while the small fish feel so secure they even clean the mouths of the larger fish. Not everyone plays fair, however, since there are small fish that mimic the cleaners and take a bite out of the larger fish, and cleaner fish are sometimes found in the stomachs of the fish they clean.

Figure 28.27
The bullhorn acacia is adapted so as to provide a home
for ants of the species *Pseudomyrmex ferruginea*. *a.* The
thorns are hollow and the ants live inside. *b.* The base of
leaves have nectaries (openings) where ants can feed.
c. Leaves have bodies at the tips that ants harvest for
larvae food.

a.

b.

c.

Figure 28.28
Cleaning symbiosis. Neon gobies and a Spanish hogfish
cleaning a Nassau grouper.

Summary

The biosphere is divided into ecosystems, units where populations interact with one another and the physical environment. All the populations make up a community. Populations with a positive rate of natural increase (r) have an increase in size for the year expressed as $I = rN$, where (N) is the number of individuals in the population. Since (N) is larger each year, so is (I). This means the population is increasing in size exponentially, and the growth curve would be a J-shaped, or exponential one.

The maximum rate of natural increase (biotic potential) for a population (r_{max}) occurs only when there is no environmental resistance. An r_{max} might be observed for a short time during exponential growth, but eventually environmental resistance causes the growth curve to become sigmoidal as the population size levels off at the carrying capacity of the environment. Sometimes populations exhibit a J-shaped curve followed by a sharp decline. Opportunistic species often colonize an area and then are cut back by the first frost, for example. Or unrestricted growth is often followed by a crash as the population greatly exceeds the environment's carrying capacity.

Another characteristic of populations is survivorship. Populations tend to have one of three types of survivorship curves depending on whether most individuals live out the normal life span, die at a constant rate regardless of age, or die early. Each population also has a particular type of age structure diagram that indicates the number of individuals in the prereproductive, reproductive, and postreproductive age group. The shape of the pyramid tells whether the population is a young expanding one, a stable one, or a dying one.

Populations seem to have a particular overall strategy for continued existence. Those that are opportunists produce many young within a short period of time and rely on rapid dispersal to new, unoccupied environments. These populations tend to have a J-shaped growth curve and to be regulated by density-independent factors. Such populations belong to species that have apparently been r-selected. Other populations produce a limited number of young that they nurture for a long time. These populations tend to be regulated by density-dependent factors and belong to species that have apparently been K-selected.

Population interactions help regulate population sizes. Competition may be intraspecific or interspecific. Intraspecific competition may be either a scramble type or a contest type. The former often results in fluctuations in population size, while the latter results in maintenance of population size at or near carrying capacity. Contest competition is seen in populations that either establish territories or have a dominance hierarchy. When populations increase in size, members sometimes emigrate and intraspecies competition is reduced. Interspecific competition occurs only for a short period of time because either one population replaces the other or the two species evolve to occupy different niches. While it may seem that similar species are occupying the same niche, actually there are slight differences.

Evidence seems to suggest that predation does tend to control the size of the prey population, but the relationship may involve other factors as well. Just as predators are adapted to capture prey, prey are adapted to escape predators. Plants have defenses and are noted especially for producing toxins that affect the metabolism or life cycle of the prey. Animals have various types of defenses, as listed in table 28.5. Some animals mimic other animals that have a defense, especially those others that are poisonous to eat.

Symbiotic relationships include parasitism, commensalism, and mutualism. Parasitism, tends to control the size of the host populations. Although many parasites require secondary hosts, the hookworm, an animal parasite, does not. In social parasitism one society exploits another. In commensalism one species is benefited but the other is unaffected. Often the host simply provides a home and/or transportation. In mutualism both species benefit. The two species are often closely adapted to one another, as are flowers and their pollinators. There are also examples of mutualistic species that live together in the same locale, such as ants that keep fungal gardens. In cleaning symbiosis, however, the relationship is transitory.

Objective Questions

1. In an ecosystem, populations of organisms interact with one another and the _____ environment.
2. If resources are plentiful, populations exercise their biotic potential and undergo _____ growth.
3. Environmental resistance causes a J-shaped growth curve to become a(n) _____-shaped curve just at the level of the carrying capacity.
4. The human population has a type of survivorship curve that indicates that individuals in the population tend to _____ .
5. Organisms that are r-strategists tend to have a _____-shaped curve, followed by a sharp decline, and are regulated by density-_____ effects.
6. Scramble competition is a form of _____ specific competition.
7. In an ecosystem, it can be determined that no two populations occupy the same _____ as dictated by the competitive exclusion principle.
8. Both competition and predation are factors that _____ population sizes.
9. _____ occurs when one species resembles another that possesses an overt antipredator defense.
10. In the symbiotic relationship known as mutualism, _____ .

Study Questions

1. Draw a sigmoidal growth curve and relate the concepts of biotic potential and environmental resistance and carrying capacity to the curve.
2. Draw a J-shaped curve and discuss why populations with these curves may suddenly decline in size.
3. Draw three different survivorship curves and tell what each curve indicates about the expected time of death.
4. Draw three different age structure diagrams and tell what each one indicates about a population.
5. What is an "r-strategist"? a "K-strategist"?
6. Name four types of density-dependent effects and explain what is meant by this term.
7. What is the competitive exclusion principle and how is it related to the concept of niche?
8. Give examples of interspecific and intraspecific competition.
9. Give examples of predator adaptations that enable them to capture prey, and of prey adaptations that enable them to escape capture.
10. Contrast the way of life of a predator with that of a parasite, and commensalism with mutualism. Give examples of parasitism, commensalism, and mutualism.

Selected Key Terms

biosphere ('bī ə ,sfir) 622
community (kə 'myü nət ē) 622
logistic curve (lō 'jis tik 'kərv) 624
carrying capacity ('kar ē iŋ kə 'pas ət ē) 624
survivorship (sər 'vī vər ,ship) 626
density independent ('den sət ē ,in də 'pen dənt) 626
density dependent ('den sət ē di 'pen dənt) 626
competition (,käm pə 'tish ən) 628
niche ('nich) 631
habitat ('hab ə ,tat) 631
predation (pri 'dā shən) 633
mimicry ('mim i krē) 636

CHAPTER 29

Ecosystems
The Flow of Energy and the Cycling of Materials

Concepts

Energy flow and the cycling of chemicals (nutrients) are two main characteristics of natural ecosystems.
  a. Mature and highly diversified natural ecosystems tend to require the same amount of solar energy and nutrient input each year.
  b. The human ecosystem, as exemplified by the human food chain, has tended to utilize increasing amounts of supplemental energy and material inputs that do not cycle.
  c. Each chemical element has its own biogeochemical cycle that includes not only the populations within an ecosystem but also segments of the physical environment.
  d. To acquire supplemental energy and materials, humans have usually tapped the reservoir portions of the biogeochemical cycles.

All of the populations in a given area, together with the physical environment, comprise an *ecosystem*. Thus, an ecosystem possesses both biotic and abiotic components. The **abiotic components** include soil, water, light, inorganic nutrients, and weather variables. The **biotic components** can be categorized as either producers or consumers. **Producers** are autotrophic organisms with the capability to carry on photosynthesis and make food for themselves and indirectly for the other populations as well. In terrestrial ecosystems, the producers are predominantly herbaceous and woody seed plants, while in freshwater and marine ecosystems the dominant producers are various species of algae (fig. 29.1).

Consumers are heterotrophic organisms that must take in preformed food. It is possible to distinguish four types of consumers, depending on their food source. *Herbivores* are primary consumers that feed directly on green plants. *Carnivores* are secondary or tertiary consumers that feed only on animals. *Omnivores* feed on both plants and animals. Thus, a caterpillar feeding on a leaf is a herbivore; a green heron feeding on a fish is a carnivore; a human being who eats both leafy green vegetables and beef is an omnivore.

The fourth type of consumer is a *decomposer* that feeds on **detritus,** the remains of plants and animals. The bacteria and fungi of decay are important detritus feeders, but so are other soil organisms, such as earthworms and various small arthropods. The importance of the latter can be demonstrated by placing leaf litter in bags with mesh too fine to allow soil animals to enter; the leaf litter will not decompose well. Small soil organisms precondition the detritus so bacteria and fungi can break it down to inorganic matter that producers can use again.

When we diagram all the biotic components of an ecosystem as in figure 29.2, it is possible to illustrate that every ecosystem is characterized by **energy flow** and **chemical cycling.** These are the two fundamental aspects of ecosystems. Both begin when producers absorb solar energy and receive inorganic nutrients from the physical environment. Thereafter, the energy content of organic food is lost as heat (a nonusable form of energy) but the original inorganic nutrients—the individual chemical elements out of which organisms are made—eventually cycle back to the producers.

It is customary to think of such areas as forests and ponds as ecosystems. But it is also proper to realize that, in a sense, human beings have created their own ecosystem comprising both the country, where most of the producers (crops) are found, and the city, where most of the final consumers (people) are found.

a.

b.

Figure 29.1
Biotic components of an ecosystem. *a.* Trees are terrestrial producers and algae are aquatic producers. *b.* A great blue heron is a consumer in an aquatic ecosystem.

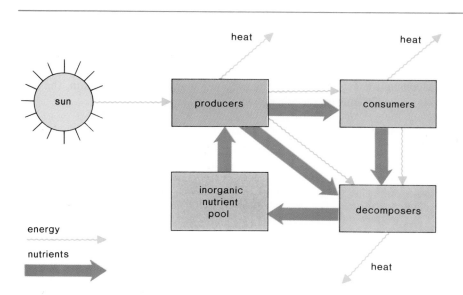

Figure 29.2
A diagram illustrating the manner in which nutrients cycle through an ecosystem. Energy does not cycle because all that is derived from the sun eventually dissipates as heat.

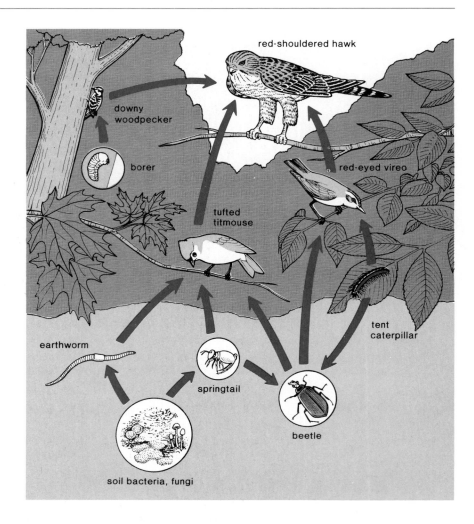

Energy Flow

Energy flow is a one-way process; that is, as the energy available in food moves through an ecosystem, it is gradually degraded to **heat** (fig. 29.2), a nonusable form of energy. This means that ecosystems are unable to function unless there is a constant energy input from an external source, such as the sun, the ultimate source of energy for our planet.

Food Chains and Food Webs

Energy flows through an ecosystem as one population feeds on another. A series of populations through which energy flows is called a **food chain.** A typical terrestrial ecosystem is shown in figure 29.3, and a typical aquatic ecosystem is shown in figure 29.4. The arrows indicate the direction of energy flow, and

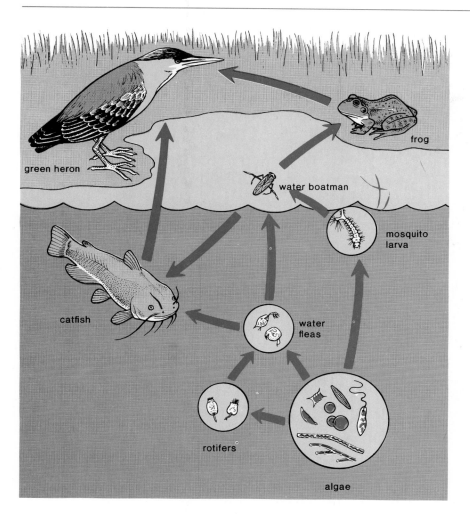

Figure 29.4
This drawing depicts populations typical of a freshwater
pond ecosystem. The arrows indicate the flow of energy.

green heron

frog

water boatman

mosquito
larva

catfish

water
fleas

rotifers

algae

thus, it is possible to decipher various food chains. Two types of food chains
can be described: grazing and detritus chains. For example, **grazing food chains**
are:

1. trees ⟶ tent caterpillars ⟶ red-eyed vireos ⟶ hawks
2. algae ⟶ water fleas ⟶ catfish ⟶ green herons

In some ecosystems (forests, rivers, and marshes) the detritus food chain
accounts for more energy flow than does the grazing food chain because most
organisms die without having been eaten. In the forest one **detritus food chain**
is:

detritus ⟶ soil bacteria ⟶ earthworms

A detritus food chain often ties in with a grazing food chain, as when earthworms are eaten by a tufted titmouse. Eventually, however, as dead organisms
decompose, all the solar energy that was taken up by the producer populations

Figure 29.5
A flow diagram illustrating the dissipation of energy in an ecosystem. Producers convert inorganic matter to organic matter after capturing a small amount of solar energy. When organic matter is used as an energy source by consumers and decomposers, the energy ultimately becomes heat. Energy, therefore, does not cycle, and ecosystems cannot exist without a continual supply of solar energy.

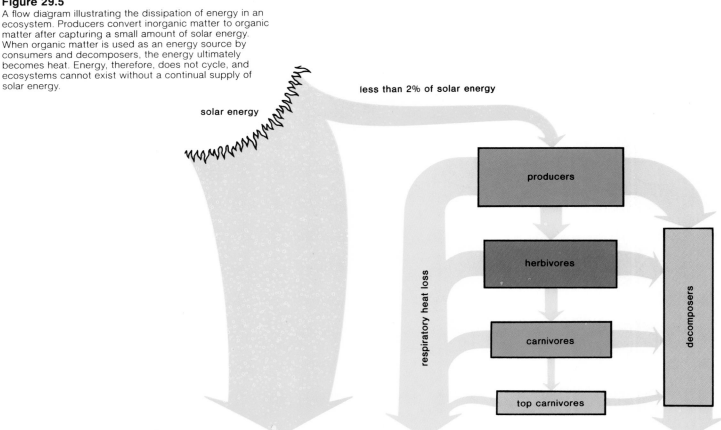

dissipates as heat. As figure 29.5 illustrates, the very small percentage of solar energy captured by producers supports the populations in an ecosystem, and eventually all this energy is returned to the atmosphere as heat. Thus, **energy does not cycle.**

Food Web

Each food chain represents just one path of energy flow through an ecosystem. Natural ecosystems have numerous food chains, each linked to others to form a complex **food web.** For example, in figure 29.3 plants are eaten by a variety of insects, and these, in turn, are eaten by several different birds, while any one of the latter may be eaten by a larger bird such as a hawk. Therefore, energy flow is better described in terms of **trophic (feeding) levels,** each one further removed from the producer population, the first level. All animals acting as primary consumers are part of a second trophic level; all animals acting as secondary consumers are part of the third level, and so on.

The trophic structure of an ecosystem can be summarized in the form of an **ecological pyramid** (fig. 29.6). The base of the pyramid represents the producer trophic level; the apex is the tertiary or some higher level consumer, and the other consumer trophic levels are in between. There are three possible kinds of pyramids. One is a *pyramid of numbers,* based on the number of

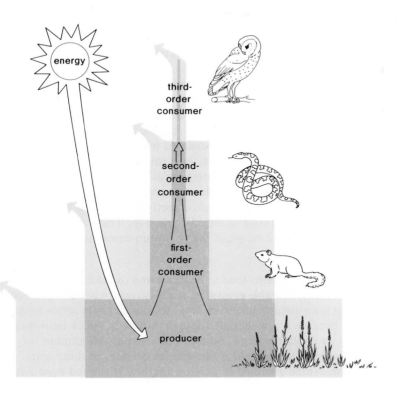

energy lost from the living system
energy retained in the living system

third-order consumer

second-order consumer

first-order consumer

producer

energy

Figure 29.6
Energy flow. At each step, an appreciable portion of energy originally trapped by the producer is lost as heat. Accordingly, organisms in each trophic level pass on less energy than they received.

organisms at each trophic level. A second is the *pyramid of biomass*. **Biomass** is the weight of living material at some particular time. To calculate the biomass, it is necessary to collect and weigh the plants in a certain area and to count as best as possible the number of different type animals. Then the total number is multiplied by the average weight per individual to give the approximate biomass at each trophic level. A third is the *pyramid of energy* (fig. 29.6), which always illustrates that each succeeding trophic level is smaller than the previous level. Less energy is found in each succeeding trophic level because of the following:

1. Of the food available, only a certain amount is captured and eaten by the next trophic level. After all, prey are adapted in many ways to escape their predators, as discussed on page 636.
2. Some of the food that is eaten cannot be digested and exits the digestive tract as waste.
3. Only a portion of the food that is digested becomes part of the organism's body. The rest is used as a source of energy.

In regard to the last point, consider that even when food molecules are used for growth and repair, energy is needed to cause these molecules to react. Synthetic reactions will not occur spontaneously, and when the body builds proteins, carbohydrates, and fats, energy in the form of ATP must be supplied. ATP is also needed by the body for such activities as muscle contraction and nerve conduction. Altogether, a significant portion of the food molecules supply the energy needed to form ATP by means of cellular respiration within mitochondria. Then, too, we must remember, as discussed on page 90, that one form of energy can never be transformed completely into another form. There is always some loss of usable energy between trophic levels for this same reason.

The light areas in figure 29.6 stand for the amount of energy that is unavailable to the next trophic level. The unavoidable loss of usable energy between feeding levels explains why food chains are relatively short—at most, four or five links—and why mice are more common than weasels, foxes, or hawks. The population sizes are appropriate to the amount of energy available at each trophic level, until finally, there is an insufficient amount of energy to support another level. The largest populations in an ecological pyramid are the producer populations at the base of the pyramid, and the smallest populations are the top predators. The population sizes of top predators are controlled by the amount of food energy available to them.

The energy considerations associated with ecological pyramids have implications for the human population. As a rule of thumb, it is generally stated that only about 10 percent of the energy available at a particular trophic level is incorporated into the tissues of the next level. This being the case, it can be reasoned that 100 lbs. of grain could directly result in 10 human lbs., but if fed to cattle, it would result in only 1 human lb. Thus, a larger human population can be sustained on grain than on grain-consuming animals.

Stability

Mature natural ecosystems tend to be stable and to exhibit the characteristics listed in table 29.1. A **stable ecosystem** is able to return to its original condition after having been disturbed in some way. **Species diversity** seems to contribute to stability because it provides alternate pathways by which individuals in the community can obtain energy and nutrients. To put this another way, a large number of species means that if one species declines or disappears, its place and function can be assumed in part by another. If the red-eyed vireo population shown in figure 29.3 declined in size, the tufted titmouse population would most likely increase in size, and in this way the hawk population would be maintained.

Generally, population sizes remain constant in mature natural ecosystems particularly due to biotic factors, such as competition and predation. Under normal circumstances, most of the solar energy utilized goes into maintenance of the ecosystem rather than into the growth of certain populations. The same amount of energy and materials are required by the system, and it is said that the **inputs** are constant. Also, because the nutrients cycle, there is no output except heat, and a natural ecosystem is independent of other ecosystems. If isolated, it could continue to exist as long as it received an adequate supply of solar energy (fig. 29.7).

Human Food Chain

When humans were hunters and gatherers, human populations, like a few native tribes today, were members of food webs, such as the one depicted in figure 29.3. With the advent of the agricultural revolution and then the industrial revolution, human populations were for the most part no longer members of such webs. Instead, humans created their own chain, which they have attempted to keep separate from natural food chains.

Whereas natural food webs are remarkable for their inner stability, the human food chain is highly unstable for the reasons listed in table 29.1. First, it may be noted that the human food chain is very short and usually has at most three links; for example,

$$\text{grain} \longrightarrow \text{cattle} \longrightarrow \text{humans}$$

Table 29.1 Ecosystems

Natural	Human
Stable	Unstable
Diverse food web	Short food chain
Population size remains constant	Population size ever increases
Cyclic (except energy)	Noncyclic
Renewable solar energy	Nonrenewable fossil fuel energy
Independent	Dependent

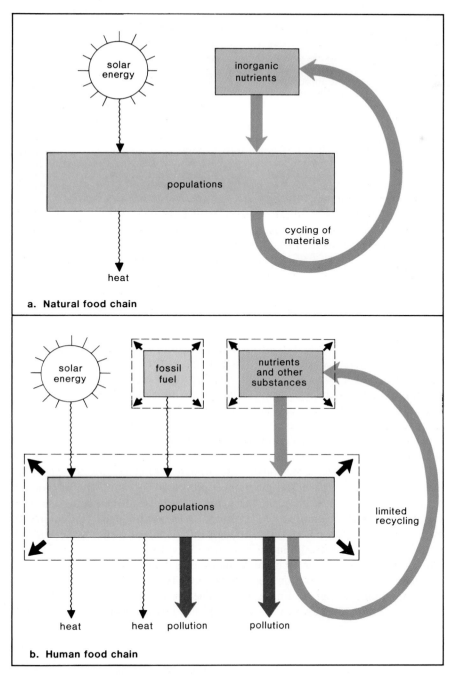

Figure 29.7
Natural food chain versus human food chain. *a.* Diagram
illustrating the cyclic nature of natural food chains and,
thus, of natural ecosystems. Notice, too, that the inputs
and outputs remain constant, as does the size of the
populations. *b.* Diagram illustrating the noncyclic nature of
the human food chain and, thus, the human ecosystem.
Notice that the use of supplemental fossil fuel energy and
of substances that do not cycle results in pollution. The
ever-increasing size of the human food chain and the
concomitant use of more artificial supplements cause an
ever greater amount of pollution.

Adding to the lack of diversity is the tendency of farmers to plant the same
hybrid variety of wheat or corn (fig. 11.6) because it gives the highest yield.
This so-called **monoculture farming** contributes to instability since one common
environmental factor, either biological or physical, can cause a crop failure in
all regions. One frequently cited example is the failure of the potato crop in
Ireland in the 1800s due to a fungal infection. A more recent example occurred in 1970 when 15 percent of the corn crop in the United States was
damaged by a corn blight. The damage could have been far worse because 80
percent of the crop planted that year was of the same susceptible hybrid variety. The fungus causing this particular corn blight is endemic to the south,

Organic Farming in the Corn Belt

A study was made of organic farming practices and results in the Midwest from 1974 to 1978. Organic farming was defined as farming without the application of conventional nitrogen, phosphorus, and potassium fertilizer, and without the application of synthetic herbicides or insecticides. Over 80 percent of the farmers in the study had previously farmed with conventional methods. For the most part, they had given up these practices because they were concerned about the health of their families and livestock and had also found that the chemicals were sometimes ineffective.

The organic farmers did not attempt to compensate for the nonuse of standard fertilizers and pesticides by adopting radically different practices from those generally used by conventional farmers. Instead, they relied more heavily on methods that most farmers have always used. For example, organic farmers used cultivation of row crops to control weeds; crop rotation, not the biological controls discussed on page 638, to combat major pests; and the growth of legumes to supply nitrogen fertility to the soil.

Table A shows that organic farming was just as profitable as conventional farming for every year in the study except 1978. The difference between the results in 1978 and the other years may have been due to growing conditions. Previously, there had been a drought, but the weather was very favorable in 1978. It would seem that when weather conditions are unfavorable, there is less advantage in using agricultural chemicals since other factors, such as soil moisture, limit yields.

Crop yields were lower on organic farms but so were operating costs. Organic farms required about two-fifths as much fossil energy to produce one dollar's worth of crop. This was probably because organic farmers did not need to use mechanized machinery to apply fertilizers, herbicides, and pesticides. Expenses were also lower for organic farmers because they were not purchasing these substances in the first place.

Their method of plowing and utilization of crop rotations resulted in about one-third less soil erosion. On the other hand, the crops had a lower protein content and a different amino acid balance indicating a

Figure 29.8
Intensive mechanized monoculture farming is commonly practiced in the United States today. Here a farmer is harvesting sorghum for silage.

but it happened to undergo a mutation that made it more lethal than before, and it spread to the cornbelt states where warm, moist weather helped it flourish. The next year resistant hybrid seed was available for planting, but this occurrence still exemplifies that planting the same type crop over a wide area produces a potentially unstable ecological situation.

The size of the populations in the human food chain are not constant. The human population has tripled since 1900, and although the rate of increase has slowed down in recent years, the overall size of the population continues to increase dramatically. More food must be produced to sustain the ever-growing human population although this population encroaches on agricultural land. A half-million hectares (one hectare = 2.471 acres) of farmland a year are taken over for other human uses, such as highways and housing. Nevertheless, food production (but not necessarily food distribution) has kept up with population increases. The newly developed hybrid plants are designed to give a high yield as long as they receive adequate **supplements** of fertilizer, pesticides, and water. The production of fertilizer and pesticides utilizes energy, and the application of these and water requires the use of mechanized farm machinery run on fossil fuel energy (fig. 29.8).

Fertilizer, pesticides, water, and fossil fuels are visible supplements to our agricultural system, but there is another supplement as well. It is said that we are **mining the soil** because many farmers are not using methods to minimize soil erosion, such as contour farming (fig. 29.9). The Department of Agriculture estimates that erosion is causing a steady drop in the productivity of land that is equivalent to the loss of 1.25 million acres per year. Even more fertilizers, pesticides, and energy supplements will be required to maintain yield.

nitrogen deficiency. But the soil showed a lack of phosphorus rather than a lack of nitrogen.

The researchers who conducted this study concluded that the present heavy reliance on chemical subsidies may be unnecessary, and that it might be well to determine how far farmers can move in the direction of reduced agricultural chemical use. They state, "The organic farmers we studied had fared reasonably well without chemical fertilizers and pesticides and without the benefit of the scientific and technical assistance routinely available to farmers following more accepted practices." A modest amount of fertilizer might have improved their results; therefore, there may be an intermediate system between conventional and organic farming which would be more attractive. This would be particularly advantageous at this time because of the worsening fossil fuel energy situation.

From W. Lockeretz, et. al., "Organic Farming in the Corn Belt," in *Science*, Vol. 211, p. 540, February 6, 1981. Copyright 1981 by the American Association for the Advancement of Science. Reprinted by permission.

Table A Economic Performance of Organic and Conventional Farms.

Year	Value of Production ($/ha)		Operating Expenses ($/ha)		Net Returns ($/ha)	
	Organic	Conventional	Organic	Conventional	Organic	Conventional
1974*	393	426	69	113	324	314
1975*	417	478	84	133	333	346
1976*	427	482	91	150	336	333
1977†	384	407	95	129	289	278
1978†	440	527	107	143	333	384

*Data from 14 organic and 14 conventional farms (7). †Data from 23 organic farms in 1977 and 19 in 1978: county-average data for conventional farms (3). The data are averaged over all cropland (including rotation hay and pasture, soil-improving crops, and crop failures).

Figure 29.9
In contour farming, crops are planted according to the lay of the land in order to reduce soil erosion. This farmer has also planted alfalfa in between the corn plants in order to replenish the nitrogen content of the soil.

Figure 29.10

In the United States, most cattle are kept in feedlots and fed grain. This means of caring for cattle requires an input of fossil fuel energy. First, supplemental fossil fuel energy must be used to grow the grain, and then fossil fuel energy must be used to make the grain available to the cattle.

Supplemental fossil fuel energy contributes not only to agricultural yields but also to animal husbandry yields. At least 50 percent of all cattle are kept in **feedlots** (fig. 29.10) where they are fed grain. Chickens are raised in a completely artificial environment where the climate is controlled, and each one has its own cage to which food is delivered on a conveyor belt. Animals raised under these conditions often have antibiotics and hormones added to their feed to increase yield.

From this discussion it is evident that the human food chain is largely noncyclic (fig. 29.7b). There is a large input of fossil fuel energy (in addition to solar energy) and other substances that do not cycle back to the producers. Unfortunately, this means there is also a large output of wastes. Excess fertilizers and pesticides from farmlands and home lawns are often transported by water and wind to aquatic ecosystems. Also, animal and human sewage are added to natural waters where the bacteria of decay are often unable to immediately break down the quantity that is disposed. Therefore, it is customary for each community to pretreat sewage in sewage treatment plants before the effluent is added to nearby bodies of water.

Thus, the human food chain cannot be isolated completely from the natural communities. In fact, it is dependent on them not only because they by themselves lend stability to the biosphere, but also because they help stabilize the human food chain by absorbing by-products, such as excess fertilizer, pesticides, and fuel breakdown products. Unfortunately, because natural communities have decreased in size as the human population has increased, the

Watering with Waste

Sprinkle wastewater onto a forest and produce clean water, fast-growing trees, and plentiful, fat earthworms.

It could be called organic forestry, or perhaps sylvan wastewater purification. By any name, a novel system is converting sewage into drinking water inexpensively in an area just 20 miles south of Atlanta.

On June 1 sprinkler heads began to rotate in 3,500 acres of hilly woodlands recently purchased by the county of Clayton. The system working at full capacity will supply almost 16 million gallons of quality drinking water to the county's 170,000 residents. And part of the expenses will be paid by the sale of timber obtained from the irrigated, fertilized forestland.

The wastewater that goes into the Clayton system comes from the homes, businesses and light industry of the county. It is pretreated to settle out the solids, or sludge, and is given just enough biological treatment, with bacteria and algae, to eliminate odors. The water is then pumped approximately 7 miles to the treatment site. There a system of 18,000 sprinklers irrigates 2,200 acres, each area receiving about 2.5 inches of water in 12 hours once a week. Water quality is monitored at 22 groundwater wells, at the creek that receives the land-purified water and at the uptake of the drinking water reservoir.

Except when a sprinkler is working, there is no obvious indication that the area is a treatment site, Nutter says. Even when a sprinkler is on, the vegetation limits the spray to a 50-foot radius. "When we first talked about spraying wastewater on

forests, people got all kinds of weird ideas about what it would look like. I think they were expecting to find toilet paper hanging in the trees," [Wade] Nutter [University of Georgia soil scientist and hydrologist] recounts. But the EPA and the local community, including such groups as the Audubon society, eventually agreed that forest wastewater treatment is an acceptable technique.

The land in Clayton County was purchased specifically for water treatment, and consequential timber growth. Loblolly pines will be sold as pulpwood. The county also is planting hardwoods that can be harvested for firewood and wood fuel on a four-year cycle. Nutter predicts that 20 percent of the operating cost of the wastewater land treatment system could be covered by timber revenue. In keeping with the EPA's interest in sites that serve multiple uses, a program of hiking and bicycle trails is also currently under consideration.

Forest wastewater treatment systems are planned for several other areas of the southeastern United States. "It is ideal for delicate areas," says Alan Fletcher, also of the University of Georgia. When asked what would limit other areas of the country from using similar land treatment systems, Nutter replied, "Just prejudice. There are very few soil and plant conditions where land treatment couldn't be used in some form."

Reprinted with permission from *Science News,* the weekly newsmagazine of science, copyright 1981 by Science Service, Inc.

natural communities are often overwhelmed. It is then that humans recognize that something is amiss and say that **pollutants**[1] have been added to the environment.

To control pollution it is necessary to control the amount of inputs and/or to control the amount and treatment of the outputs. For example, a few farmers have stopped using artificial subsidies, such as fertilizers and pesticides, to increase crop yields. One study, which is described later, indicates that this method of farming, labeled organic farming by the investigators, does have reduced yields but can be as profitable to the farmer because of the reduced costs.

Other citizens are attempting to reduce pollution by making the human food chain more cyclic. For example, the reading on the previous page describes a sewage treatment system that is now functioning outside Atlanta. Wastewater sprinkled on a forest area provides nutrients for tree growth and purifies the water at the same time. Not only is the water then fit for drinking, the trees can be harvested for human use on a four-year cycle.

This method of treating water demonstrates that it is possible to find ways to integrate the human ecosystem with the natural ecosystems. This is often called "working with nature rather than against nature." On the other hand, if we choose to continue practices that create environmental problems, we must ask ourselves if we are willing to bear the cost. As ecologists are fond of saying, "There is no free lunch." By this they mean that for every artificial benefit, there is a price. For example, if we kill off coyotes because they prey on chickens, we then have to realize that the mouse population will probably increase. In an ecosystem, as we have seen, "everything is connected to everything else."

Biogeochemical Cycles

In contrast to energy, inorganic nutrients do cycle through ecosystems. Because there is normally no input from outside an ecosystem, the same thirty or forty elements essential to life are used over and over. Since the pathways by which inorganic nutrients circulate through ecosystems involve organisms and the physical environment, they are known as **biogeochemical cycles.** For each element the cycling process (fig. 29.11) involves (1) a reservoir—that portion of the biosphere that acts as a storehouse for the element; (2) an exchange pool—that portion of the biosphere from which the producers take their nutrients; and (3) the biotic community—through which chemicals move along food chains to and from the exchange pool.

[1]Pollutant: Material added to the environment by human or natural activities that produces undesirable effects.

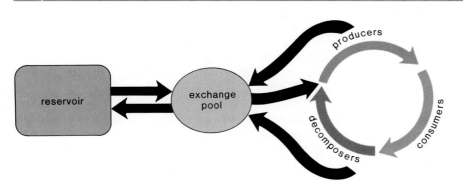

Figure 29.11
A generalized diagram for a biochemical cycle. Such cycles include a reservoir, exchange pool, and a biotic community consisting of producers, consumers, and decomposers.

Figure 29.12

Carbon cycle. Photosynthesizers take up carbon dioxide from the air or bicarbonate from the water. They and all other organisms return carbon dioxide to the environment. The carbon dioxide level is also increased when volcanoes erupt and fossil fuels are burned. Presently, the oceans are a primary reservoir for carbon dioxide because the calcium carbonate content of water forms limestone.

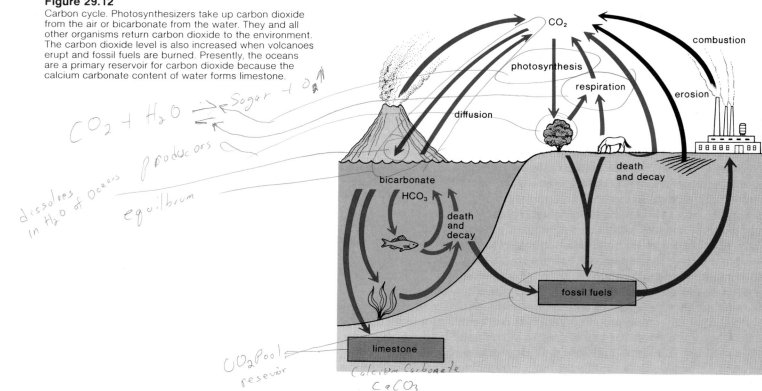

Figure 29.13

Trees are a reservoir in the carbon cycle. They act as a sink for carbon dioxide, a molecule that is incorporated into the organic compounds that make up their bodies. People are removing forests while, at the same time, adding carbon dioxide to the atmosphere. This combination of events can result only in an increased amount of CO_2 in the atmosphere.

Some biogeochemical cycles are **gaseous** while others are **sedimentary.** Elements whose main reservoirs exist in the atmosphere (or ocean) move through gaseous cycles. Four critical elements—oxygen, carbon, nitrogen, and hydrogen—are involved in mostly gaseous cycles. These four elements constitute about 97 percent of living matter. The thirty-six or so other biologically important elements follow a sedimentary cycle in which the main reservoir is the earth's crust. They initially become available as a result of rock weathering.

Carbon Cycle

Organisms in both terrestrial and aquatic ecosystems (fig. 29.12) exchange carbon dioxide with the atmosphere. On land, plants take up carbon dioxide from the air, and through photosynthesis they incorporate carbon into food that is used by autotrophs and heterotrophs alike. Then, when all organisms respire, a portion of this carbon is returned to the atmosphere as carbon dioxide.

In aquatic ecosystems the exchange of carbon dioxide with the atmosphere is indirect. Carbon dioxide from the air combines with water to give bicarbonate ions that are a source of carbon for algae that produce food for themselves and heterotrophs. Similarly, when aquatic organisms respire, the carbon dioxide they give off becomes the bicarbonate ion. The amount of bicarbonate in the water is in equilibrium with the amount in the air.

Reservoirs

Living and dead organisms contain organic carbon and serve as one of the reservoirs for the carbon cycle. The world's biota, particularly trees (fig. 29.13), contain 800 billion tons of organic carbon, and an additional 1000 to 3000 billion tons are estimated to be held in the remains of plants and animals in

the soil. Before decomposition can occur, some of these remains are subjected to physical processes that transform them into coal, oil, and natural gas. We call these materials the fossil fuels. Most of the fossil fuels were formed during the Carboniferous period, 280–350 million years ago, when an exceptionally large amount of organic matter was buried before decomposing.

Another reservoir is the inorganic carbonate that accumulates in limestone and in carbonaceous shells (fig. 29.14). Limestone particularly tends to slowly form in the oceans.

Human Influence

The activity of human beings has increased the amount of carbon dioxide in the atmosphere. Data from monitoring stations record (p. 114) an increase of 20 ppm (parts per million) in only twenty-two years. (This is equivalent to 42 billion tons of carbon.) This buildup is attributed to the burning of fossil fuels, soil erosion, and the destruction of the world's forests. When we do away with forests, we are reducing a reservoir that takes up excess carbon dioxide. At this time, the **oceans** are believed to be taking up most of the excess carbon dioxide. The burning of fossil fuels in the last twenty-two years has probably released 78 billion tons of carbon, yet the atmosphere registers an increase of only 42 billion tons.

As discussed on page 114, increased amounts of carbon dioxide in the atmosphere could possibly cause the weather to become warmer. The degree of warming could be so great as to alter the current distribution of life on earth. For example, a 3° rise in world temperatures could shift the corn- and wheat-growing belts northward, benefiting Canada and the Soviet Union.

Figure 29.15
Nitrogen cycle. Three types of bacteria are at work: nitrogen-fixing bacteria convert aerial nitrogen to a plant-usable form, nitrifying bacteria, which include both nitrite and nitrate bacteria, convert ammonia to nitrate; and the denitrifying bacteria convert nitrate back to aerial nitrogen. Humans contribute to the cycle by using aerial nitrogen to produce nitrate for fertilizers.

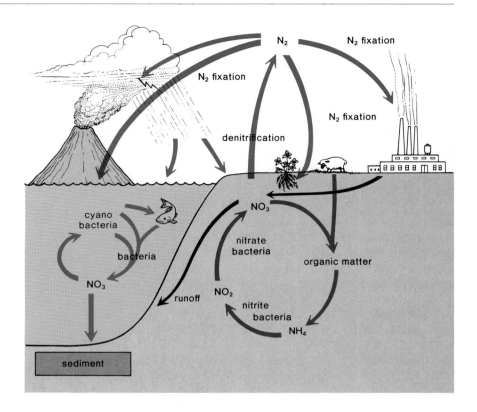

Nitrogen Cycle

Nitrogen is an abundant element. Nitrogen gas (N_2) makes up about 78 percent of the atmosphere by volume, yet nitrogen deficiency commonly limits plant growth. Unfortunately, most plants cannot incorporate N_2 into organic compounds. They must absorb nitrogen from the soil, mainly as nitrate (NO_3^-) or ammonia (NH_4^+) ions (fig. 29.15). Processes that convert nitrogen gas to usable compounds for organisms are called nitrogen-fixation processes.

Nitrogen Fixation

Atmospheric nitrogen, the *reservoir* for the nitrogen cycle, can be converted to ammonia by cyanobacteria in aquatic ecosystems and by two types of **nitrogen-fixing** bacteria in terrestrial ecosystems. One type of nitrogen-fixing bacteria is free-living in the soil, but the other type infects and lives in nodules on the roots of legumes (fig. 18.7). As figure 29.16 points out, there are at least two other ways by which nitrogen fixation occurs. Nitrogen is fixed as NO_3 in the atmosphere when cosmic radiation, meteor trails, and lightning provide the high energy needed for nitrogen to react with oxygen. Also, *humans* make a most significant contribution to the nitrogen cycle when they convert aerial nitrogen to nitrates for use in fertilizers. This industrial process requires an energy input that at least equals that of the eventual increase in crop yield. The application of fertilizers also contributes to water pollution, as discussed on page 396. Since nitrogen-fixing bacteria do not require fossil

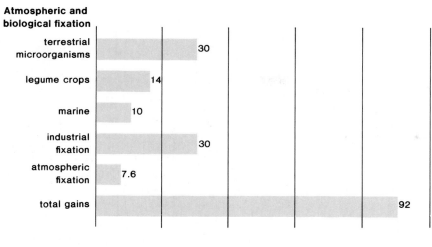

Figure 29.16
This graph estimates the amount of nitrogen being fixed and denitrifed per year. The difference in the totals represents the rate at which fixed nitrogen is accumulating in soil and water.

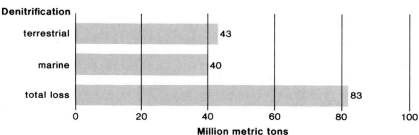

fuel energy and do not cause pollution, research is now directed toward finding a way to make all plants capable of forming nodules or, even better, through recombinant DNA technique, to possess the biochemical ability to fix nitrogen themselves.

Subcycle

Once plants absorb nitrate or ammonia, they can synthesize proteins and nucleic acids. Consumers require the resulting supply of organic nitrogen to form their own proteins. Some of this organic nitrogen quickly returns to the soil as fallen leaves, etc., or as animal wastes. The remaining organic nitrogen only returns when plants and animals die. Decomposers then break down these organic nitrogen compounds to release ammonia. Often the ammonia is further oxidized to nitrate by **nitrifying bacteria** in a process called **nitrification.** Thus, there is a subcycle among members of food chains that does not involve atmospheric nitrogen (N_2) at all.

Denitrification

Denitrification is the conversion of nitrate to atmospheric nitrogen, largely by bacterial action in both aquatic and terrestrial ecosystems. Denitrification counterbalances the process of nitrogen fixation, but not entirely. There is more nitrogen fixation, especially due to fertilizer production (fig. 29.16).

Figure 29.17

Phosphorus cycle. Normally, phosphate is in short supply. However, humans mine phosphate deposits, including guano, for inclusion in fertilizers; thus, they increase the amount of phosphorus available.

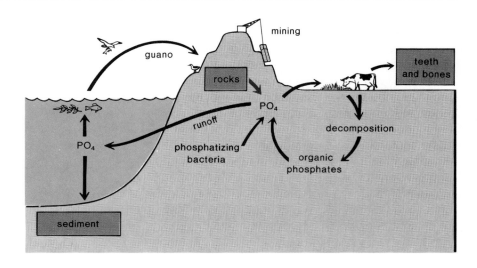

Figure 29.18

Large congregations of seabirds produce guano that can be mined for its phosphorus content. The phosphorus is used in fertilizer production.

Phosphorus Cycle

On land, plants take up phosphate ($PO_4^=$) from the soil (fig. 29.17), which also supplies other minerals, such as potassium, calcium, and magnesium. In water, floating algae have to take up these substances from the water since any amount in sediment is not available to them. Producers incorporate phosphate into a variety of molecules including lipids and nucleic acids. Animals incorporate phosphate into teeth, bones, and shells that do not decompose for very long periods of time. When decomposition does take place, most bacteria and fungi do not release phosphate directly; instead, they release organic phosphates. These are converted to inorganic phosphate by phosphatizing bacteria. Since these bacteria are not very prevalent, phosphorus is believed to be a limiting nutrient in most ecosystems.

Human Influence

For the most part, the phosphorus cycle on the land is separate from the one in the ocean, but there is one well-known exception. Seabirds, which feed on fish, deposit excrement called **guano** on sea cliffs (fig. 29.18). This contains phosphorus and other nutrients. Guano can be harvested by humans for incorporation into fertilizers. The most famous of the guano deposits are along the coasts of Ecuador and Peru, where shorebirds feed on anchovies. Overfishing of the anchovies may eventually lead to a decline in the number of seabirds and thus in guano.

Human beings also mine phosphate rock to make phosphorus available for fertilizer production, animal feed supplements, pesticides, medicines, and numerous other products. Phosphate used to be added to detergents, but this led to a great deal of foaming in natural waters, so the practice has largely been discontinued.

One well-known region where phosphorus has been mined since the late 1800s lies twenty-five miles east of Tampa, Florida. Here, the so-called Bone Valley formation was made by fossilized remains of marine animals laid down some 10 to 15 million years ago. Some 150,000 acres of this land have been strip-mined (fig. 29.19) to remove phosphorus ore, but only 30,000 of these have been reclaimed or restored to approximate their original condition. When land is strip-mined, the vegetation and soil above the deposit are removed and piled up in a manner convenient to the operation. When the land is reclaimed, the soil must be put back, preferably with the top soil uppermost, and the land must be seeded or replanted. Unless this is done, the land is subject to severe soil erosion.

Another environmental concern is that the phosphate ore is slightly radioactive; therefore, mining it poses a health threat to organisms in the area. There is also concern that this reservoir of phosphorus, the best-known reservoir in the United States, will not last more than another few decades. It would seem, then, that the best course of action would be to take all possible steps to reduce the need for phosphorus. For example, better farming practices to control soil erosion and increase the efficiency of phosphate fertilizer use would help. Also, the phosphate industry could improve their processing procedures and try to extract more phosphorus from the ore that is mined. Up to 35 percent or more of the phosphate in the original ore is lost in processing.

Figure 29.19
Phosphorus is removed from the ground by strip mining, a procedure that often leaves the land scarred. Strip miners are required by law to restore the land, but enforcement is sometimes lax.

Summary

Ecosystems contain both abiotic (physical) and biotic (living) components. The latter are arranged in food chains. Grazing food chains always begin with a producer that is capable of producing organic food, followed by a series of consumers, each of which feeds on the latter. Detritus food chains begin with dead organic matter that is consumed by decomposers, including bacteria and fungi as well as various soil animals. Eventually, all members of the food chains die and decompose. Because of this, the very same chemicals cycle through the ecosystem again and again. In contrast, energy does not cycle through an ecosystem because at each step in a food chain some energy is converted to heat until eventually all of it is dissipated.

If we consider the various food chains together, we realize that they form an intricate food web in which there are various trophic (feeding) levels. All producers are on the first level, all primary consumers are on the second level, and so forth. To illustrate that energy does not cycle, it is customary to arrange the various trophic levels to form an energy pyramid.

Then we see that each level contains less energy than the previous level. It is estimated that only 10 percent of the available energy is actually incorporated into the body tissues of the next level.

Mature natural ecosystems tend to be stable. The same amount of solar energy is utilized, and the same amount of matter cycles at all times. Each population tends to remain at a fairly constant size, and the ecosystem usually is independent of any other. In contrast, the human ecosystem tends to be unstable. The human food chain is quite short, containing at most only three links. Adding to this lack of diversity is the tendency of farmers to use only certain high-yield varieties of agricultural plants. These plants are designed to give a high yield when they receive supplements, such as fertilizers, pesticides, and water. The first two of these require energy to produce them, and all three require energy to apply them to crops. The human food chain is noncyclic, and worse yet, more fossil fuel energy and material supplements are needed each year to produce more food for a human population that continues to increase in size. The noncyclic nature of the human ecosystem results in waste substances being added to natural ecosystems. It would be beneficial for us to find ways to make the human ecosystem, including the human food chain, as cyclical as possible.

Chemical cycling through an ecosystem involves not only the biotic components of an ecosystem but also the physical environment. Each cycle involves a reservoir where the element is stored, an exchange pool from which the populations take and return nutrients, and then the populations themselves.

In general, human beings affect the carbon, nitrogen, and phosphorus cycles by withdrawing these substances from their reservoirs. For example, when human beings burn fossil fuels, they are removing carbon dioxide from a reservoir. Human beings also have learned to convert aerial nitrogen to ammonia, which allows them to remove nitrogen from the air. They also reduce the reservoir for phosphorus by mining phosphate and guano.

Objective Questions

1. Chemicals cycle through the populations of an ecosystem, but energy is said to _____ because it is all eventually dissipated as heat.
2. The first population in a food chain is always a _____ population.
3. The first carnivore population in a food chain is the _____ consumer population.
4. An ecological pyramid illustrates that each succeeding trophic level has less _____ than the previous level.
5. Natural ecosystems utilize the same amount of energy per year, but the human ecosystem utilizes an _____ _____ .

6. Each biogeochemical cycle includes (1) a _____ , (2) an exchange pool, and (3) the biotic community.
7. In the carbon cycle, when organisms _____ carbon dioxide is returned to the exchange pool.
8. Humans make a significant contribution to the nitrogen cycle when they convert aerial nitrogen to _____ for use in fertilizers.
9. During the process of denitrification, nitrates are converted to _____ .
10. Guano is the _____ of birds, which is rich in _____ .

Study Questions

1. Name four different types of consumers found in natural ecosystems.
2. Give an example of a grazing and a detritus food chain for a terrestrial and for an aquatic food chain.
3. Draw an energy pyramid and explain why such a pyramid can be used to verify that energy does not cycle.
4. Compare and contrast the basic characteristics of a natural ecosystem to the characteristics of the human food chain.
5. Discuss in detail the carbon, nitrogen, and phosphorus biogeochemical cycles. Indicate the manner in which humans disturb the equilibrium of these cycles.

Selected Key Terms

abiotic ('ā bī ,ät ik) 646
biotic (bī 'ät ik) 646
detritus (dī 'trīt əs) 646
food chain ('füd 'chān) 648
food web ('füd 'web) 650
trophic level (trō fik 'lev əl) 650
ecological pyramid (,ē kə 'läj i kəl 'pir ə ,mid) 651

biomass ('bī ō ,mas) 651
monoculture ('män ə ,kəl chər) 653
biogeochemical cycle ('bī ō 'jē ō ,kem i kəl 'sī kəl) 657
nitrification (,nī trə fə 'kā shən) 661
denitrification (dē ,nī trə fə 'kā shən) 661

Each individual ecosystem is a unit of the **ecosphere** that encompasses all the populations of the earth plus their physical environments. Within the ecosphere it is possible to distinguish major types of ecosystems, and this chapter will consider some of the most obvious types. Ecosystems are either aquatic or terrestrial, and we begin our discussion with aquatic ecosystems.

Aquatic Ecosystems

Aquatic ecosystems can be divided into two types: (1) inland or **freshwater ecosystem** and (2) oceanic or **saltwater ecosystem.** An **estuary,** however, where a river flows into the ocean, has mixed fresh and salt water, called **brackish water.** Table 30.1 lists the aquatic biomes for easy reference. In these ecosystems organisms vary according to whether they are adapted to fresh or salt water, warm or cold water, quiet or turbulent water, and the presence or absence of light. In both salt and fresh water, *floating* microscopic organisms, called **plankton,** are important components of the ecosystem. **Phytoplankton** are photosynthesizing algae that only become noticeable when they reproduce to the extent that a green scum or red tide appears on the water. **Zooplankton** are animals that feed on the phytoplankton.

Freshwater Ecosystems

Fresh water is a distillate of salt water, which is readily apparent when considering the water cycle (fig. 30.1). As the sun's rays cause seawater to evaporate, the salts are left behind. The vaporized fresh water rises into the atmosphere, cools, and falls as rain either over the ocean or over the land. A lesser amount of water also evaporates from and returns to the land. Since land lies above sea level, gravity eventually returns all fresh water to the sea, but in the meantime, it is contained within **standing waters** (lakes and ponds), **flowing waters** (streams and rivers), and **groundwater.**

When rain falls, some of the water sinks or percolates into the ground and saturates the earth to a certain level. The top of the saturation zone is called the groundwater table, or simply the **water table.** Wherever the earth contains basins or channels, water will appear to the level of the water table. The water within basins is called **lakes** and **ponds,** and the water within channels is called **streams** or **rivers.**

Sometimes groundwater is also located in a porous layer, called an **aquifer,** that lies between two sloping layers of impervious rock. Human beings often remove this water by means of specially constructed wells.

CHAPTER 30

Ecosystems of the World

Concepts

1. Each type of ecosystem has particular abiotic and biotic components.
   Aquatic ecosystems are categorized in part according to the degree of salinity; the various terrestrial ecosystems are especially determined by climate.

2. Mature (climax) ecosystems come into existence by a series of stages called succession.
   While the early stages of succession have the most growth, the final stage tends to be the most stable.

3. Ecosystems have varying degrees of productivity.
   The most productive of the ecosystems are estuaries, coral reefs, and tropical rain forests. The least productive are the open oceans and the deserts.

Table 30.1 Aquatic Ecosystems

Inland (Fresh Water)	Ocean (Salt Water)
Lakes and ponds	Ocean (Salt Water)
Rivers and streams	Oceans
Swamps, marshes, and bogs	Seashores
	Rocky
	Sandy
	Coral reef
	Estuaries and salt marshes

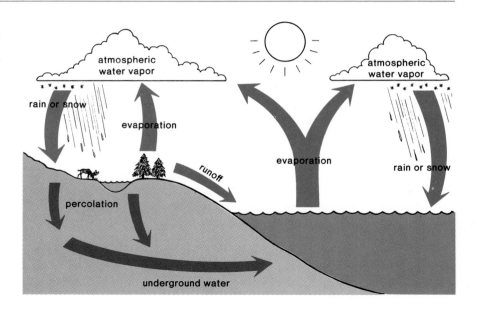

Lakes and Ponds

Lakes and ponds, which are characterized by still waters, tend to have fairly stable abiotic and biotic regions. For example, in the summer, lakes (but not ponds) in the temperate zone have three layers of water that differ in temperature (fig. 30.2). The surface layer, the **epilimnion,** is warm from solar radiation; the middle **thermocline** experiences an abrupt drop in temperature; and the **hypolimnion** is cold. These differences in temperature prevent mixing. The warmer, less dense water of the epilimnion "floats" on top of the colder, more dense water of the hypolimnion.

As the season progresses, the epilimnion becomes nutrient poor, while the hypolimnion begins to be depleted of oxygen. The phytoplankton found in the sunlit epilimnion use up nutrients as they photosynthesize. Photosynthesis releases oxygen, giving this layer a ready supply. Detritus naturally falls by gravity to the bottom of the lake, and here oxygen is used up as decomposition occurs. Decomposition releases nutrients, however.

In the fall, as the epilimnion cools, and in the spring, as it warms, a **turnover** occurs. In the fall the upper epilimnion waters become cooler than the hypolimnion waters. This causes the surface water to sink and the bottom water to rise. The fall turnover continues until the temperature is uniform throughout the lake. At this point, wind aids in the circulation of water so that mixing occurs. Eventually, oxygen and nutrients become evenly distributed.

As winter approaches, the water cools. Ice formation begins at the top, and the ice remains there because ice is less dense than cool water. Ice has an

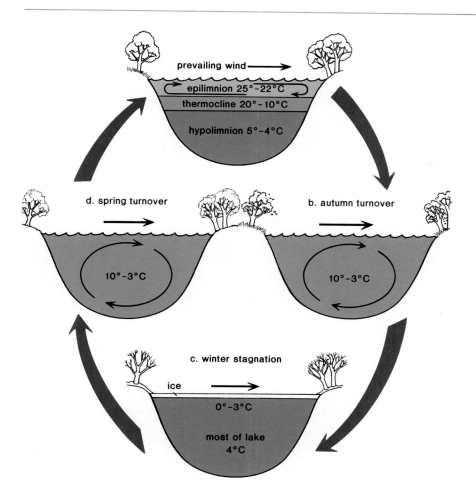

Figure 30.2
Temperature profiles of a large lake in a temperate region vary with the season. During spring and autumn turnover, the deep waters receive oxygen while the shallow waters receive nutrients.

insulating effect, preventing further cooling of the water below. This permits aquatic organisms to live through the winter in the water beneath the surface of the ice.

In the spring, as the ice melts, the cooler water on top sinks below the warmer water on the bottom. The spring turnover continues until the temperature is uniform throughout the lake. At this point, wind aids in the circulation of water as before. When the surface waters absorb solar radiation, however, **thermal stratification** occurs once more.

This vertical stratification and seasonal change of temperatures in a pond or lake basin influence the seasonal distribution of fishes and other aquatic life in the lake basin. For example, coldwater fishes move to the deeper water in summer and inhabit the upper water in winter. In the fall and spring, just after mixing occurs, phytoplankton growth is most abundant.

Figure 30.3

Life zones of a lake and habitats of organisms within a lake. *a*. Periphyton are minute organisms that cling to biotic and abiotic surfaces. *b*. Zooplankton are floating animals often found in the sunny limnetic zone where they feed on algae. *c*. Neuston are organisms that live on the surface where atmospheric oxygen is available. *d*. Nekton are free-swimming fishes and insects that often feed on the zooplankton in the limnetic zone. *e*. Benthos are the bottom-dwelling animals in the profundal zone that feed on debris that falls from above.

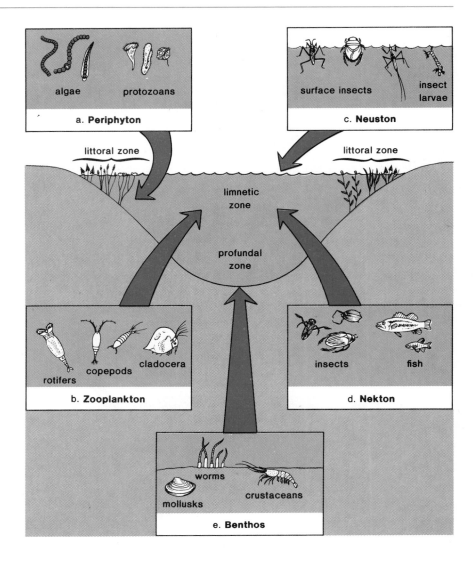

Life Zones

Lakes and ponds can be divided into three life zones: the **littoral zone** is closest to the shore; the **limnetic zone** forms the sunlit body of the lake; and the **profundal zone** is below the level of light penetration. Aquatic plants are rooted in the shallow littoral zone of a lake. The rest of the organisms are divided into five groups according to their life style (fig. 30.3).

Classification

Lakes and ponds are often classified by their nutrient status. **Oligotrophic** (nutrient-poor) **lakes** are characterized by low organic matter, low nutrient release from bottom sediments, and low levels of phytoplankton productivity. Such a lake is usually situated in a nutrient-poor **watershed,** the surrounding area that drains water into a lake. **Eutrophic** (nutrient-rich) **lakes** are characterized by high organic matter, high levels of nutrient release from bottom

a.   b.

Figure 30.4
Aquatic succession on Presque Isle, Pennsylvania. *a*. As the pond developed more sediments and the body of water became more shallow, water lilies rooted in the bottom. Cattails, rushes, and sedges began to grow at the edges where the water level was reduced to only a few inches. *b*. As the land became drier, trees began to crowd in and the remaining wet areas were invaded by meadow grasses.

sediments, high phytoplankton productivity, and often a well-developed littoral zone. Such lakes are usually situated in naturally nutrient-rich watersheds or in watersheds affected by agriculture or urban and suburban settlements. Oligotrophic lakes can become eutrophic through large inputs of nutrients. This process, called **cultural eutrophication,** accelerates the process of aquatic succession, a series of stages by which a pond and possibly a lake can fill in and disappear (fig. 30.4).

Marine Ecosystems

Marine ecosystems begin at the edges of the oceans. These include estuaries and seashores in addition to the oceans themselves.

Estuaries

An *estuary* forms where a large river flows into the ocean. A river brings fresh water into the estuary. The sea, because of the tides, brings salt water into this same area. Organisms living in an estuary must withstand constant mixing of waters and rapid changes in salinity. Not many organisms are suited to this environment, but there is an abundance of nutrients for those that are suited. An estuary acts as a **nutrient trap;** the tides bring nutrients from the sea and at the same time prevent the seaward escape of nutrients brought by the river (fig. 30.5). Therefore, estuaries are very productive (p. 685).

Only a few small fishes live permanently in an estuary, but many develop there; thus, we find larval fishes and immature fishes ever present. It has been estimated that well over half of all marine fishes develop in the protective environment of an estuary; this is why estuaries are called the **nurseries of the sea.** Shrimp and mollusks, too, use the estuary as a nursery.

Figure 30.5
An estuary is a nutrient trap. Fresh water brought by rivers flows outward toward the sea; at the same time, seawater is carried inward, bringing with it nutrients and debris. Many of these nutrients are taken up by plants in tidal marshes and mud flats, which, therefore, act as nutrient filters.

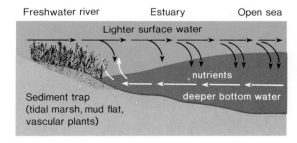

Figure 30.6
A drawing of some permanent residents of a salt marsh.

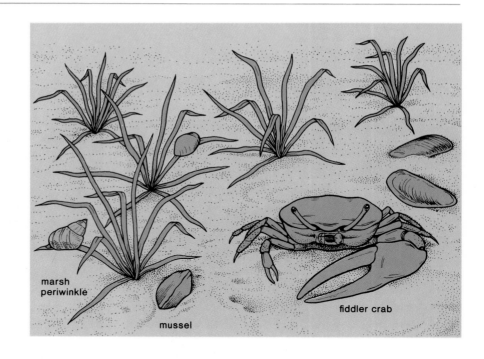

Figure 30.7
A drawing of animals that typically are found along a rocky coast in the intertidal zone.

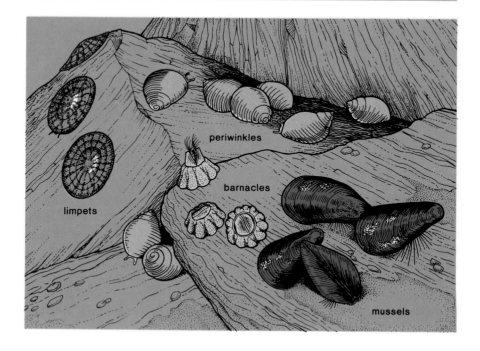

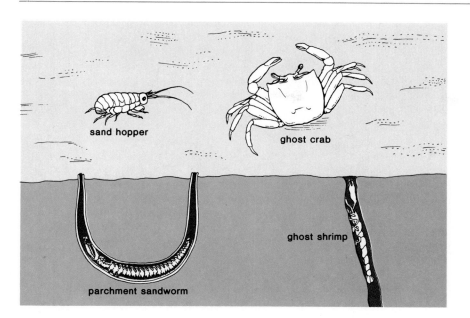

Figure 30.8
Representative animals of a sandy shore.

sand hopper

ghost crab

ghost shrimp

parchment sandworm

Salt marshes dominated by salt marsh cordgrass are often associated with estuaries. Salt marshes contribute nutrients and detrital material to estuaries in addition to being an important habitat for wildlife, a few of which are shown in figure 30.6.

Seashores

Both rocky and sandy shores are constantly bombarded by the sea as the tides roll in and out. The **rocky shore** (fig. 30.7) offers a firm substratum to which organisms can attach themselves and not be swept out to sea. Macroscopic *seaweeds,* which are the main photosynthesizers, anchor themselves to the rocks by holdfasts. Barnacles are glued to the stone by their own secretions so tightly that their calcerous outer plates remain in place even after the enclosed shrimplike animal dies. Oysters and mussels attach themselves to the rocks by filaments called **byssus threads.** Limpets are snails, just as are periwinkles. But whereas periwinkles have a coiled shell and secure themselves by hiding in crevices or under seaweeds, limpets press their single flattened cone tightly to a rock.

The shifting, unstable sands on a **sandy beach** (fig. 30.8) do not provide a suitable substratum for the attachment of organisms; therefore, nearly all the permanent residents dwell underground. They either burrow during the day and surface to feed at night, or they remain permanently within their burrows and tubes. Ghost crabs and sandhoppers (amphipods) burrow above high tide and feed at night when the tide is out. Sandworms and sand (ghost) shrimp remain within their burrows in the intertidal zone and feed on detritus whenever possible. Still lower in the beach, clams, cockles, and sand dollars are found.

Figure 30.9
Coral reef. A coral reef is a unique community of marine
organisms living within and around the skeletons. A coral
is made up of living representatives of animals called
corals. The diversity and richness of form and color are
common to the optimal environmental conditions of
tropical seas.

Figure 30.9
Coral reef. A coral reef is a unique community of marine organisms living within and around the skeletons. A coral is made up of living representatives of animals called corals. The diversity and richness of form and color are common to the optimal environmental conditions of tropical seas.

Coral Reefs

Coral reefs (fig. 30.9) are areas of biological abundance found in shallow, warm tropical waters that have a minimum temperature of 70° F. Coral reefs begin on a rocky substratum beneath the surface of the water. Their chief constituents are **stony corals,** which have a calcium carbonate (limestone) exoskeleton, and calcerous red and green algae. Corals do not usually occur individually; rather, they form colonies derived from an individual coral that has reproduced by means of budding. Corals provide a home for a microscopic dinoflagellate called **zooxanthellae.** The corals, which feed at night, and the dinoflagellate, which photosynthesize during the day, are mutualistic and share materials and nutrients.

A reef is densely populated with animal life. There are many types of small fishes (butterfly, damsel, clown, and sturgeon), all beautifully colored. In addition, the large number of crevices and caves provide shelter for filter feeders (sponges, sea squirts, and fan worms) and for scavengers (crabs and sea urchins). The barracuda and moray eel prey on these animals. Some fishes feed on the coral, but the most deadly coral predator of Pacific reefs is the **Crown-of-Thorns** starfish, which grows as large as two feet across and has from nine to twenty-one arms. Along the northeastern coast of Australia, the very existence of the Great Barrier Reef is threatened by these animals. The giant triton is their great natural foe, and some believe that the Crown-of-Thorns is proliferating because humans have killed off the triton for its handsome spiral shell.

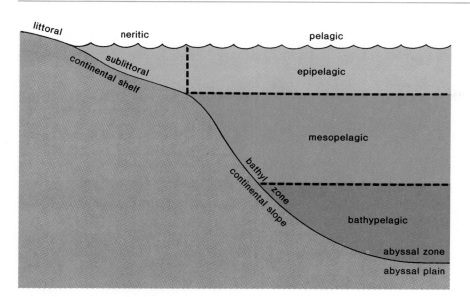

Figure 30.10
The floor of the ocean is divided into the continental shelf, the continental slope, and the abyssal plain. Organisms reside in the life zones: those of the benthos may be found in the littoral and sublittoral zones, the bathyl zone, and the abyssal zone; those of the neritic zone are found above the continental shelf; those in the pelagic zone are in the open sea.

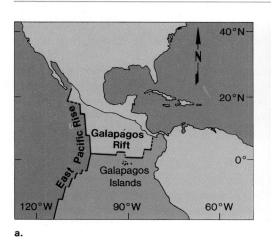

a.

b.

Figure 30.11
a. Location of the spreading zones in the East Pacific, along which deep ocean hydrothermal vents occur. The "oases" of abundant life around the vents have been given such names as Dandelions, the Rose Garden, and the Garden of Eden. b. Giant tube worms living near a deep ocean hydrothermal vent. These large worms (up to 3 m long) have no digestive tract. Their bodies contain chemosynthetic bacteria that are believed to provide them with organic molecules.

Oceans

The oceans cover approximately three-quarters of our planet. Figure 30.10 shows that the sea is at first shallow, then it becomes deep as the **continental shelf** gives way to the **continental slope** that leads to the **abyssal** (from abyss) **plains.** The flat plains are interrupted by enormous mountain chains called **oceanic ridges.** Lava flows continually out of wide rifts and forms new crustal material along the axes of the oceanic ridges; this leads to continental drift, mentioned earlier (p. 242). In some areas scientists have discovered ecosystems that seem to depend on chemosynthetic bacteria acting as producers (fig. 30.11).

Coastal Zone

The coastal zone includes the continental shelf and the waters above the shelf. Seaweed, after it is partially decomposed by bacteria, is a source of food for benthic clams, worms, and sea urchins. These are preyed on by starfish, crabs,

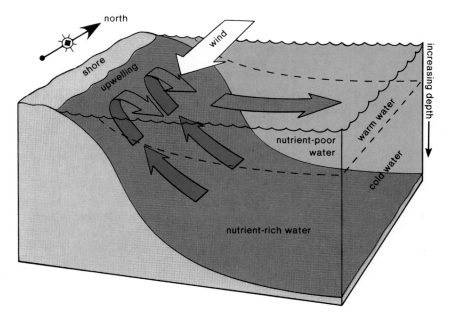

Figure 30.12
Coastal upwelling in the Northern Hemisphere. During upwelling, nutrient-rich water from below moves above nutrient-poor water on the surface. The nutrients allow producers and then consumers to dramatically increase in number. Therefore, fishing is particularly good in regions of upwelling.

Table 30.2 Ocean Life Zones

Life Zones	Location
Neritic	Shallow coastal waters
Pelagic	Deep open sea
	Epipelagic—sunlit waters
	Mesopelagic—twilight waters
	Bathypelagic—dark waters
Benthic	Sea floor
	Littoral and sublittoral— continental shelf
	Bathyl zone—continental slope
	Abyssal zone—abyssal plain

and brittle stars, all of which are eaten by bottom-dwelling fish. More importantly, the shallow sunlit waters, receiving nutrients from the sea and estuaries, produce abundant phytoplankton that grow larger than they would in the open sea. Especially in regions of **upwelling** (fig. 30.12), where surface waters are blown offshore and replaced by cold, nutrient-laden water from the deep, ample phytoplankton grow and provide food not only for zooplankton but also for small fishes. These, in turn, are food for the commercial fishes—herring, cod, and flounder. The coastal zone is more productive than the open sea (p. 685).

Open Sea

The open sea, or **pelagic zone,** is divided into the epipelagic, mesopelagic, and bathypelagic zones (fig. 30.10). Only the **epipelagic zone** is brightly lit, or euphotic; the **mesopelagic zone** is in semidarkness; and the **bathypelagic zone** is in complete darkness.

The open sea, which includes 90 percent of ocean waters, is not very productive because the surface waters are nutrient poor. Lack of nutrients in the *epipelagic zone* curtails phytoplankton growth to the extent that productivity of the open sea is equivalent to that of a terrestrial desert (chart, p. 685). Then, too, food chains tend to be longer in the ocean than on land. The small size of the *phytoplankton* and *herbivorous zooplankton* can account for this difference in food chain length, which also contributes to a low fish production. Among the phytoplankton are diatoms, dinoflagellates, and the smaller coccolithophores, which are less than a thousandth of an inch in size. They have a distinctive chalky shell and swim by moving tiny, whiplike flagella. Copepods, krill, foraminifera, and radiolarians feed on these and in turn are food for the carnivorous zooplankton, such as jellyfishes, comb jellies, wing-footed snails, sea squirts, and arrow worms.

The *nekton* (fig. 30.13) of the epipelagic zone includes herring and bluefish, which are food for the larger mackerel, tuna, and sharks. Flying fishes, which glide above the surface, are preyed upon by dolphin fishes, not to be confused with mammalian porpoises, which are also present. Whales are other mammals found in this zone. Baleen whales strain krill from the water, and the toothed sperm whales feed especially on the common squid.

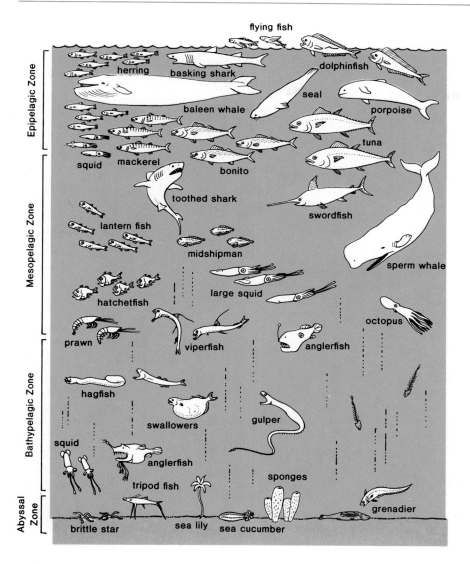

Figure 30.13
Free-swimming nekton of the epipelagic zone. In the epipelagic sunlit zone, many types of nekton populations exist because the producers are located here. Consumers are found in increased numbers in the dimly lit mesopelagic zone. Inhabitants of the dark bathypelagic zone are adapted to lack of light and to infrequent meals. The organisms in the abyssal zone are scavengers of various types.

Animals in the *mesopelagic zone* are adapted to the absence of light and tend to be translucent, red-colored, or even luminescent. Aside from luminescent jellyfishes, sea squirts, copepods, shrimp, and squid, there are luminescent carnivorous fishes, such as lantern and hatchet fishes. The *bathy-pelagic zone* is in complete darkness except for an occasional flash of bioluminescent light. Strange-looking fishes with distensible mouths and abdomens and small, tubular eyes feed on infrequent prey.

The **abyssal life zone** in the bathypelagic zone is inhabited by those animals that live in or just above the cold, dark abyssal plain. Because of the cold temperature (averaging 2° C) and the intense pressure (300 to 500 atmospheres), it was once thought that only a few specialized animals would live in the abyssal life zone. Yet a diverse assemblage of organisms has been found. Debris from the mesopelagic zone is taken in by filter feeders, such as the sea lilies that rise above the sea floor, and the clams and tubeworms that lie burrowed in the mud. Other animals, such as sea cucumbers and sea urchins, crawl around on the sea bottom, eating detritus and bacteria of decay. They, in turn, are food for predaceous brittle stars and crabs.

Figure 30.14
a. Temperature and rainfall determine the biome to a large extent. For example, rain forests are found in tropical regions where temperatures are warm and rainfall plentiful year round. Deserts, on the other hand, are found in tropical and temperate regions where rainfall is minimal. *b.* A map showing the major biomes of the world as they would be distributed if undisturbed by humans.

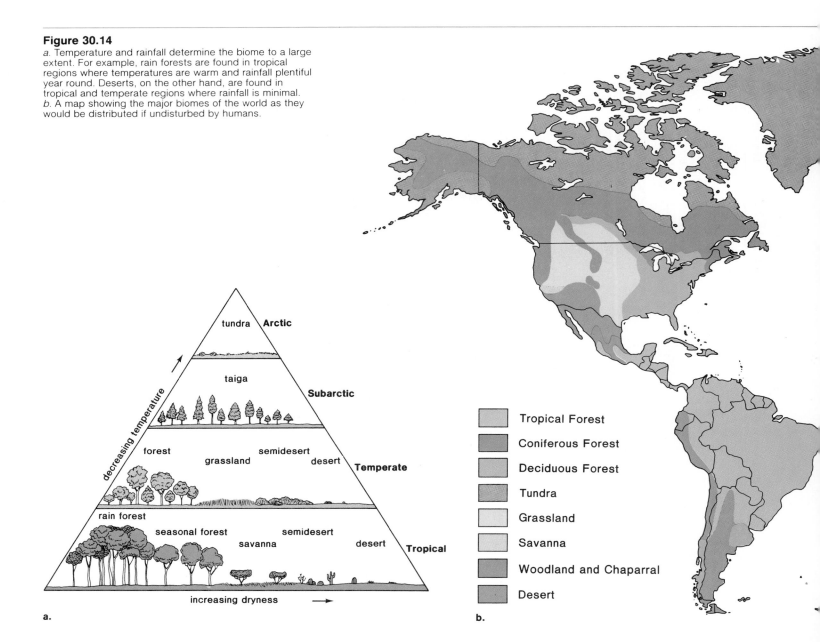

a.

b.

Terrestrial Ecosystems

Each major type of terrestrial ecosystem contains a characteristic community of populations called a **biome.** The locations of terrestrial biomes studied in this text are indicated in figure 30.14*b*. These represent the most complex biome possible for each area. Complex biomes arise by the process of **succession.** The complete process is called a **sere,** and each stage is a **seral stage.**

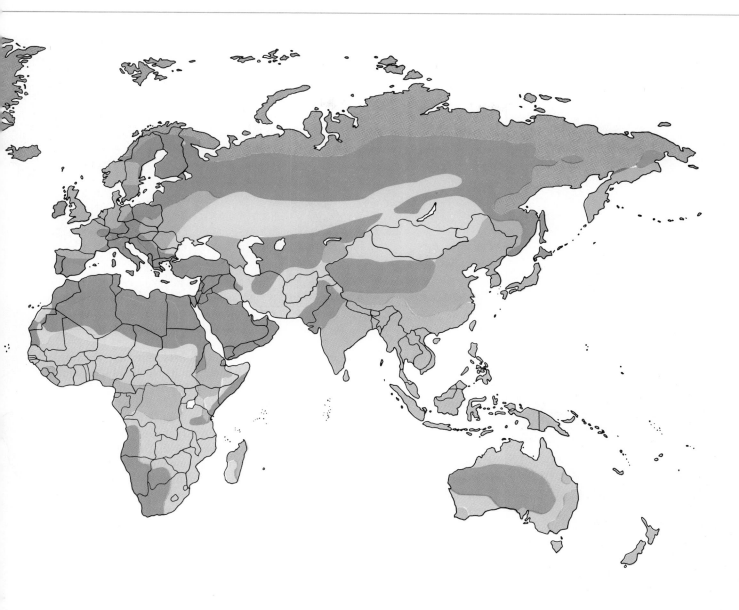

Succession

Primary succession (fig. 30.15) on land begins with bare rock. At first the rock is subjected to weathering as wind and rain act on it. Then lichens begin growing on the rock. Their acid secretions further break down the rock, and mosses and ferns take hold after a while. As soil begins to build up, longer-growing native plants begin to grow in the area, and these are eventually followed by shrubs and trees.

Figure 30.15
Primary succession includes the stages by which bare
rock becomes a climax community within a particular
biome. These photographs show a possible sequence of
events: (a.) lichens growing on bare rock; (b.) individual
plants taking hold; (c.) perennial herbs spread out over
the area.

a.

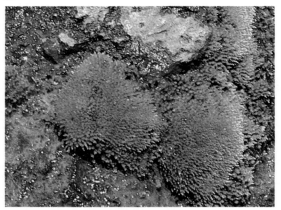

b.

c.

Figure 30.16
Secondary succession includes the stages by which
defoliated land areas, such as abandoned farmland or
strip-mined land, become a climax community again.
During secondary succession, grass and weeds are
followed by shrubs and trees and finally by a mature
climax forest.

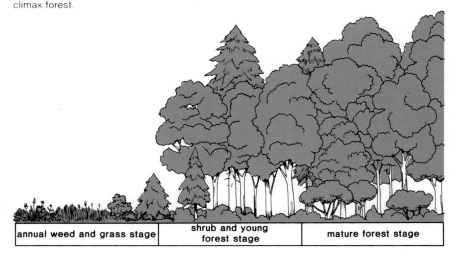

| annual weed and grass stage | shrub and young forest stage | mature forest stage |

Secondary succession (fig. 30.16) occurs in regions that have been disturbed. The stages vary from site to site, but figure 30.16 shows a possible
succession for abandoned farmland in the eastern United States. The first
community, called the pioneer community, includes plants, such as weeds, that
are able to colonize disturbed areas because of their ability to survive under
such harsh conditions as limited soil moisture and direct sunlight. This colo-
nization prepares the way for native plants of the area that most likely have
a longer growing period. Notice in figure 30.16 how one plant community replaces another until a climax community has been achieved. A different mix
of animals is associated with each of these communities. Whereas previous
communities are replaced, the final stage, a **climax community,** is able to sustain itself indefinitely. We will be studying climax communities in the next
few pages.

Climate
Physical conditions, particularly climate, determine the biome of an area.
Figure 30.14a shows that the various biomes can be related to temperature
and rainfall. Deserts are biomes with the least amount of rainfall.

Deserts
Deserts are regions of aridity with less than 20 cm of rainfall a year. True
deserts with less than 2 cm annually are infrequent, the Sahara in Africa being
the largest. **Semideserts,** however, include about one-third of all land areas.
In deserts, the days are hot because a lack of cloud cover allows the sun's rays
to penetrate easily, but the nights are cold because heat easily escapes into
the atmosphere.

Semidesert vegetation is extremely varied (fig. 30.17); annuals, herbaceous and woody perennials, deciduous and evergreen shrubs and trees all grow
here. The annuals exist most of the year as seeds and burst into flower during

a limited period of time when moisture and temperature are favorable. The entire life cycle takes place within a few weeks as foliage and then flowers develop in quick succession.

Semidesert perennials are of two types: (1) nonsucculent shrubs and trees, and (2) succulent herbaceous plants. Sagebrush, a densely branched evergreen shrub with wedge-shaped leaves, is common in the "cold" northern desert, and the creosote bush with small, well-protected leaves that may turn brown and drop off is common to the "hot" southern desert of the United States. The deciduous mesquite tree with compound leaves is a well-known tree with deep roots. Even better known, however, are the succulent leafless cacti with stems that not only store water but also carry on photosynthesis. All cacti have wide-spreading root systems that absorb great quantities of water during brief periods of rainfall. The spines provide a means of defense against desert herbivores. The prickly pear cactus has a jointed stem, while the barrel cactus has a ribbed, unjointed stem. The **saguaro** is a slow-growing giant cactus.

Like plants that live in deserts, animals must have features that allow them to survive heat and lack of water. Arthropods with a hard, external skeleton have an advantage. Millipedes, centipedes, scorpions, spiders, and numerous insects are found here. Like plants, some insects are apt to have a compressed life cycle, going from pupa to pupa within a short time since the pupa can lie inactive until it rains. Other insects burrow in the soil where a detritus food chain allows survival.

Reptiles, especially lizards and snakes, are perhaps the most characteristic group of vertebrates found in semidesert regions. Running birds, such as the ostrich in southwest Africa, the emu in Australia, and the roadrunner in the southwest United States, are numerous. Such birds as the hawk and eagle may fly in and out of the desert at will. Bats, rodents, and rabbits are numerous desert mammals. Some, such as the **kangaroo rat,** survive on metabolic water produced by respiration. Ungulates (antelopes, goats, and sheep) are able to roam far afield in search of food and water. The African **camel** is the best adapted of the ungulates; broad feet enable it to move over sand, and the reservoir of fat within the hump allows it to drink and eat infrequently. Dogs, such as the coyote and kit fox in North America and the dingo in Australia, are prominent carnivores. Lions in Africa and Asia and smaller cats, such as the mountain lion in North America, hunt in the desert.

Grasslands

Grasslands occur where rainfall is greater than 20 cm but is generally insufficient to support trees. The extensive root system of *grasses* allows them to recover quickly from drought, cold, fire, and grazing. Furthermore, their matted roots, which absorb surface water efficiently, prevent invasion by trees. Most of the grass both above and below ground dies each year. Thus, the turnover of biomass and nutrients is rapid; a complete turnover occurs in approximately three years. Because of relatively low rainfall and high evaporation rates, nutrients tend to accumulate in the lower part of the soil. At the same time, organic matter accumulates in the upper soil. The result is very fertile soil that has been exploited worldwide for agriculture.

Grasslands, which occur on all continents (fig. 30.14b), are known by various names, as indicated in table 30.3. We will discuss the tundra, an arctic grassland; the savanna, a tropical grassland; and the prairie, a temperate grassland found in the United States.

Figure 30.17
Semidesert vegetation and animal life. These organisms are adapted to retain water so that they might exist in areas with infrequent rain.

Table 30.3 Grasslands

Location	Name
Arctic Grassland	
Arctic region of all continents	Tundra
Temperate Grasslands	
North America	Prairie
South America	Pampa
Africa	Veld
Eurasia	Steppe
Tropical Grasslands	
Africa and Australia	Savanna

Tundra

The **tundra** (fig. 30.18), which means "bare mountaintops" in Finnish, is an arctic grassland that runs along the coasts and islands of the Arctic Ocean in Asia, Europe, and North America.

This northernmost biome is cold and dark much of the year. Since precipitation is only about 20 cm a year, it could possibly be considered a desert, but water frozen in the winter is plentiful in the summer because so little of it evaporates. Only the topmost layer of the earth thaws, and the **permafrost** beneath this is forever frozen. Trees are not found in the tundra for two reasons: their roots cannot penetrate the permafrost, and they cannot become anchored in the constantly shifting soil.

While the ground is covered with shortgrasses and forbs during the summer months, there are also dwarf woody shrubs and frequent patches of lichens and mosses. Only a few small animals—for example, the arctic fox; the snowshoe hare; the lemming, which resembles a rat; and a bird, the arctic tern—live in the tundra the year round, but many birds migrate there in summer. At one time, before its virtual extermination by human hunters, the musk-ox was a large year-round resident. Now only the large caribou and reindeer migrate to the tundra in the summer, and the wolves follow to prey upon them. Polar bears are common near the coast.

Plants and animals need adaptations to survive the extreme cold in the tundra. Insects require several seasons to complete their life cycle, but some plants, such as flowering shrubs, have extremely short life cycles to take advantage of the short summer season. The plants are low lying since it is warmer near the ground. Smaller animals burrow, and the large musk-ox has a thick coat and a short, squat body that conserves heat.

Savanna

The African **savanna** (fig. 30.19) is a tropical grassland that contains both trees and grasses and therefore supports populations of both browsers and grazers. The temperature is warm and there is adequate yearly rainfall, but

Figure 30.19
Vegetation and animal life of the African savanna. There are more types of grazers in the savanna than in any other biome.

a severe dry season limits the number of different types of plants. Perhaps the best known of the trees is the flat-topped **acacia tree** that sheds its leaves during a drought and never grows large because its water requirements increase as it grows.

The large, always warm African savanna supports the greatest number of different types of large herbivores of all the biomes. Elephants and giraffes are browsers. Antelopes, zebras, wildebeest, water buffalo, and rhinoceros are grazers. These are preyed upon by cheetahs and lions, whose kill is at times scavenged by hyenas and vultures.

These large animals can be active the entire year because they migrate in search of new pastures. Other animals are more active in either the wet or dry season. Most insects and birds reproduce during the wet season, while reptiles tend to reproduce during the dry season.

The savanna has been reduced in size by human encroachment, but the parts that remain, although often misused, have largely retained their natural state. "Conservation by utilization," which means the native animals are domesticated for dairy and meat products, is favored by farsighted promoters, especially since European cattle are susceptible to tsetse fly infection, a constant threat in many parts of Africa.

Prairie

When traveling from east to west across the United States, the **tallgrass prairie** gradually gives way to a **shortgrass prairie.** Although grasses dominate, they are interspersed by other herbaceous plants called **forbs.** These may catch the eye because they have colorful flowers, whereas grasses do not.

The absence of trees places a restriction on the variety of animal life. Insects abound, especially grasshoppers, crickets, leafhoppers, and spiders. Songbirds and prairie chickens sometimes feed off these, but usually these birds prefer seeds, berries, and fruits.

In contrast to the savanna, the mammalian herbivores are all grazers because there are few trees to support populations of browsers. Small mammals, such as mice, prairie dogs, and rabbits, typically burrow in the ground but usually feed above ground. Hawks, snakes, badgers, coyotes, and kit foxes capture and feed off these. The largest of the herbivores, buffalo and pronghorn antelope, had few enemies until humans killed them off. Large herds of buffalo, in the hundreds of thousands, once roamed the prairies and plains, never overgrazing the bountiful vegetation.

Forests

Forests (table 30.4) are located in all geographic zones except the arctic zone. Coniferous evergreen forests are generally associated with the subarctic zone; broad-leaved deciduous forests with the temperate zone; and broad-leaved evergreen forests with the tropical zone.

Taiga

The **taiga** is a coniferous forest extending in a broad belt across northern Eurasia and North America. The climate is characterized by cool summers, cold winters, and a growing season of about 130 days. Precipitation ranges between 40 and 100 cm per year, with much of it coming as heavy snow. The great stands of spruce, fir, and pine trees (fig. 30.20) are interrupted by many lakes and swamps. *Taiga* is Russian for "swamp land." The thickness of the forest allows little sunshine to filter through to the forest floor. As a result, there is no shrub understory; instead, there are only lichens, mosses, ferns, and some types of flowers residing close to the ground.

Table 30.4 Forests

Biome	Plants
Coniferous forest	Cone-bearing evergreen trees, such as pine and spruce No understory
Temperate deciduous forest	Broad-leaved trees, such as oak and maple Understories
Tropical rain forest	Broad-leaved evergreen trees Multilevel canopy No understory
Tropical seasonal forest	Mixed broad-leaved evergreen and deciduous Rich understories produce jungle

Figure 30.20
A coniferous forest encircles the globe in the northern temperate zone.

Productivity is good (chart, p. 685) because the *cone-bearing trees* have needlelike evergreen leaves that retain moisture and carry on photosynthesis whenever weather conditions permit. The trees acquire adequate nutrients from the soil even though it is acidic and infertile. Much of the precipitation moves through the soil, carrying with it important nutrients. However, the tree roots have a symbiotic association with fungi collectively called mycorrhizae (fig. 18.7*b*). Fungal hyphae extend into the soil and speed the transfer of nutrients from it to the plant.

Compared to other forests, the taiga has relatively few consumer species. Insects inundate the area in the summer, attack the trees, and are eaten by birds, such as woodpeckers and chickadees. Other birds, such as the crossbills and grosbeaks, extract seeds from the cones. Hawks and eagles prey on small animals, such as plentiful rodents, rabbits, and squirrels. The large herbivores, moose and bear, are apt to be found in clearings or near the water's edge where small trees, such as willows, aspens, and birches, and berry-bearing shrubs, such as blueberry and raspberry, are found. The carnivores, weasel, lynx, fox, and wolf, are especially noticeable in the summer.

Active signs of life all but disappear in the winter. Insect larvae survive in the dormant pupa stage, and many birds go south. Small animals seek shelter beneath the snow, within the ground, or in tree hollows. The snowshoe rabbit and short-tailed weasel change to their winter coats of white. Only the strongest animals survive the winter to again take advantage of the short summer season of plenty.

Other Coniferous Forests Along the coast and Piedmont region of the southern United States, wherever sandy and rocky soil does not permit a deciduous forest, there is a pine forest. On the West Coast, from California to Canada, an unusual coniferous forest has developed, dominated by hemlock, cedar, fir, and redwood. The **redwood** trees are among the largest of all American trees. Here abundant rainfall, high humidity, and moderate winter temperatures combined to favor the development of this unique forest. Coniferous trees are also found at high altitudes along the mountainous ranges of both the east and west coasts. The coniferous forests supply most of the lumber needs of the United States, and a large percentage of this lumber is from federally owned forests governed by the National Forest Service. This government agency regulates the manner and degree to which cutting can take place.

Temperate Deciduous Forests Temperate forests (fig. 28.1) are found around the world just south of the taiga. They are also found in other areas, such as parts of Japan, Australia, and South America, where there is a moderate climate with well-defined winter and summer seasons and relatively high precipitation (75 to 150 cm per year). The growing season, which lasts from the last frost of spring to the first frost of fall, ranges between 140 and 300 days. The soil is relatively fertile and mildly acidic, and litter decomposes quite rapidly.

The dominant trees are broad-leaved deciduous trees, such as oak and hickory, or beech, hemlock, and maple, which lose their leaves in the fall and regain them in the spring. The tallest of these trees form a **canopy,** an upper layer of leaves that are the first to receive sunlight. Even so, enough sunlight penetrates to provide energy for another layer of trees called **understory trees.**

Beneath these trees are shrubs that may flower in the spring before the trees have put forth their leaves. Still another layer of plant growth—mosses, lichens, and ferns—resides beneath the shrub layer. This **stratification** provides a variety of habitats for insects and birds. Ground life is also plentiful. Squirrels, cottontail rabbits, shrews, skunks, woodchucks, and chipmunks are small herbivores. These and the ground birds—turkeys, pheasants, and grouse—provide food for the carnivores. Mountain lions, bobcats, and timber wolves used to be found throughout the biome, but now only red foxes remain. The white-tail deer, a large herbivore, has increased in number of late, while the black bear, an omnivore, is all but extinct.

In contrast to the taiga, amphibians and reptiles are found in this biome because the winters are not as cold. Frogs and turtles prefer an aquatic existence, as do the mammalian beaver and muskrat.

Autumn fruits, nuts, and berries provide a supply of food for the winter, and the leaves, after turning brilliant colors, fall to the ground and contribute to the rich layer of organic matter. The minerals within the soil are washed far into the ground by the spring rains, but the deep tree roots capture these and bring them back up into the forest system again. Therefore, trees in a deciduous forest are called **nutrient pumps.**

Tropical Rain Forests The largest tropical rain forest (fig. 30.21), is found in the Amazon basin of South America, but such a forest also occurs at the equator in Africa and the Indo-Malayan region where it is always warm (between 20° and 25° C) and rain is plentiful (in excess of 200 cm per year). Therefore, these forests are very productive (chart, p. 685).

Diversity of species and luxurious growth characterize the plants of this biome. There is a triple-layered *canopy;* some of the broad-leaved evergreen trees grow to 150 feet or taller, some to 100 feet, and some to 50 feet. These tall trees often have trunks buttressed at ground level to prevent their toppling over. **Lianas,** or woody vines, which encircle the trees as they grow from tree bottom to tree top, also help strengthen the trunk. Sometimes, little light penetrates the canopy. Then there is no understory except in open areas or clearings where the presence of light may produce a thick jungle.

Although there is animal life, including many types of insects and such unique small animals, as the paca, agouti, peccaries, armadillo, and coatimundi on the ground floor, most animals have an arboreal existence. Birds, such as hummingbirds, parakeets, parrots, and toucans, are often beautifully colored. Amphibians and reptiles are also well represented by many types of snakes, lizards, and frogs. Lemurs and sloths are primitive-type primates found in the trees, but *monkeys* are more prolific. New-World monkeys such as the capuchin have a prehensile tail that helps them stay aloft; Old-World monkeys such as the rhesus do not. Monkeys feed off the fruit of the trees and are rarely carnivorous. The largest carnivores of the tropical forests are the big cats, the jaguars in South America and the leopards in the Old World.

Not only do many animals spend their entire lives in the canopy, some plants do also. *Epiphytes* are plants that grow on the tall trees but do not parasitize them. Instead, they have roots that take moisture and minerals from the air so are sometimes known as air plants. The most common types belong to the pineapple, orchid, and fern families.

Stability versus Productivity

Table A Pioneer versus Climax Community

Pioneer Community	Climax Community
Harsh environment	Most favorable environment
Biomass increasing	Biomass stable
Energy consumption inefficient	Energy consumption efficient
Some nutrient loss	Nutrient cycling
Little species diversity	High species diversity
Fluctuations common	Fluctuations do not usually occur
Little stability	High degree of stability

There is an ecological theory, based on laboratory and field data, that suggests that a biome's stability is related to its complexity. Certainly the climax, or mature, stage of succession is the most stable since it maintains itself with little change. One reason for this might very well be that the last stage is the most complex in terms of species diversity. If one species or population of a climax community is reduced in size or is eliminated, the other populations can compensate for this loss. For example, the deciduous forests in this country did not disappear when both the chestnut and elm trees succumbed to parasitic disease; the other trees simply filled in.

The early stages of succession are obviously unstable, as witnessed by the fact that they are replaced by later stages. However, investigators find that these stages show the most growth, and therefore are the most productive, because the inputs, in terms of energy and nutrients into the system, are high. But the outputs are also high because the community lacks the diversity to make full use of all the inputs. Not surprising, the producers have a high so-called P/B ratio, where P = gross primary production and B = biomass. This indicates that some of the energy derived from photosynthesis is not going into increased biomass. Some of it is being dissipated even as producer biomass is increasing.

As succession proceeds, diversity increases, and there is more efficient utilization of energy within the community. Nutrients begin to recycle. The chart compares the productivity of the producers found within the climax communities studied in this chapter. Although their productivity varies, all climax communities have a P/B ratio that approximates unity. This would not be true of the agricultural field, however.

Humans replace climax communities, which tend to have the characteristics listed in table A, with simpler communities that tend to have the same characteristics as the pioneer communities also listed in table A. Agricultural fields and managed forests show an increase in biomass and have high yields because they are provided with large inputs, such as fertilizer applications. The price paid for this growth potential, however, is a certain amount of waste and a loss of stability. We should keep in mind that stable climax communities perform many ecological services necessary to our well-being. For this reason, it would be wise to make sure that a large portion of every biome is kept in its natural state.

While we usually think of tropical forests as being nonseasonal rain forests, there are tropical forests with wet and dry seasons in India, Southeast Asia, West Africa, South and Central America, West Indies, and northern Australia. Here deciduous trees often allow enough light to pass through so that several layers of growth are possible. In addition to the animals just mentioned, certain forests also contain elephants, tigers, and hippopotomi.

Whereas the soil of a temperate deciduous forest biome is rich enough for agricultural purposes, the soil of a tropical rain forest biome is not. Nutrients are cycled directly from the litter to the plants by way of fungal roots. Productivity is high (chart, page 685) because of high temperatures, a year-long growing season, and the rapid recycling of nutrients from the litter. Of the minerals, only aluminum and iron typically remain near the surface, causing a red-colored soil known as **laterite.** When the trees are cleared, laterite bakes in the hot sun to a bricklike consistency that will not support crops. Slash-and-burn agriculture is the only type that has thus far been successful in the tropics. Trees are felled and burned, and the ashes provide enough nutrients for several harvests. Thereafter, the forest must be allowed to regrow, and a new section must be utilized for agriculture.

Mountain Biomes

When traveling from the southern to the northern hemisphere, it is possible to observe first a tropical rain forest, followed by a temperate deciduous forest, and then the taiga and tundra, in that order. A similar series of biomes could also be observed when traveling from the bottom to the top of mountains (fig. 30.22). These transitions are largely due to decreasing temperature as the altitude increases, but they are also influenced by soil conditions and rainfall.

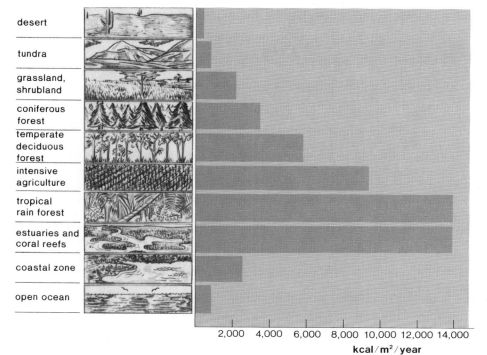

desert

tundra

grassland, shrubland

coniferous forest

temperate deciduous forest

intensive agriculture

tropical rain forest

estuaries and coral reefs

coastal zone

open ocean

2,000 4,000 6,000 8,000 10,000 12,000 14,000

kcal/m²/year

Comparison of the gross primary productivity of the world's major ecosystems. Estuaries and coral reefs are the most productive of the aquatic ecosystems. Tropical rain forests are the most productive of the terrestrial ecosystems.

From Johnson, Leland G., *Biology.* © 1983 Wm. C. Brown Publishers, Dubuque, Iowa. All Rights Reserved. Reprinted by permission.

Figure 30.22
Zones of vegetation change with altitude just as they do with latitude because vegetation is partially determined by temperature. Rainfall also determines vegetation, which is why grasslands are sometimes found at the base of mountains instead of tropical or deciduous forests.

ice

tundra

coniferous forest

deciduous forest

tropical forest

deciduous forest

coniferous forest

tundra

ice

increasing altitude

increasing latitude

Summary

Aquatic ecosystems, divided into fresh-water and saltwater systems, nearly always have a population of phytoplankton and zooplankton. Fresh water, a distillate of salt water, falls on the land and saturates the earth to the groundwater table. Rivers and streams are characterized by rapidly flowing water that gradually moves more slowly as a river approaches the ocean. The major energy source for rivers and streams is detritus.

Lakes can be divided in the summer into layers according to temperature and into life zones according to location of living organisms. Eutrophic lakes especially tend to fill in and disappear as succession occurs.

Marine ecosystems include estuaries and the seashores, both rocky and sandy, in addition to the oceans themselves. The rocky seashore, with organisms that cling to rocks, has more obvious signs of life than the sandy seashore where animals burrow in the sand. Coral reefs are spectacular communities found in shallow tropical waters.

The coastal zone, particularly wherever upwelling occurs, is more productive than the open ocean, which is comparable to a terrestrial desert. Only the epipelagic zone of the latter receives adequate sunlight to support photosynthesis, and the mesopelagic and bathypelagic zones depend on detritus from the first. Each of these zones contains organisms adapted to the conditions found there.

Terrestrial ecosystems have communities called biomes that are adapted to climate—that is, temperature and rainfall. Most deserts are high-temperature areas with less than 20 cm of rainfall a year, and the organisms found here are adapted to these conditions. Grasslands occur where rainfall is greater than 20 cm but is generally insufficient to support trees. The tundra, being a northernmost biome, supports only a limited amount of life. In contrast, the savanna, a tropical grassland, supports the greatest number of different types of large herbivores. The United States prairie has a limited variety of vegetation and animal life.

Forests require adequate rainfall. The taiga, a coniferous forest, has the least amount of rainfall. The temperate deciduous forest has trees that gain and lose their leaves because of the alternating seasons of summer and winter. Tropical forests, which include seasonal ones that experience dry and wet seasons, are the most complex and productive of all biomes.

There are a series of biomes on mountain slopes that change with altitude in the same sequence as they change when traveling from south to north. For example, a deciduous forest may precede a coniferous forest, which in turn precedes a tundralike region.

The forests are the most complex of all biomes, and it is possible to trace their development through a series of stages called succession. The first stage of succession, called the pioneer community, produces new growth but tends to be unstable. The final stage, called the climax community, produces little new growth but tends to be stable. The latter communities perform many beneficial services for humans, and their preservation is necessary for our well-being.

Objective Questions

1. In the fall, as the epilimnion of a lake cools, and in the spring as it warms, a _____ occurs.
2. Cultural eutrophication causes aquatic _____ to occur faster than it would otherwise.
3. An estuary is very productive because it acquires _____ brought both by river flow and by tidal action.
4. The chief builders of coral reefs are _____ and calcerous _____ .
5. Coasts are very productive because of _____ , a process that deposits nutrients along the coasts.
6. The neritic zone occurs along the coast, the pelagic zone is the open ocean, and the benthic zone is the _____ .
7. Each major type of terrestrial ecosystem contains a characteristic community of populations called a _____ .
8. A climax community arises due to the process of _____ , which requires a number of stages.
9. Trees are not plentiful in a desert or a grassland because of the reduced amount of _____ .
10. The tropical grassland of Africa is called the _____ .
11. Broad-leaved evergreen trees would be found in a _____ forest.
12. At the highest altitudes and highest latitudes, a _____ biome is found.

Answers to Objective Questions
1. turnover 2. succession 3. nutrients 4. stony corals, algae 5. upwelling 6. sea floor 7. biome 8. succession 9. rainfall 10. savanna 11. tropical 12. tundra

Study Questions

1. Every year the land receives a quantity of fresh water, and yet no new supply of water is added to the ecosphere. Explain.
2. Describe the temperature zones and the life zones of a lake in summer.
3. Compare a rocky seashore to a sandy seashore.
4. Describe the coastline biomes (including coral reefs) and discuss their importance to the productivity of the ocean.
5. Describe the life zones of the ocean and the organisms you would expect to find in each zone.
6. Arrange the terrestrial biomes discussed in this text in a diagram according to temperature and rainfall.
7. Describe the stages of succession by which an abandoned field becomes a forest.
8. Describe the location, climate, and populations of (1) deserts, (2) grasslands, and (3) forests.
9. Name the biomes you would expect to find when going from the base of a mountain to the top.
10. Discuss productivity and stability as they relate to succession.

Selected Key Terms

estuary ('es chə ,wer ē) 665
phytoplankton ('fīt ō ,plank tən) 665
zooplankton ('zō ,plank tən) 665
epilimnion (,ep ə 'lim nē ,än) 666
thermocline ('thər mə ,klīn) 666
hypolimnion (,hī pō 'lim nē ,än) 666
littoral zone ('lit ə rəl 'zōn) 668
limnetic zone (lim 'net ik 'zōn) 668

profundal zone (prə 'fənd əl 'zōn) 668
continental shelf (,känt ᵊn 'ent əl 'shelf) 673
continental slope (,känt ᵊn 'ent əl 'slōp) 673
biome ('bī ,ōm) 676
sere ('siər) 676
succession (sək 'sesh ən) 676

Suggested Readings for Part 11

Beddington, J. R., and May, R. M. November 1982. The harvesting of interacting species in a natural ecosystem. *Scientific American.*

Bergerud, A. T. December 1983. Prey switching in a simple ecosystem. *Scientific American.*

Borgese, E. M. March 1983. The law of the sea. *Scientific American.*

Cloud, P. September 1983. The biosphere. *Scientific American.*

Gose, J. R.; Holmes, R. T.; Likens, G. E., and Bormann, F. H. March 1978. The flow of energy in a forest ecosystem. *Scientific American.*

Kormondy, E. J. 1976. *Concepts of ecology.* 2d ed. Englewood Cliffs, N.J.: Prentice-Hall.

Moran, J. M., et al. 1980. *Introduction to environmental science.* San Francisco: W. H. Freeman.

Nebel, B. J. 1981. *Environmental science: The way the world works.* Englewood Cliffs, N.J.: Prentice-Hall.

Odum, E. P. 1983. *Basic Ecology.* New York: CBS College Publishing.

Putman, R. J., and Wratten, S. D. 1984. *Principles of Ecology.* Berkeley and Los Angeles: University of California Press.

Ricklefs, Robert E. 1983. *The Economy of Nature: A textbook in Basic Ecology.* New York: Chiron Press.

Rosenthal, G. A. January 1986. The chemical defenses of higher plants. *Scientific American.*

Smith, R. L. 1985. *Ecology and field biology.* 4th ed. New York: Harper & Row.

Whittaker, R. H. 1975. *Communities and ecosystems.* 2d ed. New York: Macmillan.

PART 12

Human Ecology

S ince 1850, the human population has expanded so rapidly that some doubt that there will be sufficient energy and food for this growth rate to continue in the future. The growth rate of the developed countries is declining, and the growth rate of the developing countries is beginning to stabilize. Even so, a large increase in population is expected in developing countries because many more young women are entering the reproductive years than are older women leaving them behind. It is suggested that energy and raw materials be used efficiently in these and all countries because many of the preferred sources are being depleted. In addition, raw materials can be cycled to assure a continued supply.

Land, air, and water pollution are common occurrences. Most countries are attempting to curtail the level of pollution in order to protect the health of the populace. All citizens should be aware of the possible ways to control pollution so that they can encourage these attempts. The human ecosystem depends on the natural ecosystems not only because they absorb pollutants, but also because they are inherently stable. Every possible step should be taken to protect natural ecosystems so as to ensure the continuance of the human ecosystem.

Los Angeles Freeway. Humans replace natural ecosystems with ones of their own design. Therefore, as more and more natural ecosystems are destroyed, the capacity of the biosphere to treat pollutants is ever decreased.

CHAPTER 31

Human Population

Concepts

1. The size of most populations in an ecosystem stays relatively constant year after year.

    In contrast, the human population has been undergoing rapid exponential growth since 1850.

2. Most populations consume approximately the same amount of resources each year.
    a. A very large human population is consuming resources and producing pollution at an unprecedented rate.
    b. Whether technology will be able to alleviate future shortages in nonrenewable resources and reduce pollution is unknown at this time.
    c. Energy is required to exploit the environment. Coal and solar energy probably will be important sources of energy in the near future.
    d. Thus far, the supply of food has kept up with increased population in most countries. A decrease in the expected population growth and the utilization of new hybrid plants may also help achieve this in the future.

The world growth rate[1] is only 1.7 percent, yet 82 million persons were added to the world population in 1986. Even if the growth rate would decline to 1.5 percent by the year 2000, approximately 89 million persons will still be added to the world population during that year. This is equivalent to the addition of a medium-sized city (250,000) every day, or the combined populations of the United Kingdom and Spain in one year. These statistics are reflective of a very large population that is presently undergoing rapid exponential growth, as can be appreciated by examining figure 31.1.

Developed and Developing Countries

The present phase of exponential growth began about 1850 at the time of the industrial revolution. In addition to European and North American countries, Russia and Japan also were industrialized. These countries, often referred to collectively as the **developed countries,** doubled their populations between 1850 and 1950. This was largely due to a decline in the death rate caused by the rise of modern medicine and improved socioeconomic conditions. The death rate decline was followed shortly by a decline in the birthrate, so the populations in the developed countries showed only modest growth between 1950 and 1975. This sequence of events (fig. 31.2) is termed the **demographic transition.**

The growth rate for the developed countries as a whole has now stabilized at about 0.6 percent. A few developed countries—Austria, Denmark, East Germany, Hungary, Sweden, West Germany—are not growing or are actually losing population. The United States has a higher growth rate (0.7 percent) than average because many people immigrate to the United States each year and a baby boom between 1947 and 1964 has resulted at this time in an unusually large number of women in the reproductive age.

Countries such as those in Africa, Asia, and Latin America are collectively known as **developing countries** because they have not as yet become industrialized. Although mortality began to decline steeply in these countries following World War II because of the importation of modern medicine from the developed countries, the birthrate did not decline to the same extent (fig. 31.2). Therefore, the populations of the developing countries began to increase dramatically. The developing countries were unable to cope adequately with such rapid population expansion, so many people in these countries are underfed, ill housed, unschooled, and living in abject poverty.

The growth rate of the developing countries did finally peak at 2.4 percent in 1960–65. Since that time, the mortality decline has slowed, and the birthrate has fallen. The growth is expected to be 1.8 percent by the end of the century. By that time, however, more than two-thirds of the world population will be in the developing countries.

The reason for a decline in the growth rate in the developing countries is not clear-cut. Previously, it was argued that decline would only be seen when socioeconomic conditions had improved. It was believed that as soon as the populace enjoyed the benefits of an industrialized society, the birthrate would decline. It is now apparent, however, that countries with the greatest decline are those with good family programs, supported by community leaders. Such programs probably can, even before socioeconomic conditions have improved, help bring about stable population sizes in the developing countries. Nevertheless, certain socioeconomic factors have also probably contributed to the

[1]Growth rate for the world is equivalent to the rate of natural increase.

Figure 31.1
Growth of the human population since the Old Stone Age.
Note the rapid increase in numbers during modern times.
The question mark signifies that the size of the population
may level off at 8, 10.5, or 14.2 billion.

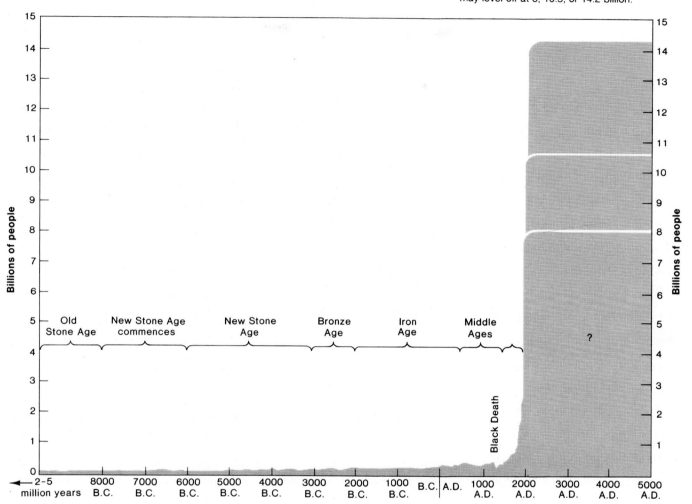

decline in the developing countries' growth rate. Relatively high Gross National Product (GNP), urbanization, low infant mortality, increased life expectancy, literacy, and education, especially of women, have all had a dampening effect on the **fertility rate**—the average number of children born to each woman.

Age Structure Comparison

Lay people are sometimes under the impression that if each couple had two children, **zero population growth** would immediately take place. However, **replacement reproduction,** as it is called, would still cause most countries to continue growth due to the age structure of the population. If there are more young women entering the reproductive years than there are older women leaving them, replacement reproduction will still result in a positive growth rate.

Figure 31.2
The demographic transition (darker color) in the
developed countries occurred in the nineteenth century,
but in the developing countries it was delayed until
recently.

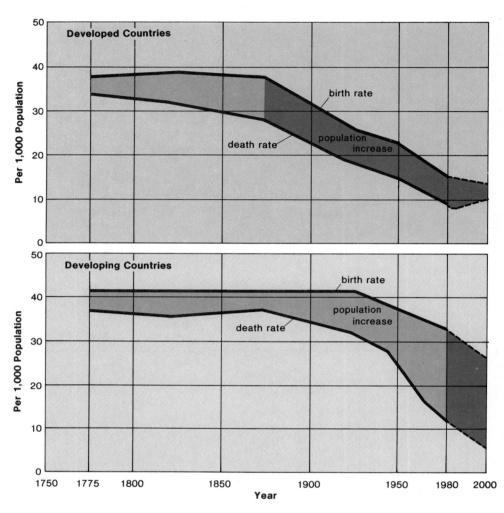

Reproduction is at or below replacement level in some twenty developed countries, including the United States. Even so, some of these countries will continue to grow modestly, in part because of the baby boom after World War II. These young women are now in their reproductive years, and even if each one has less than two children, the population will still grow. It should also be kept in mind that even the smallest growth rate can add a considerable number of individuals to a large country. For example, a rate of natural increase of 0.7 percent added over 1.6 million people to the United States population in 1982.

Whereas many developed countries are tending toward a stabilized age structure (fig. 31.3), most developing countries have a youthful profile—a large proportion of the population is below the age of fifteen. Since there are so many young women entering the reproductive years, the population will still greatly expand even after replacement reproduction is attained. The more quickly replacement reproduction is achieved, however, the sooner zero population growth will result.

The United States now has a population of over 240 million. The geographic center (the point where there are just as many persons in each direction) has moved steadily westward and recently has also moved southward. Another interesting trend is the shifting emphasis from metropolitan areas to nonmetropolitan areas. In the 1970s cities increased by 9.8 percent, but rural small towns increased by 15.8 percent.

The population size of the United States is not expected to level off any time soon for two primary reasons.

1. A baby boom between 1947 and 1964 has resulted in an unusually large number of reproductive women at this time (see the chart). Thus, although each of these women is having on the average only 1.9 children, the total number of births increased from 3.1 million a year in the mid 1970s to 3.6 million in 1980–81.
2. Many people immigrate to the United States each year. In 1981 immigration accounted for 43 percent of the annual population growth. The number of legal immigrants was about 700,000. Even though ordinarily only 20,000 legal immigrants can come from any one country, we give special permission for large numbers of political refugees to enter the United States.

In recent years, the majority of refugees have come from Latin America (e.g., Cuba) and Asia (e.g., Indonesia).

There is also substantial illegal immigration into the United States, although the exact number is not known. Estimates range from 100,000 to 500,000 or more. About 50 to 60 percent of these illegal immigrants come from Mexico, according to most estimates. There has been an effort to stem the tide of illegal immigration to the United States.

Whether the United States can ever achieve a stable population size depends on the fertility rate (the average number of children each woman bears) and the net annual immigration. If the fertility rate were kept at 1.8 and immigration limited to 500,000, the population would peak at 274 million in 2050, after which it would decline. Those who favor a curtailment of our population size point out that a fertility rate of 1.8 allows couples a great deal of freedom in deciding the number of children they will have. For example, it means that 50 percent of all couples can have two children; 30 percent can have three children; 10 percent can have one child; 5 percent can have no children; and 5 percent can have four children.

Source: Bureau of the Census

United States Population

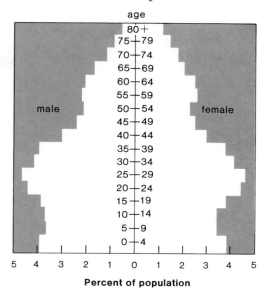

U.S. population age composition, 1985.

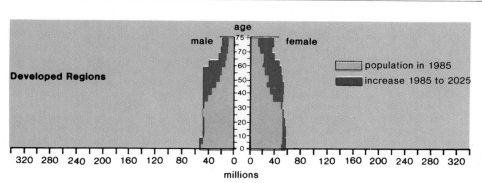

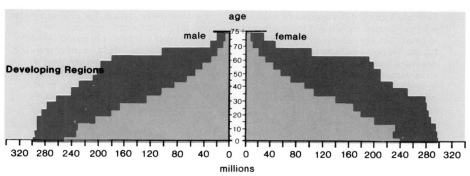

Figure 31.3
Contrasting age pyramids illustrate that the developed countries are approaching stabilization while the developing countries will expand rapidly due to their youthful profiles.
Source: Marshall Green and Robert A. Feary. World Population: The Silent Explosion. *Department of State Bulletin*, Publication 8956 (October 1978) based on U.S. Bureau of the Census Illustrative Projections of World Population to the 21st Century. Current Population Reports Special Studies.

Consequences of Population Growth

Increasing world population is putting extreme pressure on the earth's resources, environment, and social organization. While population increases in developing countries might seem to be of the gravest concern, this is not necessarily the case; each person in a developed country consumes more resources and is therefore responsible for a greater amount of pollution. Environmental

The Steady State

The human species was originally part of the natural biomes of the earth. We made the leap from hunting and gathering to farming, creating in the process much of what we call civilization. The advent of the industrial revolution removed us even farther from our original state. Today, humans most often live apart from the natural biomes (a.). The human ecosystem, consisting of people, their material goods, and their domestic plants and animals, dominates the biosphere, with an estimated 30 to 50 percent of net primary productivity being diverted and used to support the human species.

Mature natural ecosystems exist in a "steady state." The sizes of the various populations within the ecosystem remain fairly constant, while energy from the sun pours in and materials cycle continuously, so that there is little if any pollution. In contrast, the human ecosystem includes few populations (humans, domesticated plants and animals) that constantly increase in size; most energy comes from nonrenewable fossil fuels and is inefficiently used; and materials do not cycle, resulting in ever-increasing amounts of pollution. The natural ecosystems were able to absorb biodegradable pollutants until very recently, but now these cycles are overloaded. Synthetic products are an even more serious threat since they are not biodegradable and the natural ecosystems have no means of breaking them down.

The increase in world population and the consumption of nonrenewable mineral and energy resources since the industrial revolution can be represented by a J-shaped curve. It seems unlikely that this exponential increase can long continue. Critically important is the pollution caused by energy consumption and the production of goods. As long as the environment could handle the pollution fairly well, the cost was not considered. But pollution has now increased so much that it can no longer be ignored, and the once-hidden cost of pollution has now become apparent.

The ecological concept of carrying capacity suggests that there is a limit to the possible size of the human population. We do not know what that limit might be, but it seems possible that we could

overshoot it and create a population crash. The J-shaped curve would be followed by a rapid decline. However, both population and consumption of resources could level off. The human population, like other populations of the biosphere, can exist in a steady state, with no increase in number of people or in resource consumption.

A stable population size is a new experience for humans. The age structure diagram for a hypothetical stable population is shown in (b.). As you can see,

1. Over 40 percent of the people would be fairly youthful, with only about 15 percent in the senior-citizen category. Further, a combination of good health habits and advances in medical science could well mean that people would remain more youthful and more productive for a longer time than is common today.

2. There would be proportionately fewer children and teenagers than in a rapidly expanding population. This might mean a reduction in automobile accidents and certain types of crime statistically related to the teenage years.

3. There might be increased employment opportunities for women and a generally less competitive work place, since newly qualified workers would enter the job market at a more moderate rate.

4. Creativity need not be impaired. A study of Nobel Prize winners showed that the average age at which the prize-winning work was done was over thirty.

5. The quality of life for children might increase substantially, since fewer unwanted babies might be born, and the opportunity would exist for each child to receive the loving attention he or she needs.

The economy, like the population, should show no growth. For example, if the population size was to increase, the resource consumption per person should decline. In this way the amount of resource consumption would remain constant. In order to ensure that people have the goods and services they need for a comfortable, but not necessarily luxurious life, goods should have a long lifetime expectancy. Frugality is envisioned as absolutely necessary to the steady-state economy. Also, materials must be recycled so that they can be used over and over again. In

impact is measured not only in terms of population size, but also in terms of the resource used and the pollution caused by each person in the population.

$$\text{E.I.} = \frac{\text{population}}{\text{size}} \times \frac{\text{resource use}}{\text{per person}} = \frac{\text{pollution per unit}}{\text{of resource used}}$$

Therefore, there are two possible types of overpopulation. The first type is due to increased population and the second to increased resource consumption. The first type of overpopulation is more obvious in developing countries, and the second type is more obvious in developed countries.

order to provide jobs, technology could sometimes be labor intensive instead of energy intensive. The steady state would have need of very sophisticated technology, but it is hoped that the technology will not exploit the environment; instead, it is hoped that the technology will work with the environment.

Environmental preservation would be the most important consideration in a steady-state world. Renewable energy sources, such as solar energy, would play a greater role in providing energy needs. Pollution would be minimized. Ecological diversity would be maintained and overexploitation would cease. Ecological principles would serve as guidelines for specialists in all fields, thus creating a unified approach to the environment. In a steady-state world all people would strive to be aware of the environmental consequences of their actions, consciously working toward achieving balance in the ecosystems of their planet.

What would our culture be like if we had steady-state manufacturing and a steady-state population? Perhaps it would be greatly improved. Certainly there are no limits to growth in knowledge, education, art, music, scientific research, human rights, justice, and cooperative human interactions. In a steady-state world the general sense of fearful competition among peoples might diminish, allowing human compassion and creativity to prosper as never before.

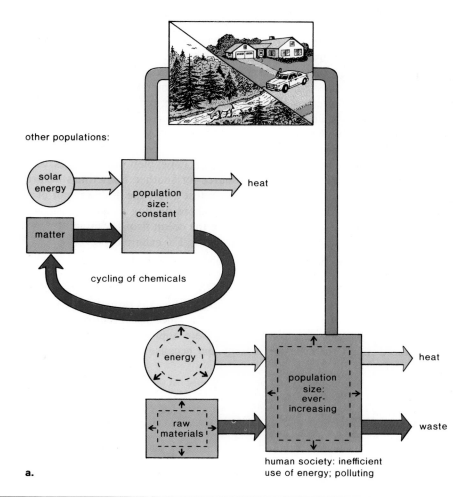

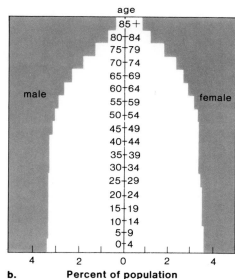

a. *In natural ecosystems, population sizes remain about the same year after year; materials cycle and energy is used efficiently. In the human ecosystem, the population size consistently increases, resulting in much pollution because of inadequate cycling of materials and inefficient use of supplemental energy. b. This diagram shows the age and sex structure of a hypothetical human population that remains the same size each year.*
b. Source: Bureau of the Census

Figure 31.4

Alternate depletion patterns for nonrenewable resources. *a.* Rapid depletion as resources are used up quickly: (*1.*) exponential consumption of a resource, followed by (*2.*) a peak, and (*3.*) decline as the resource becomes difficult to acquire. *b.* Depletion time can be extended with some recycling and less wasteful use. *c.* Efficient recycling extends depletion curve indefinitely.

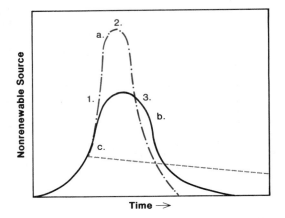

Resources

Resources, any material that is used by humans to sustain their lives or standard of living, are often categorized as nonrenewable or renewable. **Nonrenewable resources** are those with a supply that can be used up or exhausted. Fossil fuels and minerals are considered nonrenewable. Fossil fuels are certainly nonrenewable because once they have been burned, they can no longer be used as a source of energy. The classification of minerals as nonrenewable is debatable, however, because theoretically they can be recycled, and in that way the supply would never be exhausted. However, once minerals have been scattered widely, it is very difficult to retrieve them by recycling.

Renewable resources are those that can be replenished by either physical or biological means. Sunlight and rain are renewable resources, but so are crops and meat as long as care is taken not to overexploit the source of the supply.

Nonrenewable Resources

The demand for nonrenewable resources constantly increases not only because of population growth but also because per capita demand rises. There are two points of view, called the Malthusian and the Cornucopian views,[2] about resource availability under these circumstances. According to the Malthusian view, each nonrenewable resource is subject to depletion, and we are rapidly approaching the time when many if not most supplies will be exhausted. Figure 31.4 shows a **depletion curve** for any nonrenewable resource that is consumed at an increasingly rapid rate. A consumption peak is followed by a consumption decline as the resource becomes more expensive to find and process. Those who uphold the Malthusian view believe that, because of exponential consumption, finding new reserves cannot sufficiently extend a depletion curve. Nor do they believe that technology will ever overcome inevitable shortages. They admonish us, therefore, to cut back on growth, conserve resources, and recycle whenever possible to extend the depletion curve (fig. 31.4*b* and *c*).

According to the Cornucopian view, technology will constantly be able to put off the day when no further exploitation is possible. Improved technology will enable us to: (1) find new reserves, (2) exploit new reserves, and (3) substitute one mineral or energy resource for another. Sometimes we are aware of the availability of a resource but must await the development of a technology to exploit it. We now utilize offshore drilling to acquire oil; we might be able to develop the means to mine the ocean floor for minerals; and we can utilize ever poorer grades of mineral ores. For example, previously only 3 percent copper ores were mined, whereas now ores with .3 percent copper are utilized.

To exploit previously inaccessible and/or less concentrated resources, a plentiful supply of energy is required. The Malthusians do not believe that increasing amounts of energy will be available, since fossil fuel reserves are being rapidly depleted. The Cornucopians were hoping at first that nuclear energy would supply the necessary energy, but because of the many problems

[2]Malthus was an eighteenth-century economist who pointed out that since the size of a population increases geometrically and renewable resources increase only arithmetically, shortages must eventually occur. Cornucopia is a Greek word meaning horn of plenty, a symbol of everlasting abundance.

associated with nuclear energy, many are now looking forward to the development of a different energy source. They feel that, given time, a new and plentiful energy source will be found.

Malthusians and Cornucopians also view the problem of pollution differently. The Malthusians suggest that as resource consumption increases, the amount of pollution control needed to protect the environment becomes prohibitive. Previously, no heed was paid to the cost of pollution. Industries were allowed to pollute the environment, and the general public bore the cost of cleaning up after industry. Now, since controls are required by the government, the cost of pollution is added to the total cost of a product (fig. 31.5). The profit margin of the producer is thereby cut even as the consumer must pay more for the product. Unfortunately, also, greater and greater efficiency of control is required as output is increased. For example, as table 31.1 shows, if output doubles every fifteen years, pollution control efficiency must increase to 96 percent within 120 years to stay even. Since the cost of pollution control increases manyfold per level of efficiency, it may become unprofitable to continue production. It is estimated, for example, that it costs $25 per automobile to achieve a 50 percent reduction in pollution emissions, but that it will cost $400 per automobile to achieve a 95 percent reduction.

On the other hand, Cornucopians suggest that whenever strict pollution control regulations are instigated, pollution can be reduced to acceptable levels at relatively modest costs compared to the value of total output. Given reasonable periods of time, all industries and most firms will be able to accommodate themselves to the impact of the new standards. In some instances it may be possible to change industrial processes so there is no waste or to make use of a by-product that formerly was considered a waste. For example, researchers have recently reported that it will be possible to recover 150 pounds of aluminum oxide from each of 60 million tons of fly ash waste given off by coal-fired power plants in the United States.

Energy

A larger world population will require more energy in the years ahead. As figure 31.6 shows, developing nations are expected to be responsible for most of the increased demand for energy. Between 1956 and 1976, energy consumption in China increased by a factor of 14, in Latin America by a factor of 8, and in Southeast Asia by a factor of 11. This trend will continue through the year 2000 and beyond.

In contrast, the opposite trend seems to be occurring in the United States: demand is not increasing and might even diminish by almost 25 percent from 80 quads[3] in 1980 to 62 quads in 2000. Increased energy efficiency in the home, office, and industry is the primary means of conserving energy. For example, the same steam flow can be used first to generate electricity and then to supply heat to a building. This **cogeneration,** which saves about 30 percent of the energy needed to do both separately, can be practiced independently of electric utility companies by apartment complexes, small towns, and industries. Utility companies are now required by law to buy any excess electricity generated by independent sources.

[3]Quadrillion BTUs, the same amount of energy as in 8 billion gallons of gasoline.

Figure 31.5
The cost of a product includes the cost of controlling pollution due to, for example, mining, manufacturing, and distributing the product.

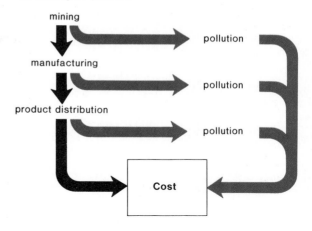

Table 31.1 Pollution Control Efficiency

Time (Years)	Pollution Level	Pollution Control Efficiency
—	One unit	—
15	Two units	50%
30	Four units	75%
60	Eight units	88%
120	Sixteen units	99%

Figure 31.6
A graph showing that the energy demands of developing countries will largely account for increased use of energy in the future.

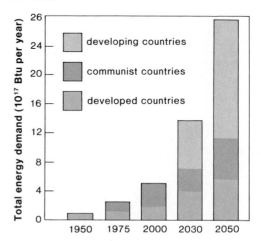

Table 31.2 Renewable and Nonrenewable Energy Sources

	Advantages	Disadvantages
Nonrenewable Fossil fuels	*Technology Well Established*	*Finite Fuel Supply*
Coal	Plentiful supply	Surface mining Air and water pollution
Petroleum	Cleaner burning	Limited supply
Natural gas	Cleanest burning	Limited supply
Nuclear		
Light water	Fuel availability	Thermal pollution Radiation pollution
Renewable	*Infinite Fuel Supply*	*Technology under Development*
Nuclear		
Breeder	Fuel availability	Radiation pollution Thermal pollution Nuclear weapons proliferation
Fusion	Fuel availability	Radiation pollution
Geothermal	Less pollution	Availability limited
Solar and wind	Nonpolluting Large and small scale possible	Noncompetitive cost
Ocean	Nonpolluting	Applicable only in certain areas
Biomass	Utilizes wastes	Air and water pollution

The energy sources currently available or being investigated are listed in table 31.2. The public is interested in development of renewable energy sources, such as solar and wind energy, but thus far the United States government has largely subsidized nuclear power.

Nonrenewable Energy Sources

The depletion curve (fig. 31.4) for petroleum is now in the decline phase, and the depletion curve for natural gas has reached its peak. Supplies of these favored fossil fuels are not expected to last more than thirty years. Petroleum became the favored fuel during the present generation (fig. 31.17) because it is easily transportable; is versatile, serving as the raw material for gasoline and many organic compounds; and is cleaner-burning than coal.

Coal is in plentiful supply, and the United States probably will depend primarily on its use to make up for decreased use of petroleum. There are environmental drawbacks to using coal in place of petroleum, however. Thousands of acres of land are strip mined, and oil shale mining in particular causes huge piles of residue having the potential to leach hazardous chemicals into underground aquifers. Coal burning adds to the air many pollutants that damage human lungs, poison lakes, and erode buildings. It is even possible that the CO_2 given off will eventually cause a global warming trend. All these concerns are discussed at length in the next chapter.

Three possible types of nuclear energy (table 31.2) can be used to generate electricity. Thus far, only light water **fission power plants,** which split ^{235}uranium, have been utilized in the United States. Approximately one hundred operating nuclear plants exist and twenty-five are under construction. Many proposed plants have been canceled, and there have been no new orders for the past several years. Some predict that the nuclear power industry is dead in this country. Possible reasons for this are numerous; for example, there is little need for new plants because demand for electricity is not increasing; it is cheaper to build and operate coal-firing plants; and the public is very much concerned about nuclear power safety.

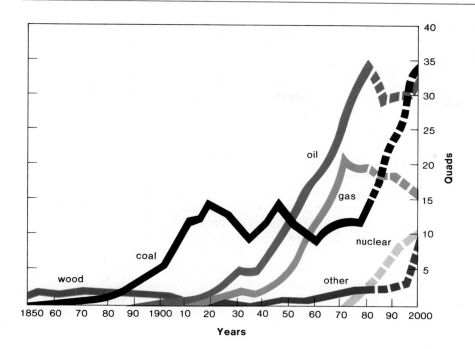

Figure 31.7
The favored energy sources change with time. Coal replaced wood and then oil replaced coal. It is possible that coal will again be more widely used in the year 2000. The term *quad* stands for Quadrillion British Thermal Unit.

In May 1986 the Chernobyl nuclear power plant in Russia suffered an accident that released so much radioactivity it triggered an alarm system in Sweden. Due to operator error, the nuclear reaction that powers the plant increased beyond control, causing a fuel meltdown. The molten fuel reacted with water, and generated an immense burst of steam that shattered the roof of the plant. The damaged reactor core and the surrounding graphite (a material that normally functions to control a nuclear reaction) began to burn, sending a plume of radioactivity into the atmosphere. Western nations were quick to point out that the West had no graphite-utilizing plants and that the Chernobyl lacked a containment tower that keeps radioactivity from entering the atmosphere should an accident occur. Even so, the drive to control the nuclear power industry was given new impetus by the Chernobyl accident, which caused several immediate deaths and may lead to hundreds more.

The public is also aware of the need to store radioactive wastes that have accumulated since the start of the nuclear age. Spent fuel rods contain low-, medium-, and high-level wastes, depending on the amount and character of the radiation they emit. High-level wastes must be stored hundreds of thousands of years before they lose their radioactivity. As yet, the government has not sanctioned any particular permanent method of disposal. Most probably, the wastes will finally be incorporated into glass or ceramic beads, packaged in metal canisters, and buried in stable salt beds, red clay deposits in the center of the ocean, or in stable rock formations. Meanwhile, some wastes have been temporarily stored above ground in tanks that on occasion have developed leaks.

Another source of radioactive waste, not often mentioned, occurs when uranium is mined. **Uranium mill tailings** are the sand that remains after uranium has been removed from mined material. The tailings contain [226]radium, which has a half-life[4] of 1,620 years. As [226]radium decays, it gives off radon gas, a substance that has been associated with the development of lung cancer.

[4]Length of time required for one-half of the radiation to dissipate.

The government was at first slow to regulate proper disposal of the tailings, and in the meantime, they were used as fill at many construction sites. This procedure has now been stopped, and mill operators are required to bury tailings beneath a thick layer of soil that is replanted with vegetation to prevent erosion of radioactive dust into the water and air.

Breeder fission reactors do not have the same environmental problems as light water nuclear power plants. They need less raw uranium, and they have little waste because they use ^{239}plutonium, a fuel that is actually generated from what would be reactor waste.

Plutonium, however, is a very toxic element that readily causes lung cancer; it is also the element used to make nuclear weapons. (The chances of a nuclear explosion are much greater with breeder than with light water reactors.) The fear of nuclear weapons proliferation, in particular, has thus far prevented the start-up of a breeder reactor in this country.

Nuclear fusion requires that two atoms, usually deuterium and tritium, be fused, and the heat needed is so great that there is no conventional container for the reaction. Scientists are now perfecting laser-beam ignition and magnetic containment for the reaction. Since the fusion reaction gives off neutrons that can change uranium to plutonium, the best use of fusion plants would be to provide fuel for breeder reactors. The latest idea is to have hybrid fusion plants that would combine both fusion and breeder fission plants. But, in any case, the fusion process is still experimental and not expected to be ready for production any time soon.

Renewable Energy Sources

Solar energy is diffuse energy that must be collected and concentrated before utilization is possible. The public, rather than the government, has provided most of the impetus for practical solutions, including solar heating of homes (fig. 31.8) and offices. **Solar collectors** placed on rooftops absorb radiant energy but do not release the resulting heat. A fluid within the solar collector heats up and is pumped to other parts of the building for space heating, cooling, or the generation of electricity. Passive systems are also possible; specially constructed glass can be used for the south wall of a building, and building materials can be designed to collect the sun's energy during the day and release it at night.

The U.S. government subsidized the building of **Solar One,** the world's largest solar-powered electrical generating facility (fig. 31.9), which began operation in 1982 at Daggett, California, in the Mojave Desert. A single boiler is located atop a large tower twenty stories high, and a large field of mirrors (called **heliostats,** which are capable of tracking the sun) reflects the sun's rays onto the tower. The water is heated to 500° C, and the steam is used to produce electricity in a conventional generator. Systems such as this require much land and cannot be placed just anywhere. For this reason, **photovoltaic (solar) cells** that produce electricity directly may be a more promising energy source. It has even been suggested that cells might be placed in orbit around the earth where they would collect intense solar energy, generate electricity, and send it back to earth via microwaves.

Solar energy is clean energy. It does not produce air or water pollution. Nor does it add additional heat to the atmosphere, since radiant energy is eventually converted to heat anyway. The problem of storage can be overcome in a number of ways, including the use of solar energy to produce hydrogen by means of hydrolysis of water. Hydrogen as either a gas or a liquid can be piped into existing pipelines and used as fuel for automobiles and airplanes. When it is burned, it forms fog, not smog (fig. 31.10).

Figure 31.8
Geohouse makes use of time-honored principles in order to stabilize internal conditions. Counterclockwise, the insets illustrate how primitive people have solved the problems of making their homes comfortable: an igloo uses a tunnel to protect the interior from inrush of cold and outrush of warm air—the geohouse uses a vestibule; pueblo villages face the winter sun for passive solar heating—the geohouse faces south; the walls of adobe houses, like the geohouse's thick walls, absorb the day's warmth to radiate it at night; a sod house of the midwest was cool in summer and warm in winter—the north side of geohouse is banked with earth; a house in the tropics is open to catch prevailing winds—the geohouse has windows and vents to ensure air exchange; an awning reduces solar radiation—roof overhangs and awnings protect the geohouse from the high summer sun.

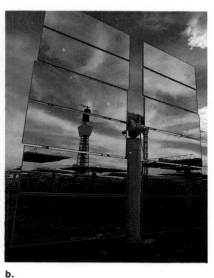

a.   b.

Figure 31.9
Solar One is the world's largest solar-powered electrical generating facility. *a.* Note the large number of heliostats surrounding a single boiler that sits atop a large tower. *b.* Close-up of a single heliostat.

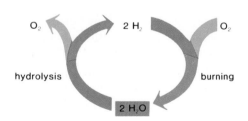

O₂ → 2 H₂ → O₂

hydrolysis   burning

2 H₂O

Figure 31.10
When hydrogen is burned, water is the product. This water can then be hydrolyzed to hydrogen again. This cycle makes hydrogen an attractive fuel.

These types of renewable energy sources have been utilized for quite some time. *Falling water* is used to produce electricity in hydroelectric plants. *Geothermal energy* is trapped heat produced by radioactive material deep beneath the surface of the earth. Water, converted to steam by this heat, may be pumped up and used to heat buildings or to generate electricity. *Wind power* provides enough force to turn vanes, blades, or propellers attached to a shaft, which, in turn, spins the motor of a generator that produces electricity. The government has allocated a small amount of money for wind research, particularly the promotion of large windmills, such as those being constructed in the windy Columbia River valley in the Pacific Northwest. Many others believe it would be best to build numerous small windmills, such as those found in the first wind farm now operating in New Hampshire.

Food

Figure 31.11*a* and *c* show that the *production of food* has increased in developed countries, and the per capita share has also increased. This means that food production has kept pace with population increase in developed countries, where the majority of people can afford the price of food and are eating better than they were in the past. Figures 31.11*b* and *d* show that food production has increased in developing countries, but the per capita share has not substantially increased because the increased food production must be divided by a large population increase. In developing countries only a segment of the population can afford the price of food and eat well; a larger portion cannot afford the price of food and do not eat adequately. (Since the per capita share is calculated by dividing the total production by the total population size, it does not necessarily mean that all persons are eating an equal amount.) The U.N. Food and Agricultural Organization (FAO) estimates that as of 1975 about 450 million people are chronically undernourished. The World Bank estimates that the figure may be as high as one billion. While some undernourished individuals live in developed countries, the majority live in developing countries. In particular, African countries, with the fastest population growth ever recorded, have had difficulty providing enough food for their populaces.

The factors we will now discuss have been identified as important considerations in regard to food production.

Factors in Food Production

Today, most **land** is already being used for one purpose or another, and it would be difficult to expand the amount of farmland. Only in the tropics (sub-Saharan Africa and the Amazon Basin of Brazil) are there still sizable tracts of land not presently utilized that have enough water to grow crops. Slash-and-burn agriculture has been traditional in this area because the soil contains so few nutrients. The jungle is cut and the debris burned, providing the nutrients to allow a few years' crops. When the soil can no longer support continued cultivation, a new area is selected for cultivation. Attempts to cultivate the land in a traditional manner have proven difficult because the soil tends to become hard and compact when exposed to the sun. If these problems could be solved, the food supply would be multiplied manyfold. Even so, many believe that tropical forests should remain as they are because they help control global pollution. (The many complications in the development of rain forests are discussed in the next chapter.)

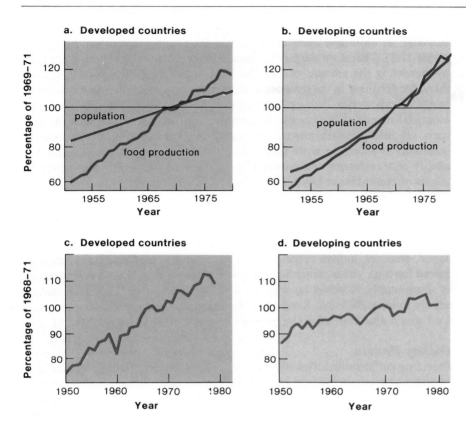

Figure 31.11
Food production. *a. and b.* Food production has increased at similar rates in the developed and developing countries over the last three decades. Unlike food production, population growth has continued at a faster rate in developing countries compared to developed countries. *c. and d.* Therefore, the developed countries have experienced a per capita gain that the developing countries have not had.

Since virtually all readily available, relatively fertile cropland is already in use, more effort should be directed toward safekeeping the land already being cultivated. Millions of acres of fertile lands are lost each year (p. 707) from encroachment by cities and suburbs, industrial development, soil erosion, and desertification. The United Nations estimates that by the end of this century twice as much land will be lost due to these factors than will be developed as farmland.

Developed countries, especially the United States, have had spectacular success in increasing yields by using monoculture agriculture. This method of agriculture along with its many environmental problems (p. 652) is now being exported to developing countries. The term **"Green Revolution"** was coined to describe the introduction and rapid spread of *high-yielding wheat and rice plants* (fig. 11.6) that, when supplied with generous amounts of fertilizer and water, will grow in warmer countries. Although Green Revolution plants require irrigation, they use water efficiently because they produce more grain per given amount of water and mature faster than traditional varieties. Research is striving to provide plants with improved internal efficiency. The new focus is on greater photosynthetic efficiency (p. 108), more efficient nutrient and water uptake, improved biological nitrogen fixation, and genetic resistance to pests and environmental stress.

Table 31.3 Corn-Yield Gains from Successive Fertilizer Applications

Nitrogen Applied	Average Gain in Corn Yield per Pound of Nitrogen Applied
First 40 pounds	27 pounds
Second 40 pounds	14 pounds
Third 40 pounds	9 pounds
Fourth 40 pounds	4 pounds
Fifth 40 pounds	1 pound

Source: U.S. Department of Agriculture. Data from Iowa, 1964.

Figure 31.12
Kwashiorkor is an illness caused by lack of protein. Characteristically, there is edema due to lack of plasma proteins. The nervous system of these children may also fail to develop, which frequently results in mental retardation.

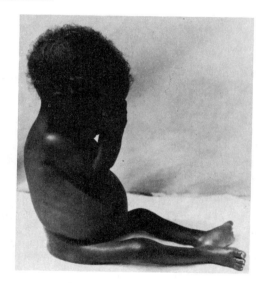

The application of **fertilizer** has contributed greatly to increased yield. When fertilizer is first applied to soil, there is a dramatic increase in yield; but later, as more and more fertilizer is used, the increase in yield drops off (table 31.3). Therefore, farmers in developed countries would profit very little by increasing the amount of fertilizer they already apply. A decrease in demand for fertilizer in the developed countries would mean that more fertilizer might be exported to developing countries. *Legumes,* you will recall, are also a natural means of increasing the nitrogen content of the soil because their roots are infected with nitrogen-fixing bacteria. Research continues on the possibility of infecting more plants with these bacteria and even on the possibility of transferring the nitrogen-fixing genes to plant cells themselves by means of recombinant DNA methods.

High agricultural yields are extremely dependent on both indirect and direct uses of **energy** as discussed in the reading on page 94. In the United States, fossil fuels have supplied the supplemental energy, as is apparent when considering that nearly all large farm machinery utilizes fossil fuel energy. In Japan, however, human labor provides much of the supplemental energy required for high yields. Since we are now entering a time when fossil fuels are in short supply, it would be wise for developing countries to rely on human energy to increase yield. Keeping people in rural areas close to the source of food would also help solve the problem of transportation and packaging.

Dietary Protein

Providing sufficient calories will not assure an adequate diet; there must also be enough protein in the food eaten. Children in the developing countries often have two forms of nutrition illness: marasmus and kwashiorkor. **Marasmus** is caused by a diet low in both calories and protein: the body is thin; the skin shriveled; the face wrinkled, with wide eyes; and the belly bloated. **Kwashiorkor** is caused by a diet that has adequate calories but is deficient in proteins; the entire body is bloated, the skin is discolored, a rash is present, and the hair has a reddish-orange tinge (fig. 31.12).

Plants are low quality, nutritionally incomplete protein sources; while they contain some amino acids, they lack others. It is possible, however, to eat a combination of plants that together contain an acceptable level of all essential amino acids. For example, wheat and beans complement each other to give a balance that is comparable to a high quality protein, such as is found in meat or cheese. Unfortunately, the new varieties of wheat and rice plants give a higher yield per acre than do beans, and farmers may prefer to plant wheat or rice instead of beans.

Most grain produced in developed countries is fed to cattle, pigs, or chickens, which then serve as a source of high quality protein. In many developing countries most grain must be consumed directly to provide calories. Grazing land is also deteriorating. Therefore, beef is not expected to supply the necessary protein. Another possibility is fish consumption. Between 1950 and 1970, fish supplied an increasing portion of the human diet until the catch averaged 18.5 kg per person annually. Since that time, although the catch has increased, the per capita share has decreased due to population growth. Overexploitation of the ocean's fisheries, and pollution of coastal waters may also eventually cause a decline in the yearly catch. Therefore, the expansion of aquaculture, the cultivation of fish, is highly desirable. Some innovative ideas, such as those described in figure 31.13, could make aquaculture an ordinary part of everyone's life, even in developed countries.

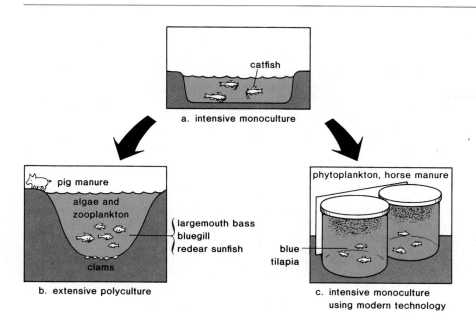

Figure 31.13
Aquaculture methods. *a*. Catfish are raised in ponds and fed commercial fish pellets. *b*. A natural detritus food web is supported by animal manure. *c*. Phytoplankton are placed in clear fiberglass cylinders that allow maximum input of solar energy. The phytoplankton flourish when fertilized, and when mature they will serve as food for fish. Methods (*b*.) and (*c*.) are more ecologically sound than method (*a*.).

- a. intensive monoculture
- b. extensive polyculture
- c. intensive monoculture using modern technology

Summary

The human population is expanding exponentially, and even though the growth rate has declined, there is a large increase in the population each year. Developed countries underwent the demographic transition between 1950 and 1975, with the result that their growth rate is now only about 0.6 percent. Developing countries are just now undergoing demographic transition, and since it was delayed, their growth rate went as high as 2.4 percent. Now it is declining slowly. Reproduction replacement will not bring about zero population growth even in developed countries because of a baby boom after World War II. In developing countries, where the average age is less than fifteen, it will be many years before reproduction replacement will mean zero population growth.

Resource consumption depends on population size and on consumption per individual. Developing countries are responsible for the first type of environmental impact and developed countries for the second type. According to the Malthusian view, we are running out of nonrenewable resources, and only conservation can help extend their depletion curves. According to the Cornucopian view, we will continue to find new ways to exploit the environment. Similarly, Malthusians do not believe pollution problems can be solved if resource consumption continues at the present high rate, but Cornucopians believe new and innovative ways to avoid increased pollution will be found.

Exploitation of the environment depends on a plentiful energy source. However, it is predicted that the United States can reduce its need for energy simply by using energy more efficiently. The demand for energy in this country has leveled off and may even decline. Among the nonrenewable sources of energy, only coal is in ready supply. The government has largely supported nuclear research but has also been instrumental in starting the first solar and wind electricity-generating plants. These two sources of renewable energy might best be used for medium- and small-scale needs. Solar ponds are recommended as a way to meet a community's heating needs, for example.

Food production has become increasingly dependent on supplements, such as fertilizer and irrigation, and on high-yield plants since the 1960s, whereas prior to this time expansion of agricultural land helped increase yields. It is predicted that developing countries will not be able to meet their need for food unless they rapidly bring population growth under control. Diets in developing countries often lack adequate protein. Grain is needed for human consumption, and grazing land is deteriorating; therefore, beef consumption will probably not increase. It is doubtful that oceanic fishing can be counted on to supply protein, for as yet, aquaculture is not sufficiently developed to do so.

Objective Questions

1. After a country has undergone the demographic transition, both the death rate and the birthrate are _____ (choose high or low).
2. In contrast to the developed countries, the developing countries are not as yet _____ .
3. The people of the developed countries have a higher standard of living and consume more _____ than do those of developing countries.
4. Those who hold a Cornucopian view concerning nonrenewable resources believe that _____ _____ .
5. For the past several years, the nonrenewable energy source of choice has been _____ .
6. The steady accumulation of _____ wastes is a deterent to the use of nuclear power.
7. Solar power can be used to hydrolyze water and the resulting _____ can be a gaseous or liquid fuel.
8. The continent least able to feed its ever-growing populace is _____ .
9. One of the chief problems in regard to a proper diet in developing countries is supplying adequate _____ .
10. Overexploitation and pollution may very well cause a _____ in the size of the yearly fish catch.

Study Questions

1. Show that a population of 10,000 persons will add more individuals with each generation even if the growth rate remains constant. Is it possible that even with a declining growth rate the same number of individuals could be added each generation?
2. Define demographic transition. When did developed countries undergo demographic transition? When did developing countries undergo demographic transition?
3. Give at least three differences between developed countries and developing countries.
4. Contrast the Malthusian view and the Cornucopian view toward nonrenewable resource supplies and concomitant pollution.
5. Draw a typical depletion curve and relate it to the consumption of fossil fuels.
6. Name three types of nuclear power and give at least one drawback for each type.
7. Name at least four types of renewable energy sources. What types of fuels might be produced from or by these sources?
8. Discuss the alternative means by which developing countries could provide enough food (calories) for their populations.
9. Discuss the alternative means by which developing countries could provide adequate protein in the diet of their people.

Selected Key Terms

zero population growth ('zē rō ,päp yə 'lā shən 'grōth) 691

depletion curve (di 'plē shən 'kərv) 696

cogeneration (,kō ,jen ə 'rā shən) 697

fission power plant ('fish ən 'paủ ər 'plant) 698

uranium mill tailings (yủ 'rā nē əm 'mil 'tā liŋz) 699

breeder fission ('brēd ər 'fish ən) 700

solar collector ('sō lər kə 'lek tər) 700

heliostat ('hē lē ə ,stat) 700

photovoltaic cell (,phō tō väl 'tā ik 'sel) 700

marasmus (mə 'raz məs) 704

kwashiorkor (,k wäsh ē 'ȯr kər) 704

As the human ecosystem becomes ever larger, it seriously encroaches on the space allotted to natural ecosystems and, at the same time, reduces the ability of the natural ecosystems to process and rid the biosphere of wastes. The wastes then accumulate and are called pollutants. These substances, added to the environment by human activities particularly, lead to undesirable effects for all living things. Human beings utilize and add pollutants to all parts of the biosphere—land, air, and water.

Land

Land Use

Human beings use land as a place to live and as a source of various renewable and nonrenewable resources. Figure 32.1 shows a categorization for land use in the United States. Federal and state governments have, thus far, set aside only a small percentage of land for parks, wilderness, and recreation areas. This is a matter of concern because only here, particularly in wilderness areas, is wildlife assured of being left undisturbed. A significant percentage of land in the United States is designated as forests, but for the most part, these lands are managed for the production of wood.

Urbanization

Cities comprise the category of land use that is most apt to increase in size and encroach on land now being used for other purposes. Whereas 20 percent of the world's population lived in cities in 1950, it is predicted that 50 percent will live in cities in 2000. New suburban areas associated with cities, a development often termed **urban sprawl,** tend to take over prime agricultural

Figure 32.1
Land usage in the United States, 1974.

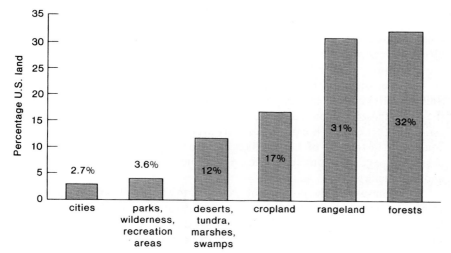

CHAPTER 32

Human Ecosystem and the Environment

Concepts

1. Human exploitation of land, air, and water resources has led to environmental problems.
   a. As land is taken over for urbanization and agriculture, natural areas decline in size. Salinization, erosion, desertification, and deforestation all reduce the quality of the land being used.
   b. Pollutants added to the air degrade the ozone shield, produce acid rain, affect the weather, and even cause indoor air pollution.
   c. Fresh water is being used maximally, and the removal of groundwater has led to subsidence, sinkholes, and saltwater intrusion. Sewage treatment plants are essential for keeping surface water clean; many organic industrial chemicals cannot be removed, however, and they pose a particular threat to the quality of water both above and below ground.

2. Human activities are causing the extinction of many different kinds of organisms.

3. In numerous ways, humans are beginning to learn how to work with nature for a better environment.

Figure 32.2
Much of the shoreline in the United States is now covered by urban sprawl. This threatens the productivity of the sea.

Figure 32.3
Huge tracts of land are disturbed each year due to strip mining.

land and wetlands after they have been drained and filled in (fig. 32.2). At one time, researchers estimate, the continental United States contained at least 127 million acres of wetlands. Today, only about 70 million acres remain, and many thousands of acres are still being lost not only to suburban and commercial development, but also to agriculture.

Cropland and Rangeland

Agriculture can use former wetland regions and also move into other areas once considered incapable of supporting it. When semidesert land is converted to agricultural land, the conversion requires many subsidies, especially irrigation. Such conversions may be temporary, however, not only because the supply of water for irrigation may be limiting, but also because of waterlogging and salinization. Water is usually channeled through furrows in a field to irrigate the land. As the water percolates down through the soil, the water table rises and may eventually reach the level of the roots. This condition, termed **waterlogging,** may cause plants to die because their roots can no longer acquire oxygen. Also, when the water table rises to within a meter or so of the surface, capillary action brings groundwater up to the surface. Both groundwater and newly applied irrigation water evaporate at the surface, causing salts to accumulate to the extent that crops can no longer be grown on the land. Salt-whitened soil is found in California, Colorado, Wyoming, and other heavily irrigated areas in the United States.

Another threat to agricultural lands and to forests, even forests located on federal lands, is surface mining (fig. 32.3) for coal and minerals. During **strip mining,** a common form of surface mining, the soil above the coal or mineral deposit is removed to expose and mine the resource. When these federal lands were first set aside as public lands, it was believed they could be used for many purposes—recreation, cattle grazing, and exploitation for resources. At that time, no one realized it would be possible to reduce an entire mountain to rubble by strip mining.

The percentage of coal obtained from strip mines has been rising steadily, increasing from about 9 percent to over 60 percent between 1940 and 1976. In 1977 a law was passed that requires strip-mining companies to reclaim the land as much as possible. Even with reclamation, however, agricultural land always is less productive, and it takes many years to reestablish a forest. In the meantime, wildlife habitats are disturbed.

Soil erosion is also causing productivity of agricultural lands to decline. **Soil erosion** occurs when wind blows and rain washes away the top soil to leave the land exposed without adequate cover. The U.S. Department of Agriculture estimates that erosion is causing a steady drop in the productivity of farmland equivalent to the loss of 1.25 million acres per year. The difference, as discussed on page 654, must be made up with more fertilizers, more pesticides, and more energy.

One answer to the problem of erosion is to adopt soil conservation measures. Unfortunately, at the present time many farmers are failing to use any form of soil conservation, and it is estimated that soil erosion is removing top soil at three times the rate at which it can be re-formed by natural means. However, there are both new and old measures that farmers can adopt. Some farmers have turned to no-tillage farming as a means to prevent soil erosion. In no-tillage farming the remains of the previous crop are left in place, and a disc is used to cut a furrow for planting new seeds. There is one drawback to this method: it requires the use of potent herbicides to prevent weed growth, and it may also require a more liberal application of pesticides, since pests thrive in the undisturbed soil and litter. For this reason, some experts prefer the older methods of preventing soil erosion, such as strip and contour farming (fig. 29.9).

The seriousness of the problem is made clear by some recent trends. People are abandoning marginal cropland and rangeland worldwide because of severe soil erosion and **desertification,** or the transformation of marginal lands to semidesert conditions. Desertification has been particularly evident along the southern edge of the Sahara in Africa, where it is estimated that 350,000 square miles of once-productive grazing land has become desert in the last fifty years. However, desertification also occurs in this country. The U.S. Bureau of Land Management, which opens up federal lands for grazing, reports that much of the rangeland it manages is in poor or bad condition, with much of its topsoil gone and with greatly reduced ability to support forage plants (fig. 32.4). As the reading on the following page explains, it is very difficult to keep a resource that is held in common from being overexploited.

Forests

Deforestation is also reducing the quality of land in the world today. This is not so great a problem in developed countries where forests are managed and a large proportion of natural hardwood forests were long ago converted to rapidly growing softwood forest plantations. However, in developing countries there can be serious problems. In tropical countries, according to a 1980 National Academy of Sciences report, in each week an area of tropical lowland forest about the size of Delaware is permanently converted to one of a variety of alternative uses, and in every year an area about the size of Great Britain is so converted. Some countries, such as Haiti, Thailand, Nigeria, and Madagascar, have already lost the majority of their forests. Others, such as Brazil, are expected to lose most of theirs by the twenty-first century.

Deforestation occurs for three main reasons. Local people are accustomed to practicing slash-and-burn agriculture. They cut down and burn the trees to provide space and inorganic nutrients to support crops. This type of farming could account for the loss of over 1 percent of tropical forests each year. The second reason for deforestation is the developed countries' demand for hardwood. Table 32.1 shows that at the present time, the majority of this wood is exported to Japan, Europe, and the United States. After the trees are cut for export, the land is sometimes used as farmland. Unfortunately, tropical soils are nutrient poor (p. 684) and typically sustain crops for only a couple of years. Severe soil erosion may eventually cause the land to become a barren wasteland. If the land is allowed to lie fallow, a secondary forest will grow back in about ten years (fig. 32.5), but it will not be as diverse as before. There is interest in establishing tree plantations on deforested land, but the capital to do so has been in short supply.

Figure 32.4
Appearance of (a.) properly grazed land and (b.) overgrazed land. Note that the properly grazed land contains mostly grass while the overgrazed land contains numerous shrubs.

a.   b.

Figure 32.5
An agroforestry experimental site in the rain forest of southern Venezuela. After the land was cut and burned, a forest succession area was established. In the background is undisturbed forest.

Table 32.1 Consumption of Tropical Hardwood Timber: Past, Present, and Projected (Million m³)

Country/Region	1960	1973	1980	1990	2000
Japan	1.5	28.9	35	38	48
United States	0.8	7.2	10	15	20
Europe	1.9	17.2	21	27	35
Total three importing regions	4.2	53.5	66	80	95
Tropical producing regions	21.0	46.5	66	117	185
Rest of world	1.0	9.0	13	18	23
Total	26.2	168.3	145	215	303

Source: FAO, Forest Resources in Asia and Far East, 1976; and S. L. Pringle, "Tropical Moist Forests in World Demand, Supply and Trade," Unasylva 23 (1976).

The Commons

The term "commons" originated in England and referred to any property or resource held in common by all citizens. Originally, it was a piece of land where all were allowed to graze their cattle.

It is not difficult to determine why a common resource is subject to abuse. The people using it are competing with one another, and therefore, there is a tendency for each to use as much of it as possible. If the common is for grazing cattle, for example, then the farmer who grazes ten cattle is better off than the farmer who grazes only nine. The difficulty is, of course, that eventually the resource is depleted and everyone loses. The rapid decline in the number of whales to the point of extinction is a modern example of overexploitation of a common resource.

The concept of the commons can also be applied to land, air, and water pollution. The industrialist who expends energy and funds not to pollute is at a disadvantage if competing companies are allowed to pollute freely. The commons can be protected only if citizens have the foresight to enact rules and regulations by which all abide. Each farmer, for example, should be allowed to graze only a certain number of cattle, depending on the number of farmers and the condition of the commons. The International Whaling Commission now sets quotas for each country, depending on the size of the whale stock.

While it may be possible for farmers grazing cattle to keep a sharp eye on one another, it is not always possible today for those using a resource to monitor each other. There is always a tendency for some to cheat, and therefore, it is necessary for a recognized authority to protect the resource and to have the right to prosecute those who do not obey the established rules and regulations. Unfortunately for whales, there is no enforcement agency. In the United States, the Environmental Protection Agency (EPA) has been charged with the task. Obviously, it is possible for the EPA either to vigorously carry out its mandate or to be lax in doing so. The attitude of the EPA can change, depending on the views of the current administration. Even so, citizens have the right to sue the EPA if it fails to uphold a law. Ultimately, then, each of us is responsible for the protection of the environment and the quality of the commons.

Finally, ranchers clear away tropical forests to make room for cattle raising to provide beef, not for the home country, but for the United States fast-food chains. Because this imported beef is cheaper, hamburgers can be sold for five cents less than if the beef was purchased in the United States. The destruction of tropical forests on this account has been termed the **"hamburger connection."**

A serious side effect to deforestation in tropical countries is the loss of biological diversity. Roughly half of the world's species are believed to live in tropical forests. The National Academy of Sciences estimates that a million species of plants and animals could disappear within twenty years from deforestation in tropical countries. Science has not even studied many of these life forms as yet.

To prevent such a loss, Dr. Ira Rubinoff, director of the Smithsonian Tropical Research Institute of Panama, has proposed a plan to preserve a portion of the world's rain forest. He calls for setting aside one thousand reserves of 250,000 acres each in the forty-nine nations that have most of the tropical forests. About forty developed nations in the temperate zone would finance this system, and an international organization would be in charge of the project's administration.

Land Pollution

Every year, the U.S. population discards billions of tons of solid wastes, much of it on land. **Solid wastes** include not only household trash but also sewage sludge, agricultural residues, mining refuse, and industrial wastes. Those solid wastes containing substances that cause human illness and sometimes even death are called **hazardous wastes.**

Household Trash

In 1920 the per capita production of waste was about 2.75 pounds per day; in 1970 about 5 pounds per day; and in 1980 about 8 pounds per day. The exponential growth of consumer products with "planned obsolescence" and the use of packaging materials to gain a competitive edge account for much of this increase.

Open dumping (fig. 32.6), sanitary landfills, or incineration have been the most common practices of disposing trash. These disposal methods have become increasingly expensive and also cause pollution problems. It would be far more satisfactory to recycle materials as much as possible and/or to use organic substances as a fuel to generate electricity. One study showed that it was possible to achieve 70 to 90 percent public participation in recycling by spending only thirty cents per household. A city the size of Washington, D.C., where 500,000 tons of waste are generated each year, could have an increase of 1,300 jobs if the community utilized solid wastes as a resource instead of throwing them away.

Hazardous Wastes

The U.S. Environmental Protection Agency, charged with the task of keeping the environment safe, estimates that in 1980 at least 57 million metric tons of the nation's total wasteload could be classified as hazardous. Hazardous wastes fall into three general categories.

1. **Heavy metals,** such as lead, mercury, cadmium, nickel, and beryllium, can accumulate in various organs, interfering with normal enzymatic actions and causing illness—including cancer.

2. **Chlorinated hydrocarbons,** also called **organochlorides,** include pesticides and numerous organic compounds, such as PCBs (polychlorinated biphenyls), in which chlorine atoms have replaced hydrogen atoms. Research has often found organochlorides to be cancer-producing in laboratory animals.

3. **Nuclear wastes** include radioactive elements that will be dangerous for thousands of years. ^{239}Plutonium must be isolated from the biosphere for 200,000 years before it loses its radioactivity.

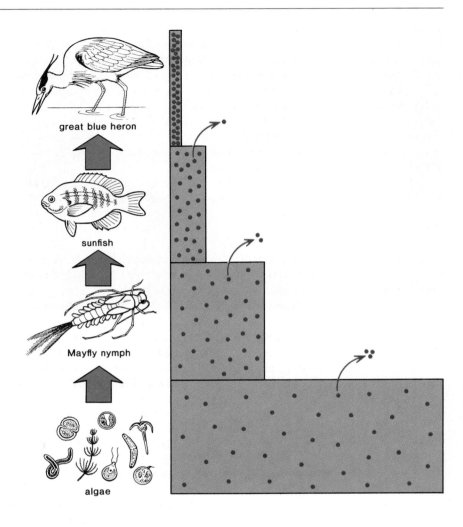

great blue heron

sunfish

Mayfly nymph

algae

Hazardous wastes are often subject to **bioaccumulation** (fig. 32.7). Bacteria in the soil do not break down these wastes, and when other organisms take them up, they remain in the body and are not excreted. Once they enter a food chain, they become more concentrated at each trophic level. Notice in figure 32.7 that the number of dots representing DDT become more concentrated as the chemical is passed along from producer to tertiary consumer. Bioaccumulation is most apt to occur in aquatic food chains; there are more trophic levels in aquatic food chains than there are in terrestrial food chains. Humans are the final consumers in both types of food chains, and in some areas, mothers' milk contains a detectable amount of DDT and PCB.

The public has become aware of hazardous dump sites (fig. 32.6) that have polluted nearby water supplies. Chemical wastes buried over a quarter century ago in Love Canal, near Niagara Falls, have seriously damaged the health of some residents there. In other places manufacturers have left thousands of drums in abandoned or uncontrolled sites where toxic chemicals are oozing out into the ground.

In an effort to prevent such occurrences in the future, Congress passed the Resource Conservation and Recovery Act of 1976, which empowers the EPA to track all significant quantities of hazardous waste from wherever it is generated to its final disposal. Stricter government regulations have encouraged industry to adopt new waste management strategies, including *reduction* (changing a manufacturing process so it does not produce hazardous by-products); *recycling* (reusing waste material); and *resource recovery* (extracting valuable material from waste). For example, in 1980 EPA regulations forced the Reliable Plating Company of Milwaukee, Wisconsin, to purchase a $30,000 ion-transfer system for chromium recovery in the company's nickel/chrome-plating operation. Although initially considered a regulatory burden, the system in its first eight months reduced by 90 percent the amount of chrome the company had to buy and cut water use from 14,000 to 200 gallons a day.

Air

The atmosphere can be divided into layers according to changes in temperature, as indicated in figure 32.8. The **troposphere,** which shows a gradual decrease in temperature, is that part of the atmosphere that concerns humans in day-to-day life. Only nitrogen, which makes up 78 percent of the gases in the troposphere, and oxygen, which makes up 21 percent, are present in appreciable amounts. The amount of carbon dioxide is quite small—only about 0.03 percent.

The temperature of the **stratosphere,** the next atmospheric layer, slowly increases. This is because the ozone layer absorbs ultraviolet radiation. **Ozone** (O_3) is formed in the following manner.

molecular oxygen   atomic oxygen
$$O_2 + energy \longrightarrow O + O$$
$$O + O_2 \longrightarrow O_3$$

The energy of the sun splits oxygen molecules. These individual oxygen atoms then combine with other oxygen molecules to give ozone. The layer of ozone in the stratosphere is called an **ozone shield.** Ozone absorbs the ultraviolet rays of the sun so they do not strike the earth. If they did penetrate the atmosphere, life on land would not be possible because living things cannot tolerate heavy doses of ultraviolet radiation. (Life in water would not be affected.) Even so, ultraviolet rays are implicated in some 200,000 to 600,000 cases of skin cancer per year in the United States.

The release of chlorine atoms into the stratosphere breaks down the ozone shield. A major source of chlorine is from chlorofluorocarbons (CFCs), such as freon, used in other countries as a propellant in aerosol cans. Governmental regulations banned the use of CFCs in aerosols in 1977, but they are still widely used as refrigerants and as industrial foaming agents. A 1982 Academy of Sciences report predicts a depletion of the ozone shield between 5 and 9 percent if CFC production continues at the 1977 rate. It is estimated that for every 1 percent decrease in ozone concentration, there could be a 2 to 10 percent increase in skin cancer.

Besides CFCs, carbon tetrachloride (CCl_4) is also a significant source of free chlorine. In addition, *nitrogen oxides* can break down ozone in a manner similar to chlorine. Production of the SST (supersonic transport plane) was halted in part because its exhaust would have put nitrogen oxides into the stratosphere. But automobile exhausts and nitrogen fertilizers are sources of nitrogen oxides that eventually could reach the stratosphere.

Figure 32.8
A graph showing variations in atmospheric temperature and ozone level in relation to altitude. Notice that as the ozone level decreases, the temperature increases, and vice versa. This is because ozone absorbs ultraviolet radiation.

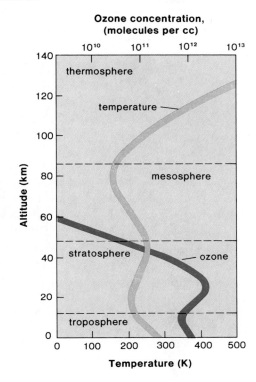

Figure 32.9
Common sources of air pollution in most cities.

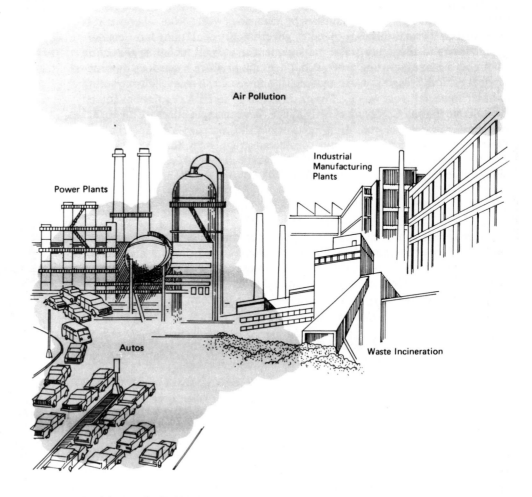

Ambient Air Pollution

Figure 32.9 shows the sources of air pollution due to the burning of fossil fuels. This combustion results in five categories of primary pollutants: carbon monoxide (CO), nitrogen oxides (NO, NO_2), sulfur oxides (SO_2, SO_3), hydrocarbons (HC), and particulates. It is obvious from figure 32.10 that modes of transportation, especially the automobile, are the main cause of carbon monoxide air pollution. This chemical combines preferentially with hemoglobin to prevent circulation of oxygen within the body, causing unconsciousness. In New York City traffic, the blood concentration of CO has been shown to reach 5.8 percent, a dangerous level when compared to the 1.5 percent that physicians consider safe. The particulates, dust and soot, can collect in the lungs, and nitrogen oxides and sulfur oxides irritate the respiratory tract. The hydrocarbons (various organic compounds) may be carcinogenic.

Unfortunately, these primary pollutants interact with one another, producing pollutants that are even more dangerous to animal and plant life.

Photochemical Smog

Photochemical smog results when two pollutants from automobile exhaust—nitrogen oxide and hydrocarbons—react with one another in the presence of sunlight to produce nitrogen dioxide (NO_2), ozone (O_3), and PAN (peroxylacetyl nitrate). **Ozone** and **PAN** are commonly referred to as oxidants.

Figure 32.10

Air pollution contains five major components: CO (carbon monoxide), HC (hydrocarbons), NO$_x$ (nitrogen oxides), particulates (solid matter) and SO$_x$ (sulfur oxides). Transportation contributes most to air pollution.

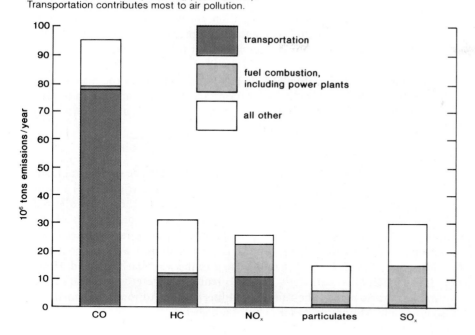

Figure 32.11

The milkweed in (a.) was exposed to ozone and appears unhealthy; the milkweed in (b.) was grown in an enclosure with filtered air and appears healthy.

a.

b.

Breathing ozone affects the respiratory and nervous systems, resulting in respiratory distress, headache, and exhaustion. These symptoms are particularly apt to appear in youngsters; therefore, Los Angeles schoolchildren must remain at rest inside the school building whenever the ozone level reaches 0.35 ppm (parts per million by weight). PAN is especially damaging to plants, resulting in leaf mottling and reduced growth (fig. 32.11).

Normally, warm air near the ground rises, so that pollutants are dispersed and carried away by air currents. Sometimes, however, air pollutants, including smog and particulates, are trapped near the earth due to **thermal inversions.** During a temperature inversion, the cold air is at ground level and the warm air is above. (This may occur when a cold front brings in cold air that settles beneath a warm layer.) Some areas surrounded by hills are particularly susceptible to the effects of a temperature inversion.

Any effect that greatly increases the level of air pollution can lead to disaster. This happened in New York City in 1963 when 200 to 400 people died and in 1966 when 168 people died. Even worse were the events in London in 1962 when 700 people died and in 1957 when 700 to 800 people died due to the effects of air pollution.

Other Pollutants

Fine carbon particles (less than 0.002 mm) caused by fossil fuel combustion are involved in air pollution. They remain suspended in the air where they absorb other pollutants, such as heavy metals (e.g., lead, hydrocarbons, sulfur, and nitrogen oxides). These particles then enter the lungs where they can increase the chances of lung cancer, emphysema, and other respiratory diseases. Diesel automobiles and trucks emit between 30 and 100 times more particulates than other vehicles, and more than 95 percent of these are fine particles.

Figure 32.12
High smokestacks disperse pollutants into the
atmosphere, causing pollution problems, particularly acid
rain, some distance away.

Happily, however, **lead** used in gasoline production has decreased from 190,000 tons in 1976 to 90,000 tons in 1980 because lead interferes with the operation of the catalytic converter, a device added to an automobile to reduce the emission of hydrocarbons. The Center for Disease Control reports that as a result of the change in gasoline composition, the blood level of lead has been reduced by an average of 36.7 percent in a representative U.S. population sample. The center believes these data show that lead should not be added to gasoline.

Asbestos is also a particulate air pollutant that enters the atmosphere from asbestos mining and milling operations and from the manufacture, disposal, and use of asbestos-containing products, such as insulation. Asbestos fibers cause a rare form of lung cancer.

Acid Rain

A side effect of air pollution is now the common occurrence of acid rain. Industrial and electrical plants built very tall smokestacks (fig. 32.12) to improve air quality in the immediate vicinity. Unfortunately, these have allowed wind currents to carry fine particles of SO_2 and NO_x to other parts of the country, where they eventually fall to the ground or where rain washes them out of the air. The pH of natural rainfall is usually slightly acidic, about 5.6, because water reacts with carbon dioxide to give the weak carbonic acid. However, fine particles that carry SO_2 and NO_x react with water to give the strong sulfuric and nitric acids. The rain now falling on the Adirondacks, northern Wisconsin, and Minnesota has a pH of 4.6 or below.

Acid rain leaches compounds out of the soil. Most affected regions in North America are not cultivated regions but areas where forests are the only possible crop. Research suggests that aluminum ions leached from the soil kill some of the coniferous trees' feeder roots while, at the same time, impairing the trees' ability to take up an adequate amount of water and nutrients.

Acid rain has also caused mountainous lakes to become crystal clear and devoid of life except for filamentous algae and fungi growing on the bottom. When a lake is first subjected to acid rain, there is no change because the acid is being buffered. But when the buffering capacity of the lake is depleted, there is a sudden downward shift in pH accompanied by an increase in dissolved metals, particularly aluminum. Aluminum ions irritate fishes' gills, causing the gills to produce a protective mucus. This initiates a process that physically erodes the gill filaments until the fish suffocate.

Acid rain also corrodes marble, metal, and stonework (fig. 32.13), an effect that is noticeable in cities. It can also leach certain elements from the soil into drinking water supplies. Water with a low pH may dissolve copper from pipes and the lead solder that is used to join pipes together.

The burning of coal supplies most of the sulfur oxides that lead to acid rain in this country. There are several ways to minimize the problem. Coal could be washed prior to burning. Low-sulfur coal could be substituted for high-sulfur coal. Devices called scrubbers could be installed in the tall stacks

a.

b.

to prevent sulfur oxide from entering the air. Now under development is a new method of burning coal that uses a mixture of coal and limestone. This technique has reduced both sulfur and nitrogen oxide emissions.

People are also addressing the problem of NO_x pollution. Continued use of the catalytic converter in automobiles is important because it reduces by about three-fourths the amount of NO_x given off. A low NO_x burner is also being developed for industrial boilers.

Weather

Scientists also believe that air pollution is affecting the weather. Some predict an eventual rise in temperature due to the **greenhouse effect,** while others predict a lowering of temperature, at least initially, due to the **refrigerator effect.**

Figure 32.14

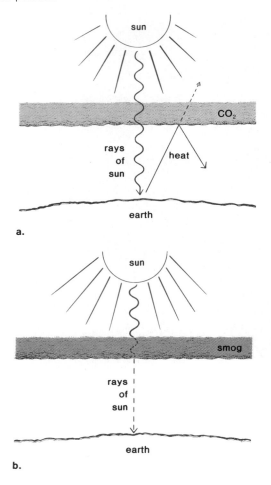

a. If the greenhouse effect holds, the earth will get warmer because heat from the sun's rays will be trapped beneath a blanket of carbon dioxide, just as heat is trapped beneath the glass of a greenhouse. *b.* If the refrigerator effect holds, the earth will get colder because the sun's rays will be unable to penetrate the particles produced by air pollution.

Greenhouse Effect

Despite its low atmospheric concentration, CO_2 plays a critical role in regulating the earth's surface temperature. CO_2 allows the sun's rays to pass through the atmosphere but absorbs and reradiates their heat back toward earth (fig. 32.14*a*). This can be compared to a greenhouse in that the glass of a greenhouse allows sunlight to pass through, then traps the resulting heat inside the structure. Since 1958 the atmospheric concentration of CO_2 has risen almost 6 percent (p. 114) and is expected to continue to increase as people burn more fossil fuels. If the present trend continues, a doubling of atmospheric CO_2 could occur sometime toward the middle of the next century. The National Academy of Sciences predicts this would cause an average annual temperature rise of 3° to 8° C.

This rise in temperature from the greenhouse effect could have a disastrous impact on agriculture and eventually on sea levels because of melting polar ice. Wind pattern changes would make the midwestern United States, the Soviet Union, and China drier. Favorable climates for the growth of crops would move north, where the soil is not so favorable for agriculture. Even though it is not certain that polar ice would melt, the sea level would rise simply because water expands when it absorbs heat. If the arctic ice should melt substantially, it is predicted that most of the world's cities would be flooded as would some of the richest farmlands. Columbia University researchers recently reported that they have observed a 35 percent decrease in the average summer area of Antarctic pack ice between 1973 and 1980.

Refrigerator Effect

Some authorities predict the world's average temperature will lower instead of rise because suspended particles in photochemical smog, aircraft exhaust, smoke from slash-and-burn agriculture, and soil erosion prevent the sun's rays from reaching the earth in the first place (fig. 32.14*b*). Most researchers believe, however, that eventually the CO_2 effect will overcome the refrigerator effect, and the average annual temperature will begin to rise.

Indoor Air Pollution

While much attention has been given to the quality of outdoor air, it has been pointed out that indoor air is often more polluted. NO_2 traced to gas combustion in stoves has been found indoors at twice the outdoor level; CO in offices, garages, and hockey rinks is routinely in excess of the EPA standard; hydrocarbons from myriad sources appear in high concentrations; and radioactive radon gas, emitted naturally from a variety of building materials, has been detected indoors at levels that exceed outdoors by factors of 2 to 20.

Because of these findings, some have questioned the advisability of tightening up residential buildings to save energy. Reduced ventilation unquestionably increases concentrations of pollutants. The EPA is now recommending that ventilation rates not be reduced below one complete air change per hour.

Water

Fresh water is required not only for domestic purposes, including drinking water, but also for crop irrigation, industrial use (both manufacturing and mineral extraction and processes), and energy production. Surface water from rivers and lakes, and groundwater from aquifers are used to meet these needs. Approximately one-half of the U.S. population depend on groundwater as their primary source of drinking water. Overall, 25 percent of the water used for all purposes comes from under the ground, including 40 percent of the water used for irrigation.

Figure 32.15
When groundwater is removed, subsidence may occur.
This leads to damaged buildings, as shown here.

Water Supply

Groundwater

The water in underground aquifers is a vast natural resource having a volume that is about 50 times the annual flow of surface water. But this water is like money in the bank; to ensure a continual and lasting supply, withdrawals cannot exceed deposits. Aquifers are recharged when rainfall and melted snow percolate into the soil. In most parts of the United States there is adequate precipitation to recharge aquifers as long as enough natural regions, particularly forests and wetlands, remain to allow this to happen. In some parts, especially the Texas high plains, central Arizona, and southern Florida, withdrawals from aquifers exceed any possibility of recharge. This is called **"groundwater mining."** In these locations the water table is dropping 7 to 8 feet per year, and residents may run out of groundwater, at least for irrigation purposes, in the next fifteen years. In the meantime, removal of water is causing land **subsidence,** a settling of the soil as it dries out. Subsidence damages canals, buildings (fig. 32.15), and underground pipes. Between June 1975 and September 1976, the Bureau of Reclamation spent $3.7 million to rehabilitate federal irrigation projects damaged by subsidence.

Withdrawal of groundwater can cause the formation of **sinkholes,** formed when underground caverns collapse because water no longer helps support the roof of the cavern. In coastal regions freshwater supply is further endangered by **saltwater intrusion.** The outflow of water from streams and aquifers usually

Figure 32.16
When groundwater is removed, the water table (height of underground water) sinks, and saltwater intrusion occurs. *a.* Normal conditions. *b.* Saltwater intrusion.

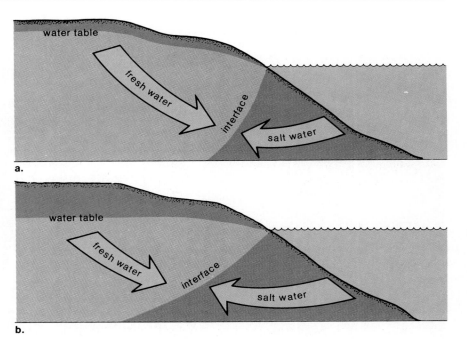

keeps them free of seawater. But as water is withdrawn, the water table can lower to the point that seawater backs up into streams and aquifers. Saltwater intrusion then reduces the supply of fresh water along the coast (fig. 32.16).

Unfortunately, areas of the country already mining groundwater are also experiencing rapid population growth. In the Tucson, Arizona, area, where the water level has dropped 110 feet in the last ten years, the population is expected to grow from 450,000 to 652,000 by 1990. Local agriculture and mining are threatened as the city requires more and more water.

Water Reservoirs

To prevent the mining of water and/or to provide a continuous supply of surface water, people in an area sometimes build a dam. The large body of water behind the dam is a type of reservoir. Dams prevent flooding during rainy seasons and store water for use during dry spells. Irrigation canals can lead from the reservoir to the countryside to provide water for agriculture. The Aswan Dam in Egypt allows farmers to harvest two, and sometimes three, crops a year instead of only one. In addition, this dam, like others, provides hydroelectric power. In this country aqueducts and reservoirs also supply drinking water to far-removed places. Los Angeles, for example, is provided with water from northern California rivers in this way.

While reservoirs have these advantages, they also have disadvantages. In arid regions 10 to 25 percent of the water held in reservoirs evaporates. Eventually, in all regions, reservoirs become unusable because they fill up with accumulated silt. A dam can also burst, causing more severe flooding than it is designed to prevent. Perhaps the greatest disadvantage of building a dam is that the ecosystems below the dam do not receive the same amount of water and silt as before.

Tampering with the water supply can affect ecosystems. The Everglades National Park (p. 30), located in southwestern Florida, is threatened because dikes and canals were built to take water from Lake Okeechobee to the

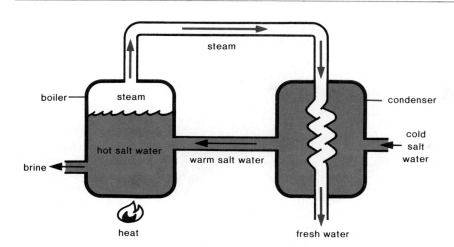

boiler
steam
steam
hot salt water
brine
heat

condenser
cold salt water
warm salt water
fresh water

Figure 32.17
Fresh water can be produced from salt water by the distillation process shown here. Notice how this process mimics the natural process by which the oceans produce fresh water (fig. 30.1).

cities along the east coast of Florida. The Everglades are a huge, extensive marsh that contains islands covered by a variety of birds, alligators, white-tail deer, bobcats, otters, black bears, and even cougars. Formerly the lake flooded in the summer, causing the glades to have a summer wet season and a winter dry season. Now the glades receive no water except by way of the canals, and it has been difficult to supply them with normal amounts of water—sometimes they receive too little and sometimes they receive too much. Without the normal fluctuation in water level, the plant life will change, and some of the animals will die because their breeding habits are attuned to the change in seasons. Whether the glades will survive has become a serious question, especially since they are also threatened by residential and industrial expansion.

Irrigation

Over 40 percent of all fresh water in the United States is used to irrigate crops grown in dry areas where sunlight is plentiful but rain water is in short supply. The San Joaquin Valley, only a small portion of California, produces more farm products than most states, but this abundance depends entirely on irrigation. The need for irrigation not only leads to the building of dams and aqueducts, the mining of underground water, and salinization, but also causes rivers to become more salty. This occurs because plants extract almost pure water from their water supply; therefore, the water that trickles down has a higher salt content than when it was first applied. As this water returns to rivers, they become increasingly salty. The Colorado River, which flows from southwestern United States into Mexico, becomes so salty that the United States has agreed to reduce the amount of salt in the river before it reaches Mexico. It will be necessary to build a desalinization plant, which will cost $12 million per year to operate. **Desalinization** (fig. 32.17) is expensive because the process is energy intensive. For this reason, desalinization is not expected to add greatly to the world's supply of fresh water at this time.

Farmers have little incentive to practice water conservation at the present time. The government subsidizes the cost of irrigation and even gives tax credits according to the amount of water a farmer uses. But this may change as the problem becomes more obvious and more costly. The problems associated with irrigation are decreased if less water is used, which makes water conservation a highly desirable goal. Drip irrigation (fig. 32.18), in which conduits deliver water to each plant root by means of small holes, is already used by some farmers to conserve water.

Figure 32.18
Drip irrigation conserves water because it is delivered directly to the roots of plants.

Figure 32.19
Human beings bring about water pollution in the many ways shown here.

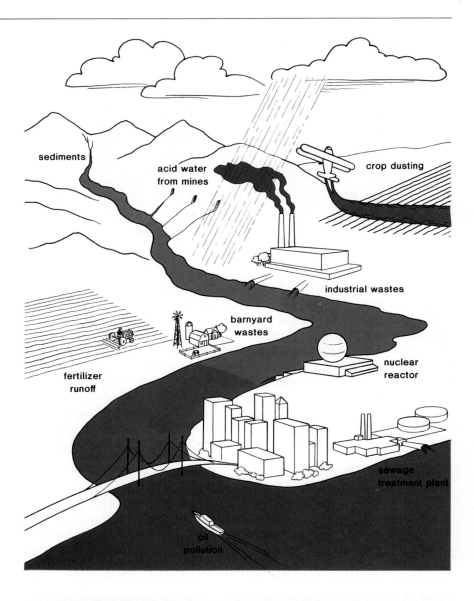

Table 32.2 Sources of Water Pollution

Leading to Cultural Eutrophication

Oxygen-demanding wastes	Biodegradable organic compounds (e.g., sewage, wastes from food processing plants, paper mills, and tanneries)
Plant nutrients	Nitrates and phosphates from detergents, fertilizers, and sewage treatment plants
Sediments	Enriched soil in water due to soil erosion
Thermal discharges	Heated water from steam-electric power plants

Health Hazards

Disease-causing agents	Bacteria and viruses from sewage (e.g., food poisoning and hepatitis)
Synthetic organic compounds	Pesticides, industrial chemicals
Inorganic chemicals and minerals	Acids from mines and air pollution; dissolved salts; heavy metals (e.g., mercury) from industry
Radioactive substances	From nuclear power plants, medical and research facilities, and nuclear weapons testing

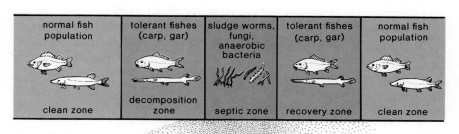

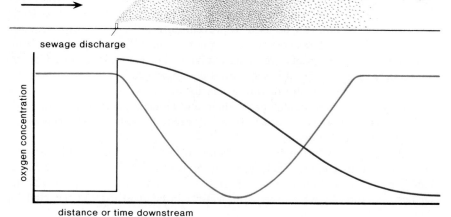

Figure 32.20
Oxygen sag curve. When raw sewage is added to a river, the water is deprived of oxygen and the normal distribution of animals is affected in the manner shown here. Notice that a plentiful amount of oxygen exists in the clean zones. After sewage is discharged, the oxygen level drops in the decomposition zone because the decay bacteria use oxygen when they break down the sewage. The septic zone contains almost no oxygen, and mostly anaerobic bacteria are at work. Once the sewage has been decomposed, the oxygen level again rises in the recovery zone.

Water Pollution

Surface Water

All sorts of pollutants enter surface waters, as depicted in figure 32.19 and listed in table 32.2. If sewage and like substances enter a river directly, biological diversity of the river is apt to suffer because bacteria and fungi of decay use up oxygen while they decompose the material. Such substances are said to have a high **BOD** (biological oxygen demand). Downstream from the point of discharge, the amount of oxygen decreases, giving a characteristic oxygen sag curve. Figure 32.20 illustrates this curve and also shows the effects upon the natural ecosystem. The portion of the river before the discharge contains a rich benthos (fig. 30.3) and a variety of fishes, including perhaps trout, pike, or bass. As oxygen level decreases, only certain fishes—carp, catfish, or gar—are present. When the oxygen level approaches zero, sludge worms and blood-worms are typical of the only forms of life that can survive. As long as minimal oxygen is available, aerobic breakdown will occur; but should oxygen disappear entirely, anaerobic breakdown gives off the odorous chemicals methane, ammonia, and hydrogen sulfide. As recovery takes place, the warm water fishes and then the cold water fishes return.

To keep high BOD organic materials from polluting rivers, municipalities have built sewage treatment plants. A **primary treatment plant** removes solids, grease, and scum before the effluent (discharge) is chlorinated to kill any harmful bacteria. A **secondary treatment plant** uses specially provided decomposers to break down any remaining organic material after primary treatment before the effluent is chlorinated and discharged. This means that instead of organic wastes, inorganic nutrients like nitrates and phosphates are added to the water. However, these nutrients also lead to water pollution, and they hasten the process of *eutrophication* that accompanies succession, or the steps by which a pond or lake becomes filled in and disappears (fig. 30.4). Therefore, these nutrients are said to contribute to *cultural eutrophication,* that portion of eutrophication caused by human activities. What happens is this: the nutrients cause algae to grow in abundance and sometimes produce algal bloom, apparent when green scum floats on the water or when there are excessive mats of filamentous algae. The death of these algae promotes the growth of a very large decomposer population. The decomposers break down the algae, but in so doing, use up oxygen. In addition, algae also consume oxygen during the night when photosynthesis is impossible. Both of these cause a decreased amount of oxygen available to fishes, ultimately causing the fishes to die. The increased amount of life and their dead remains cause the lake to be more eutrophic. A **tertiary treatment plant,** which removes nutrient molecules to prevent algal bloom, can be used to prevent eutrophication. However, since tertiary treatment plants cost twice as much as secondary plants, other ways to remove excess nutrients are being explored. For example, as described in the reading on page 656, it is sometimes possible to pass the water through a swampy area before it drains into a nearby waterway. Or the water can be used to irrigate crops and/or grow algae and aquatic plants in a man-made shallow pond. Since the latter can be used as food for animals, this represents a cyclical use of chemicals.

Municipalities often find it difficult to provide enough funds for adequate sewage treatment. Therefore, any process that eliminates the need to send sewage to local plants is highly desirable. The reading on page 727 describes a method by which the same water can be cleansed and recycled for reuse in toilets.

The building of modern sewage treatment plants has immensely helped clean up U.S. waters. In addition, large industries are no longer permitted to dump chemicals indiscriminately into water or to send their wastes to local sewage treatment plants. Therefore, many have built their own waste water treatment facilities. This two-pronged attack has helped clean up Lake Erie, which previously was said to be dying because of pollution. Even so, the presence of toxic waste is still a problem, especially since public sewage treatment plants are not designed to rid water of nonbiodegradable wastes.

Groundwater

Ordinarily, one would expect underground water to be free of pollutants because bacteria and fungi of decay found in the soil can remove most conventional contaminants before water reaches an aquifer. But it has been found that underground water is sometimes polluted with nonbiodegradable pollutants, such as organochlorides and heavy metals, and also with inorganic nitrates and chlorides. Figure 32.21 shows the ways in which pollutants can reach aquifers. Previously, industry was accustomed to running wastewater into a pit. The pollutants could then seep into the ground. Also, pollutants were injected into deep wells from which the pollutants constantly discharged. Both of these customs have been or are in the process of being phased out. However, it is very difficult for industry to find any other way to dispose of wastes. More adequately managed and controlled waste treatment plants are needed, but because citizens do not wish to live near waste treatment plants, towns are often successful in preventing their construction. In the meantime, industries are still employing less approved methods of dealing with industrial wastes.

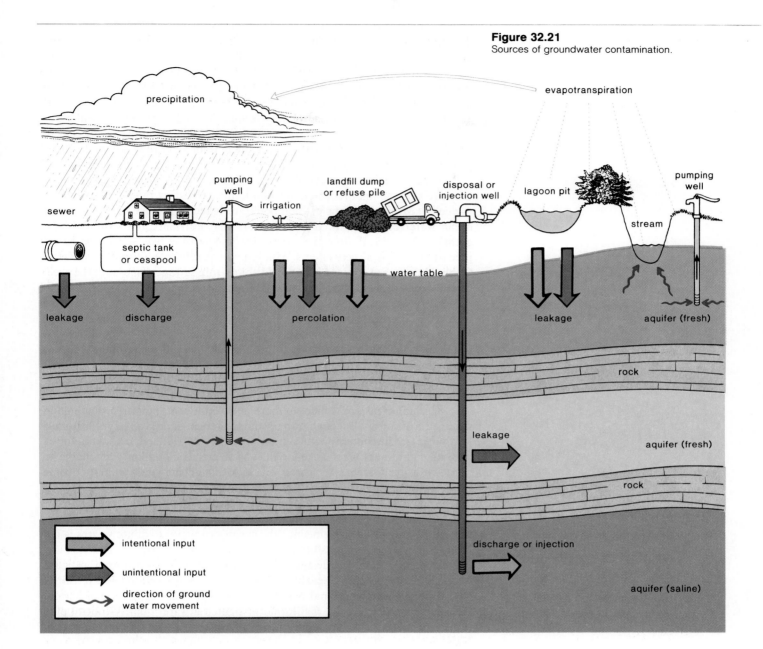

Figure 32.21
Sources of groundwater contamination.

Oceans

The coasts (p. 673), including estuaries, are the most productive portions of the marine ecosystem, but human activities severely threaten this productivity. One-third of the U.S. population now lives in cities located on the coasts, and development in these areas still continues. In numerous areas, such as along the southern Florida coast, coastal regions are bulldozed and then filled in with dredged sediments to create land for housing developments. Extensive dredging is destroying marine habitats in this country and in all parts of the globe.

Much of the world's industry is also located on the coasts because marine transportation is the cheapest, most energy-efficient mode of transporting materials from one location to another. On the Persian Gulf, Saudi Arabia is now building a new ultramodern city termed the Jubail project. It will cover an area as large as London and contain a population as numerous as that of Minneapolis. Its purpose is entirely for the development of a new petrochemical and manufacturing center.

Coastal regions are not only the immediate receptors for local pollutants, they are also the final receptors for pollutants carried by rivers that empty at the coast. Millions of tons of each pollutant listed in table 32.2 are added to ocean water each month. It now appears that nutrients and sediments, together with physical alteration from dredging and the like, will cause eutrophication of coastal ecosystems. Much of the coral reefs along the shores of Oahu, Hawaii, are covered by a putrid slime from sewage disposal in the area. On this continent, marshes are filling in with sediments and disappearing.

Toxic wastes, including heavy metals, organochlorides, and radioactive material, concentrate in marine food chains just as they do in freshwater chains. In the Baltic Sea only a few thousand gray seals remain of a population estimated at 20,000 in 1940—due to their intake of PCB-contaminated fishes. Large amounts of metal-laden wastes have accumulated in Raritan Bay, near New York, forcing termination of shellfish harvesting and reducing the yield of commercial fishing.

Coastal areas are subject to more petroleum contamination than inland waters. Large oil spills from blowouts and tanker collisions can have disastrous local effects. Oil contaminants remain in estuarine wetland sediments for as long as eight years after an oil spill, and a marsh community is unable to reestablish itself even after three years. Although it is less dramatic, some believe that routine discharges from everyday use of oil and gas represent a longer and even greater threat to the marine environment.

The buildup of wastes in the oceans could eventually lead to the destruction of the marine ecosystem since the ocean's capacity to dilute wastes is not infinite. However, it is possible to reverse the present trend.

For example, a decade ago, off the coast of southern California, pelicans were unable to breed on nearby islands, beds of giant kelp off Palos Verdes had disappeared, and animal populations in large areas of the ocean bottom were substantially altered. California instigated a water quality control plan that set limits on the type and amount of wastes that could be discharged. Now pelicans and kelp are back, and in most areas marine life seems to be in good condition.

Most cities and towns take water from a nearby source and chlorinate it or kill bacteria before it is piped to offices and homes. Used water, including drainage and sewage water, is piped to a sewage treatment plant where primary treatment removes most solids by settling, and secondary treatment, if available, removes all solids by additional bacterial digestion. Both of these sewage treatment methods release nutrients that lead to algal bloom and eutrophication. Tertiary treatment of sewage involves the removal of nutrient molecules but is very costly.

A method of sewage treatment, even if costly, that did not rely on a constant inflow of fresh water would be desirable because there are so many competing uses for fresh water. A new method that uses the same water repeatedly has been developed and is illustrated in the accompanying figure. This system processes wastes in four main stages. Sewage from toilets and wastewater from other fixtures is transferred to a chamber where denitrifying bacteria consume organic compounds using nitrate as their source of oxygen. The gaseous end products, mainly nitrogen and carbon dioxide, are vented to the atmosphere. The wastewater then flows into an aerobic digestion chamber. Here aerobic bacteria continue the digestion process and convert ammonia from urine into nitrate. Thus, the second chamber replaces nitrate that is needed as an oxygen source in the first chamber. The water is next forced through ultrafine tubular membranes that filter out bacteria and any remaining solids. Finally, the water is treated with activated carbon to improve the color and odor and with ozone to disinfect it. The water is then piped back to the toilets.

Sewage Treatment

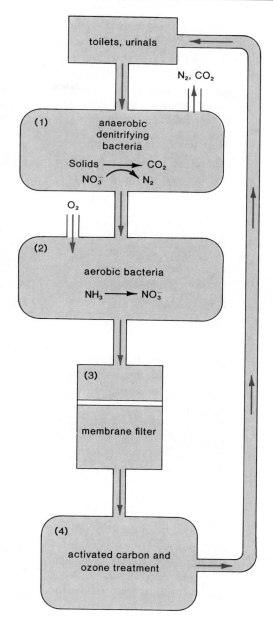

Figure 32.22
Endangered species. *a.* Bald eagles are plentiful in
Alaska, but are endangered in the contiguous states
because of pesticides, illegal shooting, and loss of habitat.
Since the banning of DDT in 1972, eagles have started to
recover. *b.* Yellow-fringed orchid, a species that is
endangered in nine states. *c.* Giant pandas survive in the
wild only in remote regions of central China. Whenever
there is a shortage of their favorite food, bamboo (being
eaten here), the giant pandas are particularly endangered.

a.

b.

c.

Biological Diversity

The continued growth of the human ecosystem with its accompanying pollution of land, air, and water threatens to bring about the extinction of many species. Until modern times more species evolved than became extinct, but now the opposite is the case, and the rate of extinction is increasing dramatically. If the current trend continues, some 20 percent of the species on earth could become extinct by 2000. Many of the plants and animals threatened with extinction are well-known organisms (fig. 32.22).

Reasons for Extinction

There are numerous reasons why the extinction rate is now increasing. We will discuss only a few of them.

Habitat Destruction

Habitat destruction is by far the biggest single cause of extinction. The two primary parts of the human ecosystem, cities and farms, both displace natural ecosystems. Human beings depend on only some twenty types of plants and even fewer types of animals to provide them with food. To make matters worse, farmers often plant exactly the same variety of corn, wheat, or rice because this variety is the one that currently produces the greatest yield. However, this custom is causing other varieties to become extinct, even ones that possess characteristics that we may need in the future to improve currently employed varieties.

Of primary concern of late is the rapid destruction of tropical rain forests to provide space for farming and animal husbandry. Most extinctions expected by 2000 will be due to rain forest destruction. Yet rain forests may serve as a storehouse of diversity from which plants and animals have migrated to populate other regions of the earth.

To prevent further extinctions because of habitat destruction, there should be a worldwide effort to preserve large areas of representative biomes, especially those rich in life forms. Such an effort has begun under the auspices of UNESCO. As of mid-1978, UNESCO has officially recognized 144 areas in thirty-five countries as part of its global network of Biosphere Reserves. This is only a beginning; even more and larger tracts of land should be set aside since the continuance of a biome requires many thousands of acres. Only then will all the various types of organisms, from the smallest to the largest, find enough room to stay in existence.

As noted in figure 32.1, the United States has thus far designated only 3 percent of its land area for parks and wilderness areas. The parks are such favorite recreational areas (even allowing the use of off-road vehicles [ORVs], which can degrade an acre of land in a 20-mile drive) that only the wilderness areas can be counted on to help preserve plants and animals.

Hunting and Collecting

Humans hunt and collect organisms for both pleasure and profit. Wealthy sportsmen now fly into otherwise unreachable Arctic regions to hunt polar bear from the air. Spotted cats (tigers, cheetahs, leopards, jaguars, etc.) are endangered because customers will pay more money for furs as these animals become more and more rare.

As is well publicized, baby harp seals are killed in great numbers because they have a white coat about four days after birth. Some animal species have been hurried toward extinction because their parts are believed, particularly by Asians, to have curative powers. Rhino horn powder is used as an aphrodisiac and as a cure for snake bites, kidney disease, and other illnesses. In an activity called **poaching,** a thousand black rhinos a year are killed illegally on preserves in Kenya—only to acquire the horn (fig. 32.23). Elephants are killed only to acquire their tusks. In 1977 total exports from Africa included ivory from approximately 60,000 elephants. In June 1978 the United States added the African elephant to its list of threatened species.

In recent years cacti have been collected for sale in the United States, Europe, Japan, and the Soviet Union. Dealers removed tens of thousands of cacti from Texas and Arizona, even from the national parks. An expedition sponsored by Japanese dealers removed all of the native cacti and other fleshy plants from an island off the Baja California coast. Several of the species taken were known only from that island and are now presumably extinct in the wild. The removal of cacti, like other keystone plants, threatens the lives of the animals that depend upon them for food and shelter.

The number of animals that were at one time legally imported into the United States for the pet industry, zoos, and medical research laboratories numbered in the hundreds of thousands. Often the means of collecting and transporting the animals caused many deaths. The mountain gorilla, found only in mountainous parts of Uganda, Zaire, and Rwanda, is the largest of the apes. Even without human intervention, 50 percent of the young die from accidents and disease. Yet when babies are captured for zoos, the parents are shot and killed. Today, there are only a few hundred gorillas in the wild. The manner of capturing chimpanzees is similar, and in many African countries the number of breeding females may already be below that required for recovery. Chimpanzees are used in medical research because they are so closely related to humans. Steps are now being taken to provide them with more normal surroundings so that they will breed in captivity. However, it would be far less expensive to allow populations to maintain themselves in the wild, with only a few chimps a year being taken for human needs.

Right now we protect plants and animals endangered by hunters, collectors, or fur and ivory traders on a species-by-species basis. The U.S. Endangered Species Act makes the importation of endangered species illegal and forbids any federal action that would destroy a species listed as imperiled. Of some 4,500 federal projects proposed since the law came into effect, only about 200 have involved potential conflicts with species preservation, and in nearly every case the conflict has been reconciled through modification in project plans. Worldwide control has also begun. A Convention on International Trade in Endangered Species of Wild Fauna and Flora *(CITIES)* was held in 1973. The convention formulated rules and regulations to which nearly fifty countries have agreed. Remaining countries are being pressured to sign this document, especially since some of these are the main importers and exporters of endangered species.

Predator Control

Local citizens are often in favor of predator control because they believe certain native animals are dangerous to humans and/or a threat to domestic livestock and wild game. The rancher himself may do the killing, or the federal government may carry it out on a large scale. In this country coyotes, wolves, foxes, mountain lions, bobcats, eagles, and hawks are among those that have been killed off as part of predator control. Unfortunately, the government has been inclined to use poisons rather than shooting predators. This kills not only predators but many other types of animals as well. Once the predators decrease in number, it becomes necessary to start poisoning their prey. In the plains, for example, coyotes were first killed off, and then prairie dogs were victimized (fig. 32.24). This meant that other animals, such as the black-footed ferret, were threatened with extinction. The end result was a much imbalanced ecosystem.

Figure 32.23
Black rhinoceros illegally killed for its horn. A thousand black rhinoceroses a year are poached in Kenya alone. This one was killed in the Masai Mara reserve.

Figure 32.24
Prairie dogs are food for prairie falcons, burrowing owls, coyotes, and black-footed ferrets. When poisoned bait is used to kill coyotes, the prairie dogs also eat it, and the other animals, including scavengers like ravens, are threatened.

Reasons for Conservation

Biome conservation and preservation of species should be the concern of all human beings for various reasons, some of which are listed here.

1. Organisms now becoming extinct may be useful to human beings in the future. While emphasis is often placed on saving vertebrates from extinction, plants and invertebrates are just as necessary to human well-being.

2. Human beings seem to have an innate need to visit, at least occasionally, a diversified natural area. As Jon Roush has stated, "When people talk about going to the country for a vacation, they do not mean simply getting out of town. They mean finding some version of diversity. A cornfield may be natural, but it is also monotonous, and hardly anyone vacations there."[1]

3. Future generations have a right to inherit a world as complete as possible. In the same vein, other living things have a right to live in their natural environment rather than as captives in zoos. To be a polar bear, the animal must live like a polar bear; to be a chimpanzee, the animal must develop and mature as its ancestors in the wild have done.

4. Finally, we must realize that the human ecosystem depends on the natural ecosystems. They not only have the capacity to absorb pollutants if we do not overwhelm them, they also contain the complexity that is necessary to stabilize the biosphere. It is in our own best interest to retain and if necessary to restore what is left of all the biomes.

[1]G. J. Roush, "On Saving Diversity," *The Nature Conservancy News* (January-February 1982), p. 8.

Summary

In the United States, as in other parts of the world, urban and suburban sprawl grows and takes over agricultural, rangeland, and natural ecosystems. Salinization, soil erosion, and strip mining all threaten to reduce the quality of the land. Desertification and deforestation are observed worldwide; loss of tropical rain forests is of great current concern.

Solid wastes, including hazardous wastes, are deposited on land. The latter, including heavy metals, organochlorides, and nuclear wastes, concentrate in food chains in which humans are a final consumer.

The atmosphere contains an ozone shield that is believed to be getting smaller due to the use of freon in aerosol cans and nitrogen oxide from exhaust. Nitrogen oxide and hydrocarbons react in sunlight to produce photochemical smog. Fine carbon particles absorb other pollutants and, for example, facilitate the distribution of nitric and sulfuric acids. These acids are the components of acid rain that is causing lakes in certain parts of the country to become devoid of fishes and other forms of wildlife. The accumulation of CO_2 may already have caused a warming trend that could affect distribution of agriculture and cause polar ice to melt, flooding some of the world's cities. Human health is also threatened by air pollution, and a new consideration of late has been indoor air pollution, which may be worse than outdoor air pollution.

Water supplies include surface water and groundwater. Groundwater supplies may run short in some parts of southwestern United States because withdrawals now exceed recharge. Dams are often built to trap surface water in reservoirs from which water can be distributed as needed. This custom has many drawbacks, including the damage that it does to natural ecosystems. Irrigation not only may cause waterlogging and salinization, it also increases the salt content of rivers.

There are two categories of water pollutants: those that are more likely to cause eutrophication and those that are more likely to cause health problems. Sewage treatment plants were designed to handle the former type, but a problem still exists. The inorganic nutrients that remain after secondary treatment cause algal bloom and fish kill that also contribute to eutrophication. These inorganic nutrients should be removed by either an artificial or a natural way before the water is returned to a river. Heavy metals and organochlorides are those hazardous wastes that are apt to pollute both surface and underground water. The coasts are the final receptors of water pollutants, and they are also subject to much development. Humanity should bear in mind that the ocean's capacity to receive pollution is finite, not infinite.

Modern times have seen an increase in the rate of extinction that is expected to accelerate. Habitat destruction, hunting and collecting, and predator control have all contributed to extinction. There are many reasons for trying to prevent extinctions, but the primary one is that the human ecosystem is dependent on natural ecosystems, and if they should cease to function, then our own existence will come to an end.

Objective Questions

1. Strip mining is a common method today of mining for _____ .
2. Desertification is the transformation of marginal lands to _____ conditions.
3. Which three geographic areas import most of the hardwood timber taken from tropical forests? _____ _____ _____
4. What are the three categories of hazardous wastes? _____ _____ _____ .
5. Hazardous wastes are often subject to _____ , a process by which they become concentrated in top consumers, including humans.
6. The chief air pollutant is _____ , largely given off by _____ .
7. The burning of _____ supplies most of the sulfur oxides that lead to acid rain in the United States.
8. Groundwater mining leads to _____ , _____ , and _____ .
9. Disposal of wastes on land leads to pollution of _____ water.
10. Three reasons for the rise in the number of animals that have become extinct in recent years are _____ , _____ , and _____ .
11. What reason do you find most compelling for maintaining the diversity of living things in the biosphere? _____ _____

Answers to Objective Questions

1. coal 2. semidesert 3. Japan, United States, and Europe 4. heavy metals, chlorinated hydrocarbons, and nuclear wastes 5. bioaccumulation 6. CO, automobiles 7. coal 8. subsidence, sinkholes, saltwater intrusion 9. ground 10. habitat destruction, hunting and collecting, predator control 11. your choice

Study Questions

1. State the ways in which land is currently used in the United States. Indicate the main ecological problems associated with each type of use.
2. What are two main categories of solid wastes? Name three types of hazardous wastes and describe the manner in which biological concentration occurs.
3. What suggestions do you have for reducing the amount of solid wastes to be disposed?
4. Describe the first two layers of the atmosphere. In what way is the ozone shield threatened?
5. What are the five categories of air pollutants caused by fossil fuel combustion? Which two are implicated in photochemical smog? in acid rain? Describe the environmental damage done by both of these.
6. What is the greenhouse effect? the refrigerator effect? How might each affect the condition of the biosphere?
7. Statisticians tell us that there is enough fresh water for human use. Give at least three indications that there must be a problem in distribution.
8. Describe the changing quality of life if high BOD material is dumped into rivers.
9. What is cultural eutrophication and what are its causes?
10. What are the main pollutants of groundwater and how did they get underground?
11. Prepare a discourse to prove to a friend that the entire ocean ecosystem is under severe stress.
12. Name at least three reasons why the extinction rate is currently increasing.
13. Give at least four reasons why it is in our own best interest to conserve biomes and to preserve biological diversity.

Selected Key Terms

strip mining ('strip 'mī niŋ) 708
desertification (di ‚zərt ə fə 'kā shən) 709
deforestation (dē ‚fȯr ə 'stā shən) 709
organochloride (ȯr ‚gan ə 'klȯr ‚īd) 711
bioaccumulation ('bī ō ə 'kyü myə 'lā shən) 712
troposphere ('trōp ə ‚sfir) 713
stratosphere ('strat ə ‚sfir) 713
thermal inversion ('thər məl in 'vər zhən) 715
groundwater mining ('grau̇n ‚dwȯt ər 'mī niŋ) 719
subsidence (səb 'sīd ənts) 719
sinkhole ('sink ‚hōl) 719
saltwater intrusion (‚sȯlt‚wȯt ər in 'trü zhən) 719
desalinization (dē ‚sal ə nə 'zā shən) 721

Suggested Readings for Part 12

Batie, S. S., and Healy, R. G. February 1983. The future of American agriculture. *Scientific American.*

Brown, L. R., et al. 1986. *State of the World: 1986.* New York: W. W. Norton and Co.

Donaldson, L. R., and Joyner, T. July 1983. The salmonid fishes as a natural livestock. *Scientific American.*

Environment. All issues of this journal contain articles covering modern ecological problems.

Gwatkin, D. R. May 1982. Life expectancy and population growth in the third world. *Scientific American.*

Keely, C. B. March 1982. Illegal migration. *Scientific American.*

Lester, R. K. March 1986. Rethenkeng nuclear power. *Scientific American.*

Martin, P. I. October 1983. Labor-intensive agriculture. *Scientific American.*

Miller, G. T. 1981. *Living in the environment.* 3d ed. Belmont, CA: Wadsworth.

Moran, J. M., et al. 1980. *Introduction to environmental science.* San Francisco: W. H. Freeman.

Nebel, B. J. 1981. *Environmental science: The way the world works.* Englewood Cliffs, NJ: Prentice-Hall.

Rasmussen, E. D. September 1982. The mechanization of agriculture. *Scientific American.*

Revelle, R. August 1982. Carbon dioxide and world climate. *Scientific American.*

Sheldon, R. P. June 1982. Phosphate rock. *Scientific American.*

Swaminathan, M. S. January 1984. Rice. *Scientific American.*

Turk, J., and Turk, A. 1984. *Environmental Science.* New York: CBS College Publishing.

Classification of Organisms

Kingdom MONERA

Unicellular procaryotic organisms lacking distinct nuclei and membrane-bound organelles. Nutrition principally by absorption, but some are photosynthetic or chemosynthetic.

    Phylum SCHIZOPHYTA Bacteria
    Phylum CYANOBACTERIA Cyanobacteria (formerly blue-green algae)

Kingdom PROTISTA

Typically unicellular, sometimes multicellular, eucaryotic organisms with distinct nuclei and organelles. Nutrition by photosynthesis, absorption, or ingestion.

    Phylum MASTIGOPHORA Flagellated protozoans
    Phylum SARCODINA Amoeboid protozoans
    Phylum CILIOPHORA Ciliated protozoans
    Phylum SPOROZOA Parasitic protozoans
    Phylum CHRYSOPHYTA Diatoms
    Phylum EUGLENOPHYTA *Euglena* and relatives
    Phylum PYRROPHYTA Dinoflagellates

Kingdom FUNGI

Multinucleate plantlike organisms lacking photosynthetic pigments. Nutrition absorptive.

    Division ZYGOMYCOTA Primitive fungi
    Division ASCOMYCOTA Sac fungi
    Division BASIDIOMYCOTA Club fungi
    Division MYXOMYCOTA Slime molds

Kingdom PLANTAE

Multicellular eucaryotic organisms with rigid cell walls and chlorophyll. Nutrition principally by photosynthesis.

    Division RHODOPHYTA Red algae
    Division PHAEOPHYTA Brown algae
    Division CHLOROPHYTA Green algae
    Division BRYOPHYTA Mosses and liverworts
    Division TRACHEOPHYTA
        Subdivision PSILOPSIDA *Psilotum*
        Subdivision LYCOPSIDA Club mosses
        Subdivision SPHENOPSIDA Horsetails
        Subdivision PTEROPSIDA
            Class FILICINEAE Ferns
            Class GYMNOSPERMAE Conifers, cycads, ginkgos
            Class ANGIOSPERMAE Flowering plants
                Subclass MONOCOTYLEDONEAE Grasses, lilies, and orchids
                Subclass DICOTYLEDONEAE Various

Kingdom ANIMALIA

Multicellular organisms without cell walls or chlorophyll. Nutrition principally ingestive with digestion in an internal cavity.

    Phylum PORIFERA Sponges
    Phylum CNIDARIA Radially symmetrical marine animals
        Class HYDROZOA *Hydra*, Portuguese man-of-war
        Class SCYPHOZOA Jellyfish
        Class ANTHOZOA Sea anemones and corals

Phylum PLATYHELMINTHES Flatworms
    Class TURBELLARIA Free-living flatworms
    Class TREMATODA Parasitic flukes
    Class CESTODA Parasitic tapeworms
Phylum NEMATODA Roundworms
Phylum ROTIFERA Rotifers
Phylum MOLLUSCA Soft-bodied, unsegmented animals
    Class AMPHINEURA Chitons
    Class MONOPLACOPHORA *Neopilina*
    Class GASTROPODA Snails and slugs
    Class PELECYPODA Clams and mussels
    Class CEPHALOPODA Squids and octopuses
Phylum ANNELIDA Segmented worms
    Class POLYCHAETA Sandworms
    Class OLIGOCHAETA Earthworms
    Class HIRUDINEA Leeches
Phylum ARTHROPODA Joint-legged animals; exoskeleton
    Class CRUSTACEA Lobsters, crabs, barnacles
    Class ARACHNIDA Spiders, scorpions, ticks
    Class CHILOPODA Centipedes
    Class DIPLOPODA Millipedes
    Class INSECTA Grasshoppers, termites, beetles
Phylum ECHINODERMATA Marine; spiny, radially symmetrical animals
    Class CRINOIDEA Sea lilies and feather stars
    Class ASTEROIDEA Sea stars
    Class OPHIUROIDEA Brittle stars
    Class ECHINOIDEA Sea urchin and sand dollar
    Class HOLOTHUROIDEA Sea cucumbers
Phylum HEMICHORDATA Acorn worms
Phylum CHORDATA Dorsal supporting rod (notochord) at some stage; dorsal
  hollow nerve cord; pharyngeal pouches or slits
    Subphylum UROCHORDATA Tunicates
    Subphylum CEPHALOCHORDATA Lancelets
    Subphylum VERTEBRATA Vertebrates
      Class AGNATHA Jawless fishes (lampreys, hagfishes)
      Class CHONDRICHTHYES Cartilaginous fishes (sharks, rays)
      Class OSTEICHTHYES Bony fishes
        Subclass Crossopterygii. Lobe-finned fishes
        Subclass Dipnoi. Lungfishes
        Subclass Actinopterygii. Ray-finned fishes
      Class AMPHIBIA Frogs, toads, salamanders
      Class REPTILIA Snakes, lizards, turtles
      Class AVES Birds
      Class MAMMALIA Mammals
        Subclass Prototheria. Egg-laying mammals
          Order MONOTREMATA Duckbilled platypus, spiny ant-
            eater
        Subclass Metatheria. Marsupial mammals
          Order MARSUPIALIA Opossums, kangaroos
        Subclass Eutheria. Placental mammals
          Order INSECTIVORA Shrews, moles
          Order CHIROPTERA Bats
          Order EDENTATA Anteaters, armadillos
          Order RODENTIA Rats, mice, squirrels
          Order LAGOMORPHA Rabbits and hares

Order CETACEA Whales, dolphins, porpoises
Order CARNIVORA Dogs, bears, weasels, cats, skunks
Order PROBOSCIDEA Elephants
Order SIRENIA Manatees
Order PERISSODACTYLA Horse, hippopotamus, zebra
Order ARTIODACTYLA Pigs, deer, cattle
Order PRIMATES Lemur, monkeys, apes, humans
    Suborder Prosimii. Lemurs, tree shrews, tarsiers, lorises, pottos
    Suborder Anthropoidea. Monkeys, apes, humans
        Superfamily Ceboidea. New world monkeys
        Superfamily Cercopithecoidea. Old world monkeys
        Superfamily Hominoidea. Apes and humans
            Family Hylobatidae. Gibbons
            Family Pongidae. Chimpanzee, gorilla, orang-utan
            Family Hominidae. *Australopithecus,** *Homo erectus,** *Homo sapiens* neanderthalis,* *Homo sapiens sapiens*

*extinct

Periodic Table of the Elements

Ia																	0
1 H 1.008	IIa											IIIa	IVa	Va	VIa	VIIa	2 He 4.00
3 Li 6.94	4 Be 9.01											5 B 10.81	6 C 12.01	7 N 14.00	8 O 15.99	9 F 18.99	10 Ne 20.18
11 Na 22.99	12 Mg 24.31	IIIb	IVb	Vb	VIb	VIIb	VIIIb			IB	IIIB	13 Al 26.98	14 Si 28.09	15 P 30.97	16 S 32.06	17 Cl 35.45	18 Ar 39.95
19 K 39.10	20 Ca 40.08	21 Sc 44.96	22 Ti 47.90	23 V 50.94	24 Cr 51.99	25 Mn 54.94	26 Fe 55.85	27 Co 58.93	28 Ni 58.71	29 Cu 63.54	30 Zn 65.37	31 Ga 69.72	32 Ge 72.59	33 As 74.92	34 Se 78.96	35 Br 79.91	36 Kr 83.80
37 Rb 85.47	38 Sr 87.62	39 Y 88.91	40 Zr 91.22	41 Nb 92.91	42 Mo 95.94	43 Tc (99)	44 Ru 101.97	45 Rh 102.91	46 Pd 106.4	47 Ag 107.87	48 Cd 112.40	49 In 114.82	50 Sn 118.69	51 Sb 121.75	52 Te 127.60	53 I 126.90	54 Xe 131.30
55 Cs 132.91	56 Ba 137.34	see below 57-71	72 Hf 178.49	73 Ta 180.95	74 W 183.85	75 Re 186.2	76 Os 190.2	77 Ir 192.2	78 Pt 195.09	79 Au 196.97	80 Hg 200.59	81 Tl 204.37	82 Pb 207.19	83 Bi 208.98	84 Po (210)	85 At (210)	86 Rn (222)
87 Fr (223)	88 Ra (226)	see below 89-103	104 Rf (261)	105 Ha (260)	106 * 263	*newly produced											

57 La 138.91	58 Ce 140.12	59 Pr 140.91	60 Nd 144.24	61 Pm (147)	62 Sm 150.35	63 Eu 151.96	64 Gd 157.25	65 Tb 158.92	66 Dy 162.50	67 Ho 164.93	68 Er 167.26	69 Tm 168.93	70 Yb 173.04	71 Lu 174.97
89 Ac (227)	90 Th 232.04	91 Pa (231)	92 U 238.03	93 Np (237)	94 Pu (242)	95 Am (243)	96 Cm (247)	97 Bk (247)	98 Cf (251)	99 Es (254)	100 Fm (253)	101 Md (256)	102 No (254)	103 Lw (257)

GLOSSARY

A

abiotic components ('ā bī ˌät ik   kəm 'pō nəntz) the nonliving components of an ecosystem   646

abscisic acid (ˌab ˌsiz ik   'as əd)   a plant hormone causing stomata closing and inhibition of growth   416

abyssal plains (ə 'bis əl   'plānz)   deep substrate of abyssal zone occurring beyond continental slope   673

acetylcholine (ə ˌset əl 'kō ˌlēn)   a neurotransmitter utilized within both the peripheral and central nervous systems   508

acid ('as əd)   a compound tending to yield hydrogen ions and to lower its pH numerically   47

acidosis (ˌas ə 'dō səs)   accumulation of acids that lowers the pH of body fluids   468

actin ('ak tən)   a muscle protein making up the thin filaments in a sarcomere; its movement shortens the sarcomere, yielding muscle contraction   546

active site ('ak tiv   ˌsīt)   that part of an enzyme molecule where the substrate fits and the chemical reaction occurs   95

active transport ('ak tiv   trans ˌpōərt)   use of a cell membrane carrier molecule to move particles from a region of lower to higher concentration; it opposes an equilibrium and requires energy   70

adaptation (ˌad ap 'tā shən)   an organism's modification in structure or function to suit the environmental condition   24

adenosine triphosphate (ə 'denə ˌsēn   trī 'fäs ˌfāt) a compound containing adenine, ribose, and three phosphates, two of which are high-energy phosphates. The breakdown of ATP to ADP and Ⓟ makes energy available for energy-requiring processes in cells   93

aerobic respiration (ˌā 'rō bik   ˌres pə 'rā shən) utilization of oxygen as the final acceptor for hydrogen atoms from a series of reactions that produce ATP   115

agranulocyte ('ā 'gran yə lō ˌsīt)   a white-blood cell type without cytoplasmic granules   448

allantois (ə 'lant ə wəs)   an extraembryonic membrane that accumulates nitrogenous wastes in the embryo of birds and reptiles, and is vestigial in mammals   608

allele (ə 'lēl)   alternative form of a gene yielding variety in a trait, i.e., brown-eyed allele and blue-eyed allele   152

allopatric (ˌal ə 'pa trik)   when species occupy different regions, with no overlap   230

alveolus (al 'vē ə ləs)   terminal microscopic, grapelike air sac found in vertebrate lungs   482

amino acid (ə 'mē ˌnō   'as əd)   organic subunits each having an amino group and an acid group, that covalently bond to produce protein molecules   59

amnion ('am nē 'än)   an extraembryonic membrane of the bird, reptile, and mammal embryo that forms an enclosing fluid-filled sac   608

amoeboid (ə 'mē ˌbȯid)   like an amoeba, particularly showing the slow creeping motility of this protozoan   270

amylase ('am ə ˌlās)   an enzyme that catalyzes chemical breakdown of starch to maltose; works in the mouth and small intestine of humans   459

anaerobic respiration (ˌan ə 'rō bik   ˌres pə 'rā shən)   utilization of some molecule other than oxygen as an acceptor for hydrogen atoms from a series of reactions that produce ATP   112

analogous structure (ə 'nal ə gəs   'strək cher) structure that functions similarly in two different type organisms   243

anaphase ('an ə ˌfāz)   stage of mitosis in which chromatid pairs separate, then each (now called a chromosome) segregates to opposite poles of the spindle   136

anemia (ə 'nē mē ə)   reduced oxygen-carrying capacity of the blood through decreased hemoglobin concentration or red blood cell count or both   444

aneuploidy ('an yü ˌplȯid ē)   a condition of extra or fewer chromosomes, i.e.,—trisomy, three chromosomes rather than two chromosomes of a particular kind   182

angiosperm ('an jē ə ˌspərm)   flowering plant, having double fertilization that results in development of a seed-bearing fruit   299

anther ('an thər)   part of stamen for pollen development   300

antheridium (ˌan thə 'rid ē əm)   male organ in nonseed plants where swimming sperm are produced   290

anthropoid ('an thrə ˌpȯid)   members of a suborder of primates including monkeys, great apes, and humans   351

antibody ('ant i ˌbäd ē)   a protein molecule usually formed naturally in an organism to combat antigens or foreign substances   450

anticodon ('an tē 'kō ˌdän)   three nucleotides on a tRNA molecule, attracted to a complementary codon on mRNA   202

appendicular skeleton (ap ən 'dik yə lər 'skəl ə tən)   part of the skeleton forming the upper extremities, shoulder girdle, lower extremities, and hip girdle   544

archegonium (ˌär ki 'gō nē əm)   female organ in nonseed plants where an egg is produced   291

archenteron ('är kent ə ˌrän)   primitive digestive tract of the gastrula in the early embryo   600

Archeozoic (är kē ə 'zō ik)   a geologic era prior to the Precambrian, about two billion years in length, having few fossils, but having some evidence of primitive life forms   238

arteriole (ar 'tir ē ˌōl)   blood vessel type distributing blood from arteries to capillaries   429

artery ('ärt ə rē)   a blood vessel that transports blood away from the heart toward arterioles   429

ascus ('as kəs)   a saclike, tubular structure producing eight spores per packet in one group of fungi (*Ascomycetes*)   277

atom ('at əm)   the smallest particle of an element that displays its properties; i.e., carbon, hydrogen, oxygen   39

ATP (ˌa ˌtē 'pē)   *see* adenosine triphosphate

atrium ('ā trē əm)   upper heart chamber, one on each side in a four-chambered heart; its pumping primes the lower ventricle   429

autonomic nervous system (,ȯt ə 'näm ik 'nər vəs 'sis təm)   a branch of the peripheral nervous system administering motor control over internal organs   513

autosome ('ȯt ə ,sōm)   any chromosome other than the sex-determining pair   133

autotrophic (,ȯt ə 'trō fik)   self-nourishing; referring to producers starting food chains that make organic molecules from inorganic nutrients   90

auxin ('ȯk sən)   plant hormone regulating growth, particularly cell elongation; also called indolacetic acid   411

axial skeleton ('ak sē əl 'skəl ət ən)   part of the skeleton forming the vertical support or axis, including the skull, rib cage, and vertebral column   543

axon ('ak ,sän)   a neuron process, conducting the impulse from cell body to synapse   502

B

bacteriophage (bak 'tir ē ə ,fāj)   a virus that parasitizes a bacterial cell as its host, destroying it by lytic action   188

balanced polymorphism ('bal ənst ,päl i 'mȯr fiz əm)   the existence in a population of two or more different forms in approximately the same proportion in each generation   227

base ('bās)   a compound tending to lower the hydrogen ion concentration in a solution and to raise its pH numerically   47

benthic ('ben thik)   pertaining to bottom-dwelling aquatic organisms   674

bilateral symmetry (,bī 'lat ə rəl 'sim ə trē)   a body having two corresponding or complementary halves   308

bile ('bīl)   a substance released from the gall-bladder to small intestine to emulsify fats prior to chemical digestion   463

binary fission ('bī nə rē 'fish ən)   splitting of a parent cell into two daughter cells; serves as an asexual form of reproduction in single-celled organisms   144

binomial name (bī 'nō mē əl 'nām)   a two-part name for kinds of organisms; devised by Linnaeus in 1758, the first name is the genus category, the second is the species category; the genus name is capitalized and both names are italicized or underlined, i.e., *Rana pipiens, Homo sapiens*   26

bioaccumulation ('bī ō ə 'kyü myə 'lā shən)   tendency of nonexcretable hazardous wastes to progressively concentrate through the links of a food chain   712

biogeochemical cycle ('bi ō ,je ō 'kem i kəl 'sī kəl)   circulating pathway of an element between the biotic and abiotic components of an ecosystem   657

biogeography (,bī ō je 'äg rə fē)   the study dealing with the geographical distribution of organisms   239

biomass (,bī ō ,mas)   amount of organic matter of species per unit area or volume   651

biome ('bī ,ōm)   a major biotic community having well-recognized life forms and a typical climax species   676

biosphere ('bī ə ,sfir)   a thin shell around the earth supporting life from a few feet into the soil to a few hundred feet above its surface   622

biotic (bī ät ik)   the living components of an ecosystem as opposed to those that are not (abiotic)   646

blastocoel (blas tə sēl)   the fluid-filled cavity of a blastula in the early embryo   600

blastopore ('blas tə ,pōr)   opening of the gastrula in the early embryo   600

blastula ('blas chə lə)   a hollow, fluid-filled ball of cells prior to gastrula formation of the early embryo   600

breeder fission ('brēd ər 'fish ən)   a nuclear reaction that produces plutonium, a nuclear power plant fuel   700

bronchiole ('brän kē ,ōl)   small tube that conducts air from a bronchus to the alveoli   482

bronchus ('brän kəs)   one of two main branches of the trachea in vertebrates that have lungs   482

bryophyte ('brī ə ,fīt)   a member of a plant group, including mosses and liverworts   289

bud ('bəd)   1. in animals, body outgrowth capable of forming an entire new organism asexually, 310; 2. in flowering plants, meristem tissue protected by leaves   373

budding ('bəd niŋ)   in animals an asexual form of reproduction whereby a new organism develops from the body of the parent   579

buffer ('bəf ər)   a substance or group of substances that tend to resist pH changes in a solution, thus stabilizing its relative acidity   48

C

canine ('kā ,nīn)   knife-shaped tooth for tearing food   459

capillary ('kap ə ,ler ē)   microscopic, plentiful blood vessel for gas and nutrient exchange with body cells   429

carbohydrate (,kär bō 'hī ,drāt)   a family of organic compounds consisting of carbon, hydrogen, and oxygen atoms; exhibits subfamilies monosaccharides, disaccharides, and polysaccharides   54

carnivore ('kär nə ,vȯr)   a secondary consumer in a food chain, that eats other animals   90

carrier ('kar ē ər)   1. a molecule that combines with a substance and actively transports it through the cell membrane, 70; 2. an individual that transmits an infectious or genetic disease   179

carrying capacity ('kar ē iŋ kə 'pas ət ē)   the maximum size of a population that can be supported by the environment in a particular locale   624

Casparian strip (,kas 'pər ə ən 'strip)   a waxy layer bordering four sides of root endodermal cells; prevents water and solute transport between adjacent cells   371

cell ('sel)   the smallest unit that displays properties of life; composed of cytoplasmic regions, possibly organelles, and surrounded by a cell membrane   7

cellular respiration ('sel yə lər ,res pə 'rā shən)   metabolic reactions that provide energy to cells including both anaerobic and aerobic processes   93

cellulose ('sel yə ,lōs)   a polysaccharide having glucose as the unit molecule; a major component of the plant cell wall   68

cell wall ('sel wȯl)   a relatively rigid structure composed mostly of polysaccharides that surrounds the cell membrane of plants, fungi, and bacteria   68

Cenozoic (,sē nə 'zō ik)   the most recent geologic era, the Age of Mammals marked by the spread of modern mammals and plants   238

central nervous system ('sen trəl 'nər vəs 'sis təm)   the brain and spinal cord in vertebrate animals   502

centriole ('sen trē ,ōl)   cell organelle, existing in pairs, that possibly organizes a mitotic spindle for chromosome movement during cell division   82

centromere ('sen trə ,mir)   a constriction where duplicates (chromatids) of a double-stranded chromosome are held together   133

cephalization (,sef ə lə 'zā shən)   having a well-recognized anterior head with concentrated nerve masses and receptors   312

cerebellum (,ser ə 'bel əm)   a brain region in back of the lower brainstem controlling muscle coordination and body equilibrium   516

cerebrum (sə 'rē brəm)   the higher forebrain; its outer gray cortex is involved in sensory and motor control, plus more abstract functions of thinking, reasoning, and memory   517

chemotaxis (,kē mō 'tak səs)   movement of an organism toward a chemical stimulus   559

chloroplast ('klōr ə ,plast)   a membrane-bounded organelle in which membranous grana contain chlorophyll and where photosynthesis takes place   100

cholinesterase (,kō lə 'nes tə ,rās)   an enzyme that inactivates acetylcholine at the synapse   508

chorion ('kor ē 'än)   an extraembryonic membrane functioning for respiratory exchange in birds and reptiles; contributes to placenta formation in mammals   608

chromatid ('krō mə təd)   a longitudinal half of a duplicated chromosome   133

chromatin ('krō mə tən)   the darkly-stained material of a cell's nucleus during interphase, it forms chromosomes prior to nuclear division   131

chromosome ('krō mə ,sōm)   a rod-shaped body in the cell nucleus visible during cell division and possessing linearly arranged genes   131

chyme ('kīm)   a highly acidic, soupy mixture of digestion products that moves from the stomach to small intestine   462

cilia ('sil ē ə)   hairlike projections having a 9 + 2 organization of microtubules that are used for locomotion by many unicellular organisms and that have various purposes in higher organisms   80

circadian rhythm (sər 'kād ē ən 'rith əm) referring to a phenomenon recurring every twenty-four hours 421

citric acid cycle (ˌsi trik 'as əd 'sī kəl) a cycle of reactions that converts acetic acid to CO₂; gives off electrons that are sent to the respiratory chain 116

class ('klas) a taxonomic category between order and phylum; a group of related orders 25

cleavage ('klē vij) earliest division of the zygote developmentally, without cytoplasmic addition or enlargement 598

clitoris ('klit ə rəs) erectile tissue of the female genitalia that is homologous to the penis 587

coacervate droplet (kō 'as ər ˌvāt 'dräp lət) mixtures of macromolecules joined in water as proven by Oparin's experiments; cell-like in properties 261

cochlea ('kō klē ə) spiral-shaped structure of the inner ear containing the receptor hair cells for hearing 539

codominance (ˌkō 'däm ə nəns) a condition in heredity whereby neither allele is dominant over the other 163

codon ('kō ˌdän) three nucleotides on an mRNA strand that is complementary of the DNA code and attracts a complementary anticodon on tRNA 198

coelocanth ('sə lə ˌkanth) a primitive, lobe-finned fish, abundant in fossils 340

coelom ('sē ləm) animal body cavity, lined by mesoderm, that lies between the digestive tract and body wall 318

coenzyme (ˌkō 'en ˌzīm) the nonprotein part of entire enzyme structure, organic in makeup, often with a vitamin as a subpart 97

cogeneration (ˌkō ˌjen ə 'rā shən) use of an energy form for a dual purpose; i.e., heat and electricity 697

cohesion-tension theory (kō 'hē zhən 'ten chən 'thē ə rē) a force that causes water molecules to cling together as they are transported through a plant by transpiration 400

collar cell ('käl ər 'sel) flagellated cell lining gastrovascular cavity of a sponge, producing internal water currents 308

colon ('kō lən) large intestine 468

colonial (kə 'lō nē əl) multicellular organism whose cells remain independent for most functions 284

colony ('käl ə nē) 1. a protistan organism composed of cells that cooperate to a degree but are not highly specialized, 270, 284; 2. unicellular or multicellular organisms found in the same locale 226

colostrum (kə 'läs trəm) watery, yellowish-white fluid produced by the breasts 591

commensalism (kə 'men sə ˌliz əm) a symbiotic relationship in which one species profits with an apparent neutral effect on the other species 641

community (kə 'myü nət ē) the interaction of many different populations in an ecosystem 622

competition (ˌkäm pə 'tish ən) interaction between two different organisms or groups of organisms that require the same resource 628

compound (käm 'paůnd) a chemical substance having two or more different elements in fixed ratio 41

concept ('kän ˌsept) a general scheme or unifying idea supported by evidence and having predictive value 14

condensation (ˌkän ˌden 'sā shən) bonding two molecules into a larger molecule with an accompanying loss of a water molecule or molecules 52

conidium (kə 'nid ē əm) an asexual spore produced by certain fungi 277

conjugation (ˌkän je 'gā shən) the transfer of genetic material from one cell to another through a cytoplasmic bridge 285

consumer (kən 'sü mər) an organism that feeds on another organism in a food chain; primary consumers eat plants, secondary consumers eat animals 29

continental shelf (ˌkänt ən 'ent əl 'shelf) that part of the ocean floor that projects from the continents 673

continental slope (ˌkänt ən 'ent əl 'slōp) ocean floor beyond continental shelf 673

contraception (ˌkän trə 'sep shən) prevention of conception or fertilization of sex cells 592

control (kən ˌtrōl) a sample that goes through all the steps of an experiment except the one being tested; a standard against which results of an experiment are checked 14

convergent evolution (kən 'vər jənt ˌev ə 'lü shən) the appearance of similar structures in different groups of organisms exposed to similar types of environments 252

cornea ('kȯr nē ə) a clear, circular outer eye layer located in front of the aqueous humor and iris 534

coronary ('kȯr ə ˌner ē) referring to the heart, particularly the local circulation that serves the heart proper 433

cortex ('kȯr ˌteks) 1. in animals, the outer layer, such as the adrenal cortex or cerebral cortex, 519; 2. in plants, the root or stem tissue between the epidermis and the central vascular tissues 371

covalent bond (ˌkō 'vā lənt 'bänd) a chemical bond in which the atoms share electrons 42

Cowper's gland (ˌkaů pərz 'gland) a small gland adding seminal fluid to moving sperm prior to its entrance into the urethra as that structure passes through the penis 585

cristae ('kris ˌtē) membranous shelves of the mitochondrion's interior formed by infoldings of its inner membrane 119

crop ('kräp) dilated foregut region in which food is stored in birds and certain invertebrates 455

crossing over ('krȯ siŋ 'ō vər) an exchange of segments between chromatids of synapsed chromosomes during meiosis 175

cuticle ('kyüt i kəl) waxy covering on upper and lower epidermis of leaves 366

cyanobacterium (ˌsī ə nō bak 'tir ē əm) formerly called blue-green alga; photosynthetic prokaryotes that contain chlorophyll and release oxygen 268

cytochrome system ('sīt ə ˌkrōm 'sis təm) a series of electron carriers that capture the energy of oxidation so that ATP synthesis is possible 117

cytokinesis (ˌsīt ō kə 'nē səs) the division of a cell into two halves, involving furrowing in animals and a cell plate in plants 137

cytokinin (ˌsīt ə 'kī nən) plant hormone that promotes cell division; development, often works in combination with auxin during organ development 413

cytoplasm ('sīt ə ˌplaz əm) the background substance of a cell 67

D

data ('dāt ə) facts or information collected through observation and/or experimentation 13

day-neutral plant ('dā 'nü trəl 'plant) a plant whose flowering is not dependent on photoperiod changes 418

deamination (ˌdē am ə 'nā shən) removal of an amino group from an amino acid; usually performed in an animal's liver 468

decomposer (ˌdē kəm 'pō zer) organisms that break down dead organic material or waste products of other organisms making inorganic nutrients available to autotrophs 29

deductive reasoning (di 'dək tiv rēz niŋ iŋ) a process of logic and reasoning, using "if . . . then" statements 13

deforestation (dē ˌfȯr ə 'stā shən) the replacement of forests with other types of biomes often due to human activities 709

deletion (di 'lē shən) a chromosome mutation that results in loss of a portion of the chromosome 181

dendrite ('den ˌdrīt) a neuron process that sends impulses toward the cell body 504

denitrification (dē ˌnī trə fə 'kā shən) the process of converting nitrogen compounds to free nitrogen, liberated in the nitrogen cycle 661

density dependent ('den sət ē də 'pen dənt) referring to factors whose intensity is dependent on population size 626

density independent ('den sət ē ˌin də 'pen dənt) referring to factors whose intensity is not dependent on population size 626

deoxyribonucleic acid (ˌdē 'äk si ˌrī bō nủ klē ik 'as əd) (DNA), an organic molecule produced from covalent bonding of nucleotide subunits, the chemical composition of genes on chromosomes 61

depletion curve (di 'plē shən 'kerv) graphical depiction of a resource's dwindling amount over time 696

depolarization (ˌdē ˌpō lə rə 'zā shən) the electrical-like change in a neuron's polarity during its transmission of an impulse 506

desalinization (dē ,sal ə nə 'zā shən) removing the salt from saltwater for human consumption 721

desertification (di zərt ə fə 'kā shən) transformation of marginal lands into semidesert conditions 709

detritus (dī 'trīt əs) falling, settled debris at the land-floor or water bed of an ecosystem that is subject to decomposer action 646

deuterostome ('düt ə rə ,stōm) a group of coelomate animals in which the second embryonic opening becomes the mouth; the first embryonic opening, the blastopore, becomes the anus 307

diaphragm ('dī ə ,fram) a dome-shaped muscle separating the thoracic cavity from the abdominal cavity in the animal body 479

diastole (dī 'as tə ,lē) relaxation period of a heart during the cardiac cycle 435

dicotyledon (,dī ,kät l 'ēd n) a flowering plant group; members show two embryonic leaves, netted-venation leaves, cylindrical vascular tissue, and other characteristics 300

differentiation (,dif ə ,ren chē 'ā shən) specialization of early embryonic cells in regard to structure and function 598

diffusion (dif 'yü zhən) the movement of particles of matter from a region of high to low concentration, tending toward an equal distribution 67

diploid ('dip ,lȯid) the condition in which cells have two of each type of chromosome; twice the number of chromosomes found in gametes that have the haploid (N) number of chromosomes 139

dominant allele ('däm ə nənt ə 'lēl) allelic form of a given gene that hides the effect of the recessive allele in a heterozygous condition 152

Down's syndrome ('daúnz 'sin ,drōm) a birth defect often due to chromosome abnormality in number, three (trisomy) of chromosome number 21 is present 183

ductus deferens ('dək təs 'def ə ,renz) a passageway conducting migrating sperm from the epididymis to the urethra 584

duplication (,dü pli 'kā shən) 1. a chromosome mutation in which a chromosome contains two groups of identical genes, 182; 2. replication of DNA 192

E

ecological pyramid (,ē kə 'läj i kəl 'pir ə ,mid) pictorial graph representing the biomass, organism number, or energy content of each trophic level in a food web from the producer to the final consumer populations 650

ecosystem ('ē kō ,sis təm) a biological community together with the associated abiotic environment 28

ectoderm ('ek tə ,dərm) outermost of an animal's three primary germ layers 312

edema (i 'dē mə) swelling due to tissue fluid accumulation in the intercellular spaces 441

element ('el ə mənt) the simplest of substances consisting of only one type of atom; i.e., carbon, hydrogen, oxygen 39

embolus ('em bə ləs) blood clot moving through a vessel 438

emigrate (,em ə 'grā shən) to move from a geographical source of reference 629

emphysema (,em fə 'zē mə) respiratory disease characterized by breathing difficulty due to a breakdown in lung alveoli 483

endocrine gland ('en də krən) a ductless gland that secretes a hormone into the bloodstream 519

endocytosis (,en də sī 'tō səs) moving particles or debris into the cell from the environment by phagocytosis (cellular eating) or pinocytosis (cellular drinking) 77

endoderm ('en də ,dərm) innermost of an animal's three primary germ layers 312

endodermis (,en də 'dər məs) internal plant root tissue, forming a single, circular array of cells around a root's vascular cylinder 366

endometrium (,en dō 'mē 'trē əm) a mucous membrane lining the inside free surface of the uterus 588

endosperm ('en də ,spərm) nutritive tissue in a seed; often triploid from a fusion of a sperm cell with two polar nucleii 362

enzyme ('en ,zīm) an organic catalyst, speeding up reaction rates in living systems 95

epicotyl ('ep i ,kät əl) plant embryo portion above the cotyledon that contributes to shoot development 364

epidermis (,ep ə 'dər məs) 1. covering tissue of roots and leaves of plants, plus stems of nonwoody organisms, 366; 2. the outermost layer of various invertebrates; also the surface skin layer of vertebrates 308

epididymis (,ep ə 'did ə məs) a structure located on top and on back of the testis to store sperm cells from the testes for maturation 584

epiglottis (,ep ə 'glät əs) a flaplike covering, hinged to the back of the larynx and capable of covering the glottis or air-tract opening 462

epilimnion (,ep ə 'lim nē ,än) the upper layer of warm, circulating water in a lake 666

epiphyte ('ep ə ,fīt) a plant that takes its nourishment from the air since its attachment to other plants gives it an aerial position 396

episiotomy (i ,piz ē 'ät ə mē) an operative incision used at the vaginal orifice to facilitate birth of a child 615

epistatic gene (,ep ə 'stat ik 'jēn) a gene that interferes with the expression of alleles that are at a different locus than the epistatic gene 165

esophagus (i 'säf ə gəs) a muscular tube for moving swallowed food from pharynx to stomach 458

estuary ('es chə ,wer ē) the terminal end of a river where freshwater and saltwater mix as they meet 665

ethologist (ē 'thäl ə jəst) one who studies animal behavior 553

eukaryotic (,yü ,kar ē 'ät ik) cells typical of organisms except bacteria and cyanobacteria having organelles and a well-defined nucleus 72

eutrophic (yu ,trō fik) pertaining to a lake with much organic matter and little dissolved oxygen 668

eutrophication (yu ,trō fə 'kā shən) the enrichment process of filling a waterway with large amounts of organic matter 669

evolution (,ev ə 'lü shən) genetic changes that occur in populations or organisms with the passage of time, often resulting in increased adaptation of organisms to the prevailing environment 17

evolutionary tree (,ev ə 'lü sh ner ē trē) a branching diagram linking ancestors and evolutionary progeny through time 246

exocytosis (,ek sō sī 'tō səs) a process by which particles or debris are expelled through a cell membrane to the extracellular environment 76

exponential (,ek spə 'nen chəl) referring to a geometrically multiplying, rapid population growth rate 623

F

facilitated transport (fə ,sil ə 'tā təd 'trans ,pōrt) particle movement, whereby a carrier molecule in the cell membrane increases the particles' diffusion rate through the membrane 70

family ('fam ə lē) a taxonomic category between the genus and order; a group of related genera 25

fatty acid ('fat ē 'as əd) an organic molecule having a long chain of carbon atoms and ending in an acidic group 56

fermentation (,fər mən 'tā shən) anaerobic breakdown of carbohydrates that results in products such as alcohol and lactic acid 112

filtration (fil 'trā shən) the forcible movement of blood components out of a blood capillary due to blood pressure; the first step in urine formation 494

fimbria ('fim brē ə) fingerlike extensions from the oviduct near the ovary 587

fission power plant ('fish ən 'paú ər 'plant) nuclear power plant that splits uranium-235 to generate energy 698

flagella (flə 'jel ə) long projections from cell membranes; produces forces for organized cell movement 80

flame cell ('flām 'sel) a flatworm cell in which a cilia tuft moves wastes into an excretory tubule 313

florigen ('flōr ə jən) a hypothetical plant hormone that may be transported to sites to induce flowering; has never been isolated 421

flower ('flaú ər) the blossom of a plant that contains the reproductive organs of higher plants 299

fluid mosaic ('flü əd mō 'zā ik) model for the cell membrane based on the changing location and pattern of protein molecules in a fluid lipid matrix 66

food chain ('füd 'chān) a succession of populations in an ecosystem linked by an energy flow and giving the order of who eats whom 648

food web ('füd 'web) a complex pattern of interlocking and crisscrossing food chains 650

fossil ('fäs əl) remains of an organism, usually plant or animal, of a past geological period; usually a petrified part, but sometimes a cast or impression 235

founder effect ('faun dər i 'fekt) the tendency for a new, small population to experience genetic drift after it has separated from an original, larger population 226

fovea centralis ('fō vē ə sen 'tral əs) a retinal region having the highest concentration of cones for color discrimination and visual activity 535

fruit ('früt) flowering plant structure consisting of one or more ripened ovaries with accessory structures 299

fruiting body ('frü tiŋ 'bäd ē) a spore-bearing structure found in certain types of fungi, such as mushrooms 275

fungus root ('fən gəs 'rüt) a root of a vascular plant covered by a fungus that has invaded it and functions symbiotically 396

G

gamete (gə 'mēt) sex cell 579

gametophyte (gə 'mēt ə ‚fīt) the gamete-producing haploid generation in the alternation of generations plant life cycle 139

ganglion ('gan glē ən) a knot or bundle of neuron cell bodies outside the central nervous system 510

gastric ('gas trik) pertaining to the stomach 462

gastrin ('gas trən) a hormone secreted by stomach cells to regulate the release of pepsin by the stomach wall 465

gastrodermis (‚gas trō 'dər məs) the epithelial layer that lines the gastrovascular cavity of coelenterates 308

gastrovascular cavity (‚gas trō 'vas kyə lər kav ə tē) a blind, branched digestive cavity that also serves a circulatory (transport) function in animals that lack a circulatory system 308

gastrula ('gas trə lə) cup-shaped early embryo with two primary germ layers, ectoderm and endoderm, enclosing a primitive digestive tract 600

gene ('jēn) the unit of heredity occupying a particular locus on the chromosome 9, 152

gene frequency ('jēn 'frē kwən sē) the percentage occurrence of each allele of a given gene in the gene pool of a population 218

gene locus ('jēn 'lō kəs) the specific location of a particular gene on a chromosome 152

gene pool ('jēn 'pül) the total of all the genes of all the individuals in a population 218

genetic drift (jə 'net ik 'drift) evolution by chance processes, as when a population becomes homozygous for one allele of a pair due to nonrandom mating 225

genetic mutation (je 'net ik mü 'tā shən) a change in the base sequence of DNA such that the sequence of amino acids is changed in a protein 210

genotype ('jē nə ‚tīp) the genes of an organism for a particular trait or traits; i.e., BB or Aa 152

genus ('jē nəs) a taxonomic category between the species and family; a group of related species 25

germ layer ('jərm 'lā ər) developmental layer of body; i.e., ectoderm, mesoderm, and endoderm 312

gibberellin ('jib ər' rel ən) plant hormone producing increased stem growth; also involved in flowering and seed germination 413

girdling ('gərd liŋ) cutting a plant part, such as the stem or trunk, circularly 401

gizzard ('giz ərd) grinding digestive compartment between crop and intestine of annelids and arthropods 326

glomerular filtrate (glə 'mer yə lər fil 'trāt) the solution found within Bowman's capsule of a kidney tubule formed by the movement of small molecules from the blood into the capsule due to the force of blood pressure 494

glottis ('glät əs) opening for airflow in the larynx 482

glucose ('glü ‚kōs) a monosaccharide, $C_6 H_{12} O_6$; a stable product of photosynthesis and the blood sugar in animals 111

glycogen ('glī kə jən) a polysaccharide, composed of branched glucose chains; the major storage carbohydrate in animals 55

glycolysis (glī 'käl ə səs) pathway of metabolism converting a sugar, usually glucose, to pyruvic acid 112

gonorrhea (‚gän ə 'rē ə) contagious venereal disease, caused by bacteria and leading to inflammation of the urogenital tract 596

gradualism ('graj ə wəl ‚iz əm) the concept that evolution and new species formation occurs over a long period with the gradual accumulation of adaptive traits 246

granulocyte ('gran yə lō ‚sīt) a white-blood cell type with cytoplasmic granules detected by chemical reagent staining 448

gravitropism (‚grav ə 'trō ‚piz əm) growth in relation to the earth's center or gravity source; roots demonstrate positive gravitropism 411

groundwater mining ('graun ‚dwort ər 'mī ninj) withdrawal of water from underground aquifers without allowing adequate time for recharging by rainwater 719

guard cell ('gärd 'sel) a bean-shaped, epidermal cell; one found on each side of a leaf stoma; their activity controls stoma size 383

guttation (‚gə 'tā shən) liberation of water droplets from the edges and tips of leaves 399

gymnosperm ('jim nə ‚spərm) a vascular plant producing naked seeds, as in conifers 296

H

habitat ('hab ə ‚tat) region of an ecosystem occupied by an organism 631

haploid ('hap ‚loid) a term referring to chromosomes occurring singularly, as in sex cells produced by meiosis 139

Hardy-Weinberg (‚härd ē 'wīn ‚bərg) a law stating that the frequency of two alleles in a population remains stable under certain assumptions such as random mating; therefore, no change or evolution occurs 218

heartwood ('härt ‚wud) central region of stem xylem that no longer functions in water transport 377

heliostat ('hē lē ə ‚stat) large-field mirrors that track the sun and reflect its energy onto a mounted boiler for solar heating 700

hemocoel ('hē mə ‚sēl) a coelomic body cavity through which blood circulates in certain invertebrates; i.e., arthropods 326

hemoglobin ('hē mə ‚glō bən) red pigment of erythrocytes for transport of oxygen 428

herbaceous (‚ər bā shəs) a plant that lacks persistent woody tissue 375

herbivore ('ər bə ‚vōr) a primary consumer in a food chain; a plant eater 90

hermaphrodite (‚hər 'maf rə ‚dīt) an animal having both male and female sex organs 313

herpes ('hər ‚pēz) a virus, one type of which causes a sexually-transmitted disease 595

heterocyst ('het ə rō ‚sist) a cell specialized for carrying out nitrogen fixation; present in certain cyanobacteria 268

heterosis (‚het ə 'rō səs) see hybrid vigor

heterospores (‚het ə rə 'spōrz) spores of two types; i.e., microspores and megaspores 296

heterotrophic (‚het ə rə 'trō fik) referring to an organism that cannot synthesize organic compounds from inorganic substances and therefore must acquire food from external sources 90

heterozygote superiority (‚het ə rō 'zī gōt sù ‚pir ē 'ór ət ē) fitness, for survival and reproduction, of individuals having the heterozygote genotype over those with either homozygous genotype 250

heterozygous (‚het ə rō 'zi gəs) alleles of a pair are different in identity, one being dominant and the other recessive 152

histone ('his tōn) a family of simple, water-soluble proteins containing a large percentage of basic amino acids; these contribute to chromosome structure 172

homeostasis (‚hō mē ō 'stā səs) a state of stability in body physiology despite any variations in the environment; i.e., relatively constant temperature, pH, blood sugar 8

homing ('hōm iŋ) returning to a home base, as a pigeon might be trained to do 560

hominoid ('häm ə ‚nóid) term for humans and apes 351

homologous pair (hō 'mäl ə gəs 'paər) pairs of chromosomes that bear the same gene loci, and synapse during prophase of the first meiotic division 133

homologous structure (hō 'mäl ə gəs 'strək chər) in evolution, fundamentally similar structures from a common ancestor 243

homozygous (ˌhō mə 'zī gəs) alleles of a pair are the same in identity, both being dominant or both being recessive 152

hormone ('hȯr ˌmōn) a chemical secreted in one part of the body that controls the activity of other parts 407

hybrid ('hī brəd) heterozygous genotype due to reproduction between two different strains 223

hybrid vigor ('hī brəd 'vig ər) a phenomenon whereby a heterozygote resulting from the crossing of two different strains is more fit than either homozygous parent 224

hydrocarbon (ˌhī drə 'kär bən) organic compound containing covalently bonded hydrogen and carbon atoms 711

hydrogen bond ('hī drə gən 'bänd) a weak bond that arises between a partially positive hydrogen and a partially negative oxygen, often on different molecules or separated by some distance 43

hydrolysis (hī 'dräl ə səs) splitting of a compound into parts in a reaction that involves addition of water, with the H⁺ ion being incorporated in one fragment and the OH⁻ ion in the other 52

hydroponics ('hī drə pän iks) the growth of plants in aqueous solutions having nutritional requirements 392

hypertension ('hī pər ˌten chən) elevated blood pressure, particularly the diastolic pressure 438

hypertonic (ˌhī pər 'tän ik) lesser water concentration, greater solute concentration than the cytoplasm of a particular cell 69

hypha ('hī fə) a filament of the vegetative body of a fungus 275

hypocotyl ('hī pə ˌkät əl) plant embryo portion below the cotyledon that contributes to development of stem 364

hypolimnion (ˌhī pō 'lim nē ˌän) a water layer extending from the thermocline to the bottom of a lake, its temperature being fairly constant and cold 666

hypothesis (hī 'päth ə səs) a supposition that is established by reasoning after consideration of available evidence and that can be tested by obtaining more data, often by experimentation 12

hypotonic (ˌhī pə 'tän ik) greater water concentration, lesser solute concentration than the cytoplasm of a particular cell 69

I

immunity (im 'yü nət ē) the ability to resist infection because of adequate production of specific antibodies 448

imprinting ('im ˌprint iŋ) a behavior pattern from early development in waterfowl; the tendency to follow the first moving object seen 563

incisor (in 'sī zər) chisel-shaped front tooth for biting and cutting food 459

independent assortment (ˌin də 'pen dənt ə 'sȯrt mənt) in heredity, referring to the independent segregation of the allelic pair relative to another pair during sex cell formation; i.e., AaBb genotype yields AB,Ab,aB,ab in gametes 157

indolacetic acid ('in ˌdōl ə ˌsēt ik 'as əd) plant growth regulator, known also as auxin 409

induction (in 'dək shən) in embryology, the ability of one body part to influence the development of another part 607

inductive reasoning (in 'dək tiv 'rēz niŋ i ŋ) a process of logic and reasoning, using specific observations to arrive at a hypothesis 12

insight ('in ˌsīt) ability to respond correctly to a new, different situation the first time it is encountered 564

interferon (ˌint ər 'fir ˌän) an antiviral agent produced by body cells 451

interphase ('int ər ˌfāz) stage of cell cycle during which DNA replication and growth are occurring and the nucleus is not actively dividing 138

inversion (in 'vər zhən) a mutation in which there is reversal of the order of genes on a chromosome by the breakage and reattachment of that fragment in opposite order 182

invertebrate (ˌin 'vərt ə brət) referring to an animal without a serial arrangement of vertebrae or a backbone 305

ionic bond (ī 'än ik 'bänd) an attraction between charged particles (ions) through their opposite charge 41

isogamete (ˌī sō gə 'mēt) a sex cell similar in shape and size to the one with which it unites 284

isomer ('ī sə mər) molecules with the same molecular formula but different structural formula and, therefore, shape 52

isotonic (ī sə 'tän ik) a solution that is equal in solute and water concentration to that of the cytoplasm of a particular cell 69

isotope (ī sə ˌtōp) forms of an element having the same atomic number but a different atomic mass due to a different number of neutrons 40

K

karyotype ('kar ē ə ˌtīp) an organized display of the homologous chromosomes cut from a photograph of the nucleus just prior to cell division 133

kingdom ('kiŋ dəm) a taxonomic category grouping related phyla (animals) or divisions (plants) 25

Klinefelter's syndrome ('klin ˌfel tərz 'sin ˌdrōm) a condition caused by the inheritance of a chromosome abnormality in number; an XXY (trisomy) condition where normally a chromosome pair exists 185

Krebs cycle ('krebz 'sī kəl) *see* citric acid cycle

kwashiorkor (ˌkwäsh ē 'ȯr kər) malnutrition from a protein-deficient diet, yielding a bloated body and discolored skin 704

L

labia major ('lā bē ə 'ma jər) the larger member of two pairs of folds lying lateral to the vaginal opening 587

labia minor ('lā bē ə 'mī nər) the smaller member of two pairs of folds lying lateral to the vaginal opening 587

lancelet ('lan slət) a marine chordate of the phylum cephalochordates 334

larva ('lär və) immature insect form between egg and pupal stage 308

larynx ('lar iŋks) voicebox, cartilage-made box for airflow between the larynx and trachea 482

lens ('lenz) a football-shaped, biconvex structure of the eye, located in back of the iris and aqueous humor that functions to focus light 534

lenticel ('lent ə 'sel) a region of loosely arrayed stem cells that allow passage of air into the stem 378

lichen ('lī kən) a symbiotic relationship between certain fungi and algae that has long been thought to be mutualistic, the fungi providing inorganic and the algae providing organic food 268

life cycle ('līf 'sī kəl) the complete life history of an organism, ranging from one stage through all following steps to that original stage 8

limnetic zone (lim 'net ik 'zōn) relating to organisms, inhabiting open sunlit waters of lakes and inland seas 668

linkage group ('liŋ kij grüp) the tendency of certain traits to be inherited together because their respective allelic pairs are on the same chromosome pair 173

lipase ('lī ˌpās) an enzyme catalyzing the chemical breakdown of fat molecules in the small intestine 463

lipid ('lip əd) a family of organic compounds that tend to be soluble in nonpolar rather than polar solvents such as water 56

littoral zone ('lit ə rəl 'zōn) pertaining to a seashore zone, especially between high and low tide lines 668

locus ('lō kəs) specific cite or location of a gene on a chromosome 152

logistic curve (lō 'jis tik 'kərv) an S-shaped curve that represents exponential growth of a population followed by a leveling off 624

long-day plant ('lȯŋ 'dā 'plant) a plant that flowers when daylight becomes longer or the dark period becomes shorter than some critical length 418

lymph ('limf) tissue fluid collected by the lymphatic system and returned to the general systemic circuit 440

M

macronutrient ('mak ‚rō 'nü trē ənt)   plant nutritional requirement; needed in relatively large amounts, such as with nitrogen, phosphorous, and potassium   391

Malpighian tubules (mal ‚pig ē ən   'tü ‚byülz)   blind, threadlike tubules for excretion of wastes near anterior end of insect hindgut   326

mammal ('mam əl)   a class of vertebrates characterized especially by the presence of hair and mammary glands   344

mantle ('mant əl)   membrane common in mollusks that encloses the visceral mass and may secrete a shell   318

mapping ('map ing)   plotting the identity, spacing, and sequence of genes along a chromosome   175

marasmus (mə 'raz məs)   malfunction condition from a low-calorie, low-protein diet that yields shriveled skin, wide eyes, and bloated belly   704

marsupial (mär 'sü pē əl)   a mammal bearing immature young nursed in a marsupium, or pouch; i.e., kangaroo, oppossum   347

matrix ('mā triks)   general term for a background substance, such as in mitochondrion's interior   119

medulla oblongata (mə 'dəl ə   'äb ‚lȯŋ gä tä)   a brain region at the base of the brainstem controlling heartbeat, blood pressure, breathing, and other vital functions   516

medusa (mi 'dü sə)   free-swimming sexual form of the coelenterate life cycle   309

meiosis (mī 'ō səs)   type of nuclear division that occurs during the production of gametes, by means of which the daughter cells receive the haploid number of chromosomes   138

meninges (‚men ən 'gēz)   several external layers encasing the brain and spinal cord   515

meristem ('mer ə ‚stem)   undifferentiated, embryonic tissue in the active growth regions of plants   364

mesoderm ('mez ə ‚dərm)   middle layer of an animal's three primary germ layers   312

mesoglea (‚mez ə 'glē ə)   a jellylike layer between the epidermis and endodermis of coelenterates   308

mesophyll tissue ('mez ə ‚fil   'tish ‚ü)   tissue in leaf's interior made of packed and loosely arrayed parenchymal cells   383

Mesozoic (‚mez ə 'zō ik)   a geologic era between the Paleozoic and Cenozoic; Age of Reptiles with Triassic, Jurassic, and Cretaceous periods   238

messenger RNA ('mes n jər)   a nucleic acid (ribonucleic acid) complementary to genetic DNA and bearing a message to direct cell protein synthesis at the ribosome   196

metabolic pool (‚met ə 'bäl ik 'pül)   metabolites that are the products of and/or the substrates of key reactions in cells allowing one type of molecule to be changed into another type, such as the conversion of carbohydrates to fats   122

metabolism (mə 'tab ə ‚liz əm)   the sum total of all chemical reactions in a cell   122

metamorphosis (‚met ə 'mȯr fə səs)   change in shape and form that an animal undergoes during development   327

metaphase ('met ə ‚fāz)   stage of mitosis during which chromosomes are lined up at the equator of the mitotic spindle   136

metastasis (mə 'tas tə səs)   the tendency of cells to wander through the body, as in cancer cells   483

micronutrient ('mī ‚krō 'nü trē ənt)   plant nutritional requirement; needed in only trace amounts as with zinc or copper   391

microsphere ('mī krə ‚sfir)   a cell-like structure that results when proteinoids are placed in water; perhaps an evolutionary precursor of the first true cell   261

microspore ('mī krə ‚spȯr)   the smaller of two spores produced by a heterosporous plant   296

migration (mī 'grā shən)   the movement of organisms to and from a geographical site   560

mimicry ('mim i krē)   superficial resemblance of one organism to one of another species; often used advantageously   636

mitochondrion (‚mīt ə 'kän drē ən)   membrane-bound cell organelle known as the powerhouse because it transforms particularly the energy of carbohydrates and fats to ATP energy   119

mitosis (mī 'tō səs)   a process in which a parent nucleus reproduces two daughter nuclei, each identical to the parent nucleus; this division is necessary for growth and development   133

molar ('mō lər)   a broad, many-cusped tooth for grinding and chewing food; after the premolar in tooth array   459

molecular clock (mə 'lek yə lər   'kläk)   the hypothesis that the number of base substitutions in DNA and/or the number of amino acid changes in a protein indicates the relative length of time that organisms have diverged from a common ancestor   246

molecule ('mäl i kyül)   a unit of a chemical substance formed by the union of two or more atoms by covalent bonding   41

monocotyledon (‚män ə ‚kät l'ēd n)   a flowering plant group; members show one embryonic leaf, parallel leaf venation, scattered vascular bundles, and other characteristics   300

monoculture farming ('män ə ‚kəl chər   färm ŋ)   farming procedure that cultivates only one species for each type of crop   653

monohybrid cross (‚män ō' hī brəd   'krȯs)   reproduction between two parents, both heterozygous for a trait; i.e., Bb × Bb   155

monotreme ('män ə ‚trēm)   an egg-laying mammal; i.e., duckbill platypus or spiny anteater   345

morphogenesis (‚mȯr f ə 'jen ə səs)   movement of early embryonic cells to establish body outline and form   598

morula ('mȯr ə lə)   mulberry-shaped mass of cells due to cleavage, prior to blastula formation during embryonic development   600

motor ('mōt ər)   referring to an output or relay of neural signals away from the central nervous system for the control of muscles   504

mutation (mü 'tā shən)   a change in inheritance due to either a chromosomal or a genetic alteration   210

mutualism ('müch ə wə ‚liz əm)   a symbiotic relationship in which both species benefit   642

mycelium (mī 'sē lē əm)   a tangled mass of hyphal filaments composing the vegetative body of a fungus   275

myocardium (‚mī ə 'kärd ē əm)   heart muscle   435

myofibril (mī ō 'fīb rəl)   a specific muscle cell organelle containing a linear arrangement of sarcomeres that shorten to produce muscle contraction   546

myosin ('mī ə sən)   a muscle protein making up the thick filaments in a sarcomere; it pulls actin to shorten the sarcomere, yielding muscle contraction   546

N

NAD (‚en ‚ā 'dē)   a coenzyme that functions as an electron acceptor or donor in varous cellular oxidation-reduction reactions   97

natural selection ('nach ə rəl   sə 'lek shən)   the guiding force of evolution caused by environmental selection of organisms most fit to reproduce, resulting in adaptation   20

nematocyst ('nem ət ə ‚sist)   threadlike fiber enclosed within the capsule of a stinging cell; when released, aids in the capture of prey   309

neotony ('nē ə tē nē)   shifting of reproductive ability to a stage of life previous to the adult stage   330

nephridium (ni 'frid ē əm)   for invertebrates, a tubular structure for excretion; its contents are released outside through a nephridiopore   490

nephron ('nef ‚rän)   microscopic kidney unit that regulates blood composition by filtration and reabsorption; over one million per human kidney   492

neritic (nə 'rit ik)   pertaining to organisms inhabiting waters above the continental shelf   674

neurotransmitter (‚nu̇r ō trans 'mit ər)   a chemical made at the ends of axons that is responsible for transmission across a synapse or neuromuscular junction   508, 549

niche ('nich)   total description of an organism's functional role in an ecosystem, from activities to reproduction   631

nitrification (‚nī trə fə 'kā shən)   in the nitrogen cycle, conversion of ammonia to nitrates by microorganisms   661

nitrogen fixation ('nī trə jen   fik 'sā shən)   a process whereby free atmospheric nitrogen is converted into compounds, such as ammonia and nitrates, usually by soil bacteria   267

node ('nōd)   region on a stem for leaf attachment   373

nondisjunction (‚nän dis 'jəŋk shən)   failure of homologous chromosomes (or chromatid pair) to separate during meiosis, yielding an abnormal chromosome number in the produced sex cell   184

norepinephrine ('nȯr ,ep ə ,nef rən) a hormone released by the adrenal medulla as a reaction to stress; also a neurotransmitter released by sympathetic neurons of the autonomic nervous system and between brain cells 508

notochord ('nōt ə ,kȯrd) a dorsal, elongated, supporting structure in chordates beneath the neural tube or spinal cord present in the embryo of all vertebrates 333

nucleolus (nü 'klē ə ləs) a dark-staining, spherical body in the cell nucleus that contains RNA 131

nucleotide ('nü klē ə ,tīd) the building block subunit of DNA and RNA consisting of a five carbon sugar bonded to a nitrogen base and phosphorus group 62

O

observation (,äb sər 'vā shən) close scrutiny of a phenomenon, a necessary part of the scientific method 12

oceanic ridges (,ō shē 'an ik rid 'jez) mountain chains found on the abyssal plain substrate 673

oligotrophic (,äl i go 'trō fik) referring to a clear-water lake lacking significant organic matter or debris 668

omnivore ('äm ni ,vȯr) a food chain organism feeding on both plant bodies (herbivore) and animals (carnivore) 454

oncogene ('ȯŋ kō jēn) a cancer-causing gene 213

ontogeny (än 'täj ə nē) embryological development, stage by stage 607

oogenesis (,ō ə 'jen ə səs) production of eggs in females by the process of meiosis and maturation 142

operant conditioning ('äp ə rənt kən 'dish ə niŋ) the association, as a result of reinforcement, of a response with a stimulus with which it was not previously associated 564

operculum (ō 'pər kyə ləm) a lid or covering over the gills or respiratory structures in some invertebrates and fishes 339

operon ('äp ə ,rän) an operator gene and the adjacent group of genes it controls 204

opportunist (,äp ər 'tü nəst) an organism that is able to flourish in a changing environmental circumstance 627

order ('ȯrd ər) a taxonomic category between the family and class; a group of related families 25

organelle (,ȯr gə 'nel) a cell region or structure in the cytoplasm having a specific function 74

organochloride (ȯr ,gan ə 'klȯr ,īd) chlorinated hydrocarbons such as pesticides and PCBs 711

organ of Corti ('ȯr gən əv 'kȯr tī) specialized region of the cochlea containing the hair cells of sound detection and discrimination 540

osmosis (äz 'mō səs) the diffusion of water through a selectively permeable membrane 67

ovary ('ōv ə rē) 1. enlarged, base portion of the pistil that eventually develops into the fruit, 300; 2. female gonad; site of oogenesis plus estrogen and progesterone production 585

oviduct ('ō və ,dəkt) tubular portion of the female reproductive tract from ovary to the uterus 587

ovipositor ('ō və 'päz ət ər) insect structure for deposition of eggs in soil 327

ovule ('äv yül) in seed plants, a structure that contains the megasporangium where meiosis occurs and the female gametophyte is produced; develops into the seed 300

oxidation (,äk sə 'dā shən) a chemical reaction that results in removal of one or more electrons from an atom, ion, or compound; oxidation of one substance occurs simultaneously with reduction of another 97

oxidative phosphorylation ('äk sə ,dāt iv fäs ,fȯr ə 'lā shən) the process of building ATP from a coupled oxidation reaction in mitochondria 117

oxyhemoglobin ('äk si 'hē mə ,glō bən) hemoglobin bound to oxygen in a loose, reversible way; formed in the lungs and dissociated at tissue cells for oxygen release 484

P

Paleozoic (,pā lē ə 'zō ik) a geologic era prior to the Mesozoic; the Age of Ancient Life marked by the rise of land animals and plants 238

parapodium (,par ə 'pōd ē əm) paired fleshy appendages from each segment of marine annelids 322

parasitism ('par ə sə ,tiz em) a symbiotic relationship in which one species benefits for growth and reproduction (parasite) to the harm of the other species (host) 638

parenchyma (pə 'ren kə mə) a relatively nonspecialized plant cell found in pith and cortex of stems, cortex of roots and spongy layer of leaves 366

parthenogenesis ('pär thə nō 'jen ə səs) development of an egg cell without fertilization 579

parturition (,pärt ə 'rish ən) passageway and delivery of a newborn organism through the terminal portion of the female reproductive tract 614

pedigree ('ped ə ,grē) referring to family tree; in genetics, a diagram indicating the phenotypes and genotypes of an individual's ancestor 179

pelagic (pə 'laj ik) the open portion of the sea 674

pepsin ('pep sən) an enzyme in the stomach that initiates chemical digestion of proteins into smaller fragments 462

peptidase ('pep tə ,dās) an enzyme that catalyzes the chemical breakdown of protein fragments (peptides) in the small intestine 464

peristalsis (,per ə 'stȯl səs) the rhythmic, wavelike contraction that moves food through the digestive tract 462

permeable ('pər mē ə bəl) allowing materials to pass through 67

petal ('pet əl) a modified, colored leaf in a flower 300

phagocytosis (,fag ə sə 'tō səs) cellular eating by engulfing within a vacuole pinched off the cell membrane 270

pharynx ('far iŋks) a common passageway (throat) for both food intake and air movement 454

phenotype ('fē nə ,tīp) the visible expression or observable result of a genotype; i.e., brown eyes, height 152

phenylketonuria (,fen l ,kēt n ur ē ə) birth defect through a recessive gene by which newborn babies lack the enzyme to properly metabolize the amino acid phenylalanine in their diet 193

pheromone ('fer ə ,mōn) substance produced and discharged into the environment by an organism that influences the behavior of another organism 559

phloem ('flō em) vascular tissue conducting organic solutes in plants; contains sieve tube cells and companion cells 292

phospholipid (,fäs fō 'lip əd) a family of derived neutral fat molecules, whereby one of the bonded fatty acids is replaced by a phosphorous-containing group 58

photoperiod ('fōt ə 'pir ē əd) relative lengths of daylight and darkness that affect the physiology and behavior of an organism 417

photosynthesis ('fōt ō 'sin thə səs) a process whereby chlorophyll-containing organisms trap sunlight energy to build a sugar from carbon dioxide and water 93

phototaxis (,fōt ō 'tak səs) movement of an organism toward a stimulus of light 559

phototropism (fō 'tä trə ,piz əm) growth in relation to sunlight; shoots display positive geotropism so that they bend toward sunlight 409

photovoltaic cell (,phō tō väl 'tā ik 'sel) a manufacture mechanism that uses sunlight to produce an electromotive force 700

phylogeny (fī 'läj ə nē) evolutionary history of a species through a study of its related ancestors 607

phylum ('fī ləm) a taxonomic category between the class and kingdom; a group of related classes 25

phytochrome ('fīt ə ,krōm) a plant photoreversible pigment; this reversion rate of two molecular forms is believed to measure the photoperiod 419

phytoplankton ('fit ō ,plank tən) microscopic, free-floating plantlike microorganisms of aquatic ecosystems 665

pistil ('pis tl) a female flower structure consisting of an ovary, style, and stigma 300

pith ('pith) soft, central, spongy portion of angiosperm stem made of parenchymal cells 377

placenta (plə 'sent ə) a structure formed from an extraembryonic membrane, the chorion and uterine tissue through which nutrient and waste exchange occur for the embryo and fetus 590

placoderm ('plak ō ,dərm) first jawed vertebrate according to the fossil record 337

plasmid ('plaz məd) a self-duplicating ring of DNA in addition to the main chromosome in the cytoplasm of bacteria 206

platelet ('plāt lət)   cell-like disks formed from fragmentation of megakaryocytes that initiate blood clotting in vertebrate blood   442

plate tectonics ('plāt tek 'tän iks)   the hypothesis that the continents lie on plates that are slowly moving in relation to one another resulting in the phenomenon of continental drift   242

pleiotropic gene (ˌplī ə 'trōp ik   'jēn)   a gene that codes for an enzyme present in many metabolic pathways and therefore has a widespread effect on the body   166

polar body ('pō lər   'bäd ē)   a nonfunctional product of oogenesis; three of the four meiotic products are of this type   142

pollination ('päl ə 'nā shən)   transfer of pollen from an anther to a stigma   296

polymorphism (ˌpäl i 'mȯr fiz əm)   the occurrence in the same habitat of two or more distinct forms of a species   228

polyp ('päl əp)   sessile form of the coelenterate life cycle that usually reproduces asexually   309

polypeptide (ˌpäl i pep ˌtīd)   two or more amino acids linked by a peptide bond   59

polyploidy ('päl i ˌplȯid ē)   a condition of an entire extra set of chromosomes in addition to the diploid set; i.e., triploid, pentaploid, etc.   182

polysaccharide (ˌpäl ə 'sak ə ˌrīd)   large carbohydrate molecule with many repetitive subunits (monosaccharides) chemically bonded together   54

polysome ('päl i ˌsōm)   a string of ribosomes, simultaneously translating different regions of the same mRNA strand during protein synthesis   201

population (ˌpäp yə 'lā shən)   a group of organisms of the same species occupying a certain area and sharing a common gene pool   218

portal ('pȯrt əl)   a circulatory pathway that begins and ends in capillaries   434

Precambrian (ˌprē 'kam brē ən)   time prior to the Cambrian period (of the Paleozoic era) during which the first life forms arose   238

predation (pri 'dā shən)   relationship whereby a predator, often larger in size, hunts a prey species for its nutrition and survival   633

preformation (ˌprē fȯr 'mā shən)   theory, incorrect, that development is but an enlargement of a miniature, preformed organism in the sex cell   598

premolar (ˌprē 'mō lər)   a broad, many-cusped tooth for grinding and chewing food, preceding the molar in tooth array   459

pressure-flow hypothesis ('presh ər   'flō hī päth ə səs)   hypothesis explaining transport through sieve tube cells of phloem, with leaves serving as a "source" and roots as a "sink" in the summer and vice versa in the spring   404

primate ('prī ˌmāt)   an animal order of mammals having hands and feet with five distinct digits, and some having an opposable thumb   349

producer (prə 'dü sər)   organisms in a food chain that make their own foods; they start the chain and are mainly green plants or other types of chlorophyll-containing organisms   29

profundal zone (prə 'fənd əl   'zōn)   pertaining to the deepest part of a body of water   668

prokaryotic (ˌprō ˌkar ē 'ät ik)   the first, primitive cells on earth, exemplified today in cyanobacteria and bacteria, lacking a defined nucleus and most organelles   72

prophase ('prō ˌfāz)   first active stage of mitosis during which chromosomes and a mitotic spindle appear   134

prosimian (prō 'sim ē ən)   the first primates; premonkey   349

prostate gland ('präs ˌtāt 'gland)   a large, donut-shaped gland beneath the bladder that adds seminal fluid to migrating sperm   585

protein (ˌprō ˌtēn)   a family of organic compounds consisting of at least carbon, hydrogen, oxygen, and nitrogen atoms, produced by covalent bonding of amino acids   59

proteinoids ('prō tēn ȯidz)   amino acid polymers perhaps like those that first formed during a chemical evolution that preceded the first cell(s)   261

prothallus (ˌprō 'thal əs)   a heart-shaped, gamete-producing plant in the fern life cycle   295

protist ('prōt əst)   an organism that belongs to the kingdom Protista that contains organisms that have characteristics of both plants and animals; for example, protozoans and some algae   270

protocell ('prōt ə 'sel)   a cell forerunner developed from cell-like microspheres   261

protostome ('prōt ō ˌstōm)   a group of coelomate animals in which the blastopore (the first embryonic opening) becomes the mouth   307

pseudocoelom (ˌsüd ə 'sē ləm)   a coelom that is not completely lined by mesoderm   315

pseudopodia (ˌsüd ə 'pōd ē ə)   false feet; temporary projections of cytoplasm serving, as in amoeba, for locomotion and engulfment of substances   270

pulmonary ('pùl mə ˌner ē)   referring to lung   432

pulse ('pəls)   vibration felt in arterial walls due to expansion of the aorta following ventricle contraction   436

Punnett square ('pən ət   'skwar)   a gridlike device that enables one to calculate the results of simple genetic crosses by lining gametic genotypes of two parents on the outside margin and their recombination in boxes inside the grid   154

pupa ('pyü pə)   insect stage between larva and winged adult; the change occurring here by metamorphosis   327

purine ('pyür ˌēn)   a type of nitrogenous base, having a double-ring structure, in a nucleotide, as in adenine and guanine   190

pyrimidine (pi 'rim ə ˌdēn)   a type of nitrogenous base, having a single-ring structure, in a nucleotide, as in cytosine, thymine, and uracil   190

pyruvic acid (pī ˌrü vik   'as əd)   the end product of glycolysis; its further fate involving fermentation or Krebs cycle incorporation depending on oxygen availability   112

R

radial symmetry ('rād ē əl   'sim ə ˌtrē)   body plan in which similar parts distribute uniformly around a central axis like spokes of a wheel   308

radula ('raj ə lə)   platelike structure equipped with rows of teeth, present in the pharynx of most mollusks   319

ramapithecene ('räm ə ˌpith ə sēn)   a fossil whose relationship to humans is in dispute since it now appears to be an ancestral orangutan   353

receptor (ri 'sep tər)   a cell, often in groups, that detects an environmental change or stimulus and initiates a nerve impulse   529

recessive allele (ri 'ses iv   ə 'lēl)   form of a gene whose effect is hidden by the dominant allele in a heterozygote   152

reduction (ri 'dək shən)   a chemical reaction that results in addition of one or more electrons to an atom, ion, or compound. Reduction of one substance occurs simultaneously with oxidation of another   97

reflex ('rē ˌfleks)   a predetermined, automatic response of an organism to a stimulus that does not require conscious intervention   560

regeneration (ri ˌjen ə 'rā shən)   reforming a lost body part from the remaining body mass by active cell division and growth   579

releaser (ri 'lē sər)   a structure, action, or sound that triggers a behavioral response   568

replication (ˌrep lə 'kā shən)   the duplication of DNA; occurs when the cell is not dividing   138, 192

repolarization (ˌrē 'pō lə rə 'zā shən)   the recovery of a neuron's polarity to the resting potential after it ceases transmitting impulses   506

resolution (ˌrez ə 'lü shən)   the ability of a microscope to form images in sharp detail; to distinguish separate particles optically   65

respiration (ˌres pə 'rā shən)   1. gas exchange, 475; 2. anaerobic respiration is the buildup of ATP during fermentation, 121; 3. aerobic respiration is the buildup of ATP by glycolysis, the Krebs cycle, and the respiratory chain   115

respiratory chain ('res pər ə ˌtȯr ē   'chān)   a series of electron carriers that begins with NAD and ends with cytochrome oxidase, the enzyme that reduces oxygen to water. As electrons are passed from one carrier to the next, energy is released and ATP is synthesized; also called electron transport system and cytochrome system   117

retina ('ret ən ə)   innermost layer of the eyeball containing the light receptors—rods and cones   535

rhizome ('rī ˌzōm)   a rootlike, underground stem   378

rhodopsin (rō 'däp sən)   a pigment contained in retinal rod cells; activated in dim light for twilight vision   536

ribonucleic acid (ˌrī bō nü klē ik   'as əd)   RNA; an organic molecule having a specific sequence of nucleotide subunits, dictated by DNA in genetic systems   61

ribosome ('rī bə ˌsōm)   cytoplasmic organelle; site of protein synthesis   200

root pressure ('rüt 'presh ər) a force generated by an osmotic gradient that serves to elevate sap through xylem a short distance 399

rumen ('rü mən) the first of four stomach chambers in certain animals, such as a cow 458

S

sac body plan ('sak 'bäd ē 'plan) a body with a digestive cavity that has only one opening, as in coelenterates and flatworms 308

salivary amylase ('sal ə ˌver ē 'am ə ˌlās) an enzyme that hydrolysizes starch to maltose and is found in saliva 459

SA node (əs ā 'nōd) mass of specialized tissue in right atrium wall of heart that initiates a heartbeat; the "pacemaker" 435

saltwater intrusion (ˌsȯlt ˌwȯt ər in 'trü zhən) the backup of ocean seawater into streams and aquifers depleted of fresh water 719

saprophytic decomposers (ˌsap rə 'fit ik ˌdē kəm 'pō zerz) heterotrophic organisms such as fungi and bacteria that live on dead organic material 267

sapwood ('sap ˌwu̇d) outer region of stem xylem, functional for transport and food storage 377

sarcomere ('sär kə ˌmir) a unit of a myofibril, many being arranged linearly along its length, that shortens to give muscle contraction 548

scientific method (ˌsī ən 'tif ik 'meth əd) a step-by-step process for discovery and generation of knowledge, ranging in steps from observation and hypothesis to theory and law 12

sclerenchyma (sklə 'ren kə mə) supportive tissue of higher plants; the nonliving cells contain thick, reinforcing walls 366

secretin (si 'krēt ən) a hormone secreted by the small intestine to regulate the release of pancreatic juice 465

seed ('sēd) a mature ovule that contains an embryo with stored food, enclosed in a protective coat 300

segmentation (ˌseg mən 'tā shən) the presence of repeating body units as is seen in the earthworm 321

segregation (ˌseg ri 'gä shən) in heredity, referring to the separation of alleles into different gametes 152

selective reabsorption (sə 'lek tiv ˌrē əb 'sȯrp shən) one of the processes involved in the formation of urine; involves the greater reabsorption of nutrient molecules compared to waste molecules from the contents of the kidney tubule into the blood 495

semicircular canals (ˌsem i sər kyə lər ka 'nalz) three half-circle-shaped canals of the inner ear, fluid-filled for registering changes for balance in motion 539

seminal vesicle (ˌsem ən əl 'ves i kəl) a gland adding seminal fluid to migrating sperm cells at the termination of their movement through the ductus deferens 585

seminiferous tubule (ˌsem ə 'nif ə rəs 'tü ˌbyül) a long, coiled structure contained within chambers of the testis where sperm are produced 582

sensory ('sen sə rē) referring to an input or relay of neural signals from receptors into the central nervous system 504

sepal ('sēp əl) a modified leaf in a flower calyx 300

sere ('siər) a series of successional changes from beginning stage to climax stage 676

seta ('sēt ə) bristlelike, chitinous projection from the annelid body 322

sex linkage ('seks 'lin kij) referring to the inheritance pattern of alleles located on the sex chromosomes, most often the X chromosome; i.e., hemophilia, color blindness 176

short-day plant ('shȯrt ˌdā 'plant) plant that flowers when daylight becomes shorter or the dark period becomes longer than some critical length 418

sickle-cell anemia ('sik əl 'sel ə 'nē m ē ə) an inherited, chronic anemia in which a large portion of the erythrocytes are sickle-shaped, depleting oxygen delivery to tissue cells, due to an abnormal hemoglobin 249

sinkhole ('sink ˌhōl) evacuated underground space, its water removed by groundwater mining, causing caverns to collapse without this water support 719

social releaser ('sō shəl ri 'lē sər) type of communication between members of the same group that evokes a social response; i.e., a color change that indicates readiness to mate 569

sociobiology (ˌsō sē ō bī 'äl ə je) a branch of biology offering genetic and evolutionary reasons for complex social behavior 573

solar collector (ˌsō lər kə 'lek tər) any manufactured structure that absorbs radiant energy 700

solute ('säl ˌyüt) the dissolved substance of a solution 68

solvent ('säl ˌvənt) the dissolving medium of a solution 68

somite ('sō ˌmīt) paired, mesodermal segments along the notochord that will develop primarily into the musculature 612

sorus ('sōr əs) a mass or cluster of sporangia found beneath leaflets of a fern frond 295

speciation (ˌspē shē 'ā shən) the process whereby a new species is produced or originates 230

species ('spē ˌshēz) a taxonomic category that is the subdivision of a genus; its members can breed successfully with each other but not with organisms of another species 25

spermatocyte (ˌspər 'mat ə 'sīt) a precursor cell, of both primary and secondary kinds, that produce a final sperm cell 583

spermatogenesis (ˌspər 'mat ə 'jen ə səs) production of sperm in males by the process of meiosis and maturation 142

spermatogonium (ˌspər 'mat ə 'gō nē əm) primordial germ cell that develops into a primary spermatocyte 582

spermatozoan (ˌspər 'mat ə 'zō ən) male sex cell or gamete; sperm 583

sphincter ('sfinj tər) a circular arrangement of muscle fibers, their contraction can close the opening of a vessel or chamber 429

spiracle ('spir i kəl) exterior opening of body wall for gas exchange in invertebrates, such as in insects 326

sporangium (spə 'ran jē əm) a specialized, caselike structure for spore production 276

spore ('spōr) 1. usually a haploid reproductive structure that develops into a haploid generation, 275; 2. in bacteria, a particularly resistant structure 266

sporophyte ('spōr ə ˌfīt) the spore-producing generation in the alternation-of-generations plant life cycle; usually a diploid organism 139

stamen ('stā mən) a pollen-producing flower structure consisting of an anther on a filament tip 300

starch ('stärch) a polysaccharide consisting of a linear bonding of the glucose, the major storage molecule of plants 55

steroid ('stiər ˌȯid) lipid soluble, biologically active molecules having four interlocking rings; examples are cholesterol, progesterone, testosterone 58

stigma ('stig mə) pistil portion that receives pollen 300

stoma ('stō mə) small opening between two guard cells on the underside of leaf epidermis; their regulated size controls the rate of gas exchange 383

stratosphere ('strat ə ˌsfir) next atmospheric layer above the earth's troposphere 713

strip mining ('strip 'mī niŋ) surface mining for a resource such as coal; exposes the mined resource to the landscape 708

style ('stīl) slender, long portion of a pistil between the stigma and the ovary 300

subsidence (səb 'sīd ənts) soil settling due to underground evacuation of water by mining 719

subspecies ('səb ˌspē shēz) subdivision of a species; different subspecies vary slightly from one another, but interbreeding is still possible 223

substrate ('səb ˌstrāt) the reactant in an enzymatic reaction; each enzyme has a specific substrate 95

succession (sək 'sesh ən) an orderly sequence of community replacement, one following the other, to an eventual climax community 676

survivorship (sər 'vī vər ˌship) usually shown graphically to depict death rates or percentage of remaining survivors of a population over time 626

symbiosis (ˌsim ˌbī ō səs) a close relationship between two different species, such as parasitism, mutualism, and commensalism 638

sympatric (ˌsim 'pa trik) when species occupy the same or overlapping regions 230

synapse ('sin ˌaps) a junction between neurons consisting of presynaptic (axon) membrane, the synaptic cleft, and the postsynaptic (usually dendrite) membrane 508

syphilis ('sif ə ləs) chronic, contagious venereal disease caused by a spirochete bacterium 596

systemic (sis 'tem ik) referring to general circulation serving overall body regions 430

systole ('sis tə ˌlē) contraction period of a heart during the cardiac cycle 435

T

taxis ('tak səs) movement of an organism toward a stimulus 559

taxonomy (tak 'sän ə mē) a system to meaningfully classify organisms into groups, based on similarities and differences, using morphology and evolution 25

teleological (ˌtel ē ə 'läj i kəl) assuming that a process is directed toward a final goal or purpose 20

telophase ('tel ə ˌfās) stage of mitosis during which the nuclei reform to give genetically identical daughter nuclei 136

territoriality (ˌter ə ˌtōr ē 'al ət ē) behavioral display by animals to closely guard a given space needed for reproduction from other intruders 570

test cross ('test 'kròs) a genetic mating in which a possible heterozygote is crossed with an organism homozygous recessive for the characteristic(s) in question in order to determine its genotype 155

testis ('tes təs) male gonad; site of spermatogenesis and testosterone production 582

testosterone (te 'stäs tə ˌrōn) male sex hormone produced from interstitial cells in the testis 583

thalamus ('thal ə məs) a lower forebrain region in vertebrates involved in crude sensory perception and screening messages intended for the higher forebrain or cerebrum 516

theory ('thē ə rē) a conceptual scheme arrived at by the scientific method 14

therapsid (thə 'rap səd) the reptilian ancestor from which the mammal evolved 343

thermal inversion ('thər məl 'in 'vər zhən) temperature inversion that traps cold air and its pollutants near the earth with the warm air above it 715

thermocline ('thər mə ˌklīn) the middle layer of water in a stratified lake that experiences a decline in temperature 666

thrombus ('thrämb bəs) stationary blood clot lodged in a vessel 438

trachea ('trā kē ə) an air tube (windpipe) of the respiratory tract in vertebrates; an air tube in insects 481

tracheid ('trā kē əd) xylem cell for water and dissolved mineral transport that is elongated and hollow with a thick, pitted cell wall 368

tracheophyte ('trā kē ə ˌfīt) vascular plants, including ferns and seed plants that have the sporophyte dominant 289

tract ('trakt) a bundle of neurons forming a transmission pathway through the brain and spinal cord 515

trait ('trāt) specific term for a distinguishing feature studied in heredity 150

transcription (trans 'krip shən) the process whereby the DNA code determines (is transcribed into) the sequence of codons in mRNA 197

translation (trans 'lā shən) the process whereby the sequence of codons in mRNA determines (is translated into) the sequence of amino acids in a polypeptide 197

translocation (ˌtrans lō 'kā shən) chromosome mutation due to the attachment of a broken chromosome fragment to a nonhomologous chromosome 182

transpiration (trans pə 'rā shən) the plant loss of water to the atmosphere, mainly through evaporation at leaf stomata 383

trichinosis (ˌtrik ə ˌnō səs) an infection that occurs when parasitic nematode forms encyst in the muscles of mammals and other vertebrates 317

trichocyst ('trik ə ˌsist) threadlike darts released by some ciliates that may help in defense against predators or in capturing prey 272

trochophore ('träk ə ˌfōr) ciliated, free-swimming larva, particularly of annelids and mollusks 320

trophic level ('trō fik 'lev əl) feeding level of one or more populations in a food web 650

troposphere ('trōp ə ˌsfir) atmospheric layer bordering earth's crust 713

trypsin ('trip sən) an enzyme catalyzing the chemical breakdown of protein molecules in the small intestine 463

tube foot ('tüb 'fut) tubular extension of an echinoderm's water vascular system ending in a sucker that aids in grasping and in locomotion 331

tube-within-a-tube body plan ('tüb with 'in a 'tüb 'bäd ē 'plan) a body with a digestive tract that has both a mouth and an anus 312

tunicate ('tü ni kət) a member of the subphylum urochordata, having the three chordate characteristics in the larval form 334

Turner's syndrome ('tər nərz 'sin ˌdrōm) a condition that results from the inheritance of an abnormality in chromosome number; an X chromosome lacks a homologous counterpart-XO 184

tympanum ('tim pə nəm) eardrum; membranous region that receives air vibrations in an auditory organ 326

typhlosole ('tif lə ˌsōl) a longitudinal enfolding of the intestinal wall of annelids that increases the area for absorption of digested food 455

U

uranium mill tailings (yu 'rā nē əm 'mil 'tā liŋz) radioactive sand remaining after uranium has been removed from mined material 699

ureter ('yur ət ər) a tubular structure conducting urine from kidney to urinary bladder 491

urethra (yu 'rē thrə) terminal structure of the urinary tract that receives urine from the bladder 491

uterus ('yüt ə rəs) pear-shaped portion of female reproductive tract that lies between the oviducts and the vagina 587

V

vagina (və 'ji nə) muscular tube leading from the uterus; the female copulatory organ and the birth canal 587

vascular cambium ('vas ˌkyə lər 'kam bē əm) embryonic tissue in the root and stem of a vascular plant that produces secondary xylem and phloem each year 377

vein ('vān) a blood vessel that transports blood toward the heart 429

vena cava ('vē nə 'kāv ə) one of two largest veins in the body; returns blood to the right atrium in a four-chambered heart 433

ventricle ('ven tri kəl) lower heart chamber, one on each side in a four-chambered heart; its pumping sends blood through an outgoing artery 429

venule ('vēn ˌyül) type of blood vessel distributing blood from capillaries to veins 430

vertebrate ('vərt ə brət) referring to an animal with a serial arrangement of vertebrae or having a backbone 305

vessel element ('ves əl 'el ə mənt) tubular, hollow xylem cell that forms a continuous pipeline for water and dissolved mineral transport 368

vestigial structure (ve 'stij əl 'strək chər) the remains of a structure that was functional in some ancestor but is no longer functional in the organism in question 246

vitamin (ˌvīt ə mən) an organic molecule that is required in small quantities for various biological processes and must be in an organism's diet because it cannot be synthesized by the organism 97

W

waggle dance ('wag əl 'dans) movement performed by worker bees to communicate distance and direction of food to other bees in the hive 569

X

xylem ('zī ləm) a vascular tissue that transports water and mineral solutes upward through the plant body 292

Z

zero population growth ('ze rō ˌpäp yə 'lā shən 'grōth) no growth in population size 691

zooplankton ('zō ˌplank tən) microscopic, free-floating animal-like microorganisms of aquatic ecosystems 665

zoospore ('zō ə ˌspōr) a motile spore produced by algae or fungi 283

zygote ('zī ˌgōt) the diploid (2N) cell formed by the union of two gametes, the product of fertilization 138

Chapter 13

13.3a: © Steven Brooke, **b:** © Sidney Fox; **page 263:** © Miryam Glickson; **13.5:** © T. J. Beveridge, University of Guelph/Biological Photo Service; **13.6a:** © Kessel, R. G., and Shih, C. Y. Scanning Electron Microscopy in Biology © 1976 Springer-Verlag, Berlin, **b:** © David Scharf/Peter Arnold, Inc., **c:** Kessel, R. G., and Shih, C. Y. Scanning Electron Microscopy in Biology. © 1976 Springer-Verlag, Berlin; **13.8a:** © R. Knauft, Biology Media/Photo Researchers, Inc., **b:** © David Hall, SPL/Photo Researchers, Inc.; **page 269(all):** © BioPhoto Associates; **13.9a:** © Biology Media/Photo Researchers, Inc.; **13.10:** © Eric Grave; **13.12:** © Kessel, R. G., and Shih, C. Y. Scanning Electron Microscopy in Biology. © 1976 Springer-Verlag, Berlin; **13.13:** © Carolina Biological Supply Company; **13.15:** © Gordon Leedale/BioPhoto Associates; **13.16(both):** © Eric Grave/Photo Researchers, Inc.; **13.19:** © Kitly Kahoot/Root Resources; **13.20a:** © William Ferguson, **b:** © Robert A. Ross; **13.21c:** © Gordon Leedale/BioPhoto Associates; **13.22:** © William E. Ferguson.

Chapter 14

14.3a: © Carolina Biological Supply Company; **14.4b:** © M. I. Walker/Photo Researchers, Inc.; **14.6a:** © J. Robert Waaland, University of Washington/Biological Photo Service; **14.7b:** © Runk/Schoenberger/Grant Heilman Photography, Inc.; **14.9:** © BioPhoto Associates/Photo Researchers, Inc.; **14.10a:** © Leonard Lee Rue III/Bruce Coleman, Inc., **b:** © Bob and Miriam Francis, **c:** © William E. Ferguson, **d:** © Kent Dannen/Photo Researchers, Inc.; **14.13:** © Field Museum of Natural History; **14.15a:** © Carolina Biological Supply Company, **b:** © BioPhoto Associates, **c:** © Carolina Biological Supply Company; **14.17:** © W. H. Hodge/Peter Arnold, Inc.; **14.18a:** © Albert Kunigh/Valan Photos, **b:** © John N. A. Lott/Biological Photo Services; **14.19:** © H. W. Elmore and R. J. Adams/Carolina Biological Supply Company; **14.21:** © Karlene Schwartz/Photo/Nats; **14.23:** © Carolina Biological Supply; **14.25a:** © Robert Lee/Photo Reseachers, Inc., **b:** © Bob Coyle; **page 301:** © Rob Sutter, North Carolina, Plant Conservation Program.

Chapter 15

15.1a: © R. J. Erwin/Photo Researchers, Inc., **b:** © James H. Carmichael/Bruce Coleman, Inc.; **15.3a:** © Bud Higdon; **15.4:** © J. Fiala; **15.8a:** © Carolina Biological Supply Company, **b:** © Ron Taylor/Bruce Coleman, Inc., **c:** © Carolina Biological Supply Company; **15.11:** © Eric Grave; **15.13b:** © Paul Nollen and Matthew Nadakavukaren; **15.14b:** © Dr. Fred Whittaker; **15.16a:** © Thomas Eisner; **15.17:** © Kirk Kreutzig/Microgaphics; **15.18:** © Markell, E. K., and Voge, M: Medical Parasitology, 4th ed. W. B. Saunders Co., 1981; **15.22a:** © Michael DiSpezio, **b:** © William E. Ferguson; **15.23b:** © F. Shiang Chia and R. Koss, University of Alberta; **15.24a:** © Michael DiSpezio, **b:** © BioPhoto Associates/NHPA; **15.26:** © Edward S. Ross; **15.27:** © Michael DiSpezio; **15.28a:** © James Carmichael/Bruce Coleman, Inc. **b:** Christopher Newbert/Bruce Coleman, Inc., **c:** © Kjell Sandved/Bruce Coleman, Inc., **d:** © William E. Ferguson, **e:** © Michael P. L. Fogden/Bruce Coleman, Inc., **f:** © William E. Ferguson; **15.29:** © David Scharf/Peter Arnold, Inc.

Chapter 16

16.1a: © Michael DiSpezio, **b:** © Robert A. Ross; **16.4a:** © Michael DiSpezio; **16.5b:** © Carolina Biological Supply Company; **16.8:** © Field Museum of Natural History, Painting by Charles R. Knight; **16.9a:** © Kenneth W. Fink/Berg and Associates, **b:** © Douglas Faulkner/Sally Faulkner Collection; **16.11(all):** © Jane Burton/Bruce Coleman, Inc.; **16.12:** Neg. # 322873 Courtesy Department of Library Services, American Museum of Natural History. **16.13a:** © Andrew Odum/Peter Arnold, Inc., **b:** © E. R. Degginger/Bruce Coleman, Inc.; **16.14a:** © Wolfgang Bayer; **16.16:** Courtesy Department of Library Services, American Museum of Natural History; **16.17:** © Conrad Kitsz/ANIMALS, ANIMALS, INC.; **16.18a:** © Graham Pizzey/Bruce Coleman, Inc., **b:** © Dallas Heaton/UniPhoto; **16.19a:** © Edward S. Ross, **b:** © Bob Coyle; **16.21:** © Rod Williams/Bruce Coleman, Inc.; **16.22:** © W. H. Muller/Peter Arnold, Inc.; **16.23a:** © Tom McHugh/Photo Researchers, Inc., **b:** © Ulrich Nebelsiek/Peter Arnold, Inc., **c:** © Dr. Helmut Albrecht/Bruce Coleman, Inc.; **16.24:** © Russ Kinne/Photo Researchers, Inc.; **16.30:** Courtesy Department of Library Services, American Museum of Natural History.

Chapter 17

17.9c, 17.11: © Carolina Biological Supply Company; **17.13(all):** © Gordon Leedale/BioPhoto Associates; **17.14(left):** © Al Bussewitz; **17.15a:** © Charles Havel/CPH Photography; **17.16a:** © Carolina Biological Supply Company, **b,c:** Runk/Schoenberger/Grant Heilman Photography, Inc.; **17.18a–17.20:** © Carolina Biological Supply Company; **17.21:** © Walter Hodge; **page 381:** © Earl Roberge/Photo Researchers, Inc.; **17.22b:** © J. H. Troughton and F. B. Sampson; **17.23a,b:** © Carolina Biological Supply Company, **c:** © Fred Bavendam/Peter Arnold, Inc., **d:** © Carolina Biological Supply Company; **17.25a,b:** © A. J. Belling/Photo Researchers, Inc., **c:** © BioPhoto Associates/Photo Researchers, Inc.; **17.26(both):** © W. H. Hodge/Peter Arnold, Inc.

Chapter 18

18.1(all): © E. J. Hewitt/Long Ashton Research Station, University of Bristol, England; **18.4:** © J. H. Troughton and L. Donaldson; **18.7a:** © Gordon Leedale/BioPhoto Associates, **b:** Carolina Biological Supply Company, **c(both):** © Donald Marx, USDA Forest Service; **18.8:** © Steven Kaufman/Peter Arnold, Inc.; **18.9a:** © Dr. B. A. Meylan, **b:** © J. H. Troughton and L. Donaldson; **18.10:** © W. H. Hodge/Peter Arnold, Inc.; **18.13b:** © BioPhoto Associates/Photo Researchers, Inc., **c:** © Walter Eschrich and Eberhard Fritz; **18.15b:** © J. H. Troughton and F. B. Sampsin/John Wiley and Sons.

Chapter 19

19.1a: © John D. Cunningham, **b:** © W. A. Banaszewski/Visuals Unlimited, **c:** © Dr. William M. Harlow/Photo Researchers, Inc.; **19.2:** © USDA; **page 409:** © Courtesy of R. J. Weaver; **19.7a:** © William E. Ferguson, **b:** © Bob Coyle; **19.8:** © Robert E. Lyons/Visuals Unlimited; **19.9(all):** © Kiem Tran Thanh Van and her colleagues; **19.10:** © Runk/Schoenberger/Grant Heilman Photography Inc.; **19.12a:** © David Newman/Visuals Unlimited, **b:** © William J. Weber/Visuals Unlimited; **19.16a:** © David M. Stone/Photo/Nats; **19.18:** © Frank B. Salisbury.

Chapter 20

20.1a: © Eric Grave/Photo Researchers, Inc., **b:** © Carolina Biological Supply Company, **c:** © Michael DiSpezio; **20.5:** © Kessel, R. G. and Kardon, R. H.: Tissues and Organs: A Text-Atlas of Scanning Electron Microscopy. © 1979 by W. H. Freeman and Company; **page 430:** © Bettmann News Photos; **20.10a:** © Manfred Kage/Peter Arnold; Inc., **b:** Dr. Paul Heidger; **20.11b:** © Yokochi, C., and Rohen J. W.: Photographic Anatomy of the Human Body, 2nd ed, **c:** Igakushoin, Ltd. 1978; **20.15:** © Edwin A. Reschke; **20.16:** © Reproduced with permission from "The Morphology of Human Blood Cells" by L. W. Diggs, Dorothy Sturm, and Ann Bell © 1984 Abbott Laboratories; **20.18:** "Scanning Electron Micrograph of an Erthocyte Enmeshed", Bernstein, E. and Kairimen, E., Vol 173, Photo, 27 August 1971; **20.20a(both):** Stuart I. Fox; **20.22:** © Jean-Paul Revel.

Chapter 21

21.1: © Karl Ammann/Bruce Coleman, Inc.; **21.5:** © C. R. Wyttenbach, University of Kansas/Biological Photo Service, Inc.; **21.7:** © Carol Hughes/Bruce Coleman, Inc.; **21.14, 21.15(both):** Kessel, R. G., and Kardon, R. H.: Tissues and Organs: A Text-Atlas of Scanning Electron Microscopy. © 1979 by W. H. Freeman and Company; **page 466:** © Martin Rotker/Taurus Photos, Inc.; **21.19:** © J. Somers/Taurus Photos, Inc.; **21.20 a,b:** © WHO, **c,d:** © Centers for Disease Control, Atlanta, GA.; **page 472:** © Bob Coyle.

Chapter 22

22.1a: © Douglas Faulkner/Sally Faulkner Collection, **b:** © James H. Carmichael, **c:** © Tom J. Ulrich/Visuals Unlimited; **page 483:** © Martin M. Rotker/Taurus Photos, Inc.; **22.13:** © Kessel, R. G. and Kardon, R. H.: Tissues and Organs: A Text-Atlas of Scanning Electron Microscopy. © 1979 by W. H. Freeman and Company; **22.16:** © Landrum Shettles; **22.27:** © R. B. Wilson, Eppiley Institute for Research in Cancer.

Chapter 23

23.6: © Linda Bartlett, © 1981; **23.11:** © Randolph E. Perkins; **23.27:** © Gary R. Zahm/DRK Photo, Inc.; **23.30:** © Lester V. Bergman and Associates, Inc.; **23.32:** © Edwin A. Reschke

Chapter 24

page 532: © Michael Borque/Valan Photos; **24.6b:** © Dr. T. Kuwabara; **24.7(both):** © William Vandivert; **24.8:** © T. Norman Tait/BioPhoto Associates; **page 541(both):** © Robert S. Preston/Courtesy of Professor J. E. Hawkins, Kresge; **24.13:** © Joe McDonald/Bruce Coleman, Inc.; **24.15(both), 24.18(all):** © Edwin A. Reschke; **24.21d:** © John D. Cunningham, **e:** © Dr. H. E. Huxley.

Chapter 25

25.1: © Dick Hanley/Photo Researchers, Inc.; **25.3:** © Wayne Lankinen/DRK Photo; **25.4:** © David G. Allen; **page 556:** © H. Armstrong Roberts, Inc.; **25.7:** © The American Museum of Natural History; **25.8(both):** © Fredrick Webster; **25.9:** © Bob Coyle; **25.11:** © Dwight Kuhn; **25.12:** © Thomas McAvoy, Life Magazine © 1955 Time, Inc.; **25.13:** © Russ Kinne/Photo Researchers, Inc.; **25.14:** © Nina Leen, Life Magazine © 1964 Time, Inc.; **page 567:** © Susan Kuklin; **25.18:** © Herbert Parsons/Photo/Nats; **25.20:** © Susan Kuklin/Science Source/Photo Researchers, Inc.; **25.21:** © Tom McHugh/Photo Researchers, Inc.; **25.22:** © W. Perry Conway; **25.23:** © Philip Hart/ANIMALS, ANIMALS; **25.26:** © John D. Cunningham.

Chapter 26

26.2: © John Shaw/Bruce Coleman, Inc.; **26.3:** © Dieter Blum/Peter Arnold, Inc.; **26.5(all):** © Jane Burton/Bruce Coleman, Inc.; **26.12:** © Dr. Landrum Shettles; **26.13a:** © Dr. William Byrd, **b:** © Dr. Mia Tegner, Scripps Institute of Oceanography; **26.19:** © Charles Lightdale/Photo Researchers, Inc.

Chapter 27

Page 606: © Isao Kubo; **27.11b:** © Carnegie Institute of Embryology; **27.15:** © Claude Edelmann, Petit Format et Giugoz from the book First Days of Life/Black Star; **27.18:** © Eve Arnold/Magnum Photos, Inc.

Chapter 28

28.1a: © Larry West, **b:** © John Shaw/Bruce Coleman, Inc., **c:** © Larry West; **28.11:** © Irven DeVore/Anthro-photos; **28.19a:** © H. S. Wesenberg and G. A. Antipa; **28.20:** © Ronald Austing/Photo Researchers, Inc.; **28.21:** © Michael P. L. Fagden/Bruce Coleman, Inc.; **28.22(both):** © Carolina Biological Supply Company; **page 639:** © Australian Information Service; **28.23:** © Hunter, G. W., Swartzwelder, J. C., and Clyde, D. F.: Tropical Medicine 5th ed: W. B. Saunders Company, 1976; **28.25:** © C. B. Frith/Bruce Coleman, Inc.; **28.26:** © Robert Lee/Photo Researchers, Inc.; **28.27(all):** © Dr. Daniel Jantzen; **28.28:** © Douglas Faulkner/Sally Faulkner Collection.

Chapter 29

29.1a: © Michael G. Galbridge, **b:** © Leonard Lee Rue III/Bruce Coleman, Inc.; **29.8:** © William H. Allen Jr.; **29.9, 29.10:** © Bob Coyle; **page 656:** © Pat Smith, University of Georgia Agriculture Experimental Stations; **29.13:** © John Shaw/Bruce Coleman, Inc.; **29.14:** © Paolo Koch/Photo Researchers, Inc.; **29.18:** © Dr. M. P. Kahl/Bruce Coleman, Inc.; **29.19:** "Debate over an Essential Resource," Carter, L. J. Science, Vol. 209 pp. 372–376, photo, 18 July 1980.

Chapter 30

30.4(both): © Edward J. Kormondy; **30.9:** © Douglas Faulkner/Sally Faulkner Collection; **30.11b:** © Jack Donnelly, Woods Hole Oceangraphic Institution; **30.15a:** © Stephen J. Krasemann/Peter Arnold, Inc., **b:** © Richard Ferguson/William Ferguson, **c:** © Mary Thacher/Photo Researchers, Inc.; **30.17:** © W. H. Hodge/Peter Arnold, Inc.; **30.18:** © Stephen J. Krasemann/Peter Arnold, Inc.; **30.19:** © C. Gans, University of Michigan/Biological Photo Service; **30.20:** © Norman Owen Tomalin/Bruce Coleman, Inc.; **30.21:** © Richard Estes/Photo Researchers, Inc.

Chapter 31

31.9(both): © Georg Gerster/Photo Researchers, Inc.; **31.12:** © UNICEF.

Chapter 32

32.2: © Wendell Mentzen, 1980; **32.3:** © U.S.D.A. Photo; **32.4:** © USDA Soil Conservation Service; **32.5:** © Carl Jordon, Institute of Ecology, University of Georgia; **32.6:** © Tim McCabe/Soil Conservation Service; **32.11(both):** © Peter Arnold, Inc.; **32.13a:** © G. Gscheidle/Peter Arnold, Inc.; **b:** © Field Museum of Natural History, **b:** © Don and Pat Valenti; **32.15, 32.18:** © USDA Soil Conservation Service; **32.22a:** © Jerg Kroener, **b:** © John Neel, **c:** © Norman Myers/Bruce Coleman, Inc.; **32.23:** © Karl Ammann/Bruce Coleman, Inc.

Line Art

Chapter 2

2.7: Adapted from W. M. Levi, The Pigeon (Sumter, S.C.: Levi Publishing Company, 1957). **2.10 a, b:** From BIOLOGICAL SCIENCE: An Inquiry into Life, BSCS Yellow Version, Fourth Edition. Copyright © 1980 Biological Sciences Curriculum Study. Reprinted by permission.

Chapter 3

3.8 a, b: From Johnson, Leland G., Biology. © 1983 Wm. C. Brown Publishers, Dubuque, Iowa. All Rights Reserved. Reprinted by permission. **3.10 a, b:** From Johnson, Leland G., Biology. © 1983 Wm. C. Brown Publishers, Dubuque, Iowa. All Rights Reserved. Reprinted by permission.

Chapter 4

4.8: From Johnson, Leland G., Biology. © 1983 Wm. C. Brown Publishers, Dubuque, Iowa. All Rights Reserved. Reprinted by permission. **4.9 a, b:** From Johnson, Leland G., Biology. © 1983 Wm. C. Brown Publishers, Dubuque, Iowa. All Rights Reserved. Reprinted by permission.

Chapter 5

5.13: From Johnson, Leland G., Biology. © 1983 Wm. C. Brown Publishers, Dubuque, Iowa. All Rights Reserved. Reprinted by permission.

Chapter 6

6.2: From Johnson, Leland G., Biology. © 1983 Wm. C. Brown Publishers, Dubuque, Iowa. All Rights Reserved. Reprinted by permission. **6.10:** From Johnson, Leland G., Biology. © 1983 Wm. C. Brown Publishers, Dubuque, Iowa. All Rights Reserved. Reprinted by permission. **6.13:** From Johnson, Leland G., Biology. © 1983 Wm. C. Brown Publishers, Dubuque, Iowa. All Rights Reserved. Reprinted by permission. **Excerpt, p. 121** (ATP Formation in Mitochondria): From Johnson, Leland G., Biology. © 1983 Wm. C. Brown Publishers, Dubuque, Iowa. All Rights Reserved. Reprinted by permission.

Chapter 8

8.7: From Stern, Kingsley R., Introductory Plant Biology, Third Edition. Copyright © 1979, 1982, 1985 Wm. C. Brown Publishers, Dubuque, Iowa. All Rights Reserved. Reprinted by permission.

Chapter 9

9.12: From Johnson, Leland G., Biology. © 1983 Wm. C. Brown Publishers, Dubuque, Iowa. All Rights Reserved. Reprinted by permission.

Chapter 11

11.5: From Douglas J. Futuyma, Evolutionary Biology, 1978. Reprinted by permission from Sinauer Associates, Inc. **11.8:** From Gardner and Snustad, Principles of Genetics, p. 556. Copyright John Wiley and Sons, Inc., New York, N.Y. Reprinted by permission. **11.15:** Reprinted by permission from North American Trees, Third Edition by Robert Preston, Jr., © 1976 by The Iowa State University Press, 2121 South State Avenue, Ames, Iowa 50010.

Chapter 12

12.5: From Johnson, Leland G., Biology. © 1983 Wm. C. Brown Publishers, Dubuque, Iowa. All Rights Reserved. Reprinted by permission. **12.8 a–c:** From Johnson, Leland G., Biology. © 1983 Wm. C. Brown Publishers, Dubuque, Iowa. All Rights Reserved. Reprinted by permission. **12.14:** Source: General Zoology, 6th edition, by Storer. © McGraw-Hill Book Company, New York, N.Y. **12.18 a, b:** From Johnson, Leland G., Biology. © 1983 Wm. C. Brown Publishers, Dubuque, Iowa. All Rights Reserved. Reprinted by permission. **12.23:** From Volpe, E. Peter, Understanding Evolution, 4th ed. © 1967, 1970, 1977, 1981 Wm. C. Brown Publishers, Dubuque, Iowa. All Rights Reserved. Reprinted by permission.

Chapter 13

13.11: From Cox, F. E. G. and Vickerman, K., (1967), The Protozoa, pp. 15, 25, 46. Houghton-Mifflin. Reprinted by permission of the author. **13.14:** From Cox, F. E. G. and Vickerman, K., (1967), The Protozoa, pp. 15, 25, 46. Houghton-Mifflin. Reprinted by permission of the author. **13.21:** From Stern, Kingsley R., Introductory Plant Biology, 3d ed. © 1979, 1982, 1985 Wm. C. Brown Publishers, Dubuque, Iowa. All Rights Reserved. Reprinted by permission.

Chapter 14

14.3 b: Copyright © 1974 Kendall/Hunt Publishing Company. **14.5:** Copyright © 1974 Kendall/Hunt Publishing Company. **14.8 a, b:** From Johnson, Leland G., Biology. © 1983 Wm. C. Brown Publishers, Dubuque, Iowa. All Rights Reserved. Reprinted by permission.

Chapter 16

16.14 b: From An Introduction to Embryology, 5th ed., by B. I. Balinsky and B. C. Fabian. Copyright © 1981 by CBS College Publishing. Reprinted by permission of CBS College Publishing. **16.26:** From Fossil Man by Michael H. Day. Copyright © 1970 by Grosset and Dunlap, Inc. Copyright © 1969 by The Hamlyn Group. Used by permission of Grosset and Dunlap, Inc. **16.27:** From Fossil Man by Michael H. Day. Copyright © 1970 by Grosset and Dunlap, Inc. Copyright © 1969 by The Hamlyn Group. Used by permission of Grosset and Dunlap, Inc.

Chapter 17

17.9 a, b: From MacMillan Science Company, Inc., Chicago, Illinois. **17.10:** From Johnson, Leland G., Biology. © 1983 Wm. C. Brown Publishers, Dubuque, Iowa. All Rights Reserved. Reprinted by permission. **17.14:** From Johnson, Leland G., Biology. © 1983 Wm. C. Brown Publishers, Dubuque, Iowa. All Rights Reserved. Reprinted by permission. **17.15:** From Johnson, Leland G., Biology. © 1983 Wm. C. Brown Publishers, Dubuque, Iowa. All Rights Reserved. Reprinted by permission. **17.17:** Copyright © 1974 Kendall/Hunt Publishing Company. **17.18:** Copyright © 1974 Kendall/Hunt Publishing Company. **17.19:** From Johnson, Leland G., Biology. © 1983 Wm. C. Brown Publishers, Dubuque, Iowa. All Rights Reserved. Reprinted by permission.

Chapter 18

18.2: From *The Physiology of Plants,* edited and translated by A. J. Ewart. © 1904 Oxford University Press. Reprinted by permission. **18.11 *a–c:*** From Johnson, Leland G., *Biology.* © 1983 Wm. C. Brown Publishers, Dubuque, Iowa. All Rights Reserved. Reprinted by permission. **18.15 *a:*** Reproduced from *The Journal of Cell Biology,* 1968, Vol. #38, p. 298 by copyright permission of The Rockefeller University Press.

Chapter 19

19.5: From Johnson, Leland G., *Biology.* © 1983 Wm. C. Brown Publishers, Dubuque, Iowa. All Rights Reserved. Reprinted by permission.

Chapter 20

20.3 *a, b:* From Johnson, Leland G., *Biology.* © 1983 Wm. C. Brown Publishers, Dubuque, Iowa. All Rights Reserved. Reprinted by permission. **20.6 *a–c:*** From Johnson, Leland G., *Biology.* © 1983 Wm. C. Brown Publishers, Dubuque, Iowa. All Rights Reserved. Reprinted by permission. **20.7:** From Johnson, Leland G., *Biology.* © 1983 Wm. C. Brown Publishers, Dubuque, Iowa. All Rights Reserved. Reprinted by permission. **20.8:** From Van De Graaff, Kent M., *Human Anatomy.* © 1984 Wm. C. Brown Publishers, Dubuque, Iowa. All Rights Reserved. Reprinted by permission. **20.9:** From Hole, John W., Jr., *Human Anatomy and Physiology,* 3d ed. © 1981, 1984 Wm. C. Brown Publishers, Dubuque, Iowa. All Rights Reserved. Reprinted by permission. **20.11 *a:*** From Hole, John W., Jr., *Human Anatomy and Physiology,* 3d ed. © 1984 Wm. C. Brown Publishers, Dubuque, Iowa. All Rights Reserved. Reprinted by permission. **20.12:** From Johnson, Leland G., *Biology.* © 1983 Wm. C. Brown Publishers, Dubuque, Iowa. All Rights Reserved. Reprinted by permission.

Chapter 21

21.2 *a:* From Johnson, Leland G., *Biology.* © 1983 Wm. C. Brown Publishers, Dubuque, Iowa. All Rights Reserved. Reprinted by permission. **21.8 *a, b:*** From Johnson, Leland G., *Biology.* © 1983 Wm. C. Brown Publishers, Dubuque, Iowa. All Rights Reserved. Reprinted by permission. **21.9:** From Johnson, Leland G., *Biology.* © 1983 Wm. C. Brown Publishers, Dubuque, Iowa. All Rights Reserved. Reprinted by permission. **21.18:** From Hole, John W., Jr., *Human Anatomy and Physiology,* 3d ed. © 1981, 1984 Wm. C. Brown Publishers, Dubuque, Iowa. All Rights Reserved. Reprinted by permission. **21.21:** © 1984 by Consumer's Union of U.S., Inc., Mt. Vernon, New York: based on tests for *Consumer Reports,* October 1984.

Chapter 22

22.6 *a, b:* From Johnson, Leland G., *Biology.* © 1983 Wm. C. Brown Publishers, Dubuque, Iowa. All Rights Reserved. Reprinted by permission. **22.12:** From Van De Graaff, Kent M., *Human Anatomy.* © 1984 Wm. C. Brown Publishers, Dubuque, Iowa. All Rights Reserved. Reprinted by permission. **22.14:** From "The Quaternary Structure of Hemoglobin," adapted by permission from R. E. Dickerson and I. Geis, *The Structure and Action of Proteins.* Benjamin/Cummings, Menlo Park, CA., 1969. Copyright 1969 by Dickerson and Geis. **22.15:** From *Human Physiology* by A. J. Vander, et. al. © 1970 McGraw-Hill Book Company, New York, N.Y. Reprinted by permission. **22.16 *b:*** From Johnson, Leland G., *Biology.* © 1983 Wm. C. Brown Publishers, Dubuque, Iowa. All Rights Reserved. Reprinted by permission. **22.17:** From Johnson, Leland G., *Biology.* © 1983 Wm. C. Brown Publishers, Dubuque, Iowa. All Rights Reserved. Reprinted by permission. **22.19:** From *An Introduction to Embryology,* 5th ed., by B. I. Balinsky. Copyright © 1981 by CBS College Publishing. Reprinted by permission of CBS College Publishing. **22.23:** From Johnson, Leland G., *Biology.* © 1983 Wm. C. Brown Publishers, Dubuque, Iowa. All Rights Reserved. Reprinted by permission.

Chapter 23

23.10 *b, c:* From Hole, John W., Jr., *Human Anatomy and Physiology,* 3d ed. © 1978, 1981, 1984 Wm. C. Brown Publishers, Dubuque, Iowa. All Rights Reserved. Reprinted by permission. **23.13 *b, c:*** From Hole, John W., Jr., *Human Anatomy and Physiology,* 3d ed. © 1984 Wm. C. Brown Publishers, Dubuque, Iowa. All Rights Reserved. Reprinted by permission. **23.16:** From Hole, John W., Jr., *Human Anatomy and Physiology,* 3d ed. © 1978, 1981, 1984 Wm. C. Brown Publishers, Dubuque, Iowa. All Rights Reserved. Reprinted by permission. **23.18:** From Johnson, Leland G., *Biology.* © 1983 Wm. C. Brown Publishers, Dubuque, Iowa. All Rights Reserved. Reprinted by permission. **23.19:** From Hole, John W., Jr., *Human Anatomy and Physiology,* 3d ed. © 1981, 1984 Wm. C. Brown Publishers, Dubuque, Iowa. All Rights Reserved. Reprinted by permission.

Chapter 24

24.1: From Nelson, G. E., *Fundamental Concepts of Biology.* © 1974 by John Wiley and Sons, Inc. Reprinted by permission of John Wiley and Sons, Inc. **24.2:** From Hole, John W., Jr., *Human Anatomy and Physiology,* 3d ed. © 1984 Wm. C. Brown Publishers, Dubuque, Iowa. All Rights Reserved. Reprinted by permission. **24.3:** From Hole, John W., Jr., *Human Anatomy and Physiology,* 3d ed. © 1981, 1984 Wm. C. Brown Publishers, Dubuque, Iowa. All Rights Reserved. Reprinted by permission. **24.5:** From Hole, John W., Jr., *Human Anatomy and Physiology,* 3d ed. © 1981, 1984 Wm. C. Brown Publishers, Dubuque, Iowa. All Rights Reserved. Reprinted by permission. **24.11:** From Hole, John W., Jr., *Human Anatomy and Physiology,* 3d ed. © 1984 Wm. C. Brown Publishers, Dubuque, Iowa. All Rights Reserved. Reprinted by permission. **24.13:** From Hole, John W., Jr., *Human Anatomy and Physiology,* 3d ed. © 1978, 1981, 1984 Wm. C. Brown Publishers, Dubuque, Iowa. All Rights Reserved. Reprinted by permission. **24.16:** From Hole, John W., Jr., *Human Anatomy and Physiology,* 3d ed. © 1981, 1984 Wm. C. Brown Publishers, Dubuque, Iowa. All Rights Reserved. Reprinted by permission. **24.18 *a–c:*** From Hole, John W., Jr., *Human Anatomy and Physiology,* 3d ed. © 1978, 1981, 1984 Wm. C. Brown Publishers, Dubuque, Iowa. All Rights Reserved. Reprinted by permission. **24.23:** From *Nerve, Muscle and Synapse,* by Bernard Katz. © 1966 McGraw-Hill Book Company, New York, N.Y. Reprinted by permission.

Chapter 25

25.13: From *Biology: The Behavioral View,* by Suthers, R. A., and R. A. Gallant. Copyright © 1973 John Wiley and Sons, Inc. **25.16:** From *Biology Today,* Second Edition, by David L. Kirk. Copyright © 1972, 1975 by Random House, Inc. Reprinted by permission of CRM Books, a Division of Random House, Inc. **25.24 *a:*** From Helena Curtis, *Biology,* Fourth Edition, Worth Publishers, New York, 1983, page 933. Reprinted by permission. **25.24 *b:*** From Curtis and Barnes, *Invitation to Biology,* Fourth Edition, Worth Publishers, New York, 1985, page 546. Reprinted by permission.

Chapter 27

27.13 (*left*): From Hole, John W., Jr., *Human Anatomy and Physiology,* 3d ed. © 1984 Wm. C. Brown Publishers, Dubuque, Iowa. All Rights Reserved. Reprinted by permission. **27.13 (*right*):** From Hole, John W., Jr., *Human Anatomy and Physiology,* 3d ed. © 1978, 1981, 1984 Wm. C. Brown Publishers, Dubuque, Iowa. All Rights Reserved. Reprinted by permission. **27.16 *a–c:*** Courtesy Carnation Company.

Chapter 28

28.2: From Johnson, Leland G., *Biology.* © 1983 Wm. C. Brown Publishers, Dubuque, Iowa. All Rights Reserved. Reprinted by permission. **28.3:** From *Laboratory Studies in Biology: Observations and Their Implications,* by Chester A. Lawson, Ralph W. Lewis, Mary Alice Burmester, and Garrett Hardin. Copyright © 1955 by W. H. Freeman and Company. Reprinted by permission. **28.5:** After C. Kabat and D. R. Thompson, 1963, *Wisconsin Quail 1834–1962: Population Dynamics and Habitat Management,* Tech. Bull. No. 30, Wisconsin Conservation Department, Madison, WI. **28.6:** From "The Rise and Fall of a Reindeer Herd," by V. B. Scheffer, in *Scientific Monthly,* Vol. 73, December 1951, pp. 356–362. Copyright 1951 by the American Association for the Advancement of Science. Reprinted by permission. **28.7:** From Johnson, Leland G., *Biology.* © 1983 Wm. C. Brown Publishers, Dubuque, Iowa. All Rights Reserved. Reprinted by permission. **28.9:** Source: Fig. 15.4 "Oscillations in the numbers of adult sheep . . ." (p. 480) in *Ecology and Field Biology,* 3rd edition by Robert Leo Smith (after Nickolson). Copyright © 1980 by Robert Leo Smith. Reprinted by permission of Harper and Row, Publishers, Inc. **28.14 *a–c:*** From Kelly and McGrath: *Biology: Evolution and Adaptation to the Environment.* Copyright © 1975 Houghton-Mifflin Company. Used by permission. **28.15 *a–e:*** From "Population Ecology of Some Warblers in Northeastern Coniferous Forests," by R. H. MacArthur, *Ecology,* 39, pp. 599–619. Copyright © by the Ecological Society of America. Reprinted by permission. **28.16:** From "Physiological Ecology of Three Codominant Successional Annuals," by N. K. Wieland and F. A. Bazzaz, *Ecology,* 56, 681–88. Copyright © by the Ecological Society of America. Reprinted by permission. **28.18:** From *Fundamentals of Ecology,* 3rd edition, by Eugene P. Odum. Copyright © 1971 by W. B. Saunders Company. Reprinted by permission of Holt, Rinehart and Winston, CBS College Publishing.

Chapter 29

29.6: From Volpe, E. Peter, *Biology and Human Concerns,* 3d ed. © 1975, 1979, 1983 Wm. C. Brown Publishers, Dubuque, Iowa. All Rights Reserved. Reprinted by permission.

Chapter 30

30.5: From Correll, D., "Estuarine Productivity," in *BioScience,* Vol. No. 281, p. 649. Copyright © 1978 by the American Institute of Biological Sciences. Reprinted by permission. **30.11 *a:*** From Johnson, Leland G., *Biology.* © 1983 Wm. C. Brown Publishers, Dubuque, Iowa. All Rights Reserved. Reprinted by permission.

Chapter 31

31.2: From Elaine M. Murphy, *World Population: Toward the Next Century,* Population Reference Bureau, Washington, D.C., 1971 and other sources. **31.4:** From *Fundamentals of Ecology,* 3rd edition, by Eugene P. Odum. Copyright © 1971 by W. B. Saunders Company. Reprinted by permission of Holt, Rinehart and Winston, CBS College Publishing. **31.6:** Reprinted with permission from *Chemical and Engineering News,* June 21, 1982, p. 7. Copyright 1982 American Chemical Society. **31.8:** Painting by William H. Bond. © National Geographic Society **31.11 *a–d:*** From "The World Food Situation and Global Grain Prospects," by T. N. Barr, in *Science,* Vol. 214, October 4, 1981, pp. 1087–1095. Copyright 1981 by the American Association for the Advancement of Science. Reprinted by permission.

Chapter 32

32.9: Drawing by Emily Bookhultz.

H

Habitat, 631
  destruction, 728
Haeckel, Ernst, 622
Hagfishes, 337
Hair, 344
Hair cell, 538, 539
Hales, Stephen, 400
Hamburger connection, 710
Hammer, 538, 539
Hammer, K. C., 419
Hand, 352
Haploid gametophyte generation, 287
Haploid number, 139
Hard-shelled eggs, 344
Hardwood trees, 300
Hardy, G. H., 218
Hardy-Weinberg Law, 218–21
Hare, 634
Harelip, 167
Harvey, William, 430
Hatchet foot, 319
Haversian systems, 542
Hawk, 343, 633
Hazardous wastes, 710, 711–13
HCG, 590, 611
Head, of molecule, 58
Hearing, 529, 538, 539, 540–41
Heart
    human, 432, 433, 434–37
    valves of, 436, 437
Heartbeat, 435–37
Heart disease, 483
Heartwood, 377
Heat, 529, 648
Heavy metals, 711
Heliostats, 700, 701
Helix, double-, model, 191
Heme, 484
Heme group, 61
Hemocoel, 326, 428
Hemoglobin, 156, 428, 442, 480, 485
    molecule, 61
    sickle-cell, 195
    types of, 486
Hemophilia, 177, 178, 208, 445
Henle, loop of. See Loop of Henle
Hepatica, 622
Hepatic portal system, 434, 467
Hepatic portal vein, 434
Hepatitis, 466
Herbaceous dicot stem, 374
Herbaceous plants, 300
Herbaceous stems, 375
Herbivores, 90, 98, 454, 646
    mammalian, 458
Herbivorous zooplankton, 674
Hereditary material, 188–89
Heredity
    code of, 197–98
    in genetic expression, 167
    and hypertension, 439
Herman, Louis, 566
Hermaphrodite, 313, 320, 579

Heron, 636, 647
Herpes, 594, 595–96
Herring Gull's World, 554
Herrnstein, R. J., 566
Hershey, A. D., 189
Heterocysts, 268
Heterosis, 224
Heterospores, 296
Heterotrophic fermenter, 263
Heterotrophic organisms, 90, 92
Heterotrophs, 78
Heterozygous, 152
Hierarchy, dominance, 570, 630
Hindbrain, 516
Histamines, 448
Histones, 172
H.M.S. *Beagle*, 17, 18, 25
Hodgkins, A. L., 506
Hodgkin's disease, 301
Hodos (scientist), 567
Holstein cattle, 164, 165
Homeostasis, 8, 127, 475
Homing, 560
Hominid, 355–56
Hominid ancestor, 352–55
Hominoid, 351–55
Hominoid ancestor, 351–52
Homo erectus, 355, 356–57
Homo habilis, 355, 356, 357
Homologous chromosomes, 170
Homologous pair, 133
Homologous structures, 243
Homo sapiens, 355, 357–58
Homo sapiens neanderthal, 357
Homo sapiens sapiens, 358
Homozygous, 152
Honeybees, 568–69
Hooke, Robert, 65
Hookworm, 318, 639
Hoppe, Peter, 602
Hormone, follicle stimulating. See FSH
Hormone-receptor complex, 520
Hormone regulation, 584
Hormones
    adrenocorticotropic. See ACTH
    antiuretic. See ADH
    and blood, 590
    cellular activity of, 521
    corticotropin releasing. See CRH
    egg-laying. See ELH
    female sex, 586
    and flowering, 421
    gonadotropin releasing. See GnRH
    growth. See GH
    human, 521
    and insect development, 606
    interstitial cell stimulating. See ICSH
    luteinizing. See LH
    mammalian, 523
    as messengers, 520–21
    parathyroid, 525
    in plants, 407, 409–15, 416
    plasma levels, 588
    releasing, 524
    thyroid releasing. See TRH
    thyroid stimulating. See TSH

Horse, 459
    evolution of, 253
Horseweed, 629
Hot, 529
Housekeeping system, 515
Hrdy, Sarah Blaffer, 591
Human, 356–58
    blood, 442–47
    circulation, 432–40
    circulatory system, 432
        classification of, 348
    development, 609–13
    digestive system, 460
    ear, 538, 539
    ecology, 689
    ecosystem, 29–30, 652
        and environment, 707–32
    evolution, 349–58
    eye, 534–35, 537
    food chain, 652–57
    gas exchange system, 481–86
    genetic diseases, 156–57
    glands, 521
    heart, 432, 433, 434–37
    hormones, 521
    kidney, 491–96
    nervous system, 502–4
    population, 690–706
    reproduction, 582–91
    respiratory system, 481
    skeleton, 543–44
    skin, 537
Human chorionic gonadotropin. *See* HCG
Hummingbird, 343, 642
Hunting, 728–29
Huntington's disease, 207
Huxley, A. F., 506, 546
Huxley, H. E., 546
Hybrid, 223
Hybrid vigor, 224
Hydra, 310, 476, 502, 503, 579
Hydrocarbons, 711
Hydrochloric acid, 462
Hydrocortisone, 525
Hydrogen, 39, 40, 42, 52, 57, 104, 390, 701
Hydrogen bonding, 43, 59
Hydroid, 310–11
Hydrolysis, 52, 53
Hydroponics, 392
Hydrostatic skeleton, 318, 321, 542
Hydrothermal vents, 673
Hyena, 556
Hypertension, 438–39
Hypertonic solution, 68, 69
Hypertonic urine, 495–96
Hypha, 275
Hypocotyl, 364
Hypolimnion, 666
Hypothalamus, 516, 522, 523, 524, 584
Hypothesis, 12
Hypotonic solution, 68, 69
H zone, 548

I

IAA, 409
I band, 548
Ice, 44, 46
ICSH, 584
Iguana, 19
Illmensee, Karl, 602
Immunity, 448–51
  side effects, 451
Immunization, 596
Implantation, 590, 609
Impotency, 584
Imprinting, 563
Inborn errors of metabolism, 193–94
Incisors, 459
Inclusive fitness, 573
Incomplete dominance, 163–64
Incomplete gut, 454–56
Incomplete metamorphosis, 327
Incomplete proteins, 469
Incomplete ventilation, 480
Incurrent siphon, 457
Incus, 539
Independent Assortment, Law of. *See* Law of
    Independent Assortment
Indian mallow, 633
Indolacetic acid. *See* IAA
Indoor air pollution, 718
Inducer, 204, 603
Inducible operon model, 204
Induction, 604–7
Inductive reasoning, 12
Industrial melanism, 250–51
Industrial Revolution, 358
Infection, defense against, 447–51
Infectious diseases, caused by bacteria, 267
Inferior vena cava, 433
Infertility, 592–94
Inflammatory reaction, 448
Influence, sex, 181
Ingram, V. M., 195
Inhalation, 479
Inheritance
    alteration of genetic, 207
    intermediate, 164
    multitrait, 157–62
    patterns of, 148–69
    polygenic, 166–67
    single-trait, 150–57
Inhibitors, plant growth, 415
Initiation, 200
Innate behavior, 559–62
Inner cell mass, 610
Inputs, 652
Insecta, class, 326–27
Insectivora, order, 347
Insects, 8, 326–27
    and bacteria, 248–49
    development, 606
    eyes, 533, 537
    respiration in, 478
Insight, 564
Insight learning, 564
Instinctive behavior, 559
Insulin, 208, 526
Integration, and coordination in animals, 501
Intercalated discs, 435, 546

Interferon, 451
Intermediate forms, 246
Intermediate inheritance, 164
Internal environment, 426
Internal skeleton, 308, 330, 335
Internal stimuli, 565
International Whaling Commission, 710
Interneuron, 504
Internodes, 373
Interoceptors, 529
Interphase, 134, 138
Interspecific competition, 628, 630–31
Interstitial cells, 583, 584
Interstitial cell stimulating hormone. *See* ICSH
Intertidal zone, 670
Intestinal juice, 464
Intestines, 455, 457, 458 460
    large, 468
    small, 462–66
Intragene, 210
Intraspecific competition, 628–30
Introns, 210
Invagination, 478, 600, 606
Inversion, 182
    thermal, 715
Invertebrates, 305, 426–28
    age of, 335
    and behavior, 559
    lower, 307
In vitro, 206
In vitro fertilization, 593–94
In vivo, 206
Involuntary muscles, 545
Ion, 41, 48, 395
Ionic bonds, 41
Ionic concentration, relative, 489
Iris, 378, 534
Iron, 39, 62, 390, 484
Irrigation, 721–22
Isle Royale, 625, 634
Islets of Langerhans, 525
Isogametes, 284
Isolation, 230, 231, 232
Isomer, 52, 54
Isotonic fluids, 488
Isotonic solution, 68
Isotope, 40–41
IUD, 592
Ivanowsky, Dimitri, 269

J

Jackrabbit, 241
Jacob, Francois, 204
Jagendorf, Andre, 104
Japanese beetle, 638
Jaundice, 466
Jaws, 337
Jellyfish, 309, 311
Johanson, Donald, 356
Jointed appendages, 324
Joly (scientist), 400
J-shaped curve, 623, 625
Jumping genes, 210
Jurassic period, 238, 282, 335
Juvenile-onset diabetes, 526

K

Kamil, Alan, 567
Kangaroo rat, 679
Karyotype, 133, 183
    preparation, 132
Ketoacidosis, diabetic, 526
Kettlewell, H. B. D., 250
Kidney, 66
    artificial, 497
    failure, 496–97
    human, 491–96
    structure, 492–94
Kingdom, 25
    Fungi, 275–79
    Protista, 270–75
    See also Classification of organisms
Kingdom Monera, 265–69
Kingfisher, 38
Kinins, 448
Kin recognition, 562–63
Kin selection, 573
Kit, 9
Klamath weed, 638
Klinefelter's syndrome, 185
Knuckle-walking, 352, 354
Koala bear, 345
Köhler, Georges, 451
Kohlrabi, 22
Krebs cycle, 113, 115, 116–17, 118, 119, 122
K-selected species, 627
K-strategist species, 627
Kurosawa (scientist), 413
Kwashiorkor, 470, 704

L

Labial palps, 457
Labia major, 587
Labor, 614–15
Lactase, 95
Lacteal, 440, 441, 464
Lactic acid, 112, 113
    buildup, 548
Lactose, 95
Ladder-type nervous system, 312
Ladybugs, 638
Lakes, 665
    classification of, 668–69
    and ponds, 666–67
Lamarck, Jean Baptiste, 20, 24
Lamellae, 475, 477
    stroma, 100
Laminaria, 287
Lampreys, 337
Lancelet, 334
Land
    adaptation to, 288–89, 387
    in food production, 702
    pollution, 710
    use, 707–10
Langerhans, islets of. *See* Islets of Langerhans
Lansteiner, Karl, 446
Large intestine, 468
Larva, 9, 308, 581, 628

Neutron, 39
Neutrophil, 448
Newt, 340
New World monkeys, 351
Niacin, 470
Nibonuclease, 95
Niche, 29, 631
Nicholson, A. H., 628
Nicotinamide adenine dinucleotide phosphate.
　　See NADP
Nicotinamide adenine dinucleotide. *See* NAD
Night vision, 536
Nitrification, 661
Nitrifying bacteria, 661
Nitrogen, 39, 390
Nitrogen cycle, 660–61
Nitrogen fixation, 660–61
Nitrogen-fixing bacteria, 267
Nitrogenous wastes, 487–88
Nitrogen oxides, 713
N number, 139
Nodes, 373
　　of heart, 434
　　lymph, 440, 441
Nodes of Ranvier, 505, 507
Noise, 541
Nonbreeders, 574
Noncyclic photophosphorylation, 102
Nondisjunction, 184
Nonpolar bond, 42
Nonrenewable energy sources, 698–700
Nonrenewable resources, 696–97
Nonteleological mechanism, 20
Noradrenalin, 525
Norepinephrine, 508, 525
Nose, 481
Notochord, 333, 604
Nuclear envelope, 73
Nuclear fusion, 700
Nuclear power plant, Chernobyl, 699
Nuclear wastes, 711
Nucleases, 465
Nucleic acids, 53, 61–62
　　See also specific acid
Nucleoid, 265
Nucleolus, 72, 73, 74, 131
Nucleoplasm, 131
Nucleosome, 172
Nucleotides, 53, 62, 190
　　and plant products, 108
Nucleus, 39, 73, 74, 131–33
　　and DNA and RNA, 74
Nurseries of the sea, 669
Nutrient trap, 669
Nutrition, 267–68, 468–73
　　and digestion, 454–74
　　in plants, 390–406
Nuts, 622
Nutter, Wade, 656

O

Oats, 410
Obesity, and hypertension, 439
Obligate anaerobe, 268
Obligate parasite, 189
O'Brien, Stephen, 230
Observations, 12
Occipital lobe, 519
Ocean, 659, 673, 725–27
Oceanic ridges, 673
Oedogonium, 285
Old World monkeys, 351
Olfactory cells, 529, 531
Oligochaeta, class, 323
Oligotrophic lakes, 668
Omnivores, 454, 646
Oncogene, 212, 213
One gene-one enzyme hypothesis, 195
One gene-one polypeptide hypothesis, 195
Ontogeny, 607
Oocyte, 586
Oogenesis, 142
Oparin, A. I., 261
Open circulatory system, 319, 426–28
Open sea, 674–75
Operant conditioning, 564
Operator, 204
Operculum, 339, 475, 477
Operon, 204
Ophrys, 246
Opium, 387
Opportunistic species, 627
Opsin, 536
Optic nerve, 535
Oral contraception, 592
Orangutan, 351
Orbital, 39, 40
Orchid, 246, 728
Order
　　Artiodactyla, 347
　　Carnivora, 347
　　Cetacea, 347
　　Chiroptera, 347
　　Insectivora, 347
　　Perissodactyla, 347
　　Primates, 347
　　Rotentia, 347
　　Marsupialia, 347
　　Monotremata, 345
　　See also Classification of organisms
Ordovician period, 238, 282, 335
Organelles
　　and energy, 77–79
　　eukaryotic, 74
Organic evolution, 263–64
Organic farming, 654–55
Organisms, classification of. *See* Classification of
　　organisms
Organization, of life, 6–7
Organizer, 603
Organochlorides, 711

Organ of Corti, 540, 541
Organs
　　accessory, 587
　　excretory, 490–91
　　vegetative, 366–68
Orgasm
　　in females, 587
　　in male, 585
　　reasons for, 590–91
Origin of Species, 25, 217
Oscillatoria, 27
Oscilloscope, 506, 507
Osculum, 308
Osmosis, 67–70
　　in animal cells, 68
　　local, 71
　　in plant cells, 68–69
Osmotic concentration, 497
Osmotic pressure, 68, 444
Osmotic regulation, 488–90
Osteichthyes, class, 337–40
Osterich, 18, 19
Ostrich, 343
Otolith, 539
Ovarian cycle, 586, 589
Ovary, 300, 386, 522, 523, 585, 586
Oviduct, 585, 586, 587, 609
Ovipositor, 327
Ovulation, 586, 598
Ovule, 149
Ovum, 586
Owl, 343, 571
Oxidation, 97, 116
　　glucose, 112–15
Oxidative phosphorylation, 120
Oxygen, 39, 40, 42, 52, 390
　　in blood, 431, 476, 477
　　and respiration, 480
Oxygen debt, 548
Oxygen sag curve, 723
Oxyhemoglobin, 484, 485
Oxytocin, 521, 523, 524
Ozone, 714, 715
Ozone shield, 713

P

Pacemaker, 435
Pacinian corpuscle, 538
Pain, 529
Paleozoic era, 238, 260, 264, 282, 335
Palisade layer, 383
Palps, labial, 457
PAN, 714, 715
Pancreas, 522, 523, 526
Pancreatic amylase, 463, 465
Pancreatic juice, 463
Panda, 728
Pangaea, 242
Pangenesis, 143

R

Rabbit, 503
Raccoon, 556
Race, and hypertension, 439
Radial symmetry, 308, 309, 312, 330
Radiation, 181
    adaptive, 251–52
Radicle, 365
Radioactive tracers, in plants, 402
Radium, 699
Radula, 319, 458
Rain, acid, 49, 716–17
Rainfall, world, 676
Ramapithecus, 353
Rangeland, 708–9
Ranvier, nodes of. *See* Nodes of Ranvier
Rat, 679
Rate of natural increase, 623
Raup, David, 246
Ray-finned fish, 338–39
Ray, 338
Reabsorption, 495
Reaction
    condensation, 52, 53
    light and dark, 101
    transition, 116
Reaction center, 102
Reasoning, 12, 13, 564
Recapitulates, 607
Receptable, 300, 385
Receptors, 529, 530, 537
    touch, 536
Receptor sites, 448, 508, 509
Recessive allele, 152
Recognition, kin, 562–63
Recombinant DNA, 206
Recombination, 224
Recycling, 713
Red algae, 288
Red blood cells, 442, 448
Redirection, 572
Reduced hemoglobin, 485
Reduction, 97–98
Redwood trees, 682
Reef, coral, 672
Reflex arc, 512
Reflexes, 560–62
    simple, 512
Refractory period, 585
Refrigerator effect, 114, 717, 718
Regeneration, 579
Regulators, plant growth, 408–9
Regulatory genes, 204–5
Reindeer, 625
Reinforcement, positive, 564
Relative ionic concentration, 489
Releaser effect, 568
Releasing hormones, 524
Releasing mechanisms, 560
Remoras, 641

Renewable energy sources, 700–702
Renewable resources, 696
Repetitive DNA, 211
Replacement reproduction, 691
Replication, 133
    DNA, 191–93
Repolarization, 506
Repressible operon model, 204
Repressor, 204
Reproduction, 8–9
    animal, 577, 578–97
    asexual, 283, 578–79
    cell, 130–47
    control of, 592–94
    human, 582–91
    patterns of, 578–81
    sexual, 283, 579–80
    vegetative, 379
Reproductive behavior, 573
Reproductive isolation, 230, 231, 232
Reproductive structures, 289
Reproductive systems
    female, 585–91
    male, 582–85
Reptiles, 341–43
    mammallike, 343
Reptilia, class, 341–43
RER. *See* Rough endoplasmic reticulum
Reservoirs, 658–59, 720–21
Resistance, environmental, 624
Resolution, 65
Resource Conservation and Recovery Act, 713
Resource partitioning, 633
Resources, 696
    nonrenewable, 696–97
    renewable, 696
Respiration, 475–86
    aerobic, 115–18
    anaerobic, 112, 121–22
    cellular, 92
        and photosynthesis, 123–27
    circulation, 118
Respiratory center, 482
Respiratory chain, 117
Respiratory pigment, 428
Respiratory system, human, 481
Response, 7
    approach and avoidance, 572
Resting potential, 506, 507
Retina, 535
Retinal, 536
Retrovirus, 207, 208, 213, 269
Reward, 564
"R" group, 59, 61, 122
Rhea, 18, 19
Rh factor, 447
    and birth defects, 613
Rhinoceros, 729
Rhizobium, 396
Rhizoid, 276
Rhizome, 378

Rhizophora mangle, 26
Rhizopus, 276
Rhodophyta, division, 288
Rhodopsin, 536
RhoGam, and birth defects, 613
Rhynia major, 292
Rhythm method, 592
Rib cage, 479
Riboflavin, 470
Ribonucleic acid. *See* RNA
Ribose, 62
Ribosomal ribonucleic acid. *See* rRNA
Ribosome, 72, 73, 74, 75, 200
Rickets, 471
Rilling, Mark, 566, 567
Ringdove, 565
Rings, in dicots, 375
Ritualization, 569, 570
Rivers, 665
RNA, 95
    and cell division, 131, 136
    and chromosomes, 172
    and gene therapy, 209
    and nucleic acids, 61–62
    and nucleotides, 196
    and nucleus, 74
    and retroviruses, 213
    and viruses, 269
Robin, 561
Rock, 236
Rocky shore, 671
Rodentia, order, 347
Rods, 535
Root, 289, 393, 397
    fungus, 396
Root adaptations, 372
Root apex, 364
Root apical meristem, 364
Root cap, 369
Root hair cycle, 370
Root hairs, 397
Root pressure, 399
Root system, 366, 369–72
Root systems, 372
Rose, 27
Rotation, 615
Rough endoplasmic reticulum, 72, 73, 74, 75, 76
    See also Endoplasmic reticulum
Roundworm, 315–18
Rouse sarcoma, 212
Roush, Jon, 730
Roux, Wilhelm, 598, 599, 603, 604
Royal Society, 25
rRNA, 200
    and cell division, 131
R-selected species, 627
R-strategist species, 627
Rubinoff, Ira, 710
Rumen, 56, 458
Runners, 378, 379

Y

Y chromosomes, active, 180
Yeast, 112, 277
Yolk, 488, 598, 599
Yolk plug, 601
Yolk sac, 608, 610

Z

Zea mays, 375, 383
Zeatin, 413
Zebra, 126, 241
Zero population growth, 691
Zinc, 39
Z line, 548
Zones
    of cell division, 369
    coastal, 673, 674, 675
    of elongation, 370
    life, 668, 675
    of maturation, 370
    of vegetation, 685
Zooflagellates, 272
Zooplankton, 665, 674
Zoospore, 283, 284, 285, 286
Zooxanthellae, 672
Zygomycota, division, 276
Zygospore, 276
Zygote, 138, 142, 283, 284, 285, 286, 331, 593, 609